Sixth Edition

Complete Solutions Guide

Chemistry

Steven S. Zumdahl
Susan Arena Zumdahl

Thomas J. Hummel
Steven S. Zumdahl
Susan Arena Zumdahl

University of Illinois at Urbana-Champaign

HOUGHTON MIFFLIN COMPANY   Boston   New York

Editor-in-Chief: Charles Hartford
Executive Editor: Richard Stratton
Development Editor: Sara Wise
Senior Project Editor: Cathy Brooks
Editorial Assistant: Rosemary Mack
Senior Marketing Manager: Katherine Greig

Cover credit: Masaaki Kazama/Photonica

Printed in the U.S.A.

ISBN: 0-618-22168-9

23456789 – CRS – 06 05 04 03

TABLE OF CONTENTS

TO THE STUDENT: HOW TO USE THIS GUIDE

Solutions to all of the end of chapter questions and exercises are in this manual. This "Solutions Guide" can be very valuable if you use it properly. The way NOT to use it is to look at an exercise in the book and then immediately check the solution, often saying to yourself, "That's easy, I can do it." Developing problem solving skills takes practice. Don't look up a solution to a problem until you have tried to work it on your own. If you are completely stuck, see if you can find a similar problem in the Sample Exercises in the chapter. Only look up the solution as a last resort. If you do this for a problem, look for a similar problem in the end of chapter exercises and try working it. The more problems you do, the easier chemistry becomes. It is also in your self interest to try to work as many problems as possible. Most exams that you will take in chemistry will involve a lot of problem solving. If you have worked several problems similar to the ones on an exam, you will do much better than if the exam is the first time you try to solve a particular type of problem. No matter how much you read and study the text, or how well you think you understand the material, you don't really understand it until you have taken the information in the text and applied the principles to problem solving. You will make mistakes, but the good students learn from their mistakes.

In this manual we have worked problems as in the textbook. We have shown intermediate answers to the correct number of significant figures and used the rounded answer in later calculations. Thus, some of your answers may differ slightly from ours. When we have not followed this convention, we have usually noted this in the solution. The most common exception is when working with the natural logarithm (ln) function, where we usually carried extra significant figures in order to reduce round-off error. In addition, we tried to use constants and conversion factors reported to at least one more significant figure as compared to numbers given in the problem. The practice of carrying one extra significant figure in constants helps minimize round-off error.

We are grateful to Claire O. Szoke for her outstanding effort in preparing the manuscript for this manual. We also thank Marjorie M. Needham and Linda C. Bush for their careful and thorough accuracy review of the Solutions Guide.

TJH
SSZ
SAZ

CHAPTER ONE

CHEMICAL FOUNDATIONS

Questions

16. a. Law versus theory: A law is a concise statement or equation that summarizes observed behavior. A theory is a set of hypotheses that gives an overall explanation of some phenomenon. A law summarizes what happens; a theory (or model) attempts to explain why it happens.

    b. Theory versus experiment: A theory is an explanation of why things behave the way they do, while an experiment is the process of observing that behavior. Theories attempt to explain the results of experiments and are, in turn, tested by further experiments.

    c. Qualitative versus quantitative: A qualitative observation only describes a quality while a quantitative observation attaches a number to the observation. Examples: Qualitative observations: The water was hot to the touch. Mercury was found in the drinking water. Quantitative observations: The temperature of the water was 62°C. The concentration of mercury in the drinking water was 1.5 ppm.

    d. Hypothesis versus theory: Both are explanations of experimental observation. A theory is a set of hypotheses that has been tested over time and found to still be valid, with (perhaps) some modifications.

17. No, it is useful whenever a systematic approach of observation and hypothesis testing can be used.

18. a. No   b. Yes   c. Yes

    Only statements b and c can be determined from experiment.

19. Volume readings are estimated to one decimal place past the markings on the glassware. The assumed uncertainty is ±1 in the estimated digit. For glassware a, the volume would be estimated to the tenths place since the markings are to the ones place. A sample reading would be 4.2 with an uncertainty of ±0.1. This reading has two significant figures. For glassware b, 10.52 ±0.01 would be a sample reading and the uncertainty; this reading has four significant figures. For glassware c, 18 ±1 would be a sample reading and uncertainty, with the reading having two significant figures.

20. Accuracy: How close a measurement or series of measurements are to an accepted or true value. Precision: How close a series of measurements of the same thing are to each other. The results, average = 14.91 ± 0.03% are precise (close to each other) but are not accurate (not close to the true value).

21.   Chemical changes involve the making and breaking of chemical forces (bonds). Physical changes do not. The identity of a substance changes after a chemical change, but not after a physical change.

22.   Many techniques of chemical analysis need to be performed on relatively pure materials. Thus, a separation step often is necessary to remove materials that will interfere with the analytical measurement.

Exercises

Significant Figures and Unit Conversions

23.   a.   inexact   b.   exact   c.   exact

For c, $\dfrac{36 \text{ in}}{\text{yd}} \times \dfrac{2.54 \text{ cm}}{\text{in}} \times \dfrac{1 \text{ m}}{100 \text{ cm}} = \dfrac{0.9144 \text{ m}}{\text{yd}}$ (All conversion factors used are exact.)

    d.   inexact; Although this number appears to be exact, it probably isn't. The announced attendance may be tickets sold but not the number who were actually in the stadium. Some people who paid may not have gone, some may leave early or arrive late, some may sneak in without paying, etc.

    e.   exact   f.   inexact

24.   a.   exact   b.   inexact; 0.9144 m/yd is exact (see Exercise 1.23c)
                                           Thus, there are 1/0.9144 = 1.093613 ... yd/m.

    c.   exact   d.   inexact (π has an infinite number of decimal places.)

25.   a. $\underline{12}$; 2 significant figures (S.F.); Nonzero integers always count as significant figures.

    b. $\underline{1098}$; 4 S.F.; Captive zeros always count as significant figures.

    c. $\underline{2001}$; 4 S.F.   d. $\underline{2.001} \times 10^3$; 4 S.F.

    e.   0.0000$\underline{101}$; 3 S.F.; Leading zeros never count as significant figures.

    f. $\underline{1.01} \times 10^{-5}$; 3 S.F.

    g. $\underline{1000.}$; 4 S.F.; Trailing zeros are only counted as significant figures if the number contains a decimal point.

    h. $\underline{22.04030}$; 7 S.F.; The trailing zero is a significant figure since the number contains a decimal point.

26.   a. $\underline{1}00$; 1 S.F.   b. $\underline{1.0} \times 10^2$; 2 S.F.

       c. $\underline{1.00} \times 10^3$; 3 S.F.   d. $\underline{100.}$; 3 S.F.

       e. $0.00\underline{48}$; 2 S.F.   f. $0.00\underline{480}$; 3 S.F.

       g. $\underline{4.80} \times 10^{-3}$; 3 S.F.   h. $\underline{4.800} \times 10^{-3}$; 4 S.F.

27.   a. 3.13×10^2 b. 3.13×10^{-4} c. 3.13×10^7

       d. 3.13×10^{-1} e. 3.13×10^{-2}

28.   a. 5×10^2 b. 4.8×10^2 c. 4.80×10^2 d. 4.800×10^2

29.   For addition and/or subtraction, the result has the same number of decimal places as the number in the calculation with the fewest decimal places. When the result is rounded to the correct number of significant figures, the last significant figure stays the same if the number after this significant figure is less than 5 and increases by one if the number is greater than or equal to 5.

       a. $97.381 + 4.2502 + 0.99195 = 102.62315 = 102.623$; Since 97.381 has only three decimal places, the result should only have three decimal places.

       b. $171.5 + 72.915 - 8.23 = \underline{236.1}85 = 236.2$

       c. $1.00914 + 0.87104 + 1.2012 = \underline{3.0813}8 = 3.0814$

       d. $21.901 - 13.21 - 4.0215 = \underline{4.66}95 = 4.67$

30.   For multiplication and/or division, the result has the same number of significant figures as the number in the calculation with the fewest significant figures.

       a. $\dfrac{0.102 \times 0.0821 \times 273}{1.01} = \underline{2.26}35 = 2.26$

       b. $0.14 \times 6.022 \times 10^{23} = \underline{8.4}31 \times 10^{22} = 8.4 \times 10^{22}$; Since 0.14 only has two significant figures, the result should only have two significant figures.

       c. $4.0 \times 10^4 \times 5.021 \times 10^{-3} \times 7.34993 \times 10^2 = \underline{1.4}76 \times 10^5 = 1.5 \times 10^5$

       d. $\dfrac{2.00 \times 10^6}{3.00 \times 10^{-7}} = \underline{6.66}67 \times 10^{12} = 6.67 \times 10^{12}$

31.   a. 467; The difference of $25.27 - 24.16 = 1.11$ has only three significant figures. The answer will only have three significant figures since we have a four significant figure number multiplied by a five significant figure number multiplied by a three significant figure number. For this problem and for subsequent problems, the addition/subtraction rule must be applied separately from the multiplication/division rule.

b.   0.24; The difference of 8.925 - 8.904 = 0.021 has only 2 significant figures. When a two significant figure number is divided by a four significant figure number, the result is reported to two significant figures (division rule).

c. $(9.04 - 8.23 + 21.954 + 81.0) \div 3.1416 = 103.8 \div 3.1416 = 33.04$

Here, apply the addition/subtraction rule first; then apply the multiplication/division rule to arrive at the four significant figure answer. We will generally round off at intermediate steps in order to show the correct number of significant figures. However, you should round off at the end of all the mathematical operations in order to avoid round-off error. Make sure you keep track of the correct number of significant figures during intermediate steps, but round off at the end.

d. $\dfrac{9.2 \times 100.65}{8.321 + 4.026} = \dfrac{9.2 \times 100.65}{12.347} = 75$

e. $0.1654 + 2.07 - 2.114 = 0.12$

Uncertainty begins to appear in the second decimal place. Numbers were added as written and the answer was rounded off to 2 decimal places at the end. If you round to 2 decimal places and add you get 0.13. Always round off at the end of the operation to avoid round-off error.

f. $8.27(4.987 - 4.962) = 8.27(0.025) = 0.21$

g. $\dfrac{9.5 + 4.1 + 2.8 + 3.175}{4} = \dfrac{19.6}{4} = 4.90 = 4.9$

Uncertainty appears in the first decimal place. The average of several numbers can only be as precise as the least precise number. Averages can be exceptions to the significant figure rules.

h. $\dfrac{9.025 - 9.024}{9.025} \times 100 = \dfrac{0.001}{9.025} \times 100 = 0.01$

32.   a. $6.022 \times 10^{23} \times 1.05 \times 10^{2} = 6.32 \times 10^{25}$

b. $\dfrac{6.6262 \times 10^{-34} \times 2.998 \times 10^{8}}{2.54 \times 10^{-9}} = 7.82 \times 10^{-17}$

c. $1.285 \times 10^{-2} + 1.24 \times 10^{-3} + 1.879 \times 10^{-1}$

$= 0.1285 \times 10^{-1} + 0.0124 \times 10^{-1} + 1.879 \times 10^{-1} = 2.020 \times 10^{-1}$

When the exponents are different, it is easiest to apply the addition/subtraction rule when all numbers are based on the same power of 10.

d. $1.285 \times 10^{-2} - 1.24 \times 10^{-3} = 1.285 \times 10^{-2} - 0.124 \times 10^{-2} = 1.161 \times 10^{-2}$

e. $\dfrac{(1.00866 - 1.00728)}{6.02205 \times 10^{23}} = \dfrac{0.00138}{6.02205 \times 10^{23}} = 2.29 \times 10^{-27}$

f. $\dfrac{9.875 \times 10^2 - 9.795 \times 10^2}{9.875 \times 10^2} \times 100 = \dfrac{0.080 \times 10^2}{9.875 \times 10^2} \times 100 = 8.1 \times 10^{-1}$

g. $\dfrac{9.42 \times 10^2 + 8.234 \times 10^2 + 1.625 \times 10^3}{3}$

$$= \dfrac{0.942 \times 10^3 + 0.8234 \times 10^3 + 1.625 \times 10^3}{3} = 1.130 \times 10^3$$

33.   a. $8.43 \text{ cm} \times \dfrac{1 \text{ m}}{100 \text{ cm}} \times \dfrac{1000 \text{ mm}}{\text{m}} = 84.3 \text{ mm}$ b. $2.41 \times 10^2 \text{ cm} \times \dfrac{1 \text{ m}}{100 \text{ cm}} = 2.41 \text{ m}$

c. $294.5 \text{ nm} \times \dfrac{1 \text{ m}}{1 \times 10^9 \text{ nm}} \times \dfrac{100 \text{ cm}}{\text{m}} = 2.945 \times 10^{-5} \text{ cm}$ d. $1.445 \times 10^4 \text{ m} \times \dfrac{1 \text{ km}}{1000 \text{ m}} = 14.45 \text{ km}$

e. $235.3 \text{ m} \times \dfrac{1000 \text{ mm}}{\text{m}} = 2.353 \times 10^5 \text{ mm}$

f. $903.3 \text{ nm} \times \dfrac{1 \text{ m}}{1 \times 10^9 \text{ nm}} \times \dfrac{1 \times 10^6 \text{ } \mu\text{m}}{\text{m}} = 0.9033 \text{ } \mu\text{m}$

34.   a. $1 \text{ Tg} \times \dfrac{1 \times 10^{12} \text{ g}}{\text{Tg}} \times \dfrac{1 \text{ kg}}{1000 \text{ g}} = 1 \times 10^9 \text{ kg}$

b. $6.50 \times 10^2 \text{ Tm} \times \dfrac{1 \times 10^{12} \text{ m}}{\text{Tm}} \times \dfrac{1 \times 10^9 \text{ nm}}{\text{m}} = 6.50 \times 10^{23} \text{ nm}$

c. $25 \text{ fg} \times \dfrac{1 \text{ g}}{1 \times 10^{15} \text{ fg}} \times \dfrac{1 \text{ kg}}{1000 \text{ g}} = 25 \times 10^{-18} \text{ kg} = 2.5 \times 10^{-17} \text{ kg}$

d. $8.0 \text{ dm}^3 \times \dfrac{1 \text{ L}}{\text{dm}^3} = 8.0 \text{ L}$ $(1 \text{ L} = 1 \text{ dm}^3 = 1000 \text{ cm}^3 = 1000 \text{ mL})$

e. $1 \text{ mL} \times \dfrac{1 \text{ L}}{1000 \text{ mL}} \times \dfrac{1 \times 10^6 \text{ } \mu\text{L}}{\text{L}} = 1 \times 10^3 \text{ } \mu\text{L}$

f. $1 \text{ } \mu\text{g} \times \dfrac{1 \text{ g}}{1 \times 10^6 \text{ } \mu\text{g}} \times \dfrac{1 \times 10^{12} \text{ pg}}{\text{g}} = 1 \times 10^6 \text{ pg}$

35.   a. Appropriate conversion factors are found in Appendix 6. In general, the number of significant figures we use in the conversion factors will be one more than the number of significant figures from the numbers given in the problem. This is usually sufficient to avoid round-off error.

$$3.91 \text{ kg} \times \frac{1 \text{ lb}}{0.4536 \text{ kg}} = 8.62 \text{ lb}; \quad 0.62 \text{ lb} \times \frac{16 \text{ oz}}{\text{lb}} = 9.9 \text{ oz}$$

Baby's weight = 8 lb and 9.9 oz or to the nearest ounce, 8 lb and 10 oz.

$$51.4 \text{ cm} \times \frac{1 \text{ in}}{2.54 \text{ cm}} = 20.2 \text{ in} \approx 20\ 1/4 \text{ in} = \text{baby's height}$$

b. $25{,}000 \text{ mi} \times \dfrac{1.61 \text{ km}}{\text{mi}} = 4.0 \times 10^4 \text{ km}; \quad 4.0 \times 10^4 \text{ km} \times \dfrac{1000 \text{ m}}{\text{km}} = 4.0 \times 10^7 \text{ m}$

c. $V = 1 \times w \times h = 1.0 \text{ m} \times \left(5.6 \text{ cm} \times \dfrac{1 \text{ m}}{100 \text{ cm}} \right) \times \left(2.1 \text{ dm} \times \dfrac{1 \text{ m}}{10 \text{ dm}} \right) = 1.2 \times 10^{-2} \text{ m}^3$

$$1.2 \times 10^{-2} \text{ m}^3 \times \left(\frac{10 \text{ dm}}{\text{m}} \right)^3 \times \frac{1 \text{ L}}{\text{dm}^3} = 12 \text{ L}$$

$$12 \text{ L} \times \frac{1000 \text{ cm}^3}{\text{L}} \times \left(\frac{1 \text{ in}}{2.54 \text{ cm}} \right)^3 = 730 \text{ in}^3; \quad 730 \text{ in}^3 \times \left(\frac{1 \text{ ft}}{12 \text{ in}} \right)^3 = 0.42 \text{ ft}^3$$

36.   a. $908 \text{ oz} \times \dfrac{1 \text{ lb}}{16 \text{ oz}} \times \dfrac{0.4536 \text{ kg}}{\text{lb}} = 25.7 \text{ kg}$

b. $12.8 \text{ L} \times \dfrac{1 \text{ qt}}{0.9463 \text{ L}} \times \dfrac{1 \text{ gal}}{4 \text{ qt}} = 3.38 \text{ gal}$

c. $125 \text{ mL} \times \dfrac{1 \text{ L}}{1000 \text{ mL}} \times \dfrac{1 \text{ qt}}{0.9463 \text{ L}} = 0.132 \text{ qt}$

d. $2.89 \text{ gal} \times \dfrac{4 \text{ qt}}{1 \text{ gal}} \times \dfrac{1 \text{ L}}{1.057 \text{ qt}} \times \dfrac{1000 \text{ mL}}{1 \text{ L}} = 1.09 \times 10^4 \text{ mL}$

e. $4.48 \text{ lb} \times \dfrac{453.6 \text{ g}}{1 \text{ lb}} = 2.03 \times 10^3 \text{ g}$

f. $550 \text{ mL} \times \dfrac{1 \text{ L}}{1000 \text{ mL}} \times \dfrac{1.06 \text{ qt}}{\text{L}} = 0.58 \text{ qt}$

37.   a. $1.25 \text{ mi} \times \dfrac{8 \text{ furlongs}}{\text{mi}} = 10.0 \text{ furlongs}; \quad 10.0 \text{ furlongs} \times \dfrac{40 \text{ rods}}{\text{furlong}} = 4.00 \times 10^2 \text{ rods}$

$$4.00 \times 10^2 \text{ rods} \times \frac{5.5 \text{ yd}}{\text{rod}} \times \frac{36 \text{ in}}{\text{yd}} \times \frac{2.54 \text{ cm}}{\text{in}} \times \frac{1 \text{ m}}{100 \text{ cm}} = 2.01 \times 10^3 \text{ m}$$

$$2.01 \times 10^3 \text{ m} \times \frac{1 \text{ km}}{1000 \text{ m}} = 2.01 \text{ km}$$

b. Let's assume we know this distance to ± 1 yard. First convert 26 miles to yards.

$$26 \text{ mi} \times \frac{5280 \text{ ft}}{\text{mi}} \times \frac{1 \text{ yd}}{3 \text{ ft}} = 45{,}760.\ \text{yd}$$

$$26 \text{ mi} + 385 \text{ yd} = 45,760. \text{ yd} + 385 \text{ yd} = 46,145 \text{ yards}$$

$$46,145 \text{ yard} \times \frac{1 \text{ rod}}{5.5 \text{ yd}} = 8390.0 \text{ rods}; \quad 8390.0 \text{ rods} \times \frac{1 \text{ furlong}}{40 \text{ rods}} = 209.75 \text{ furlongs}$$

$$46,145 \text{ yard} \times \frac{36 \text{ in}}{\text{yd}} \times \frac{2.54 \text{ cm}}{\text{in}} \times \frac{1 \text{ m}}{100 \text{ cm}} = 42,195 \text{ m}; \quad 42,195 \text{ m} \times \frac{1 \text{ km}}{1000 \text{ m}} = 42.195 \text{ km}$$

38.   a. $1 \text{ ha} \times \dfrac{10,000 \text{ m}^2}{\text{ha}} \times \left(\dfrac{1 \text{ km}}{1000 \text{ m}}\right)^2 = 1 \times 10^{-2} \text{ km}^2$

   b. $5.5 \text{ acre} \times \dfrac{160 \text{ rod}^2}{\text{acre}} \times \left(\dfrac{5.5 \text{ yd}}{\text{rod}} \times \dfrac{36 \text{ in}}{\text{yd}} \times \dfrac{2.54 \text{ cm}}{\text{in}} \times \dfrac{1 \text{ m}}{100 \text{ cm}}\right)^2 = 2.2 \times 10^4 \text{ m}^2$

$$2.2 \times 10^4 \text{ m}^2 \times \frac{1 \text{ ha}}{1 \times 10^4 \text{ m}^2} = 2.2 \text{ ha}; \quad 2.2 \times 10^4 \text{ m}^2 \times \left(\frac{1 \text{ km}}{1000 \text{ m}}\right)^2 = 0.022 \text{ km}^2$$

   c.   Area of lot $= 120 \text{ ft} \times 75 \text{ ft} = 9.0 \times 10^3 \text{ ft}^2$

$$9.0 \times 10^3 \text{ ft}^2 \times \left(\frac{1 \text{ yd}}{3 \text{ ft}} \times \frac{1 \text{ rod}}{5.5 \text{ yd}}\right)^2 \times \frac{1 \text{ acre}}{160 \text{ rod}^2} = 0.21 \text{ acre}; \quad \frac{\$6,500}{0.21 \text{ acre}} = \frac{\$31,000}{\text{acre}}$$

We can use our result from (b) to get the conversion factor between acres and ha (5.5 acre = 2.2 ha.). Thus, 1 ha = 2.5 acre.

$$0.21 \text{ acre} \times \frac{1 \text{ ha}}{2.5 \text{ acre}} = 0.084 \text{ ha}; \quad \text{The price is: } \frac{\$6,500}{0.084 \text{ ha}} = \frac{\$77,000}{\text{ha}}$$

39.   a. $1 \text{ troy lb} \times \dfrac{12 \text{ troy oz}}{\text{troy lb}} \times \dfrac{20 \text{ pw}}{\text{troy oz}} \times \dfrac{24 \text{ grains}}{\text{pw}} \times \dfrac{0.0648 \text{ g}}{\text{grain}} \times \dfrac{1 \text{ kg}}{1000 \text{ g}} = 0.373 \text{ kg}$

$$1 \text{ troy lb} = 0.373 \text{ kg} \times \frac{2.205 \text{ lb}}{\text{kg}} = 0.822 \text{ lb}$$

   b. $1 \text{ troy oz} \times \dfrac{20 \text{ pw}}{\text{troy oz}} \times \dfrac{24 \text{ grains}}{\text{pw}} \times \dfrac{0.0648 \text{ g}}{\text{grain}} = 31.1 \text{ g}$

$$1 \text{ troy oz} = 31.1 \text{ g} \times \frac{1 \text{ carat}}{0.200 \text{ g}} = 156 \text{ carats}$$

   c. $1 \text{ troy lb} = 0.373 \text{ kg}; \quad 0.373 \text{ kg} \times \dfrac{1000 \text{ g}}{\text{kg}} \times \dfrac{1 \text{ cm}^3}{19.3 \text{ g}} = 19.3 \text{ cm}^3$

40.   a. $1 \text{ grain ap} \times \dfrac{1 \text{ scruple}}{20 \text{ grain ap}} \times \dfrac{1 \text{ dram ap}}{3 \text{ scruples}} \times \dfrac{3.888 \text{ g}}{\text{dram ap}} = 0.06480 \text{ g}$

From the previous question, we are given that 1 grain troy = 0.0648 g = 1 grain ap. So, the two are the same.

       b. $1 \text{ oz ap} \times \dfrac{8 \text{ dram ap}}{\text{oz ap}} \times \dfrac{3.888 \text{ g}}{\text{dram ap}} \times \dfrac{1 \text{ oz troy*}}{31.1 \text{ g}} = 1.00 \text{ oz troy}$ *See Exercise 39b.

       c. $5.00 \times 10^2 \text{ mg} \times \dfrac{1 \text{ g}}{1000 \text{ mg}} \times \dfrac{1 \text{ dram ap}}{3.888 \text{ g}} \times \dfrac{3 \text{ scruples}}{\text{dram ap}} = 0.386 \text{ scruple}$

             $0.386 \text{ scruple} \times \dfrac{20 \text{ grains ap}}{\text{scruple}} = 7.72 \text{ grains ap}$

       d. $1 \text{ scruple} \times \dfrac{1 \text{ dram ap}}{3 \text{ scruples}} \times \dfrac{3.888 \text{ g}}{\text{dram ap}} = 1.296 \text{ g}$

41. $1.71 \text{ warp factor} = \left(5.00 \times \dfrac{3.00 \times 10^8 \text{ m}}{\text{s}} \right) \times \dfrac{1.094 \text{ yd}}{\text{m}} \times \dfrac{60 \text{ s}}{\text{min}} \times \dfrac{60 \text{ min}}{\text{hr}}$

$\times \dfrac{1 \text{ knot}}{2000 \text{ yd/hr}} = 2.95 \times 10^9 \text{ knots}$

42. $\dfrac{100. \text{ m}}{9.79 \text{ s}} = 10.2 \text{ m/s}; \quad \dfrac{100. \text{ m}}{9.79 \text{ s}} \times \dfrac{1 \text{ km}}{1000 \text{ m}} \times \dfrac{60 \text{ s}}{\text{min}} \times \dfrac{60 \text{ min}}{\text{hr}} = 36.8 \text{ km/hr}$

       $\dfrac{100. \text{ m}}{9.79 \text{ s}} \times \dfrac{1.0936 \text{ yd}}{\text{m}} \times \dfrac{3 \text{ ft}}{\text{yd}} = 33.5 \text{ ft/s}; \quad \dfrac{33.5 \text{ ft}}{\text{s}} \times \dfrac{1 \text{ mi}}{5280 \text{ ft}} \times \dfrac{60 \text{ s}}{\text{min}} \times \dfrac{60 \text{ min}}{\text{hr}} = 22.8 \text{ mi/hr}$

       $1.00 \times 10^2 \text{ yd} \times \dfrac{1 \text{ m}}{1.0936 \text{ yd}} \times \dfrac{9.79 \text{ s}}{100. \text{ m}} = 8.95 \text{ s}$

43. $\dfrac{14 \text{ km}}{\text{L}} \times \dfrac{1 \text{ mi}}{1.61 \text{ km}} \times \dfrac{3.79 \text{ L}}{\text{gal}} = 33 \text{ mi/gal};$ The spouse's car has the better gas mileage.

44. $112 \text{ km} \times \dfrac{0.6214 \text{ mi}}{\text{km}} \times \dfrac{1 \text{ hr}}{65 \text{ mi}} = 1.1 \text{ hr} = 1 \text{ hr and 6 min}$

       $112 \text{ km} \times \dfrac{0.6214 \text{ mi}}{\text{km}} \times \dfrac{1 \text{ gal}}{28 \text{ mi}} \times \dfrac{3.785 \text{ L}}{\text{gal}} = 9.4 \text{ L of gasoline}$

45. $1.00 \text{ lb} \times \dfrac{0.4536 \text{ kg}}{\text{lb}} \times \dfrac{4.00 \text{ euros}}{\text{kg}} \times \dfrac{\$1.00}{1.14 \text{ euros}} = \$1.59$

46. $1.5 \text{ teaspoons} \times \dfrac{80. \text{ mg acet}}{0.50 \text{ teaspoon}} = 240 \text{ mg acetaminophen}$

$\dfrac{240 \text{ mg acet}}{24 \text{ lb}} \times \dfrac{1 \text{ lb}}{0.454 \text{ kg}} = 22 \text{ mg acetaminophen/kg}$

$\dfrac{240 \text{ mg acet}}{35 \text{ lb}} \times \dfrac{1 \text{ lb}}{0.454 \text{ kg}} = 15 \text{ mg acetaminophen/kg}$

The range is from 15 mg to 22 mg acetaminophen per kg of body weight.

Temperature

47. $T_C = \dfrac{5}{9}(T_F - 32) = \dfrac{5}{9}(102.5 - 32) = 39.2\,°C; \ T_K = T_C + 273.2 = 312.4 \text{ K (Note: 32 is exact)}$

48. $T_C = \dfrac{5}{9}(74 - 32) = 23\,°C; \ T_K = 23 + 273 = 296 \text{ K}$

49.   a. $T_F = \dfrac{9}{5} \times T_C + 32 = \dfrac{9}{5} \times 78.1\,°C + 32 = 173\,°F$

$T_K = T_C + 273.2 = 78.1 + 273.2 = 351.3 \text{ K}$

b. $T_F = \dfrac{9}{5} \times (-25) + 32 = -13\,°F; \ T_K = -25 + 273 = 248 \text{ K}$

c. $T_F = \dfrac{9}{5} \times (-273) + 32 = -459\,°F; \ T_K = -273 + 273 = 0 \text{ K}$

d. $T_F = \dfrac{9}{5} \times 801 + 32 = 1470\,°F; \ T_K = 801 + 273 = 1074 \text{ K}$

50.   a. $T_C = T_K - 273 = 233 - 273 = -40.\,°C$

$T_F = \dfrac{9}{5} \times T_C + 32 = \dfrac{9}{5} \times (-40.) + 32 = -40.\,°F$

b. $T_C = 4 - 273 = -269\,°C; \ T_F = \dfrac{9}{5} \times (-269) + 32 = -452\,°F$

c. $T_C = 298 - 273 = 25\,°C; \ T_F = \dfrac{9}{5} \times 25 + 32 = 77\,°F$

d. $T_C = 3680 - 273 = 3410\,°C; \ T_F = \dfrac{9}{5} \times 3410 + 32 = 6170\,°F$

51. We can do this two ways. One way is to calculate the high and low temperature and get the uncertainty from the range. $20.6°C \pm 0.1°C$ means the temperature can range from $20.5°C$ to $20.7°C$.

$$T_F = \frac{9}{5} \times T_c + 32 \leftarrow \text{(exact)}; \quad T_F = \frac{9}{5} \times 20.6 + 32 = 69.1°F$$

$$T_F \text{(min)} = \frac{9}{5} \times 20.5 + 32 = 68.9°F; \quad T_F \text{(max)} = \frac{9}{5} \times 20.7 + 32 = 69.3°F$$

So the temperature ranges from $68.9°F$ to $69.3°F$ which we can express as $69.1 \pm 0.2°F$.

An alternative way is to treat the uncertainty and the temperature in $°C$ separately.

$$T_F = \frac{9}{5} \times T_C + 32 = \frac{9}{5} \times 20.6 + 32 = 69.1°F; \quad \pm 0.1°C \times \frac{9°F}{5°C} = \pm 0.18°F \approx \pm 0.2°F$$

Combining the two calculations: $T_F = 69.1 \pm 0.2°F$

52. $96.1°F \pm 0.2°F$; First, convert $96.1°F$ to $°C$. $T_C = \frac{5}{9}(T_F - 32) = \frac{5}{9}(96.1 - 32) = 35.6°C$

A change in temperature of $9°F$ is equal to a change in temperature of $5°C$. So the uncertainty is:

$$\pm 0.2°F \times \frac{5°C}{9°F} = \pm 0.1°C. \text{ Thus, } 96.1 \pm 0.2°F = 35.6 \pm 0.1°C$$

Density

53. $$\frac{2.70 \text{ g}}{\text{cm}^3} \times \frac{1 \text{ kg}}{1000 \text{ g}} \times \left(\frac{100 \text{ cm}}{\text{m}}\right)^3 = \frac{2.70 \times 10^3 \text{ kg}}{\text{m}^3}$$

$$\frac{2.70 \text{ g}}{\text{cm}^3} \times \frac{1 \text{ lb}}{453.6 \text{ g}} \times \left(\frac{2.54 \text{ cm}}{\text{in}}\right)^3 \times \left(\frac{12 \text{ in}}{\text{ft}}\right)^3 = \frac{169 \text{ lb}}{\text{ft}^3}$$

54. $$\text{mass} = 350 \text{ lb} \times \frac{453.6 \text{ g}}{\text{lb}} = 1.6 \times 10^5 \text{ g}; \quad V = 1.2 \times 10^4 \text{ in}^3 \times \left(\frac{2.54 \text{ cm}}{\text{in}}\right)^3 = 2.0 \times 10^5 \text{ cm}^3$$

$$\text{density} = \frac{\text{mass}}{\text{volume}} = \frac{1.6 \times 10^5 \text{ g}}{2.0 \times 10^5 \text{ cm}^3} = 0.80 \text{ g/cm}^3$$

Since the material has a density less than water, it will float in water.

55. $$V = \frac{4}{3}\pi r^3 = \frac{4}{3} \times 3.14 \times \left(7.0 \times 10^5 \text{ km} \times \frac{1000 \text{ m}}{\text{km}} \times \frac{100 \text{ cm}}{\text{m}}\right)^3 = 1.4 \times 10^{33} \text{ cm}^3$$

$$\text{density} = \frac{\text{mass}}{\text{volume}} = \frac{2 \times 10^{36} \text{ kg} \times \frac{1000 \text{ g}}{\text{kg}}}{1.4 \times 10^{33} \text{ cm}^3} = 1.4 \times 10^6 \text{ g/cm}^3 = 1 \times 10^6 \text{ g/cm}^3$$

56. $V = 1 \times w \times h = 2.9 \text{ cm} \times 3.5 \text{ cm} \times 10.0 \text{ cm} = 1.0 \times 10^2 \text{ cm}^3$

$d = \text{density} = \dfrac{615.0 \text{ g}}{1.0 \times 10^2 \text{ cm}^3} = \dfrac{6.2 \text{ g}}{\text{cm}^3}$

57. $5.0 \text{ carat} \times \dfrac{0.200 \text{ g}}{\text{carat}} \times \dfrac{1 \text{ cm}^3}{3.51 \text{ g}} = 0.28 \text{ cm}^3$

58. $2.8 \text{ mL} \times \dfrac{1 \text{ cm}^3}{\text{mL}} \times \dfrac{3.51 \text{ g}}{\text{cm}^3} \times \dfrac{1 \text{ carat}}{0.200 \text{ g}} = 49 \text{ carats}$

59. $V = 21.6 \text{ mL} - 12.7 \text{ mL} = 8.9 \text{ mL}; \quad \text{density} = \dfrac{33.42 \text{ g}}{8.9 \text{ mL}} = 3.8 \text{ g/mL} = 3.8 \text{ g/cm}^3$

60. $5.25 \text{ g} \times \dfrac{1 \text{ cm}^3}{10.5 \text{ g}} = 0.500 \text{ cm}^3 = 0.500 \text{ mL}$

The volume in the cylinder will rise to 11.7 mL (11.2 mL + 0.500 mL = 11.7 mL).

61.   a. Both have the same mass of 1.0 kg.

b. 1.0 mL of mercury; Mercury has a greater density than water. Note: 1 mL = 1 cm³

$1.0 \text{ mL} \times \dfrac{13.6 \text{ g}}{\text{mL}} = 14 \text{ g of mercury}; \quad 1.0 \text{ mL} \times \dfrac{0.998 \text{ g}}{\text{mL}} = 1.0 \text{ g of water}$

c. Same; Both represent 19.3 g of substance.

$19.3 \text{ mL} \times \dfrac{0.9982 \text{ g}}{\text{mL}} = 19.3 \text{ g of water}; \quad 1.00 \text{ mL} \times \dfrac{19.32 \text{ g}}{\text{mL}} = 19.3 \text{ g of gold}$

d. 1.0 L of benzene (880 g vs 670 g)

$75 \text{ mL} \times \dfrac{8.96 \text{ g}}{\text{mL}} = 670 \text{ g of copper}; \quad 1.0 \text{ L} \times \dfrac{1000 \text{ mL}}{\text{L}} \times \dfrac{0.880 \text{ g}}{\text{mL}} = 880 \text{ g of benzene}$

62.   a. 1.0 kg feather; Feathers are less dense than lead.

b. 100 g water since water is less dense than gold.   c. Same; Both volumes are 1.0 L.

63. $V = 1.00 \times 10^3 \text{ g} \times \dfrac{1 \text{ cm}^3}{22.57 \text{ g}} = 44.3 \text{ cm}^3$

$44.3 \text{ cm}^3 = 1 \times w \times h = 4.00 \text{ cm} \times 4.00 \text{ cm} \times h, \quad h = 2.77 \text{ cm}$

64. $V = 22 \text{ g} \times \dfrac{1 \text{ cm}^3}{8.96 \text{ g}} = 2.5 \text{ cm}^3$; $V = \pi r^2 \times l$ where $l = $ length of the wire

$$2.5 \text{ cm}^3 = \pi \times \left(\dfrac{0.25 \text{ mm}}{2} \right)^2 \times \left(\dfrac{1 \text{ cm}}{10 \text{ mm}} \right)^2 \times l, \; l = 5.1 \times 10^3 \text{ cm} = 170 \text{ ft}$$

Classification and Separation of Matter

65.   Solid: own volume, own shape, does not flow; Liquid: own volume, takes shape of container, flows; Gas: takes volume and shape of container, flows

66.   Homogeneous: Having visibly indistinguishable parts (the same throughout). Heterogeneous: Having visibly distinguishable parts (not uniform throughout).

a.   heterogeneous (Due to mulch, water, roots, etc., which can all be present.)

b.   heterogeneous: There is usually a fair amount of particulate matter present in the atmosphere (dirt, smog) in addition to condensed water (rain, clouds). However, an atmosphere consisting of only clean air can be considered homogeneous.

c.   heterogeneous (due to bubbles)   d.   homogeneous

e.   homogeneous   f.   homogeneous

67.   A gas has molecules that are very far apart from each other while a solid or liquid has molecules that are very close together. An element has the same type of atom, whereas a compound contains two or more different elements. Picture i represents an element that exists as two atoms bonded together (like H_2 or O_2 or N_2). Picture iv represents a compound (like CO, NO, or HF). Pictures iii and iv contain representations of elements that exist as individual atoms (like Ar, Ne, or He).

a.   Picture iv represents a gaseous compound. Note that pictures ii and iii also contain a gaseous compound, but they also both have a gaseous element present.

b.   Picture vi represents a mixture of two gaseous elements.

c.   Picture v represents a solid element.

d.   Pictures ii and iii both represent a mixture of a gaseous element and a gaseous compound.

68.   a. pure   b. mixture   c. mixture   d. pure   e. mixture (copper and zinc)

f. pure   g. mixture   h. mixture   i. pure

Iron and uranium are elements. Water and table salt are compounds. Water is H_2O and table salt is NaCl. Compounds are composed of two or more elements.

69.   A physical change is a change in the state of a substance (solid, liquid and gas are the three states of matter); a physical change does not change the chemical composition of the substance. A chemical change is a change in which a given substance is converted into another substance having a different formula (composition).

    a.   Vaporization refers to a liquid converting to a gas, so this is a physical change. The formula (composition) of the moth ball does not change.

    b.   This is a chemical change since hydrofluoric acid (HF) is reacting with glass (SiO_2) to form new compounds which wash away.

    c.   This is a physical change since all that is happening is the conversion of liquid alcohol to gaseous alcohol. The alcohol formula (C_2H_5OH) does not change.

    d.   This is a chemical change since the acid is reacting with cotton to form new compounds.

70.   a.   Distillation separates components of a mixture, so the orange liquid is a mixture (has an average color of the yellow liquid and the red solid). Distillation utilizes boiling point differences to separate out the components of a mixture. Distillation is a physical change since the components of the mixture do not become different compounds or elements.

    b.   Decomposition is a type of chemical reaction. The crystalline solid is a compound, and decomposition is a chemical change where new substances are formed.

    c.   Tea is a mixture of tea compounds dissolved in water. The process of mixing sugar into tea is a physical change. Sugar doesn't react with the tea compounds, it just makes the solution sweeter.

Additional Exercises

71. $1\ \mu\text{mol} \times \dfrac{1\ \text{mol}}{1 \times 10^6\ \mu\text{mol}} \times \dfrac{6.02 \times 10^{23}\ \text{atoms}}{\text{mol}} = 6.02 \times 10^{17}$ atoms of helium

    1.25×10^{20} atoms $\times \dfrac{1\ \text{mol}}{6.02 \times 10^{23}\ \text{atoms}} = 2.08 \times 10^{-4}$ mol helium

72. $126\ \text{gal} \times \dfrac{4\ \text{qt}}{\text{gal}} \times \dfrac{1\ \text{L}}{1.057\ \text{qt}} = 477\ \text{L}$

73.   Total volume $= \left(200.\ \text{m} \times \dfrac{100\ \text{cm}}{\text{m}}\right) \times \left(300.\ \text{m} \times \dfrac{100\ \text{cm}}{\text{m}}\right) \times 4.0\ \text{cm} = 2.4 \times 10^9\ \text{cm}^3$

    Vol. of topsoil covered by 1 bag $= \left[10.\ \text{ft}^2 \times \left(\dfrac{12\ \text{in}}{\text{ft}}\right)^2 \times \left(\dfrac{2.54\ \text{cm}}{\text{in}}\right)^2\right] \times \left(1.0\ \text{in} \times \dfrac{2.54\ \text{cm}}{\text{in}}\right)$

$$= 2.4 \times 10^4\ \text{cm}^3$$

    $2.4 \times 10^9\ \text{cm}^3 \times \dfrac{1\ \text{bag}}{2.4 \times 10^4\ \text{cm}^3} = 1.0 \times 10^5$ bags topsoil

74.   a.   No; If the volumes were the same, then the gold idol would have a much greater mass because gold is much more dense than sand.

b.   Mass $= 1.0\ L \times \dfrac{1000\ cm^3}{L} \times \dfrac{19.32\ g}{cm^3} \times \dfrac{1\ kg}{1000\ g} = 19.32\ kg\ (= 42.59\ lb)$

It wouldn't be easy to play catch with the idol since it would have a mass of over 40 pounds.

75.   Volume of lake $= 100\ mi^2 \times \left(\dfrac{5280\ ft}{mi}\right)^2 \times 20\ ft = 6 \times 10^{10}\ ft^3$

$6 \times 10^{10}\ ft^3 \times \left(\dfrac{12\ in}{ft} \times \dfrac{2.54\ cm}{in}\right)^3 \times \dfrac{1\ mL}{cm^3} \times \dfrac{0.4\ \mu g}{mL} = 7 \times 10^{14}\ \mu g\ mercury$

$7 \times 10^{14}\ \mu g \times \dfrac{1\ g}{10^6\ \mu g} \times \dfrac{1\ kg}{10^3\ g} = 7 \times 10^5\ kg\ of\ mercury$

76. $mass_{benzene} = 58.80\ g - 25.00\ g = 33.80\ g;\ \ V_{benzene} = 33.80\ g \times \dfrac{1\ cm^3}{0.880\ g} = 38.4\ cm^3$

$V_{solid} = 50.0\ cm^3 - 38.4\ cm^3 = 11.6\ cm^3;\ \ density = \dfrac{25.00\ g}{11.6\ cm^3} = 2.16\ g/cm^3$

77.   a.   Volume × density = mass; the orange block is more dense. Since mass (orange) > mass (blue) and since volume (orange) < volume (blue), the density of the orange block must be greater to account for the larger mass of the orange block.

b.   Which block is more dense cannot be determined. Since mass (orange) > mass (blue) and since volume (orange) > volume (blue), the density of the orange block may or may not be larger than the blue block. If the blue block is more dense, its density cannot be so large that its mass is larger than the orange block's mass.

c.   The blue block is more dense. Since mass (blue) = mass (orange) and since volume (blue) < volume (orange), the density of the blue block must be larger in order to equate the masses.

d.   The blue block is more dense. Since mass (blue) > mass (orange) and since the volumes are equal, the density of the blue block must be larger in order to give the blue block the larger mass.

78.   Circumference $= c = 2\pi r;\ \ V = \dfrac{4\pi r^3}{3} = \dfrac{4\pi}{3}\left(\dfrac{c}{2\pi}\right)^3 = \dfrac{c^3}{6\pi^2}$

Largest density $= \dfrac{5.25\ oz}{\dfrac{(9.00\ in)^3}{6\pi^2}} = \dfrac{5.25\ oz}{12.3\ in^3} = \dfrac{0.427\ oz}{in^3}$

$$\text{Smallest density} = \frac{5.00 \text{ oz}}{\dfrac{(9.25 \text{ in})^3}{6\pi^2}} = \frac{5.00 \text{ oz}}{13.4 \text{ in}^3} = \frac{0.373 \text{ oz}}{\text{in}^3}$$

Maximum range is: $\dfrac{(0.373 - 0.427) \text{ oz}}{\text{in}^3}$ or 0.40 ± 0.03 oz/in³   (Uncertainty in 2nd decimal place.)

79. $V = V_{final} - V_{initial}$; $d = \dfrac{28.90 \text{ g}}{9.8 \text{ cm}^3 - 6.4 \text{ cm}^3} = \dfrac{28.90 \text{ g}}{3.4 \text{ cm}^3} = 8.5 \text{ g/cm}^3$

$d_{max} = \dfrac{mass_{max}}{V_{min}}$;   We get V_{min} from 9.7 cm³ - 6.5 cm³ = 3.2 cm³.

$d_{max} = \dfrac{28.93 \text{ g}}{3.2 \text{ cm}^3} = \dfrac{9.0 \text{ g}}{\text{cm}^3}$; $d_{min} = \dfrac{mass_{min}}{V_{max}} = \dfrac{28.87 \text{ g}}{9.9 \text{ cm}^3 - 6.3 \text{ cm}^3} = \dfrac{8.0 \text{ g}}{\text{cm}^3}$

The density is: 8.5 ± 0.5 g/cm³.

Challenge Problems

80.   a. $\dfrac{2.70 - 2.64}{2.70} \times 100 = 2\%$ b. $\dfrac{|16.12 - 16.48|}{16.12} \times 100 = 2.2\%$

c. $\dfrac{1.000 - 0.9981}{1.000} \times 100 = \dfrac{0.002}{1.000} \times 100 = 0.2\%$

81.   In subtraction, the result gets smaller but the uncertainties add. If the two numbers are very close together, the uncertainty may be larger than the result. For example, let us assume we want to take the difference of the following two measured quantities: 999,999 ± 2 and 999,996 ± 2. The difference is 3 ± 4. Because of the uncertainty, subtracting two similar numbers is bad practice.

82.   a.   At some point in 1982, the composition of the metal used in minting pennies was changed since the mass changed during this year (assuming the volume of the pennies were constant).

b.   It should be expressed as 3.08 ± 0.05 g. The uncertainty in the second decimal place will swamp any effect of the next decimal places.

83.   Heavy pennies (old): mean mass = 3.08 ± 0.05 g

Light pennies (new): mean mass = $\dfrac{(2.467 + 2.545 + 2.518)}{3} = 2.51 \pm 0.04$ g

Since we are assuming that the volume is additive, let's calculate the volume of 100. g of each type of penny then calculate the density of the alloy. For 100. g of the old pennies, 95 g will be Cu and 5 g will be Zn.

$$V = 95 \text{ g Cu} \times \frac{1 \text{ cm}^3}{8.96 \text{ g}} + 5 \text{ g Zn} \times \frac{1 \text{ cm}^3}{7.14 \text{ g}} = 11.3 \text{ cm}^3 \text{ (carrying one extra sig. fig.)}$$

$$\text{Density of old pennies} = \frac{100. \text{ g}}{11.3 \text{ cm}^3} = 8.8 \text{ g/cm}^3$$

For 100. g of new pennies, 97.6 g will be Zn and 2.4 g will be copper.

$$V = 2.4 \text{ g Cu} \times \frac{1 \text{ cm}^3}{8.96 \text{ g}} + 97.6 \text{ g Zn} \times \frac{1 \text{ cm}^3}{7.14 \text{ g}} = 13.94 \text{ cm}^3 \text{ (carrying one extra sig. fig.)}$$

$$\text{Density of new pennies} = \frac{100. \text{ g}}{13.94 \text{ cm}^3} = 7.17 \text{ g/cm}^3$$

Since $d = \dfrac{\text{mass}}{\text{volume}}$ and since the volume of both types of pennies are assumed equal, then:

$$\frac{d_{new}}{d_{old}} = \frac{\text{mass}_{new}}{\text{mass}_{old}} = \frac{7.17 \text{ g/cm}^3}{8.8 \text{ g/cm}^3} = 0.81$$

The calculated average mass ratio is: $\dfrac{\text{mass}_{new}}{\text{mass}_{old}} = \dfrac{2.51 \text{ g}}{3.08 \text{ g}} = 0.815$

To the first two decimal places, the ratios are the same. If the assumptions are correct, then we can reasonably conclude that the difference in mass is accounted for by the difference in alloy used.

84. a.

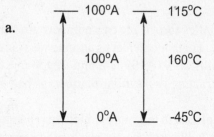

A change in temperature of 160°C equals a change in temperature of 100°A.

So, $\dfrac{160 \text{ °C}}{100 \text{ °C}}$ is our unit conversion for a degree change in temperature.

At the freezing point: 0°A = -45°C

Combining the two pieces of information:

$$T_A = (T_C + 45°C) \times \frac{100°A}{160°C} = (T_C + 45°C) \times \frac{5°A}{8°C} \text{ or } T_C = T_A \times \frac{8°C}{5°A} - 45°C$$

b. $T_C = (T_F - 32) \times \dfrac{5}{9}; \quad T_C = T_A \times \dfrac{8}{5} - 45 = (T_F - 32) \times \dfrac{5}{9}$

$$T_F - 32 = \frac{9}{5} \times \left[T_A \times \frac{8}{5} - 45 \right] = T_A \times \frac{72}{25} - 81, \quad T_F = T_A \times \frac{72°F}{25°A} - 49°F$$

c. $T_C = T_A \times \dfrac{8}{5} - 45$ and $T_C = T_A$; So, $T_C = T_C \times \dfrac{8}{5} - 45$, $\dfrac{3 \times T_C}{5} = 45$, $T_C = 75°C = 75°A$

d. $T_C = 86°A \times \dfrac{8°C}{5°A} - 45°C = 93°C$; $T_F = 86°A \times \dfrac{72°F}{25°A} - 49°F = 199°F = 2.0 \times 10^2°F$

e. $T_A = (45°C + 45°C) \times \dfrac{5°A}{8°C} = 56°A$

85.   a. One possibility is that rope B is not attached to anything and rope A and rope C are connected via a pair of pulleys and/or gears.

       b. Try to pull rope B out of the box. Measure the distance moved by C for a given movement of A. Hold either A or C firmly while pulling on the other.

86.   The bubbles of gas is air in the sand that is escaping; methanol and sand are not reacting. We will assume that the mass of trapped air is insignificant.

mass of dry sand = 37.3488 g - 22.8317 g = 14.5171 g

mass of methanol = 45.2613 g - 37.3488 g = 7.9125 g

Volume of sand particles (air absent) = volume of sand and methanol - volume of methanol

Volume of sand particles (air absent) = 17.6 mL - 10.00 mL = 7.6 mL

density of dry sand (air present) = $\dfrac{14.5171\,g}{10.0\,mL}$ = 1.45 g/mL

density of methanol = $\dfrac{7.9125\,g}{10.00\,mL}$ = 0.7913 g/mL

density of sand particles (air absent) = $\dfrac{14.5171\,g}{7.6\,mL}$ = 1.9 g/mL

CHAPTER TWO

ATOMS, MOLECULES, AND IONS

Questions

15. a. Atoms have mass and are neither destroyed nor created by chemical reactions. Therefore, mass is neither created nor destroyed by chemical reactions. Mass is conserved.

   b. The composition of a substance depends on the number and kinds of atoms that form it.

   c. Compounds of the same elements differ only in the numbers of atoms of the elements forming them, i.e., NO, N_2O, NO_2.

16. Some elements exist as molecular substances. That is, hydrogen normally exists as H_2 molecules, not single hydrogen atoms. The same is true for N_2, O_2, F_2, Cl_2, etc.

17. Deflection of cathode rays by magnetic and electric fields led to the conclusion that they were negatively charged. The cathode ray was produced at the negative electrode and repelled by the negative pole of the applied electric field.

18. J. J. Thomson discovered electrons. Henri Becquerel discovered radioactivity. Lord Rutherford proposed the nuclear model of the atom. Dalton's original model proposed that atoms were indivisible particles (that is, atoms had no internal structure). Thomson and Becquerel discovered subatomic particles, and Rutherford's model attempted to describe the internal structure of the atom composed of these subatomic particles. In addition, the existence of isotopes, atoms of the same element but with different mass, had to be included in the model.

19. The proton and neutron have similar mass with the mass of the neutron slightly larger than that of the proton. Each of these particles has a mass approximately 1800 times greater than that of an electron. The combination of the protons and the neutrons in the nucleus makes up the bulk of the mass of an atom, but the electrons make the greatest contribution to the chemical properties of the atom.

20. If the plum pudding model were correct (a diffuse positive charge with electrons scattered throughout), then alpha particles should have traveled through the thin foil with very minor deflections in their path. This was not the case as a few of the alpha particles were deflected at very large angles. Rutherford reasoned that the large deflections of these alpha particles could be caused only by a center of concentrated positive charge that contains most of the atom's mass (the nuclear model of the atom).

18

21.   The atomic number of an element is equal to the number of protons in the nucleus of an atom of that element. The mass number is the sum of the number of protons plus neutrons in the nucleus. The atomic mass is the actual mass of a particular isotope (including electrons). As we will see in Chapter Three, the average mass of an atom is taken from a measurement made on a large number of atoms. The average atomic mass value is listed in the periodic table.

22.   A family is a set of elements in the same vertical column. A family is also called a group. A period is a set of elements in the same horizontal row.

23.   A compound will always contain the same numbers (and types) of atoms. A given amount of hydrogen will react only with a specific amount of oxygen. Any excess oxygen will remain unreacted.

24.   The halogens have a high affinity for electrons, and one important way they react is to form anions of the type X^-. The alkali metals tend to give up electrons easily and in most of their compounds exist as M^+ cations. Note: These two very reactive groups are only one electron away (in the periodic table) from the least reactive family of elements, the noble gases.

Exercises

Development of the Atomic Theory

25.   a. The composition of a substance depends on the numbers of atoms of each element making up the compound (i.e., on the formula of the compound) and not on the composition of the mixture from which it was formed.

   b. Avogadro's hypothesis implies that volume ratios are equal to molecule ratios at constant temperature and pressure. $H_2(g) + Cl_2(g) \rightarrow 2 HCl(g)$. From the balanced equation (2 molecules of HCl are produced per molecule of H_2 or Cl_2 reacted), the volume of HCl produced will be twice the volume of H_2 (or Cl_2) reacted.

26.   From Avogadro's hypothesis, volume ratios are equal to molecule ratios at constant temperature and pressure. Therefore, we can write a balanced equation using the volume data, $Cl_2 + 3 F_2 \rightarrow 2 X$. Two molecules of X contain 6 atoms of F and two atoms of Cl. The formula of X is ClF_3 for a balanced equation.

27. $\dfrac{1.188}{1.188} = 1.000; \quad \dfrac{2.375}{1.188} = 1.999; \quad \dfrac{3.563}{1.188} = 2.999$

   The masses of fluorine are simple ratios of whole numbers to each other, 1:2:3.

28.   Hydrazine: 1.44×10^{-1} g H/g N;   Ammonia: 2.16×10^{-1} g H/g N

   Hydrogen azide: 2.40×10^{-2} g H/g N

Let's try all of the ratios:

$$\frac{0.144}{0.0240} = 6.00; \quad \frac{0.216}{0.0240} = 9.00; \quad \frac{0.216}{0.144} = 1.50 = \frac{3}{2}$$

All the masses of hydrogen in these three compounds can be expressed as simple whole number ratios. The g H/g N in hydrazine, ammonia, and hydrogen azide are in the ratios 6:9:1.

29.   To get the atomic mass of H to be 1.00, we divide the mass of hydrogen that reacts with 1.00 g of oxygen by 0.126, i.e., $\frac{0.126}{0.126} = 1.00$. To get Na, Mg and O on the same scale, we do the same division.

Na: $\frac{2.875}{0.126} = 22.8$; Mg: $\frac{1.500}{0.126} = 11.9$; O: $\frac{1.00}{0.126} = 7.94$

	H	O	Na	Mg
Relative Value	1.00	7.94	22.8	11.9
Accepted Value	1.008	16.00	22.99	24.31

The atomic masses of O and Mg are incorrect; the atomic masses of H and Na are close to the values in the periodic table. Something must be wrong about the assumed formulas of the compounds. It turns out the correct formulas are H_2O, Na_2O, and MgO. The smaller discrepancies result from the error in the atomic mass of H.

30.   If the formula is InO, then one atomic mass of In would combine with one atomic mass of O, or:

$$\frac{A}{16.00} = \frac{4.784 \text{ g In}}{1.000 \text{ g O}}, \quad A = \text{atomic mass of In} = 76.54$$

If the formula is In_2O_3, then two times the atomic mass of In will combine with three times the atomic mass of O, or:

$$\frac{2A}{(3)16.00} = \frac{4.784 \text{ g In}}{1.000 \text{ g O}}, \quad A = \text{atomic mass of In} = 114.8$$

The latter number is the atomic mass of In used in the modern periodic table.

The Nature of the Atom

31.   Density of hydrogen nucleus (contains one proton only):

$$V_{nucleus} = \frac{4}{3} \pi r^3 = \frac{4}{3} (3.14) (5 \times 10^{-14} \text{ cm})^3 = 5 \times 10^{-40} \text{ cm}^3$$

$$d = \frac{1.67 \times 10^{-24} \text{ g}}{5 \times 10^{-40} \text{ cm}^3} = 3 \times 10^{15} \text{ g/cm}^3$$

Density of H-atom (contains one proton and one electron):

$$V_{atom} = \frac{4}{3}(3.14)(1 \times 10^{-8} \text{ cm})^3 = 4 \times 10^{-24} \text{ cm}^3$$

$$d = \frac{1.67 \times 10^{-24} + 9 \times 10^{-28} \text{ g}}{4 \times 10^{-24} \text{ cm}^3} = 0.4 \text{ g/cm}^3$$

32. Since electrons move about the nucleus at an average distance of about 1×10^{-8} cm, then the diameter of an atom is about 2×10^{-8} cm. Let's set up a ratio:

$$\frac{\text{diameter of nucleus}}{\text{diameter of atom}} = \frac{1 \text{ mm}}{\text{diameter of model}} = \frac{1 \times 10^{-13} \text{ cm}}{2 \times 10^{-8} \text{ cm}}, \text{ Solving:}$$

diameter of model $= 2 \times 10^5$ mm $= 200$ m

33. 5.93×10^{-18} C $\times \dfrac{1 \text{ electron charge}}{1.602 \times 10^{-19} \text{ C}} = 37$ negative (electron) charges on the oil drop

34. First, divide all charges by the smallest quantity, 6.40×10^{-13}.

$$\frac{2.56 \times 10^{-12}}{6.40 \times 10^{-13}} = 4.00; \quad \frac{7.68}{0.640} = 12.00; \quad \frac{3.84}{0.640} = 6.00$$

Since all charges are whole number multiples of 6.40×10^{-13} zirkombs, then the charge on one electron could be 6.40×10^{-13} zirkombs. However, 6.40×10^{-13} zirkombs could be the charge of two electrons (or three electrons, etc.). All one can conclude is that the charge of an electron is 6.40×10^{-13} zirkombs or an integer fraction of 6.40×10^{-13} zirkombs.

35. sodium - Na; beryllium - Be; manganese - Mn; chromium - Cr; uranium - U

36. fluorine - F; chlorine - Cl; bromine - Br; sulfur - S; oxygen - O; phosphorus - P

37. Sn - tin; Pt - platinum; Co - cobalt; Ni - nickel; Mg - magnesium; Ba - barium; K - potassium

38. As - arsenic; I - iodine; Xe - xenon; He - helium; C - carbon; Si - silicon

39. The noble gases are He, Ne, Ar, Kr, Xe, and Rn (helium, neon, argon, krypton, xenon, and radon). Radon has only radioactive isotopes. In the periodic table, the whole number enclosed in parentheses is the mass number of the longest-lived isotope of the element.

40. promethium (Pm) and technetium (Tc)

41. a. Eight; Li to Ne   b. Eight; Na to Ar

     c. Eighteen; K to Kr   d. Five; N, P, As, Sb, Bi

42. a. Six; Be, Mg, Ca, Sr, Ba, Ra   b. Five; O, S, Se, Te, Po

c. Four; Ni, Pd, Pt, Uun   d. Six; He, Ne, Ar, Kr, Xe, Rn

43. a. $^{238}_{94}$Pu: 94 protons, 238 - 94 = 144 neutrons   b. $^{65}_{29}$Cu: 29 protons, 65 - 29 = 36 neutrons

c. $^{52}_{24}$Cr: 24 protons, 28 neutrons   d. $^{4}_{2}$He: 2 protons, 2 neutrons

e. $^{60}_{27}$Co: 27 protons, 33 neutrons   f. $^{54}_{24}$Cr: 24 protons, 30 neutrons

44. a. $^{79}_{35}$Br: 35 protons, 79 - 35 = 44 neutrons. Since the charge of the atom is neutral, the number of protons = the number of electrons = 35.

b. $^{81}_{35}$Br: 35 protons, 46 neutrons, 35 electrons

c. $^{239}_{94}$Pu: 94 protons, 145 neutrons, 94 electrons

d. $^{133}_{55}$Cs: 55 protons, 78 neutrons, 55 electrons

e. $^{3}_{1}$H: 1 proton, 2 neutrons, 1 electron

f. $^{56}_{26}$Fe: 26 protons, 30 neutrons, 26 electrons

45. a. Element #5 is boron. $^{12}_{5}$B   b. Z = 7; A = 7 + 8 = 15; $^{15}_{7}$N

c. Z = 17; A = 17 + 18 = 35; $^{35}_{17}$Cl   d. A = 92 + 143 = 235; $^{235}_{92}$U

e. Z = 6; A = 14; $^{14}_{6}$C   f. Z = 15; A = 31; $^{31}_{15}$P

46. a. Element 8 is oxygen. A = mass number = 9 + 8 = 17; $^{17}_{8}$O

b. Chlorine is element 17. $^{37}_{17}$Cl   c. Cobalt is element 27. $^{60}_{27}$Co

d. Z = 26; A = 26 + 31 + 57; $^{57}_{26}$Fe   e. Iodine is element 53. $^{131}_{53}$I

f. Lithium is element 3. $^{7}_{3}$Li

47. Atomic number = 63 (Eu); Charge = +63 - 60 = +3; Mass number = 63 + 88 = 151; Symbol: $^{151}_{63}$Eu^{3+}

Atomic number = 50 (Sn); Mass number = 50 + 68 = 118; Net charge = +50 - 48 = +2; The symbol is $^{118}_{50}$Sn^{2+}.

48.   Atomic number = 16 (S); Charge = +16 - 18 = -2; Mass number = 16 + 18 = 34; Symbol: $^{34}_{16}S^{2-}$

Atomic number = 16 (S); Charge = +16 - 18 = -2; Mass number = 16 + 16 = 32; Symbol: $^{32}_{16}S^{2-}$

49.

Symbol	Number of protons in nucleus	Number of neutrons in nucleus	Number of electrons	Net charge
$^{75}_{33}As^{3+}$	33	42	30	3+
$^{128}_{52}Te^{2-}$	52	76	54	2-
$^{32}_{16}S$	16	16	16	0
$^{204}_{81}Tl^{+}$	81	123	80	1+
$^{195}_{78}Pt$	78	117	78	0

50.

Symbol	Number of protons in nucleus	Number of neutrons in nucleus	Number of electrons	Net charge
$^{238}_{92}U$	92	146	92	0
$^{40}_{20}Ca^{2+}$	20	20	18	2+
$^{51}_{23}V^{3+}$	23	28	20	3+
$^{89}_{39}Y$	39	50	39	0
$^{79}_{35}Br^{-}$	35	44	36	1-
$^{31}_{15}P^{3-}$	15	16	18	3-

51.   Metals: Mg, Ti, Au, Bi, Ge, Eu, Am.   Nonmetals: Si, B, At, Rn, Br.

52.   Si, Ge, B, At. The elements at the boundary between the metals and the nonmetals are: B, Si, Ge, As, Sb, Te, Po, At. Aluminum has mostly properties of metals.

53.   a and d. A group is a vertical column of elements in the periodic table. Elements in the same family (group) have similar chemical properties.

54.   a. transition metals   b. alkaline earth metals
       c. alkali metals   d. noble gases
       e. halogens

55.   Carbon is a nonmetal. Silicon and germanium are metalloids. Tin and lead are metals. Thus, metallic character increases as one goes down a family in the periodic table.

56.   The metallic character decreases from left to right.

57.   Metals lose electrons to form cations, and nonmetals gain electrons to form anions. Group 1A, 2A and 3A metals form stable +1,+2 and +3 charged cations, respectively. Group 5A, 6A and 7A nonmetals form -3, -2 and -1 charged anions, respectively.

       a. Lose 1 e⁻ to form Na⁺.   b. Lose 2 e⁻ to form Sr²⁺.   c. Lose 2 e⁻ to form Ba²⁺.

       d. Gain 1 e⁻ to form I⁻.   e. Lose 3 e⁻ to form Al³⁺.   f. Gain 2 e⁻ to form S²⁻.

58.   a. Lose 2 e⁻ to form Ra²⁺.   b. Lose 3 e⁻ to form In³⁺.   c. Gain 3 e⁻ to form P³⁻.

       d. Gain 2 e⁻ to form Te²⁻.   e. Gain 1 e⁻ to form Br⁻.   f. Lose 1 e⁻ to form Rb⁺.

Nomenclature

59.   a. sodium chloride   b. rubidium oxide

       c. calcium sulfide   d. aluminum iodide

60.   a. mercury(I) oxide   b. iron(III) bromide

       c. cobalt(II) sulfide   d. titanium(IV) chloride

61.   a. chromium(VI) oxide   b. chromium(III) oxide   c. aluminum oxide

       d. sodium hydride   e. calcium bromide

       f. zinc chloride (Zinc only forms +2 ions so no Roman numerals are needed for zinc compounds.)

62.   a. cesium fluoride   b. lithium nitride

       c. silver sulfide (Silver only forms +1 ions so no Roman numerals are needed.)

       d. manganese(IV) oxide   e. titanium(IV) oxide   f. strontium phosphide

63.   a. potassium perchlorate   b. calcium phosphate
       c. aluminum sulfate   d. lead(II) nitrate

64.   a. barium sulfite   b. sodium nitrite
       c. potassium permanganate   d. potassium dichromate

65.   a. nitrogen triiodide   b. sulfur difluoride
       c. phosphorus trichloride   d. dinitrogen tetrafluoride

66.   a. dinitrogen tetroxide   b. iodine trichloride
       c. sulfur dioxide   d. diphosphorus pentasulfide

67.   a. copper(I) iodide   b. copper(II) iodide   c. cobalt(II) iodide
       d. sodium carbonate   e. sodium hydrogen carbonate or sodium bicarbonate
       f. tetrasulfur tetranitride   g. sulfur hexafluoride   h. sodium hypochlorite
       i. barium chromate   j. ammonium nitrate

68.   a. acetic acid   b. ammonium nitrite   c. cobalt(III) sulfide
       d. iodine monochloride   e. lead(II) phosphate   f. potassium iodate
       g. sulfuric acid   h. strontium nitride   i. aluminum sulfite
       j. tin(IV) oxide   k. sodium chromate   l. hypochlorous acid

69.   a. CsBr   b. $BaSO_4$ c. NH_4Cl d. ClO
       e. $SiCl_4$ f. ClF_3 g. BeO   h. MgF_2

70.   a. SF_2 b. SF_6 c. NaH_2PO_4
       d. Li_3N e. $Cr_2(CO_3)_3$ f. SnF_2
       g. $NH_4C_2H_3O_2$ h. NH_4HSO_4 i. $Co(NO_3)_3$
       j. Hg_2Cl_2; Mercury(I) exists as Hg_2^{2+} ions.   k. $KClO_3$ l. NaH

71.   a. Na_2O b. Na_2O_2 c. KCN
       d. $Cu(NO_3)_2$ e. $SeBr_4$ f. PbS
       g. PbS_2 h. CuCl   i. GaAs (Predict Ga^{3+} and As^{3-} ions.)
       j. CdSe (Cadmium only forms +2 charged ions in compounds.)
       k. ZnS (Zinc only forms +2 charged ions in compounds.)
       l. HNO_2 m. P_2O_5

72.   a. $(NH_4)_2HPO_4$ b. Hg_2S c. SiO_2

        d. Na_2SO_3 e. $Al(HSO_4)_3$ f. NCl_3

        g. HBr   h. $HBrO_2$ i. $HBrO_4$

        j. KHS   k. CaI_2 l. $CsClO_4$

Additional Exercises

73.   Yes, 1.0 g H would react with 37.0 g ^{37}Cl and 1.0 g H would react with 35.0 g ^{35}Cl.

        No, the mass ratio of H/Cl would always be 1 g H/37 g Cl for ^{37}Cl and 1 g H/35 g Cl for ^{35}Cl.
        As long as we had pure ^{37}Cl or pure ^{35}Cl, the above ratios will always hold. If we have a mixture
        (such as the natural abundance of chlorine), the ratio will also be constant as long as the composition
        of the mixture of the two isotopes does not change.

74. $^{98}_{43}Tc$: 43 protons and 55 neutrons; $^{99}_{43}Tc$: 43 protons and 56 neutrons

        Tc is in the same family as Mn. We would expect Tc to have properties similar to Mn.

        permanganate: MnO_4^-; pertechnetate: TcO_4^-; ammonium pertechnetate: NH_4TcO_4

75.   a. nitric acid, HNO_3 b. perchloric acid, $HClO_4$ c. acetic acid, $HC_2H_3O_2$

        d. sulfuric acid, H_2SO_4 e. phosphoric acid, H_3PO_4

76.   a. Fe^{2+}: 26 protons (Fe is element 26.); protons - electrons = charge, 26 -2 = 24 electrons; FeO is
            the formula since the oxide ion has a -2 charge.

        b. Fe^{3+}: 26 protons; 23 electrons; Fe_2O_3 c. Ba^{2+}: 56 protons; 54 electrons; BaO

        d. Cs^+: 55 protons; 54 electrons; Cs_2O e. S^{2-}: 16 protons; 18 electrons; Al_2S_3

        f. P^{3-}: 15 protons; 18 electrons; AlP   g. Br^-: 35 protons; 36 electrons; $AlBr_3$

        h. N^{3-}: 7 protons; 10 electrons; AlN

77.   a. $Pb(C_2H_3O_2)_2$: lead(II) acetate   b. $CuSO_4$: copper(II) sulfate

        c. CaO: calcium oxide   d. $MgSO_4$: magnesium sulfate

        e. $Mg(OH)_2$: magnesium hydroxide   f. $CaSO_4$: calcium sulfate

        g. N_2O: dinitrogen monoxide or nitrous oxide

78.   a.   This is element 52, tellurium. Te forms stable -2 charged ions (like other oxygen family members).

       b.   Rubidium. Rb, element 37, forms stable +1 charged ions.

       c.   Argon. Ar is element 18.

       d.   Astatine. At is element 85.

79.   A chemical formula gives the actual number and kind of atoms in a compound. In all cases, 12 hydrogen atoms are present. For example:

$$4 \text{ molecules } H_3PO_4 \times \frac{3 \text{ atoms H}}{\text{molecule } H_3PO_4} = 12 \text{ atoms H}$$

80.   From the XBr_2 formula, the charge on element X is +2. Therefore, the element has 88 protons, which identifies it as radium, Ra. 230 - 88 = 142 neutrons

81.   a. Ca^{2+} and N^{3-}: Ca_3N_2, calcium nitride   b. K^+ and O^{2-}: K_2O, potassium oxide

       c. Rb^+ and F^-: RbF, rubidium fluoride   d. Mg^{2+} and S^{2-}: MgS, magnesium sulfide

       e. Ba^{2+} and I^-: BaI_2, barium iodide   f. Al^{3+} and Se^{2-}: Al_2Se_3, aluminum selenide

       g. Cs^+ and P^{3-}: Cs_3P, cesium phosphide

       h. In^{3+} and Br^-: $InBr_3$, indium(III) bromide. In also forms In^+ ions, but one would predict In^{3+} ions from its position in the periodic table.

82.   These compounds are similar to phosphate (PO_4^{3-}) compounds. Na_3AsO_4 contains Na^+ ions and AsO_4^{3-} ions. The name would be sodium arsenate. H_3AsO_4 is analogous to phosphoric acid, H_3PO_4. H_3AsO_4 would be arsenic acid. $Mg_3(SbO_4)_2$ contains Mg^{2+} ions and SbO_4^{3-} ions, and the name would be magnesium antimonate.

Challenge Problems

83.   Copper(Cu), silver (Ag) and gold(Au) make up the coinage metals.

84.   Avogadro proposed that equal volumes of gases (at constant temperature and pressure) contain equal numbers of molecules. In terms of balanced equations, Avogadro's hypothesis implies that volume ratios will be identical to molecule ratios. Assuming one molecule of octane reacting, then 1 molecule of C_xH_y produces 8 molecules of CO_2 and 9 molecules of H_2O. $C_xH_y + O_2 \rightarrow 8 \ CO_2 + 9 \ H_2O$. Since all the carbon in octane ends up as carbon in CO_2, then octane contains 8 atoms of C. Similarly, all hydrogen in octane ends up as hydrogen in H_2O, so one molecule of octane contains $9 \times 2 = 18$ atoms of H. Octane formula = C_8H_{18} and the ratio of C:H = 8:18 or 4:9.

85.   Compound I: $\dfrac{14.0 \text{ g R}}{3.00 \text{ g Q}} = \dfrac{4.67 \text{ g R}}{1.00 \text{ g Q}}$; Compound II: $\dfrac{7.00 \text{ g R}}{4.50 \text{ g Q}} = \dfrac{1.56 \text{ g R}}{1.00 \text{ g Q}}$

The ratio of the masses of R that combine with 1.00 g Q is: $\dfrac{4.67}{1.56} = 2.99 \approx 3$

As expected from the law of multiple proportions, this ratio is a small whole number.

Since Compound I contains three times the mass of R per gram of Q as compared to Compound II (RQ), then the formula of Compound I should be R_3Q.

86.   The alchemists were incorrect. The solid residue must have come from the flask.

87.   a.   Both compounds have C_2H_6O as the formula. Because they have the same formula, their mass percent composition will be identical. However, these are different compounds with different properties since the atoms are bonded together differently. These compounds are called isomers of each other.

       b.   When wood burns, most of the solid material in wood is converted to gases, which escape. The gases produced are most likely CO_2 and H_2O.

       c.   The atom is not an indivisible particle, but is instead composed of other smaller particles, e.g., electrons, neutrons, protons.

       d.   The two hydride samples contain different isotopes of either hydrogen and/or lithium. Although the compounds are composed of different isotopes, their properties are similar because different isotopes of the same element have similar properties (except, of course, their mass).

CHAPTER THREE

STOICHIOMETRY

Questions

18. The two major isotopes of boron are ^{10}B and ^{11}B. The listed mass of 10.81 is the average mass of a very large number of boron atoms.

19. The molecular formula tells us the actual number of atoms of each element in a molecule (or formula unit) of a compound. The empirical formula tells only the simplest whole number ratio of atoms of each element in a molecule. The molecular formula is a whole number multiple of the empirical formula. If that multiplier is one, the molecular and empirical formulas are the same. For example, both the molecular and empirical formulas of water are H_2O. They are the same. For hydrogen peroxide, the empirical formula is OH; the molecular formula is H_2O_2.

20. Side reactions may occur. For example, in the combustion of CH_4 (methane) to CO_2 and H_2O, some CO is also formed. Also, some reactions only go part way to completion and reach a state of equilibrium where both reactants and products are present (see Ch. 13).

Exercises

Atomic Masses and the Mass Spectrometer

21. A = atomic mass = 0.7899(23.9850 amu) + 0.1000(24.9858 amu) + 0.1101(25.9826 amu)

    A = 18.95 amu + 2.499 amu + 2.861 amu = 24.31 amu

22. A = 0.0140(203.973) + 0.2410(205.9745) + 0.2210(206.9759) + 0.5240(207.9766)

    A = 2.86 + 49.64 + 45.74 + 109.0 = 207.2 amu; From the periodic table, the element is Pb.

23. Let x = % of ^{151}Eu and y = % of ^{153}Eu, then $x + y = 100$ and $y = 100 - x$.

    $$151.96 = \frac{x(150.9196) + (100 - x)(152.9209)}{100}$$

    $15196 = 150.9196\,x + 15292.09 - 152.9209\,x,\ \ -96 = -2.0013\,x$

    $x = 48\%$; 48% ^{151}Eu and $100 - 48 = 52\%$ ^{153}Eu

24.   Let A = mass of ^{185}Re: 186.207 = 0.6260(186.956) + 0.3740(A), 186.207 - 117.0 = 0.3740(A)

A = $\dfrac{69.2}{0.3740}$ = 185 amu (A = 184.95 amu without rounding to proper significant figures.)

25.   There are three peaks in the mass spectrum, each 2 mass units apart. This is consistent with two isotopes, differing in mass by two mass units. The peak at 157.84 corresponds to a Br_2 molecule composed of two atoms of the lighter isotope. This isotope has mass equal to 157.84/2 or 78.92. This corresponds to ^{79}Br. The second isotope is ^{81}Br with mass equal to 161.84/2 = 80.92. The peaks in the mass spectrum correspond to $^{79}Br_2$, $^{79}Br^{81}Br$, and $^{81}Br_2$ in order of increasing mass. The intensities of the highest and lowest mass tell us the two isotopes are present in about equal abundance. The actual abundance is 50.69% ^{79}Br and 49.31% ^{81}Br. The calculation of the abundance from the mass spectrum is beyond the scope of this text.

26.   GaAs can be either ^{69}GaAs or ^{71}GaAs. The mass spectrum for GaAs will have 2 peaks at 144 (= 69 + 75) and 146 (= 71 + 75) with intensities in the ratio of 60:40 or 3:2.

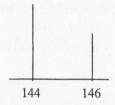

144   146

Ga$_2$As$_2$ can be ^{69}Ga$_2$As$_2$, ^{69}Ga^{71}GaAs$_2$, or ^{71}Ga$_2$As$_2$. The mass spectrum will have 3 peaks at 288, 290, and 292 with intensities in the ratio of 36:48:16 or 9:12:4. We get this ratio from the following probability table:

	^{69}Ga (0.60)	^{71}Ga (0.40)
^{69}Ga (0.60)	0.36	0.24
^{71}Ga (0.40)	0.24	0.16

288   290   292

Moles and Molar Masses

27.   When more than one conversion factor is necessary to determine the answer, we will usually put all the conversion factors into one calculation instead of determining intermediate answers. This method reduces round-off error and is a time saver.

$$500. \text{ atoms Fe} \times \frac{1 \text{ mol Fe}}{6.022 \times 10^{23} \text{ atoms Fe}} \times \frac{55.85 \text{ g Fe}}{\text{mol Fe}} = 4.64 \times 10^{-20} \text{ g Fe}$$

28. $$500.0 \text{ g Fe} \times \frac{1 \text{ mol Fe}}{55.85 \text{ g Fe}} = 8.953 \text{ mol Fe}$$

$$8.953 \text{ mol Fe} \times \frac{6.022 \times 10^{23} \text{ atoms Fe}}{\text{mol Fe}} = 5.391 \times 10^{24} \text{ atoms Fe}$$

29. $$1.00 \text{ carat} \times \frac{0.200 \text{ g C}}{\text{carat}} \times \frac{1 \text{ mol C}}{12.01 \text{ g C}} \times \frac{6.022 \times 10^{23} \text{ atoms C}}{\text{mol C}} = 1.00 \times 10^{22} \text{ atoms C}$$

30. $$5.0 \times 10^{21} \text{ atoms C} \times \frac{1 \text{ mol C}}{6.022 \times 10^{23} \text{ atoms C}} = 8.3 \times 10^{-3} \text{ mol C}$$

$$8.3 \times 10^{-3} \text{ mol C} \times \frac{12.01 \text{ g C}}{\text{mol C}} = 0.10 \text{ g C}$$

31. Al_2O_3: $2(26.98) + 3(16.00) = 101.96$ g/mol

Na_3AlF_6: $3(22.99) + 1(26.98) + 6(19.00) = 209.95$ g/mol

32.   HFC - 134a, CH_2FCF_3: $2(12.01) + 2(1.008) + 4(19.00) = 102.04$ g/mol

HCFC-124, $CHClFCF_3$: $2(12.01) + 1(1.008) + 1(35.45) + 4(19.00) = 136.48$ g/mol

33.   a.   The formula is NH_3. 14.01 g/mol $+ 3(1.008$ g/mol$) = 17.03$ g/mol

b.   The formula is N_2H_4. $2(14.01) + 4(1.008) = 32.05$ g/mol

c. $(NH_4)_2Cr_2O_7$: $2(14.01) + 8(1.008) + 2(52.00) + 7(16.00) = 252.08$ g/mol

34.   a.   The formula is P_4O_6. $4(30.97$ g/mol$) + 6(16.00$ g/mol$) = 219.88$ g/mol

b. $Ca_3(PO_4)_2$: $3(40.08) + 2(30.97) + 8(16.00) = 310.18$ g/mol

c. Na_2HPO_4: $2(22.99) + 1(1.008) + 1(30.97) + 4(16.00) = 141.96$ g/mol

35.   a. $1.00 \text{ g NH}_3 \times \dfrac{1 \text{ mol NH}_3}{17.03 \text{ g NH}_3} = 0.0587 \text{ mol NH}_3$

       b. $1.00 \text{ g N}_2\text{H}_4 \times \dfrac{1 \text{ mol N}_2\text{H}_4}{32.05 \text{ g N}_2\text{H}_4} = 0.0312 \text{ mol N}_2\text{H}_4$

       c. $1.00 \text{ g (NH}_4)_2\text{Cr}_2\text{O}_7 \times \dfrac{1 \text{ mol (NH}_4)_2\text{Cr}_2\text{O}_7}{252.08 \text{ g (NH}_4)_2\text{Cr}_2\text{O}_7} = 3.97 \times 10^{-3} \text{ mol (NH}_4)_2\text{Cr}_2\text{O}_7$

36.   a. $1.00 \text{ g P}_4\text{O}_6 \times \dfrac{1 \text{ mol P}_4\text{O}_6}{219.88 \text{ g}} = 4.55 \times 10^{-3} \text{ mol P}_4\text{O}_6$

       b. $1.00 \text{ g Ca}_3(\text{PO}_4)_2 \times \dfrac{1 \text{ mol Ca}_3(\text{PO}_4)_2}{310.18 \text{ g}} = 3.22 \times 10^{-3} \text{ mol Ca}_3(\text{PO}_4)_2$

       c. $1.00 \text{ g Na}_2\text{HPO}_4 \times \dfrac{1 \text{ mol Na}_2\text{HPO}_4}{141.96 \text{ g}} = 7.04 \times 10^{-3} \text{ mol Na}_2\text{HPO}_4$

37.   a. $5.00 \text{ mol NH}_3 \times \dfrac{17.03 \text{ g NH}_4}{\text{mol NH}_3} = 85.2 \text{ g NH}_3$

       b. $5.00 \text{ mol N}_2\text{H}_4 \times \dfrac{32.05 \text{ g N}_2\text{H}_4}{\text{mol N}_2\text{H}_4} = 160. \text{ g N}_2\text{H}_4$

       c. $5.00 \text{ mol (NH}_4)_2\text{Cr}_2\text{O}_7 \times \dfrac{252.08 \text{ g (NH}_4)_2\text{Cr}_2\text{O}_7}{\text{mol (NH}_4)_2\text{Cr}_2\text{O}_7} = 1260 \text{ g (NH}_4)_2\text{Cr}_2\text{O}_7$

38.   a. $5.00 \text{ mol P}_4\text{O}_6 \times \dfrac{219.88 \text{ g}}{\text{mol P}_4\text{O}_6} = 1.10 \times 10^3 \text{ g P}_4\text{O}_6$

       b. $5.00 \text{ mol Ca}_3(\text{PO}_4)_2 \times \dfrac{310.18 \text{ g}}{\text{mol Ca}_3(\text{PO}_4)_2} = 1.55 \times 10^3 \text{ g Ca}_3(\text{PO}_4)_2$

       c. $5.00 \text{ mol Na}_2\text{HPO}_4 \times \dfrac{141.96 \text{ g}}{\text{mol Na}_2\text{HPO}_4} = 7.10 \times 10^2 \text{ g Na}_2\text{HPO}_4$

39.   Chemical formulas give atom ratios as well as mol ratios.

       a. $5.00 \text{ mol NH}_3 \times \dfrac{1 \text{ mol N}}{\text{mol NH}_3} \times \dfrac{14.01 \text{ g N}}{\text{mol N}} = 70.1 \text{ g N}$

       b. $5.00 \text{ mol N}_2\text{H}_4 \times \dfrac{2 \text{ mol N}}{\text{mol N}_2\text{H}_4} \times \dfrac{14.01 \text{ g N}}{\text{mol N}} = 140. \text{ g N}$

       c. $5.00 \text{ mol (NH}_4)_2\text{Cr}_2\text{O}_7 \times \dfrac{2 \text{ mol N}}{\text{mol (NH}_4)_2\text{Cr}_2\text{O}_7} \times \dfrac{14.01 \text{ g N}}{\text{mol N}} = 140. \text{ g N}$

40.   a. $5.00 \text{ mol } P_4O_6 \times \dfrac{4 \text{ mol P}}{\text{mol } P_4O_6} \times \dfrac{30.97 \text{ g P}}{\text{mol P}} = 619 \text{ g P}$

      b. $5.00 \text{ mol } Ca_3(PO_4)_2 \times \dfrac{2 \text{ mol P}}{\text{mol } Ca_3(PO_4)_2} \times \dfrac{30.97 \text{ g P}}{\text{mol P}} = 310. \text{ g P}$

      c. $5.00 \text{ mol } Na_2HPO_4 \times \dfrac{1 \text{ mol P}}{\text{mol } Na_2HPO_4} \times \dfrac{30.97 \text{ g P}}{\text{mol P}} = 155 \text{ g P}$

41.   a. $1.00 \text{ g } NH_3 \times \dfrac{1 \text{ mol } NH_3}{17.03 \text{ g } NH_3} \times \dfrac{6.022 \times 10^{23} \text{ molecules } NH_3}{\text{mol } NH_3} = 3.54 \times 10^{22} \text{ molecules } NH_3$

      b. $1.00 \text{ g } N_2H_4 \times \dfrac{1 \text{ mol } N_2H_4}{32.05 \text{ g } N_2H_4} \times \dfrac{6.022 \times 10^{23} \text{ molecules } N_2H_4}{\text{mol } N_2H_4} = 1.88 \times 10^{22} \text{ molecules } N_2H_4$

      c. $1.00 \text{ g } (NH_4)_2Cr_2O_7 \times \dfrac{1 \text{ mol } (NH_4)_2Cr_2O_7}{252.08 \text{ g } (NH_4)_2Cr_2O_7} \times \dfrac{6.022 \times 10^{23} \text{ formula units } (NH_4)_2Cr_2O_7}{\text{mol } (NH_4)_2Cr_2O_7}$

$$= 2.39 \times 10^{21} \text{ formula units } (NH_4)_2Cr_2O_7$$

42.   a. $1.00 \text{ g } P_4O_6 \times \dfrac{1 \text{ mol } P_4O_6}{219.88 \text{ g}} \times \dfrac{6.022 \times 10^{23} \text{ molecules}}{\text{mol } P_4O_6} = 2.74 \times 10^{21} \text{ molecules } P_4O_6$

      b. $1.00 \text{ g } Ca_3(PO_4)_2 \times \dfrac{1 \text{ mol } Ca_3(PO_4)_2}{310.18 \text{ g}} \times \dfrac{6.022 \times 10^{23} \text{ formula units}}{\text{mol } Ca_3(PO_4)_2}$

$$= 1.94 \times 10^{21} \text{ formula units } Ca_3(PO_4)_2$$

      c. $1.00 \text{ g } Na_2HPO_4 \times \dfrac{\text{mol } Na_2HPO_4}{141.96 \text{ g}} \times \dfrac{6.022 \times 10^{23} \text{ formula units}}{\text{mol } Na_2HPO_4}$

$$= 4.24 \times 10^{21} \text{ formula units } Na_2HPO_4$$

43.   Using answers from Exercise 41:

      a. $3.54 \times 10^{22} \text{ molecules } NH_3 \times \dfrac{1 \text{ atom N}}{\text{molecule } NH_3} = 3.54 \times 10^{22} \text{ atoms N}$

      b. $1.88 \times 10^{22} \text{ molecules } N_2H_4 \times \dfrac{2 \text{ atoms N}}{\text{molecule } N_2H_4} = 3.76 \times 10^{22} \text{ atoms N}$

      c. $2.39 \times 10^{21} \text{ formula units } (NH_4)_2Cr_2O_7 \times \dfrac{2 \text{ atoms N}}{\text{formula unit } (NH_4)_2Cr_2O_7} = 4.78 \times 10^{21} \text{ atoms N}$

44.   Using answers from Exercise 42:

   a. 2.74×10^{21} molecules $P_4O_6 \times \dfrac{4 \text{ atoms P}}{\text{molecule } P_4O_6} = 1.10 \times 10^{22}$ atoms P

   b. 1.94×10^{21} formula units $Ca_3(PO_4)_2 \times \dfrac{2 \text{ atoms P}}{\text{formula unit } Ca_3(PO_4)_2} = 3.88 \times 10^{21}$ atoms P

   c. 4.24×10^{21} formula units $Na_2HPO_4 \times \dfrac{1 \text{ atom P}}{\text{formula unit } Na_2HPO_4} = 4.24 \times 10^{21}$ atoms P

45.   Molar mass of $C_6H_8O_6 = 6(12.01) + 8(1.008) + 6(16.00) = 176.12$ g/mol

   $500.0 \text{ mg} \times \dfrac{1 \text{ g}}{1000 \text{ mg}} \times \dfrac{1 \text{ mol}}{176.12 \text{ g}} = 2.839 \times 10^{-3}$ mol

   $2.839 \times 10^{-3} \text{ mol} \times \dfrac{6.022 \times 10^{23} \text{ molecules}}{\text{mol}} = 1.710 \times 10^{21}$ molecules

46.   a. $9(12.01) + 8(1.008) + 4(16.00) = 180.15$ g/mol

   b. $500. \text{ mg} \times \dfrac{1 \text{ g}}{1000 \text{ mg}} \times \dfrac{1 \text{ mol}}{180.15 \text{ g}} = 2.78 \times 10^{-3}$ mol

   $2.78 \times 10^{-3} \text{ mol} \times \dfrac{6.022 \times 10^{23} \text{ molecules}}{\text{mol}} = 1.67 \times 10^{21}$ molecules

47.   a. 2.49×10^{20} molecules $CO \times \dfrac{1 \text{ mol CO}}{6.022 \times 10^{23} \text{ molecules CO}} = 4.13 \times 10^{-4}$ mol CO

   b. $15.0 \text{ g } CuSO_4 \times \dfrac{1 \text{ mol } CuSO_4}{159.62 \text{ g } CuSO_4} = 9.40 \times 10^{-2}$ mol $CuSO_4$

   c. $100 \text{ molecules } H_2SO_4 \times \dfrac{1 \text{ mol } H_2SO_4}{6.022 \times 10^{23} \text{ molecules } H_2SO_4} = 1.661 \times 10^{-22}$ mol H_2SO_4

   d. $6.210 \text{ mg } K_2O \times \dfrac{1 \text{ g}}{1000 \text{ mg}} \times \dfrac{1 \text{ mol } K_2O}{94.20 \text{ g } K_2O} = 6.592 \times 10^{-5}$ mol K_2O

48.   a. $150.0 \text{ g } Fe_2O_3 \times \dfrac{1 \text{ mol}}{159.70 \text{ g}} = 0.9393$ mol Fe_2O_3

   b. $10.0 \text{ mg } NO_2 \times \dfrac{1 \text{ g}}{1000 \text{ mg}} \times \dfrac{1 \text{ mol}}{46.01 \text{ g}} = 2.17 \times 10^{-4}$ mol NO_2

   c. $1.5 \times 10^{16} \text{ molecules } BF_3 \times \dfrac{1 \text{ mol}}{6.02 \times 10^{23} \text{ molecules}} = 2.5 \times 10^{-8}$ mol BF_3

49. a. $1.27 \text{ mmol CO}_2 \times \dfrac{1 \text{ mol}}{1000 \text{ mmol}} \times \dfrac{44.01 \text{ g CO}_2}{\text{mol CO}_2} = 5.59 \times 10^{-2} \text{ g CO}_2$

b. $2.00 \times 10^{22} \text{ molecules NCl}_3 \times \dfrac{1 \text{ mol NCl}_3}{6.022 \times 10^{23} \text{ molecules}} \times \dfrac{120.36 \text{ g}}{\text{mol NCl}_3} = 4.00 \text{ g NCl}_3$

c. $0.00451 \text{ mol (NH}_4)_2\text{CO}_3 \times \dfrac{96.09 \text{ g (NH}_4)_2\text{CO}_3}{\text{mol (NH}_4)_2\text{CO}_3} = 0.433 \text{ g (NH}_4)_2\text{CO}_3$

d. $1 \text{ molecule N}_2 \times \dfrac{1 \text{ mol N}_2}{6.022 \times 10^{23} \text{ molecules N}_2} \times \dfrac{28.02 \text{ g N}_2}{\text{mol N}_2} = 4.653 \times 10^{-23} \text{ g N}_2$

e. $62.7 \text{ mol CuSO}_4 \times \dfrac{159.62 \text{ g CuSO}_4}{\text{mol CuSO}_4} = 1.00 \times 10^4 \text{ g CuSO}_4$

50. a. A chemical formula gives atom ratios as well as mole ratios. We will use both ideas to show how these conversion factors can be used.

Molar mass of $C_2H_5O_2N = 2(12.01) + 5(1.008) + 2(16.00) + 14.01 = 75.07 \text{ g/mol}$

$5.00 \text{ g C}_2\text{H}_5\text{O}_2\text{N} \times \dfrac{1 \text{ mol C}_2\text{H}_5\text{O}_2\text{N}}{75.07 \text{ g C}_2\text{H}_5\text{O}_2\text{N}} \times \dfrac{6.022 \times 10^{23} \text{ molecules C}_2\text{H}_5\text{O}_2\text{N}}{\text{mol C}_2\text{H}_5\text{O}_2\text{N}}$

$\times \dfrac{1 \text{ atom N}}{\text{molecule C}_2\text{H}_5\text{O}_2\text{N}} = 4.01 \times 10^{22} \text{ atoms N}$

b. Molar mass of $Mg_3N_2 = 3(24.31) + 2(14.01) = 100.95 \text{ g/mol}$

$5.00 \text{ g Mg}_3\text{N}_2 \times \dfrac{1 \text{ mol Mg}_3\text{N}_2}{100.95 \text{ g Mg}_3\text{N}_2} \times \dfrac{6.022 \times 10^{23} \text{ formula units Mg}_3\text{N}_2}{\text{mol Mg}_3\text{N}_2} \times \dfrac{2 \text{ atoms N}}{\text{mol Mg}_3\text{N}_2}$

$= 5.97 \times 10^{22} \text{ atoms N}$

c. Molar mass of $Ca(NO_3)_2 = 40.08 + 2(14.01) + 6(16.00) = 164.10 \text{ g/mol}$

$5.00 \text{ g Ca(NO}_3)_2 \times \dfrac{1 \text{ mol Ca(NO}_3)_2}{164.10 \text{ g Ca(NO}_3)_2} \times \dfrac{2 \text{ mol N}}{\text{mol Ca(NO}_3)_2} \times \dfrac{6.022 \times 10^{23} \text{ atoms N}}{\text{mol N}}$

$= 3.67 \times 10^{22} \text{ atoms N}$

d. Molar mass of $N_2O_4 = 2(14.01) + 4(16.00) = 92.02 \text{ g/mol}$

$5.00 \text{ g N}_2\text{O}_4 \times \dfrac{1 \text{ mol N}_2\text{O}_4}{92.02 \text{ g N}_2\text{O}_4} \times \dfrac{2 \text{ mol N}}{\text{mol N}_2\text{O}_4} \times \dfrac{6.022 \times 10^{23} \text{ atoms N}}{\text{mol N}} = 6.54 \times 10^{22} \text{ atoms N}$

51.   a. $14 \text{ mol C} \left(\dfrac{12.01 \text{ g}}{\text{mol C}} \right) + 18 \text{ mol H} \left(\dfrac{1.008 \text{ g}}{\text{mol H}} \right) + 2 \text{ mol N} \left(\dfrac{14.01 \text{ g}}{\text{mol N}} \right) + 5 \text{ mol O} \left(\dfrac{16.00 \text{ g}}{\text{mol O}} \right)$

$$= 294.30 \text{ g/mol}$$

   b. $10.0 \text{ g aspartame} \times \dfrac{1 \text{ mol}}{294.30 \text{ g}} = 3.40 \times 10^{-2} \text{ mol}$

   c. $1.56 \text{ mol} \times \dfrac{294.30 \text{ g}}{\text{mol}} = 459 \text{ g}$

   d. $5.0 \text{ mg} \times \dfrac{1 \text{ g}}{1000 \text{ mg}} \times \dfrac{1 \text{ mol}}{294.30 \text{ g}} \times \dfrac{6.02 \times 10^{23} \text{ molecules}}{\text{mol}} = 1.0 \times 10^{19} \text{ molecules}$

   e.   The chemical formula tells us that 1 molecule of aspartame contains two atoms of N. The chemical formula also says that 1 mol of aspartame contains two mol of N.

$$1.2 \text{ g aspartame} \times \dfrac{1 \text{ mol aspartame}}{294.30 \text{ g aspartame}} \times \dfrac{2 \text{ mol N}}{\text{mol aspartame}} \times \dfrac{6.02 \times 10^{23} \text{ atoms N}}{\text{mol N}}$$

$$= 4.9 \times 10^{21} \text{ atoms of nitrogen}$$

   f. $1.0 \times 10^{9} \text{ molecules} \times \dfrac{1 \text{ mol}}{6.02 \times 10^{23} \text{ molecules}} \times \dfrac{294.30 \text{ g}}{\text{mol}} = 4.9 \times 10^{-13} \text{ g or } 490 \text{ fg}$

   g. $1 \text{ molecule aspartame} \times \dfrac{1 \text{ mol}}{6.022 \times 10^{23} \text{ molecules}} \times \dfrac{294.30 \text{ g}}{\text{mol}} = 4.887 \times 10^{-22} \text{ g}$

52.   a. $2(12.01) + 3(1.008) + 3(35.45) + 2(16.00) = 165.39 \text{ g/mol}$

   b. $500.0 \text{ g} \times \dfrac{1 \text{ mol}}{165.39 \text{ g}} = 3.023 \text{ mol}$ c. $2.0 \times 10^{-2} \text{ mol} \times \dfrac{165.39 \text{ g}}{\text{mol}} = 3.3 \text{ g}$

   d. $5.0 \text{ g C}_2\text{H}_3\text{Cl}_3\text{O}_2 \times \dfrac{1 \text{ mol}}{165.39 \text{ g}} \times \dfrac{6.02 \times 10^{23} \text{ molecules}}{\text{mol}} \times \dfrac{3 \text{ atoms Cl}}{\text{molecule}}$

$$= 5.5 \times 10^{22} \text{ atoms of chlorine}$$

   e. $1.0 \text{ g Cl} \times \dfrac{1 \text{ mol Cl}}{35.45 \text{ g}} \times \dfrac{1 \text{ mol C}_2\text{H}_3\text{Cl}_3\text{O}_2}{3 \text{ mol Cl}} \times \dfrac{165.39 \text{ g C}_2\text{H}_3\text{Cl}_3\text{O}_2}{\text{mol C}_2\text{H}_3\text{Cl}_3\text{O}_2} = 1.6 \text{ g chloral hydrate}$

   f. $500 \text{ molecules} \times \dfrac{1 \text{ mol}}{6.022 \times 10^{23} \text{ molecules}} \times \dfrac{165.39 \text{ g}}{\text{mol}} = 1.373 \times 10^{-19} \text{ g}$

Percent Composition

53.   In 1 mole of $YBa_2Cu_3O_7$, there are 1 mole of Y, 2 moles of Ba, 3 moles of Cu and 7 moles of O.

Molar mass $= 1 \text{ mol Y} \left(\dfrac{88.91 \text{ g Y}}{\text{mol Y}} \right) + 2 \text{ mol Ba} \left(\dfrac{137.3 \text{ g Ba}}{\text{mol Ba}} \right)$

$$+ 3 \text{ mol Cu} \left(\dfrac{63.55 \text{ g Cu}}{\text{mol Cu}} \right) + 7 \text{ mol O} \left(\dfrac{16.00 \text{ g O}}{\text{mol O}} \right)$$

Molar mass $= 88.91 + 274.6 + 190.65 + 112.00 = 666.2 \text{ g/mol}$

$\%Y = \dfrac{88.91 \text{ g}}{666.2 \text{ g}} \times 100 = 13.35\% \text{ Y}; \quad \%Ba = \dfrac{274.6 \text{ g}}{666.2 \text{ g}} \times 100 = 41.22\% \text{ Ba}$

$\%Cu = \dfrac{190.65 \text{ g}}{666.2 \text{ g}} \times 100 = 28.62\% \text{ Cu}; \quad \%O = \dfrac{112.00 \text{ g}}{666.2 \text{ g}} \times 100 = 16.81\% \text{ O}$

54.   a. $C_3H_4O_2$: Molar mass $= 3(12.01) + 4(1.008) + 2(16.00) = 36.03 + 4.032 + 32.00 = 72.06 \text{ g/mol}$

$\%C = \dfrac{36.03 \text{ g C}}{72.06 \text{ g compound}} \times 100 = 50.00\% \text{ C}; \quad \%H = \dfrac{4.032 \text{ g H}}{72.06 \text{ g compound}} \times 100 = 5.595\% \text{ H}$

$\%O = 100.00 - (50.00 + 5.595) = 44.41\% \text{ O} \text{ or } \%O = \dfrac{32.00 \text{ g}}{72.06 \text{ g}} \times 100 = 44.41\% \text{ O}$

b. $C_4H_6O_2$: Molar mass $= 4(12.01) + 6(1.008) + 2(16.00) = 48.04 + 6.048 + 32.00 = 86.09 \text{ g/mol}$

$\%C = \dfrac{48.04 \text{ g}}{86.09 \text{ g}} \times 100 = 55.80\% \text{ C}; \quad \%H = \dfrac{6.048 \text{ g}}{86.09 \text{ g}} \times 100 = 7.025\% \text{ H}$

$\%O = 100.00 - (55.80 + 7.025) = 37.18\% \text{ O}$

c. C_3H_3N: Molar mass $= 3(12.01) + 3(1.008) + 1(14.01) = 36.03 + 3.024 + 14.01 = 53.06 \text{ g/mol}$

$\%C = \dfrac{36.03 \text{ g}}{53.06 \text{ g}} \times 100 = 67.90\% \text{ C}; \quad \%H = \dfrac{3.024 \text{ g}}{53.06 \text{ g}} \times 100 = 5.699\% \text{ H}$

$\%N = \dfrac{14.01 \text{ g}}{53.06 \text{ g}} \times 100 = 26.40\% \text{ N} \text{ or } \%N = 100.00 - (67.90 + 5.699) = 26.40\% \text{ N}$

55.   a.   NO: $\%N = \dfrac{14.01 \text{ g N}}{30.01 \text{ g NO}} \times 100 = 46.68\% \text{ N}$

b. NO_2: $\%N = \dfrac{14.01 \text{ g N}}{46.01 \text{ g NO}_2} \times 100 = 30.45\% \text{ N}$

c. N_2O_4: $\%N = \dfrac{28.02 \text{ g N}}{92.02 \text{ g N}_2O_4} \times 100 = 30.45\% \text{ N}$

d. N_2O: $\%N = \dfrac{28.02 \text{ g N}}{44.02 \text{ g N}_2\text{O}} \times 100 = 63.65\% \text{ N}$

The order from lowest to highest mass percentage of nitrogen is: $NO_2 = N_2O_4 < NO < N_2O$.

56. $C_8H_{10}N_4O_2$: molar mass = $8(12.01) + 10(1.008) + 4(14.01) + 2(16.00) = 194.20$ g/mol

$\%C = \dfrac{8(12.01) \text{ g C}}{194.20 \text{ g C}_8\text{H}_{10}\text{N}_4\text{O}_2} \times 100 = \dfrac{96.08}{194.20} \times 100 = 49.47\% \text{ C}$

$C_{12}H_{22}O_{11}$: molar mass = $12(12.01) + 22(1.008) + 11(16.00) = 342.30$ g/mol

$\%C = \dfrac{12(12.01) \text{ g C}}{342.30 \text{ g C}_{12}\text{H}_{22}\text{O}_{11}} \times 100 = 42.10\% \text{ C}$

C_2H_5OH: molar mass = $2(12.01) + 6(1.008) + 1(16.00) = 46.07$ g/mol

$\%C = \dfrac{2(12.01) \text{ g C}}{46.07 \text{ g C}_2\text{H}_5\text{OH}} \times 100 = 52.14\% \text{ C}$

The order from lowest to highest mass percentage of carbon is: sucrose ($C_{12}H_{22}O_{11}$) < caffeine ($C_8H_{10}N_4O_2$) < ethanol (C_2H_5OH)

57.   There are many valid methods to solve this problem. We will assume 100.00 g of compound; then determine from the information in the problem how many mol of compound equals 100.00 g of compound. From this information, we can determine the mass of one mol of compound (the molar mass) by setting up a ratio. Assuming 100.00 g cyanocobalamin:

mol cyanocobalamin $= 4.34 \text{ g Co} \times \dfrac{1 \text{ mol Co}}{58.93 \text{ g Co}} \times \dfrac{1 \text{ mol cyanocobalamin}}{\text{mol Co}}$

$= 7.36 \times 10^{-2} \text{ mol cyanocobalamin}$

$\dfrac{x \text{ g cyanocobalamin}}{1 \text{ mol cyanocobalamin}} = \dfrac{100.00 \text{ g}}{7.36 \times 10^{-2} \text{ mol}}$, $x = $ molar mass $= 1360$ g/mol

58.   There are 0.390 g Cu for every 100.00 g of fungal laccase. Assuming 100.00 g fungal laccase:

mol fungal laccase $= 0.390 \text{ g Cu} \times \dfrac{1 \text{ mol Cu}}{63.55 \text{ g Cu}} \times \dfrac{1 \text{ mol fungal laccase}}{4 \text{ mol Cu}} = 1.53 \times 10^{-3} \text{ mol}$

$\dfrac{x \text{ g fungal laccase}}{1 \text{ mol fungal laccase}} = \dfrac{100.00 \text{ g}}{1.53 \times 10^{-3} \text{ mol}}$, $x = $ molar mass $= 6.54 \times 10^4$ g/mol

Empirical and Molecular Formulas

59.   a.   Molar mass of CH_2O = 1 mol C $\left(\dfrac{12.01 \text{ g}}{\text{mol C}} \right)$ + 2 mol H $\left(\dfrac{1.008 \text{ g H}}{\text{mol H}} \right)$

$$+ 1 \text{ mol O} \left(\dfrac{16.00 \text{ g}}{\text{mol O}} \right) = 30.03 \text{ g/mol}$$

$\%C = \dfrac{12.01 \text{ g C}}{30.03 \text{ g } CH_2O} \times 100 = 39.99\% \text{ C};\ \ \%H = \dfrac{2.016 \text{ g H}}{30.03 \text{ g } CH_2O} \times 100 = 6.713\% \text{ H}$

$\%O = \dfrac{16.00 \text{ g O}}{30.03 \text{ g } CH_2O} \times 100 = 53.28\% \text{ O}\quad$ or $\%O = 100.00 - (39.99 + 6.713) = 53.30\%$

   b.   Molar Mass of $C_6H_{12}O_6$ = 6(12.01) + 12(1.008) + 6(16.00) = 180.16 g/mol

$\%C = \dfrac{72.06 \text{ g C}}{180.16 \text{ g } C_6H_{12}O_6} \times 100 = 40.00\%;\ \ \%H = \dfrac{12(1.008) \text{ g}}{180.16 \text{ g}} \times 100 = 6.714\%$

$\%O = 100.00 - (40.00 + 6.714) = 53.29\%$

   c.   Molar mass of $HC_2H_3O_2$ = 2(12.01) + 4(1.008) + 2(16.00) = 60.05 g/mol

$\%C = \dfrac{24.02 \text{ g}}{60.05 \text{ g}} \times 100 = 40.00\%;\quad \%H = \dfrac{4.032 \text{ g}}{60.05 \text{ g}} \times 100 = 6.714\%$

$\%O = 100.00 - (40.00 + 6.714) = 53.29\%$

60.   All three compounds have the same empirical formula, CH_2O, and different molecular formulas. The composition of all three in mass percent is also the same (within rounding differences). Therefore, elemental analysis will give us only the empirical formula.

61.   a.   The molecular formula is N_2O_4. The smallest whole number ratio of the atoms (the empirical formula) is NO_2.

   b.   Molecular formula: C_3H_6; empirical formula = CH_2

   c.   Molecular formula: P_4O_{10}; empirical formula = P_2O_5

   d.   Molecular formula: $C_6H_{12}O_6$; empirical formula = CH_2O

62.   a.   SNH: Empirical formula mass = 32.07 + 14.01 + 1.008 = 47.09 g

$\dfrac{188.35}{47.09} = 4.000$; So the molecular formula is $(SNH)_4$ or $S_4N_4H_4$.

b. $NPCl_2$: Empirical formula mass $= 14.01 + 30.97 + 2(35.45) = 115.88$ g/mol

$$\frac{347.64}{115.88} = 3.0000; \text{ Molecular formula is } (NPCl_2)_3 \text{ or } N_3P_3Cl_6.$$

c. CoC_4O_4: $58.93 + 4(12.01) + 4(16.00) = 170.97$ g/mol

$$\frac{341.94}{170.97} = 2.0000; \text{ Molecular formula: } Co_2C_8O_8$$

d.   SN: $32.07 + 14.01 = 46.08$ g/mol; $\dfrac{184.32}{46.08} = 4.000$; Molecular formula: S_4N_4

63.   Out of 100.0 g of the pigment, there are:

$$59.9 \text{ g Ti} \times \frac{1 \text{ mol Ti}}{47.88 \text{ g Ti}} = 1.25 \text{ mol Ti}; \quad 40.1 \text{ g O} \times \frac{1 \text{ mol O}}{16.00 \text{ g O}} = 2.51 \text{ mol O}$$

Empirical formula $= TiO_2$ since mol O to mol Ti are in a 2:1 mol ratio (2.51/1.25 = 2.01).

64.   Out of 100.00 g of adrenaline, there are:

$$56.79 \text{ g C} \times \frac{1 \text{ mol C}}{12.01 \text{ g C}} = 4.729 \text{ mol C}; \quad 6.56 \text{ g H} \times \frac{1 \text{ mol H}}{1.008 \text{ g H}} = 6.51 \text{ mol H}$$

$$28.37 \text{ g O} \times \frac{1 \text{ mol O}}{16.00 \text{ g O}} = 1.773 \text{ mol O}; \quad 8.28 \text{ g N} \times \frac{1 \text{ mol N}}{14.01 \text{ g N}} = 0.591 \text{ mol N}$$

Dividing each mol value by the smallest number:

$$\frac{4.729}{0.591} = 8.00; \quad \frac{6.51}{0.591} = 11.0; \quad \frac{1.773}{0.591} = 3.00; \quad \frac{0.591}{0.591} = 1.00$$

This gives adrenaline an empirical formula of $C_8H_{11}O_3N$.

65.   Compound I: mass O $= 0.6498$ g Hg_xO_y - 0.6018 g Hg $= 0.0480$ g O

$$0.6018 \text{ g Hg} \times \frac{1 \text{ mol Hg}}{200.6 \text{ g Hg}} = 3.000 \times 10^{-3} \text{ mol Hg}$$

$$0.0480 \text{ g O} \times \frac{1 \text{ mol O}}{16.00 \text{ g O}} = 3.00 \times 10^{-3} \text{ mol O}$$

The mol ratio between Hg and O is 1:1, so the empirical formula of compound I is HgO.

Compound II: mass Hg $= 0.4172$ g Hg_xO_y - 0.016 g O $= 0.401$ g Hg

$$0.401 \text{ g Hg} \times \frac{1 \text{ mol Hg}}{200.6 \text{ g Hg}} = 2.00 \times 10^{-3} \text{ mol Hg}; \quad 0.016 \text{ g O} \times \frac{1 \text{ mol O}}{16.00 \text{ g O}} = 1.0 \times 10^{-3} \text{ mol O}$$

The mol ratio between Hg and O is 2:1, so the empirical formula is Hg_2O.

66. $1.121 \text{ g N} \times \dfrac{1 \text{ mol N}}{14.01 \text{ g N}} = 8.001 \times 10^{-2} \text{ mol N}; \quad 0.161 \text{ g H} \times \dfrac{1 \text{ mol H}}{1.008 \text{ g H}} = 1.60 \times 10^{-1} \text{ mol H}$

$0.480 \text{ g C} \times \dfrac{1 \text{ mol C}}{12.01 \text{ g C}} = 4.00 \times 10^{-2} \text{ mol C}; \quad 0.640 \text{ g O} \times \dfrac{1 \text{ mol O}}{16.00 \text{ g O}} = 4.00 \times 10^{-2} \text{ mol O}$

Dividing all mol values by the smallest number:

$\dfrac{8.001 \times 10^{-2}}{4.00 \times 10^{-2}} = 2.00; \quad \dfrac{1.60 \times 10^{-1}}{4.00 \times 10^{-2}} = 4.00; \quad \dfrac{4.00 \times 10^{-2}}{4.00 \times 10^{-2}} = 1.00$

Empirical formula = N_2H_4CO

67.   Out of 100.0 g compound: $30.4 \text{ g N} \times \dfrac{1 \text{ mol N}}{14.01 \text{ g N}} = 2.17 \text{ mol N}$

%O = 100.0 - 30.4 = 69.6% O; $69.6 \text{ g O} \times \dfrac{1 \text{ mol O}}{16.00 \text{ g O}} = 4.35 \text{ mol O}$

$\dfrac{2.17}{2.17} = 1.00; \quad \dfrac{4.35}{2.17} = 2.00;$ Empirical formula is NO_2.

The empirical formula mass of $NO_2 \approx 14 + 2(16) = 46$ g/mol.

$\dfrac{92 \text{ g}}{46 \text{ g}} = 2.0;$ Therefore, the molecular formula is N_2O_4.

68.   Out of 100.0 g, there are:

$69.6 \text{ g S} \times \dfrac{1 \text{ mol S}}{32.07 \text{ g S}} = 2.17 \text{ mol S}; \quad 30.4 \text{ g N} \times \dfrac{1 \text{ mol N}}{14.01 \text{ g N}} = 2.17 \text{ mol N}$

Empirical formula is SN since mol values are in a 1:1 mol ratio.

The empirical formula mass of SN is ~ 46 g. Since $\dfrac{184}{46} = 4.0$, the molecular formula is S_4N_4.

69.   Assuming 100.00 g of compound (mass hydrogen = 100.00 g - 49.31 g C - 43.79 g O = 6.90 g H):

$49.31 \text{ g C} \times \dfrac{1 \text{ mol C}}{12.01 \text{ g C}} = 4.106 \text{ mol C}; \quad 6.90 \text{ g H} \times \dfrac{1 \text{ mol H}}{1.008 \text{ g H}} = 6.85 \text{ mol H}$

$43.79 \text{ g O} \times \dfrac{1 \text{ mol O}}{16.00 \text{ g O}} = 2.737 \text{ mol O}$

Dividing all mole values by 2.737 gives:

$\dfrac{4.106}{2.737} = 1.500; \quad \dfrac{6.85}{2.737} = 2.50; \quad \dfrac{2.737}{2.737} = 1.000$

Since a whole number ratio is required, the empirical formula is $C_3H_5O_2$.

The empirical formula mass is: $3(12.01) + 5(1.008) + 2(16.00) = 73.07$ g/mol

$$\frac{\text{molar mass}}{\text{empirical formula mass}} = \frac{146.1}{73.07} = 1.999; \quad \text{molecular formula} = (C_3H_5O_2)_2 = C_6H_{10}O_4$$

70.   Assuming 100.00 g of compound (mass oxygen = 100.00 g - 41.39 g C - 3.47 g H = 55.14 g O):

$$41.39 \text{ g C} \times \frac{1 \text{ mol C}}{12.01 \text{ g C}} = 3.446 \text{ mol C}; \quad 3.47 \text{ g H} \times \frac{1 \text{ mol H}}{1.008 \text{ g H}} = 3.44 \text{ mol H}$$

$$55.14 \text{ g O} \times \frac{1 \text{ mol O}}{16.00 \text{ g O}} = 3.446 \text{ mol O}$$

All are the same mol values so the empirical formula is CHO. The empirical formula mass is $12.01 + 1.008 + 16.00 = 29.02$ g/mol.

$$\text{molar mass} = \frac{15.0 \text{ g}}{0.129 \text{ mol}} = 116 \text{ g/mol}$$

$$\frac{\text{molar mass}}{\text{empirical mas}} = \frac{116}{29.02} = 4.00; \quad \text{molecular formula} = (CHO)_4 = C_4H_4O_4$$

71.   When combustion data are given, it is assumed that all the carbon in the compound ends up as carbon in CO_2 and all the hydrogen in the compound ends up as hydrogen in H_2O. In the sample of propane combusted, the moles of C and H are:

$$\text{mol C} = 2.641 \text{ g CO}_2 \times \frac{1 \text{ mol CO}_2}{44.01 \text{ g CO}_2} \times \frac{1 \text{ mol C}}{\text{mol CO}_2} = 0.06001 \text{ mol C}$$

$$\text{mol H} = 1.442 \text{ g H}_2\text{O} \times \frac{1 \text{ mol H}_2\text{O}}{18.02 \text{ g H}_2\text{O}} \times \frac{2 \text{ mol H}}{\text{mol H}_2\text{O}} = 0.1600 \text{ mol H}$$

$$\frac{\text{mol H}}{\text{mol C}} = \frac{0.1600}{0.06001} = 2.666$$

Multiplying this ratio by three gives the empirical formula of C_3H_8.

72.   This compound contains nitrogen, and one way to determine the amount of nitrogen in the compound is to calculate composition by mass percent. We assume that all of the carbon in 33.5 mg CO_2 came from the 35.0 mg of compound and all of the hydrogen in 41.1 mg H_2O came from the 35.0 mg of compound.

$$3.35 \times 10^{-2} \text{ g CO}_2 \times \frac{1 \text{ mol CO}_2}{44.01 \text{ g CO}_2} \times \frac{1 \text{ mol C}}{\text{mol CO}_2} \times \frac{12.01 \text{ g C}}{\text{mol C}} = 9.14 \times 10^{-3} \text{ g C}$$

$$\%C = \frac{9.14 \times 10^{-3} \text{ g C}}{3.50 \times 10^{-2} \text{ g compound}} \times 100 = 26.1\% \text{ C}$$

$$4.11 \times 10^{-2} \text{ g H}_2\text{O} \times \frac{1 \text{ mol H}_2\text{O}}{18.02 \text{ g H}_2\text{O}} \times \frac{2 \text{ mol H}}{\text{mol H}_2\text{O}} \times \frac{1.008 \text{ g H}}{\text{mol H}} = 4.60 \times 10^{-3} \text{ g H}$$

$$\%\text{H} = \frac{4.60 \times 10^{-3} \text{ g H}}{3.50 \times 10^{-2} \text{ g compound}} \times 100 = 13.1\% \text{ H}$$

The mass percent of nitrogen is obtained by difference:

$$\%\text{N} = 100.0 - (26.1 + 13.1) = 60.8\% \text{ N}$$

Now perform the empirical formula determination by first assuming 100.0 g of compound. Out of 100.0 g of compound, there are:

$$26.1 \text{ g C} \times \frac{1 \text{ mol C}}{12.01 \text{ g C}} = 2.17 \text{ mol C}; \quad 13.1 \text{ g H} \times \frac{1 \text{ mol H}}{1.008 \text{ g H}} = 13.0 \text{ mol H}$$

$$60.8 \text{ g N} \times \frac{1 \text{ mol N}}{14.01 \text{ g N}} = 4.34 \text{ mol N}$$

Dividing all mol values by 2.17 gives: $\frac{2.17}{2.17} = 1.00$; $\frac{13.0}{2.17} = 5.99$; $\frac{4.34}{2.17} = 2.00$

The empirical formula is CH_6N_2.

73. The combustion data allow determination of the amount of hydrogen in cumene. One way to determine the amount of carbon in cumene is to determine the mass percent of hydrogen in the compound from the data in the problem; then determine the mass percent of carbon by difference (100.0 - mass %H = mass %C).

$$42.8 \text{ mg H}_2\text{O} \times \frac{1 \text{ g}}{1000 \text{ mg}} \times \frac{2.016 \text{ g H}}{18.02 \text{ g H}_2\text{O}} \times \frac{1000 \text{ mg}}{\text{g}} = 4.79 \text{ mg H}$$

$$\%\text{H} = \frac{4.79 \text{ mg H}}{47.6 \text{ mg cumene}} \times 100 = 10.1\% \text{ H}; \quad \%\text{C} = 100.0 - 10.1 = 89.9\% \text{ C}$$

Now solve this empirical formula problem. Out of 100.0 g cumene, we have:

$$89.9 \text{ g C} \times \frac{1 \text{ mol C}}{12.01 \text{ g C}} = 7.49 \text{ mol C}; \quad 10.1 \text{ g H} \times \frac{1 \text{ mol H}}{1.008 \text{ g H}} = 10.0 \text{ mol H}$$

$\frac{10.0}{7.49} = 1.34 \approx \frac{4}{3}$, i.e., mol H to mol C are in a 4:3 ratio. Empirical formula = C_3H_4

Empirical formula mass $\approx 3(12) + 4(1) = 40$ g/mol

The molecular formula is $(C_3H_4)_3$ or C_9H_{12} since the molar mass will be between 115 and 125 g/mol (molar mass $\approx 3 \times 40$ g/mol = 120 g/mol).

74.   First, we will determine composition by mass percent:

$$16.01 \text{ mg } CO_2 \times \frac{1 \text{ g}}{1000 \text{ mg}} \times \frac{12.01 \text{ g C}}{44.01 \text{ g } CO_2} \times \frac{1000 \text{ mg}}{\text{g}} = 4.369 \text{ mg C}$$

$$\%C = \frac{4.369 \text{ mg C}}{10.68 \text{ mg compound}} \times 100 = 40.91\% \text{ C}$$

$$4.37 \text{ mg } H_2O \times \frac{1 \text{ g}}{1000 \text{ mg}} \times \frac{2.016 \text{ g H}}{18.02 \text{ g } H_2O} \times \frac{1000 \text{ mg}}{\text{g}} = 0.489 \text{ mg H}$$

$$\%H = \frac{0.489 \text{ mg}}{10.68 \text{ mg}} \times 100 = 4.58\% \text{ H}; \quad \%O = 100.00 - (40.91 + 4.58) = 54.51\% \text{ O}$$

So, in 100.00 g of the compound, we have:

$$40.91 \text{ g C} \times \frac{1 \text{ mol C}}{12.01 \text{ g C}} = 3.406 \text{ mol C}; \quad 4.58 \text{ g H} \times \frac{1 \text{ mol H}}{1.008 \text{ g H}} = 4.54 \text{ mol H}$$

$$54.51 \text{ g O} \times \frac{1 \text{ mol O}}{16.00 \text{ g O}} = 3.407 \text{ mol O}$$

Dividing by the smallest number: $\frac{4.54}{3.406} = 1.33 = \frac{4}{3}$; the empirical formula is $C_3H_4O_3$.

The empirical formula mass of $C_3H_4O_3$ is $\approx 3(12) + 4(1) + 3(16) = 88$ g.

Since $\frac{176.1}{88} = 2.0$, then the molecular formula is $C_6H_8O_6$.

Balancing Chemical Equations

75.   When balancing reactions, start with elements that appear in only one of the reactants and one of the products, then go on to balance the remaining elements.

a. $Fe + O_2 \rightarrow Fe_2O_3$. Balancing Fe first, then O, gives: $2 \text{ Fe} + 3/2 \text{ } O_2 \rightarrow Fe_2O_3$. The best balanced equation contains the smallest whole numbers. To convert to whole numbers, multiply each coefficient by two, which gives: $4 \text{ Fe(s)} + 3 \text{ } O_2(g) \rightarrow 2 \text{ } Fe_2O_3(s)$

b. $Ca + H_2O \rightarrow Ca(OH)_2 + H_2$; Calcium is already balanced, so concentrate on oxygen next. Balancing O gives: $Ca(s) + 2 \text{ } H_2O(l) \rightarrow Ca(OH)_2(aq) + H_2(g)$. The equation is balanced. Note: Hydrogen is the most difficult element to balance since it appears in both products. It is generally easiest to save these atoms for last when balancing an equation.

c. $Ba(OH)_2 + H_2SO_4 \rightarrow BaSO_4 + H_2O$; Ba and S are already balanced. There are 6 O atoms on the reactant side and, in order to get 6 O atoms on the product side, we will need 2 H_2O molecules. The balanced equation is: $Ba(OH)_2(aq) + H_2SO_4(aq) \rightarrow BaSO_4(s) + 2 \text{ } H_2O(l)$.

76.   a. $C_6H_{12}O_6(s) + O_2(g) \rightarrow CO_2(g) + H_2O(g)$

Balance C atoms: $C_6H_{12}O_6 + O_2 \rightarrow 6\ CO_2 + H_2O$

Balance H atoms: $C_6H_{12}O_6 + O_2 \rightarrow 6\ CO_2 + 6\ H_2O$

Lastly, balance O atoms: $C_6H_{12}O_6(s) + 6\ O_2(g) \rightarrow 6\ CO_2(g) + 6\ H_2O(g)$

The equation is balanced.

b. $Fe_2S_3(s) + HCl(g) \rightarrow FeCl_3(s) + H_2S(g)$

Balance Fe atoms: $Fe_2S_3 + HCl \rightarrow 2\ FeCl_3 + H_2S$

Balance S atoms: $Fe_2S_3 + HCl \rightarrow 2\ FeCl_3 + 3\ H_2S$

There are 6 H and 6 Cl on right, so balance with 6 HCl on left:

$Fe_2S_3(s) + 6\ HCl(g) \rightarrow 2\ FeCl_3(s) + 3\ H_2S(g)$. Equation is balanced.

c. $CS_2(l) + NH_3(g) \rightarrow H_2S(g) + NH_4SCN(s)$

C and S balanced; balance N:

$CS_2 + 2\ NH_3 \rightarrow H_2S + NH_4SCN$

H is also balanced. So: $CS_2(l) + 2\ NH_3(g) \rightarrow H_2S(g) + NH_4SCN(s)$

77.   a. $Cu(s) + 2\ AgNO_3(aq) \rightarrow 2\ Ag(s) + Cu(NO_3)_2(aq)$

b. $Zn(s) + 2\ HCl(aq) \rightarrow ZnCl_2(aq) + H_2(g)$

c. $Au_2S_3(s) + 3\ H_2(g) \rightarrow 2\ Au(s) + 3\ H_2S(g)$

78.   a. $3\ Ca(OH)_2(aq) + 2\ H_3PO_4(aq) \rightarrow 6\ H_2O(l) + Ca_3(PO_4)_2(s)$

b. $Al(OH)_3(s) + 3\ HCl(aq) \rightarrow AlCl_3(aq) + 3\ H_2O(l)$

c. $2\ AgNO_3(aq) + H_2SO_4(aq) \rightarrow Ag_2SO_4(s) + 2\ HNO_3(aq)$

79.   a.   The formulas of the reactants and products are $C_6H_6(l) + O_2(g) \rightarrow CO_2(g) + H_2O(g)$.
To balance this combustion reaction, notice that all of the carbon in C_6H_6 has to end up as carbon in CO_2 and all of the hydrogen in C_6H_6 has to end up as hydrogen in H_2O. To balance C and H, we need 6 CO_2 molecules and 3 H_2O molecules for every 1 molecule of C_6H_6. We do oxygen last. Since we have 15 oxygen atoms in 6 CO_2 molecules and 3 H_2O molecules, we need 15/2 O_2 molecules in order to have 15 oxygen atoms on the reactant side.

$C_6H_6(l) + \dfrac{15}{2} O_2(g) \rightarrow 6\ CO_2(g) + 3\ H_2O(g);$ Multiply by two to give whole numbers.

$2\ C_6H_6(l) + 15\ O_2(g) \rightarrow 12\ CO_2(g) + 6\ H_2O(g)$

b. The formulas of the reactants and products are $C_4H_{10}(g) + O_2(g) \rightarrow CO_2(g) + H_2O(g)$.

$C_4H_{10}(g) + \dfrac{13}{2} O_2(g) \rightarrow 4\ CO_2(g) + 5\ H_2O(g);$ Multiply by two to give whole numbers.

$2\ C_4H_{10}(g) + 13\ O_2(g) \rightarrow 8\ CO_2(g) + 10\ H_2O(g)$

c. $C_{12}H_{22}O_{11}(s) + 12\ O_2(g) \rightarrow 12\ CO_2(g) + 11\ H_2O(g)$

d. $2\ Fe(s) + \dfrac{3}{2} O_2(g) \rightarrow Fe_2O_3(s);$ For whole numbers: $4\ Fe(s) + 3\ O_2(g) \rightarrow 2\ Fe_2O_3(s)$

e. $2\ FeO(s) + \dfrac{1}{2} O_2(g) \rightarrow Fe_2O_3(s);$ For whole numbers, multiply by two.

$4\ FeO(s) + O_2(g) \rightarrow 2\ Fe_2O_3(s)$

80.   a. $16\ Cr(s) + 3\ S_8(s) \rightarrow 8\ Cr_2S_3(s)$

b. $2\ NaHCO_3(s) \rightarrow Na_2CO_3(s) + CO_2(g) + H_2O(g)$

c. $2\ KClO_3(s) \rightarrow 2\ KCl(s) + 3\ O_2(g)$

d. $2\ Eu(s) + 6\ HF(g) \rightarrow 2\ EuF_3(s) + 3\ H_2(g)$

81.   a. $SiO_2(s) + C(s) \rightarrow Si(s) + CO(g)$

Balance oxygen atoms: $SiO_2 + C \rightarrow Si + 2\ CO$

Balance carbon atoms: $SiO_2(s) + 2\ C(s) \rightarrow Si(s) + 2\ CO(g)$

b. $SiCl_4(l) + Mg(s) \rightarrow Si(s) + MgCl_2(s)$

Balance Cl atoms: $SiCl_4 + Mg \rightarrow Si + 2\ MgCl_2$

Balance Mg atoms: $SiCl_4(l) + 2\ Mg(s) \rightarrow Si(s) + 2\ MgCl_2(s)$

c. $Na_2SiF_6(s) + Na(s) \rightarrow Si(s) + NaF(s)$

Balance F atoms: $Na_2SiF_6 + Na \rightarrow Si + 6\ NaF$

Balance Na atoms: $Na_2SiF_6(s) + 4\ Na(s) \rightarrow Si(s) + 6\ NaF(s)$

82.   Unbalanced equation:

$$CaF_2 \cdot 3Ca_3(PO_4)_2(s) + H_2SO_4(aq) \rightarrow H_3PO_4(aq) + HF(aq) + CaSO_4 \cdot 2H_2O(s)$$

Balancing Ca^{2+}, F^-, and PO_4^{3-}:

$$CaF_2 \cdot 3Ca_3(PO_4)_2(s) + H_2SO_4(aq) \rightarrow 6\ H_3PO_4(aq) + 2\ HF(aq) + 10\ CaSO_4 \cdot 2H_2O(s)$$

On the right-hand side there are 20 extra hydrogen atoms, 10 extra sulfates, and 20 extra water molecules. We can balance the hydrogen and sulfate with 10 sulfuric acid molecules. The extra waters came from the water in the sulfuric acid solution. The balanced equation is:

$$CaF_2 \cdot 3Ca_3(PO_4)_2(s) + 10\ H_2SO_4(aq) + 20\ H_2O(l) \rightarrow 6\ H_3PO_4(aq) + 2\ HF(aq) + 10\ CaSO_4 \cdot 2H_2O(s)$$

83. $C_{12}H_{22}O_{11}(aq) + H_2O(l) \rightarrow 4\ C_2H_5OH(aq) + 4\ CO_2(g)$

84. $CaSiO_3(s) + 6\ HF(aq) \rightarrow CaF_2(aq) + SiF_4(g) + 3\ H_2O(l)$

Reaction Stoichiometry

85.   The stepwise method to solve stoichiometry problems is outlined in the text. Instead of calculating intermediate answers for each step, we will combine conversion factors into one calculation. This practice reduces round-off error and saves time.

The balanced reaction is: $(NH_4)_2Cr_2O_7(s) \rightarrow Cr_2O_3(s) + N_2(g) + 4\ H_2O(g)$

$$10.8\text{ g }(NH_4)_2Cr_2O_7 \times \frac{1\text{ mol }(NH_4)_2Cr_2O_7}{252.08\text{ g}} = 4.28 \times 10^{-2}\text{ mol }(NH_4)_2Cr_2O_7$$

$$4.28 \times 10^{-2}\text{ mol }(NH_4)_2Cr_2O_7 \times \frac{1\text{ mol }Cr_2O_3}{\text{mol }(NH_4)_2Cr_2O_7} \times \frac{152.00\text{ g }Cr_2O_3}{\text{mol }Cr_2O_3} = 6.51\text{ g }Cr_2O_3$$

$$4.28 \times 10^{-2}\text{ mol }(NH_4)_2Cr_2O_7 \times \frac{1\text{ mol }N_2}{\text{mol }(NH_4)_2Cr_2O_7} \times \frac{28.02\text{ g }N_2}{\text{mol }N_2} = 1.20\text{ g }N_2$$

$$4.28 \times 10^{-2}\text{ mol }(NH_4)_2Cr_2O_7 \times \frac{4\text{ mol }H_2O}{\text{mol }(NH_4)_2Cr_2O_7} \times \frac{18.02\text{ g }H_2O}{\text{mol }H_2O} = 3.09\text{ g }H_2O$$

86. $Fe_2O_3(s) + 2\ Al(s) \rightarrow 2\ Fe(l) + Al_2O_3(s)$

$$15.0\text{ g Fe} \times \frac{1\text{ mol Fe}}{55.85\text{ g Fe}} = 0.269\text{ mol Fe};\ \ 0.269\text{ mol Fe} \times \frac{2\text{ mol Al}}{2\text{ mol Fe}} \times \frac{26.98\text{ g Al}}{\text{mol Al}} = 7.26\text{ g Al}$$

$$0.269\text{ mol Fe} \times \frac{1\text{ mol }Fe_2O_3}{2\text{ mol Fe}} \times \frac{159.70\text{ g }Fe_2O_3}{\text{mol }Fe_2O_3} = 21.5\text{ g }Fe_2O_3$$

$$0.269\text{ mol Fe} \times \frac{1\text{ mol }Al_2O_3}{2\text{ mol Fe}} \times \frac{101.96\text{ g }Al_2O_3}{\text{mol }Al_2O_3} = 13.7\text{ g }Al_2O_3$$

87. $1.000 \text{ kg Al} \times \dfrac{1000 \text{ g Al}}{\text{kg Al}} \times \dfrac{1 \text{ mol Al}}{26.98 \text{ g Al}} \times \dfrac{3 \text{ mol NH}_4\text{ClO}_4}{3 \text{ mol Al}} \times \dfrac{117.49 \text{ g NH}_4\text{ClO}_4}{\text{mol NH}_4\text{ClO}_4} = 4355 \text{ g}$

88. a. $\text{Ba(OH)}_2\bullet 8\text{H}_2\text{O(s)} + 2 \text{ NH}_4\text{SCN(s)} \rightarrow \text{Ba(SCN)}_2\text{(s)} + 10 \text{ H}_2\text{O(l)} + 2 \text{ NH}_3\text{(g)}$

     b. $6.5 \text{ g Ba(OH)}_2\bullet 8\text{H}_2\text{O} \times \dfrac{1 \text{ mol Ba(OH)}_2\bullet 8\text{H}_2\text{O}}{315.4 \text{ g}} = 0.0206 \text{ mol} = 0.021 \text{ mol}$

     $0.021 \text{ mol Ba(OH)}_2\bullet 8\text{H}_2\text{O} \times \dfrac{2 \text{ mol NH}_4\text{SCN}}{1 \text{ mol Ba(OH)}_2\bullet 8\text{H}_2\text{O}} \times \dfrac{76.13 \text{ g NH}_4\text{SCN}}{\text{mol NH}_4\text{SCN}} = 3.2 \text{ g NH}_4\text{SCN}$

89. $1.0 \text{ ton CuO} \times \dfrac{907 \text{ kg}}{\text{ton}} \times \dfrac{1000 \text{ g}}{\text{kg}} \times \dfrac{1 \text{ mol CuO}}{79.55 \text{ g CuO}} \times \dfrac{1 \text{ mol C}}{2 \text{ mol CuO}} \times \dfrac{12.01 \text{ g C}}{\text{mol C}} \times \dfrac{100. \text{ g coke}}{95 \text{ g C}}$

     $$= 7.2 \times 10^4 \text{ g coke}$$

90. $1.0 \times 10^4 \text{ kg waste} \times \dfrac{3.0 \text{ kg NH}_4^+}{100 \text{ kg waste}} \times \dfrac{1000 \text{ g}}{\text{kg}} \times \dfrac{1 \text{ mol NH}_4^+}{18.04 \text{ g NH}_4^+} \times \dfrac{1 \text{ mol C}_5\text{H}_7\text{O}_2\text{N}}{55 \text{ mol NH}_4^+}$

     $\times \dfrac{113.12 \text{ g C}_5\text{H}_7\text{O}_2\text{N}}{\text{mol C}_5\text{H}_7\text{O}_2\text{N}} = 3.4 \times 10^4 \text{ g tissue if all NH}_4^+ \text{ converted}$

Since only 95% of the NH_4^+ ions react:

     mass of tissue = $(0.95) (3.4 \times 10^4 \text{ g}) = 3.2 \times 10^4 \text{ g}$ or 32 kg bacterial tissue

91. a. Molar mass = $195.1 + 2(14.01) + 6(1.008) + 2(35.45) = 300.1 \text{ g/mol}$

     $\% \text{ Pt} = \dfrac{195.1 \text{ g}}{300.1 \text{ g}} \times 100 = 65.01\% \text{ Pt}; \quad \% \text{ N} = \dfrac{28.02 \text{ g}}{300.1 \text{ g}} \times 100 = 9.337\% \text{ N}$

     $\% \text{ H} = \dfrac{6.048 \text{ g}}{300.1 \text{ g}} \times 100 = 2.015\% \text{ H}; \quad \% \text{ Cl} = \dfrac{70.90 \text{ g}}{300.1 \text{ g}} \times 100 = 23.63\% \text{ Cl}$

     65.01% Pt; 9.337% N; 2.015% H; 23.63% Cl

     b. $100. \text{ g K}_2\text{PtCl}_4 \times \dfrac{1 \text{ mol K}_2\text{PtCl}_4}{415.1 \text{ g K}_2\text{PtCl}_4} \times \dfrac{1 \text{ mol Pt(NH}_3)_2\text{Cl}_2}{\text{mol K}_2\text{PtCl}_4} \times \dfrac{300.1 \text{ g Pt(NH}_3)_2\text{Cl}_2}{\text{mol Pt(NH}_3)_2\text{Cl}_2}$

     $$= 72.3 \text{ g Pt(NH}_3)_2\text{Cl}_2$$

     $100. \text{ g K}_2\text{PtCl}_4 \times \dfrac{1 \text{ mol K}_2\text{PtCl}_4}{415.1 \text{ g K}_2\text{PtCl}_4} \times \dfrac{2 \text{ mol KCl}}{\text{mol K}_2\text{PtCl}_4} \times \dfrac{74.55 \text{ g KCl}}{\text{mol KCl}} = 35.9 \text{ g KCl}$

92. a. $1.00 \times 10^2 \text{ g C}_7\text{H}_6\text{O}_3 \times \dfrac{1 \text{ mol C}_7\text{H}_6\text{O}_3}{138.12 \text{ g C}_7\text{H}_6\text{O}_3} \times \dfrac{1 \text{ mol C}_4\text{H}_6\text{O}_3}{1 \text{ mol C}_7\text{H}_6\text{O}_3} \times \dfrac{102.09 \text{ g C}_4\text{H}_6\text{O}_3}{\text{mol C}_4\text{H}_6\text{O}_3} = 73.9 \text{ g C}_4\text{H}_6\text{O}_3$

b. $1.00 \times 10^2 \text{ g C}_7\text{H}_6\text{O}_3 \times \dfrac{1 \text{ mol C}_7\text{H}_6\text{O}_3}{138.12 \text{ g C}_7\text{H}_6\text{O}_3} \times \dfrac{1 \text{ mol C}_9\text{H}_8\text{O}_4}{1 \text{ mol C}_7\text{H}_6\text{O}_3} \times \dfrac{180.15 \text{ g C}_9\text{H}_8\text{O}_4}{\text{mol C}_9\text{H}_8\text{O}_4}$

$= 1.30 \times 10^2$ g aspirin

Limiting Reactants and Percent Yield

93.   a. $\text{Mg(s)} + \text{I}_2\text{(s)} \rightarrow \text{MgI}_2\text{(s)}$

From the balanced equation, 100 molecules of I_2 reacts completely with 100 atoms of Mg. We have a stoichiometric mixture. Neither is limiting.

b.   150 atoms Mg $\times \dfrac{1 \text{ molecule I}_2}{1 \text{ atom Mg}} = 150$ molecules I_2 needed

We need 150 molecules I_2 to react completely with 150 atoms Mg; we only have 100 molecules I_2. Therefore, I_2 is limiting.

c.   200 atoms Mg $\times \dfrac{1 \text{ molecule I}_2}{1 \text{ atom Mg}} = 200$ molecules I_2; Mg is limiting since 300 molecules I_2 are present.

d.   0.16 mol Mg $\times \dfrac{1 \text{ mol I}_2}{1 \text{ mol Mg}} = 0.16$ mol I_2; Mg is limiting since 0.25 mol I_2 are present.

e.   0.14 mol Mg $\times \dfrac{1 \text{ mol I}_2}{1 \text{ mol Mg}} = 0.14$ mol I_2 needed; Stoichiometric mixture. Neither is limiting.

f.   0.12 mol Mg $\times \dfrac{1 \text{ mol I}_2}{1 \text{ mol Mg}} = 0.12$ mol I_2 needed; I_2 is limiting since only 0.08 mol I_2 are present.

g.   6.078 g Mg $\times \dfrac{1 \text{ mol Mg}}{24.31 \text{ g Mg}} \times \dfrac{1 \text{ mol I}_2}{1 \text{ mol Mg}} \times \dfrac{253.8 \text{ g I}_2}{\text{mol I}_2} = 63.46$ g I_2

Stoichiometric mixture. Neither is limiting.

h.   1.00 g Mg $\times \dfrac{1 \text{ mol Mg}}{24.31 \text{ g Mg}} \times \dfrac{1 \text{ mol I}_2}{1 \text{ mol Mg}} \times \dfrac{253.8 \text{ g I}_2}{\text{mol I}_2} = 10.4$ g I_2

10.4 g I_2 needed, but we only have 2.00 g. I_2 is limiting.

i.   From h above, we calculated that 10.4 g I_2 will react completely with 1.00 g Mg. We have 20.00 g I_2. I_2 is in excess. Mg is limiting.

94. $2 \text{ H}_2\text{(g)} + \text{O}_2\text{(g)} \rightarrow 2 \text{ H}_2\text{O(g)}$

a.   50 molecules $\text{H}_2 \times \dfrac{1 \text{ molecule O}_2}{2 \text{ molecules H}_2} = 25$ molecules O_2

Stoichiometric mixture. Neither is limiting.

b. $100 \text{ molecules } H_2 \times \dfrac{1 \text{ molecule } O_2}{2 \text{ molecules } H_2} = 50 \text{ molecules } O_2$; O_2 is limiting since only 40 molecules O_2 are present.

c.   From b, 50 molecules of O_2 will react completely with 100 molecules of H_2. We have 100 molecules (an excess) of O_2. So, H_2 is limiting.

d. $0.50 \text{ mol } H_2 \times \dfrac{1 \text{ mol } O_2}{2 \text{ mol } H_2} = 0.25 \text{ mol } O_2$; H_2 is limiting since 0.75 mol O_2 are present.

e. $0.80 \text{ mol } H_2 \times \dfrac{1 \text{ mol } O_2}{2 \text{ mol } H_2} = 0.40 \text{ mol } O_2$; H_2 is limiting since 0.75 mol O_2 are present.

f. $1.0 \text{ g } H_2 \times \dfrac{1 \text{ mol } H_2}{2.016 \text{ g } H_2} \times \dfrac{1 \text{ mol } O_2}{2 \text{ mol } H_2} = 0.25 \text{ mol } O_2$

Stoichiometric mixture, neither is limiting.

g. $5.00 \text{ g } H_2 \times \dfrac{1 \text{ mol } H_2}{2.016 \text{ g } H_2} \times \dfrac{1 \text{ mol } O_2}{2 \text{ mol } H_2} \times \dfrac{32.00 \text{ g } O_2}{\text{mol } O_2} = 39.7 \text{ g } O_2$; H_2 is limiting since 56.00 g O_2 are present.

95.   a. $10.0 \text{ g Hg} \times \dfrac{1 \text{ mol Hg}}{200.6 \text{ g Hg}} = 4.99 \times 10^{-2} \text{ mol Hg}$

$9.00 \text{ g } Br_2 \times \dfrac{1 \text{ mol } Br_2}{159.80 \text{ g } Br_2} = 5.63 \times 10^{-2} \text{ mol } Br_2$

The required mol ratio from the balanced equation is 1 mol Br_2 to 1 mol Hg. The actual mol ratio is:

$$\dfrac{5.63 \times 10^{-2} \text{ mol Br}}{4.99 \times 10^{-2} \text{ mol Hg}} = 1.13$$

This is higher than the required ratio, so Hg is the limiting reagent.

$4.99 \times 10^{-2} \text{ mol Hg} \times \dfrac{1 \text{ mol } HgBr_2}{\text{mol Hg}} \times \dfrac{360.4 \text{ g } HgBr_2}{\text{mol } HgBr_2} = 18.0 \text{ g } HgBr_2 \text{ produced}$

$4.99 \times 10^{-2} \text{ mol Hg} \times \dfrac{1 \text{ mol } Br_2}{1 \text{ mol Hg}} \times \dfrac{159.80 \text{ g } Br_2}{\text{mol } Br_2} = 7.97 \text{ g } Br_2 \text{ reacted}$

excess Br_2 = 9.00 g Br_2 - 7.97 g Br_2 = 1.03 g Br_2

b. $5.00 \text{ mL Hg} \times \dfrac{13.6 \text{ g Hg}}{\text{mL Hg}} \times \dfrac{1 \text{ mol Hg}}{200.6 \text{ g Hg}} = 0.339 \text{ mol Hg}$

$5.00 \text{ mL } Br_2 \times \dfrac{3.10 \text{ g } Br_2}{\text{mL } Br_2} \times \dfrac{1 \text{ mol } Br_2}{159.80 \text{ g } Br_2} = 0.0970 \text{ mol } Br_2$

Br_2 is limiting since the actual moles of Br_2 present is well below the required 1:1 mol ratio.

$$0.0970 \text{ mol } Br_2 \times \frac{1 \text{ mol } HgBr_2}{\text{mol } Br_2} \times \frac{360.4 \text{ g } HgBr_2}{\text{mol } HgBr_2} = 35.0 \text{ g } HgBr_2 \text{ produced}$$

96. $$1.50 \text{ g } BaO_2 \times \frac{1 \text{ mol } BaO_2}{169.3 \text{ g } BaO_2} = 8.86 \times 10^{-3} \text{ mol } BaO_2$$

$$25.0 \text{ mL} \times \frac{0.0272 \text{ g } HCl}{\text{mL}} \times \frac{1 \text{ mol } HCl}{36.46 \text{ g } HCl} = 1.87 \times 10^{-2} \text{ mol } HCl$$

The required mol ratio from the balanced reaction is 2 mol HCl to 1 mol BaO_2. The actual ratio is:

$$\frac{1.87 \times 10^{-2} \text{ mol } HCl}{8.86 \times 10^{-3} \text{ mol } BaO_2} = 2.11$$

Since the actual mol ratio is larger than the required mol ratio, the denominator (BaO_2) is the limiting reagent.

$$8.86 \times 10^{-3} \text{ mol } BaO_2 \times \frac{1 \text{ mol } H_2O_2}{\text{mol } BaO_2} \times \frac{34.02 \text{ g } H_2O_2}{\text{mol } H_2O_2} = 0.301 \text{ g } H_2O_2$$

The amount of HCl reacted is:

$$8.86 \times 10^{-3} \text{ mol } BaO_2 \times \frac{2 \text{ mol } HCl}{\text{mol } BaO_2} = 1.77 \times 10^{-2} \text{ mol } HCl$$

excess mol HCl = 1.87×10^{-2} mol - 1.77×10^{-2} mol = 1.0×10^{-3} mol HCl

mass of excess HCl = 1.0×10^{-3} mol HCl $\times \dfrac{36.46 \text{ g } HCl}{\text{mol } HCl} = 3.6 \times 10^{-2}$ g HCl

97. $Ca_3(PO_4)_2 + 3 H_2SO_4 \rightarrow 3 CaSO_4 + 2 H_3PO_4$

$$1.0 \times 10^3 \text{ g } Ca_3(PO_4)_2 \times \frac{1 \text{ mol } Ca_3(PO_4)_2}{310.18 \text{ g } Ca_3(PO_4)_2} = 3.2 \text{ mol } Ca_3(PO_4)_2$$

$$1.0 \times 10^3 \text{ g conc. } H_2SO_4 \times \frac{98 \text{ g } H_2SO_4}{100 \text{ g conc. } H_2SO_4} \times \frac{1 \text{ mol } H_2SO_4}{98.09 \text{ g } H_2SO_4} = 10. \text{ mol } H_2SO_4$$

The required mol ratio from the balanced equation is 3 mol H_2SO_4 to 1 mol $Ca_3(PO_4)_2$. The actual ratio is: $\dfrac{10. \text{ mol } H_2SO_4}{3.2 \text{ mol } Ca_3(PO_4)_2} = 3.1$

This is higher than the required mol ratio, so $Ca_3(PO_4)_2$ is the limiting reagent.

$$3.2 \text{ mol } Ca_3(PO_4)_2 \times \frac{3 \text{ mol } CaSO_4}{\text{mol } Ca_3(PO_4)_2} \times \frac{136.15 \text{ g } CaSO_4}{\text{mol } CaSO_4} = 1300 \text{ g } CaSO_4 \text{ produced}$$

$$3.2 \text{ mol } Ca_3(PO_4)_2 \times \frac{2 \text{ mol } H_3PO_4}{\text{mol } Ca_3(PO_4)_2} \times \frac{97.99 \text{ g } H_3PO_4}{\text{mol } H_3PO_4} = 630 \text{ g } H_3PO_4 \text{ produced}$$

98.   An alternative method to solve limiting reagent problems is to assume each reactant is limiting and calculate how much product could be produced from each reactant. The reactant that produces the smallest amount of product will run out first and is the limiting reagent.

$$5.00 \times 10^6 \text{ g NH}_3 \times \frac{1 \text{ mol NH}_3}{17.03 \text{ g NH}_3} \times \frac{2 \text{ mol HCN}}{2 \text{ mol NH}_3} = 2.94 \times 10^5 \text{ mol HCN}$$

$$5.00 \times 10^6 \text{ g O}_2 \times \frac{1 \text{ mol O}_2}{32.00 \text{ g O}_2} \times \frac{2 \text{ mol HCN}}{3 \text{ mol O}_2} = 1.04 \times 10^5 \text{ mol HCN}$$

$$5.00 \times 10^6 \text{ g CH}_4 \times \frac{1 \text{ mol CH}_4}{16.04 \text{ g CH}_4} \times \frac{2 \text{ mol HCN}}{2 \text{ mol CH}_4} = 3.12 \times 10^5 \text{ mol HCN}$$

O_2 is limiting since it produces the smallest amount of HCN. Although more product could be produced from NH_3 and CH_4, only enough O_2 is present to produce 1.04×10^5 mol HCN. The mass of HCN produced is:

$$1.04 \times 10^5 \text{ mol HCN} \times \frac{27.03 \text{ g HCN}}{\text{mol HCN}} = 2.81 \times 10^6 \text{ g HCN}$$

$$5.00 \times 10^6 \text{ g O}_2 \times \frac{1 \text{ mol O}_2}{32.00 \text{ g O}_2} \times \frac{6 \text{ mol H}_2\text{O}}{3 \text{ mol O}_2} \times \frac{18.02 \text{ g H}_2\text{O}}{1 \text{ mol H}_2\text{O}} = 5.63 \times 10^6 \text{ g H}_2\text{O}$$

99. $C_2H_6(g) + Cl_2(g) \rightarrow C_2H_5Cl(g) + HCl(g)$

$$300. \text{ g C}_2\text{H}_6 \times \frac{1 \text{ mol C}_2\text{H}_6}{30.07 \text{ g C}_2\text{H}_6} = 9.98 \text{ mol C}_2\text{H}_6; \quad 650. \text{ g Cl}_2 \times \frac{1 \text{ mol Cl}_2}{70.90 \text{ g Cl}_2} = 9.17 \text{ mol Cl}_2$$

The balanced equation requires a 1:1 mol ratio between reactants. 9.17 mol of C_2H_6 will react with all of the Cl_2 present (9.17 mol). Since 9.98 mol C_2H_6 is present, Cl_2 is the limiting reagent.

The theoretical yield of C_2H_5Cl is:

$$9.17 \text{ mol Cl}_2 \times \frac{1 \text{ mol C}_2\text{H}_5\text{Cl}}{\text{mol Cl}_2} \times \frac{64.51 \text{ g C}_2\text{H}_5\text{Cl}}{\text{mol C}_2\text{H}_5\text{Cl}} = 592 \text{ g C}_2\text{H}_5\text{Cl}$$

$$\text{Percent yield} = \frac{\text{actual}}{\text{theoretical}} \times 100 = \frac{490. \text{ g}}{592 \text{ g}} \times 100 = 82.8\%$$

100. $C_7H_6O_3 + C_4H_6O_3 \rightarrow C_9H_8O_4 + HC_2H_3O_2$

$$1.50 \text{ g C}_7\text{H}_6\text{O}_3 \times \frac{1 \text{ mol C}_7\text{H}_6\text{O}_3}{138.12 \text{ g C}_7\text{H}_6\text{O}_3} = 1.09 \times 10^{-2} \text{ mol C}_7\text{H}_6\text{O}_3$$

$$2.00 \text{ g C}_4\text{H}_6\text{O}_3 \times \frac{1 \text{ mol C}_4\text{H}_6\text{O}_3}{102.09 \text{ g C}_4\text{H}_6\text{O}_3} = 1.96 \times 10^{-2} \text{ mol C}_4\text{H}_6\text{O}_3$$

$C_7H_6O_3$ is the limiting reagent since the actual moles of $C_7H_6O_3$ are below the required 1:1 mol ratio. The theoretical yield of aspirin is:

$$1.09 \times 10^{-2} \text{ mol } C_7H_6O_3 \times \frac{1 \text{ mol } C_9H_8O_4}{\text{mol } C_7H_6O_3} \times \frac{180.15 \text{ g } C_9H_8O_4}{\text{mol } C_9H_8O_4} = 1.96 \text{ g } C_9H_8O_4$$

$$\% \text{ yield} = \frac{1.50 \text{ g}}{1.96 \text{ g}} \times 100 = 76.5\%$$

101. $2.50 \text{ metric tons } Cu_3FeS_3 \times \dfrac{1000 \text{ kg}}{\text{metric ton}} \times \dfrac{1000 \text{ g}}{\text{kg}} \times \dfrac{1 \text{ mol } Cu_3FeS_3}{342.71 \text{ g}} \times \dfrac{3 \text{ mol Cu}}{1 \text{ mol } Cu_3FeS_3} \times \dfrac{63.55 \text{ g}}{\text{mol Cu}}$

$$= 1.39 \times 10^6 \text{ g Cu (theoretical)}$$

$$1.39 \times 10^6 \text{ g Cu (theoretical)} \times \frac{86.3 \text{ g Cu (actual)}}{100. \text{ g Cu (theoretical)}} = 1.20 \times 10^6 \text{ g Cu} = 1.20 \times 10^3 \text{ kg Cu}$$

$$= 1.20 \text{ metric tons Cu (actual)}$$

102. $P_4(s) + 6 F_2(g) \rightarrow 4 PF_3(g)$; The theoretical yield of PF_3 is:

$$120. \text{ g } PF_3 \text{ (actual)} \times \frac{100.0 \text{ g } PF_3 \text{ (theoretical)}}{78.1 \text{ g } PF_3 \text{ (actual)}} = 154 \text{ g } PF_3 \text{ (theoretical)}$$

$$154 \text{ g } PF_3 \times \frac{1 \text{ mol } PF_3}{87.97 \text{ g } PF_3} \times \frac{6 \text{ mol } F_2}{4 \text{ mol } PF_3} \times \frac{38.00 \text{ g } F_2}{\text{mol } F_2} = 99.8 \text{ g } F_2$$

99.8 g F_2 are needed to produce an actual PF_3 yield of 78.1%.

Additional Exercises

103. $\dfrac{9.123 \times 10^{-23} \text{ g}}{\text{atom}} \times \dfrac{6.022 \times 10^{23} \text{ atom}}{\text{mol}} = \dfrac{54.94 \text{ g}}{\text{mol}}$

The atomic mass is 54.94 amu. From the periodic table, the element is manganese (Mn).

104. In one hour, the 1000. kg of wet cereal produced contains 580 kg H_2O and 420 kg of cereal. We want the final product to contain 20.% H_2O. Let x = mass of H_2O in final product.

$$\frac{x}{420 + x} = 0.20, \ x = 84 + 0.20 \, x, \ x = 105 \approx 110 \text{ kg } H_2O$$

The amount of water to be removed is 580 - 110 = 470 kg/hr.

105. Empirical formula mass = 12.01 + 1.008 = 13.02 g/mol; Since 104.14/13.02 = 7.998 ≈ 8, the molecular formula for styrene is $(CH)_8 = C_8H_8$.

$$2.00 \text{ g } C_8H_8 \times \frac{1 \text{ mol } C_8H_8}{104.14 \text{ g } C_8H_8} \times \frac{8 \text{ mol H}}{\text{mol } C_8H_8} \times \frac{6.022 \times 10^{23} \text{ atoms H}}{\text{mol H}} = 9.25 \times 10^{22} \text{ atoms H}$$

106. $41.98 \text{ mg } CO_2 \times \dfrac{12.01 \text{ mg C}}{44.01 \text{ mg } CO_2} = 11.46 \text{ mg C}; \quad \%C = \dfrac{11.46 \text{ mg}}{19.81 \text{ mg}} \times 100 = 57.85\% \text{ C}$

$6.45 \text{ mg } H_2O \times \dfrac{2.016 \text{ mg H}}{18.02 \text{ mg } H_2O} = 0.722 \text{ mg H}; \quad \%H = \dfrac{0.722 \text{ mg}}{19.81 \text{ mg}} \times 100 = 3.64\% \text{ H}$

$\%O = 100.00 - (57.85 + 3.64) = 38.51\% \text{ O}$

Out of 100.00 g terephthalic acid, there are:

$57.85 \text{ g C} \times \dfrac{1 \text{ mol C}}{12.01 \text{ g C}} = 4.817 \text{ mol C}; \quad 3.64 \text{ g H} \times \dfrac{1 \text{ mol H}}{1.008 \text{ g H}} = 3.61 \text{ mol H}$

$38.51 \text{ g O} \times \dfrac{1 \text{ mol O}}{16.00 \text{ g O}} = 2.407 \text{ mol O}$

$\dfrac{4.817}{2.407} = 2.001; \quad \dfrac{3.61}{2.407} = 1.50; \quad \dfrac{2.407}{2.407} = 1.000$

C:H:O mol ratio is 2:1.5:1 or 4:3:2. Empirical formula: $C_4H_3O_2$

Mass of $C_4H_3O_2 \approx 4(12) + 3(1) + 2(16) = 83$

Molar mass $= \dfrac{41.5 \text{ g}}{0.250 \text{ mol}} = 166 \text{ g/mol}; \quad \dfrac{166}{83} = 2; \quad$ Molecular formula: $C_8H_6O_4$

107. $17.3 \text{ g H} \times \dfrac{1 \text{ mol H}}{1.008 \text{ g H}} = 17.2 \text{ mol H}; \quad 82.7 \text{ g C} \times \dfrac{1 \text{ mol C}}{12.01 \text{ g C}} = 6.89 \text{ mol C}$

$\dfrac{17.2}{6.89} = 2.50; \quad$ The empirical formula is C_2H_5.

The empirical formula mass is ~29 g, so two times the empirical formula would put the compound in the correct range of the molar mass. Molecular formula $= (C_2H_5)_2 = C_4H_{10}$

$2.59 \times 10^{23} \text{ atoms H} \times \dfrac{1 \text{ molecule } C_4H_{10}}{10 \text{ atoms H}} \times \dfrac{1 \text{ mol } C_4H_{10}}{6.022 \times 10^{23} \text{ molecules}} = 4.30 \times 10^{-2} \text{ mol } C_4H_{10}$

$4.30 \times 10^{-2} \text{ mol } C_4H_{10} \times \dfrac{58.12 \text{ g}}{\text{mol } C_4H_{10}} = 2.50 \text{ g } C_4H_{10}$

108.   Assuming 100.00 g E_3H_8:

$\text{mol E} = 8.73 \text{ g H} \times \dfrac{1 \text{ mol H}}{1.008 \text{ g H}} \times \dfrac{3 \text{ mol E}}{8 \text{ mol H}} = 3.25 \text{ mol E}$

$\dfrac{x \text{ g E}}{1 \text{ mol E}} = \dfrac{91.27 \text{ g E}}{3.25 \text{ mol E}}, \quad x = \text{molar mass of E} = 28.1 \text{ g/mol}; \quad \text{atomic mass of E} = 28.1 \text{ amu}$

109.   Mass of H_2O = 0.755 g $CuSO_4 \cdot xH_2O$ - 0.483 g $CuSO_4$ = 0.272 g H_2O

$$0.483 \text{ g } CuSO_4 \times \frac{1 \text{ mol } CuSO_4}{159.62 \text{ g } CuSO_4} = 0.00303 \text{ mol } CuSO_4$$

$$0.272 \text{ g } H_2O \times \frac{1 \text{ mol } H_2O}{18.02 \text{ g } H_2O} = 0.0151 \text{ mol } H_2O$$

$$\frac{0.0151 \text{ mol } H_2O}{0.00303 \text{ mol } CuSO_4} = \frac{4.98 \text{ mol } H_2O}{1 \text{ mol } CuSO_4}; \text{ Compound formula} = CuSO_4 \cdot 5H_2O, \, x = 5$$

110.   a.   Only acrylonitrile contains nitrogen. If we have 100.00 g of polymer:

$$8.80 \text{ g N} \times \frac{1 \text{ mol } C_3H_3N}{14.01 \text{ g N}} \times \frac{53.06 \text{ g } C_3H_3N}{1 \text{ mol } C_3H_3N} = 33.3 \text{ g } C_3H_3N$$

$$\% \, C_3H_3N = \frac{33.3 \text{ g } C_3H_3N}{100.00 \text{ g polymer}} = 33.3\% \, C_3H_3N$$

Only butadiene in the polymer reacts with Br_2:

$$0.605 \text{ g } Br_2 \times \frac{1 \text{ mol } Br_2}{159.80 \text{ g } Br_2} \times \frac{1 \text{ mol } C_4H_6}{\text{mol } Br_2} \times \frac{54.09 \text{ g } C_4H_6}{\text{mol } C_4H_6} = 0.205 \text{ g } C_4H_6$$

$$\% \, C_4H_6 = \frac{0.205 \text{ g}}{1.20 \text{ g}} \times 100 = 17.1\% \, C_4H_6$$

b.   If we have 100.0 g of polymer:

$$33.3 \text{ g } C_3H_3N \times \frac{1 \text{ mol } C_3H_3N}{53.06 \text{ g}} = 0.628 \text{ mol } C_3H_3N$$

$$17.1 \text{ g } C_4H_6 \times \frac{1 \text{ mol } C_4H_6}{54.09 \text{ g } C_4H_6} = 0.316 \text{ mol } C_4H_6$$

$$49.6 \text{ g } C_8H_8 \times \frac{1 \text{ mol } C_8H_8}{104.14 \text{ g } C_8H_8} = 0.476 \text{ mol } C_8H_8$$

Dividing by 0.316: $\dfrac{0.628}{0.316} = 1.99$; $\dfrac{0.316}{0.316} = 1.00$; $\dfrac{0.476}{0.316} = 1.51$

This is close to a mol ratio of 4:2:3. Thus, there are 4 acrylonitrile to 2 butadiene to 3 styrene molecules in the polymer or $(A_4B_2S_3)_n$.

111. $1.20 \text{ g CO}_2 \times \dfrac{1 \text{ mol CO}_2}{44.01 \text{ g}} \times \dfrac{1 \text{ mol C}}{\text{mol CO}_2} \times \dfrac{1 \text{ mol C}_{24}\text{H}_{30}\text{N}_3\text{O}}{24 \text{ mol C}} \times \dfrac{376.51 \text{ g}}{\text{mol C}_{24}\text{H}_{30}\text{N}_3\text{O}} = 0.428 \text{ g C}_{24}\text{H}_{30}\text{N}_3\text{O}$

$\dfrac{0.428 \text{ g C}_{24}\text{H}_{30}\text{N}_3\text{O}}{1.00 \text{ g sample}} \times 100 = 42.8\% \text{ C}_{24}\text{H}_{30}\text{N}_3\text{O}$

112.   a. $\text{CH}_4(g) + 4 \text{ S}(s) \rightarrow \text{CS}_2(l) + 2 \text{ H}_2\text{S}(g)$ or $2 \text{ CH}_4(g) + \text{S}_8(s) \rightarrow 2 \text{ CS}_2(l) + 4 \text{ H}_2\text{S}(g)$

b. $120. \text{ g CH}_4 \times \dfrac{1 \text{ mol CH}_4}{16.04 \text{ g CH}_4} = 7.48 \text{ mol CH}_4$; $120. \text{ g S} \times \dfrac{1 \text{ mol S}}{32.07 \text{ g S}} = 3.74 \text{ mol S}$

The required S to CH_4 mol ratio is 4:1. The actual S to CH_4 mol ratio is:

$\dfrac{3.74 \text{ mol S}}{7.48 \text{ mol CH}_4} = 0.500$

This is well below the required ratio so sulfur is the limiting reagent.

The theoretical yield of CS_2 is: $3.74 \text{ mol S} \times \dfrac{1 \text{ mol CS}_2}{4 \text{ mol S}} \times \dfrac{76.15 \text{ g CS}_2}{\text{mol CS}_2} = 71.2 \text{ g CS}_2$

The same amount of CS_2 would be produced using the balanced equation with S_8.

113. $453 \text{ g Fe} \times \dfrac{1 \text{ mol Fe}}{55.85 \text{ g Fe}} \times \dfrac{1 \text{ mol Fe}_2\text{O}_3}{2 \text{ mol Fe}} \times \dfrac{159.70 \text{ g Fe}_2\text{O}_3}{\text{mol Fe}_2\text{O}_3} = 648 \text{ g Fe}_2\text{O}_3$

mass % $\text{Fe}_2\text{O}_3 = \dfrac{648 \text{ g Fe}_2\text{O}_3}{752 \text{ g ore}} \times 100 = 86.2\%$

114.   a.   Mass of Zn in alloy $= 0.0985 \text{ g ZnCl}_2 \times \dfrac{65.38 \text{ g Zn}}{136.28 \text{ g ZnCl}_2} = 0.0473 \text{ g Zn}$

%Zn $= \dfrac{0.0473 \text{ g Zn}}{0.5065 \text{ g brass}} \times 100 = 9.34\% \text{ Zn}$; %Cu $= 100.00 - 9.34 = 90.66\% \text{ Cu}$

b.   The Cu remains unreacted. After filtering, washing, and drying, the mass of the unreacted copper could be measured.

115.   Assuming one mol of vitamin A (286.4 g Vitamin A):

mol C $= 286.4 \text{ g Vitamin A} \times \dfrac{0.8386 \text{ g C}}{\text{g Vitamin A}} \times \dfrac{1 \text{ mol C}}{12.01 \text{ g C}} = 20.00 \text{ mol C}$

mol H $= 286.4 \text{ g Vitamin A} \times \dfrac{0.1056 \text{ g H}}{\text{g Vitamin A}} \times \dfrac{1 \text{ mol H}}{1.008 \text{ g H}} = 30.00 \text{ mol H}$

Since one mol of Vitamin A contains 20 mol C and 30 mol H, the molecular formula of Vitamin A is $C_{20}H_{30}E$. To determine E, let's calculate the molar mass of E.

$$286.4 \text{ g} = 20(12.01) + 30(1.008) + \text{molar mass E},\ \text{molar mass E} = 16.0 \text{ g/mol}$$

From the periodic table, E = oxygen and the molecular formula of Vitamin A is $C_{20}H_{30}O$.

Challenge Problems

116. $\dfrac{^{85}\text{Rb}}{^{87}\text{Rb}} = 2.591$; Assuming 100 atoms, let x = number of ^{85}Rb atoms and $100 - x$ = number of ^{87}Rb atoms.

$$\frac{x}{100 - x} = 2.591,\ x = 259.1 - 2.591\,x,\ x = \frac{259.1}{3.591} = 72.15\% \ ^{85}\text{Rb}$$

$$0.7215\,(84.9117) + 0.2785\,(A) = 85.4678,\ A = \frac{85.4678 - 61.26}{0.2785} = 86.92 \text{ amu} = \text{atomic mass of } ^{87}\text{Rb}$$

117.   First, we will determine composition in mass percent. We assume all the carbon in the 0.213 g CO_2 came from 0.157 g of the compound and that all the hydrogen in the 0.0310 g H_2O came from the 0.157 g of the compound.

$$0.213 \text{ g } CO_2 \times \frac{12.01 \text{ g C}}{44.01 \text{ g } CO_2} = 0.0581 \text{ g C};\ \%C = \frac{0.0581 \text{ g C}}{0.157 \text{ g compound}} \times 100 = 37.0\% \text{ C}$$

$$0.0310 \text{ g } H_2O \times \frac{2.016 \text{ g H}}{18.02 \text{ g } H_2O} = 3.47 \times 10^{-3} \text{ g H};\ \%H = \frac{3.47 \times 10^{-3} \text{ g}}{0.157 \text{ g}} = 2.21\% \text{ H}$$

We get %N from the second experiment:

$$0.0230 \text{ g } NH_3 \times \frac{14.01 \text{ g N}}{17.03 \text{ g } NH_3} = 1.89 \times 10^{-2} \text{ g N}$$

$$\%N = \frac{1.89 \times 10^{-2} \text{ g}}{0.103 \text{ g}} \times 100 = 18.3\% \text{ N}$$

The mass percent of oxygen is obtained by difference:

$$\%O = 100.00 - (37.0 + 2.21 + 18.3) = 42.5\%$$

So out of 100.00 g of compound, there are:

$$37.0 \text{ g C} \times \frac{1 \text{ mol C}}{12.01 \text{ g C}} = 3.08 \text{ mol C};\ \ 2.21 \text{ g H} \times \frac{1 \text{ mol H}}{1.008 \text{ g H}} = 2.19 \text{ mol H}$$

$$18.3 \text{ g N} \times \frac{1 \text{ mol N}}{14.01 \text{ g N}} = 1.31 \text{ mol N};\ \ 42.5 \text{ g O} \times \frac{1 \text{ mol O}}{16.00 \text{ g O}} = 2.66 \text{ mol O}$$

The last, and often the hardest part, is to find simple whole number ratios. Divide all mole values by the smallest number:

$$\frac{3.08}{1.31} = 2.35; \quad \frac{2.19}{1.31} = 1.67; \quad \frac{1.31}{1.31} = 1.00; \quad \frac{2.66}{1.31} = 2.03$$

Multiplying all these ratios by 3 gives an empirical formula of $C_7H_5N_3O_6$.

118. $1.0 \times 10^6 \text{ kg HNO}_3 \times \dfrac{1000 \text{ g HNO}_3}{\text{kg HNO}_3} \times \dfrac{1 \text{ mol HNO}_3}{63.02 \text{ g HNO}_3} = 1.6 \times 10^7 \text{ mol HNO}_3$

We need to get the relationship between moles of HNO_3 and moles of NH_3. We have to use all 3 equations.

$$\frac{2 \text{ mol HNO}_3}{3 \text{ mol NO}_2} \times \frac{2 \text{ mol NO}_2}{2 \text{ mol NO}} \times \frac{4 \text{ mol NO}}{4 \text{ mol NH}_3} = \frac{16 \text{ mol HNO}_3}{24 \text{ mol NH}_3}$$

Thus, we can produce 16 mol HNO_3 for every 24 mol NH_3 we begin with:

$$1.6 \times 10^7 \text{ mol HNO}_3 \times \frac{24 \text{ mol NH}_3}{16 \text{ mol HNO}_3} \times \frac{17.03 \text{ g NH}_3}{\text{mol NH}_3} = 4.1 \times 10^8 \text{ g or } 4.1 \times 10^5 \text{ kg}$$

This is an oversimplified answer. In practice, the NO produced in the third step is recycled back continuously into the process in the second step. If this is taken into consideration, then the conversion factor between mol NH_3 and mol HNO_3 turns out to be 1:1, i.e., 1 mol of NH_3 produces 1 mol of HNO_3. Taking into consideration that NO is recycled back gives an answer of $2.7 \times 10^5 \text{ kg NH}_3$ reacted.

119.   Total mass of copper used:

$$10{,}000 \text{ boards} \times \frac{(8.0 \text{ cm} \times 16.0 \text{ cm} \times 0.060 \text{ cm})}{\text{board}} \times \frac{8.96 \text{ g}}{\text{cm}^3} = 6.9 \times 10^5 \text{ g Cu}$$

Amount of Cu removed = $0.80 \times 6.9 \times 10^5 \text{ g} = 5.5 \times 10^5 \text{ g Cu}$

$$5.5 \times 10^5 \text{ g Cu} \times \frac{1 \text{ mol Cu}}{63.55 \text{ g Cu}} \times \frac{1 \text{ mol Cu(NH}_3)_4\text{Cl}_2}{\text{mol Cu}} \times \frac{202.59 \text{ g Cu(NH}_3)_4\text{Cl}_2}{\text{mol Cu(NH}_3)_4\text{Cl}_2}$$
$$= 1.8 \times 10^6 \text{ g Cu(NH}_3)_4\text{Cl}_2$$

$$5.5 \times 10^5 \text{ g Cu} \times \frac{1 \text{ mol Cu}}{63.55 \text{ g Cu}} \times \frac{4 \text{ mol NH}_3}{\text{mol Cu}} \times \frac{17.03 \text{ g NH}_3}{\text{mol NH}_3} = 5.9 \times 10^5 \text{ g NH}_3$$

120.   a.   From the reaction stoichiometry we would expect to produce 4 mol of acetaminophen for every 4 mol of $C_6H_5O_3N$ reacted. The actual yield is 3 moles of acetaminophen compared to a theoretical yield of 4 moles of acetaminophen. Solving for percent yield by mass (where M = molar mass acetaminophen):

$$\% \text{ yield} = \frac{3 \text{ mol} \times M}{4 \text{ mol} \times M} \times 100 = 75\%$$

b.   The product of the percent yields of the individual steps must equal the overall yield, 75%.

(0.87) (0.98) (x) = 0.75, x = 0.88; Step III has a % yield = 88%.

121.   10.00 g XCl_2 + excess $Cl_2 \rightarrow$ 12.55 g XCl_4; 2.55 g Cl reacted with XCl_2 to form XCl_4. XCl_4 contains 2.55 g Cl and 10.00 g XCl_2. From mol ratios, 10.00 g XCl_2 must also contain 2.55 g Cl; mass X in XCl_2 = 10.00 - 2.55 = 7.45 g X.

$$2.55 \text{ g Cl} \times \frac{1 \text{ mol Cl}}{35.45 \text{ g Cl}} \times \frac{1 \text{ mol } XCl_2}{2 \text{ mol Cl}} \times \frac{1 \text{ mol X}}{\text{mol } XCl_2} = 3.60 \times 10^{-2} \text{ mol X}$$

So, 3.60×10^{-2} mol X must equal 7.45 g X. The molar mass of X is:

$$\frac{7.45 \text{ g X}}{3.60 \times 10^{-2} \text{ mol X}} = \frac{207 \text{ g}}{\text{mol X}}; \text{ Atomic mass} = 207 \text{ amu so X is Pb.}$$

122.   4.000 g $M_2S_3 \rightarrow$ 3.723 g MO_2

There must be twice as many mol of MO_2 as mol of M_2S_3 in order to balance M in the reaction. Setting up an equation for 2 mol MO_2 = mol M_2S_3 where A = molar mass M:

$$2\left(\frac{4.000 \text{ g}}{2 \text{ A} + 3(32.07)}\right) = \frac{3.723 \text{ g}}{\text{A} + 2(16.00)}, \quad \frac{8.000}{2 \text{ A} + 96.21} = \frac{3.723}{\text{A} + 32.00}$$

8.000 A + 256.0 = 7.446 A + 358.2, 0.554 A = 102.2, A = 184 g/mol; atomic mass = 184 amu

123.   Consider the case of aluminum plus oxygen. Aluminum forms Al^{3+} ions; oxygen forms O^{2-} anions. The simplest compound of the two elements is Al_2O_3. Similarly, we would expect the formula of any group 6A element with Al to be Al_2X_3. Assuming this, out of 100.00 g of compound there are 18.56 g Al and 81.44 g of the unknown element, X. Let's use this information to determine the molar mass of X which will allow us to identify X from the periodic table.

$$18.56 \text{ g Al} \times \frac{1 \text{ mol Al}}{26.98 \text{ g Al}} \times \frac{3 \text{ mol X}}{2 \text{ mol Al}} = 1.032 \text{ mol X}$$

81.44 g of X must contain 1.032 mol of X.

The molar mass of X = $\dfrac{81.44 \text{ g X}}{1.032 \text{ mol X}}$ = 78.91 g/mol X.

From the periodic table, the unknown element is selenium and the formula is Al_2Se_3.

124. $NaCl(aq) + Ag^+(aq) \rightarrow AgCl(s);\ \ KCl(aq) + Ag^+(aq) \rightarrow AgCl(s)$

$$8.5904 \text{ g AgCl} \times \frac{1 \text{ mol AgCl}}{143.4 \text{ g AgCl}} \times \frac{1 \text{ mol Cl}^-}{1 \text{ mol AgCl}} = 5.991 \times 10^{-2} \text{ mol Cl}^-$$

The molar masses of NaCl and KCl are 58.44 and 74.55 g/mol, respectively. Let x = g NaCl and y = g KCl:

$$x + y = 4.000 \text{ g and } \frac{x}{58.44} + \frac{y}{74.55} = 5.991 \times 10^{-2} \text{ total mol Cl}^- \text{ or } 1.276\,x + y = 4.466$$

Solving using simultaneous equations:

$$
\begin{array}{rl}
1.276\,x + y = & 4.466 \\
-x - y = & -4.000 \\
\hline
0.276\,x \quad\quad = & 0.466, \quad x = 1.69 \text{ g NaCl and } y = 2.31 \text{ g KCl}
\end{array}
$$

$$\% \text{ NaCl} = \frac{1.69 \text{ g}}{4.000 \text{ g}} \times 100 = 42.3\% \text{ NaCl};\ \ \% \text{ KCl} = 57.7\%$$

125.   The balanced equations are:

$$4 \text{ NH}_3(g) + 5 \text{ O}_2(g) \rightarrow 4 \text{ NO}(g) + 6 \text{ H}_2\text{O}(g) \text{ and } 4 \text{ NH}_3(g) + 7 \text{ O}_2(g) \rightarrow 4 \text{ NO}_2(g) + 6 \text{ H}_2\text{O}(g)$$

Let 4x = number of mol of NO formed, and let 4y = number of mol of NO_2 formed. Then:

$$4x \text{ NH}_3 + 5x \text{ O}_2 \rightarrow 4x \text{ NO} + 6x \text{ H}_2\text{O} \text{ and } 4y \text{ NH}_3 + 7y \text{ O}_2 \rightarrow 4y \text{ NO}_2 + 6y \text{ H}_2\text{O}$$

All the NH_3 reacted, so 4x + 4y = 2.00. 10.00 - 6.75 = 3.25 mol O_2 reacted, so 5x + 7y = 3.25.

Solving by the method of simultaneous equations:

$$
\begin{array}{rl}
20\,x + 28\,y = & 13.0 \\
-20\,x - 20\,y = & -10.0 \\
\hline
8\,y = & 3.0, \quad y = 0.38;\ \ 4x + 4 \times 0.38 = 2.00,\ \ x = 0.12
\end{array}
$$

mol NO = 4x = 4 × 0.12 = 0.48 mol NO formed

CHAPTER FOUR

TYPES OF CHEMICAL REACTIONS AND SOLUTION STOICHIOMETRY

Questions

9. "Slightly soluble" refers to substances that dissolve only to a small extent. A slightly soluble salt may still dissociate completely to ions and, hence, be a strong electrolyte. An example of such a substance is $Mg(OH)_2$. It is a strong electrolyte, but not very soluble. A weak electrolyte is a substance that doesn't dissociate completely to produce ions. A weak electrolyte may be very soluble in water, or it may not be very soluble. Acetic acid is an example of a weak electrolyte that is very soluble in water.

10. Measure the electrical conductivity of a solution and compare it to the conductivity of a solution of equal concentration of a strong electrolyte.

Exercises

Aqueous Solutions: Strong and Weak Electrolytes

11. a. $NaBr(s) \rightarrow Na^+(aq) + Br^-(aq)$

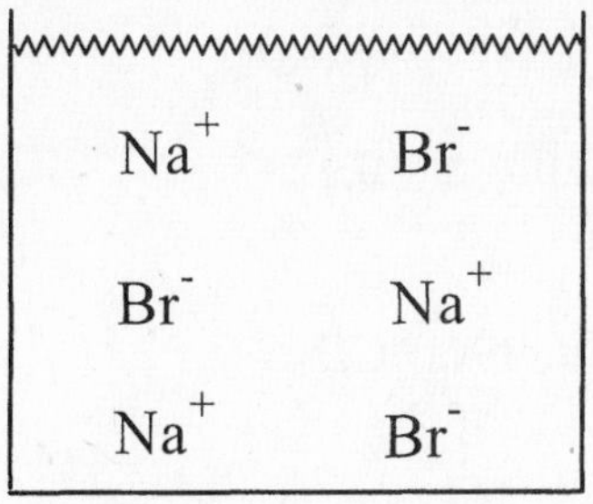

b. $MgCl_2(s) \rightarrow Mg^{2+}(aq) + 2\,Cl^-(aq)$

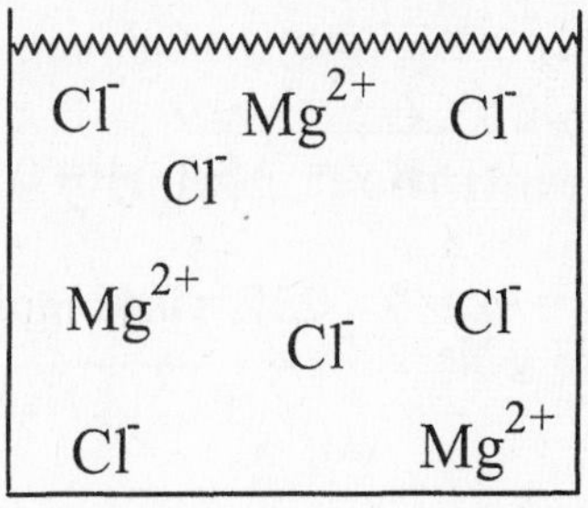

Your drawing should show equal numbers of Na^+ and Br^- ions.

Your drawing should show twice the number of Cl^- ions as Mg^{2+} ions.

c. $Al(NO_3)_3(s) \rightarrow Al^{3+}(aq) + 3\,NO_3^-(aq)$ d. $(NH_4)_2SO_4(s) \rightarrow 2\,NH_4^+(aq) + SO_4^{2-}(aq)$

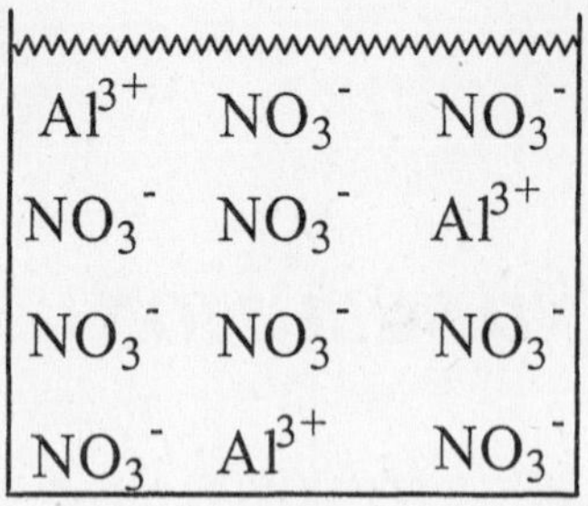

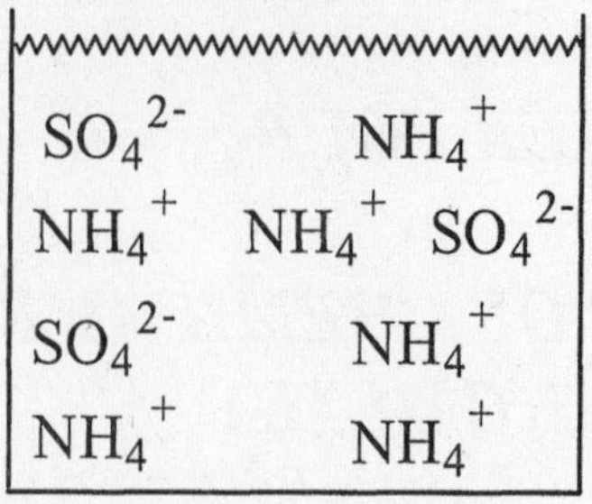

Al^{3+}	NO_3^-	NO_3^-
NO_3^-	NO_3^-	Al^{3+}
NO_3^-	NO_3^-	NO_3^-
NO_3^-	Al^{3+}	NO_3^-

SO_4^{2-}	NH_4^+	
NH_4^+	NH_4^+	SO_4^{2-}
SO_4^{2-}	NH_4^+	
NH_4^+	NH_4^+	

For e-i, your drawings should show equal numbers of the cations and anions present as each salt is a 1:1 salt. The ions present are listed in the following dissolution reactions.

e. $NaOH(s) \rightarrow Na^+(aq) + OH^-(aq)$ f. $FeSO_4(s) \rightarrow Fe^{2+}(aq) + SO_4^{2-}(aq)$

g. $KMnO_4(s) \rightarrow K^+(aq) + MnO_4^-(aq)$ h. $HClO_4(aq) \rightarrow H^+(aq) + ClO_4^-(aq)$

i. $NH_4C_2H_3O_2(s) \rightarrow NH_4^+(aq) + C_2H_3O_2^-(aq)$

12. a. $Ba(NO_3)_2(aq) \rightarrow Ba^{2+}(aq) + 2\,NO_3^-(aq)$; Picture iv represents the Ba^{2+} and NO_3^- ions present in $Ba(NO_3)_2(aq)$.

b. $NaCl(aq) \rightarrow Na^+(aq) + Cl^-(aq)$; Picture ii represents $NaCl(aq)$.

c. $K_2CO_3(aq) \rightarrow 2\,K^+(aq) + CO_3^{2-}(aq)$; Picture iii represents $K_2CO_3(aq)$.

d. $MgSO_4(aq) \rightarrow Mg^{2+}(aq) + SO_4^{2-}(aq)$; Picture i represents $MgSO_4(aq)$.

13. $CaCl_2(s) \rightarrow Ca^{2+}(aq) + 2\,Cl^-(aq)$

14. $MgSO_4(s) \rightarrow Mg^{2+}(aq) + SO_4^{2-}(aq)$; $NH_4NO_3(s) \rightarrow NH_4^+(aq) + NO_3^-(aq)$

Solution Concentration: Molarity

15. a. $5.623 \text{ g NaHCO}_3 \times \dfrac{1 \text{ mol NaHCO}_3}{84.01 \text{ g NaHCO}_3} = 6.693 \times 10^{-2} \text{ mol NaHCO}_3$

$$M = \frac{6.693 \times 10^{-2} \text{ mol}}{250.0 \text{ mL}} \times \frac{1000 \text{ mL}}{\text{L}} = 0.2677\ M \text{ NaHCO}_3$$

b. $0.1846 \text{ g K}_2Cr_2O_7 \times \dfrac{1 \text{ mol K}_2Cr_2O_7}{294.20 \text{ g K}_2Cr_2O_7} = 6.275 \times 10^{-4} \text{ mol K}_2Cr_2O_7$

$$M = \frac{6.275 \times 10^{-4} \text{ mol}}{500.0 \times 10^{-3} \text{ L}} = 1.255 \times 10^{-3}\ M \text{ K}_2Cr_2O_7$$

c. $0.1025 \text{ g Cu} \times \dfrac{1 \text{ mol Cu}}{63.55 \text{ g Cu}} = 1.613 \times 10^{-3} \text{ mol Cu} = 1.613 \times 10^{-3} \text{ mol Cu}^{2+}$

$$M = \dfrac{1.613 \times 10^{-3} \text{ mol Cu}^{2+}}{200.0 \text{ mL}} \times \dfrac{1000 \text{ mL}}{\text{L}} = 8.065 \times 10^{-3} \; M \text{ Cu}^{2+}$$

16. $75.0 \text{ mL} \times \dfrac{0.79 \text{ g}}{\text{mL}} \times \dfrac{1 \text{ mol}}{46.07 \text{ g}} = 1.3 \text{ mol C}_2\text{H}_5\text{OH}; \; \text{Molarity} = \dfrac{1.3 \text{ mol}}{0.250 \text{ L}} = 5.2 \; M \text{ C}_2\text{H}_5\text{OH}$

17.   a. $\text{CaCl}_2(s) \rightarrow \text{Ca}^{2+}(aq) + 2 \text{ Cl}^-(aq); \; M_{\text{Ca}^{2+}} = 0.15 \; M; \; M_{\text{Cl}^-} = 2(0.15) = 0.30 \; M$

b. $\text{Al(NO}_3)_3(s) \rightarrow \text{Al}^{3+}(aq) + 3 \text{ NO}_3^-(aq); \; M_{\text{Al}^{3+}} = 0.26 \; M; \; M_{\text{NO}_3^-} = 3(0.26) = 0.78 \; M$

c. $\text{K}_2\text{Cr}_2\text{O}_7(s) \rightarrow 2 \text{ K}^+(aq) + \text{Cr}_2\text{O}_7^{2-}(aq); \; M_{\text{K}^+} = 2(0.25) = 0.50 \; M; \; M_{\text{Cr}_2\text{O}_7^{2-}} = 0.25 \; M$

d. $\text{Al}_2(\text{SO}_4)_3(s) \rightarrow 2 \text{ Al}^{3+}(aq) + 3 \text{ SO}_4^{2-}(aq)$

$$M_{\text{Al}^{3+}} = \dfrac{2.0 \times 10^{-3} \text{ mol Al}_2(\text{SO}_4)_3}{\text{L}} \times \dfrac{2 \text{ mol Al}^{3+}}{\text{mol Al}_2(\text{SO}_4)_3} = 4.0 \times 10^{-3} \; M$$

$$M_{\text{SO}_4^{2-}} = \dfrac{2.0 \times 10^{-3} \text{ mol Al}_2(\text{SO}_4)_3}{\text{L}} \times \dfrac{3 \text{ mol SO}_4^{2-}}{\text{mol Al}_2(\text{SO}_4)_3} = 6.0 \times 10^{-3} \; M$$

18.   a. $M_{\text{Ca(NO}_3)_2} = \dfrac{0.100 \text{ mol Ca(NO}_3)_2}{0.100 \text{ L}} = 1.00 \; M$

$\text{Ca(NO}_3)_2(s) \rightarrow \text{Ca}^{2+}(aq) + 2 \text{ NO}_3^-(aq); \; M_{\text{Ca}^{2+}} = 1.00 \; M; \; M_{\text{NO}_3^-} = 2(1.00) = 2.00 \; M$

b. $M_{\text{Na}_2\text{SO}_4} = \dfrac{2.5 \text{ mol Na}_2\text{SO}_4}{1.25 \text{ L}} = 2.0 \; M$

$\text{Na}_2\text{SO}_4(s) \rightarrow 2 \text{ Na}^+(aq) + \text{SO}_4^{2-}(aq); \; M_{\text{Na}^+} = 2(2.0) = 4.0 \; M; \; M_{\text{SO}_4^{2-}} = 2.0 \; M$

c. $5.00 \text{ g NH}_4\text{Cl} \times \dfrac{1 \text{ mol NH}_4\text{Cl}}{53.49 \text{ g NH}_4\text{Cl}} = 0.0935 \text{ mol NH}_4\text{Cl}$

$$M_{\text{NH}_4\text{Cl}} = \dfrac{0.0935 \text{ mol NH}_4\text{Cl}}{0.5000 \text{ L}} = 0.187 M$$

$\text{NH}_4\text{Cl}(s) \rightarrow \text{NH}_4^+(aq) + \text{Cl}^-(aq); \; M_{\text{NH}_4^+} = M_{\text{Cl}^-} = 0.187 \; M$

d.　$1.00 \text{ g K}_3\text{PO}_4 \times \dfrac{1 \text{ mol K}_3\text{PO}_4}{212.27 \text{ g}} = 4.71 \times 10^{-3} \text{ mol K}_3\text{PO}_4$

$M_{\text{K}_3\text{PO}_4} = \dfrac{4.71 \times 10^{-3} \text{ mol}}{0.2500 \text{ L}} = 0.0188 \text{ } M$

$\text{K}_3\text{PO}_4(s) \rightarrow 3 \text{ K}^+(aq) + \text{PO}_4^{3-}(aq);\ M_{\text{K}^+} = 3(0.0188) = 0.0564 \text{ } M;\ M_{\text{PO}_4^{3-}} = 0.0188 \text{ } M$

19.　mol solute = volume (L) $\times$ molarity $\left(\dfrac{\text{mol}}{\text{L}} \right)$;　$\text{AlCl}_3(s) \rightarrow \text{Al}^{3+}(aq) + 3 \text{ Cl}^-(aq)$

mol Cl$^-$ = $0.1000 \text{ L} \times \dfrac{0.30 \text{ mol AlCl}_3}{\text{L}} \times \dfrac{3 \text{ mol Cl}^-}{\text{mol AlCl}_3} = 9.0 \times 10^{-2}$ mol Cl$^-$

$\text{MgCl}_2(s) \rightarrow \text{Mg}^{2+}(aq) + 2 \text{ Cl}^-(aq)$

mol Cl$^-$ = $0.0500 \text{ L} \times \dfrac{0.60 \text{ mol MgCl}_2}{\text{L}} \times \dfrac{2 \text{ mol Cl}^-}{\text{mol MgCl}_2} = 6.0 \times 10^{-2}$ mol Cl$^-$

$\text{NaCl}(s) \rightarrow \text{Na}^+(aq) + \text{Cl}^-(aq)$

mol Cl$^-$ = $0.2000 \text{ L} \times \dfrac{0.40 \text{ mol NaCl}}{\text{L}} \times \dfrac{1 \text{ mol Cl}^-}{\text{mol NaCl}} = 8.0 \times 10^{-2}$ mol Cl$^-$

100.0 mL of 0.30 M AlCl$_3$ contains the most moles of Cl$^-$ ions.

20.　$\text{NaOH}(s) \rightarrow \text{Na}^+(aq) + \text{OH}^-(aq)$,　2 total mol of ions (1 mol Na$^+$ and 1 mol Cl$^-$) per mol NaOH.

$0.1000 \text{ L} \times \dfrac{0.100 \text{ mol NaOH}}{\text{L}} \times \dfrac{2 \text{ mol ions}}{\text{mol NaOH}} = 2.0 \times 10^{-2}$ mol ions

$\text{BaCl}_2(s) \rightarrow \text{Ba}^{2+}(aq) + 2 \text{ Cl}^-(aq)$,　3 total mol of ions per mol BaCl$_2$.

$0.0500 \text{ L} \times \dfrac{0.200 \text{ mol}}{\text{L}} \times \dfrac{3 \text{ mol ions}}{\text{mol BaCl}_2} = 3.0 \times 10^{-2}$ mol ions

$\text{Na}_3\text{PO}_4(s) \rightarrow 3 \text{ Na}^+(aq) + \text{PO}_4^{3-}(aq)$,　4 total mol of ions per mol Na$_3$PO$_4$.

$0.0750 \text{ L} \times \dfrac{0.150 \text{ mol Na}_3\text{PO}_4}{\text{L}} \times \dfrac{4 \text{ mol ions}}{\text{mol Na}_3\text{PO}_4} = 4.50 \times 10^{-2}$ mol ions

75.0 mL of 0.150 M Na$_3$PO$_4$ contains the largest number of ions.

21.   Molar mass of $NaHCO_3 = 22.99 + 1.008 + 12.01 + 3(16.00) = 84.01$ g/mol

Volume = 0.350 g $NaHCO_3 \times \dfrac{1 \text{ mol } NaHCO_3}{84.01 \text{ g } NaHCO_3} \times \dfrac{1 \text{ L}}{0.100 \text{ mol } NaHCO_3} = 0.0417 \text{ L} = 41.7$ mL

41.7 mL of 0.100 M $NaHCO_3$ contains 0.350 g $NaHCO_3$.

22.   Molar mass of NaOH = $22.99 + 16.00 + 1.008 = 40.00$ g/mol

Mass NaOH = $0.2500 \text{ L} \times \dfrac{0.400 \text{ mol NaOH}}{\text{L}} \times \dfrac{40.00 \text{ g NaOH}}{\text{mol NaOH}} = 4.00$ g NaOH

23.   a. $2.00 \text{ L} \times \dfrac{0.250 \text{ mol NaOH}}{\text{L}} \times \dfrac{40.00 \text{ g NaOH}}{\text{mol}} = 20.0$ g NaOH

Place 20.0 g NaOH in a 2 L volumetric flask; add water to dissolve the NaOH, and fill to the mark with water, mixing several times along the way.

b. $2.00 \text{ L} \times \dfrac{0.250 \text{ mol NaOH}}{\text{L}} \times \dfrac{1 \text{ L stock}}{1.00 \text{ mol NaOH}} = 0.500$ L

Add 500. mL of 1.00 M NaOH stock solution to a 2 L volumetric flask; fill to the mark with water, mixing several times along the way.

c. $2.00 \text{ L} \times \dfrac{0.100 \text{ mol } K_2CrO_4}{\text{L}} \times \dfrac{194.20 \text{ g } K_2CrO_4}{\text{mol } K_2CrO_4} = 38.8$ g K_2CrO_4

Similar to the solution made in part a, instead using 38.8 g K_2CrO_4.

d. $2.00 \text{ L} \times \dfrac{0.100 \text{ mol } K_2CrO_4}{\text{L}} \times \dfrac{1 \text{ L stock}}{1.75 \text{ mol } K_2CrO_4} = 0.114$ L

Similar to the solution made in part b, instead using 114 mL of the 1.75 M K_2CrO_4 stock solution.

24.   a. $1.00 \text{ L solution} \times \dfrac{0.50 \text{ mol } H_2SO_4}{\text{L}} = 0.50$ mol H_2SO_4

$0.50 \text{ mol } H_2SO_4 \times \dfrac{1 \text{ L}}{18 \text{ mol } H_2SO_4} = 2.8 \times 10^{-2}$ L conc. H_2SO_4 or 28 mL

Dilute 28 mL of concentrated H_2SO_4 to a total volume of 1.00 L with water.

b.   We will need 0.50 mol HCl.

$0.50 \text{ mol HCl} \times \dfrac{1 \text{ L}}{12 \text{ mol HCl}} = 4.2 \times 10^{-2} \text{ L} = 42$ mL

Dilute 42 mL of concentrated HCl to a final volume of 1.00 L.

c. We need 0.50 mol $NiCl_2$.

$$0.50 \text{ mol } NiCl_2 \times \frac{1 \text{ mol } NiCl_2 \cdot 6H_2O}{\text{mol } NiCl_2} \times \frac{237.69 \text{ g } NiCl_2 \cdot 6H_2O}{\text{mol } NiCl_2 \cdot 6H_2O} = 118.8 \text{ g } NiCl_2 \cdot 6H_2O \approx 120 \text{ g}$$

Dissolve 120 g $NiCl_2 \cdot 6H_2O$ in water, and add water until the total volume of the solution is 1.00 L.

d. $1.00 \text{ L} \times \dfrac{0.50 \text{ mol } HNO_3}{\text{L}} = 0.50 \text{ mol } HNO_3$

$$0.50 \text{ mol } HNO_3 \times \frac{1 \text{ L}}{16 \text{ mol } HNO_3} = 0.031 \text{ L} = 31 \text{ mL}$$

Dissolve 31 mL of concentrated reagent in water. Dilute to a total volume of 1.00 L.

e. We need 0.50 mol Na_2CO_3.

$$0.50 \text{ mol } Na_2CO_3 \times \frac{105.99 \text{ g } Na_2CO_3}{\text{mol}} = 53 \text{ g } Na_2CO_3$$

Dissolve 53 g Na_2CO_3 in water, dilute to 1.00 L.

25. $10.8 \text{ g } (NH_4)_2SO_4 \times \dfrac{1 \text{ mol}}{132.15 \text{ g}} = 8.17 \times 10^{-2} \text{ mol } (NH_4)_2SO_4$

$$\text{Molarity} = \frac{8.17 \times 10^{-2} \text{ mol}}{100.0 \text{ mL}} \times \frac{1000 \text{ mL}}{\text{L}} = 0.817 \text{ } M \text{ } (NH_4)_2SO_4$$

Moles of $(NH_4)_2SO_4$ in final solution

$$10.00 \times 10^{-3} \text{ L} \times \frac{0.817 \text{ mol}}{\text{L}} = 8.17 \times 10^{-3} \text{ mol}$$

$$\text{Molarity of final solution} = \frac{8.17 \times 10^{-3} \text{ mol}}{(10.00 + 50.00) \text{ mL}} \times \frac{1000 \text{ mL}}{\text{L}} = 0.136 \text{ } M \text{ } (NH_4)_2SO_4$$

$(NH_4)_2SO_4(s) \rightarrow 2 \text{ } NH_4^+(aq) + SO_4^{2-}(aq); \text{ } M_{NH_4^+} = 2(0.136) = 0.272 \text{ } M; \text{ } M_{SO_4^{2-}} = 0.136 \text{ } M$

26. $\text{mol } Na_2CO_3 = 0.0700 \text{ L} \times \dfrac{3.0 \text{ mol } Na_2CO_3}{\text{L}} = 0.21 \text{ mol } Na_2CO_3$

$Na_2CO_3(s) \rightarrow 2 \text{ } Na^+(aq) + CO_3^{2-}(aq); \text{ mol } Na^+ = 2(0.21) = 0.42 \text{ mol}$

$\text{mol } NaHCO_3 = 0.0300 \text{ L} \times \dfrac{1.0 \text{ mol } NaHCO_3}{\text{L}} = 0.030 \text{ mol } NaHCO_3$

$NaHCO_3(s) \rightarrow Na^+(aq) + HCO_3^-(aq); \text{ mol } Na^+ = 0.030 \text{ mol}$

$$M_{Na^+} = \frac{\text{total mol } Na^+}{\text{total volume}} = \frac{0.42 \text{ mol} + 0.030 \text{ mol}}{0.0700 \text{ L} + 0.0300 \text{ L}} = \frac{0.45 \text{ mol}}{0.1000 \text{ L}} = 4.5 \text{ } M \text{ } Na^+$$

27.   Stock solution $= \dfrac{10.0 \text{ mg}}{500.0 \text{ mL}} = \dfrac{10.0 \times 10^{-3} \text{ g}}{500.0 \text{ mL}} = \dfrac{2.00 \times 10^{-5} \text{ g steroid}}{\text{mL}}$

$100.0 \times 10^{-6} \text{ L stock} \times \dfrac{1000 \text{ mL}}{\text{L}} \times \dfrac{2.00 \times 10^{-5} \text{ g steroid}}{\text{mL}} = 2.00 \times 10^{-6} \text{ g steroid}$

This is diluted to a final volume of 100.0 mL.

$\dfrac{2.00 \times 10^{-6} \text{ g steroid}}{100.0 \text{ mL}} \times \dfrac{1000 \text{ mL}}{\text{L}} \times \dfrac{1 \text{ mol steroid}}{336.43 \text{ g steroid}} = 5.94 \times 10^{-8} \ M \text{ steroid}$

28.   Stock solution:

$1.584 \text{ g Mn}^{2+} \times \dfrac{1 \text{ mol Mn}^{2+}}{54.94 \text{ g Mn}^{2+}} = 2.883 \times 10^{-2} \text{ mol Mn}^{2+}; \ \dfrac{2.883 \times 10^{-2} \text{ mol Mn}^{2+}}{1.000 \text{ L}} = 2.883 \times 10^{-2} \ M$

Solution A contains:

$50.00 \text{ mL} \times \dfrac{1 \text{ L}}{1000 \text{ mL}} \times \dfrac{2.883 \times 10^{-2} \text{ mol}}{\text{L}} = 1.442 \times 10^{-3} \text{ mol Mn}^{2+}$

$\text{Molarity} = \dfrac{1.442 \times 10^{-3} \text{ mol}}{1000.0 \text{ mL}} \times \dfrac{1000 \text{ mL}}{\text{L}} = 1.442 \times 10^{-3} \ M$

Solution B contains:

$10.0 \text{ mL} \times \dfrac{1 \text{ L}}{1000 \text{ mL}} \times \dfrac{1.442 \times 10^{-3} \text{ mol}}{\text{L}} = 1.442 \times 10^{-5} \text{ mol Mn}^{2+}$

$\text{Molarity} = \dfrac{1.442 \times 10^{-5} \text{ mol}}{0.2500 \text{ L}} = 5.768 \times 10^{-5} \ M$

Solution C contains:

$10.00 \times 10^{-3} \text{ L} \times \dfrac{5.768 \times 10^{-5} \text{ mol}}{\text{L}} = 5.768 \times 10^{-7} \text{ mol Mn}^{2+}$

$\text{Molarity} = \dfrac{5.768 \times 10^{-7} \text{ mol}}{0.5000 \text{ L}} = 1.154 \times 10^{-6} \ M$

Precipitation Reactions

29.   In these reactions, soluble ionic compounds are mixed together. To predict the precipitate, switch the anions and cations in the two reactant compounds to predict possible products; then use the solubility rules in Table 4.1 to predict if any of these possible products are insoluble (are the precipitate).

   a.   Possible products $= BaSO_4$ and $NaCl$; precipitate $= BaSO_4(s)$
   b.   Possible products $= PbCl_2$ and KNO_3; precipitate $= PbCl_2(s)$
   c.   Possible products $= Ag_3PO_4$ and $NaNO_3$; precipitate $= Ag_3PO_4(s)$
   d.   Possible products $= NaNO_3$ and $Fe(OH)_3$; precipitate $= Fe(OH)_3(s)$

30.   a. Possible products = $FeCl_2$ and K_2SO_4; Both salts are soluble so no precipitate forms.
       b. Possible products = $Al(OH)_3$ and $Ba(NO_3)_2$; precipitate = $Al(OH)_3(s)$
       c. Possible products = $CaSO_4$ and NaCl; precipitate = $CaSO_4(s)$
       d. Possible products = KNO_3 and NiS; precipitate = NiS(s)

31.   For the following answers, the balanced molecular equation is first, followed by the complete ionic equation, then the net ionic equation.

       a. $BaCl_2(aq) + Na_2SO_4(aq) \rightarrow BaSO_4(s) + 2\ NaCl(aq)$

           $Ba^{2+}(aq) + 2\ Cl^-(aq) + 2\ Na^+(aq) + SO_4^{2-}(aq) \rightarrow BaSO_4(s) + 2\ Na^+(aq) + 2\ Cl^-(aq)$

           $Ba^{2+}(aq) + SO_4^{2-}(aq) \rightarrow BaSO_4(s)$

       b. $Pb(NO_3)_2(aq) + 2\ KCl(aq) \rightarrow PbCl_2(s) + 2\ KNO_3(aq)$

           $Pb^{2+}(aq) + 2\ NO_3^-(aq) + 2\ K^+(aq) + 2\ Cl^-(aq) \rightarrow PbCl_2(s) + 2\ K^+(aq) + 2\ NO_3^-(aq)$

           $Pb^{2+}(aq) + 2\ Cl^-(aq) \rightarrow PbCl_2(s)$

       c. $3\ AgNO_3(aq) + Na_3PO_4(aq) \rightarrow Ag_3PO_4(s) + 3\ NaNO_3(aq)$

           $3\ Ag^+(aq) + 3\ NO_3^-(aq) + 3\ Na^+(aq) + PO_4^{3-}(aq) \rightarrow Ag_3PO_4(s) + 3\ Na^+(aq) + 3\ NO_3^-(aq)$

           $3\ Ag^+(aq) + PO_4^{3-}(aq) \rightarrow Ag_3PO_4(s)$

       d. $3\ NaOH(aq) + Fe(NO_3)_3(aq) \rightarrow Fe(OH)_3(s) + 3\ NaNO_3(aq)$

           $3\ Na^+(aq) + 3\ OH^-(aq) + Fe^{3+}(aq) + 3\ NO_3^-(aq) \rightarrow Fe(OH)_3(s) + 3\ Na^+(aq) + 3\ NO_3^-(aq)$

           $Fe^{3+}(aq) + 3\ OH^-(aq) \rightarrow Fe(OH)_3(s)$

32.   a. No reaction occurs since all possible products are soluble salts.

       b. $2\ Al(NO_3)_3(aq) + 3\ Ba(OH)_2(aq) \rightarrow 2\ Al(OH)_3(s) + 3\ Ba(NO_3)_2(aq)$

           $2\ Al^{3+}(aq) + 6\ NO_3^-(aq) + 3\ Ba^{2+}(aq) + 6\ OH^-(aq) \rightarrow 2\ Al(OH)_3(s) + 3\ Ba^{2+}(aq) + 6\ NO_3^-(aq)$

           $Al^{3+}(aq) + 3\ OH^-(aq) \rightarrow Al(OH)_3(s)$

       c. $CaCl_2(aq) + Na_2SO_4(aq) \rightarrow CaSO_4(s) + 2\ NaCl(aq)$

           $Ca^{2+}(aq) + 2\ Cl^-(aq) + 2\ Na^+(aq) + SO_4^{2-}(aq) \rightarrow CaSO_4(s) + 2\ Na^+(aq) + 2\ Cl^-(aq)$

           $Ca^{2+}(aq) + SO_4^{2-}(aq) \rightarrow CaSO_4(s)$

d. $K_2S(aq) + Ni(NO_3)_2(aq) \rightarrow 2\ KNO_3(aq) + NiS(s)$

$2\ K^+(aq) + S^{2-}(aq) + Ni^{2+}(aq) + 2\ NO_3^-(aq) \rightarrow 2\ K^+(aq) + 2\ NO_3^-(aq) + NiS(s)$

$Ni^{2+}(aq) + S^{2-}(aq) \rightarrow NiS(s)$

33. a. When $CuSO_4(aq)$ is added to $Na_2S(aq)$, the precipitate that forms is $CuS(s)$. Therefore, Na^+ (the grey spheres) and SO_4^{2-} (the blueish-green spheres) are the spectator ions.

$CuSO_4(aq) + Na_2S(aq) \rightarrow CuS(s) + Na_2SO_4(aq);\ Cu^{2+}(aq) + S^{2-}(aq) \rightarrow CuS(s)$

b. When $CoCl_2(aq)$ is added to $NaOH(aq)$, the precipitate that forms is $Co(OH)_2(s)$. Therefore, Na^+ (the grey spheres) and Cl^- (the green spheres) are the spectator ions.

$CoCl_2(aq) + 2\ NaOH(aq) \rightarrow Co(OH)_2(s) + 2\ NaCl(aq);\ Co^{2+}(aq) + 2\ OH^-(aq) \rightarrow Co(OH)_2(s)$

c. When $AgNO_3(aq)$ is added to $KI(aq)$, the precipitate that forms is $AgI(s)$. Therefore, K^+ (the red spheres) and NO_3^- (the blue spheres) are the spectator ions.

$AgNO_3(aq) + KI(aq) \rightarrow AgI(s) + KNO_3(aq);\ Ag^+(aq) + I^-(aq) \rightarrow AgI(s)$

34. There are many acceptable choices for spectator ions. We will generally choose Na^+ and NO_3^- as the spectator ions because sodium salts and nitrate salts are usually soluble in water.

a. $Fe(NO_3)_3(aq) + 3\ NaOH(aq) \rightarrow Fe(OH)_3(s) + 3\ NaNO_3(aq)$

b. $Hg_2(NO_3)_2(aq) + 2\ NaCl(aq) \rightarrow Hg_2Cl_2(s) + 2\ NaNO_3(aq)$

c. $Pb(NO_3)_2(aq) + Na_2SO_4(aq) \rightarrow PbSO_4(s) + 2\ NaNO_3(aq)$

d. $BaCl_2(aq) + Na_2CrO_4(aq) \rightarrow BaCrO_4(s) + 2\ NaCl(aq)$

35. a. $(NH_4)_2SO_4(aq) + Ba(NO_3)_2(aq) \rightarrow 2\ NH_4NO_3(aq) + BaSO_4(s)$

$Ba^{2+}(aq) + SO_4^{2-}(aq) \rightarrow BaSO_4(s)$

b. $Pb(NO_3)_2(aq) + 2\ NaCl(aq) \rightarrow PbCl_2(s) + 2\ NaNO_3(aq)$

$Pb^{2+}(aq) + 2\ Cl^-(aq) \rightarrow PbCl_2(s)$

c. Potassium phosphate and sodium nitrate are both soluble in water. No reaction occurs.

d. No reaction occurs since all possible products are soluble.

e. $CuCl_2(aq) + 2\ NaOH(aq) \rightarrow Cu(OH)_2(s) + 2\ NaCl(aq)$

$Cu^{2+}(aq) + 2\ OH^-(aq) \rightarrow Cu(OH)_2(s)$

36.   a. $CrCl_3(aq) + 3\ NaOH(aq) \rightarrow Cr(OH)_3(s) + 3\ NaCl(aq)$

$Cr^{3+}(aq) + 3\ OH^-(aq) \rightarrow Cr(OH)_3(s)$

b. $2\ AgNO_3(aq) + (NH_4)_2CO_3(aq) \rightarrow Ag_2CO_3(s) + 2\ NH_4NO_3(aq)$

$2\ Ag^+(aq) + CO_3^{2-}(aq) \rightarrow Ag_2CO_3(s)$

c. $CuSO_4(aq) + Hg_2(NO_3)_2(aq) \rightarrow Cu(NO_3)_2(aq) + Hg_2SO_4(s)$

$Hg_2^{2+}(aq) + SO_4^{2-}(aq) \rightarrow Hg_2SO_4(s)$

d.   No reaction occurs since all possible products (SrI_2 and KNO_3) are soluble.

37.   Three possibilities are:

Addition of K_2SO_4 solution to give a white ppt. of $PbSO_4$. Addition of NaCl solution to give a white ppt. of $PbCl_2$. Addition of K_2CrO_4 solution to give a bright yellow ppt. of $PbCrO_4$.

38.   Since no precipitates formed upon addition of NaCl or Na_2SO_4, we can conclude that Hg_2^{2+} and Ba^{2+} are not present in the sample since Hg_2Cl_2 and $BaSO_4$ are insoluble salts. However, Mn^{2+} may be present since Mn^{2+} does not form a precipitate with either NaCl or Na_2SO_4. Since a precipitate formed with NaOH, the solution must contain Mn^{2+} because it forms a precipitate with OH^- [$Mn(OH)_2(s)$].

39. $2\ AgNO_3(aq) + Na_2CrO_4(aq) \rightarrow Ag_2CrO_4(s) + 2\ NaNO_3(aq)$

$$0.0750\ L \times \frac{0.100\ mol\ AgNO_3}{L} \times \frac{1\ mol\ Na_2CrO_4}{2\ mol\ AgNO_3} \times \frac{161.98\ g\ Na_2CrO_4}{mol\ Na_2CrO_4} = 0.607\ g\ Na_2CrO_4$$

40. $2\ Na_3PO_4(aq) + 3\ Pb(NO_3)_2(aq) \rightarrow Pb_3(PO_4)_2(s) + 6\ NaNO_3(aq)$

$$0.1500\ L \times \frac{0.250\ mol\ Pb(NO_3)_2}{L} \times \frac{2\ mol\ Na_3PO_4}{3\ mol\ Pb(NO_3)_2} \times \frac{1\ L\ Na_3PO_4}{0.100\ mol\ Na_3PO_4} = 0.250\ L$$

$$= 250.\ mL\ Na_3PO_4$$

41. $Al(NO_3)_3(aq) + 3\ KOH(aq) \rightarrow Al(OH)_3(s) + 3\ KNO_3(aq)$

$$0.0500\ L \times \frac{0.200\ mol\ Al(NO_3)_3}{L} = 0.0100\ mol\ Al(NO_3)_3$$

$$0.2000\ L \times \frac{0.100\ mol\ KOH}{L} = 0.0200\ mol\ KOH$$

From the balanced equation, 3 mol of KOH are required to react with 1 mol of $Al(NO_3)_3$ (3:1 mol ratio). The actual KOH to $Al(NO_3)_3$ mol ratio present is $0.0200/0.0100 = 2$ (2:1). Since the actual mol ratio present is less than the required mol ratio, KOH is the limiting reagent.

$$0.0200 \text{ mol KOH} \times \frac{1 \text{ mol Al(OH)}_3}{3 \text{ mol KOH}} \times \frac{78.00 \text{ g Al(OH)}_3}{\text{mol Al(OH)}_3} = 0.520 \text{ g Al(OH)}_3$$

42.   The balanced equation is: $3 \text{ BaCl}_2(aq) + \text{Fe}_2(\text{SO}_4)_3(aq) \rightarrow 3 \text{ BaSO}_4(s) + 2 \text{ FeCl}_3(aq)$

$$100.0 \text{ mL BaCl}_2 \times \frac{1 \text{ L}}{1000 \text{ mL}} \times \frac{0.100 \text{ mol BaCl}_2}{\text{L}} = 1.00 \times 10^{-2} \text{ mol BaCl}_2$$

$$100.0 \text{ mL Fe}_2(\text{SO}_4)_3 \times \frac{1 \text{ L}}{1000 \text{ mL}} \times \frac{0.100 \text{ mol Fe}_2(\text{SO}_4)_3}{\text{L Fe}_2(\text{SO}_4)_3} = 1.00 \times 10^{-2} \text{ mol Fe}_2(\text{SO}_4)_3$$

The required mol BaCl_2 to mol $\text{Fe}_2(\text{SO}_4)_3$ ratio from the balanced reaction is 3:1. The actual mol ratio is 0.0100/0.0100 = 1 (1:1). This is well below the required mol ratio, so BaCl_2 is the limiting reagent.

$$0.0100 \text{ mol BaCl}_2 \times \frac{3 \text{ mol BaSO}_4}{3 \text{ mol BaCl}_2} \times \frac{233.4 \text{ g BaSO}_4}{\text{mol BaSO}_4} = 2.33 \text{ g BaSO}_4$$

43. $2 \text{ AgNO}_3(aq) + \text{CaCl}_2(aq) \rightarrow 2 \text{ AgCl}(s) + \text{Ca(NO}_3)_2(aq)$

$$\text{mol AgNO}_3 = 0.1000 \text{ L} \times \frac{0.20 \text{ mol AgNO}_3}{\text{L}} = 0.020 \text{ mol AgNO}_3$$

$$\text{mol CaCl}_2 = 0.1000 \text{ L} \times \frac{0.15 \text{ mol CaCl}_2}{\text{L}} = 0.015 \text{ mol CaCl}_2$$

The required mol AgNO_3 to mol CaCl_2 ratio is 2:1 (from the balanced equation). The actual mol ratio present is 0.020/0.015 = 1.3 (1.3:1). Therefore, AgNO_3 is the limiting reagent.

$$\text{mass AgCl} = 0.020 \text{ mol AgNO}_3 \times \frac{1 \text{ mol AgCl}}{1 \text{ mol AgNO}_3} \times \frac{143.4 \text{ g AgCl}}{\text{mol AgCl}} = 2.9 \text{ g AgCl}$$

The net ionic equation is: $\text{Ag}^+(aq) + \text{Cl}^-(aq) \rightarrow \text{AgCl}(s)$. The ions remaining in solution are the unreacted Cl^- ions and the spectator ions, NO_3^- and Ca^{2+} (all Ag^+ is used up in forming AgCl). The mol of each ion present initially (before reaction) can be easily determined from the mol of each reactant. 0.020 mol AgNO_3 dissolves to form 0.020 mol Ag^+ and 0.020 mol NO_3^-. 0.015 mol CaCl_2 dissolves to form 0.015 mol Ca^{2+} and 2(0.015) = 0.030 mol Cl^-.

mol unreacted Cl^- = 0.030 mol Cl^- initially - 0.020 mol Cl^- reacted = 0.010 mol Cl^- unreacted

$$M_{\text{Cl}^-} = \frac{0.010 \text{ mol Cl}^-}{\text{total volume}} = \frac{0.010 \text{ mol Cl}^-}{0.1000 \text{ L} + 0.1000 \text{ L}} = 0.050 \ M \ \text{Cl}^-$$

The molarity of the spectator ions are:

$$M_{NO_3^-} = \frac{0.020 \text{ mol } NO_3^-}{0.2000 \text{ L}} = 0.10 \ M \ NO_3^-; \quad M_{Ca^{2+}} = \frac{0.015 \text{ mol } Ca^{2+}}{0.2000 \text{ L}} = 0.075 \ M \ Ca^{2+}$$

44.   a. $Cu(NO_3)_2(aq) + 2 \ KOH(aq) \rightarrow Cu(OH)_2(s) + 2 \ KNO_3(aq)$

Solution A contains $2.00 \text{ L} \times 2.00 \text{ mol/L} = 4.00 \text{ mol } Cu(NO_3)_2$ and solution B contains $2.00 \text{ L} \times 3.00 \text{ mol/L} = 6.00 \text{ mol } KOH$. Lets assume in our picture that we have 4 formula units of $Cu(NO_3)_2$ (4 Cu^{2+} ions and 8 NO_3^- ions) and 6 formula units of KOH (6 K^+ ions and 6 OH^- ions). With 4 Cu^{2+} ions and 6 OH^- ions present, then OH^- is limiting. One Cu^{2+} ion remains as 3 $Cu(OH)_2(s)$ formula units form as precipitate. The following drawing summarizes the ions that remain in solution and the relative amount of precipitate that forms. Note that K^+ and NO_3^- ions are spectator ions. In the drawing, V_1 is the volume of solution A or B and V_2 is the volume of the combined solutions with $V_2 = 2 \ V_1$. The drawing exaggerates the amount of precipitate that would actually form.

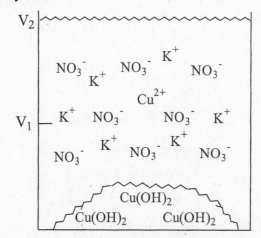

b.   The spectator ion concentrations will be one-half of the original spectator ion concentrations in the individual beakers because the volume was doubled. Or using moles, $M_{K^+} = \dfrac{6.00 \text{ mol } K^+}{4.00 \text{ L}}$ $= 1.50 \ M$ and $M_{NO_3^-} = \dfrac{8.00 \text{ mol } NO_3^-}{4.00 \text{ L}} = 2.00 \ M$. The concentration of OH^- ions will be zero since OH^- is the limiting reagent. From the drawing, the number of Cu^{2+} ions will decrease by a factor of four as the precipitate forms. Since the volume of solution doubled, the concentration of Cu^{2+} ions will decrease by a factor of eight after the two beakers are mixed:

$$M_{Cu^{2+}} = 2.00 \left(\frac{1}{8} \right) = 0.250 \ M$$

Alternately, one could certainly use moles to solve for $M_{Cu^{2+}}$:

$$\text{mol } Cu^{2+} \text{ reacted} = 2.00 \text{ L} \times \frac{3.00 \text{ mol } OH^-}{L} \times \frac{1 \text{ mol } Cu^{2+}}{2 \text{ mol } OH^-} = 3.00 \text{ mol } Cu^{2+} \text{ reacted}$$

$$\text{mol } Cu^{2+} \text{ present initially} = 2.00 \text{ L} \times \frac{2.00 \text{ mol } Cu^{2+}}{\text{L}} = 4.00 \text{ mol } Cu^{2+} \text{ present initially}$$

$$\text{excess } Cu^{2+} \text{ present after reaction} = 4.00 \text{ mol} - 3.00 \text{ mol} = 1.00 \text{ mol } Cu^{2+} \text{ excess}$$

$$M_{Cu^{2+}} = \frac{1.00 \text{ mol } Cu^{2+}}{2.00 \text{ L} + 2.00 \text{ L}} = 0.250 \ M$$

$$\text{mass of precipitate} = 6.00 \text{ mol KOH} \times \frac{1 \text{ mol } Cu(OH)_2}{2 \text{ mol KOH}} \times \frac{97.57 \text{ g } Cu(OH)_2}{\text{mol } Cu(OH)_2} = 293 \text{ g } Cu(OH)_2$$

Acid-Base Reactions

45.   All the bases in this problem are ionic compounds containing OH⁻. The acids are either strong or
weak electrolytes. The best way to determine if an acid is a strong or weak electrolyte is to
memorize all the strong electrolytes (strong acids). Any other acid you encounter that is not a strong
acid will be a weak electrolyte (a weak acid) and the formula should be left unaltered in the complete
ionic and net ionic equations. The strong acids to recognize are HCl, HBr, HI, HNO_3, $HClO_4$ and
H_2SO_4. For the answers below, the order of the equations are molecular, complete ionic and net
ionic.

a. $2 \ HClO_4(aq) + Mg(OH_2)(s) \rightarrow 2 \ H_2O(l) + Mg(ClO_4)_2(aq)$

$2 \ H^+(aq) + 2 \ ClO_4^-(aq) + Mg(OH)_2(s) \rightarrow 2 \ H_2O(l) + Mg^{2+}(aq) + 2 \ ClO_4^-(aq)$

$2 \ H^+(aq) + Mg(OH)_2(s) \rightarrow 2 \ H_2O(l) + Mg^{2+}(aq)$

b. $HCN(aq) + NaOH(aq) \rightarrow H_2O(l) + NaCN(aq)$

$HCN(aq) + Na^+(aq) + OH^-(aq) \rightarrow H_2O(l) + Na^+(aq) + CN^-(aq)$

$HCN(aq) + OH^-(aq) \rightarrow H_2O(l) + CN^-(aq)$

c. $HCl(aq) + NaOH(aq) \rightarrow H_2O(l) + NaCl(aq)$

$H^+(aq) + Cl^-(aq) + Na^+(aq) + OH^-(aq) \rightarrow H_2O(l) + Na^+(aq) + Cl^-(aq)$

$H^+(aq) + OH^-(aq) \rightarrow H_2O(l)$

46.   a. $3 \ HNO_3(aq) + Al(OH)_3(s) \rightarrow 3 \ H_2O(l) + Al(NO_3)_3(aq)$

$3 \ H^+(aq) + 3 \ NO_3^-(aq) + Al(OH)_3(s) \rightarrow 3 \ H_2O(l) + Al^{3+}(aq) + 3 \ NO_3^-(aq)$

$3 \ H^+(aq) + Al(OH)_3(s) \rightarrow 3 \ H_2O(l) + Al^{3+}(aq)$

b. $HC_2H_3O_2(aq) + KOH(aq) \rightarrow H_2O(l) + KC_2H_3O_2(aq)$

$HC_2H_3O_2(aq) + K^+(aq) + OH^-(aq) \rightarrow H_2O(l) + K^+(aq) + C_2H_3O_2^-(aq)$

$HC_2H_3O_2(aq) + OH^-(aq) \rightarrow H_2O(l) + C_2H_3O_2^-(aq)$

c. $Ca(OH)_2(aq) + 2\ HCl(aq) \rightarrow 2\ H_2O(l) + CaCl_2(aq)$

$Ca^{2+}(aq) + 2\ OH^-(aq) + 2\ H^+(aq) + 2\ Cl^-(aq) \rightarrow 2\ H_2O(l) + Ca^{2+}(aq) + 2\ Cl^-(aq)$

$2\ H^+(aq) + 2\ OH^-(aq) \rightarrow 2\ H_2O(l)$ or $H^+(aq) + OH^-(aq) \rightarrow H_2O(l)$

47.   All the acids in this problem are strong electrolytes. The acids to recognize as strong electrolytes are HCl, HBr, HI, HNO_3, $HClO_4$ and H_2SO_4.

a. $KOH(aq) + HNO_3(aq) \rightarrow H_2O(l) + KNO_3(aq)$

$K^+(aq) + OH^-(aq) + H^+(aq) + NO_3^-(aq) \rightarrow H_2O(l) + K^+(aq) + NO_3^-(aq)$

$OH^-(aq) + H^+(aq) \rightarrow H_2O(l)$

b. $Ba(OH)_2(aq) + 2\ HCl(aq) \rightarrow 2\ H_2O(l) + BaCl_2(aq)$

$Ba^{2+}(aq) + 2\ OH^-(aq) + 2\ H^+(aq) + 2\ Cl^-(aq) \rightarrow 2\ H_2O(l) + Ba^{2+}(aq) + 2\ Cl^-(aq)$

$2\ OH^-(aq) + 2\ H^+(aq) \rightarrow 2\ H_2O(l)$ or $OH^-(aq) + H^+(aq) \rightarrow H_2O(l)$

c. $3\ HClO_4(aq) + Fe(OH)_3(s) \rightarrow 3\ H_2O(l) + Fe(ClO_4)_3(aq)$

$3\ H^+(aq) + 3\ ClO_4^-(aq) + Fe(OH)_3(s) \rightarrow 3\ H_2O(l) + Fe^{3+}(aq) + 3\ ClO_4^-(aq)$

$3\ H^+(aq) + Fe(OH)_3(s) \rightarrow 3\ H_2O(l) + Fe^{3+}(aq)$

48.   a. $AgOH(s) + HBr(aq) \rightarrow AgBr(s) + H_2O(l)$

$AgOH(s) + H^+(aq) + Br^-(aq) \rightarrow AgBr(s) + H_2O(l)$

$AgOH(s) + H^+(aq) + Br^-(aq) \rightarrow AgBr(s) + H_2O(l)$

b. $Sr(OH)_2(aq) + 2\ HI(aq) \rightarrow 2\ H_2O(l) + SrI_2(aq)$

$Sr^{2+}(aq) + 2\ OH^-(aq) + 2\ H^+(aq) + 2\ I^-(aq) \rightarrow 2\ H_2O(l) + Sr^{2+}(aq) + 2\ I^-(aq)$

$2\ OH^-(aq) + 2\ H^+(aq) \rightarrow 2\ H_2O(l)$ or $OH^-(aq) + H^+(aq) \rightarrow H_2O(l)$

c. $Cr(OH)_3(s) + 3\ HNO_3(aq) \rightarrow 3\ H_2O(l) + Cr(NO_3)_3(aq)$

$Cr(OH)_3(s) + 3\ H^+(aq) + 3\ NO_3^-(aq) \rightarrow 3\ H_2O(l) + Cr^{3+}(aq) + 3\ NO_3^-(aq)$

$Cr(OH)_3(s) + 3\ H^+(aq) \rightarrow 3\ H_2O(l) + Cr^{3+}(aq)$

49.   If we begin with 50.00 mL of 0.200 M NaOH, then:

$$50.00 \times 10^{-3} \text{ L} \times \frac{0.200 \text{ mol}}{\text{L}} = 1.00 \times 10^{-2} \text{ mol NaOH is to be neutralized.}$$

a. $NaOH(aq) + HCl(aq) \rightarrow NaCl(aq) + H_2O(l)$

$$1.00 \times 10^{-2} \text{ mol NaOH} \times \frac{1 \text{ mol HCl}}{\text{mol NaOH}} \times \frac{1 \text{ L}}{0.100 \text{ mol}} = 0.100 \text{ L or } 100. \text{ mL}$$

b. $HNO_3(aq) + NaOH(aq) \rightarrow H_2O(l) + NaNO_3(aq)$

$$1.00 \times 10^{-2} \text{ mol NaOH} \times \frac{1 \text{ mol HNO}_3}{\text{mol NaOH}} \times \frac{1 \text{ L}}{0.150 \text{ mol HNO}_3} = 6.67 \times 10^{-2} \text{ L or } 66.7 \text{ mL}$$

c. $HC_2H_3O_2(aq) + NaOH(aq) \rightarrow H_2O(l) + NaC_2H_3O_2(aq)$

$$1.00 \times 10^{-2} \text{ mol NaOH} \times \frac{1 \text{ mol HC}_2H_3O_2}{\text{mol NaOH}} \times \frac{1 \text{ L}}{0.200 \text{ mol HC}_2H_3O_2} = 5.00 \times 10^{-2} \text{ L or } 50.0 \text{ mL}$$

50.   We begin with 25.00 mL of 0.200 M HCl or 25.00×10^{-3} L $\times$ 0.200 mol/L $= 5.00 \times 10^{-3}$ mol HCl.

a. $HCl(aq) + NaOH(aq) \rightarrow H_2O(l) + NaCl(aq)$

$$5.00 \times 10^{-3} \text{ mol HCl} \times \frac{1 \text{ mol NaOH}}{\text{mol HCl}} \times \frac{1 \text{ L}}{0.100 \text{ mol NaOH}} = 5.00 \times 10^{-2} \text{ L or } 50.0 \text{ mL}$$

b. $2 \text{ HCl}(aq) + Ba(OH)_2(aq) \rightarrow 2 \text{ H}_2O(l) + BaCl_2(aq)$

$$5.00 \times 10^{-3} \text{ mol HCl} \times \frac{1 \text{ mol Ba(OH)}_2}{2 \text{ mol HCl}} \times \frac{1 \text{ L}}{0.0500 \text{ mol Ba(OH)}_2} = 5.00 \times 10^{-2} \text{ L} = 50.0 \text{ mL}$$

c. $HCl(aq) + KOH(aq) \rightarrow H_2O(l) + KCl(aq)$

$$5.00 \times 10^{-3} \text{ mol HCl} \times \frac{1 \text{ mol KOH}}{\text{mol HCl}} \times \frac{1 \text{ L}}{0.250 \text{ mol KOH}} = 2.00 \times 10^{-2} \text{ L or } 20.0 \text{ mL}$$

51. $HNO_3(aq) + NaOH(aq) \rightarrow NaNO_3(aq) + H_2O(l)$

$$15.0 \text{ g NaOH} \times \frac{1 \text{ mol NaOH}}{40.00 \text{ g}} = 0.375 \text{ mol NaOH}$$

$$0.1500 \text{ L} \times \frac{0.250 \text{ mol HNO}_3}{\text{L}} = 0.0375 \text{ mol HNO}_3$$

We have added more moles of NaOH than mol of HNO_3 present. Since NaOH and HNO_3 react in a 1:1 mol ratio, NaOH is in excess and the solution will be basic. The ions present after reaction will be the excess OH^- ions and the spectator ions, Na^+ and NO_3^-. The moles of ions present initially are:

mol NaOH = mol Na^+ = mol OH^- = 0.375 mol

mol HNO_3 = mol H^+ = mol NO_3^- = 0.0375 mol

The net ionic reaction occurring is: $H^+(aq) + OH^-(aq) \rightarrow H_2O(l)$

The mol of excess OH^- remaining after reaction will be the initial mol of OH^- minus the amount of OH^- neutralized by reaction with H^+:

mol excess OH^- = 0.375 mol - 0.0375 mol = 0.338 mol OH^- excess

The concentration of ions present is:

$$M_{OH^-} = \frac{mol\ OH^-\ excess}{volume} = \frac{0.338\ mol\ OH^-}{0.1500\ L} = 2.25\ M\ OH^-$$

$$M_{NO_3^-} = \frac{0.0375\ mol\ NO_3^-}{0.1500\ L} = 0.250\ M\ NO_3^-;\quad M_{Na^+} = \frac{0.375\ mol}{0.1500\ L} = 2.50\ M\ Na^+$$

52. $Ba(OH)_2(aq) + 2\ HCl(aq) \rightarrow BaCl_2(aq) + 2\ H_2O(l);\ H^+(aq) + OH^-(aq) \rightarrow H_2O(l)$

$$75.0 \times 10^{-3}\ L \times \frac{0.250\ mol\ HCl}{L} = 1.88 \times 10^{-2}\ mol\ HCl = 1.88 \times 10^{-2}\ mol\ H^+ + 1.88 \times 10^{-2}\ mol\ Cl^-$$

$$225.0 \times 10^{-3}\ L \times \frac{0.0550\ mol\ Ba(OH)_2}{L} = 1.24 \times 10^{-2}\ mol\ Ba(OH)_2 = 1.24 \times 10^{-2}\ mol\ Ba^{2+}$$

$$+ 2.48 \times 10^{-2}\ mol\ OH^-$$

The net ionic equation requires a 1:1 mol ratio between OH^- and H^+. The actual mol OH^- to mol H^+ ratio is greater than 1:1 so OH^- is in excess.

Since 1.88×10^{-2} mol OH^- will be neutralized by the H^+, we have $(2.48 - 1.88) \times 10^{-2}$ = 0.60×10^{-2} mol OH^- remaining in excess.

$$M_{OH^-} = \frac{mol\ OH^-\ excess}{total\ volume} = \frac{6.0 \times 10^{-3}\ mol\ OH^-}{0.0750\ L + 0.2250\ L} = 2.0 \times 10^{-2}\ M\ OH^-$$

53. $HCl(aq) + NaOH(aq) \rightarrow H_2O(l) + NaCl(aq)$

$$24.16 \times 10^{-3}\ L\ NaOH \times \frac{0.106\ mol\ NaOH}{L\ NaOH} \times \frac{1\ mol\ HCl}{mol\ NaOH} = 2.56 \times 10^{-3}\ mol\ HCl$$

$$Molarity\ of\ HCl = \frac{2.56 \times 10^{-3}\ mol}{25.00 \times 10^{-3}\ L} = 0.102\ M\ HCl$$

54. $2 HNO_3(aq) + Ca(OH)_2(aq) \rightarrow 2 H_2O(l) + Ca(NO_3)_2(aq)$

$$35.00 \times 10^{-3} \text{ L HNO}_3 \times \frac{0.0500 \text{ mol HNO}_3}{\text{L HNO}_3} \times \frac{1 \text{ mol Ca(OH)}_2}{2 \text{ mol HNO}_3} \times \frac{1 \text{ L Ca(OH)}_2}{0.0200 \text{ mol Ca(OH)}_2}$$

$$= 0.0438 \text{ L} = 43.8 \text{ mL Ca(OH)}_2$$

55.   Since KHP is a monoprotic acid, the reaction is: $NaOH(aq) + KHP(aq) \rightarrow H_2O(l) + NaKP(aq)$

Mass KHP $= 0.02046 \text{ L NaOH} \times \dfrac{0.1000 \text{ mol NaOH}}{\text{L NaOH}} \times \dfrac{1 \text{ mol KHP}}{\text{mol NaOH}} \times \dfrac{204.22 \text{ g KHP}}{\text{mol KHP}} = 0.4178 \text{ g KHP}$

56. $NaOH(aq) + KHP(aq) \rightarrow NaKP(aq) + H_2O(l)$

$0.1082 \text{ g KHP} \times \dfrac{1 \text{ mol KHP}}{204.22 \text{ g KHP}} \times \dfrac{1 \text{ mol NaOH}}{\text{mol KHP}} = 5.298 \times 10^{-4} \text{ mol NaOH}$

There are 5.298×10^{-4} mol of sodium hydroxide in 34.67 mL of solution. Therefore, the concentration of sodium hydroxide is:

$$\frac{5.298 \times 10^{-4} \text{ mol}}{34.67 \times 10^{-3} \text{ L}} = 1.528 \times 10^{-2} \; M \text{ NaOH}$$

Oxidation-Reduction Reactions

57.   Apply rules in Table 4.2.

a. $KMnO_4$ is composed of K^+ and MnO_4^- ions. Assign oxygen a value of -2, which gives manganese a +7 oxidation state since the sum of oxidation states for all atoms in MnO_4^- must equal the -1 charge on MnO_4^-. K, +1; O, -2; Mn, +7.

b.   Assign O a -2 oxidation state, which gives nickel a +4 oxidation state. Ni, +4; O, -2.

c. $K_4Fe(CN)_6$ is composed of K^+ cations and $Fe(CN)_6^{4-}$ anions. $Fe(CN)_6^{4-}$ is composed of iron and CN^- anions. For an overall anion charge of -4, iron must have a +2 oxidation state.

d. $(NH_4)_2HPO_4$ is made of NH_4^+ cations and HPO_4^{2-} anions. Assign +1 as the oxidation state of H and -2 as the oxidation state of O. In NH_4^+, $x + 4(+1) = +1$, $x = -3 =$ oxidation state of N. In HPO_4^{2-}, $+1 + y + 4(-2) = -2$, $y = +5 =$ oxidation state of P.

e.   O, -2; P, +3   f.   O, -2; Fe, + 8/3

g.   O, -2; F, -1; Xe, +6   h.   F, -1; S, +4

i.   O, -2; C, +2   j.   Na, +1; O, -2; C, +3

58.   a. UO_2^{2+}: O, -2; For U, $x + 2(-2) = +2$, $x = \underline{+6}$

        b. As_2O_3: O, -2; For As, $2(x) + 3(-2) = 0$, $x = \underline{+3}$

        c. $NaBiO_3$: Na, +1; O, -2; For Bi, $+1 + x + 3(-2) = 0$, $x = \underline{+5}$

        d. As_4: As, 0

        e. $HAsO_2$: assign H = +1 and O = -2; For As, $+1 + x + 2(-2) = 0$; $x = \underline{+3}$

        f. $Mg_2P_2O_7$: Composed of Mg^{2+} ions and $P_2O_7^{4-}$ ions. Oxidation states are:

                Mg, +2; O, -2; P, +5

        g. $Na_2S_2O_3$: Composed of Na^+ ions and $S_2O_3^{2-}$ ions.   Na, +1; O, -2; S, +2

        h. Hg_2Cl_2: Hg, +1; Cl, -1

        i. $Ca(NO_3)_2$: Composed of Ca^{2+} ions and NO_3^- ions. Ca, +2; O, -2; N, +5

59.   a.   HBr: H, +1; Br, -1

        b.   HOBr: H, +1; O, -2; For Br, $+1 + 1(-2) + x = 0$, $x = +1$

        c. Br_2: Br, 0

        d. $HBrO_4$: H, +1; O, -2; For Br, $+1 + 4(-2) + x = 0$, $x = +7$

        e. BrF_3: F, -1; For Br, $x + 3(-1) = 0$, $x = +3$

60.   a.   -3   b.   -3   c. $2(x) + 4(+1) = 0$, $x = -2$
        d.   +2   e.   +1   f.   +4
        g.   +3   h.   +5   i.   0

61.   To determine if the reaction is an oxidation-reduction reaction, assign oxidation numbers. If the oxidation numbers change for some elements, the reaction is a redox reaction. If the oxidation numbers do not change, the reaction is not a redox reaction. In redox reactions, the species oxidized (called the reducing agent) shows an increase in oxidation numbers and the species reduced (called the oxidizing agent) shows a decrease in oxidation numbers.

	Redox?	Oxidizing Agent	Reducing Agent	Substance Oxidized	Substance Reduced
a.	Yes	O_2	CH_4	CH_4 (C)	O_2 (O)
b.	Yes	HCl	Zn	Zn	HCl (H)
c.	No	-	-	-	-
d.	Yes	O_3	NO	NO (N)	O_3 (O)
e.	Yes	H_2O_2	H_2O_2	H_2O_2 (O)	H_2O_2 (O)
f.	Yes	CuCl	CuCl	CuCl (Cu)	CuCl (Cu)

In c, no oxidation numbers change from reactants to products.

62.

	Redox?	Oxidizing Agent	Reducing Agent	Substance Oxidized	Substance Reduced
a.	Yes	Ag^+	Cu	Cu	Ag^+
b.	No	-	-	-	-
c.	No		-	-	-
d.	Yes	$SiCl_4$	Mg	Mg	$SiCl_4$ (Si)
e.	No		-	-	-

In b, c, and e, no oxidation numbers change.

63.   Use the method of half-reactions described in Section 4.10 of the text to balance these redox reactions. The first step always is to separate the reaction into the two half-reactions, then balance each half-reaction separately.

a. $Zn \rightarrow Zn^{2+} + 2\ e^-$ $2e^- + 2\ HCl \rightarrow H_2 + 2\ Cl^-$

Adding the two balanced half-reactions, $Zn(s) + 2\ HCl(aq) \rightarrow H_2(g) + Zn^{2+}(aq) + 2\ Cl^-(aq)$

b. $3\ I^- \rightarrow I_3^- + 2e^-$ $ClO^- \rightarrow Cl^-$
          $2e^- + 2H^+ + ClO^- \rightarrow Cl^- + H_2O$

Adding the two balanced half-reactions so electrons cancel:

   $3\ I^-(aq) + 2\ H^+(aq) + ClO^-(aq) \rightarrow I_3^-(aq) + Cl^-(aq) + H_2O(l)$

c. $As_2O_3 \rightarrow H_3AsO_4$ $NO_3^- \rightarrow NO + 2\ H_2O$
      $As_2O_3 \rightarrow 2\ H_3AsO_4$ $4\ H^+ + NO_3^- \rightarrow NO + 2\ H_2O$
      Left 3 - O; Right 8 - O $(3\ e^- + 4\ H^+ + NO_3^- \rightarrow NO + 2\ H_2O) \times 4$
      Right hand side has 5 extra O.
      Balance the oxygen atoms first using H_2O, then balance H using H^+, and finally balance charge using electrons.

      $(5\ H_2O + As_2O_3 \rightarrow 2\ H_3AsO_4 + 4\ H^+ + 4\ e^-) \times 3$

Common factor is a transfer of 12 e⁻. Add half-reactions so electrons cancel.

$$12 \text{ e}^- + 16 \text{ H}^+ + 4 \text{ NO}_3^- \rightarrow 4 \text{ NO} + 8 \text{ H}_2\text{O}$$
$$15 \text{ H}_2\text{O} + 3 \text{ As}_2\text{O}_3 \rightarrow 6 \text{ H}_3\text{AsO}_4 + 12 \text{ H}^+ + 12 \text{ e}^-$$

$$7 \text{ H}_2\text{O(l)} + 4 \text{ H}^+\text{(aq)} + 3 \text{ As}_2\text{O}_3\text{(s)} + 4 \text{ NO}_3^-\text{(aq)} \rightarrow 4 \text{ NO(g)} + 6 \text{ H}_3\text{AsO}_4\text{(aq)}$$

d. $(2 \text{ Br}^- \rightarrow \text{Br}_2 + 2 \text{ e}^-) \times 5$ $\qquad MnO_4^- \rightarrow Mn^{2+} + 4 \text{ H}_2\text{O}$
$\qquad\qquad\qquad (5 \text{ e}^- + 8 \text{ H}^+ + \text{MnO}_4^- \rightarrow \text{Mn}^{2+} + 4 \text{ H}_2\text{O}) \times 2$

Common factor is a transfer of 10 e⁻.

$$10 \text{ Br}^- \rightarrow 5 \text{ Br}_2 + 10 \text{ e}^-$$
$$10 \text{ e}^- + 16 \text{ H}^+ + 2 \text{ MnO}_4^- \rightarrow 2 \text{ Mn}^{2+} + 8 \text{ H}_2\text{O}$$

$$16 \text{ H}^+\text{(aq)} + 2 \text{ MnO}_4^-\text{(aq)} + 10 \text{ Br}^-\text{(aq)} \rightarrow 5 \text{ Br}_2\text{(l)} + 2 \text{ Mn}^{2+}\text{(aq)} + 8 \text{ H}_2\text{O(l)}$$

e. $\text{CH}_3\text{OH} \rightarrow \text{CH}_2\text{O}$ $\qquad\qquad\qquad Cr_2O_7^{2-} \rightarrow Cr^{3+}$
$(\text{CH}_3\text{OH} \rightarrow \text{CH}_2\text{O} + 2 \text{ H}^+ + 2 \text{ e}^-) \times 3$ $\qquad 14 \text{ H}^+ + \text{Cr}_2\text{O}_7^{2-} \rightarrow 2 \text{ Cr}^{3+} + 7 \text{ H}_2\text{O}$
$\qquad\qquad\qquad\qquad\qquad\qquad 6 \text{ e}^- + 14 \text{ H}^+ + \text{Cr}_2\text{O}_7^{2-} \rightarrow 2 \text{ Cr}^{3+} + 7 \text{ H}_2\text{O}$

Common factor is a transfer of 6 e⁻.

$$3 \text{ CH}_3\text{OH} \rightarrow 3 \text{ CH}_2\text{O} + 6 \text{ H}^+ + 6 \text{ e}^-$$
$$6 \text{ e}^- + 14 \text{ H}^+ + \text{Cr}_2\text{O}_7^{2-} \rightarrow 2 \text{ Cr}^{3+} + 7 \text{ H}_2\text{O}$$

$$8 \text{ H}^+\text{(aq)} + 3 \text{ CH}_3\text{OH(aq)} + \text{Cr}_2\text{O}_7^{2-}\text{(aq)} \rightarrow 2 \text{ Cr}^{3+}\text{(aq)} + 3 \text{ CH}_2\text{O(aq)} + 7 \text{ H}_2\text{O(l)}$$

64. a. $(\text{Cu} \rightarrow \text{Cu}^{2+} + 2 \text{ e}^-) \times 3$ $\qquad\qquad NO_3^- \rightarrow NO + 2 \text{ H}_2\text{O}$
$\qquad\qquad\qquad\qquad (3 \text{ e}^- + 4 \text{ H}^+ + \text{NO}_3^- \rightarrow \text{NO} + 2 \text{ H}_2\text{O}) \times 2$

Adding the two balanced half-reactions so electrons cancel:

$$3 \text{ Cu} \rightarrow 3 \text{ Cu}^{2+} + 6 \text{ e}^-$$
$$6 \text{ e}^- + 8 \text{ H}^+ + 2 \text{ NO}_3^- \rightarrow 2 \text{ NO} + 4 \text{ H}_2\text{O}$$

$$3 \text{ Cu(s)} + 8 \text{ H}^+\text{(aq)} + 2 \text{ NO}_3^-\text{(aq)} \rightarrow 3 \text{ Cu}^{2+}\text{(aq)} + 2 \text{ NO(g)} + 4 \text{ H}_2\text{O(l)}$$

b. $(2 \text{ Cl}^- \rightarrow \text{Cl}_2 + 2 \text{ e}^-) \times 3$ $\qquad\qquad Cr_2O_7^{2-} \rightarrow 2 \text{ Cr}^{3+} + 7 \text{ H}_2\text{O}$
$\qquad\qquad\qquad\qquad 6 \text{ e}^- + 14 \text{ H}^+ + \text{Cr}_2\text{O}_7^{2-} \rightarrow 2 \text{ Cr}^{3+} + 7 \text{ H}_2\text{O}$

Add the two half-reactions with six electrons transferred:

$$6 \; Cl^- \rightarrow 3 \; Cl_2 + 6 \; e^-$$
$$6 \; e^- + 14 \; H^+ + Cr_2O_7^{2-} \rightarrow 2 \; Cr^{3+} + 7 \; H_2O$$

$$14 \; H^+(aq) + Cr_2O_7^{2-}(aq) + 6 \; Cl^-(aq) \rightarrow 3 \; Cl_2(g) + 2 \; Cr^{3+}(aq) + 7 \; H_2O(l)$$

c. $Pb \rightarrow PbSO_4$ $PbO_2 \rightarrow PbSO_4$

$Pb + H_2SO_4 \rightarrow PbSO_4 + 2 \; H^+$ $PbO_2 + H_2SO_4 \rightarrow PbSO_4 + 2 \; H_2O$

$Pb + H_2SO_4 \rightarrow PbSO_4 + 2 \; H^+ + 2 \; e^-$ $2 \; e^- + 2 \; H^+ + PbO_2 + H_2SO_4 \rightarrow PbSO_4 + 2 \; H_2O$

Add the two half-reactions with two electrons transferred:

$$2 \; e^- + 2 \; H^+ + PbO_2 + H_2SO_4 \rightarrow PbSO_4 + 2 \; H_2O$$
$$Pb + H_2SO_4 \rightarrow PbSO_4 + 2 \; H^+ + 2 \; e^-$$

$$Pb(s) + 2 \; H_2SO_4(aq) + PbO_2(s) \rightarrow 2 \; PbSO_4(s) + 2 \; H_2O(l)$$

This is the reaction that occurs in an automobile lead-storage battery.

d. $Mn^{2+} \rightarrow MnO_4^-$
$(4 \; H_2O + Mn^{2+} \rightarrow MnO_4^- + 8 \; H^+ + 5 \; e^-) \times 2$

$$NaBiO_3 \rightarrow Bi^{3+} + Na^+$$
$$6 \; H^+ + NaBiO_3 \rightarrow Bi^{3+} + Na^+ + 3 \; H_2O$$
$$(2 \; e^- + 6 \; H^+ + NaBiO_3 \rightarrow Bi^{3+} + Na^+ + 3 \; H_2O) \times 5$$

$$8 \; H_2O + 2 \; Mn^{2+} \rightarrow 2 \; MnO_4^- + 16 \; H^+ + 10 \; e^-$$
$$10 \; e^- + 30 \; H^+ + 5 \; NaBiO_3 \rightarrow 5 \; Bi^{3+} + 5 \; Na^+ + 15 \; H_2O$$

$$8 \; H_2O + 30 \; H^+ + 2 \; Mn^{2+} + 5 \; NaBiO_3 \rightarrow 2 \; MnO_4^- + 5 \; Bi^{3+} + 5 \; Na^+ + 15 \; H_2O + 16 \; H^+$$

Simplifying :

$$14 \; H^+(aq) + 2 \; Mn^{2+}(aq) + 5 \; NaBiO_3(s) \rightarrow 2 \; MnO_4^-(aq) + 5 \; Bi^{3+}(aq) + 5 \; Na^+(aq) + 7 \; H_2O(l)$$

e. $H_3AsO_4 \rightarrow AsH_3$ $(Zn \rightarrow Zn^{2+} + 2 \; e^-) \times 4$
$H_3AsO_4 \rightarrow AsH_3 + 4 \; H_2O$
$8 \; e^- + 8 \; H^+ + H_3AsO_4 \rightarrow AsH_3 + 4 \; H_2O$

$$8 \; e^- + 8 \; H^+ + H_3AsO_4 \rightarrow AsH_3 + 4 \; H_2O$$
$$4 \; Zn \rightarrow 4 \; Zn^{2+} + 8 \; e^-$$

$$8 \; H^+(aq) + H_3AsO_4(aq) + 4 \; Zn(s) \rightarrow 4 \; Zn^{2+}(aq) + AsH_3(g) + 4 \; H_2O(l)$$

65.   Use the same method as with acidic solutions. After the final balanced equation, convert H^+ to OH^- as described in section 4.10 of the text. The extra step involves converting H^+ into H_2O by adding equal moles of OH^- to each side of the reaction. This converts the reaction to a basic solution while keeping it balanced.

a.
$$Al \rightarrow Al(OH)_4^-$$
$$4\,H_2O + Al \rightarrow Al(OH)_4^- + 4\,H^+$$
$$4\,H_2O + Al \rightarrow Al(OH)_4^- + 4\,H^+ + 3\,e^-$$

$$MnO_4^- \rightarrow MnO_2$$
$$3\,e^- + 4\,H^+ + MnO_4^- \rightarrow MnO_2 + 2\,H_2O$$

$$4\,H_2O + Al \rightarrow Al(OH)_4^- + 4\,H^+ + 3\,e^-$$
$$3\,e^- + 4\,H^+ + MnO_4^- \rightarrow MnO_2 + 2\,H_2O$$
_____
$$2\,H_2O(l) + Al(s) + MnO_4^-(aq) \rightarrow Al(OH)_4^-(aq) + MnO_2(s)$$

Since H^+ doesn't appear in the final balanced reaction, we are done.

b.
$$Cl_2 \rightarrow Cl^-$$
$$2\,e^- + Cl_2 \rightarrow 2\,Cl^-$$

$$Cl_2 \rightarrow OCl^-$$
$$2\,H_2O + Cl_2 \rightarrow 2\,OCl^- + 4\,H^+ + 2\,e^-$$

$$2\,e^- + Cl_2 \rightarrow 2\,Cl^-$$
$$2\,H_2O + Cl_2 \rightarrow 2\,OCl^- + 4\,H^+ + 2\,e^-$$
_____
$$2\,H_2O + 2\,Cl_2 \rightarrow 2\,Cl^- + 2\,OCl^- + 4\,H^+$$

Now convert to a basic solution. Add $4\,OH^-$ to both sides of the equation. The $4\,OH^-$ will react with the $4\,H^+$ on the product side to give $4\,H_2O$. After this step, cancel identical species on both sides ($2\,H_2O$). Applying these steps gives: $4\,OH^- + 2\,Cl_2 \rightarrow 2\,Cl^- + 2\,OCl^- + 2\,H_2O$, which can be further simplified to:

$$2\,OH^-(aq) + Cl_2(g) \rightarrow Cl^-(aq) + OCl^-(aq) + H_2O(l)$$

c.
$$NO_2^- \rightarrow NH_3$$
$$6\,e^- + 7\,H^+ + NO_2^- \rightarrow NH_3 + 2\,H_2O$$

$$Al \rightarrow AlO_2^-$$
$$(2\,H_2O + Al \rightarrow AlO_2^- + 4\,H^+ + 3\,e^-) \times 2$$

Common factor is a transfer of $6\,e^-$.

$$6e^- + 7\,H^+ + NO_2^- \rightarrow NH_3 + 2\,H_2O$$
$$4\,H_2O + 2\,Al \rightarrow 2\,AlO_2^- + 8\,H^+ + 6\,e^-$$
_____
$$OH^- + 2\,H_2O + NO_2^- + 2\,Al \rightarrow NH_3 + 2\,AlO_2^- + H^+ + OH^-$$

Reducing gives: $OH^-(aq) + H_2O(l) + NO_2^-(aq) + 2\,Al(s) \rightarrow NH_3(g) + 2\,AlO_2^-(aq)$

66. a. $Cr \rightarrow Cr(OH)_3$ $CrO_4^{2-} \rightarrow Cr(OH)_3$
          $3 H_2O + Cr \rightarrow Cr(OH)_3 + 3 H^+ + 3 e^-$ $3 e^- + 5 H^+ + CrO_4^{2-} \rightarrow Cr(OH)_3 + H_2O$

$$3 H_2O + Cr \rightarrow Cr(OH)_3 + 3 H^+ + 3 e^-$$
$$3 e^- + 5 H^+ + CrO_4^{2-} \rightarrow Cr(OH)_3 + H_2O$$

$$\overline{2 OH^- + 2 H^+ + 2 H_2O + Cr + CrO_4^{2-} \rightarrow 2 Cr(OH)_3 + 2 OH^-}$$

Two OH^- were added above to each side to convert to a basic solution. The two OH^- react with the 2 H^+ on the reactant side to produce 2 H_2O. The overall balanced equation is:

$$4 H_2O(l) + Cr(s) + CrO_4^{2-}(aq) \rightarrow 2 Cr(OH)_3(s) + 2 OH^-(aq)$$

  b. $S^{2-} \rightarrow S$ $MnO_4^- \rightarrow MnS$
       $(S^{2-} \rightarrow S + 2 e^-) \times 5$ $MnO_4^- + S^{2-} \rightarrow MnS$
                                                   $(5 e^- + 8 H^+ + MnO_4^- + S^{2-} \rightarrow MnS + 4 H_2O) \times 2$

Common factor is a transfer of 10 e^-.

$$5 S^{2-} \rightarrow 5 S + 10 e^-$$
$$10 e^- + 16 H^+ + 2 MnO_4^- + 2 S^{2-} \rightarrow 2 MnS + 8 H_2O$$

$$\overline{16 OH^- + 16 H^+ + 7 S^{2-} + 2 MnO_4^- \rightarrow 5 S + 2 MnS + 8 H_2O + 16 OH^-}$$

$$16 H_2O + 7 S^{2-} + 2 MnO_4^- \rightarrow 5 S + 2 MnS + 8 H_2O + 16 OH^-$$

Reducing gives: $8 H_2O(l) + 7 S^{2-}(aq) + 2 MnO_4^-(aq) \rightarrow 5 S(s) + 2 MnS(s) + 16 OH^-(aq)$

  c. $CN^- \rightarrow CNO^-$
       $(H_2O + CN^- \rightarrow CNO^- + 2 H^+ + 2 e^-) \times 3$
                                                   $MnO_4^- \rightarrow MnO_2$
                                                   $(3 e^- + 4 H^+ + MnO_4^- \rightarrow MnO_2 + 2 H_2O) \times 2$

Common factor is a transfer of 6 electrons.

$$3 H_2O + 3 CN^- \rightarrow 3 CNO^- + 6 H^+ + 6 e^-$$
$$6 e^- + 8 H^+ + 2 MnO_4^- \rightarrow 2 MnO_2 + 4 H_2O$$

$$\overline{2 OH^- + 2 H^+ + 3 CN^- + 2 MnO_4^- \rightarrow 3 CNO^- + 2 MnO_2 + H_2O + 2 OH^-}$$

Reducing gives:

$$H_2O(l) + 3 CN^-(aq) + 2 MnO_4^-(aq) \rightarrow 3 CNO^-(aq) + 2 MnO_2(s) + 2 OH^-(aq)$$

67. $NaCl + H_2SO_4 + MnO_2 \rightarrow Na_2SO_4 + MnCl_2 + Cl_2 + H_2O$

We could balance this reaction by the half-reaction method or by inspection. Let's try inspection. To balance Cl^-, we need 4 NaCl:

$4 NaCl + H_2SO_4 + MnO_2 \rightarrow Na_2SO_4 + MnCl_2 + Cl_2 + H_2O$

Balance the Na^+ and SO_4^{2-} ions next:

$4 NaCl + 2 H_2SO_4 + MnO_2 \rightarrow 2 Na_2SO_4 + MnCl_2 + Cl_2 + H_2O$

On the left side: 4-H and 10-O; On the right side: 8-O not counting H_2O

We need 2 H_2O on the right side to balance H and O:

$4 NaCl(aq) + 2 H_2SO_4(aq) + MnO_2(s) \rightarrow 2 Na_2SO_4(aq) + MnCl_2(aq) + Cl_2(g) + 2 H_2O(l)$

68. $Au + HNO_3 + HCl \rightarrow AuCl_4^- + NO$

Only deal with ions that are reacting (omit H^+): $Au + NO_3^- + Cl^- \rightarrow AuCl_4^- + NO$

The balanced half-reactions are:

$Au + 4 Cl^- \rightarrow AuCl_4^- + 3 e^-$ $3 e^- + 4 H^+ + NO_3^- \rightarrow NO + 2 H_2O$

Adding the two balanced half-reactions:

$Au(s) + 4 Cl^-(aq) + 4 H^+(aq) + NO_3^-(aq) \rightarrow AuCl_4^-(aq) + NO(g) + 2 H_2O(l)$

Additional Exercises

69. $0.100 \text{ g Ca} \times \dfrac{1 \text{ mol Ca}}{40.08 \text{ g Ca}} \times \dfrac{1 \text{ mol Ca(OH)}_2}{\text{mol Ca}} \times \dfrac{2 \text{ mol OH}^-}{\text{mol Ca(OH)}_2} = 4.99 \times 10^{-3} \text{ mol OH}^-$

Molarity $= \dfrac{4.99 \times 10^{-3} \text{ mol}}{450. \times 10^{-3} \text{ L}} = 1.11 \times 10^{-2} \, M \text{ OH}^-$

70.   mol $CaCl_2$ present $= 0.230 \text{ L CaCl}_2 \times \dfrac{0.275 \text{ mol CaCl}_2}{\text{L CaCl}_2} = 6.33 \times 10^{-2} \text{ mol CaCl}_2$

The volume of $CaCl_2$ solution after evaporation is:

$6.33 \times 10^{-2} \text{ mol CaCl}_2 \times \dfrac{1 \text{ L CaCl}_2}{1.10 \text{ mol CaCl}_2} = 5.75 \times 10^{-2} \text{ L} = 57.5 \text{ mL CaCl}_2$

Volume H_2O evaporated = 230. mL - 57.5 mL = 173 mL H_2O evaporated

71.   There are other possible correct choices for the following answers. We have listed only three possible reactants in each case.

a. $AgNO_3$, $Pb(NO_3)_2$, and $Hg_2(NO_3)_2$ would form precipitates with the Cl^- ion.
$Ag^+(aq) + Cl^-(aq) \rightarrow AgCl(s)$; $Pb^{2+}(aq) + 2\ Cl^-(aq) \rightarrow PbCl_2(s)$;
$Hg_2^{2+}(aq) + 2\ Cl^-(aq) \rightarrow Hg_2Cl_2(s)$

b. Na_2SO_4, Na_2CO_3, and Na_3PO_4 would form precipitates with the Ca^{2+} ion.
$Ca^{2+}(aq) + SO_4^{2-}(aq) \rightarrow CaSO_4(s)$; $Ca^{2+} + CO_3^{2-}(aq) \rightarrow CaCO_3(s)$
$3\ Ca^{2+}(aq) + 2\ PO_4^{3-}(aq) \rightarrow Ca_3(PO_4)_2(s)$

c. $NaOH$, Na_2S, and Na_2CO_3 would form precipitates with the Fe^{3+} ion.
$Fe^{3+}(aq) + 3\ OH^-(aq) \rightarrow Fe(OH)_3(s)$; $2\ Fe^{3+}(aq) + 3\ S^{2-}(aq) \rightarrow Fe_2S_3(s)$;
$2\ Fe^{3+}(aq) + 3\ CO_3^{2-}(aq) \rightarrow Fe_2(CO_3)_3(s)$

d. $BaCl_2$, $Pb(NO_3)_2$, and $Ca(NO_3)_2$ would form precipitates with the SO_4^{2-} ion.
$Ba^{2+}(aq) + SO_4^{2-}(aq) \rightarrow BaSO_4(s)$; $Pb^{2+}(aq) + SO_4^{2-}(aq) \rightarrow PbSO_4(s)$;
$Ca^{2+}(aq) + SO_4^{2-}(aq) \rightarrow CaSO_4(s)$

e. Na_2SO_4, $NaCl$, and NaI would form precipitates with the Hg_2^{2+} ion.
$Hg_2^{2+}(aq) + SO_4^{2-}(aq) \rightarrow Hg_2SO_4(s)$; $Hg_2^{2+}(aq) + 2\ Cl^-(aq) \rightarrow Hg_2Cl_2(s)$;
$Hg_2^{2+}(aq) + 2\ I^-(aq) \rightarrow Hg_2I_2(s)$

f. $NaBr$, Na_2CrO_4, and Na_3PO_4 would form precipitates with the Ag^+ ion.
$Ag^+(aq) + Br^-(aq) \rightarrow AgBr(s)$; $2\ Ag^+(aq) + CrO_4^{2-}(aq) \rightarrow Ag_2CrO_4(s)$;
$3\ Ag^+(aq) + PO_4^{3-}(aq) \rightarrow Ag_3PO_4(s)$

72.   a. $MgCl_2(aq) + 2\ AgNO_3(aq) \rightarrow 2\ AgCl(s) + Mg(NO_3)_2(aq)$

$$0.641\ g\ AgCl \times \frac{1\ mol\ AgCl}{143.4\ g\ AgCl} \times \frac{1\ mol\ MgCl_2}{2\ mol\ AgCl} \times \frac{95.21\ g}{mol\ MgCl_2} = 0.213\ g\ MgCl_2$$

$$\frac{0.213\ g\ MgCl_2}{1.50\ g\ mixture} \times 100 = 14.2\%\ MgCl_2$$

b. $$0.213\ g\ MgCl_2 \times \frac{1\ mol\ MgCl_2}{95.21\ g} \times \frac{2\ mol\ AgNO_3}{mol\ MgCl_2} \times \frac{1\ L}{0.500\ mol\ AgNO_3} \times \frac{1000\ mL}{1\ L}$$

$$= 8.95\ mL\ AgNO_3$$

73. $$1.00\ L \times \frac{0.200\ mol\ Na_2S_2O_3}{L} \times \frac{1\ mol\ AgBr}{2\ mol\ Na_2S_2O_3} \times \frac{187.8\ g\ AgBr}{mol\ AgBr} = 18.8\ g\ AgBr$$

74. $M_2SO_4(aq) + CaCl_2(aq) \rightarrow CaSO_4(s) + 2\ MCl(aq)$

$$1.36\ g\ CaSO_4 \times \frac{1\ mol\ CaSO_4}{136.15\ g\ CaSO_4} \times \frac{1\ mol\ M_2SO_4}{mol\ CaSO_4} = 9.99 \times 10^{-3}\ mol\ M_2SO_4$$

From the problem, 1.42 g M_2SO_4 was reacted so:

$$1.42 \text{ g } M_2SO_4 = 9.99 \times 10^{-3} \text{ mol } M_2SO_4, \quad \text{molar mass} = \frac{1.42 \text{ g } M_2SO_4}{9.99 \times 10^{-3} \text{ mol } M_2SO_4} = 142 \text{ g/mol}$$

142 amu = 2(atomic mass M) + 32.07 + 4(16.00), atomic mass M = 23 amu

From periodic table, M = Na(sodium).

75.   All the sulfur in $BaSO_4$ came from the saccharin. The conversion from $BaSO_4$ to saccharin utilizes the molar masses of each.

$$0.5032 \text{ g } BaSO_4 \times \frac{32.07 \text{ g S}}{233.4 \text{ g } BaSO_4} \times \frac{183.19 \text{ g saccharin}}{32.07 \text{ g S}} = 0.3949 \text{ g saccharin}$$

$$\frac{\text{Avg. mass}}{\text{Tablet}} = \frac{0.3949 \text{ g}}{10 \text{ tablets}} = \frac{3.949 \times 10^{-2} \text{ g}}{\text{tablet}} = \frac{39.49 \text{ mg}}{\text{tablet}}$$

$$\text{Avg. mass \%} = \frac{0.3949 \text{ g saccharin}}{0.5894 \text{ g}} \times 100 = 67.00\% \text{ saccharin by mass}$$

76.   a. $Fe^{3+}(aq) + 3 \text{ OH}^-(aq) \rightarrow Fe(OH)_3(s)$

$Fe(OH)_3$: 55.85 + 3(16.00) + 3(1.008) = 106.87 g/mol

$$0.107 \text{ g } Fe(OH)_3 \times \frac{55.85 \text{ g Fe}}{106.87 \text{ g } Fe(OH)_3} = 0.0559 \text{ g Fe}$$

   b. $Fe(NO_3)_3$: 55.85 + 3(14.01) + 9(16.00) = 241.86 g/mol

$$0.0559 \text{ g Fe} \times \frac{241.86 \text{ g } Fe(NO_3)_3}{55.85 \text{ g Fe}} = 0.242 \text{ g } Fe(NO_3)_3$$

   c. Mass % $Fe(NO_3)_3 = \frac{0.242 \text{ g}}{0.456 \text{ g}} \times 100 = 53.1\%$

77. $Cr(NO_3)_3(aq) + 3 \text{ NaOH}(aq) \rightarrow Cr(OH)_3(s) + 3 \text{ NaNO}_3(aq)$

$$\text{mol NaOH used} = 2.06 \text{ g } Cr(OH)_3 \times \frac{1 \text{ mol } Cr(OH)_3}{103.02 \text{ g}} \times \frac{3 \text{ mol NaOH}}{1 \text{ mol } Cr(OH)_3} = 6.00 \times 10^{-2} \text{ mol NaOH}$$
to form precipitate

$NaOH(aq) + HCl(aq) \rightarrow NaCl(aq) + H_2O(l)$

$$\text{mol NaOH used} = 0.1000 \text{ L} \times \frac{0.400 \text{ mol HCl}}{\text{L}} \times \frac{1 \text{ mol NaOH}}{\text{mol HCl}} = 4.00 \times 10^{-2} \text{ mol NaOH}$$
to react with HCl

$$M_{NaOH} = \frac{\text{mol NaOH}}{\text{volume}} = \frac{6.00 \times 10^{-2} \text{ mol} + 4.00 \times 10^{-2} \text{ mol}}{0.0500 \text{ L}} = 2.00 \text{ } M \text{ NaOH}$$

78.   a. Perchloric acid reacted with potassium hydroxide is a possibility.

$HClO_4(aq) + KOH(aq) \rightarrow H_2O(l) + KClO_4(aq)$

b. Nitric acid reacted with cesium hydroxide is a possibility.

$HNO_3(aq) + CsOH(aq) \rightarrow H_2O(l) + CsNO_3(aq)$

c. Hydroiodic acid reacted with calcium hydroxide is a possibility.

$2\ HI(aq) + Ca(OH)_2(aq) \rightarrow 2\ H_2O(l) + CaI_2(aq)$

79. $HC_2H_3O_2(aq) + NaOH(aq) \rightarrow H_2O(l) + NaC_2H_3O_2(aq)$

a. $16.58 \times 10^{-3}\ \text{L soln} \times \dfrac{0.5062\ \text{mol NaOH}}{\text{L soln}} \times \dfrac{1\ \text{mol acetic acid}}{\text{mol NaOH}} = 8.393 \times 10^{-3}\ \text{mol acetic acid}$

Concentration of acetic acid $= \dfrac{8.393 \times 10^{-3}\ \text{mol}}{0.01000\ \text{L}} = 0.8393\ M$

b. If we have 1.000 L of solution: total mass $= 1000.\ \text{mL} \times \dfrac{1.006\ \text{g}}{\text{mL}} = 1006\ \text{g}$

Mass of $HC_2H_3O_2 = 0.8393\ \text{mol} \times \dfrac{60.05\ \text{g}}{\text{mol}} = 50.40\ \text{g}$

Mass % acetic acid $= \dfrac{50.40\ \text{g}}{1006\ \text{g}} \times 100 = 5.010\%$

80. $Mg(s) + 2\ HCl(aq) \rightarrow MgCl_2(aq) + H_2(g)$

$3.00\ \text{g Mg} \times \dfrac{1\ \text{mol Mg}}{24.31\ \text{g Mg}} \times \dfrac{2\ \text{mol HCl}}{\text{mol Mg}} \times \dfrac{1\ \text{L HCl}}{5.0\ \text{mol HCl}} = 0.0494\ \text{L} = 49.4\ \text{mL HCl}$

81.   Let HA = unknown acid; $HA(aq) + NaOH(aq) \rightarrow NaA(aq) + H_2O(l)$

mol HA present $= 0.0250\ \text{L} \times \dfrac{0.500\ \text{mol NaOH}}{\text{L}} \times \dfrac{1\ \text{mol HA}}{1\ \text{mol NaOH}} = 0.0125\ \text{mol HA}$

$\dfrac{x\ \text{g HA}}{\text{mol HA}} = \dfrac{2.20\ \text{g HA}}{0.0125\ \text{mol HA}}$, x = molar mass of HA = 176 g/mol

Empirical formula weight $\approx 3(12) + 4(1) + 3(16) = 88$ g/mol

Since 176/88 = 2.0, the molecular formula is $(C_3H_4O_3)_2 = C_6H_8O_6$.

82.   a. $Al(s) + 3\ HCl(aq) \rightarrow AlCl_3(aq) + 3/2\ H_2(g)$ or $2\ Al(s) + 6\ HCl(aq) \rightarrow 2\ AlCl_3(aq) + 3\ H_2(g)$

Hydrogen is reduced (goes from +1 oxidation state to 0 oxidation state) and aluminum Al is oxidized (0 $\rightarrow$ +3).

b. Balancing S is most complicated since sulfur is in both products. Balance C and H first then worry about S.

$$CH_4(g) + 4\ S(s) \rightarrow CS_2(l) + 2\ H_2S(g)$$

Sulfur is reduced (0 → -2) and carbon is oxidized (-4 → +4).

c. Balance C and H first, then balance O.

$$C_3H_8(g) + 5\ O_2(g) \rightarrow 3\ CO_2(g) + 4\ H_2O(l)$$

Oxygen is reduced (0 → -2) and carbon is oxidized (-8/3 → +4).

d. Although this reaction is mass balanced, it is not charge balanced. We need 2 mol of silver on each side to balance the charge.

$$Cu(s) + 2\ Ag^+(aq) \rightarrow 2\ Ag(s) + Cu^{2+}(aq)$$

Silver is reduced (+1 → 0) and copper is oxidized (0 → +2).

83. $Mn + HNO_3 \rightarrow Mn^{2+} + NO_2$

$$Mn \rightarrow Mn^{2+} + 2\ e^- \qquad\qquad HNO_3 \rightarrow NO_2$$
$$HNO_3 \rightarrow NO_2 + H_2O$$
$$(e^- + H^+ + HNO_3 \rightarrow NO_2 + H_2O) \times 2$$

$$Mn \rightarrow Mn^{2+} + 2\ e^-$$
$$2\ e^- + 2\ H^+ + 2\ HNO_3 \rightarrow 2\ NO_2 + 2\ H_2O$$
$$\overline{\rule{6cm}{0.4pt}}$$
$$2\ H^+(aq) + Mn(s) + 2\ HNO_3(aq) \rightarrow Mn^{2+}(aq) + 2\ NO_2(g) + 2\ H_2O(l)$$

$Mn^{2+} + IO_4^- \rightarrow MnO_4^- + IO_3^-$

$$(4\ H_2O + Mn^{2+} \rightarrow MnO_4^- + 8\ H^+ + 5\ e^-) \times 2 \qquad\qquad (2\ e^- + 2\ H^+ + IO_4^- \rightarrow IO_3^- + H_2O) \times 5$$

$$8\ H_2O + 2\ Mn^{2+} \rightarrow 2\ MnO_4^- + 16\ H^+ + 10\ e^-$$
$$10\ e^- + 10\ H^+ + 5\ IO_4^- \rightarrow 5\ IO_3^- + 5\ H_2O$$
$$\overline{\rule{6cm}{0.4pt}}$$
$$3\ H_2O(l) + 2\ Mn^{2+}(aq) + 5\ IO_4^-(aq) \rightarrow 2\ MnO_4^-(aq) + 5\ IO_3^-(aq) + 6\ H^+(aq)$$

Challenge Problems

84.   a. 5.0 ppb Hg in water $= \dfrac{5.0\ \text{ng Hg}}{\text{g soln}} = \dfrac{5.0 \times 10^{-9}\ \text{g Hg}}{\text{mL soln}}$

$$\frac{5.0 \times 10^{-9}\ \text{g Hg}}{\text{mL soln}} \times \frac{1\ \text{mol Hg}}{200.6\ \text{g Hg}} \times \frac{1000\ \text{mL}}{\text{L}} = 2.5 \times 10^{-8}\ M\ \text{Hg}$$

b. $\dfrac{1.0 \times 10^{-9} \text{ g CHCl}_3}{\text{mL}} \times \dfrac{1 \text{ mol CHCl}_3}{119.37 \text{ g CHCl}_3} \times \dfrac{1000 \text{ mL}}{\text{L}} = 8.4 \times 10^{-9} \ M \text{ CHCl}_3$

c. $10.0 \text{ ppm As} = \dfrac{10.0 \ \mu\text{g As}}{\text{g soln}} = \dfrac{10.0 \times 10^{-6} \text{ g As}}{\text{mL soln}}$

$\dfrac{10.0 \times 10^{-6} \text{ g As}}{\text{mL soln}} \times \dfrac{1 \text{ mol As}}{74.92 \text{ g As}} \times \dfrac{1000 \text{ mL}}{\text{L}} = 1.33 \times 10^{-4} \ M \text{ As}$

d. $\dfrac{0.10 \times 10^{-6} \text{ g DDT}}{\text{mL}} \times \dfrac{1 \text{ mol DDT}}{354.46 \text{ g DDT}} \times \dfrac{1000 \text{ mL}}{\text{L}} = 2.8 \times 10^{-7} \ M \text{ DDT}$

85. a. $0.308 \text{ g AgCl} \times \dfrac{35.45 \text{ g Cl}}{143.4 \text{ g AgCl}} = 0.0761 \text{ g Cl}; \ \%\text{Cl} = \dfrac{0.0761 \text{ g}}{0.256 \text{ g}} \times 100 = 29.7\% \text{ Cl}$

Cobalt(III) oxide, Co_2O_3: $2(58.93) + 3(16.00) = 165.86$ g/mol

$0.145 \text{ g Co}_2\text{O}_3 \times \dfrac{117.86 \text{ g Co}}{165.86 \text{ g Co}_2\text{O}_3} = 0.103 \text{ g Co}; \ \%\text{Co} = \dfrac{0.103 \text{ g}}{0.416 \text{ g}} \times 100 = 24.8\% \text{ Co}$

The remainder, $100.0 - (29.7 + 24.8) = 45.5\%$, is water.

Assuming 100.0 g of compound:

$45.5 \text{ g H}_2\text{O} \times \dfrac{2.016 \text{ g H}}{18.02 \text{ g H}_2\text{O}} = 5.09 \text{ g H}; \ \%\text{H} = \dfrac{5.09 \text{ g H}}{100.0 \text{ g compound}} \times 100 = 5.09\% \text{ H}$

$45.5 \text{ g H}_2\text{O} \times \dfrac{16.00 \text{ g O}}{18.02 \text{ g H}_2\text{O}} = 40.4 \text{ g O}; \ \%\text{O} = \dfrac{40.4 \text{ g O}}{100.0 \text{ g compound}} \times 100 = 40.4\% \text{ O}$

The mass percent composition is 24.8% Co, 29.7% Cl, 5.09% H and 40.4% O.

b. Out of 100.0 g of compound, there are:

$24.8 \text{ g Co} \times \dfrac{1 \text{ mol}}{58.93 \text{ g Co}} = 0.421 \text{ mol Co}; \ 29.7 \text{ g Cl} \times \dfrac{1 \text{ mol}}{35.45 \text{ g Cl}} = 0.838 \text{ mol Cl}$

$5.09 \text{ g H} \times \dfrac{1 \text{ mol}}{1.008 \text{ g H}} = 5.05 \text{ mol H}; \ 40.4 \text{ g O} \times \dfrac{1 \text{ mol}}{16.00 \text{ g O}} = 2.53 \text{ mol O}$

Dividing all results by 0.421, we get $CoCl_2 \cdot 6H_2O$.

c. $CoCl_2 \cdot 6H_2O(aq) + 2 \ AgNO_3(aq) \rightarrow 2 \ AgCl(s) + Co(NO_3)_2(aq) + 6 \ H_2O(l)$

$CoCl_2 \cdot 6H_2O(aq) + 2 \ NaOH(aq) \rightarrow Co(OH)_2(s) + 2 \ NaCl(aq) + 6 \ H_2O(l)$

$Co(OH)_2 \rightarrow Co_2O_3$ This is an oxidation-reduction reaction. Thus, we also need to include an oxidizing agent. The obvious choice is O_2.

$4 \ Co(OH)_2(s) + O_2(g) \rightarrow 2 \ Co_2O_3(s) + 4 \ H_2O(l)$

86.   a. $C_{12}H_{10-n}Cl_n + n\ Ag^+ \rightarrow n\ AgCl$; molar mass (AgCl) = 143.4 g/mol

molar mass (PCB) = 12(12.01) + (10-n) (1.008) + n(35.45) = 154.20 + 34.44 n

Since n mol AgCl are produced for every 1 mol PCB reacted, then n(143.4) grams of AgCl will be produced for every (154.20 + 34.44 n) grams of PCB reacted.

$$\frac{\text{mass of AgCl}}{\text{mass of PCB}} = \frac{143.4\ n}{154.20 + 34.44\ n} \quad \text{or} \quad \text{mass}_{AgCl} (154.20 + 34.44\ n) = \text{mass}_{PCB} (143.4\ n)$$

b.   0.4791 (154.20 + 34.44 n) = 0.1947 (143.4 n), 73.88 + 16.50 n = 27.92 n

73.88 = 11.42 n, n = 6.469

87.   a. $2\ AgNO_3(aq)\ +\ K_2CrO_4(aq)\ \rightarrow\ Ag_2CrO_4(s) + 2\ KNO_3(aq)$

Molar mass: 169.9 g/mol   194.20 g/mol   331.8 g/mol

The molar mass of Ag_2CrO_4 is 331.8 g/mol, so one mol of precipitate was formed.

We have equal masses of $AgNO_3$ and K_2CrO_4. Since the molar mass of $AgNO_3$ is less than that of K_2CrO_4, then we have more mol of $AgNO_3$ present. However, we will not have twice the mol of $AgNO_3$ present as compared to K_2CrO_4 as required by the balanced reaction; this is because the molar mass of $AgNO_3$ is no where near one-half the molar mass of K_2CrO_4. Therefore, $AgNO_3$ is limiting.

$$\text{mass}\ AgNO_3 = 1.000\ \text{mol}\ Ag_2CrO_4 \times \frac{2\ \text{mol}\ AgNO_3}{\text{mol}\ Ag_2CrO_4} \times \frac{169.9\ g}{\text{mol}\ AgNO_3} = 339.8\ g\ AgNO_3$$

Since equal masses of reactants are present, then 339.8 g K_2CrO_4 were present initially.

$$M_{K^+} = \frac{\text{mol}\ K^+}{\text{total volume}} = \frac{339.8\ g\ K_2CrO_4 \times \dfrac{1\ \text{mol}\ K_2CrO_4}{194.20\ g} \times \dfrac{2\ \text{mol}\ K^+}{\text{mol}\ K_2CrO_4}}{0.5000\ L} = 7.000\ M\ K^+$$

b.   mol CrO_4^{2-} present initially $= 339.8\ g\ K_2CrO_4 \times \dfrac{1\ \text{mol}\ K_2CrO_4}{194.20\ g} \times \dfrac{1\ \text{mol}\ CrO_4^{2-}}{\text{mol}\ K_2CrO_4} = 1.750\ \text{mol}\ CrO_4^{2-}$

mol CrO_4^{2-} in precipitate $= 1.000\ \text{mol}\ Ag_2CrO_4 \times \dfrac{1\ \text{mol}\ CrO_4^{2-}}{1\ \text{mol}\ Ag_2CrO_4} = 1.000\ \text{mol}\ CrO_4^{2-}$

$$M_{CrO_4^{2-}} = \frac{\text{excess mol}\ CrO_4^{2-}}{\text{total volume}} = \frac{1.750\ \text{mol} - 1.000\ \text{mol}}{0.5000\ L + 0.5000\ L} = \frac{0.750\ \text{mol}}{1.0000\ L} = 0.750\ M$$

88.   Molar masses: KCl, $39.10 + 35.45 = 74.55$ g/mol; KBr, $39.10 + 79.90 = 119.00$ g/mol

AgCl, $107.9 + 35.45 = 143.4$ g/mol; AgBr, $107.9 + 79.90 = 187.8$ g/mol

Let x = number of moles of KCl in mixture and y = number of moles of KBr in mixture. Since $Ag^+ + Cl^- \rightarrow AgCl$ and $Ag^+ + Br^- \rightarrow AgBr$, then x = moles AgCl and y = moles AgBr.

Setting up two equations:

$$0.1024 \text{ g} = 74.55\,x + 119.0\,y \text{ and } 0.1889 \text{ g} = 143.4\,x + 187.8\,y$$

Multiply the first equation by $\dfrac{187.8}{119.0}$, and subtract from the second.

$$\begin{aligned} 0.1889 &= 143.4\,x + 187.8\,y \\ -0.1616 &= -117.7\,x - 187.8\,y \\ \hline 0.0273 &= 25.7\,x, \qquad x = 1.06 \times 10^{-3} \text{ mol KCl} \end{aligned}$$

$$1.06 \times 10^{-3} \text{ mol KCl} \times \frac{74.55 \text{ g KCl}}{\text{mol KCl}} = 0.0790 \text{ g KCl}$$

$$\% \text{ KCl} = \frac{0.0790 \text{ g}}{0.1024 \text{ g}} \times 100 = 77.1\%, \quad \% \text{ KBr} = 100.0 - 77.1 = 22.9\%$$

89. $0.298 \text{ g BaSO}_4 \times \dfrac{96.07 \text{ g SO}_4^{2-}}{233.4 \text{ g BaSO}_4} = 0.123 \text{ g SO}_4^{2-}$; $\% \text{ sulfate} = \dfrac{0.123 \text{ g SO}_4^{2-}}{0.205 \text{ g}} = 60.0\%$

Assume we have 100.0 g of the mixture of Na_2SO_4 and K_2SO_4. There are:

$$60.0 \text{ g SO}_4^{2-} \times \frac{1 \text{ mol}}{96.07 \text{ g}} = 0.625 \text{ mol SO}_4^{2-}$$

There must be $2 \times 0.625 = 1.25$ mol of +1 cations to balance the -2 charge of SO_4^{2-}.

Let x = number of moles of K^+ and y = number of moles of Na^+; then $x + y = 1.25$.

The total mass of Na^+ and K^+ must be 40.0 g in the assumed 100.0 g of mixture. Setting up an equation:

$$x \text{ mol K}^+ \times \frac{39.10 \text{ g}}{\text{mol}} + y \text{ mol Na}^+ \times \frac{22.99 \text{ g}}{\text{mol}} = 40.0 \text{ g}$$

So, we have two equations with two unknowns: $x + y = 1.25$ and $39.10\,x + 22.99\,y = 40.0$

Since $x = 1.25 - y$, then $39.10(1.25 - y) + 22.99\,y = 40.0$

$48.9 - 39.10\,y + 22.99\,y = 40.0$, $-16.11\,y = -8.9$

$y = 0.55$ mol Na^+ and $x = 1.25 - 0.55 = 0.70$ mol K^+

Therefore:

$$0.70 \text{ mol K}^+ \times \frac{1 \text{ mol K}_2\text{SO}_4}{2 \text{ mol K}^+} = 0.35 \text{ mol K}_2\text{SO}_4; \quad 0.35 \text{ mol K}_2\text{SO}_4 \times \frac{174.27 \text{ g}}{\text{mol}} = 61 \text{ g K}_2\text{SO}_4$$

Since we assumed 100.0 g, the mixture is 61% K_2SO_4 and 39% Na_2SO_4.

90.   a. $H_3PO_4(aq) + 3 \text{ NaOH}(aq) \rightarrow 3 H_2O(l) + Na_3PO_4(aq)$

         b. $3 H_2SO_4(aq) + 2 \text{ Al(OH)}_3(s) \rightarrow 6 H_2O(l) + Al_2(SO_4)_3(aq)$

         c. $H_2Se(aq) + Ba(OH)_2(aq) \rightarrow 2 H_2O(l) + BaSe(s)$

         d. $H_2C_2O_4(aq) + 2 \text{ NaOH}(aq) \rightarrow 2 H_2O(l) + Na_2C_2O_4(aq)$

91. $2 H_3PO_4(aq) + 3 \text{ Ba(OH)}_2(aq) \rightarrow 6 H_2O(l) + Ba_3(PO_4)_2(s)$

$$0.01420 \text{ L} \times \frac{0.141 \text{ mol H}_3\text{PO}_4}{\text{L}} \times \frac{3 \text{ mol Ba(OH)}_2}{2 \text{ mol H}_3\text{PO}_4} \times \frac{1 \text{ L Ba(OH)}_2}{0.0521 \text{ mol Ba(OH)}_2} = 0.0576 \text{ L}$$

$$= 57.6 \text{ mL Ba(OH)}_2$$

92. $35.08 \text{ mL NaOH} \times \dfrac{1 \text{ L}}{1000 \text{ mL}} \times \dfrac{2.12 \text{ mol NaOH}}{\text{L NaOH}} \times \dfrac{1 \text{ mol H}_2\text{SO}_4}{2 \text{ mol NaOH}} = 3.72 \times 10^{-2} \text{ mol H}_2\text{SO}_4$

$$\text{Molarity} = \frac{3.72 \times 10^{-2} \text{ mol}}{10.00 \text{ mL}} \times \frac{1000 \text{ mL}}{\text{L}} = 3.72 \text{ } M \text{ H}_2\text{SO}_4$$

93.   a. $MgO(s) + 2 \text{ HCl}(aq) \rightarrow MgCl_2(aq) + H_2O(l)$

         $Mg(OH)_2(s) + 2 \text{ HCl}(aq) \rightarrow MgCl_2(aq) + 2 H_2O(l)$

         $Al(OH)_3(s) + 3 \text{ HCl}(aq) \rightarrow AlCl_3(aq) + 3 H_2O(l)$

         b.   Let's calculate the number of moles of HCl neutralized per gram of substance. We can get these directly from the balanced equations and the molar masses of the substances.

$$\frac{2 \text{ mol HCl}}{\text{mol MgO}} \times \frac{1 \text{ mol MgO}}{40.31 \text{ g MgO}} = \frac{4.962 \times 10^{-2} \text{ mol HCl}}{\text{g MgO}}$$

$$\frac{2 \text{ mol HCl}}{\text{mol Mg(OH)}_2} \times \frac{1 \text{ mol Mg(OH)}_2}{58.33 \text{ g Mg(OH)}_2} = \frac{3.429 \times 10^{-2} \text{ mol HCl}}{\text{g Mg(OH)}_2}$$

$$\frac{3 \text{ mol HCl}}{\text{mol Al(OH)}_3} \times \frac{1 \text{ mol Al(OH)}_3}{78.00 \text{ g Al(OH)}_3} = \frac{3.846 \times 10^{-2} \text{ mol HCl}}{\text{g Al(OH)}_3}$$

Therefore, one gram of magnesium oxide would neutralize the most 0.10 M HCl.

94.   We get the empirical formula from the elemental analysis. Out of 100.00 g carminic acid there are:

$$53.66 \text{ g C} \times \frac{1 \text{ mol C}}{12.01 \text{ g C}} = 4.468 \text{ mol C}; \quad 4.09 \text{ g H} \times \frac{1 \text{ mol H}}{1.008 \text{ g H}} = 4.06 \text{ mol H}$$

$$42.25 \text{ g O} \times \frac{1 \text{ mol O}}{16.00 \text{ g O}} = 2.641 \text{ mol O}$$

Dividing the moles by the smallest number gives:

$$\frac{4.468}{2.641} = 1.692; \quad \frac{4.06}{2.641} = 1.54$$

These numbers don't give obvious mol ratios. Let's determine the mol C to mol H ratio.

$$\frac{4.468}{4.06} = 1.10 = \frac{11}{10}$$

So let's try $\frac{4.06}{10} = 0.406$ as a common factor: $\quad \frac{4.468}{0.406} = 11.0; \quad \frac{4.06}{0.406} = 10.0; \quad \frac{2.641}{0.406} = 6.50$

Therefore, $C_{22}H_{20}O_{13}$ is the empirical formula.

We can get molar mass from the titration data.

$$18.02 \times 10^{-3} \text{ L soln} \times \frac{0.0406 \text{ mol NaOH}}{\text{L soln}} \times \frac{1 \text{ mol carminic acid}}{\text{mol NaOH}} = 7.32 \times 10^{-4} \text{ mol carminic acid}$$

$$\text{Molar mass} = \frac{0.3602 \text{ g}}{7.32 \times 10^{-4} \text{ mol}} = \frac{492 \text{ g}}{\text{mol}}$$

The empirical formula mass of $C_{22}H_{20}O_{13} \approx 22(12) + 20(1) + 13(16) = 492$ g.

Therefore, the molecular formula of carminic acid is also $C_{22}H_{20}O_{13}$.

95. $\text{mol C}_6\text{H}_8\text{O}_7 = 0.250 \text{ g C}_6\text{H}_8\text{O}_7 \times \frac{1 \text{ mol C}_6\text{H}_8\text{O}_7}{192.12 \text{ g C}_6\text{H}_8\text{O}_7} = 1.30 \times 10^{-3} \text{ mol C}_6\text{H}_8\text{O}_7$

Let H_xA represent citric acid where x is the number of acidic hydrogens. The balanced neutralization reaction is:

$$H_xA(aq) + x \text{ OH}^-(aq) \rightarrow x \text{ H}_2\text{O}(l) + A^{x-}(aq)$$

$$\text{mol OH}^- \text{ reacted} = 0.0372 \text{ L} \times \frac{0.105 \text{ mol OH}^-}{\text{L}} = 3.91 \times 10^{-3} \text{ mol OH}^-$$

$$x = \frac{\text{mol OH}^-}{\text{mol citric acid}} = \frac{3.91 \times 10^{-3} \text{ mol}}{1.30 \times 10^{-3} \text{ mol}} = 3.01$$

Therefore, the general acid formula for citric acid is H_3A, meaning that citric acid has three acidic hydrogens per citric acid molecule (citric acid is a triprotic acid).

96.   a. $HCl(aq)$ dissociates to $H^+(aq) + Cl^-(aq)$. For simplicity let's use H^+ and Cl^- separately.

$$H^+ \rightarrow H_2 \qquad\qquad\qquad\qquad\qquad\qquad Fe \rightarrow HFeCl_4$$
$$(2\ H^+ + 2\ e^- \rightarrow H_2) \times 3 \qquad\qquad\qquad (H^+ + 4\ Cl^- + Fe \rightarrow HFeCl_4 + 3\ e^-) \times 2$$

$$6\ H^+ + 6\ e^- \rightarrow 3\ H_2$$
$$2\ H^+ + 8\ Cl^- + 2\ Fe \rightarrow 2\ HFeCl_4 + 6\ e^-$$

$$\overline{8\ H^+ + 8\ Cl^- + 2\ Fe \rightarrow 2\ HFeCl_4 + 3\ H_2}$$

or $8\ HCl(aq) + 2\ Fe(s) \rightarrow 2\ HFeCl_4(aq) + 3\ H_2(g)$

b.

$$IO_3^- \rightarrow I_3^- \qquad\qquad\qquad\qquad\qquad\qquad I^- \rightarrow I_3^-$$
$$3\ IO_3^- \rightarrow I_3^- \qquad\qquad\qquad\qquad\qquad (3\ I^- \rightarrow I_3^- + 2\ e^-) \times 8$$
$$3\ IO_3^- \rightarrow I_3^- + 9\ H_2O$$
$$16\ e^- + 18\ H^+ + 3\ IO_3^- \rightarrow I_3^- + 9\ H_2O$$

$$16\ e^- + 18\ H^+ + 3\ IO_3^- \rightarrow I_3^- + 9\ H_2O$$
$$24\ I^- \rightarrow 8\ I_3^- + 16\ e^-$$

$$\overline{18\ H^+ + 24\ I^- + 3\ IO_3^- \rightarrow 9\ I_3^- + 9\ H_2O}$$

Reducing: $6\ H^+(aq) + 8\ I^-(aq) + IO_3^-(aq) \rightarrow 3\ I_3^-(aq) + 3\ H_2O(l)$

c. $(Ce^{4+} + e^- \rightarrow Ce^{3+}) \times 97$

$$Cr(NCS)_6^{4-} \rightarrow Cr^{3+} + NO_3^- + CO_2 + SO_4^{2-}$$
$$54\ H_2O + Cr(NCS)_6^{4-} \rightarrow Cr^{3+} + 6\ NO_3^- + 6\ CO_2 + 6\ SO_4^{2-} + 108\ H^+$$

Charge on left –4. Charge on right $= +3 + 6(-1) + 6(-2) + 108(+1) = +93$. Add 97 e^- to the right, then add the two balanced half-reactions with a common factor of 97 e^- transferred.

$$54\ H_2O + Cr(NCS)_6^{4-} \rightarrow Cr^{3+} + 6\ NO_3^- + 6\ CO_2 + 6\ SO_4^{2-} + 108\ H^+ + 97\ e^-$$
$$97\ e^- + 97\ Ce^{4+} \rightarrow 97\ Ce^{3+}$$

$$\overline{97\ Ce^{4+}(aq) + 54\ H_2O(l) + Cr(NCS)_6^{4-}(aq) \rightarrow 97\ Ce^{3+}(aq) + Cr^{3+}(aq) + 6\ NO_3^-(aq) + 6\ CO_2(g)}$$

$$+ 6\ SO_4^{2-}(aq) + 108\ H^+(aq)$$

This is very complicated. A check of the net charge is a good check to see if the equation is balanced. Left: charge $= 97(+4) - 4 = +384$. Right: charge $= 97(+3) + 3 + 6(-1) + 6(-2) + 108(+1) = +384$.

d. $CrI_3 \rightarrow CrO_4^{2-} + IO_4^-$ $Cl_2 \rightarrow Cl^-$

$(16\ H_2O + CrI_3 \rightarrow CrO_4^{2-} + 3\ IO_4^- + 32\ H^+ + 27\ e^-) \times 2$ $(2\ e^- + Cl_2 \rightarrow 2\ Cl^-) \times 27$

Common factor is a transfer of 54 e^-.

$$54\ e^- + 27\ Cl_2 \rightarrow 54\ Cl^-$$
$$32\ H_2O + 2\ CrI_3 \rightarrow 2\ CrO_4^{2-} + 6\ IO_4^- + 64\ H^+ + 54\ e^-$$

$$32\ H_2O + 2\ CrI_3 + 27\ Cl_2 \rightarrow 54\ Cl^- + 2\ CrO_4^{2-} + 6\ IO_4^- + 64\ H^+$$

Add 64 OH^- to both sides and convert 64 H^+ into 64 H_2O.

$$64\ OH^- + 32\ H_2O + 2\ CrI_3 + 27\ Cl_2 \rightarrow 54\ Cl^- + 2\ CrO_4^{2-} + 6\ IO_4^- + 64\ H_2O$$

Reducing gives:

$$64\ OH^-(aq) + 2\ CrI_3(s) + 27\ Cl_2(g) \rightarrow 54\ Cl^-(aq) + 2\ CrO_4^{2-}(aq) + 6\ IO_4^-(aq) + 32\ H_2O(l)$$

e. $Ce^{4+} \rightarrow Ce(OH)_3$

$(e^- + 3\ H_2O + Ce^{4+} \rightarrow Ce(OH)_3 + 3\ H^+) \times 61$

$$Fe(CN)_6^{4-} \rightarrow Fe(OH)_3 + CO_3^{2-} + NO_3^-$$
$$Fe(CN)_6^{4-} \rightarrow Fe(OH)_3 + 6\ CO_3^{2-} + 6\ NO_3^-$$

There are 39 extra O atoms on right. Add 39 H_2O to left, then add 75 H^+ to right to balance H^+.

$$39\ H_2O + Fe(CN)_6^{4-} \rightarrow Fe(OH)_3 + 6\ CO_3^{2-} + 6\ NO_3^- + 75\ H^+$$
net charge = -4   net charge = +57

Add 61 e^- to the right; then add the two balanced half-reactions with a common factor of 61 e^- transferred.

$$39\ H_2O + Fe(CN)_6^{4-} \rightarrow Fe(OH)_3 + 6\ CO_3^{2-} + 6\ NO_3^- + 75\ H^+ + 61\ e^-$$
$$61\ e^- + 183\ H_2O + 61\ Ce^{4+} \rightarrow 61\ Ce(OH)_3 + 183\ H^+$$

$$222\ H_2O + Fe(CN)_6^{4-} + 61\ Ce^{4+} \rightarrow 61\ Ce(OH)_3 + Fe(OH)_3 + 6\ CO_3^{2-} + 6\ NO_3^- + 258\ H^+$$

Adding 258 OH^- to each side then reducing gives:

$$258\ OH^-(aq) + Fe(CN)_6^{4-}(aq) + 61\ Ce^{4+}(aq) \rightarrow 61\ Ce(OH)_3(s) + Fe(OH)_3(s)$$
$$+ 6\ CO_3^{2-}(aq) + 6\ NO_3^-(aq) + 36\ H_2O(l)$$

f. $Fe(OH)_2 \rightarrow Fe(OH)_3$ $H_2O_2 \rightarrow H_2O$
        $(H_2O + Fe(OH)_2 \rightarrow Fe(OH)_3 + H^+ + e^-) \times 2$ $2\,e^- + 2\,H^+ + H_2O_2 \rightarrow 2\,H_2O$

$$2\,H_2O + 2\,Fe(OH)_2 \rightarrow 2\,Fe(OH)_3 + 2\,H^+ + 2\,e^-$$
$$2\,e^- + 2\,H^+ + H_2O_2 \rightarrow 2\,H_2O$$

$$2\,H_2O + 2\,H^+ + 2\,Fe(OH)_2 + H_2O_2 \rightarrow 2\,Fe(OH)_3 + 2\,H_2O + 2\,H^+$$

Reducing gives: $2\,Fe(OH)_2(s) + H_2O_2(aq) \rightarrow 2\,Fe(OH)_3(s)$

97.   The amount of KHP used $= 0.4016\ g \times \dfrac{1\ mol}{204.22\ g} = 1.967 \times 10^{-3}$ mol KHP

Since one mole of NaOH reacts completely with one mole of KHP, the NaOH solution contains 1.967×10^{-3} mol NaOH.

Molarity of NaOH $= \dfrac{1.967 \times 10^{-3}\ mol}{25.06 \times 10^{-3}\ L} = \dfrac{7.849 \times 10^{-2}\ mol}{L}$

Maximum molarity $= \dfrac{1.967 \times 10^{-3}\ mol}{25.01 \times 10^{-3}\ L} = \dfrac{7.865 \times 10^{-2}\ mol}{L}$

Minimum molarity $= \dfrac{1.967 \times 10^{-3}\ mol}{25.11 \times 10^{-3}\ L} = \dfrac{7.834 \times 10^{-2}\ mol}{L}$

We can express this as $0.07849 \pm 0.00016\ M$. An alternative is to express the molarity as $0.0785 \pm 0.0002\ M$. The second way shows the actual number of significant figures in the molarity. The advantage of the first method is that it shows that we made all of our individual measurements to four significant figures.

CHAPTER FIVE

GASES

Questions

16. a. Heating the can will increase the pressure of the gas inside the can, $P \propto T$, V and n constant. As the pressure increases, it may be enough to rupture the can.

    b. As you draw a vacuum in your mouth, atmospheric pressure pushing on the surface of the liquid forces the liquid up the straw.

    c. The external atmospheric pressure pushes on the can. Since there is no opposing pressure from the air inside, the can collapses.

    d. How "hard" the tennis ball is depends on the difference between the pressure of the air inside the tennis ball and atmospheric pressure. A "sea level" ball will be much "harder" at high altitude since the external pressure is lower at high altitude. A "high altitude" ball will be "soft" at sea level.

17. $PV = nRT = $ constant at constant n and T. At two sets of conditions, $P_1V_1 = $ constant $= P_2V_2$.

    $P_1V_1 = P_2V_2$ (Boyle's law).

    $\dfrac{V}{T} = \dfrac{nR}{P} = $ constant at constant n and P. At two sets of conditions, $\dfrac{V_1}{T_1} = $ constant $= \dfrac{V_2}{T_2}$.

    $\dfrac{V_1}{T_1} = \dfrac{V_2}{T_2}$ (Charles's law)

18. Boyle's law: $P \propto 1/V$ at constant n and T

    In the kinetic molecular theory (kmt), P is proportional to the collision frequency which is proportional to 1/V. As the volume increases there will be fewer collisions per unit area with the walls of the container and pressure will decrease (Boyle's law).

    Charles's law: $V \propto T$ at constant n and P

    When a gas is heated to a higher temperature, the speeds of the gas molecules increase and thus hit the walls of the container more often and with more force. In order to keep the pressure constant, the volume of the container must increase (this increases surface area which decreases the number of collisions per unit area which decreases the pressure). Therefore, volume and temperature are directly related at constant n and P (Charles's law).

19.   The kinetic molecular theory assumes that gas particles do not exert forces on each other and that gas particles are volumeless. Real gas particles do exert attractive forces for each other, and real gas particles do have volumes. A gas behaves most ideally at low pressures and high temperatures. The effect of attractive forces is minimized at high temperatures since the gas particles are moving very rapidly. At low pressure, the container volume is relatively large (P and V are inversely related) so the volume of the container taken up by the gas particles is negligible.

20.   Molecules in the condensed phases (liquids and solids) are very close together. Molecules in the gaseous phase are very far apart. A sample of gas is mostly empty space. Therefore, one would expect 1 mol of $H_2O(g)$ to occupy a huge volume as compared to 1 mol of $H_2O(l)$.

21.   Method 1: molar mass $= \dfrac{dRT}{P}$

Determine the density of a gas at a measurable temperature and pressure, then use the above equation to determine the molar mass.

Method 2: $\dfrac{\text{effusion rate for gas 1}}{\text{effusion rate for gas 2}} = \sqrt{\dfrac{(\text{molar mass})_2}{(\text{molar mass})_1}}$

Determine the effusion rate of the unknown gas relative to some known gas; then use Graham's law of effusion (the above equation) to determine the molar mass.

22.   a.   At constant temperature, the average kinetic energy of the He gas sample will equal the average kinetic energy of the Cl_2 gas sample. In order for the average kinetic energy to be the same, the smaller He atoms must move at a faster average velocity as compared to Cl_2. Therefore, plot A, with the slower average velocity, would be for the Cl_2 sample, and plot B would be for the He sample. Note the average velocity in each plot is a little past the top peak.

        b.   As temperature increases, the average velocity of a gas will increase. Plot A would be for $O_2(g)$ at 273 K and plot B, with the faster average velocity, would be for $O_2(g)$ at 1273 K.

              Since a gas behaves more ideally at higher temperatures, $O_2(g)$ at 1273 K would behave most ideally.

23.   Rigid container (constant volume): As reactants are converted to products, the mol of gas particles present decrease by one-half. As n decreases, the pressure will decrease (by one-half). Density is the mass per unit volume. Mass is conserved in a chemical reaction, so the density of the gas will not change since mass and volume do not change.

        Flexible container (constant pressure): Pressure is constant since the container changes volume in order to keep a constant pressure. As the mol of gas particles decrease by a factor of 2, the volume of the container will decrease (by one-half). We have the same mass of gas in a smaller volume, so the gas density will increase (is doubled).

24.   a.   Containers ii, iv, vi, and viii have volumes twice that of containers i, iii, v, and vii. Containers iii, iv, vii, and viii have twice the number of molecules (mol) present as compared to containers i, ii, v, and vi. The container with the lowest pressure will be the one which has the fewest mol of gas present in the largest volume (containers ii and vi both have the lowest P). The smallest container with the most mol of gas present will have the highest pressure (containers iii and vii both have the highest P). All the other containers (i, iv, v and viii) will have the same pressure between the two extremes. The order is: ii = vi < i = iv = v = viii < iii = vii.

b.   All have the same average kinetic energy since the temperature is the same in each container. Only the temperature determines the average kinetic energy.

c.   The least dense gas will be container ii since it has the fewest of the lighter Ne atoms present in the largest volume. Container vii has the most dense gas since the largest number of the heavier Ar atoms are present in the smallest volume. To figure out the ordering for the other containers, we will calculate the relative density of each. In the table below, m_1 equals the mass of Ne in container i, V_1 equals the volume of container i, and d_1 equals the density of the gas in container i.

Container	i	ii	iii	iv	v	vi	vii	viii
mass, volume	m_1, V_1	$m_1, 2V_1$	$2m_1, V_1$	$2m_1, 2V_1$	$2m_1, V_1$	$2m_1, 2V_1$	$4m_1, V_1$	$4m_1, 2V_1$
density $\left(\dfrac{\text{mass}}{\text{volume}}\right)$	$\dfrac{m_1}{V_1}=d_1$	$\dfrac{m_1}{2V_1}=\dfrac{1}{2}d_1$	$\dfrac{2m_1}{V_1}=2d_1$	$\dfrac{2m_1}{2V_1}=d_1$	$\dfrac{2m_1}{V_1}=2d_1$	$\dfrac{2m_1}{2V_1}=d_1$	$\dfrac{4m_1}{V_1}=4d_1$	$\dfrac{4m_1}{2V_1}=2d_1$

From the table, the order of gas density is: ii < i = iv = vi < iii = v = viii < vii

d. $\mu_{rms} = (3\,RT/M)^{1/2}$; the root mean square velocity only depends on the temperature and the molar mass. Since T is constant, the heavier argon molecules will have the slower root mean square velocity as compared to the neon molecules. The order is:
v = vi = vii = viii < i = ii = iii = iv.

Exercises

Pressure

25.   a. $4.8 \text{ atm} \times \dfrac{760 \text{ mm Hg}}{\text{atm}} = 3.6 \times 10^3 \text{ mm Hg}$; b. $3.6 \times 10^3 \text{ mm Hg} \times \dfrac{1 \text{ torr}}{\text{mm Hg}} = 3.6 \times 10^3 \text{ torr}$

c. $4.8 \text{ atm} \times \dfrac{1.013 \times 10^5 \text{ Pa}}{\text{atm}} = 4.9 \times 10^5 \text{ Pa}$;   d. $4.8 \text{ atm} \times \dfrac{14.7 \text{ psi}}{\text{atm}} = 71 \text{ psi}$

26.   a. $2200 \text{ psi} \times \dfrac{1 \text{ atm}}{14.7 \text{ psi}} = 150 \text{ atm}$;   b. $150 \text{ atm} \times \dfrac{1.013 \times 10^5 \text{ Pa}}{\text{atm}} \times \dfrac{1 \text{ MPa}}{1 \times 10^6 \text{ Pa}} = 15 \text{ MPa}$

c. $150 \text{ atm} \times \dfrac{760 \text{ torr}}{\text{atm}} = 1.1 \times 10^5 \text{ torr}$

27. $6.5 \text{ cm} \times \dfrac{10 \text{ mm}}{\text{cm}} = 65 \text{ mm Hg or } 65 \text{ torr}; \quad 65 \text{ torr} \times \dfrac{1 \text{ atm}}{760 \text{ torr}} = 8.6 \times 10^{-2} \text{ atm}$

$8.6 \times 10^{-2} \text{ atm} = \dfrac{1.013 \times 10^5 \text{ Pa}}{\text{atm}} = 8.7 \times 10^3 \text{ Pa}$

28. $20.0 \text{ in Hg} \times \dfrac{2.54 \text{ cm}}{\text{in}} \times \dfrac{10 \text{ mm}}{\text{cm}} = 508 \text{ mm Hg} = 508 \text{ torr}; \quad 508 \text{ torr} \times \dfrac{1 \text{ atm}}{760 \text{ torr}} = 0.668 \text{ atm}$

29.   If the levels of Hg in each arm of the manometer are equal, the pressure in the flask is equal to atmospheric pressure. When they are unequal, the difference in height in mm will be equal to the difference in pressure in mm Hg between the flask and the atmosphere. Which level is higher will tell us whether the pressure in the flask is less than or greater than atmospheric.

   a. $P_{flask} < P_{atm}; P_{flask} = 760. - 118 = 642 \text{ mm Hg} = 642 \text{ torr}; \quad 642 \text{ torr} \times \dfrac{1 \text{ atm}}{760 \text{ torr}} = 0.845 \text{ atm}$

$0.845 \text{ atm} \times \dfrac{1.013 \times 10^5 \text{ Pa}}{\text{atm}} = 8.56 \times 10^4 \text{ Pa}$

   b. $P_{flask} > P_{atm}; P_{flask} = 760. \text{ torr} + 215 \text{ torr} = 975 \text{ torr}; \quad 975 \text{ torr} \times \dfrac{1 \text{ atm}}{760 \text{ torr}} = 1.28 \text{ atm}$

$1.28 \text{ atm} \times \dfrac{1.013 \times 10^5 \text{ Pa}}{\text{atm}} = 1.30 \times 10^5 \text{ Pa}$

   c. $P_{flask} = 635 - 118 = 517 \text{ torr}; \quad P_{flask} = 635 + 215 = 850. \text{ torr}$

30.   a.   The pressure is proportional to the mass of the fluid. The mass is proportional to the volume of the column of fluid (or to the height of the column assuming the area of the column of fluid is constant).

$d = \dfrac{\text{mass}}{\text{volume}};$ In this case, the volume of silicon oil will be the same as the volume of Hg in Exercise 5.29.

$V = \dfrac{m}{d}; \quad V_{Hg} = V_{oil}, \quad \dfrac{m_{Hg}}{d_{Hg}} = \dfrac{m_{oil}}{d_{oil}}, \quad m_{oil} = \dfrac{m_{Hg} d_{oil}}{d_{Hg}}$

Since P is proportional to the mass of liquid:

$P_{oil} = P_{Hg}\left(\dfrac{d_{oil}}{d_{Hg}}\right) = P_{Hg}\left(\dfrac{1.30}{13.6}\right) = 0.0956 \, P_{Hg}$

This conversion applies only to the column of liquid.

P_{flask} = 760. torr - (118 × 0.0956) torr = 760. - 11.3 = 749 torr

$$749 \text{ torr} \times \frac{1 \text{ atm}}{760 \text{ torr}} = 0.986 \text{ atm}; \quad 0.986 \text{ atm} \times \frac{1.013 \times 10^5 \text{ Pa}}{\text{atm}} = 9.99 \times 10^4 \text{ Pa}$$

P_{flask} = 760. torr + (215 × 0.0956) torr = 760. + 20.6 = 781 torr

$$781 \text{ torr} \times \frac{1 \text{ atm}}{760 \text{ torr}} = 1.03 \text{ atm}; \quad 1.03 \text{ atm} \times \frac{1.013 \times 10^5 \text{ Pa}}{\text{atm}} = 1.04 \times 10^5 \text{ Pa}$$

b. If we are measuring the same pressure, the height of the silicon oil column would be 13.6 ÷ 1.30 = 10.5 times the height of a mercury column. The advantage of using a less dense fluid than mercury is in measuring small pressures. The quantity measured (length) will be larger for the less dense fluid. Thus, the measurement will be more precise.

Gas Laws

31. From Boyle's law, $P_1V_1 = P_2V_2$ at constant n and T.

$$P_2 = \frac{P_1V_1}{V_2} = \frac{5.20 \text{ atm} \times 0.400 \text{ L}}{2.14 \text{ L}} = 0.972 \text{ atm}$$

As expected, as the volume increased, the pressure decreased.

32. The pressure exerted on the balloon is constant and the moles of gas present is constant. From Charles's law, $V_1/T_1 = V_2/T_2$ at constant P and n.

$$V_2 = \frac{V_1T_2}{T_1} = \frac{700. \text{ mL} \times 100. \text{ K}}{(273.2 + 20.0) \text{ K}} = 239 \text{ mL}$$

As expected, as the temperature decreased, the volume decreased.

33. From Avogadro's law, $V_1/n_1 = V_2/n_2$ at constant T and P.

$$V_2 = \frac{V_1n_2}{n_1} = \frac{11.2 \text{ L} \times 2.00 \text{ mol}}{0.500 \text{ mol}} = 44.8 \text{ L}$$

As expected, as the mol of gas present increases, volume increases.

34. As NO_2 is converted completely into N_2O_4, the mol of gas present will decrease by one-half (from the 2:1 mol ratio in the balanced equation). Using Avogadro's law,

$$\frac{V_1}{n_1} = \frac{V_2}{n_2}, \quad V_2 = V_1 \times \frac{n_2}{n_1} = 25.0 \text{ mL} \times \frac{1}{2} = 12.5 \text{ mL}$$

$N_2O_4(g)$ will occupy one-half the original volume of $NO_2(g)$. This is expected since the mol of gas present decrease by one-half when NO_2 is converted into N_2O_4.

35.   a. $PV = nRT$, $V = \dfrac{nRT}{P} = \dfrac{2.00 \text{ mol} \times \dfrac{0.08206 \text{ L atm}}{\text{mol K}} \times (155 + 273) \text{ K}}{5.00 \text{ atm}} = 14.0 \text{ L}$

       b. $PV = nRT$, $n = \dfrac{PV}{RT} = \dfrac{0.300 \text{ atm} \times 2.00 \text{ L}}{\dfrac{0.08206 \text{ L atm}}{\text{mol K}} \times 155 \text{ K}} = 4.72 \times 10^{-2} \text{ mol}$

       c. $PV = nRT$, $T = \dfrac{PV}{nR} = \dfrac{4.47 \text{ atm} \times 25.0 \text{ L}}{2.01 \text{ mol} \times \dfrac{0.08206 \text{ L atm}}{\text{mol K}}} = 678 \text{ K} = 405°C$

       d. $PV = nRT$, $P = \dfrac{nRT}{V} = \dfrac{10.5 \text{ mol} \times \dfrac{0.08206 \text{ L atm}}{\text{mol K}} \times (273 + 75) \text{ K}}{2.25 \text{ L}} = 133 \text{ atm}$

36.   a. $P = 7.74 \times 10^3 \text{ Pa} \times \dfrac{1 \text{ atm}}{1.013 \times 10^5 \text{ Pa}} = 0.0764 \text{ atm}$; $T = 25 + 273 = 298 \text{ K}$

             $PV = nRT$, $n = \dfrac{PV}{RT} = \dfrac{0.0764 \text{ atm} \times 0.0122 \text{ L}}{\dfrac{0.08206 \text{ L atm}}{\text{mol K}} \times 298 \text{ K}} = 3.81 \times 10^{-5} \text{ mol}$

       b. $PV = nRT$, $P = \dfrac{nRT}{V} = \dfrac{0.421 \text{ mol} \times \dfrac{0.08206 \text{ L atm}}{\text{mol K}} \times 223 \text{ K}}{0.0430 \text{ L}} = 179 \text{ atm}$

       c. $V = \dfrac{nRT}{P} = \dfrac{4.4 \times 10^{-2} \text{ mol} \times \dfrac{0.08206 \text{ L atm}}{\text{mol K}} \times (331 + 273) \text{ K}}{455 \text{ torr} \times \dfrac{1 \text{ atm}}{760 \text{ torr}}} = 3.6 \text{ L}$

       d. $T = \dfrac{PV}{nR} = \dfrac{\left(745 \text{ mm Hg} \times \dfrac{1 \text{ atm}}{760 \text{ mm Hg}}\right) \times 11.2 \text{ L}}{0.401 \text{ mol} \times \dfrac{0.08206 \text{ L atm}}{\text{mol K}}} = 334 \text{ K} = 61°C$

37. $n = \dfrac{PV}{RT} = \dfrac{135 \text{ atm} \times 200.0 \text{ L}}{0.08206 \dfrac{\text{L atm}}{\text{mol K}} \times (273 + 24) \text{ K}} = 1.11 \times 10^3 \text{ mol}$

For He: $1.11 \times 10^3 \text{ mol} \times \dfrac{4.003 \text{ g He}}{\text{mol}} = 4.44 \times 10^3 \text{ g He}$

For H_2: $1.11 \times 10^3 \text{ mol} \times \dfrac{2.016 \text{ g He}}{\text{mol}} = 2.24 \times 10^3 \text{ g H}_2$

38. $P = \dfrac{nRT}{V} = \dfrac{\left(0.60\ g \times \dfrac{1\ mol}{32.00\ g}\right) \times \dfrac{0.08206\ L\ atm}{mol\ K} \times (273 + 22)\ K}{5.0\ L} = 0.091\ atm$

39.   a. $PV = nRT$; $175\ g\ Ar \times \dfrac{1\ mol\ Ar}{39.95\ g\ Ar} = 4.38\ mol\ Ar$

$T = \dfrac{PV}{nR} = \dfrac{10.0\ atm \times 2.50\ L}{4.38\ mol \times \dfrac{0.08206\ L\ atm}{mol\ K}} = 69.6\ K$

b. $PV = nRT$, $P = \dfrac{nRT}{V} = \dfrac{4.38\ mol \times \dfrac{0.08206\ L\ atm}{mol\ K} \times 225\ K}{2.50\ L} = 32.3\ atm$

40. $0.050\ mL \times \dfrac{1.149\ g}{mL} \times \dfrac{1\ mol\ O_2}{32.00\ g} = 1.8 \times 10^{-3}\ mol\ O_2$

$V = \dfrac{nRT}{P} = \dfrac{1.8 \times 10^{-3}\ mol \times \dfrac{0.08206\ L\ atm}{mol\ K} \times 310.\ K}{1.0\ atm} = 4.6\ 10^{-2}\ L = 46\ mL$

41.   At constant n and T, $PV = nRT =$ constant, $P_1V_1 = P_2V_2$; At sea level, $P = 1.00\ atm = 760.\ mm\ Hg$.

$V_2 = \dfrac{P_1V_1}{P_2} = \dfrac{760.\ mm\ Hg \times 2.0\ L}{500.\ mm\ Hg} = 3.0\ L$

The balloon will burst at this pressure since the volume must expand beyond the 2.5 L limit of the balloon.

Note: To solve this problem, we did not have to convert the pressure units into atm; the units of mm Hg canceled each other. In general, only convert units if you have to. Whenever the gas constant R is not used to solve a problem, pressure and volume units must only be consistent, and not necessarily in units of atm and L. The exception is temperature as T must <u>always</u> be converted to the Kelvin scale.

42. $PV = nRT$, n is constant. $\dfrac{PV}{T} = nR =$ constant, $\dfrac{P_1V_1}{T_1} = \dfrac{P_2V_2}{T_2}$

$V_2 = 1.040\ V_1$, so $\dfrac{V_1}{V_2} = \dfrac{1.000}{1.040}$

$P_2 = \dfrac{P_1V_1T_2}{V_2T_1} = 100.\ psi \times \dfrac{1.000}{1.040} \times \dfrac{(273 + 58)\ K}{(273 + 19)\ K} = 109\ psi$

43. $PV = nRT$, V and n constant, so $\dfrac{P}{T} = \dfrac{nR}{V}$ = constant and $\dfrac{P_1}{T_1} = \dfrac{P_2}{T_2}$.

$P_2 = \dfrac{P_1 T_2}{T_1} = 13.7\ \text{MPa} \times \dfrac{(273 + 450.)\ \text{K}}{(273 + 23)\ \text{K}} = 33.5\ \text{MPa}$

44.   a.   At constant n and V, $\dfrac{P_1}{T_1} = \dfrac{P_2}{T_2}$, $P_2 = \dfrac{P_1 T_2}{T_1} = 40.0\ \text{atm} \times \dfrac{318\ \text{K}}{273\ \text{K}} = 46.6\ \text{atm}$

       b. $\dfrac{P_1}{T_1} = \dfrac{P_2}{T_2}$, $T_2 = \dfrac{T_1 P_2}{P_1} = 273\ \text{K} \times \dfrac{150.\ \text{atm}}{40.0\ \text{atm}} = 1.02 \times 10^3\ \text{K}$

       c. $T_2 = \dfrac{T_1 P_2}{P_1} = 273\ \text{K} \times \dfrac{25.0\ \text{atm}}{40.0\ \text{atm}} = 171\ \text{K}$

45. $PV = nRT$, n constant; $\dfrac{PV}{T} = nR$ = constant, $\dfrac{P_1 V_1}{T_1} = \dfrac{P_2 V_2}{T_2}$

$P_2 = \dfrac{P_1 V_1 T_2}{V_2 T_1} = 710.\ \text{torr} \times \dfrac{5.0 \times 10^2\ \text{mL}}{25\ \text{mL}} \times \dfrac{(273 + 820.)\ \text{K}}{(273 + 30.)\ \text{K}} = 5.1 \times 10^4\ \text{torr}$

46. $PV = nRT$, V constant; $\dfrac{nT}{P} = \dfrac{V}{R}$ = constant, $\dfrac{n_1 T_1}{P_1} = \dfrac{n_2 T_2}{P_2}$; mol × molar mass = mass

$\dfrac{n_1\,(\text{molar mass})\,T_1}{P_1} = \dfrac{n_2\,(\text{molar mass})\,T_2}{P_2}$, $\dfrac{\text{mass}_1 \times T_1}{P_1} = \dfrac{\text{mass}_2 \times T_2}{P_2}$

$\text{mass}_2 = \dfrac{\text{mass}_1 \times T_1 P_2}{T_2 P_1} = \dfrac{1.00 \times 10^3\ \text{g} \times 291\ \text{K} \times 650.\ \text{psi}}{299\ \text{K} \times 2050.\ \text{psi}} = 309\ \text{g Ar remains}$

47. $PV = nRT$, n is constant. $\dfrac{PV}{T} = nR$ = constant, $\dfrac{P_1 V_1}{T_1} = \dfrac{P_2 V_2}{T_2}$, $V_2 = \dfrac{V_1 P_1 T_2}{P_2 T_1}$

$V_2 = 1.00\ \text{L} \times \dfrac{760.\ \text{torr}}{220.\ \text{torr}} \times \dfrac{(273 - 31)\ \text{K}}{(273 + 23)\ \text{K}} = 2.82\ \text{L};\ \Delta V = 2.82 - 1.00 = 1.82\ \text{L}$

48. $PV = nRT$, P is constant. $\dfrac{nT}{V} = \dfrac{P}{R}$ = constant, $\dfrac{n_1 T_1}{V_1} = \dfrac{n_2 T_2}{V_2}$

$\dfrac{n_2}{n_1} = \dfrac{T_1 V_2}{T_2 V_1} = \dfrac{294\ \text{K}}{335\ \text{K}} \times \dfrac{4.20 \times 10^3\ \text{m}^3}{4.00 \times 10^3\ \text{m}^3} = 0.921$

Gas Density, Molar Mass, and Reaction Stoichiometry

49.   STP: T = 273 K and P = 1.00 atm; $n = \dfrac{PV}{RT} = \dfrac{1.00\ \text{atm} \times 1.5\ \text{L}}{\dfrac{0.08206\ \text{L atm}}{\text{mol K}} \times 273\ \text{K}} = 6.7 \times 10^{-2}$ mol He

Or we can use the fact that at STP, 1 mol of an ideal gas occupies 22.42 L.

$1.5\ \text{L} \times \dfrac{1\ \text{mol He}}{22.42\ \text{L}} = 6.7 \times 10^{-2}$ mol He; 6.7×10^{-2} mol He $\times \dfrac{4.003\ \text{g He}}{\text{mol He}} = 0.27$ g He

50. $CO_2(s) \rightarrow CO_2(g)$; $4.00\ \text{g CO}_2 \times \dfrac{1\ \text{mol CO}_2}{44.01\ \text{g CO}_2} = 9.09 \times 10^{-2}$ mol CO_2

At STP, the molar volume of a gas is 22.42 L. 9.09×10^{-2} mol $CO_2 \times \dfrac{22.42\ \text{L}}{\text{mol CO}_2} = 2.04$ L

51. $C_6H_{12}O_6(s) + 6\ O_2(g) \rightarrow 6\ CO_2(g) + 6\ H_2O(g)$

$5.00\ \text{g C}_6\text{H}_{12}\text{O}_6 \times \dfrac{1\ \text{mol C}_6\text{H}_{12}\text{O}_6}{180.16\ \text{g}} \times \dfrac{6\ \text{mol O}_2}{\text{mol C}_6\text{H}_{12}\text{O}_6} = 0.167$ mol O_2

$V = \dfrac{nRT}{P} = \dfrac{0.167\ \text{mol} \times 0.08206\ \dfrac{\text{L atm}}{\text{mol K}} \times 301\ \text{K}}{0.976\ \text{atm}} = 4.23$ L O_2

Since T and P are constant, the volume of each gas will be directly proportional to the mol of gas present. The balanced equation says that equal mol of CO_2 and H_2O will be produced as mol of O_2 reacted. So the volumes of CO_2 and H_2O produced will equal the volume of O_2 reacted.

$V_{CO_2} = V_{H_2O} = V_{O_2} = 4.23$ L

52.   Since the solution is 50.0% H_2O_2 by mass, the mass of H_2O_2 decomposed is 125/2 = 62.5 g.

$62.5\ \text{g H}_2\text{O}_2 \times \dfrac{1\ \text{mol H}_2\text{O}_2}{34.02\ \text{g H}_2\text{O}_2} \times \dfrac{1\ \text{mol O}_2}{2\ \text{mol H}_2\text{O}_2} = 0.919$ mol O_2

$V = \dfrac{nRT}{P} = \dfrac{0.919\ \text{mol} \times 0.08206\ \dfrac{\text{L atm}}{\text{mol K}} \times 300.\ \text{K}}{746\ \text{torr} \times \dfrac{1\ \text{atm}}{760\ \text{torr}}} = 23.0$ L O_2

53.

$$n_{H_2} = \frac{PV}{RT} = \frac{1.0 \text{ atm} \times \left[4800 \text{ m}^3 \times \left(\frac{100 \text{ cm}}{\text{m}}\right)^3 \times \frac{1 \text{ L}}{1000 \text{ cm}^3}\right]}{\frac{0.08206 \text{ L atm}}{\text{mol K}} \times 273 \text{ K}} = 2.1 \times 10^5 \text{ mol}$$

2.1×10^5 mol H_2 are in the balloon. This is 80.% of the total amount of H_2 that had to be generated:

0.80 (total mol H_2) $= 2.1 \times 10^5$, total mol $H_2 = 2.6 \times 10^5$ mol H_2

$$2.6 \times 10^5 \text{ mol } H_2 \times \frac{1 \text{ mol Fe}}{\text{mol } H_2} \times \frac{55.85 \text{ g Fe}}{\text{mol Fe}} = 1.5 \times 10^7 \text{ g Fe}$$

$$2.6 \times 10^5 \text{ mol } H_2 \times \frac{1 \text{ mol } H_2SO_4}{\text{mol } H_2} \times \frac{98.09 \text{ g } H_2SO_4}{\text{mol } H_2SO_4} \times \frac{100 \text{ g reagent}}{98 \text{ g } H_2SO_4} = 2.6 \times 10^7 \text{ g of 98\% sulfuric acid}$$

54. $2 NaN_3(s) \rightarrow 2 Na(s) + 3 N_2(g)$

$$n_{N_2} = \frac{PV}{RT} = \frac{1.00 \text{ atm} \times 70.0 \text{ L}}{\frac{0.08206 \text{ L atm}}{\text{mol K}} \times 273 \text{ K}} = 3.12 \text{ mol } N_2 \text{ needed to fill air bag.}$$

$$\text{mass } NaN_3 \text{ reacted} = 3.12 \text{ mol } N_2 \times \frac{2 \text{ mol } NaN_3}{3 \text{ mol } N_2} \times \frac{65.02 \text{ g } NaN_3}{\text{mol } NaN_3} = 135 \text{ g } NaN_3$$

55. $CH_3OH + 3/2 \ O_2 \rightarrow CO_2 + 2 \ H_2O$ or $2 \ CH_3OH(l) + 3 \ O_2(g) \rightarrow 2 \ CO_2(g) + 4 \ H_2O(g)$

$$50.0 \text{ mL} \times \frac{0.850 \text{ g}}{\text{mL}} \times \frac{1 \text{ mol}}{32.04 \text{ g}} = 1.33 \text{ mol } CH_3OH(l) \text{ available}$$

$$n_{O_2} = \frac{PV}{RT} = \frac{2.00 \text{ atm} \times 22.8 \text{ L}}{\frac{0.08206 \text{ L atm}}{\text{mol K}} \times 300. \text{ K}} = 1.85 \text{ mol } O_2 \text{ available}$$

$$1.33 \text{ mol } CH_3OH \times \frac{3 \text{ mol } O_2}{2 \text{ mol } CH_3OH} = 2.00 \text{ mol } O_2$$

2.00 mol O_2 are required to react completely with all of the CH_3OH available. We only have 1.85 mol O_2, so O_2 is limiting.

$$1.85 \text{ mol } O_2 \times \frac{4 \text{ mol } H_2O}{3 \text{ mol } O_2} = 2.47 \text{ mol } H_2O$$

56.   For ammonia (in one minute):

$$n_{NH_3} = \frac{PV}{RT} = \frac{90. \text{ atm} \times 500. \text{ L}}{\frac{0.08206 \text{ L atm}}{\text{mol K}} \times 496 \text{ K}} = 1.1 \times 10^3 \text{ mol } NH_3$$

NH_3 flows into the reactor at a rate of 1.1×10^3 mol/min.

For CO_2 (in one minute):

$$n_{CO_2} = \frac{PV}{RT} = \frac{45 \text{ atm} \times 600. \text{ L}}{\dfrac{0.08206 \text{ L atm}}{\text{mol K}} \times 496 \text{ K}} = 6.6 \times 10^2 \text{ mol } CO_2$$

CO_2 flows into the reactor at 6.6×10^2 mol/min.

To react completely with 1.1×10^3 mol NH_3/min, we need:

$$\frac{1.1 \times 10^3 \text{ mol } NH_3}{\text{min}} \times \frac{1 \text{ mol } CO_2}{2 \text{ mol } NH_3} = 5.5 \times 10^2 \text{ mol } CO_2/\text{min}$$

Since 660 mol CO_2/min are present, ammonia is the limiting reagent.

$$\frac{1.1 \times 10^3 \text{ mol } NH_3}{\text{min}} \times \frac{1 \text{ mol urea}}{2 \text{ mol } NH_3} \times \frac{60.06 \text{ g urea}}{\text{mol urea}} = 3.3 \times 10^4 \text{ g urea/min}$$

57.   a. $CH_4(g) + NH_3(g) + O_2(g) \rightarrow HCN(g) + H_2O(g)$; Balancing H first, then O, gives:

$$CH_4 + NH_3 + \frac{3}{2}O_2 \rightarrow HCN + 3 H_2O \text{ or } 2 CH_4(g) + 2 NH_3(g) + 3 O_2(g) \rightarrow 2 HCN(g) + 6 H_2O(g)$$

b. $PV = nRT$, T and P constant; $\dfrac{V_1}{n_1} = \dfrac{V_2}{n_2}$, $\dfrac{V_1}{V_2} = \dfrac{n_1}{n_2}$

Since the volumes are all measured at constant T and P, the volumes of gas present are directly proportional to the mol of gas present (Avogadro's law). Because Avogadro's law applies, the balanced reaction gives mol relationships as well as volume relationships. Therefore, 2 L of CH_4, 2 L of NH_3 and 3 L of O_2 are required by the balanced equation for the production of 2 L of HCN. The actual volume ratio is 20.0 L CH_4:20.0 L NH_3:20.0 L O_2 (or 1:1:1). The volume of O_2 required to react with all of the CH_4 and NH_3 present is 20.0 L ×(3/2) = 30.0 L. Since only 20.0 L of O_2 are present, O_2 is the limiting reagent. The volume of HCN produced is:

$$20.0 \text{ L } O_2 \times \frac{2 \text{ L HCN}}{3 \text{ L } O_2} = 13.3 \text{ L HCN}$$

58.   Since P and T are constant, V and n are directly proportional. The balanced equation requires 2 L of H_2 to react with 1 L of CO (2:1 volume ratio due to 2:1 mol ratio in balanced equation). The actual volume ratio present in one minute is 16.0 L/25.0 L = 0.640 (0.640:1). Since the actual volume ratio present is smaller than the required volume ratio, H_2 is the limiting reactant. The volume of CH_3OH produced at STP will be one-half the volume of H_2 reacted due to the 1:2 mol ratio in the balanced equation. In one minute, 16.0 L/2 = 8.00 L CH_3OH are produced (theoretical yield).

$$n_{CH_3OH} = \frac{PV}{RT} = \frac{1.00 \text{ atm} \times 8.00 \text{ L}}{\frac{0.08206 \text{ L atm}}{\text{mol K}} \times 273 \text{ K}} = 0.357 \text{ mol CH}_3\text{OH}\ \text{ in one minute}$$

$$0.357 \text{ mol CH}_3\text{OH} \times \frac{32.04 \text{ g CH}_3\text{OH}}{\text{mol CH}_3\text{OH}} = 11.4 \text{ g CH}_3\text{OH (theoretical yield per minute)}$$

$$\% \text{ yield} = \frac{\text{actual yield}}{\text{theoretical yield}} \times 100 = \frac{5.30 \text{ g}}{11.4 \text{ g}} \times 100 = 46.5\% \text{ yield}$$

59.   One of the equations developed in the text to determine molar mass is:

$$\text{molar mass} = \frac{dRT}{P} \text{ where } d = \text{density in units of g/L}$$

$$\text{molar mass} = \frac{1.65 \text{ g/L} \times \frac{0.08206 \text{ L atm}}{\text{mol K}} \times (273 + 27) \text{ K}}{734 \text{ torr} \times \frac{1 \text{ atm}}{760 \text{ torr}}} = 42.1 \text{ g/mol}$$

The empirical formula mass of $CH_2 = 12.01 + 2(1.008) = 14.03$ g/mol.

$$\frac{42.1}{14.03} = 3.00; \text{ Molecular formula} = C_3H_6$$

60. $P \times (\text{molar mass}) = dRT, d = \dfrac{\text{mass}}{\text{volume}}, P \times (\text{molar mass}) = \dfrac{\text{mass}}{V} \times RT$

$$\text{Molar mass} = M = \frac{\text{mass} \times RT}{PV} = \frac{0.800 \text{ g} \times \frac{0.08206 \text{ L atm}}{\text{mol K}} \times 373 \text{ K}}{(750. \text{ torr} \times \frac{1 \text{ atm}}{760 \text{ torr}}) \times 0.256 \text{ L}} = 96.9 \text{ g/mol}$$

Mass of CHCl $\approx 12.0 + 1.0 + 35.5 = 48.5; \ \dfrac{96.9}{48.5} = 2.00; \text{ Molecular formula is } C_2H_2Cl_2.$

61. $P \times (\text{molar mass}) = dRT, d = \text{density} = \dfrac{P \times (\text{molar mass})}{RT}$

For $SiCl_4$, molar mass = M = 28.09 + 4(35.45) = 169.89 g/mol

$$d = \frac{(758 \text{ torr} \times \frac{1 \text{ atm}}{760 \text{ torr}}) \times \frac{169.89 \text{ g}}{\text{mol}}}{\frac{0.08206 \text{ L atm}}{\text{mol K}} \times 358 \text{ K}} = 5.77 \text{ g/L for SiCl}_4$$

For SiHCl$_3$, molar mass = M = 28.09 + 1.008 + 3(35.45) = 135.45 g/mol

$$d = \frac{PM}{RT} = \frac{(758 \text{ torr} \times \frac{1\text{ atm}}{760\text{ torr}}) \times \frac{135.45\text{ g}}{\text{mol}}}{\frac{0.08206\text{ L atm}}{\text{mol K}} \times 358\text{ K}} = 4.60 \text{ g/L for SiHCl}_3$$

62. $$d_{UF_6} = \frac{P \times (\text{molar mass})}{RT} = \frac{\left(745 \text{ torr} \times \frac{1\text{ atm}}{760\text{ torr}}\right) \times 352.0\text{ g/mol}}{\frac{0.08206\text{ L atm}}{\text{mol K}} \times 333\text{ K}} = 12.6 \text{ g/L}$$

Partial Pressure

63. $$P_{CO_2} = \frac{nRT}{V} = \frac{\left(7.8 \text{ g} \times \frac{1\text{ mol}}{44.01\text{ g}}\right) \times \frac{0.08206\text{ L atm}}{\text{mol K}} \times 300.\text{ K}}{4.0\text{ L}} = 1.1 \text{ atm}$$

With air present, the partial pressure of CO$_2$ will still be 1.1 atm. The total pressure will be the sum of the partial pressures, $P_{total} = P_{CO_2} + P_{air}$.

$$P_{total} = 1.1 \text{ atm} + \left(740 \text{ torr} \times \frac{1\text{ atm}}{760\text{ torr}} \right) = 1.1 + 0.97 = 2.1 \text{ atm}$$

64. $$n_{H_2} = 1.00 \text{ g H}_2 \times \frac{1 \text{ mol H}_2}{2.016 \text{ g H}_2} = 0.496 \text{ mol H}_2; \quad n_{He} = 1.00 \text{ g He} \times \frac{1 \text{ mol He}}{4.003 \text{ g He}} = 0.250 \text{ mol He}$$

$$P_{H_2} = \frac{n_{H_2} \times RT}{V} = \frac{0.496 \text{ mol} \times \frac{0.08206\text{ L atm}}{\text{mol K}} \times (273 + 27)\text{ K}}{1.00\text{ L}} = 12.2 \text{ atm}$$

$$P_{He} = \frac{n_{He} \times RT}{V} = 6.15 \text{ atm}; \quad P_{total} = P_{H_2} + P_{He} = 12.2 \text{ atm} + 6.15 \text{ atm} = 18.4 \text{ atm}$$

65. Use the relationship $P_1V_1 = P_2V_2$ for each gas, since T and n for each gas is constant.

For H$_2$: $P_2 = \frac{P_1 V_1}{V_2} = 475 \text{ torr} \times \frac{2.00\text{ L}}{3.00\text{ L}} = 317 \text{ torr}$

For N$_2$: $P_2 = 0.200 \text{ atm} \times \frac{1.00\text{ L}}{3.00\text{ L}} = 0.0667 \text{ atm}$; $0.0667 \text{ atm} \times \frac{760\text{ torr}}{\text{atm}} = 50.7 \text{ torr}$

$$P_{total} = P_{H_2} + P_{N_2} = 317 + 50.7 = 368 \text{ torr}$$

66. For H_2: $P_2 = \dfrac{P_1 V_1}{V_2} = 360.\ \text{torr} \times \dfrac{2.00\ \text{L}}{3.00\ \text{L}} = 240.\ \text{torr}$

$P_{TOT} = P_{H_2} + P_{N_2},\ \ P_{N_2} = P_{TOT} - P_{H_2} = 320.\ \text{torr} - 240.\ \text{torr} = 80.\ \text{torr}$

For N_2: $P_1 = \dfrac{P_2 V_2}{V_1} = 80.\ \text{torr} \times \dfrac{3.00\ \text{L}}{1.00\ \text{L}} = 240\ \text{torr}$

67. a. mol fraction CH_4 = $\chi_{CH_4} = \dfrac{P_{CH_4}}{P_{total}} = \dfrac{0.175\ \text{atm}}{0.175\ \text{atm} + 0.250\ \text{atm}} = 0.412;\ \ \chi_{O_2} = 1.000 - 0.412 = 0.588$

b. $PV = nRT,\ \ n_{total} = \dfrac{P_{total} \times V}{RT} = \dfrac{0.425\ \text{atm} \times 10.5\ \text{L}}{0.08206\ \dfrac{\text{L atm}}{\text{mol K}} \times 338\ \text{K}} = 0.161\ \text{mol}$

c. $\chi_{CH_4} = \dfrac{n_{CH_4}}{n_{total}},\ \ n_{CH_4} = \chi_{CH_4} \times n_{total} = 0.412 \times 0.161\ \text{mol} = 6.63 \times 10^{-2}\ \text{mol } CH_4$

$6.63 \times 10^{-2}\ \text{mol } CH_4 \times \dfrac{16.04\ \text{g } CH_4}{\text{mol } CH_4} = 1.06\ \text{g } CH_4$

$n_{O_2} = 0.588 \times 0.161\ \text{mol} = 9.47 \times 10^{-2}\ \text{mol } O_2;\ \ 9.47 \times 10^{-2}\ \text{mol } O_2 \times \dfrac{32.00\ \text{g } O_2}{\text{mol } O_2} = 3.03\ \text{g } O_2$

68. If we had 100.0 g of the gas, we would have 50.0 g He and 50.0 g Xe.

$\chi_{He} = \dfrac{n_{He}}{n_{He} + n_{Xe}} = \dfrac{\dfrac{50.0\ \text{g}}{4.003\ \text{g/mol}}}{\dfrac{50.0\ \text{g}}{4.003\ \text{g/mol}} + \dfrac{50.0\ \text{g}}{131.3\ \text{g/mol}}} = \dfrac{12.5\ \text{mol He}}{12.5\ \text{mol He} + 0.381\ \text{mol Xe}} = 0.970$

$P_{He} = \chi_{He} P_{total} = 0.970 \times 600.\ \text{torr} = 582\ \text{torr};\ \ P_{Xe} = 600. - 582 = 18\ \text{torr}$

69. $P_{TOT} = P_{H_2} + P_{H_2O},\ \ 1.032\ \text{atm} = P_{H_2} + 32\ \text{torr} \times \dfrac{1\ \text{atm}}{760\ \text{torr}},\ \ P_{H_2} = 1.032 - 0.042 = 0.990\ \text{atm}$

$n_{H_2} = \dfrac{P_{H_2} V}{RT} = \dfrac{0.990\ \text{atm} \times 0.240\ \text{L}}{0.08206\ \dfrac{\text{L atm}}{\text{mol K}} \times 303\ \text{K}} = 9.56 \times 10^{-3}\ \text{mol } H_2$

$9.56 \times 10^{-3}\ \text{mol } H_2 \times \dfrac{1\ \text{mol Zn}}{\text{mol } H_2} \times \dfrac{65.38\ \text{g Zn}}{\text{mol Zn}} = 0.625\ \text{g Zn}$

70. To calculate the volume of gas, we can use P_{total} and n_{total} ($V = n_{tot}RT/P_{tot}$) or we can use P_{He} and n_{He} ($V = n_{He}RT/P_{He}$). Since n_{H_2O} is unknown, we will use P_{He} and n_{He}.

$P_{He} + P_{H_2O} = 1.00$ atm $= 760.$ torr $= P_{He} + 23.8$ torr, $P_{He} = 736$ torr

$n_{He} = 0.586$ g $\times \dfrac{1 \text{ mol}}{4.003 \text{ g}} = 0.146$ mol He

$$V = \frac{n_{He}RT}{P_{He}} = \frac{0.146 \text{ mol} \times \dfrac{0.08206 \text{ L atm}}{\text{mol K}} \times 298 \text{ K}}{736 \text{ torr} \times \dfrac{1 \text{ atm}}{760 \text{ torr}}} = 3.69 \text{ L}$$

71. $2 \text{ NaClO}_3(s) \rightarrow 2 \text{ NaCl}(s) + 3 \text{ O}_2(g)$

$P_{total} = P_{O_2} + P_{H_2O}$, $P_{O_2} = P_{total} - P_{H_2O} = 734$ torr $- 19.8$ torr $= 714$ torr

$$n_{O_2} = \frac{P_{O_2} \times V}{RT} = \frac{\left(714 \text{ torr} \times \dfrac{1 \text{ atm}}{760 \text{ torr}}\right) \times 0.0572 \text{ L}}{\dfrac{0.08206 \text{ L atm}}{\text{mol K}} \times (273 + 22) \text{ K}} = 2.22 \times 10^{-3} \text{ mol O}_2$$

Mass $NaClO_3$ decomposed $= 2.22 \times 10^{-3}$ mol $O_2 \times \dfrac{2 \text{ mol NaClO}_3}{3 \text{ mol O}_2} \times \dfrac{106.44 \text{ g NaClO}_3}{\text{mol NaClO}_3} = 0.158$ g $NaClO_3$

Mass % $NaClO_3 = \dfrac{0.158 \text{ g}}{0.8765 \text{ g}} \times 100 = 18.0\%$

72.   10.10 atm - 7.62 atm = 2.48 atm is the pressure of the amount of F_2 reacted.

$PV = nRT$, V and T are constant. $\dfrac{P}{n} = $ constant, $\dfrac{P_1}{n_1} = \dfrac{P_2}{n_2}$ or $\dfrac{P_1}{P_2} = \dfrac{n_1}{n_2}$

$\dfrac{\text{moles F}_2 \text{ reacted}}{\text{moles Xe reacted}} = \dfrac{2.48 \text{ atm}}{1.24 \text{ atm}} = 2.00$; So $Xe + 2 F_2 \rightarrow XeF_4$

Kinetic Molecular Theory and Real Gases

73. $(KE)_{avg} = (3/2)$ RT; At 273 K: $(KE)_{avg} = \dfrac{3}{2} \times \dfrac{8.3145 \text{ J}}{\text{mol K}} \times 273 \text{ K} = 3.40 \times 10^3$ J/mol

At 546 K: $(KE)_{avg} = \dfrac{3}{2} \times \dfrac{8.3145 \text{ J}}{\text{mol K}} \times 546 \text{ K} = 6.81 \times 10^3$ J/mol

74. $(KE)_{avg} = (3/2)$ RT. Since the kinetic energy depends only on temperature, CH_4 (Exercise 5.73) and N_2 at the same temperature will have the same average kinetic energy. So, for N_2, the average kinetic energy is 3.40×10^3 J/mol (at 273 K) and 6.81×10^3 J/mol (at 546 K).

75. $u_{rms} = \left(\dfrac{3RT}{M} \right)^{1/2}$, where $R = \dfrac{8.3145 \text{ J}}{\text{mol K}}$ and M = molar mass in kg = 1.604×10^{-2} kg/mol for CH_4

For CH_4 at 273 K: $u_{rms} = \left(\dfrac{\dfrac{3 \times 8.3145 \text{ J}}{\text{mol K}} \times 273 \text{ K}}{1.604 \times 10^{-2} \text{ kg/mol}} \right)^{1/2} = 652$ m/s

Similarly u_{rms} for CH_4 at 546 K is 921 m/s.

For N_2 at 273 K: $u_{rms} = \left(\dfrac{\dfrac{3 \times 8.3145 \text{ J}}{\text{mol K}} \times 273 \text{ K}}{2.802 \times 10^{-2} \text{ kg/mol}} \right)^{1/2} = 493$ m/s

Similarly for N_2 at 546 K, $u_{rms} = 697$ m/s.

76. $u_{rms} = \left(\dfrac{3RT}{M} \right)^{1/2}$; $\dfrac{u_{UF_6}}{u_{He}} = \dfrac{\left(\dfrac{3RT_{UF_6}}{M_{UF_6}} \right)^{1/2}}{\left(\dfrac{3RT_{He}}{M_{He}} \right)^{1/2}} = \left(\dfrac{M_{He}T_{UF_6}}{M_{UF_6}T_{He}} \right)^{1/2}$

We want the root mean square velocities to be equal, and this occurs when $M_{He}T_{UF_6} = M_{UF_6}T_{He}$.   The ratio of the temperatures is:

$$\frac{T_{UF_6}}{T_{He}} = \frac{M_{UF_6}}{M_{He}} = \frac{352.0}{4.003} = 87.93$$

The heavier UF_6 molecules would need a temperature 87.93 times that of the He atoms in order for the root mean square velocities to be equal.

77. $KE_{ave} = (3/2) RT$ and $KE = (1/2) mv^2$; As the temperature increases, the average kinetic energy of the gas sample will increase.   The average kinetic energy increases because the increased temperature results in an increase in the average velocity of the gas molecules.

78.

	a	b	c	d
avg. KE	inc	dec	same ($KE \propto T$)	same
avg. velocity	inc	dec	same ($\frac{1}{2} mv^2 = KE \propto T$)	same
coll. freq wall	inc	dec	inc	inc

Average kinetic energy and average velocity depend on T. As T increases, both average kinetic energy and average velocity increase. At constant T, both average kinetic energy and average velocity are constant. The collision frequency is proportional to the average velocity (as velocity increases it takes less time to move to the next collision) and to the quantity n/V (as molecules per volume increase, collision frequency increases).

79.   a.   They will all have the same average kinetic energy since they are all at the same temperature.

      b.   Flask C; H_2 has the smallest molar mass. At constant T, the lightest molecules are the fastest (on the average). This must be true in order for the average kinetic energies to be constant.

80.   a.   All the gases have the same average kinetic energy since they are all at the same temperature.

      b.   At constant T, the lighter the gas molecule, the faster the average velocity.

           Xe (131.3 g/mol) < Cl_2 (70.90 g/mol) < O_2 (32.00 g/mol) < H_2 (2.016 g/mol)
           slowest   fastest

      c.   At constant T, the lighter H_2 molecules have a faster average velocity than the heavier O_2 molecules. As temperature increases, the average velocity of the gas molecules increases. Separate samples of H_2 and O_2 can only have the same average velocities if the temperature of the O_2 sample is greater than the temperature of the H_2 sample.

81.   Graham's law of effusion:

      $$\frac{Rate_1}{Rate_2} = \left(\frac{M_2}{M_1}\right)^{1/2} \text{ where M = molar mass; } \frac{31.50}{30.50} = \left(\frac{32.00}{M}\right)^{1/2} = 1.033$$

      $\dfrac{32.00}{M} = 1.067$, so M = 29.99 g/mol; Of the choices, the gas would be NO, nitrogen monoxide.

82. $\dfrac{Rate_1}{Rate_2} = \left(\dfrac{M_2}{M_1}\right)^{1/2}$; $Rate_1 = \dfrac{24.0 \text{ mL}}{min}$, $Rate_2 = \dfrac{47.8 \text{ mL}}{min}$, $M_2 = \dfrac{16.04 \text{ g}}{mol}$ and $M_1 = ?$

      $\dfrac{24.0}{47.8} = \left(\dfrac{16.04}{M_1}\right)^{1/2} = 0.502$, $16.04 = (0.502)^2 \times M_1$, $M_1 = \dfrac{16.04}{0.252} = \dfrac{63.7 \text{ g}}{mol}$

83. $\dfrac{Rate_1}{Rate_2} = \left(\dfrac{M_2}{M_1}\right)^{1/2}$, $\dfrac{Rate\ (^{12}C\ ^{17}O)}{Rate\ (^{12}C\ ^{18}O)} = \left(\dfrac{30.0}{29.0}\right)^{1/2} = 1.02$; $\dfrac{Rate\ (^{12}C\ ^{16}O)}{Rate\ (^{12}C\ ^{18}O)} = \left(\dfrac{30.0}{28.0}\right)^{1/2} = 1.04$

      The relative rates of effusion of $^{12}C^{16}O$: $^{12}C^{17}O$: $^{12}C^{18}O$ are 1.04: 1.02: 1.00.

      Advantage: CO_2 isn't as toxic as CO.

      Major disadvantages of using CO_2 instead of CO:

1. Can get a mixture of oxygen isotopes in CO_2.

2. Some species, e.g., $^{12}C^{16}O^{18}O$ and $^{12}C^{17}O_2$, would effuse (gaseously diffuse) at about the same rate since the masses are about equal. Thus, some species cannot be separated from each other.

84. $\dfrac{Rate_1}{Rate_2} = \left(\dfrac{M_2}{M_1}\right)^{1/2}$ where M = molar mass; Let Gas (1) = He, Gas (2) = Cl_2

$\dfrac{\dfrac{1.0\ L}{4.5\ min}}{\dfrac{1.0\ L}{t}} = \left(\dfrac{70.90}{4.003}\right)^{1/2},\quad \dfrac{t}{4.5\ min} = 4.209,\ \ t = 19\ min$

85.   a. $P = \dfrac{nRT}{V} = \dfrac{0.5000\ mol \times \dfrac{0.08206\ L\ atm}{mol\ K} \times (25.0 + 273.2)\ K}{1.0000\ L} = 12.24\ atm$

   b. $\left[P + a\left(\dfrac{n}{V}\right)^2\right] \times (V - nb) = nRT$; For N_2: a = 1.39 atm L^2/mol^2 and b = 0.0391 L/mol

$\left[P + 1.39\left(\dfrac{0.5000}{1.0000}\right)^2 atm\right] \times (1.0000\ L - 0.5000 \times 0.0391\ L) = 12.24\ L\ atm$

(P + 0.348 atm) × (0.9805 L) = 12.24 L atm

$P = \dfrac{12.24\ L\ atm}{0.9805\ L} - 0.348\ atm = 12.48 - 0.348 = 12.13\ atm$

   c.   The ideal gas law is high by 0.11 atm or $\dfrac{0.11}{12.13} \times 100 = 0.91\%$.

86.   a. $P = \dfrac{nRT}{V} = \dfrac{0.5000\ mol \times \dfrac{0.08206\ L\ atm}{mol\ K} \times 298.2\ K}{10.000\ L} = 1.224\ atm$

   b. $\left[P + a\left(\dfrac{n}{V}\right)^2\right] \times (V - nb) = nRT$; For N_2: a = 1.39 atm L^2/mol^2 and b = 0.0391 L/mol

$\left[P + 1.39\left(\dfrac{0.5000}{10.000}\right)^2 atm\right] \times (10.000\ L - 0.5000 \times 0.0391\ L) = 12.24\ L\ atm$

$(P + 0.00348 \text{ atm}) \times (10.000 \text{ L} - 0.0196 \text{ L}) = 12.24 \text{ L atm}$

$$P + 0.00348 \text{ atm} = \frac{12.24 \text{ L atm}}{9.980 \text{ L}} = 1.226 \text{ atm}, \quad P = 1.226 - 0.00348 = 1.223 \text{ atm}$$

c.   The results agree to ± 0.001 atm (0.08%).

d.   In Exercise 5.85, the pressure is relatively high and there is a significant disagreement. In 5.86, the pressure is around 1 atm, and both gas law equations show better agreement. The ideal gas law is valid at relatively low pressures.

Atmospheric Chemistry

87. $\chi_{NO} = 5 \times 10^{-7}$ from Table 5.4. $P_{NO} = \chi_{NO} \times P_{total} = 5 \times 10^{-7} \times 1.0 \text{ atm} = 5 \times 10^{-7}$ atm

$$PV = nRT, \quad \frac{n}{V} = \frac{P}{RT} = \frac{5 \times 10^{-7} \text{ atm}}{\dfrac{0.08206 \text{ L atm}}{\text{mol K}} \times 273 \text{ K}} = 2 \times 10^{-8} \text{ mol NO/L}$$

$$\frac{2 \times 10^{-8} \text{ mol}}{\text{L}} \times \frac{1 \text{ L}}{1000 \text{ cm}^3} \times \frac{6.022 \times 10^{23} \text{ molecules}}{\text{mol}} = 1 \times 10^{13} \text{ molecules NO/cm}^3$$

88. $\chi_{He} = 5.24 \times 10^{-6}$ from Table 5.4. $P_{He} = \chi_{He} \times P_{total} = 5.24 \times 10^{-6} \times 1.0 \text{ atm} = 5.2 \times 10^{-6}$ atm

$$\frac{n}{V} = \frac{P}{RT} = \frac{5.2 \times 10^{-6} \text{ atm}}{\dfrac{0.08206 \text{ L atm}}{\text{mol K}} \times 298 \text{ K}} = 2.1 \times 10^{-7} \text{ mol He/L}$$

$$\frac{2 \times 10^{-7} \text{ mol}}{\text{L}} \times \frac{1 \text{ L}}{1000 \text{ cm}^3} \times \frac{6.022 \times 10^{23} \text{ atoms}}{\text{mol}} = 1.2 \times 10^{14} \text{ atoms He/cm}^3$$

89.   At 100. km, $T \approx -75°C$ and $P \approx 10^{-4.5} \approx 3 \times 10^{-5}$ atm.

$$PV = nRT, \quad \frac{PV}{T} = nR = \text{constant}, \quad \frac{P_1 V_1}{T_1} = \frac{P_2 V_2}{T_2}$$

$$V_2 = \frac{V_1 P_1 T_2}{P_2 T_1} = \frac{10.0 \text{ L} \times (3 \times 10^{-5} \text{ atm}) \times 273 \text{ K}}{1.0 \text{ atm} \times 198 \text{ K}} = 4 \times 10^{-4} \text{ L} = 0.4 \text{ mL}$$

90.   At 15 km, $T \approx -50°C$ and $P = 0.1$ atm. Use $\dfrac{P_1 V_1}{T_1} = \dfrac{P_2 V_2}{T_2}$ since n is constant.

$$V_2 = \frac{V_1 P_1 T_2}{P_2 T_1} = \frac{1.0 \text{ L} \times 1.00 \text{ atm} \times 223 \text{ K}}{0.1 \text{ atm} \times 298 \text{ K}} = 7 \text{ L}$$

91. $N_2(g) + O_2(g) \rightarrow 2\ NO(g)$, automobile combustion or formed by lightning

       $2\ NO(g) + O_2(g) \rightarrow 2\ NO_2(g)$, reaction with atmospheric O_2

       $2\ NO_2(g) + H_2O(l) \rightarrow HNO_3(aq) + HNO_2(aq)$, reaction with atmospheric H_2O

       $S(s) + O_2(g) \rightarrow SO_2(g)$, combustion of coal

       $2\ SO_2(g) + O_2(g) \rightarrow 2SO_3(g)$, reaction with atmospheric O_2

       $H_2O(l) + SO_3(g) \rightarrow H_2SO_4(aq)$, reaction with atmospheric H_2O

92. $2\ HNO_3(aq) + CaCO_3(s) \rightarrow Ca(NO_3)_2(aq) + H_2O(l) + CO_2(g)$

       $H_2SO_4(aq) + CaCO_3(s) \rightarrow CaSO_4(aq) + H_2O(l) + CO_2(g)$

Additional Exercises

93.   a. $PV = nRT$

           $PV = \text{Constant}$

       b. $PV = nRT$

           $P = \left(\dfrac{nR}{V}\right) \times T = \text{Const} \times T$

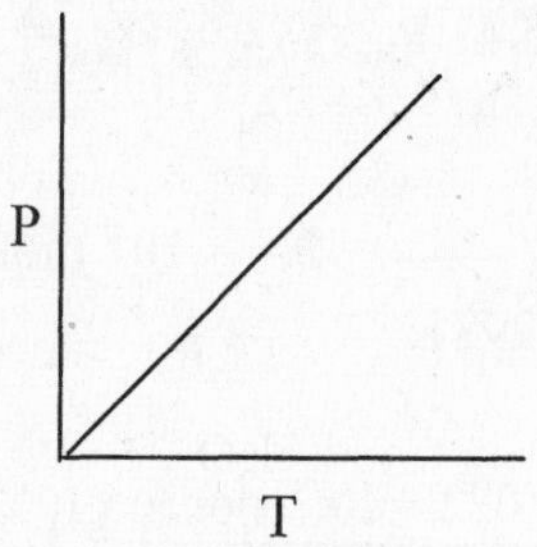

       c. $PV = nRT$

           $T = \left(\dfrac{P}{nR}\right) \times V = \text{Const} \times V$

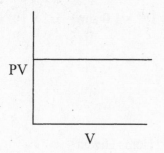

       d. $PV = nRT$

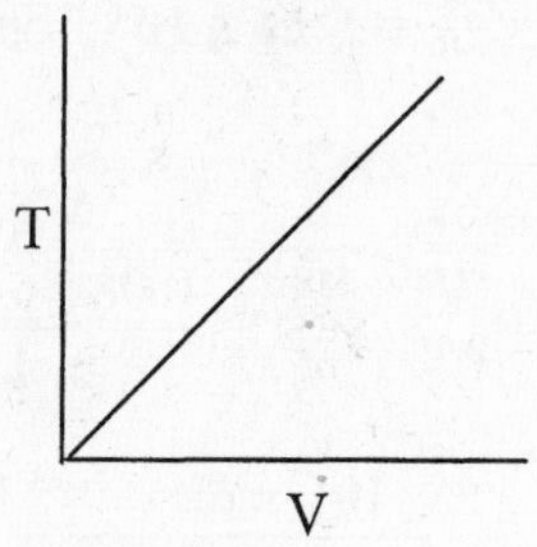

           $PV = \text{Constant}$

       e. $P = \dfrac{nRT}{V} = \dfrac{\text{Constant}}{V}$

           $P = \text{Constant} \times \dfrac{1}{V}$

       f. $PV = nRT$

           $\dfrac{PV}{T} = nR = \text{Constant}$

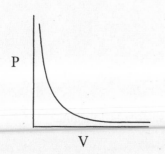
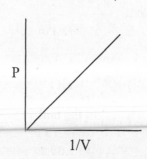
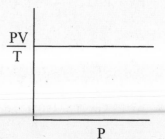

94.   At constant T and P, Avogadro's law applies; that is, equal volumes contain equal moles of molecules. In terms of balanced equations, we can say that mol ratios and volume ratios between the various reactants and products will be equal to each other. $Br_2 + 3 F_2 \rightarrow 2 X$; Two moles of X must contain two moles of Br and 6 moles of F; X must have the formula BrF_3.

95. $Mn(s) + x\ HCl(g) \rightarrow MnCl_x(s) + \dfrac{x}{2}\ H_2(g)$

$$n_{H_2} = \dfrac{PV}{RT} = \dfrac{0.951\ \text{atm} \times 3.22\ \text{L}}{\dfrac{0.08206\ \text{L atm}}{\text{mol K}} \times 373\ \text{K}} = 0.100\ \text{mol}\ H_2$$

mol Cl in compound = mol HCl = $0.100\ \text{mol}\ H_2 \times \dfrac{x\ \text{mol Cl}}{\dfrac{x}{2}\ \text{mol}\ H_2} = 0.200\ \text{mol Cl}$

$$\dfrac{\text{mol Cl}}{\text{mol Mn}} = \dfrac{0.200\ \text{mol Cl}}{2.747\ \text{g Mn} \times \dfrac{1\ \text{mol Mn}}{54.94\ \text{g Mn}}} = \dfrac{0.200\ \text{mol Cl}}{0.05000\ \text{mol Mn}} = 4.00$$

The formula of compound is $MnCl_4$.

96. $2 H_2(g) + O_2(g) \rightarrow 2 H_2O(g)$; Since P and T are constant, volume ratios will equal mol ratios ($V_f/V_i = n_f/n_i$). Let x = mol H_2 = mol O_2 present initially. H_2 will be limiting since a 2:1 H_2 to O_2 mol ratio is required by the balanced equation, but only a 1:1 mol ratio is present. Therefore, no H_2 will be present after the reaction goes to completion. However, excess $O_2(g)$ will be present as well as the $H_2O(g)$ produced.

mol O_2 reacted = x mol $H_2 \times \dfrac{1\ \text{mol}\ O_2}{2\ \text{mol}\ H_2} = x/2\ \text{mol}\ O_2$

mol O_2 remaining = x mol O_2 initially - $x/2$ mol O_2 reacted = $x/2$ mol O_2

mol H_2O produced = x mol $H_2 \times \dfrac{2\ \text{mol}\ H_2O}{2\ \text{mol}\ H_2} = x\ \text{mol}\ H_2O$

Total mol gas initially = x mol H_2 + x mol O_2 = $2x$

Total mol gas after reaction = $x/2$ mol O_2 + x mol H_2O = $1.5\ x$

$\dfrac{n_f}{n_i} = \dfrac{V_f}{V_i} = \dfrac{1.5\ x}{2\ x} = \dfrac{1.5}{2} = 0.75$; $V_f/V_i = 0.75$:1 or 3:4

97.   We will apply Boyle's law to solve. $PV = nRT$ = contstant, $P_1V_1 = P_2V_2$

Let condition (1) correspond to He from the tank that can be used to fill balloons. We must leave 1.0 atm of He in the tank, so P_1 = 200. atm - 1.00 = 199 atm and V_1 = 15.0 L. Condition (2) will correspond to the filled balloons with P_2 = 1.00 atm and V_2 = N(2.00 L) where N is the number of filled balloons, each at a volume of 2.00 L.

199 atm × 15.0 L = 1.00 atm × N(2.00 L), N = 1492.5; We can't fill 0.5 of a balloon, so N = 1492 balloons or to 3 significant figures, 1490 balloons.

98.   mol of He removed $= \dfrac{PV}{RT} = \dfrac{1.00 \text{ atm} \times 1.75 \times 10^{-3} \text{ L}}{\dfrac{0.08206 \text{ L atm}}{\text{mol K}} \times 298 \text{ K}} = 7.16 \times 10^{-5}$ mol

In the original flask, 7.16×10^{-5} mol of He exerted a partial pressure of 1.960 - 1.710 = 0.250 atm.

$V = \dfrac{nRT}{P} = \dfrac{7.16 \times 10^{-5} \text{ mol} \times 0.08206 \times 298 \text{ K}}{0.250 \text{ atm}} = 7.00 \times 10^{-3}$ L = 7.00 mL

99.   For O_2, n and T are constant, so $P_1V_1 = P_2V_2$.

$P_1 = \dfrac{P_2V_2}{V_1} = 785 \text{ torr} \times \dfrac{1.94 \text{ L}}{2.00 \text{ L}} = 761 \text{ torr} = P_{O_2}$

$P_{tot} = P_{O_2} + P_{H_2O},\ P_{H_2O} = 785 - 761 = 24$ torr

100. $4\ C_3H_5N_3O_9(s) \rightarrow 12\ CO_2(g) + 6\ N_2(g) + 10\ H_2O(g) + O_2(g)$; For every 4 mol of nitroglycerin reacted, 12 + 6 + 10 + 1 = 29 mol of gas are produced.

mol gas produced $= 25.0 \text{ g } C_3H_5N_3O_9 \times \dfrac{1 \text{ mol } C_3H_5N_3O_9}{227.10 \text{ g}} \times \dfrac{29 \text{ mol gas}}{4 \text{ mol } C_3H_5N_3O_9} = 0.798$ mol

$P_{tot} = \dfrac{n_{tot}RT}{V} = \dfrac{0.798 \text{ mol} \times 0.08206 \dfrac{\text{L atm}}{\text{mol K}} \times 773 \text{ K}}{10.0 \text{ L}} = 5.06$ atm

101. $1.00 \times 10^3 \text{ kg Mo} \times \dfrac{1000 \text{ g}}{\text{kg}} \times \dfrac{1 \text{ mol Mo}}{95.94 \text{ g Mo}} = 1.04 \times 10^4$ mol Mo

$1.04 \times 10^4 \text{ mol Mo} \times \dfrac{1 \text{ mol MoO}_3}{\text{mol Mo}} \times \dfrac{7/2 \text{ mol O}_2}{\text{mol MoO}_3} = 3.64 \times 10^4 \text{ mol O}_2$

$V_{O_2} = \dfrac{n_{O_2}RT}{P} = \dfrac{3.64 \times 10^4 \text{ mol} \times \dfrac{0.08206 \text{ L atm}}{\text{mol K}} \times 290. \text{ K}}{1.00 \text{ atm}} = 8.66 \times 10^5 \text{ L of } O_2$

$8.66 \times 10^5 \text{ L } O_2 \times \dfrac{100 \text{ L air}}{21 \text{ L } O_2} = 4.1 \times 10^6$ L air

$$1.04 \times 10^4 \text{ mol Mo} \times \frac{3 \text{ mol H}_2}{\text{mol Mo}} = 3.12 \times 10^4 \text{ mol H}_2$$

$$V_{H_2} = \frac{3.12 \times 10^4 \text{ mol} \times \dfrac{0.08206 \text{ L atm}}{\text{mol K}} \times 290. \text{ K}}{1.00 \text{ atm}} = 7.42 \times 10^5 \text{ L of H}_2$$

102.   For NH_3: $P_2 = \dfrac{P_1 V_1}{V_2} = 0.500 \text{ atm} \times \dfrac{2.00 \text{ L}}{3.00 \text{ L}} = 0.333 \text{ atm}$

For O_2: $P_2 = \dfrac{P_1 V_1}{V_2} = 1.50 \text{ atm} \times \dfrac{1.00 \text{ L}}{3.00 \text{ L}} = 0.500 \text{ atm}$

After the stopcock is opened, V and T will be constant, so $P \propto n$. The balanced equation requires:

$$\frac{n_{O_2}}{n_{NH_3}} = \frac{P_{O_2}}{P_{NH_3}} = \frac{5}{4} = 1.25$$

The actual ratio present is: $\dfrac{P_{O_2}}{P_{NH_3}} = \dfrac{0.500 \text{ atm}}{0.333 \text{ atm}} = 1.50$

The actual ratio is larger than the required ratio, so NH_3 in the denominator is limiting. Since equal mol of NO will be produced as NH_3 reacted, the partial pressure of NO produced is 0.333 atm (the same as P_{NH_3} reacted).

103. $750. \text{ mL juice} \times \dfrac{12 \text{ mL C}_2\text{H}_5\text{OH}}{100 \text{ mL juice}} = 90. \text{ mL C}_2\text{H}_5\text{OH present}$

$90. \text{ mL C}_2\text{H}_5\text{OH} \times \dfrac{0.79 \text{ g C}_2\text{H}_5\text{OH}}{\text{mL C}_2\text{H}_5\text{OH}} \times \dfrac{1 \text{ mol C}_2\text{H}_5\text{OH}}{46.07 \text{ g C}_2\text{H}_5\text{OH}} \times \dfrac{2 \text{ mol CO}_2}{2 \text{ mol C}_2\text{H}_5\text{OH}} = 1.5 \text{ mol CO}_2$

The CO_2 will occupy (825 - 750. =) 75 mL not occupied by the liquid (headspace).

$$P_{CO_2} = \frac{n_{CO_2} \times RT}{V} = \frac{1.5 \text{ mol} \times \dfrac{0.08206 \text{ L atm}}{\text{mol K}} \times 298 \text{ K}}{75 \times 10^{-3} \text{ L}} = 490 \text{ atm}$$

Actually, enough CO_2 will dissolve in the wine to lower the pressure of CO_2 to a much more reasonable value.

104. $PV = nRT$, V and T are constant. $\dfrac{P_1}{n_1} = \dfrac{P_2}{n_2}$ or $\dfrac{P_1}{P_2} = \dfrac{n_1}{n_2}$

When V and T are constant, pressure is directly proportional to moles of gas present, and pressure ratios are identical to mol ratios.

At 25°C: $2 H_2(g) + O_2(g) \rightarrow 2 H_2O(l)$; $H_2O(l)$ is produced.

The balanced equation requires 2 mol H_2 for every mol O_2 reacted. The same ratio (2:1) holds true for pressure units. The actual pressure ratio present is 2 atm H_2 to 3 atm O_2, well below the required 2:1 ratio. Therefore, H_2 is the limiting reactant. The only gas present at 25°C after the reaction goes to completion will be the excess O_2.

$$P_{O_2} \text{(reacted)} = 2.00 \text{ atm } H_2 \times \frac{1 \text{ atm } O_2}{2 \text{ atm } H_2} = 1.00 \text{ atm } O_2$$

$$P_{O_2} \text{(excess)} = P_{O_2} \text{(initially)} - P_{O_2} \text{(reacted)} = 3.00 \text{ atm} - 1.00 \text{ atm} = 2.00 \text{ atm } O_2 = P_{total}$$

At 125°C: $2 H_2(g) + O_2(g) \rightarrow 2 H_2O(g)$; $H_2O(g)$ is produced.

The major difference in the problem is that gaseous water is now a product, which will increase the total pressure.

$$P_{H_2O} \text{(produced)} = 2.00 \text{ atm } H_2 \times \frac{2 \text{ atm } H_2O}{2 \text{ atm } H_2} = 2.00 \text{ atm } H_2O$$

$$P_{total} = P_{O_2} \text{(excess)} + P_{H_2O} \text{(produced)} = 2.00 \text{ atm } O_2 + 2.00 \text{ atm } H_2O = 4.00 \text{ atm}$$

105.   If Be^{3+}, the formula is $Be(C_5H_7O_2)_3$ and $M \approx 13.5 + 15(12) + 21(1) + 6(16) = 311$ g/mol.

If Be^{2+}, the formula is $Be(C_5H_7O_2)_2$ and $M \approx 9.0 + 10(12) + 14(1) + 4(16) = 207$ g/mol.

Data Set I ($M = dRT/P$ and $d = $ mass/V):

$$M = \frac{\text{mass} \times RT}{PV} = \frac{0.2022 \text{ g} \times \dfrac{0.08206 \text{ L atm}}{\text{mol K}} \times 286 \text{ K}}{(765.2 \text{ torr} \times \dfrac{1 \text{ atm}}{760 \text{ torr}}) \times 22.6 \times 10^{-3} \text{ L}} = 209 \text{ g/mol}$$

Data Set II:

$$M = \frac{\text{mass} \times RT}{PV} = \frac{0.2224 \text{ g} \times \dfrac{0.08206 \text{ L atm}}{\text{mol K}} \times 290. \text{ K}}{(764.6 \text{ torr} \times \dfrac{1 \text{ atm}}{760 \text{ torr}}) \times 26.0 \times 10^{-3} \text{ L}} = 202 \text{ g/mol}$$

These results are close to the expected value of 207 g/mol for $Be(C_5H_7O_2)_2$. Thus, we conclude from these data that beryllium is a divalent element with an atomic mass of 9.0 amu.

106.   Out of 100.00 g compounds, there are:

$$58.51 \text{ g C} \times \frac{1 \text{ mol C}}{12.01 \text{ g C}} = 4.872 \text{ mol C}; \quad \frac{4.872}{2.435} = 2.001$$

$$7.37 \text{ g H} \times \frac{1 \text{ mol H}}{1.008 \text{ g H}} = 7.31 \text{ mol H}; \quad \frac{7.31}{2.435} = 3.00$$

$$34.12 \text{ g N} \times \frac{1 \text{ mol N}}{14.01 \text{ g N}} = 2.435 \text{ mol N}; \quad \frac{2.435}{2.435} = 1.000$$

Empirical formula: C_2H_3N

$$\frac{\text{Rate}_1}{\text{Rate}_2} = \left(\frac{M_2}{M_1}\right)^{1/2}; \quad \text{Let Gas (1) = He;} \quad 3.20 = \left(\frac{M_2}{4.003}\right)^{1/2}, \quad M_2 = 41.0 \text{ g/mol}$$

Empirical formula mass of $C_2H_3N \approx 2(12.0) + 3(1.0) + 1(14.0) = 41.0$. So, the molecular formula is also C_2H_3N.

107. $P_{total} = P_{N_2} + P_{H_2O}$, $P_{N_2} = 726 \text{ torr} - 23.8 \text{ torr} = 702 \text{ torr} \times \dfrac{1 \text{ atm}}{760 \text{ torr}} = 0.924 \text{ atm}$

$$PV = nRT, \quad n_{N_2} = \frac{P_{N_2} \times V}{RT} = \frac{0.924 \text{ atm} \times 31.8 \times 10^{-3} \text{ L}}{\dfrac{0.08206 \text{ L atm}}{\text{mol K}} \times 298 \text{ K}} = 1.20 \times 10^{-3} \text{ mol N}_2$$

Mass of N in compound $= 1.20 \times 10^{-3} \text{ mol} \times \dfrac{28.02 \text{ g N}_2}{\text{mol}} = 3.36 \times 10^{-2} \text{ g}$

$\% \text{ N} = \dfrac{3.36 \times 10^{-2} \text{ g}}{0.253 \text{ g}} \times 100 = 13.3\% \text{ N}$

108. $0.2766 \text{ g CO}_2 \times \dfrac{12.01 \text{ g C}}{44.01 \text{ g CO}_2} = 7.548 \times 10^{-2} \text{ g C}; \quad \% \text{ C} = \dfrac{7.548 \times 10^{-2} \text{ g}}{0.1023 \text{ g}} \times 100 = 73.78\% \text{ C}$

$0.0991 \text{ g H}_2\text{O} \times \dfrac{2.016 \text{ g H}}{18.02 \text{ g H}_2\text{O}} = 1.11 \times 10^{-2} \text{ g H}; \quad \% \text{ H} = \dfrac{1.11 \times 10^{-2} \text{ g}}{0.1023 \text{ g}} \times 100 = 10.9\% \text{ H}$

$$PV = nRT, \quad n_{N_2} = \frac{PV}{RT} = \frac{1.00 \text{ atm} \times 27.6 \times 10^{-3} \text{ L}}{\dfrac{0.08206 \text{ L atm}}{\text{mol K}} \times 273 \text{ K}} = 1.23 \times 10^{-3} \text{ mol N}_2$$

$1.23 \times 10^{-3} \text{ mol N}_2 \times \dfrac{28.02 \text{ g N}_2}{\text{mol N}_2} = 3.45 \times 10^{-2} \text{ g nitrogen}$

$$\% N = \frac{3.45 \times 10^{-2}\,g}{0.4831\,g} \times 100 = 7.14\%\ N$$

$$\% O = 100.00 - (73.78 + 10.9 + 7.14) = 8.2\%\ O$$

Out of 100.00 g of compound, there are:

$$73.78\ g\ C \times \frac{1\ mol}{12.01\ g} = 6.143\ mol\ C;\ \ 7.14\ g\ N \times \frac{1\ mol}{14.01\ g} = 0.510\ mol\ N$$

$$10.9\ g\ H \times \frac{1\ mol}{1.008\ g} = 10.8\ mol\ H;\ \ 8.2\ g\ O \times \frac{1\ mol}{16.00\ g} = 0.51\ mol\ O$$

Dividing all values by 0.51 gives an empirical formula of $C_{12}H_{21}NO$.

$$\text{Molar mass} = \frac{dRT}{P} = \frac{\dfrac{4.02\ g}{L} \times \dfrac{0.08206\ L\ atm}{mol\ K} \times 400.\ K}{256\ torr \times \dfrac{1\ atm}{760\ torr}} = 392\ g/mol$$

Empirical formula mass of $C_{12}H_{21}NO \approx 195\ g/mol$ and $\dfrac{392}{195} \approx 2$.

Thus, the molecular formula is $C_{24}H_{42}N_2O_2$.

109.   At constant T, the lighter the gas molecules, the faster the average velocity. Therefore, the pressure will increase initially because the lighter H_2 molecules will effuse into container A faster than air will escape. However, the pressures will eventually equalize once the gases have had time to mix thoroughly.

110.   The van der Waals' constant b is a measure of the size of the molecule. Thus, C_3H_8 should have the largest value of b since it has the largest molar mass (size).

111.   The values of a are: H_2, $\dfrac{0.244\ L^2\ atm}{mol^2}$; CO_2, 3.59; N_2, 1.39; CH_4, 2.25

Since a is a measure of interparticle attractions, the attractions are greatest for CO_2.

Challenge Problems

112. $PV = nRT$, V and T are constant. $\dfrac{P_1}{n_1} = \dfrac{P_2}{n_2},\ \dfrac{P_2}{P_1} = \dfrac{n_2}{n_1}$.

We will do this limiting reagent problem using an alternative method. Let's calculate the partial pressure of C_3H_3N that can be produced from each of the starting materials assuming each reactant is limiting. The reactant that produces the smallest amount of product will run out first and is the limiting reagent.

$$P_{C_3H_3N} = 0.500 \text{ MPa } C_3H_6 \times \frac{2 \text{ MPa } C_3H_3N}{2 \text{ MPa } C_3H_6} = 0.500 \text{ MPa if } C_3H_6 \text{ is limiting.}$$

$$P_{C_3H_3N} = 0.800 \text{ MPa } NH_3 \times \frac{2 \text{ MPa } C_3H_3N}{2 \text{ MPa } NH_3} = 0.800 \text{ MPa if } NH_3 \text{ is limiting.}$$

$$P_{C_3H_3N} = 1.500 \text{ MPa } O_2 \times \frac{2 \text{ MPa } C_3H_3N}{3 \text{ MPa } O_2} = 1.000 \text{ MPa if } O_2 \text{ is limiting.}$$

Thus, C_3H_6 is limiting. Although more product could be produced from NH_3 and O_2, there is only enough C_3H_6 to produce 0.500 MPa of C_3H_3N. The partial pressure of C_3H_3N after the reaction is:

$$0.500 \times 10^6 \text{ Pa} \times \frac{1 \text{ atm}}{1.013 \times 10^5 \text{ Pa}} = 4.94 \text{ atm}$$

$$n = \frac{PV}{RT} = \frac{4.94 \text{ atm} \times 150. \text{ L}}{\frac{0.08206 \text{ L atm}}{\text{mol K}} \times 298 \text{ K}} = 30.3 \text{ mol } C_3H_3N$$

$$30.3 \text{ mol} \times \frac{53.06 \text{ g}}{\text{mol}} = 1.61 \times 10^3 \text{ g } C_3H_3N \text{ can be produced.}$$

113. $BaO(s) + CO_2(g) \rightarrow BaCO_3(s);\ CaO(s) + CO_2(g) \rightarrow CaCO_3(s)$

$$n_i = \frac{P_iV}{RT} = \text{initial moles of } CO_2 = \frac{\frac{750.}{760} \text{ atm} \times 1.50 \text{ L}}{\frac{0.08206 \text{ L atm}}{\text{mol K}} \times 303.2 \text{ K}} = 0.0595 \text{ mol } CO_2$$

$$n_f = \frac{P_fV}{RT} = \text{final moles of } CO_2 = \frac{\frac{230.}{760} \text{ atm} \times 1.50 \text{ L}}{\frac{0.08206 \text{ L atm}}{\text{mol K}} \times 303.2 \text{ K}} = 0.0182 \text{ mol } CO_2$$

0.0595 - 0.0182 = 0.0413 mol CO_2 reacted.

Since each metal reacts 1:1 with CO_2, the mixture contains 0.0413 mol of BaO and CaO. The molar masses of BaO and CaO are 153.3 g/mol and 56.08 g/mol, respectively.

Let x = g BaO and y = g CaO, so:

$$x + y = 5.14 \text{ g and } \frac{x}{153.3} + \frac{y}{56.08} = 0.0413 \text{ mol}$$

Solving by simultaneous equations:

$$x + 2.734\,y = 6.33$$
$$\underline{-x -y = -5.14}$$
$$ 1.734\,y = 1.19$$

$y = 0.686$ g CaO and $5.14 - y = x = 4.45$ g BaO

% BaO $= \dfrac{4.45 \text{ g BaO}}{5.14 \text{ g}} \times 100 = 86.6\%$ BaO; % CaO $= 100.0 - 86.6 = 13.4\%$ CaO

114. $Cr(s) + 3\ HCl(aq) \rightarrow CrCl_3(aq) + 3/2\ H_2(g)$; $Zn(s) + 2\ HCl(aq) \rightarrow ZnCl_2(aq) + H_2(g)$

mol H_2 produced $= n = \dfrac{PV}{RT} = \dfrac{\left(750.\text{ torr} \times \dfrac{1\text{ atm}}{760\text{ torr}}\right) \times 0.225\text{ L}}{\dfrac{0.08206\text{ L atm}}{\text{mol K}} \times (273 + 27)\text{ K}} = 9.02 \times 10^{-3}$ mol H_2

9.02×10^{-3} mol $H_2 =$ mol H_2 from Cr reaction + mol H_2 from Zn reaction

From the balanced equation: 9.02×10^{-3} mol $H_2 =$ mol Cr $\times\ (3/2) +$ mol Zn $\times\ 1$

Let $x =$ mass of Cr and $y =$ mass of Zn, then:

$$x + y = 0.362 \text{ g and } 9.02 \times 10^{-3} = \frac{1.5\,x}{52.00} + \frac{y}{65.38}$$

We have two equations and two unknowns. Solving by simultaneous equations:

$$9.02 \times 10^{-3} = 0.02885\,x + 0.01530\,y$$
$$\underline{-0.01530 \times 0.362 = -0.01530\,x - 0.01530\,y}$$
$$3.48 \times 10^{-3} = 0.01355\,x \qquad\qquad x = \text{mass Cr} = \frac{3.48 \times 10^{-3}}{0.01355} = 0.257 \text{ g}$$

$y =$ mass Zn $= 0.362$ g $- 0.257$ g $= 0.105$ g Zn; mass % Zn $= \dfrac{0.105 \text{ g}}{0.362 \text{ g}} \times 100 = 29.0\%$ Zn

115.   a.   The reaction is: $CH_4(g) + 2\ O_2(g) \rightarrow CO_2(g) + 2\ H_2O(g)$

$PV = nRT,\ \dfrac{PV}{n} = RT = \text{constant},\ \dfrac{P_{CH_4} V_{CH_4}}{n_{CH_4}} = \dfrac{P_{air} V_{air}}{n_{air}}$

The balanced equation requires 2 mol O_2 for every mol of CH_4 that reacts. For three times as much oxygen, we would need 6 mol O_2 per mol of CH_4 reacted ($n_{O_2} = 6\ n_{CH_4}$). Air is 21% mol percent O_2, so $n_{O_2} = 0.21\ n_{air}$. Therefore, the mol of air we would need to delivery the excess O_2 are:

$$n_{O_2} = 0.21\ n_{air} = 6\ n_{CH_4},\quad n_{air} = 29\ n_{CH_4},\quad \frac{n_{air}}{n_{CH_4}} = 29$$

In one minute:

$$V_{air} = V_{CH_4} \times \frac{n_{air}}{n_{CH_4}} \times \frac{P_{CH_4}}{P_{air}} = 200.\ L \times 29 \times \frac{1.50\ atm}{1.00\ atm} = 8.7 \times 10^3\ L\ air/min$$

b. If x moles of CH_4 were reacted, then $6x$ mol O_2 were added, producing $0.950\,x$ mol CO_2 and $0.050\,x$ mol of CO. In addition, $2x$ mol H_2O must be produced to balance the hydrogens.

$$CH_4(g) + 2\ O_2(g) \rightarrow CO_2(g) + 2\ H_2O(g); \quad CH_4(g) + 3/2\ O_2(g) \rightarrow CO(g) + 2\ H_2O(g)$$

Amount O_2 reacted:

$$0.950\ x\ mol\ CO_2 \times \frac{2\ mol\ O_2}{mol\ CO_2} = 1.90\ x\ mol\ O_2$$

$$0.050\ x\ mol\ CO \times \frac{1.5\ mol\ O_2}{mol\ CO} = 0.075\ x\ mol\ O_2$$

Amount of O_2 left in reaction mixture $= 6.00\,x - 1.90\,x - 0.075\,x = 4.03\,x$ mol O_2

Amount of N_2 $= 6.00\ x\ mol\ O_2 \times \frac{79\ mol\ N_2}{21\ mol\ O_2} = 22.6\ x \approx 23\ x\ mol\ N_2$

The reaction mixture contains:

$$0.950\ x\ mol\ CO_2 + 0.050\ x\ mol\ CO + 4.03\ x\ mol\ O_2 + 2.00\ x\ mol\ H_2O$$

$$+ 23\ x\ mol\ N_2 = 30.\ x\ total\ mol\ of\ gas$$

$$\chi_{CO} = \frac{0.050\ x}{30.\ x} = 0.0017; \quad \chi_{CO_2} = \frac{0.950\ x}{30.\ x} = 0.032; \quad \chi_{O_2} = \frac{4.03\ x}{30.\ x} = 0.13;$$

$$\chi_{H_2O} = \frac{2.00\ x}{30.\ x} = 0.067; \quad \chi_{N_2} = \frac{23\ x}{30.\ x} = 0.77$$

116. The reactions are:

$$C(s) + 1/2\ O_2(g) \rightarrow CO(g) \text{ and } C(s) + O_2(g) \rightarrow CO_2(g)$$

$$PV = nRT, \quad P = n\left(\frac{RT}{V}\right) = n\ (constant)$$

Since the pressure has increased by 17.0%, the number of moles of gas has also increased by 17.0%.

$$n_{final} = 1.170\ n_{initial} = 1.170\ (5.00) = 5.85\ mol\ gas = n_{O_2} + n_{CO} + n_{CO_2}$$

$$n_{CO} + n_{CO_2} = 5.00 \quad (balancing\ moles\ of\ C)$$

$$n_{O_2} + n_{CO} + n_{CO_2} = 5.85$$
$$-(n_{CO} + n_{CO_2} = 5.00)$$
$$\overline{\qquad n_{O_2} \qquad = 0.85}$$

If all C was converted to CO_2, no O_2 would be left. If all C was converted to CO, we would get 5 mol CO and 2.5 mol excess O_2 in the reaction mixture. In the final mixture: $n_{CO} = 2n_{O_2}$

$$n_{CO} = 2n_{O_2} = 1.70 \text{ mol CO};\ \ 1.70 + n_{CO_2} = 5.00,\ \ n_{CO_2} = 3.30 \text{ mol } CO_2$$

$$\chi_{CO} = \frac{1.70}{5.85} = 0.291; \qquad \chi_{CO_2} = \frac{3.30}{5.85} = 0.564; \qquad \chi_{O_2} = \frac{0.85}{5.85} = 0.145 \approx 0.15$$

117.   a.   Volume of hot air: $V = \dfrac{4}{3}\pi r^3 = \dfrac{4}{3}\pi(2.50 \text{ m})^3 = 65.4 \text{ m}^3$

   (Note: radius = diameter/2 = 5.00/2 = 2.50 m)

$$65.4 \text{ m}^3 \times \left(\frac{10 \text{ dm}}{\text{m}}\right)^3 \times \frac{1 \text{ L}}{\text{dm}^3} = 6.54 \times 10^4 \text{ L}$$

$$n = \frac{PV}{RT} = \frac{\left(745 \text{ torr} \times \dfrac{1 \text{ atm}}{760 \text{ torr}}\right) \times 6.54 \times 10^4 \text{ L}}{\dfrac{0.08206 \text{ L atm}}{\text{mol K}} \times (273 + 65) \text{ K}} = 2.31 \times 10^3 \text{ mol air}$$

Mass of hot air $= 2.31 \times 10^3 \text{ mol} \times \dfrac{29.0 \text{ g}}{\text{mol}} = 6.70 \times 10^4 \text{ g}$

Mass of air displaced:

$$n = \frac{PV}{RT} = \frac{\dfrac{745}{760} \text{ atm} \times 6.54 \times 10^4 \text{ L}}{\dfrac{0.08206 \text{ L atm}}{\text{mol K}} \times (273 + 21) \text{ K}} = 2.66 \times 10^3 \text{ mol air}$$

Mass $= 2.66 \times 10^3 \text{ mol} \times \dfrac{29.0 \text{ g}}{\text{mol}} = 7.71 \times 10^4 \text{ g of air displaced}$

Lift $= 7.71 \times 10^4 \text{ g} - 6.70 \times 10^4 \text{ g} = 1.01 \times 10^4 \text{ g}$

   b.   Mass of air displaced is the same, 7.71×10^4 g. Moles of He in balloon will be the same as moles of air displaced, 2.66×10^3 mol, since P, V and T are the same.

Mass of He $= 2.66 \times 10^3 \text{ mol} \times \dfrac{4.003 \text{ g}}{\text{mol}} = 1.06 \times 10^4 \text{ g}$

Lift $= 7.71 \times 10^4 \text{g} - 1.06 \times 10^4 \text{ g} = 6.65 \times 10^4 \text{ g}$

c.   Mass of hot air:

$$n = \frac{PV}{RT} = \frac{\dfrac{630.}{760}\,\text{atm} \times 6.54 \times 10^4\,\text{L}}{\dfrac{0.08206\,\text{L atm}}{\text{mol K}} \times 338\,\text{K}} = 1.95 \times 10^3\,\text{mol air}$$

$$1.95 \times 10^3\,\text{mol} \times \frac{29.0\,\text{g}}{\text{mol}} = 5.66 \times 10^4\,\text{g of hot air}$$

Mass of air displaced:

$$n = \frac{PV}{RT} = \frac{\dfrac{630.}{760}\,\text{atm} \times 6.54 \times 10^4\,\text{L}}{\dfrac{0.08206\,\text{L atm}}{\text{mol K}} \times 294\,\text{K}} = 2.25 \times 10^3\,\text{mol air}$$

$$2.25 \times 10^3\,\text{mol} \times \frac{29.0\,\text{g}}{\text{mol}} = 6.53 \times 10^4\,\text{g of air displaced}$$

Lift = 6.53×10^4 g - 5.66×10^4 g = 8.7×10^3 g

118. $\left(P + \dfrac{an^2}{V^2}\right) \times (V-nb) = nRT,\ \ PV + \dfrac{an^2V}{V^2} - nbP - \dfrac{an^3b}{V^2} = nRT,\ \ PV + \dfrac{an^2}{V} - nbP - \dfrac{an^3b}{V^2} = nRT$

At low P and high T, the molar volume of a gas will be relatively large. The an^2/V and an^3b/V^2 terms become negligible because V is large. Since nb is the actual volume of the gas molecules themselves, then nb << V and the -nbP term is negligible compared to PV. Thus PV = nRT.

119. a.   If we have 1.0×10^6 L of air, then there are 3.0×10^2 L of CO.

$$P_{CO} = \chi_{CO} \times P_{total};\ \ \chi_{CO} = \frac{V_{CO}}{V_{total}} \text{ since } V \propto n;\ \ P_{CO} = \frac{3.0 \times 10^2\,\text{L}}{1.0 \times 10^6\,\text{L}} \times 628\,\text{torr} = 0.19\,\text{torr}$$

b. $n_{CO} = \dfrac{P_{CO} \times V}{RT};$ Assuming 1.0 cm^3 of air = 1.0 mL = 1.0×10^{-3} L:

$$n_{CO} = \frac{\dfrac{0.19}{760}\,\text{atm} \times 1.0 \times 10^{-3}\,\text{L}}{\dfrac{0.08206\,\text{L atm}}{\text{mol K}} \times 273\,\text{K}} = 1.1 \times 10^{-8}\,\text{mol CO}$$

$$1.1 \times 10^{-8}\,\text{mol} \times \frac{6.022 \times 10^{23}\,\text{molecules}}{\text{mol}} = 6.6 \times 10^{15}\,\text{molecules CO in the 1.0 cm}^3\text{ of air}$$

120.   a.   Initially, $P_{N_2} = P_{H_2} = 1.00$ atm and the total pressure is 2.00 atm ($P_{tot} = P_{N_2} + P_{H_2}$). The total pressure after reaction will also be 2.00 atm since we have a constant pressure container. Since V and T are constant before the reaction takes place, there must be equal moles of N_2 and H_2 present initially. Let $x = $ mol $N_2 = $ mol H_2 that are present initially. From the balanced equation, $N_2(g) + 3\ H_2(g) \rightarrow 2\ NH_3(g)$, H_2 will be limiting since three times as many mol of H_2 are required to react as compared to mol of N_2.

After the reaction occurs, none of the H_2 remains (it is the limiting reagent).

$$\text{mol } NH_3 \text{ produced} = x \text{ mol } H_2 \times \frac{2 \text{ mol } NH_3}{3 \text{ mol } H_2} = 2x/3$$

$$\text{mol } N_2 \text{ reacted} = x \text{ mol } H_2 \times \frac{1 \text{ mol } N_2}{3 \text{ mol } H_2} = x/3$$

mol N_2 remaining = x mol N_2 present initially - $x/3$ mol N_2 reacted = $2x/3$ mol N_2 remaining

After the reaction goes to completion, equal mol of $N_2(g)$ and $NH_3(g)$ are present ($2x/3$). Since equal mol are present, then the partial pressure of each gas must be equal ($P_{N_2} = P_{NH_3}$).

$P_{tot} = 2.00$ atm $= P_{N_2} + P_{NH_3}$; Solving: $P_{N_2} = 1.00$ atm $= P_{NH_3}$

   b. $V \propto n$ since P and T are constant. The mol of gas present initially are:

$$n_{N_2} + n_{H_2} = x + x = 2x \text{ mol}$$

After reaction, the mol of gas present are:

$$n_{N_2} + n_{NH_3} = \frac{2x}{3} + \frac{2x}{3} = 4x/3 \text{ mol}$$

$$\frac{V_{after}}{V_{initial}} = \frac{n_{after}}{n_{initial}} = \frac{4x/3}{2x} = \frac{2}{3}$$

The volume of the container will be two-thirds the original volume so:

$V = 2/3(15.0 \text{ L}) = 10.0 \text{ L}$

CHAPTER SIX

THERMOCHEMISTRY

Questions

9. A coffee-cup calorimeter is at constant (atmospheric) pressure. The heat released or gained at constant pressure is ΔH. A bomb calorimeter is at constant volume. The heat released or gained at constant volume is ΔE.

10. Plot a represents an exothermic reaction. In an exothermic process, the bonds in the product molecules are stronger (on average) than those in the reactant molecules. The net result is that the quantity of energy $\Delta(PE)$ is transferred to the surroundings as heat when reactants are converted to products.

    For an endothermic process, energy flows into the system as heat to increase the potential energy of the system. In an endothermic process, the products have higher potential energy (weaker bonds on average) than the reactants.

11. The specific heat capacities are: 0.89 J/g•°C (Al) and 0.45 J/g•°C (Fe)
    Al would be the better choice. It has a higher heat capacity and a lower density than Fe. Using Al, the same amount of heat could be dissipated by a smaller mass, keeping the mass of the amplifier down.

12. $H_2O(l) \rightarrow H_2O(g)$; Heat must be added to boil water so q is positive. Since a certain quantity of $H_2O(g)$ occupies a much larger volume than the same quantity of $H_2O(l)$, an expansion will occur as $H_2O(l)$ is converted to $H_2O(g)$. Therefore, w will be negative for this process, i.e., the system does work on the surroundings when the expansion occurs.

13. A state function is a function whose change depends only on the initial and final states and not on how one got from the initial to the final state. If H and E were not state functions, the law of conservation of energy (first law) would not be true.

14. If Hess's law were not true it would be possible to create energy by reversing a reaction using a different series of steps. This violates the law of conservation of energy (first law). Thus, Hess's law is another statement of the law of conservation of energy.

15. In order to compare values of ΔH to each other, a common reference (or zero) point must be chosen. The definition of ΔH_f° establishes the pure elements in their standard states as that common reference point.

16.   Advantages: H_2 burns cleanly (less pollution) and gives a lot of energy per gram of fuel.

       Disadvantages: Expense and storage.

Exercises

Potential and Kinetic Energy

17. $KE = \frac{1}{2}mv^2$; Convert mass and velocity to SI units. $1 J = \frac{1 \text{ kg m}^2}{s^2}$

       $Mass = 5.25 \text{ oz} \times \frac{1 \text{ lb}}{16 \text{ oz}} \times \frac{1 \text{ kg}}{2.205 \text{ lb}} = 0.149 \text{ kg}$

       $Velocity = \frac{1.0 \times 10^2 \text{ mi}}{hr} \times \frac{1 \text{ hr}}{60 \text{ min}} \times \frac{1 \text{ min}}{60 \text{ s}} \times \frac{1760 \text{ yd}}{\text{mi}} \times \frac{1 \text{ m}}{1.094 \text{ yd}} = \frac{45 \text{ m}}{s}$

       $KE = \frac{1}{2}mv^2 = \frac{1}{2} \times 0.149 \text{ kg} \times \left(\frac{45 \text{ m}}{s}\right)^2 = 150 \text{ J}$

18. $KE = \frac{1}{2}mv^2 = \frac{1}{2} \times \left(1.0 \times 10^{-5} \text{ g} \times \frac{1 \text{ kg}}{1000 \text{ g}}\right) \times \left(\frac{2.0 \times 10^5 \text{ cm}}{\text{sec}} \times \frac{1 \text{ m}}{100 \text{ cm}}\right)^2 = 2.0 \times 10^{-2} \text{ J}$

19. $KE = \frac{1}{2}mv^2 = \frac{1}{2} \times 2.0 \text{ kg} \times \left(\frac{1.0 \text{ m}}{s}\right)^2 = 1.0 \text{ J}$; $KE = \frac{1}{2}mv^2 = \frac{1}{2} \times 1.0 \text{ kg} \times \left(\frac{2.0 \text{ m}}{s}\right)^2 = 2.0 \text{ J}$

       The 1.0 kg object with a velocity of 2.0 m/s has the greater kinetic energy.

20.   Ball A: $PE = mgz = 2.00 \text{ kg} \times \frac{9.80 \text{ m}}{s^2} \times 10.0 \text{ m} = \frac{196 \text{ kg m}^2}{s^2} = 196 \text{ J}$

       At Point I: All of this energy is transferred to Ball B. All of B's energy is kinetic energy at this
              point. $E_{total} = KE = 196 \text{ J}$. At point II, the sum of the total energy will equal 196 J.

       At Point II: $PE = mgz = 4.00 \text{ kg} \times \frac{9.80 \text{ m}}{s^2} \times 3.00 \text{ m} = 118 \text{ J}$

              $KE = E_{total} - PE = 196 \text{ J} - 118 \text{ J} = 78 \text{ J}$

Heat and Work

21.   a. $\Delta E = q + w = 51 \text{ kJ} + (-15 \text{ kJ}) = 36 \text{ kJ}$

       b. $\Delta E = 100. \text{ kJ} + (-65 \text{ kJ}) = 35 \text{ kJ}$ c. $\Delta E = -65 + (-20.) = -85 \text{ kJ}$

       d.   When the system delivers work to the surroundings, $w < 0$. This is the case in all these
              examples, a, b and c.

22.   a. $\Delta E = q + w = -47\text{ kJ} + 88\text{ kJ} = 41\text{ kJ}$

b. $\Delta E = 82 + 47 = 129\text{ kJ}$ c. $\Delta E = 47 + 0 = 47\text{ kJ}$

d.   When the surroundings deliver work to the system, $w > 0$. This is the case for a and b.

23. $\Delta E = q + w = -125 + 104 = -21\text{ kJ}$

24.   Step 1: $\Delta E_1 = q + w = 72\text{ J} + 35\text{ J} = 107\text{ J}$;   Step 2: $\Delta E_2 = 35\text{ J} - 72\text{ J} = -37\text{ J}$

$\Delta E_{overall} = \Delta E_1 + \Delta E_2 = 107\text{ J} - 37\text{ J} = 70.\text{ J}$

25. $w = -P\Delta V = -P \times (V_f - V_i) = -2.0\text{ atm} \times (5.0 \times 10^{-3}\text{ L} - 5.0\text{ L}) = -2.0\text{ atm} \times (-5.0\text{ L}) = 10.\text{ L atm}$

We can also calculate the work in Joules.

$1\text{ atm} = 1.013 \times 10^5\text{ Pa} = 1.013 \times 10^5 \dfrac{\text{kg}}{\text{m s}^2}$; $1\text{ L} = 1000\text{ cm}^3 = 1 \times 10^{-3}\text{ m}^3$

$1\text{ L atm} = 1 \times 10^{-3}\text{ m}^3 \times 1.013 \times 10^5 \dfrac{\text{kg}}{\text{m s}^2} = 101.3 \dfrac{\text{kg m}^2}{\text{s}^2} = 101.3\text{ J}$

$w = 10.\text{ L atm} \times \dfrac{101.3\text{ J}}{\text{L atm}} = 1013\text{ J} = 1.0 \times 10^3\text{ J}$

26.   In this problem $q = w = -950.\text{ J}$

$-950.\text{ J} \times \dfrac{1\text{ L atm}}{101.3\text{ J}} = -9.38\text{ L atm of work done by the gases.}$

$w = -P\Delta V,\ -9.38\text{ L atm} = \dfrac{-650.}{760}\text{ atm} \times (V_f - 0.040\text{ L}),\ V_f - 0.040 = 11.0\text{ L},\ V_f = 11.0\text{ L}$

27. $q = \text{molar heat capacity} \times \text{mol} \times \Delta T = \dfrac{20.8\text{ J}}{°\text{C mol}} \times 39.1\text{ mol} \times (38.0 - 0.0)\,°\text{C} = 30{,}900\text{ J} = 30.9\text{ kJ}$

$w = -P\Delta V = -1.00\text{ atm} \times (998\text{ L} - 876\text{ L}) = -122\text{ L atm} \times \dfrac{101.3\text{ J}}{\text{L atm}} = -12{,}400\text{ J} = -12.4\text{ kJ}$

$\Delta E = q + w = 30.9\text{ kJ} + (-12.4\text{ kJ}) = 18.5\text{ kJ}$

28. $H_2O(g) \rightarrow H_2O(l)$; $\Delta E = q + w$; $q = -40.66\text{ kJ}$; $w = -P\Delta V$

Volume of 1 mol $H_2O(l) = 1\text{ mol } H_2O(l) \times \dfrac{18.02\text{ g}}{\text{mol}} \times \dfrac{1\text{cm}^3}{0.996\text{ g}} = 18.1\text{ cm}^3 = 18.1\text{ mL}$

$w = -P\Delta V = -1.00\text{ atm} \times (0.0181\text{ L} - 30.6\text{ L}) = 30.6\text{ L atm} \times \dfrac{101.3\text{ J}}{\text{L atm}} = 3.10 \times 10^3\text{ J} = 3.10\text{ kJ}$

$\Delta E = q + w = -40.66\text{ kJ} + 3.10\text{ kJ} = -37.56\text{ kJ}$

Properties of Enthalpy

29.   This is an endothermic reaction so heat must be absorbed in order to convert reactants into products. The high temperature environment of internal combustion engines provides the heat.

30.   One should try to cool the reaction mixture or provide some means of removing heat since the reaction is very exothermic (heat is released). The $H_2SO_4(aq)$ will get very hot and possibly boil unless cooling is provided.

31.   a.   Heat is absorbed from the water (it gets colder) as KBr dissolves, so this is an endothermic process.

       b.   Heat is released as CH_4 is burned, so this is an exothermic process.

       c.   Heat is released to the water (it gets hot) as H_2SO_4 is added, so this is an exothermic process.

       d.   Heat must be added (absorbed) to boil water, so this is an endothermic process.

32.   a.   The combustion of gasoline releases heat, so this is an exothermic process.

       b. $H_2O(g) \rightarrow H_2O(l)$; Heat is released when water vaper condenses, so this is an exothermic process.

       c.   To convert a solid to a gas, heat must be absorbed, so this is an endothermic process.

       d.   Heat must be added (absorbed) in order to break a bond, so this is an endothermic process.

33. $4\ Fe(s) + 3\ O_2(g) \rightarrow 2\ Fe_2O_3(s)$ $\Delta H = -1652$ kJ; Note that 1652 kJ of heat are released when 4 mol Fe react with 3 mol O_2 to produce 2 mol Fe_2O_3.

       a. $4.00\ \text{mol Fe} \times \dfrac{-1652\ \text{kJ}}{4\ \text{mol Fe}} = -1650$ kJ heat released

       b. $1.00\ \text{ml Fe}_2O_3 \times \dfrac{-1652\ \text{kJ}}{2\ \text{mol Fe}_2O_3} = -826$ kJ heat released

       c. $1.00\ \text{g Fe} \times \dfrac{1\ \text{mol Fe}}{55.85\ \text{g}} \times \dfrac{-1652\ \text{kJ}}{4\ \text{mol Fe}} = -7.39$ kJ heat released

       d. $10.0\ \text{g Fe} \times \dfrac{1\ \text{mol Fe}}{55.85\ \text{g}} = 0.179$ mol Fe; $2.00\ \text{g O}_2 \times \dfrac{1\ \text{mol O}_2}{32.00\ \text{g}} = 0.0625$ mol O_2

       0.179 mol Fe/0.0625 mol O_2 = 2.86; The balanced equation requires a 4 mol Fe/3 mol O_2 = 1.33 mol ratio. O_2 is limiting since the actual mol Fe/mol O_2 ratio is less than the required mol ratio.

       $0.0625\ \text{mol O}_2 \times \dfrac{-1652\ \text{kJ}}{3\ \text{mol O}_2} = -34.4$ kJ heat released

34.   a. $1.00 \text{ mol } H_2O \times \dfrac{-572 \text{ kJ}}{2 \text{ mol } H_2O} = -286$ kJ heat released

b. $4.03 \text{ g } H_2 \times \dfrac{1 \text{ mol } H_2}{2.016 \text{ g } H_2} \times \dfrac{-572 \text{ kJ}}{2 \text{ mol } H_2} = -572$ kJ heat released

c. $186 \text{ g } O_2 \times \dfrac{1 \text{ mol } O_2}{32.00 \text{ g } O_2} \times \dfrac{-572 \text{ kJ}}{\text{mol } O_2} = -3320$ kJ heat released

d. $n_{H_2} = \dfrac{PV}{RT} = \dfrac{1.0 \text{ atm} \times 2.0 \times 10^8 \text{ L}}{\dfrac{0.08206 \text{ L atm}}{\text{mol K}} \times 298 \text{ K}} = 8.2 \times 10^6$ mol H_2

$8.2 \times 10^6 \text{ mol } H_2 \times \dfrac{-572 \text{ kJ}}{2 \text{ mol } H_2} = -2.3 \times 10^9$ kJ heat released

35.   From Sample Exercise 6.3, $q = 1.3 \times 10^8$ J. Since the heat transfer process is only 60.%

efficient, the total energy required is: $1.3 \times 10^8 \text{ J} \times \dfrac{100. \text{ J}}{60. \text{ J}} = 2.2 \times 10^8$ J

mass $C_3H_8 = 2.2 \times 10^8 \text{ J} \times \dfrac{1 \text{ mol } C_3H_8}{2221 \times 10^3 \text{ J}} \times \dfrac{44.09 \text{ g } C_3H_8}{\text{mol } C_3H_8} = 4.4 \times 10^3$ g C_3H_8

36.   a. $1.00 \text{ g } CH_4 \times \dfrac{1 \text{ mol } CH_4}{16.04 \text{ g } CH_4} \times \dfrac{-891 \text{ kJ}}{\text{mol } CH_4} = -55.5$ kJ

b. $PV = nRT, \ n = \dfrac{PV}{RT} = \dfrac{\dfrac{740.}{760} \text{ atm} \times 1.00 \times 10^3 \text{ L}}{\dfrac{0.08206 \text{ L atm}}{\text{mol K}} \times 298 \text{ K}} = 39.8$ mol

$39.8 \text{ mol} \times \dfrac{-891 \text{ kJ}}{\text{mol}} = -3.55 \times 10^4$ kJ

Calorimetry and Heat Capacity

37.   Specific heat capacity is defined as the amount of heat necessary to raise the temperature of one gram of substance by one degree Celsius. Therefore, $H_2O(l)$ with the largest heat capacity value requires the largest amount of heat for this process. The amount of heat for $H_2O(l)$ is:

energy $= s \times m \times \Delta T = \dfrac{4.18 \text{ J}}{\text{g }^\circ C} \times 25.0 \text{ g} \times (37.0^\circ C - 15.0^\circ C) = 2.30 \times 10^3$ J

The largest temperature change when a certain amount of energy is added to a certain mass of substance will occur for the substance with the smallest specific heat capacity. This is Hg(l), and the temperature change for this process is:

$$\Delta T = \frac{energy}{s \times m} = \frac{10.7 \text{ kJ} \times \frac{1000 \text{ J}}{\text{kJ}}}{\frac{0.14 \text{ J}}{\text{g}\,^\circ\text{C}} \times 550.\text{ g}} = 140\,^\circ\text{C}$$

38.   a. s = specific heat capacity = $\dfrac{0.24 \text{ J}}{\text{g}\,^\circ\text{C}} = \dfrac{0.24 \text{ J}}{\text{g K}}$ since $\Delta T(\text{K}) = \Delta T(^\circ\text{C})$.

energy = $s \times m \times \Delta T = \dfrac{0.24 \text{ J}}{\text{g}\,^\circ\text{C}} \times 150.0 \text{ g} \times (298 \text{ K} - 273 \text{ K}) = 9.0 \times 10^2 \text{ J}$

b.   molar heat capacity = $\dfrac{0.24 \text{ J}}{\text{g}\,^\circ\text{C}} \times \dfrac{107.9 \text{ g Ag}}{\text{mol Ag}} = \dfrac{26 \text{ J}}{\text{mol}\,^\circ\text{C}}$

c. $1250 \text{ J} = \dfrac{0.24 \text{ J}}{\text{g}\,^\circ\text{C}} \times m \times (15.2\,^\circ\text{C} - 12.0\,^\circ\text{C})$, $m = \dfrac{1250}{0.24 \times 3.2} = 1.6 \times 10^3 \text{ g Ag}$

39.   The units for specific heat capacity (s) are J/g•°C. $s = \dfrac{78.2 \text{ J}}{45.6 \text{ g} \times 13.3\,^\circ\text{C}} = \dfrac{0.129 \text{ J}}{\text{g}\,^\circ\text{C}}$

Molar heat capacity = $\dfrac{0.129 \text{ J}}{\text{g}\,^\circ\text{C}} \times \dfrac{207.2 \text{ g}}{\text{mol Pb}} = \dfrac{26.7 \text{ J}}{\text{mol}\,^\circ\text{C}}$

40. $s = \dfrac{585 \text{ J}}{125.6 \text{ g} \times (53.5 - 20.0)\,^\circ\text{C}} = 0.139 \text{ J/g•}^\circ\text{C}$

Molar heat capacity = $\dfrac{0.139 \text{ J}}{\text{g}\,^\circ\text{C}} \times \dfrac{200.6 \text{ g}}{\text{mol Hg}} = \dfrac{27.9 \text{ J}}{\text{mol}\,^\circ\text{C}}$

41. | Heat loss by hot water | = | Heat gain by cooler water |

The magnitude of heat loss and heat gain are equal in calorimetry problems. The only difference is the sign (positive or negative). To avoid sign errors, keep all quantities positive and, if necessary, deduce the correct signs at the end of the problem. Water has a specific heat capacity = s = 4.18 J/°C•g = 4.18 J/K•g (ΔT in °C = ΔT in K).

Heat loss by hot water = $s \times m \times \Delta T = \dfrac{4.18 \text{ J}}{\text{g K}} \times 50.0 \text{ g} \times (330.\text{ K} - T_f)$

Heat gain by cooler water = $\dfrac{4.18 \text{ J}}{\text{g K}} \times 30.0 \text{ g} \times (T_f - 280.\text{ K})$; Heat loss = Heat gain, so:

$\dfrac{209 \text{ J}}{\text{K}} \times (330.\text{ K} - T_f) = \dfrac{125 \text{ J}}{\text{K}} \times (T_f - 280.\text{ K})$, $6.90 \times 10^4 - 209 T_f = 125 T_f - 3.50 \times 10^4$

$334 T_f = 1.040 \times 10^5$, $T_f = 311 \text{ K}$

Note that the final temperature is closer to the temperature of the more massive hot water, which is as it should be.

42.   Heat loss by Al + heat loss by Fe = heat gain by water; Keeping all quantities positive to avoid sign error:

$$\frac{0.89\,J}{g\,°C} \times 5.00\ g\ Al \times (100.0°C - T_f) + \frac{0.45\,J}{g\,°C} \times 10.00\ g\ Fe \times (100.0 - T_f)$$

$$= \frac{4.18\,J}{g\,°C} \times 97.3\ g\ H_2O \times (T_f - 22.0°C)$$

4.5(100.0 - T_f) + 4.5(100.0 - T_f) = 407(T_f - 22.0),   450 - 4.5 T_f + 450 - 4.5 T_f = 407 T_f - 8950

416 T_f = 9850, T_f = 23.7°C

43.   Heat gained by water = heat loss by metal = s × m × ΔT where s = specific heat capacity.

Heat gain = $\dfrac{4.18\,J}{g\,°C}$ × 150.0 g × (18.3°C - 15.0°C) = 2100 J

A common error in calorimetry problems is sign errors. Keeping all quantities positive helps eliminate sign errors.

heat loss = 2100 J = s × 150.0 g × (75.0°C - 18.3°C),   s = $\dfrac{2100\,J}{150.0\ g \times 56.7°C}$ = 0.25 J/g•°C

44.   Heat gain by water = heat loss by Cu; Keeping all quantities positive to avoid sign errors:

$$\frac{4.18\,J}{g\,°C} \times m \times (24.9°C - 22.3°C) = \frac{0.20\,J}{g\,°C} \times 110.\ g\ Cu \times (82.4°C - 24.9°C)$$

11 m = 1300,   m = 120 g H_2O

45. 50.0×10^{-3} L × 0.100 mol/L = 5.00×10^{-3} mol of both $AgNO_3$ and HCl are reacted. Thus, 5.00×10^{-3} mol of AgCl will be produced since there is a 1:1 mol ratio between reactants.

Heat lost by chemicals = Heat gained by solution

Heat gain = $\dfrac{4.18\,J}{g\,°C}$ × 100.0 g × (23.40 - 22.60)°C = 330 J

Heat loss = 330 J; This is the heat evolved (exothermic reaction) when 5.00×10^{-3} mol of AgCl is produced. So q = -330 J and ΔH (heat per mol AgCl formed) is negative with a value of:

$$\Delta H = \frac{-330\,J}{5.00 \times 10^{-3}\ mol} \times \frac{1\ kJ}{1000\ J} = -66\ kJ/mol$$

Note: Sign errors are common with calorimetry problems. However, the correct sign for ΔH can easily be determined from the ΔT data, i.e., if ΔT of the solution increases, then the reaction is exothermic since heat was released, and if ΔT of the solution decreases, then the reaction is endothermic since the reaction absorbed heat from the water. For calorimetry problems, keep all quantities positive until the end of the calculation, then decide the sign for ΔH. This will help eliminate sign errors.

46. $NH_4NO_3(s) \rightarrow NH_4^+(aq) + NO_3^-(aq)$ $\Delta H = ?$; mass of solution $= 75.0$ g $+ 1.60$ g $= 76.6$ g

Heat lost by solution = Heat gained as NH_4NO_3 dissolves. To help eliminate sign errors, we will keep all quantities positive (q and ΔT), then deduce the correct sign for ΔH at the end of the problem. Here, since temperature decreases as NH_4NO_3 dissolves, heat is absorbed as NH_4NO_3 dissolves, so it is an endothermic process (ΔH is positive).

Heat lost by solution $= \dfrac{4.18 \text{ J}}{\text{g}^\circ\text{C}} \times 76.6 \text{ g} \times (25.00 - 23.34)^\circ\text{C} = 532 \text{ J} = $ heat gained as NH_4NO_3 dissolves

$$\Delta H = \frac{532 \text{ J}}{1.60 \text{ g } NH_4NO_3} \times \frac{80.05 \text{ g } NH_4NO_3}{\text{mol } NH_4NO_3} \times \frac{1 \text{ kJ}}{1000 \text{ J}} = 26.6 \text{ kJ/mol } NH_4NO_3 \text{ dissolving}$$

47.   Since ΔH is exothermic, the temperature of the solution will increase as $CaCl_2(s)$ dissolves. Keeping all quantities positive:

Heat loss as $CaCl_2$ dissolves $= 11.0 \text{ g } CaCl_2 \times \dfrac{1 \text{ mol } CaCl_2}{110.98 \text{ g } CaCl_2} \times \dfrac{81.5 \text{ kJ}}{\text{mol } CaCl_2} = 8.08 \text{ kJ}$

Heat gained by solution $= 8.08 \times 10^3 \text{ J} = \dfrac{4.18 \text{ J}}{\text{g }^\circ\text{C}} \times (125 + 11.0) \text{ g} \times (T_f - 25.0^\circ\text{C})$

$T_f - 25.0^\circ\text{C} = \dfrac{8.08 \times 10^3}{4.18 \times 136} = 14.2^\circ\text{C}$, $T_f = 14.2^\circ\text{C} + 25.0^\circ\text{C} = 39.2^\circ\text{C}$

48. $0.100 \text{ L} \times \dfrac{0.500 \text{ mol HCl}}{\text{L}} = 5.00 \times 10^{-2} \text{ mol HCl}$

$0.300 \text{ L} \times \dfrac{0.100 \text{ mol Ba(OH)}_2}{\text{L}} = 3.00 \times 10^{-2} \text{ mol Ba(OH)}_2$

To react with all the HCl present, $5.00 \times 10^{-2}/2 = 2.50 \times 10^{-2}$ mol $Ba(OH)_2$ are required. Since 3.00×10^{-2} mol $Ba(OH)_2$ are present, HCl is the limiting reactant.

$5.00 \times 10^{-2} \text{ mol HCl} \times \dfrac{118 \text{ kJ}}{2 \text{ mol HCl}} = 2.95 \text{ kJ}$ of heat is evolved by reaction.

Heat gained by solution $= 2.95 \times 10^3 \text{ J} = \dfrac{4.18 \text{ J}}{\text{g }^\circ\text{C}} \times 400.0 \text{ g} \times \Delta T$

$\Delta T = 1.76^\circ\text{C} = T_f - T_i = T_f - 25.0^\circ\text{C}$, $T_f = 26.8^\circ\text{C}$

49.   First, we need to get the heat capacity of the calorimeter from the combustion of benzoic acid.

Heat lost by combustion = Heat gained by calorimeter

Heat loss $= 0.1584 \text{ g} \times \dfrac{26.42 \text{ kJ}}{\text{g}} = 4.185 \text{ kJ}$

Heat gain $= 4.185 \text{ kJ} = C_{cal} \times \Delta T$, $C_{cal} = \dfrac{4.185 \text{ kJ}}{2.54^\circ\text{C}} = 1.65 \text{ kJ/}^\circ\text{C}$

Now we can calculate the heat of combustion of vanillin. Heat loss = Heat gain

Heat gain by calorimeter = $\dfrac{1.65\,\text{kJ}}{^\circ\text{C}} \times 3.25\,^\circ\text{C} = 5.36\,\text{kJ}$

Heat loss = 5.36 kJ, which is the heat evolved by the combustion of the vanillin.

$\Delta E_{comb} = \dfrac{-5.36\,\text{kJ}}{0.2130\,\text{g}} = -25.2\,\text{kJ/g};\quad \Delta E_{comb} = \dfrac{-25.2\,\text{kJ}}{\text{g}} \times \dfrac{152.14\,\text{g}}{\text{mol}} = -3830\,\text{kJ/mol}$

50.   Heat gain by calorimeter = $\dfrac{1.56\,\text{kJ}}{^\circ\text{C}} \times 3.2\,^\circ\text{C} = 5.0\,\text{kJ}$ = heat loss by quinone

Heat loss = 5.0 kJ, which is the heat evolved (exothermic reaction) by the combustion of 0.1964 g of quinone.

$\Delta E_{comb} = \dfrac{-5.0\,\text{kJ}}{0.1964\,\text{g}} = -25\,\text{kJ/g};\qquad \Delta E_{comb} = \dfrac{-25\,\text{kJ}}{\text{g}} \times \dfrac{108.09\,\text{g}}{\text{mol}} = -2700\,\text{kJ/mol}$

Hess's Law

51.   Information given:

$$C(s) + O_2(g) \rightarrow CO_2(g) \qquad\qquad \Delta H = -393.7\,\text{kJ}$$
$$CO(g) + 1/2\,O_2(g) \rightarrow CO_2(g) \qquad \Delta H = -283.3\,\text{kJ}$$

Using Hess's Law:

$$2\,C(s) + 2\,O_2(g) \rightarrow 2\,CO_2(g) \qquad\qquad \Delta H_1 = 2(-393.7\,\text{kJ})$$
$$2\,CO_2(g) \rightarrow 2\,CO(g) + O_2(g) \qquad\qquad \Delta H_2 = -2(-283.3\,\text{kJ})$$

$$2\,C(s) + O_2(g) \rightarrow 2\,CO(g) \qquad\qquad \Delta H = \Delta H_1 + \Delta H_2 = -220.8\,\text{kJ}$$

Note: The enthalpy change for a reaction that is reversed is the negative quantity of the enthalpy change for the original reaction. If the coefficients in a balanced reaction are multiplied by an integer, the value of ΔH is multiplied by the same integer.

52. $C_4H_4(g) + 5\,O_2(g) \rightarrow 4\,CO_2(g) + 2\,H_2O(l) \qquad \Delta H_{comb} = -2341\,\text{kJ}$
   $C_4H_8(g) + 6\,O_2(g) \rightarrow 4\,CO_2(g) + 4\,H_2O(l) \qquad \Delta H_{comb} = -2755\,\text{kJ}$
   $H_2(g) + 1/2\,O_2(g) \rightarrow H_2O(l) \qquad\qquad\qquad\qquad \Delta H_{comb} = -286\,\text{kJ}$

By convention, $H_2O(l)$ is produced when enthalpies of combustion are given and, since per mol quantities are given, the combustion reaction refers to 1 mol of that quantity reacting with $O_2(g)$.

Using Hess's Law to solve:

$$C_4H_4(g) + 5\ O_2(g) \rightarrow 4\ CO_2(g) + 2\ H_2O(l) \qquad \Delta H_1 = -2341\ kJ$$
$$4\ CO_2(g) + 4\ H_2O(l) \rightarrow C_4H_8(g) + 6\ O_2(g) \qquad \Delta H_2 = -(2755\ kJ)$$
$$2\ H_2(g) + O_2(g) \rightarrow 2\ H_2O(l) \qquad \Delta H_3 = 2(-286\ kJ)$$

$$C_4H_4(g) + 2\ H_2(g) \rightarrow C_4H_8(g) \qquad \Delta H = \Delta H_1 + \Delta H_2 + \Delta H_3 = -158\ kJ$$

53.

$$2\ NO_2 \rightarrow N_2 + 2\ O_2 \qquad \Delta H = -(67.7\ kJ)$$
$$N_2 + 2\ O_2 \rightarrow N_2O_4 \qquad \Delta H = 9.7\ kJ$$

$$2\ NO_2(g) \rightarrow N_2O_4(g) \qquad \Delta H = -58.0\ kJ$$

54.

$$2\ N_2(g) + 6\ H_2(g) \rightarrow 4\ NH_3(g) \qquad \Delta H = -4(46\ kJ)$$
$$6\ H_2O(g) \rightarrow 6\ H_2(g) + 3\ O_2(g) \qquad \Delta H = -3(-484\ kJ)$$

$$2\ N_2(g) + 6\ H_2O(g) \rightarrow 3\ O_2(g) + 4\ NH_3(g) \qquad \Delta H = 1268\ kJ$$

No, since the reaction is very endothermic (requires a lot of heat), it would not be a practical way of making ammonia due to the high energy costs required.

55.

$$NO + O_3 \rightarrow NO_2 + O_2 \qquad \Delta H = -199\ kJ$$
$$3/2\ O_2 \rightarrow O_3 \qquad \Delta H = -1/2(-427\ kJ)$$
$$O \rightarrow 1/2\ O_2 \qquad \Delta H = -1/2(495\ kJ)$$

$$NO(g) + O(g) \rightarrow NO_2(g) \qquad \Delta H = -233\ kJ$$

56.

$$C_6H_4(OH)_2 \rightarrow C_6H_4O_2 + H_2 \qquad \Delta H = 177.4\ kJ$$
$$H_2O_2 \rightarrow H_2 + O_2 \qquad \Delta H = -(191.2\ kJ)$$
$$2\ H_2 + O_2 \rightarrow 2\ H_2O(g) \qquad \Delta H = 2(-241.8\ kJ)$$
$$2\ H_2O(g) \rightarrow 2\ H_2O(l) \qquad \Delta H = 2(-43.8\ kJ)$$

$$C_6H_4(OH)_2(aq) + H_2O_2(aq) \rightarrow C_6H_4O_2(aq) + 2\ H_2O(l) \qquad \Delta H = -202.6\ kJ$$

57.

$$4\ HNO_3 \rightarrow 2\ N_2O_5 + 2\ H_2O \qquad \Delta H = -2(-76.6\ kJ)$$
$$2\ N_2 + 6\ O_2 + 2\ H_2 \rightarrow 4\ HNO_3 \qquad \Delta H = 4(-174.1\ kJ)$$
$$2\ H_2O \rightarrow 2\ H_2 + O_2 \qquad \Delta H = -2(-285.8\ kJ)$$

$$2\ N_2(g) + 5\ O_2(g) \rightarrow 2\ N_2O_5(g) \qquad \Delta H = 28.4\ kJ$$

58.

$$P_4O_{10} \rightarrow P_4 + 5\ O_2 \qquad \Delta H = -(-2967.3\ kJ)$$
$$10\ PCl_3 + 5\ O_2 \rightarrow 10\ Cl_3PO \qquad \Delta H = 10(-285.7\ kJ)$$
$$6\ PCl_5 \rightarrow 6\ PCl_3 + 6\ Cl_2 \qquad \Delta H = -6(-84.2\ kJ)$$
$$P_4 + 6\ Cl_2 \rightarrow 4\ PCl_3 \qquad \Delta H = -1225.6\ kJ$$

$$P_4O_{10}(s) + 6\ PCl_5(g) \rightarrow 10\ Cl_3PO(g) \qquad \Delta H = -610.1\ kJ$$

Standard Enthalpies of Formation

59.   The change in enthalpy that accompanies the formation of one mole of a compound from its elements, with all substances in their standard states, is the standard enthalpy of formation for a compound. The reactions that refer to ΔH_f° are:

$$Na(s) + 1/2\ Cl_2(g) \rightarrow NaCl(s);\ \ H_2(g) + 1/2\ O_2(g) \rightarrow H_2O(l)$$

$$6\ C(graphite,\ s) + 6\ H_2(g) + 3\ O_2(g) \rightarrow C_6H_{12}O_6(s);\ \ Pb(s) + S(rhombic,\ s) + 2\ O_2(g)$$
$$\rightarrow PbSO_4(s)$$

60.   a.   aluminum oxide = Al_2O_3; $2\ Al(s) + 3/2\ O_2(g) \rightarrow Al_2O_3(s)$

   b. $C_2H_5OH(l) + 3\ O_2(g) \rightarrow 2\ CO_2(g) + 3\ H_2O(l)$

   c. $NaOH(aq) + HCl(aq) \rightarrow H_2O(l) + NaCl(aq)$

   d. $2\ C(graphite,\ s) + 3/2\ H_2(g) + 1/2\ Cl_2(g) \rightarrow C_2H_3Cl(g)$

   e. $C_6H_6(l) + 15/2\ O_2(g) \rightarrow 6\ CO_2(g) + 3\ H_2O(l)$

   Note: ΔH_{comb} values assume one mol of compound combusted.

   f. $NH_4Br(s) \rightarrow NH_4^+(aq) + Br^-(aq)$

61.   In general: $\Delta H^\circ = \Sigma n_p \Delta H_{f,\ products}^\circ - \Sigma n_r \Delta H_{f,\ reactants}^\circ$ and all elements in their standard state have $\Delta H_f^\circ = 0$ by definition.

   a.   The balanced equation is: $2\ NH_3(g) + 3\ O_2(g) + 2\ CH_4(g) \rightarrow 2\ HCN(g) + 6\ H_2O(g)$

   $$\Delta H^\circ = [2\ mol\ HCN \times \Delta H_{f\ HCN}^\circ + 6\ mol\ H_2O(g) \times \Delta H_{f\ H_2O}^\circ]$$

   $$- [2\ mol\ NH_3 \times \Delta H_{f\ NH_3}^\circ + 2\ mol\ CH_4 \times \Delta H_{f\ CH_4}^\circ]$$

   $$\Delta H^\circ = [2(135.1) + 6(-242)] - [2(-46) + 2(-75)] = -940.\ kJ$$

   b. $Ca_3(PO_4)_2(s) + 3\ H_2SO_4(l) \rightarrow 3\ CaSO_4(s) + 2\ H_3PO_4(l)$

   $$\Delta H^\circ = \left[3\ mol\ CaSO_4 \left(\frac{-1433\ kJ}{mol} \right) + 2\ mol\ H_3PO_4(l) \left(\frac{-1267\ kJ}{mol} \right) \right]$$

   $$- \left[1\ mol\ Ca_3(PO_4)_2 \left(\frac{-4126\ kJ}{mol} \right) + 3\ mol\ H_2SO_4(l) \left(\frac{-814\ kJ}{mol} \right) \right]$$

   $$\Delta H^\circ = -6833\ kJ - (-6568\ kJ) = -265\ kJ$$

c. $NH_3(g) + HCl(g) \rightarrow NH_4Cl(s)$

$$\Delta H° = [1 \text{ mol } NH_4Cl \times \Delta H°_{f\,NH_4Cl}] - [1 \text{ mol } NH_3 \times \Delta H°_{f\,NH_3} + 1 \text{ mol } HCl \times \Delta H°_{f\,HCl}]$$

$$\Delta H° = \left[1 \text{ mol} \left(\frac{-314 \text{ kJ}}{\text{mol}} \right) \right] - \left[1 \text{ mol} \left(\frac{-46 \text{ kJ}}{\text{mol}} \right) + 1 \text{ mol} \left(\frac{-92 \text{ kJ}}{\text{mol}} \right) \right]$$

$$\Delta H° = -314 \text{ kJ} + 138 \text{ kJ} = -176 \text{ kJ}$$

62.   a.   The balanced equation is: $C_2H_5OH(l) + 3 O_2(g) \rightarrow 2 CO_2(g) + 3 H_2O(g)$

$$\Delta H° = \left[2 \text{ mol} \left(\frac{-393.5 \text{ kJ}}{\text{mol}} \right) + 3 \text{ mol} \left(\frac{-242 \text{ kJ}}{\text{mol}} \right) \right] - \left[1 \text{ mol} \left(\frac{-278 \text{ kJ}}{\text{mol}} \right) \right]$$

$$= -1513 \text{ kJ} - (-278 \text{ kJ}) = -1235 \text{ kJ}$$

b. $SiCl_4(l) + 2 H_2O(l) \rightarrow SiO_2(s) + 4 HCl(aq)$

Since $HCl(aq)$ is $H^+(aq) + Cl^-(aq)$, then $\Delta H°_f = 0 - 167 = -167$ kJ/mol.

$$\Delta H° = \left[4 \text{ mol} \left(\frac{-167 \text{ kJ}}{\text{mol}} \right) + 1 \text{ mol} \left(\frac{-911 \text{ kJ}}{\text{mol}} \right) \right] - \left[1 \text{ mol} \left(\frac{-687 \text{ kJ}}{\text{mol}} \right) + 2 \text{ mol} \left(\frac{-286 \text{ kJ}}{\text{mol}} \right) \right]$$

$$\Delta H° = -1579 \text{ kJ} - (-1259 \text{ kJ}) = -320. \text{ kJ}$$

c. $MgO(s) + H_2O(l) \rightarrow Mg(OH)_2(s)$

$$\Delta H° = \left[1 \text{ mol} \left(\frac{-925 \text{ kJ}}{\text{mol}} \right) \right] - \left[1 \text{ mol} \left(\frac{-602 \text{ kJ}}{\text{mol}} \right) + 1 \text{ mol} \left(\frac{-286 \text{ kJ}}{\text{mol}} \right) \right]$$

$$\Delta H° = -925 \text{ kJ} - (-888 \text{ kJ}) = -37 \text{ kJ}$$

63.   a. $4 NH_3(g) + 5 O_2(g) \rightarrow 4 NO(g) + 6 H_2O(g)$; $\Delta H° = \Sigma n_p \Delta H°_{f\,products} - \Sigma n_r \Delta H°_{f\,reactants}$

$$\Delta H° = \left[4 \text{ mol} \left(\frac{90. \text{ kJ}}{\text{mol}} \right) + 6 \text{ mol} \left(\frac{-242 \text{ kJ}}{\text{mol}} \right) \right] - \left[4 \text{ mol} \left(\frac{-46 \text{ kJ}}{\text{mol}} \right) \right] = -908 \text{ kJ}$$

$2 NO(g) + O_2(g) \rightarrow 2 NO_2(g)$

$$\Delta H° = \left[2 \text{ mol} \left(\frac{34 \text{ kJ}}{\text{mol}} \right) \right] - \left[2 \text{ mol} \left(\frac{90. \text{ kJ}}{\text{mol}} \right) \right] = -112 \text{ kJ}$$

$$3 \ NO_2(g) + H_2O(l) \rightarrow 2 \ HNO_3(aq) + NO(g)$$

$$\Delta H^{\circ} = \left[2 \ mol \left(\frac{-207 \ kJ}{mol} \right) + 1 \ mol \left(\frac{90. \ kJ}{mol} \right) \right]$$

$$- \left[3 \ mol \left(\frac{34 \ kJ}{mol} \right) + 1 \ mol \left(\frac{-286 \ kJ}{mol} \right) \right] = \text{-140. kJ}$$

Note: All ΔH_f° values are assumed $\pm$ 1 kJ.

b. $12 \ NH_3(g) + 15 \ O_2(g) \rightarrow 12 \ NO(g) + 18 \ H_2O(g)$
      $12 \ NO(g) + 6 \ O_2(g) \rightarrow 12 \ NO_2(g)$
      $12 \ NO_2(g) + 4 \ H_2O(l) \rightarrow 8 \ HNO_3(aq) + 4 \ NO(g)$
      $4 \ H_2O(g) \rightarrow 4 \ H_2O(l)$

$$\overline{12 \ NH_3(g) + 21 \ O_2(g) \rightarrow 8 \ HNO_3(aq) + 4 \ NO(g) + 14 \ H_2O(g)}$$

The overall reaction is exothermic since each step is exothermic.

64. $4 \ Na(s) + O_2(g) \rightarrow 2 \ Na_2O(s), \quad \Delta H^{\circ} = 2 \ mol \left(\frac{-416 \ kJ}{mol} \right) = \text{-832 kJ}$

$$2 \ Na(s) + 2 \ H_2O(l) \rightarrow 2 \ NaOH(aq) + H_2(g)$$

$$\Delta H^{\circ} = \left[2 \ mol \left(\frac{-470. \ kJ}{mol} \right) \right] - \left[2 \ mol \left(\frac{-286 \ kJ}{mol} \right) \right] = \text{-368 kJ}$$

$$2 \ Na(s) + CO_2(g) \rightarrow Na_2O(s) + CO(g)$$

$$\Delta H^{\circ} = \left[1 \ mol \left(\frac{-416 \ kJ}{mol} \right) + 1 \ mol \left(\frac{-110.5 \ kJ}{mol} \right) \right] - \left[1 \ mol \left(\frac{-393.5 \ kJ}{mol} \right) \right] = \text{-133 kJ}$$

In both cases, sodium metal reacts with the "extinguishing agent." Both reactions are exothermic and each reaction produces a flammable gas, H_2 and CO, respectively.

65. $3 \ Al(s) + 3 \ NH_4ClO_4(s) \rightarrow Al_2O_3(s) + AlCl_3(s) + 3 \ NO(g) + 6 \ H_2O(g)$

$$\Delta H^{\circ} = \left[6 \ mol \left(\frac{-242 \ kJ}{mol} \right) + 3 \ mol \left(\frac{90. \ kJ}{mol} \right) + 1 \ mol \left(\frac{-704 \ kJ}{mol} \right) + 1 \ mol \left(\frac{-1676 \ kJ}{mol} \right) \right]$$

$$- \left[3 \ mol \left(\frac{-295 \ kJ}{mol} \right) \right] = \text{-2677 kJ}$$

66. $5 N_2O_4(l) + 4 N_2H_3CH_3(l) \rightarrow 12 H_2O(g) + 9 N_2(g) + 4 CO_2(g)$

$$\Delta H^\circ = \left[12 \text{ mol} \left(\frac{-242 \text{ kJ}}{\text{mol}} \right) + 4 \text{ mol} \left(\frac{-393.5 \text{ kJ}}{\text{mol}} \right) \right]$$

$$- \left[5 \text{ mol} \left(\frac{-20. \text{ kJ}}{\text{mol}} \right) + 4 \text{ mol} \left(\frac{54 \text{ kJ}}{\text{mol}} \right) \right] = -4594 \text{ kJ}$$

67. $2 ClF_3(g) + 2 NH_3(g) \rightarrow N_2(g) + 6 HF(g) + Cl_2(g)$ $\Delta H^\circ = -1196$ kJ

$$\Delta H^\circ = [6 \; \Delta H^\circ_{f \, HF}] - [2 \; \Delta H^\circ_{f \, ClF_3} + 2 \; \Delta H^\circ_{f \, NH_3}]$$

$$-1196 \text{ kJ} = 6 \text{ mol} \left(\frac{-271 \text{ kJ}}{\text{mol}} \right) - 2 \; \Delta H^\circ_{f \, ClF_3} - 2 \text{ mol} \left(\frac{-46 \text{ kJ}}{\text{mol}} \right)$$

$$-1196 \text{ kJ} = -1626 \text{ kJ} - 2 \; \Delta H^\circ_{f \, ClF_3} + 92 \text{ kJ}, \; \Delta H^\circ_{f \, ClF_3} = \frac{(-1626 + 92 + 1196) \text{ kJ}}{2 \text{ mol}} = \frac{-169 \text{ kJ}}{\text{mol}}$$

68. $C_2H_4(g) + 3 O_2(g) \rightarrow 2 CO_2(g) + 2 H_2O(l)$ $\Delta H^\circ = -1411.1$ kJ

$$\Delta H^\circ = -1411.1 \text{ kJ} = 2(-393.5) \text{ kJ} + 2(-285.8) \text{ kJ} - \Delta H^\circ_{f \, C_2H_4}$$

$$-1411.1 \text{ kJ} = -1358.6 \text{ kJ} - \Delta H^\circ_{f \, C_2H_4}, \; \Delta H^\circ_{f \, C_2H_4} = 52.5 \text{ kJ/mol}$$

Energy Consumption and Sources

69. $C_2H_5OH(l) + 3 O_2(g) \rightarrow 2 CO_2(g) + 3 H_2O(l)$

$$\Delta H^\circ = [2 \, (-393.5 \text{ kJ}) + 3(-286 \text{ kJ})] - (-278 \text{ kJ}) = -1367 \text{ kJ/mol ethanol}$$

$$\frac{-1367 \text{ kJ}}{\text{mol}} \times \frac{1 \text{ mol}}{46.07 \text{ g}} = -29.67 \text{ kJ/g}$$

70. $CO(g) + 2 H_2(g) \rightarrow CH_3OH(l), \; \Delta H^\circ = -239 \text{ kJ} - (-110.5 \text{ kJ}) = -129$ kJ

71. $C_3H_8(g) + 5 O_2(g) \rightarrow 3 CO_2(g) + 4 H_2O(l)$

$$\Delta H^\circ = [3(-393.5 \text{ kJ}) + 4(-286 \text{ kJ})] - [-104 \text{ kJ}] = -2221 \text{ kJ/mol } C_3H_8$$

$$\frac{-2221 \text{ kJ}}{\text{mol}} \times \frac{1 \text{ mol}}{44.09 \text{ g}} = \frac{-50.37 \text{ kJ}}{\text{g}} \text{ vs. } -47.7 \text{ kJ/g for octane (Sample Exercise 6.11)}$$

The fuel values are very close. An advantage of propane is that it burns more cleanly. The boiling point of propane is -42°C. Thus, it is more difficult to store propane and there are extra safety hazards associated with using high pressure compressed gas tanks.

72.   Since 1 mol of $C_2H_2(g)$ and 1 mol of $C_4H_{10}(g)$ have equivalent volumes at the same T and P, then:

$$\frac{\text{enthalpy of combustion per volume of } C_2H_2}{\text{enthalpy of combustion per volume of } C_4H_{10}} = \frac{\text{enthalpy of combustion per mol } C_2H_2}{\text{enthalpy of combustion per mol } C_4H_{10}}$$

$$\frac{\text{enthalpy of combustion per volume of } C_2H_2}{\text{enthalpy of combustion per volume of } C_4H_{10}} = \frac{\dfrac{-49.9 \text{ kJ}}{g\,C_2H_2} \times \dfrac{26.04 \text{ g } C_2H_2}{\text{mol } C_2H_2}}{\dfrac{-49.5 \text{ kJ}}{g\,C_4H_{10}} \times \dfrac{58.12 \text{ g } C_4H_{10}}{\text{mol } C_4H_{10}}} = 0.452$$

Almost twice the volume of acetylene is needed to furnish the same energy as a given volume of butane.

73.   The molar volume of a gas at STP is 22.42 L (from Chapter 5).

$$4.19 \times 10^6 \text{ kJ} \times \frac{1 \text{ mol } CH_4}{891 \text{ kJ}} \times \frac{22.42 \text{ L } CH_4}{\text{mol } CH_4} = 1.05 \times 10^5 \text{ L } CH_4$$

74.   Mass of H_2O = $1.00 \text{ gal} \times \dfrac{3.785 \text{ L}}{\text{gal}} \times \dfrac{1000 \text{ mL}}{L} \times \dfrac{1.00 \text{ g}}{mL} = 3790 \text{ g } H_2O$

Energy required (theoretical) = $s \times m \times \Delta T = \dfrac{4.18 \text{ J}}{g\,°C} \times 3790 \text{ g} \times 10.0 \text{ °C} = 1.58 \times 10^5 \text{ J}$

For an actual (80.0% efficient) process, more than this quantity of energy is needed since heat is always lost in any transfer of energy. The energy required is:

$$1.58 \times 10^5 \text{ J} \times \frac{100. \text{ J}}{80.0 \text{ J}} = 1.98 \times 10^5 \text{ J}$$

Mass of C_2H_2 = $1.98 \times 10^5 \text{ J} \times \dfrac{1 \text{ mol } C_2H_2}{1300. \times 10^3 \text{ J}} \times \dfrac{26.04 \text{ g } C_2H_2}{\text{mol } C_2H_2} = 3.97 \text{ g } C_2H_2$

Additional Exercises

75.   a. $2\,SO_2(g) + O_2(g) \rightarrow 2\,SO_3(g)$ (w = -PΔV); Since the volume of the piston apparatus decreased as reactants were converted to products, w is positive (w > 0).

b. $COCl_2(g) \rightarrow CO(g) + Cl_2(g)$; Since the volume increased, w is negative (w < 0).

c. $N_2(g) + O_2(g) \rightarrow 2\,NO(g)$; Since the volume did not change, no PV work is done (w = 0).

In order to predict the sign of w for a reaction, compare the coefficients of all the product gases in the balanced equation to the coefficients of all the reactant gases. When a balanced reaction has more mol of product gases than mol of reactant gases (as in b), the reaction will expand in volume (ΔV positive), and the system does work on the surroundings. When a balanced reaction has a decrease in the mol of gas from reactants to products (as in a), the reaction will contract in volume (ΔV negative), and the surroundings will do compression work on the system. When there is no change in the mol of gas from reactants to products (as in c), $\Delta V = 0$ and $w = 0$.

76. $w = -P\Delta V$; $\Delta n = $ mol gaseous products - mol gaseous reactants. Only gases can do PV work (we ignore solids and liquids). When a balanced reaction has more mol of product gases than mol of reactant gases (Δn positive), the reaction will expand in volume (ΔV positive) and the system will do work on the surroundings. For example, in reaction c, $\Delta n = 2 - 0 = 2$ mol, and this reaction would do expansion work against the surroundings. When a balanced reaction has a decrease in the mol of gas from reactants to products (Δn negative), the reaction will contract in volume (ΔV negative) and the surroundings will do compression work on the system, e.g., reaction a where $\Delta n = 0 - 1 = -1$. When there is no change in the mol of gas from reactants to products, $\Delta V = 0$ and $w = 0$, e.g., reaction b where $\Delta n = 2 - 2 = 0$.

When $\Delta V > 0$ ($\Delta n > 0$), then $w < 0$ and system does work on the surroundings (c and e).

When $\Delta V < 0$ ($\Delta n < 0$), then $w > 0$ and the surroundings do work on the system (a and d).

When $\Delta V = 0$ ($\Delta n = 0$), then $w = 0$ (b).

77. $\Delta E_{overall} = \Delta E_{step\,1} + \Delta E_{step\,2}$; This is a cyclic process which means that the overall initial state and final state are the same. Since ΔE is a state function, $\Delta E_{overall} = 0$ and $\Delta E_{step\,1} = -\Delta E_{step\,2}$.

$\Delta E_{step\,1} = q + w = 45\ J + (-10.\ J) = 35\ J$

$\Delta E_{step\,2} = -\Delta E_{step\,1} = -35\ J = q + w,\ -35\ J = -60\ J + w,\ w = 25\ J$

78. $w = -P\Delta V$; We need the final volume of the gas. Since T and n are constant, $P_1 V_1 = P_2 V_2$.

$$V_2 = \frac{V_1 P_1}{P_2} = \frac{10.0\,L\,(15.0\ atm)}{2.00\ atm} = 75.0\ L$$

$w = -P\Delta V = -2.00\ atm\,(75.0\ L - 10.0\ L) = -130.\ L\ atm \times \dfrac{101.3\ J}{L\ atm} \times \dfrac{1\ kJ}{1000\ J} = -13.2\ kJ = work$

79.   Heat loss by hot water = heat gain by cold water; Keeping all quantities positive to avoid sign errors:

$$\frac{4.18\ J}{g\,^{\circ}C} \times m_{hot} \times (55.0\,^{\circ}C - 37.0\,^{\circ}C) = \frac{4.18\ J}{g\,^{\circ}C} \times 90.0\ g \times (37.0\,^{\circ}C - 22.0\,^{\circ}C)$$

$$m_{hot} = \frac{90.0\ g \times 15.0\,^{\circ}C}{18.0\,^{\circ}C} = 75.0\ g\ hot\ water\ needed$$

80. $2 K(s) + 2 H_2O(l) \rightarrow 2 KOH(aq) + H_2(g)$, $\Delta H° = 2(-481 \text{ kJ}) - 2(-286 \text{ kJ}) = -390. \text{ kJ}$

$5.00 \text{ g K} \times \dfrac{1 \text{ mol K}}{39.10 \text{ g K}} \times \dfrac{-390. \text{ kJ}}{2 \text{ mol K}} = -24.9$ kJ of heat released upon reaction of 5.00 g of potassium.

$24{,}900 \text{ J} = \dfrac{4.18 \text{ J}}{\text{g °C}} \times (1.00 \times 10^3 \text{ g}) \times \Delta T$, $\Delta T = \dfrac{24{,}900}{4.18 \times 1.00 \times 10^3} = 5.96°C$

Final temperature = 24.0 + 5.96 = 30.0°C

81. $HCl(aq) + NaOH(aq) \rightarrow H_2O(l) + NaCl(aq)$ $\Delta H = -56$ kJ

$0.2000 \text{ L} \times \dfrac{0.400 \text{ mol HCl}}{\text{L}} = 8.00 \times 10^{-2}$ mol HCl

$0.1500 \text{ L} \times \dfrac{0.500 \text{ mol NaOH}}{\text{L}} = 7.50 \times 10^{-2}$ mol NaOH

Since the balanced reaction requires a 1:1 mol ratio between HCl and NaOH, and since fewer mol of NaOH are actually present as compared to HCl, then NaOH is the limiting reagent.

Heat released = 7.50×10^{-2} mol NaOH $\times \dfrac{-56 \text{ kJ}}{\text{mol NaOH}} = -4.2$ kJ heat released

82.   The specific heat of water is 4.18 J/g•°C, which is equal to 4.18 kJ/kg•°C.

We have 1.00 kg of H_2O, so: $1.00 \text{ kg} \times \dfrac{4.18 \text{ kJ}}{\text{kg °C}} = 4.18$ kJ/°C

This is the portion of the heat capacity that can be attributed to H_2O.

Total heat capacity = $C_{cal} + C_{H_2O}$, $C_{cal} = 10.84 - 4.18 = 6.66$ kJ/°C

83.   Heat released = $1.056 \text{ g} \times 26.42 \text{ kJ/g} = 27.90$ kJ = Heat gain by water and calorimeter

Heat gain = 27.90 kJ = $\dfrac{4.18 \text{ kJ}}{\text{kg °C}} \times 0.987 \text{ kg} \times \Delta T + \dfrac{6.66 \text{ kJ}}{°C} \times \Delta T$

$27.90 = (4.13 + 6.66) \Delta T = 10.79 \Delta T$, $\Delta T = 2.586°C$

$2.586°C = T_f - 23.32°C$, $T_f = 25.91°C$

84.   To avoid fractions, let's first calculate ΔH for the reaction:

$$6\ FeO(s) + 6\ CO(g) \rightarrow 6\ Fe(s) + 6\ CO_2(g)$$

$$
\begin{array}{ll}
6\ FeO + 2\ CO_2 \rightarrow 2\ Fe_3O_4 + 2\ CO & \Delta H^\circ = -2(18\ kJ) \\
2\ Fe_3O_4 + CO_2 \rightarrow 3\ Fe_2O_3 + CO & \Delta H^\circ = -(-39\ kJ) \\
\underline{3\ Fe_2O_3 + 9\ CO \rightarrow 6\ Fe + 9\ CO_2} & \underline{\Delta H^\circ = 3(-23\ kJ)}
\end{array}
$$

$$6\ FeO(s) + 6\ CO(g) \rightarrow 6\ Fe(s) + 6\ CO_2(g) \qquad \Delta H^\circ = -66\ kJ$$

So for: $FeO(s) + CO(g) \rightarrow Fe(s) + CO_2(g) \qquad \Delta H^\circ = \dfrac{-66\ kJ}{6} = -11\ kJ$

85.   a. $\Delta H^\circ = 3\ mol\ (227\ kJ/mol) - 1\ mol\ (49\ kJ/mol) = 632\ kJ$

       b.   Since $3\ C_2H_2(g)$ is higher in energy than $C_6H_6(l)$, acetylene will release more energy per gram when burned in air.

86. $$Cu(s) + 1/2\ O_2(g) \rightarrow CuO(s) \qquad \Delta H^\circ = \Delta H^\circ_{f,CuO(s)}$$

$$
\begin{array}{ll}
Cu(s) + CuO(s) \rightarrow Cu_2O(s) & \Delta H^\circ = -(11\ kJ) \\
\underline{Cu_2O(s) + 1/2\ O_2(g) \rightarrow 2\ CuO(s)} & \underline{\Delta H^\circ = -1/2(288\ kJ)}
\end{array}
$$

$$Cu(s) + 1/2\ O_2(g) \rightarrow CuO(s) \qquad \Delta H^\circ = -155\ kJ = \Delta H^\circ_{f,CuO(s)}$$

87.   a. $C_2H_4(g) + O_3(g) \rightarrow CH_3CHO(g) + O_2(g), \qquad \Delta H^\circ = -166\ kJ - [143\ kJ + 52\ kJ] = -361\ kJ$

       b. $O_3(g) + NO(g) \rightarrow NO_2(g) + O_2(g), \qquad \Delta H^\circ = 34\ kJ - [90.\ kJ + 143\ kJ] = -199\ kJ$

       c. $SO_3(g) + H_2O(l) \rightarrow H_2SO_4(aq), \qquad \Delta H^\circ = -909\ kJ - [-396\ kJ + (-286\ kJ)] = -227\ kJ$

       d. $2\ NO(g) + O_2(g) \rightarrow 2\ NO_2(g), \qquad \Delta H^\circ = 2(34)\ kJ - 2(90.)\ kJ = -112\ kJ$

Challenge Problems

88.   Only when there is a volume change can PV work be done. In pathway 1 (steps 1+2), only the first step does PV work (step 2 has a constant volume of 30.0 L). In pathway 2 (steps 3+4), only step 4 does PV work (step 3 has a constant volume of 10.0 L).

Pathway 1: $w = -P\Delta V = -2.00\ atm\ (30.0\ L - 10.0\ L) = -40.0\ L\ atm \times \dfrac{101.3\ J}{L\ atm} = -4.05 \times 10^3\ J$

Pathway 2: $w = -P\Delta V = -1.00\ atm\ (30.0\ L - 10.0\ L) = -20.0\ L\ atm\ \dfrac{101.3\ J}{L\ atm} = -2.03 \times 10^3\ J$

Note: The sign is (-) because the system is doing work on the surroundings (an expansion).

We get a different value of work for the two pathways which both have the same initial and final states. Since w depends on the pathway, work cannot be a state function.

89. a. $C_{12}H_{22}O_{11}(s) + 12\ O_2(g) \rightarrow 12\ CO_2(g) + 11\ H_2O(l)$

b. A bomb calorimeter is at constant volume, so heat released $= q_v = \Delta E$:

$$\Delta E = \frac{-24.00\ kJ}{1.46\ g} \times \frac{342.30\ g}{mol} = -5630\ kJ/mol\ C_{12}H_{22}O_{11}$$

c. Since $PV = nRT$, then $P\Delta V = RT\Delta n$ where $\Delta n =$ mol gaseous products - mol gaseous reactants.

$$\Delta H = \Delta E + P\Delta V = \Delta E + RT\Delta n$$

For this reaction, $\Delta n = 12 - 12 = 0$, so $\Delta H = \Delta E = -5630\ kJ/mol$.

90. Energy needed $= \dfrac{20. \times 10^3\ g\ C_{12}H_{22}O_{11}}{hr} \times \dfrac{1\ mol\ C_{12}H_{22}O_{11}}{342.30\ g\ C_{12}H_{22}O_{11}} \times \dfrac{5640\ kJ}{mol} = 3.3 \times 10^5\ kJ/hr$

Energy from sun $= 1.0\ kW/m^2 = 1000\ W/m^2 = \dfrac{1000\ J}{s\ m^2} = \dfrac{1.0\ kJ}{s\ m^2}$

$$10,000\ m^2 \times \frac{1.0\ kJ}{s\ m^2} \times \frac{60\ s}{min} \times \frac{60\ min}{hr} = 3.6 \times 10^7\ kJ/hr$$

% efficiency $= \dfrac{\text{Energy used per hour}}{\text{Total energy per hour}} \times 100 = \dfrac{3.3 \times 10^5\ kJ}{3.6 \times 10^7\ kJ} \times 100 = 0.92\%$

91. Energy used in 8.0 hours $= 40.\ kWh = \dfrac{40.\ kJ\ h}{s} \times \dfrac{3600\ s}{h} = 1.4 \times 10^5\ kJ$

Energy from the sun in 8.0 hours $= \dfrac{1.0\ kJ}{s\ m^2} \times \dfrac{60\ s}{min} \times \dfrac{60\ min}{h} \times 8.0\ h = 2.9 \times 10^4\ kJ/m^2$

Only 13% of the sunlight is converted into electricity:

$0.13 \times (2.9 \times 10^4\ kJ/m^2) \times \text{Area} = 1.4 \times 10^5\ kJ,\quad \text{Area} = 37\ m^2$

92. a. $2\ HNO_3(aq) + Na_2CO_3(s) \rightarrow 2\ NaNO_3(aq) + H_2O(l) + CO_2(g)$

$\Delta H° = [2(-467\ kJ) + (-286\ kJ) + (-393.5\ kJ)] - [2(-207\ kJ) + (-1131\ kJ)] = -69\ kJ$

2.0×10^4 gallons $\times \dfrac{4\ qt}{gal} \times \dfrac{946\ mL}{qt} \times \dfrac{1.42\ g}{mL} = 1.1 \times 10^8$ g of concentrated nitric acid solution

1.1×10^8 g solution $\times \dfrac{70.0\ g\ HNO_3}{100.0\ g\ solution} = 7.7 \times 10^7\ g\ HNO_3$

$7.7 \times 10^7\ g\ HNO_3 \times \dfrac{1\ mol}{63.02\ g} \times \dfrac{1\ mol\ Na_2CO_3}{2\ mol\ HNO_3} \times \dfrac{105.99\ g\ Na_2CO_3}{mol\ Na_2CO_3} = 6.5 \times 10^7\ g\ Na_2CO_3$

There are $(7.7 \times 10^7/63.02)$ mol of HNO_3 from the previous calculation. There are 69 kJ of heat evolved for every two moles of nitric acid neutralized. Combining these two results:

$$7.7 \times 10^7 \text{ g } HNO_3 \times \frac{1 \text{ mol } HNO_3}{63.02 \text{ g } HNO_3} \times \frac{-69 \text{ kJ}}{2 \text{ mol } HNO_3} = -4.2 \times 10^7 \text{ kJ}$$

b. They feared the heat generated by the neutralization reaction would vaporize the unreacted nitric acid, causing widespread airborne contamination.

93. $400 \text{ kcal} \times \dfrac{4.18 \text{ kJ}}{\text{kcal}} = 1.67 \times 10^3 \text{ kJ} \approx 2 \times 10^3 \text{ kJ}$

$$PE = mgz = \left(180 \text{ lb} \times \frac{1 \text{ kg}}{2.205 \text{ lb}} \right) \times \frac{9.80 \text{ m}}{s^2} \times \left(8 \text{ in} \times \frac{2.54 \text{ cm}}{\text{in}} \times \frac{1 \text{ m}}{100 \text{ cm}} \right) = 160 \text{ J} \approx 200 \text{ J}$$

200 J of energy are needed to climb one step. The total number of steps to climb are:

$$2 \times 10^6 \text{ J} \times \frac{1 \text{ step}}{200 \text{ J}} = 1 \times 10^4 \text{ steps}$$

94. $H_2(g) + 1/2\ O_2(g) \rightarrow H_2O(1)\ \ \Delta H = \Delta H^{\circ}_{f,\ H_2O(l)} = -285.8 \text{ kJ}$

$w = -P\Delta V$; Since $PV = nRT$, then at constant T and P, $P\Delta V = RT\Delta n$ where Δn = mol gaseous products - mol gaseous reactants.

For: $2\ H_2O(1) \rightarrow 2\ H_2(g) + O_2(g)$, $\Delta H = -2(-285.8 \text{ kJ}) = 571.6 \text{ kJ}$ and $\Delta n = 3 - 0 = 3$

$\Delta H = \Delta E + P\Delta V = \Delta E + RT\Delta n$, $\Delta E = \Delta H - RT\Delta n = 571.6 \times 10^3 \text{ J} - 8.3145 \text{ J/mol} \bullet \text{K } (298 \text{ K})(3 \text{ mol})$

$\Delta E = 5.716 \times 10^5 \text{ J} - 7430 \text{ J} = 5.642 \times 10^5 \text{ J} = 564.2 \text{ kJ}$

CHAPTER SEVEN

ATOMIC STRUCTURE AND PERIODICITY

Questions

15.   Planck's discovery that heated bodies give off only certain frequencies of light and Einstein's study of the photoelectric effect.

16.   When something is quantized, it can only have certain discrete values. In the Bohr model of the H-atom, the energy of the electron is quantized.

17.   Only very small particles with a tiny mass exhibit wave and particle properties, e.g., an electron. Some evidence supporting the wave properties of matter are:

   1) Electrons can be diffracted like light.

   2) The electron microscope uses electrons in a fashion similar to the way in which light is used in a light microscope.

18.   a. A discrete bundle of light energy.

   b. A number describing a discrete energy state of an electron.

   c. The lowest energy state of the electron(s) in an atom or ion.

   d. An allowed energy state that is higher in energy than the ground state.

19.   n: Gives the energy (it completely specifies the energy only for the H-atom or ions with one electron) and the relative size of the orbitals.

   ℓ: Gives the type (shape) of orbital.

   m_ℓ: Gives information about the direction in which the orbital is pointing.

20.   The 2p orbitals differ from each other in the direction in which they point in space. The 2p and 3p orbitals differ from each other in their size, energy and number of nodes.

21.   A nodal surface in an atomic orbital is a surface in which the probability of finding an electron is zero.

22.   The electrostatic energy of repulsion, from Coulomb's Law, will be of the form Q^2/r where Q is the charge of the electron and r is the distance between the two electrons. From the Heisenberg uncertainty principle, we cannot know precisely the position of each electron. Thus, we cannot know precisely the distance between the electrons nor the value of the electrostatic repulsions.

23.   No, the spin is a convenient model. Since we cannot locate or "see" the electron, we cannot see if it is spinning.

24.   There is a higher probability of finding the 4s electron very close to the nucleus than the 3d electron.

25.   The outermost electrons are the valence electrons. When atoms interact with each other, it will be the outermost electrons that are involved in these interactions. In addition, how tightly the nucleus holds these outermost electrons determines atomic size, ionization energy and other properties of atoms.

        Elements in the same group have similar valence electron configurations and, as a result, have similar chemical properties.

26.   If one more electron is added to a half-filled subshell, electron-electron repulsions will increase since two electrons must now occupy the same atomic orbital. This may slightly decrease the stability of the atom. Hence, half-filled subshells minimize electron-electron repulsions.

27.   Across a period, the positive charge from the nucleus increases as protons are added. The number of electrons also increase, but these outer electrons do not completely shield the increasing nuclear charge from each other. The general result is that the outer electrons are more strongly bound as one goes across a period which results in larger ionization energies (and smaller size).

        Aluminum is out of order because the electrons in the filled 3s orbital shield some of the nuclear charge from the 3p electron. Hence, the 3p electron is less tightly bound than a 3s electron resulting in a lower ionization energy for aluminum as compared to magnesium. The ionization energy of sulfur is lower than phosphorus because of the extra electron-electron repulsions in the doubly occupied sulfur 3p orbital. These added repulsions, which are not present in phosphorus, make it slightly easier to remove an electron from sulfur as compared to phosphorus.

28.   As successive electrons are removed, the net positive charge on the resultant ion increases. This increase in positive charge binds the remaining electrons more firmly, and the ionization energy increases.

        The electron configuration for Si is $1s^2 2s^2 2p^6 3s^2 3p^2$. There is a large jump in ionization energy when going from the removal of valence electrons to the removal of core electrons. For silicon, this occurs when the fifth electron is removed since we go from the valence electrons in $n = 3$ to the core electrons in $n = 2$. There should be another big jump when the thirteenth electron is removed, i.e., when the 1s electrons are removed.

29.   Ionization energy: $P(g) \rightarrow P^+(g) + e^-$;   electron affinity: $P(g) + e^- \rightarrow P^-(g)$

30.   Size decreases from left to right and increases going down the periodic table. So, going one element right and one element down would result in a similar size for the two elements diagonal to each other. The ionization energies will be similar for the diagonal elements since the periodic trends also oppose each other. Electron affinities are harder to predict, but atoms with similar size and ionization energy should also have similar electron affinities.

31.   The valence electrons are strongly attracted to the nucleus for elements with large ionization energies. One would expect these species to readily accept another electron and have very exothermic electron affinities. The noble gases are an exception; they have a large IE but have an endothermic EA. Noble gases have a stable arrangement of electrons. Adding an electron disrupts this stable arrangement, resulting in unfavorable electron affinities.

32.   Electron-electron repulsions become more important when we try to add electrons to an atom. From the standpoint of electron-electron repulsions, larger atoms would have more favorable (more exothermic) electron affinities. Considering only electron-nucleus attractions, smaller atoms would be expected to have the more favorable (more exothermic) EA's. These trends are exactly the opposite of each other. Thus, the overall variation in EA is not as great as ionization energy in which attractions to the nucleus dominate.

33.   For hydrogen and one-electron ions (hydrogenlike ions), all atomic orbitals with the same n value have the same energy. For polyatomic atoms/ions, the energy of the atomic orbitals also depends on ℓ. Since there are more nondegenerate energy levels for polyatomic atoms/ions as compared to hydrogen, there are many more possible electronic transitions resulting in more complicated line spectra.

34.   Each element has a characteristic spectrum. Thus, the presence of the characteristic spectral lines of an element confirms its presence in any particular sample.

35.   Yes, the maximum number of unpaired electrons in any configuration corresponds to a minimum in electron-electron repulsions.

36.   The electron is no longer part of that atom. The proton and electron are completely separated.

37.   Ionization energy is for removal of the electron from the atom in the gas phase. The work function is for the removal of an electron from the solid.

       $M(g) \rightarrow M^+(g) + e^-$ ionization energy; $M(s) \rightarrow M^+(s) + e^-$ work function

38. Li^+ ions will be the smallest of the alkali metal cations and will be most strongly attracted to the water molecules.

Exercises

Light and Matter

39. 1×10^{-9} m = 1 nm; $\nu = \dfrac{c}{\lambda} = \dfrac{2.998 \times 10^8 \text{ m/s}}{780. \times 10^{-9} \text{ m}} = 3.84 \times 10^{14} \text{ s}^{-1}$

40.   99.5 MHz = 99.5×10^6 Hz = 99.5×10^6 s^{-1}; $\lambda = \dfrac{c}{\nu} = \dfrac{2.998 \times 10^8 \text{ m/s}}{99.5 \times 10^6 \text{ s}^{-1}} = 3.01$ m

41. $\nu = \dfrac{c}{\lambda} = \dfrac{3.00 \times 10^8 \text{ m/s}}{1.0 \times 10^{-2} \text{ m}} = 3.0 \times 10^{10} \text{ s}^{-1}$

$E = h\nu = 6.63 \times 10^{-34} \text{ J s} \times 3.0 \times 10^{10} \text{ s}^{-1} = 2.0 \times 10^{-23}$ J/photon

$\dfrac{2.0 \times 10^{-23} \text{ J}}{\text{photon}} \times \dfrac{6.02 \times 10^{23} \text{ photons}}{\text{mol}} = 12$ J/mol

42. $E = h\nu = \dfrac{hc}{\lambda} = \dfrac{6.63 \times 10^{-34} \text{ J s} \times 3.00 \times 10^8 \text{ m/s}}{25 \text{ nm} \times \dfrac{1 \text{ m}}{1 \times 10^9 \text{ nm}}} = 8.0 \times 10^{-18}$ J/photon

$\dfrac{8.0 \times 10^{-18} \text{ J}}{\text{photon}} \times \dfrac{6.02 \times 10^{23} \text{ photons}}{\text{mol}} = 4.8 \times 10^6$ J/mol

43.   The wavelength is the distance between consecutive wave peaks. Wave a shows 4 wavelengths and wave b shows 8 wavelengths.

Wave a: $\lambda = \dfrac{1.6 \times 10^{-3} \text{ m}}{4} = 4.0 \times 10^{-4}$ m

Wave b: $\lambda = \dfrac{1.6 \times 10^{-3} \text{ m}}{8} = 2.0 \times 10^{-4}$ m

Wave a has the longer wavelength. Since frequency and photon energy are both inversely proportional to wavelength, then wave b will have the higher frequency and larger photon energy since it has the shorter wavelength.

$\nu = \dfrac{c}{\lambda} = \dfrac{3.00 \times 10^8 \text{ m/s}}{2.0 \times 10^{-4} \text{ m}} = 1.5 \times 10^{12} \text{ s}^{-1}$

$E = \dfrac{hc}{\lambda} = \dfrac{6.63 \times 10^{-34} \text{ J s} \times 3.00 \times 10^8 \text{ m/s}}{2.0 \times 10^{-4} \text{ m}} = 9.9 \times 10^{-22}$ J

Since both waves are examples of electromagnetic radiation, then both waves travel at the same speed, c, the speed of light. From Figure 7.2 of the text, both of these waves represent infrared electromagnetic radiation.

44.   Referencing figure 7.2 of the text, 2.12×10^{-10} m electromagnetic radiation is X-rays.

$$\lambda = \frac{c}{\nu} = \frac{2.9979 \times 10^8 \text{ m/s}}{107.1 \times 10^6 \text{ s}^{-1}} = 2.799 \text{ m}$$

From the wavelength calculated above, 107.1 MHz electromagnetic radiation is FM radiowaves.

$$\lambda = \frac{hc}{E} = \frac{6.626 \times 10^{-34} \text{ J s} \times 2.998 \times 10^8 \text{ m/s}}{3.97 \times 10^{-19} \text{ J}} = 5.00 \times 10^{-7} \text{ m}$$

The 3.97×10^{-19} J/photon electromagnetic radiation is visible (green) light.

The photon energy and frequency order will be the exact opposite of the wavelength ordering because E and ν are both inversely related to λ. From the calculated wavelengths above, the order of photon energy and frequency is:

FM radiowaves < visible (green) light < X-rays
longest λ shortest λ
lowest ν highest ν
smallest E   largest E

45.   The energy needed to remove a single electron is:

$$\frac{279.7 \text{ kJ}}{\text{mol}} \times \frac{1 \text{ mol}}{6.0221 \times 10^{23}} = 4.645 \times 10^{-22} \text{ kJ} = 4.645 \times 10^{-19} \text{ J}$$

$$E = \frac{hc}{\lambda}, \ \lambda = \frac{hc}{E} = \frac{6.6261 \times 10^{-34} \text{ J s} \times 2.9979 \times 10^8 \text{ m/s}}{4.645 \times 10^{-19} \text{ J}} = 4.277 \times 10^{-7} \text{ m} = 427.7 \text{ nm}$$

46. $$\frac{208.4 \text{ kJ}}{\text{mol}} \times \frac{1 \text{ mol}}{6.0221 \times 10^{23}} = 3.461 \times 10^{-22} \text{ kJ} = 3.461 \times 10^{-19} \text{ J to remove one electron}$$

$$E = \frac{hc}{\lambda}, \ \lambda = \frac{hc}{E} = \frac{6.6261 \times 10^{-34} \text{ J s} \times 2.9979 \times 10^8 \text{ m/s}}{3.461 \times 10^{-19} \text{ J}} = 5.739 \times 10^{-7} \text{ m} = 573.9 \text{ nm}$$

47.   a.   90.% of speed of light = $0.90 \times 3.00 \times 10^8$ m/s = 2.7×10^8 m/s

$$\lambda = \frac{h}{mv}, \ \lambda = \frac{6.63 \times 10^{-34} \text{ J s}}{1.67 \times 10^{-27} \text{ kg} \times 2.7 \times 10^8 \text{ m/s}} = 1.5 \times 10^{-15} \text{ m} = 1.5 \times 10^{-6} \text{ nm}$$

Note: For units to come out, the mass must be in kg since $1 \text{ J} = \frac{1 \text{ kg m}^2}{\text{s}^2}$.

b. $$\lambda = \frac{h}{mv} = \frac{6.63 \times 10^{-34} \text{ J s}}{0.15 \text{ kg} \times 10.0 \text{ m/s}} = 4.4 \times 10^{-34} \text{ m} = 4.4 \times 10^{-25} \text{ nm}$$

This number is so small that it is essentially zero. We cannot detect a wavelength this small. The meaning of this number is that we do not have to consider the wave properties of large objects.

48. a. $\lambda = \dfrac{h}{mv} = \dfrac{6.626 \times 10^{-34} \text{ J s}}{1.675 \times 10^{-27} \text{ kg} \times (0.0100 \times 2.998 \times 10^8 \text{ m/s})} = 1.32 \times 10^{-13} \text{ m}$

   b. $\lambda = \dfrac{h}{mv}$, $v = \dfrac{h}{\lambda m} = \dfrac{6.626 \times 10^{-34} \text{ J s}}{75 \times 10^{-12} \text{ m} \times 1.675 \times 10^{-27} \text{ kg}} = 5.3 \times 10^3 \text{ m/s}$

49. $\lambda = \dfrac{h}{mv} = \dfrac{6.63 \times 10^{-34} \text{ J s}}{9.11 \times 10^{-31} \text{ kg} \times (1.0 \times 10^{-3} \times 3.00 \times 10^8 \text{ m/s})} = 2.4 \times 10^{-9} \text{ m} = 2.4 \text{ nm}$

50. $\lambda = \dfrac{h}{mv}$, $v = \dfrac{h}{\lambda m}$; For $\lambda = 1.0 \times 10^2$ nm $= 1.0 \times 10^{-7}$ m:

   $v = \dfrac{6.63 \times 10^{-34} \text{ J s}}{9.11 \times 10^{-31} \text{ kg} \times 1.0 \times 10^{-7} \text{ m}} = 7.3 \times 10^3 \text{ m/s}$

   For $\lambda = 1.0$ nm $= 1.0 \times 10^{-9}$ m: $v = \dfrac{6.63 \times 10^{-34} \text{ J s}}{9.11 \times 10^{-31} \text{ kg} \times 1.0 \times 10^{-9} \text{ m}} = 7.3 \times 10^5 \text{ m/s}$

Hydrogen Atom: The Bohr Model

51. For the H atom (Z = 1): $E_n = -2.178 \times 10^{-18} \text{ J}/n^2$; For a spectral transition, $\Delta E = E_f - E_i$:

$$\Delta E = -2.178 \times 10^{-18} \text{ J} \left(\dfrac{1}{n_f^2} - \dfrac{1}{n_i^2} \right)$$

where n_i and n_f are the levels of the initial and final states, respectively. A positive value of ΔE always corresponds to an absorption of light, and a negative value of ΔE always corresponds to an emission of light.

   a. $\Delta E = -2.178 \times 10^{-18} \text{ J} \left(\dfrac{1}{2^2} - \dfrac{1}{3^2} \right) = -2.178 \times 10^{-18} \text{ J} \left(\dfrac{1}{4} - \dfrac{1}{9} \right)$

   $\Delta E = -2.178 \times 10^{-18} \text{ J} \times (0.2500 - 0.1111) = -3.025 \times 10^{-19} \text{ J}$

   The photon of light must have precisely this energy (3.025×10^{-19} J).

   $|\Delta E| = E_{photon} = h\nu = \dfrac{hc}{\lambda}$ or $\lambda = \dfrac{hc}{|\Delta E|} = \dfrac{6.6261 \times 10^{-34} \text{ J s} \times 2.9979 \times 10^8 \text{ m/s}}{3.025 \times 10^{-19} \text{ J}}$

   $= 6.567 \times 10^{-7} \text{ m} = 656.7 \text{ nm}$

   From Figure 7.2, this is visible electromagnetic radiation (red light).

b. $\Delta E = -2.178 \times 10^{-18} \text{ J} \left(\dfrac{1}{2^2} - \dfrac{1}{4^2} \right) = -4.084 \times 10^{-19} \text{ J}$

$$\lambda = \frac{hc}{|\Delta E|} = \frac{6.6261 \times 10^{-34} \text{ J s} \times 2.9979 \times 10^8 \text{ m/s}}{4.084 \times 10^{-19} \text{ J}} = 4.864 \times 10^{-7} \text{ m} = 486.4 \text{ nm}$$

This is visible electromagnetic radiation (green-blue light).

c. $\Delta E = -2.178 \times 10^{-18} \text{ J} \left(\dfrac{1}{1^2} - \dfrac{1}{2^2} \right) = -1.634 \times 10^{-18} \text{ J}$

$$\lambda = \frac{6.6261 \times 10^{-34} \text{ J s} \times 2.9979 \times 10^8 \text{ m/s}}{1.634 \times 10^{-18} \text{ J}} = 1.216 \times 10^{-7} \text{ m} = 121.6 \text{ nm}$$

This is ultraviolet electromagnetic radiation.

52.   a. $\Delta E = -2.178 \times 10^{-18} \text{ J} \left(\dfrac{1}{3^2} - \dfrac{1}{4^2} \right) = -1.059 \times 10^{-19} \text{ J}$

$$\lambda = \frac{hc}{|\Delta E|} = \frac{6.6261 \times 10^{-34} \text{ J s} \times 2.9979 \times 10^8 \text{ m/s}}{1.059 \times 10^{-19} \text{ J}} = 1.876 \times 10^{-6} \text{ m} = 1876 \text{ nm}$$

From Figure 7.2, this is infrared electromagnetic radiation.

b. $\Delta E = -2.178 \times 10^{-18} \text{ J} \left(\dfrac{1}{4^2} - \dfrac{1}{5^2} \right) = -4.901 \times 10^{-20} \text{ J}$

$$\lambda = \frac{hc}{|\Delta E|} = \frac{6.6261 \times 10^{-34} \text{ J s} \times 2.9979 \times 10^8 \text{ m/s}}{4.901 \times 10^{-20} \text{ J}} = \lambda = 4.053 \times 10^{-6} \text{ m} = 4053 \text{ nm (infrared)}$$

c. $\Delta E = -2.178 \times 10^{-18} \text{ J} \left(\dfrac{1}{3^2} - \dfrac{1}{5^2} \right) = -1.549 \times 10^{-19} \text{ J}$

$$\lambda = \frac{hc}{|\Delta E|} = \frac{6.6261 \times 10^{-34} \text{ J s} \times 2.9979 \times 10^8 \text{ m/s}}{1.549 \times 10^{-19} \text{ J}} = 1.282 \times 10^{-6} \text{ m} = 1282 \text{ nm (infrared)}$$

53.

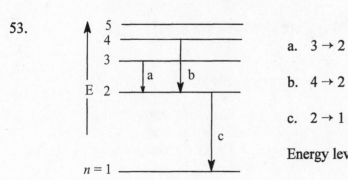

a. $3 \rightarrow 2$

b. $4 \rightarrow 2$

c. $2 \rightarrow 1$

Energy levels are not to scale.

54.

```
     5 ──────────────
     4 ──── b↓ ──────
     3 ──── ↓a ──── ↓c ──
                              a.  $4 \to 3$
  E  2 ──────────────
                              b.  $5 \to 4$
     │
     n = 1 ──────────────    c.  $5 \to 3$;  Energy levels are not to scale.
```

55. $\Delta E = -2.178 \times 10^{-18} \text{ J} \left(\dfrac{1}{n_f^2} - \dfrac{1}{n_i^2} \right) = -2.178 \; 10^{-18} \text{ J} \left(\dfrac{1}{5^2} - \dfrac{1}{1^2} \right) = 2.091 \times 10^{-18} \text{ J} = E_{photon}$

$\lambda = \dfrac{hc}{E} = \dfrac{6.6261 \times 10^{-34} \text{ J s} \times 2.9979 \times 10^8 \text{ m/s}}{2.091 \times 10^{-18} \text{ J}} = 9.500 \times 10^{-8} \text{ m} = 95.00 \text{ nm}$

Since wavelength and energy are inversely related, visible light ($\lambda \approx 400 - 700$ nm) is not energetic enough to excite an electron in hydrogen from $n = 1$ to $n = 5$.

$\Delta E = -2.178 \times 10^{-18} \text{ J} \left(\dfrac{1}{6^2} - \dfrac{1}{2^2} \right) = 4.840 \times 10^{-19} \text{ J}$

$\lambda = \dfrac{hc}{E} = \dfrac{6.6261 \times 10^{-34} \text{ J s} \times 2.9979 \times 10^8 \text{ m/s}}{4.840 \times 10^{-19} \text{ J}} = 4.104 \times 10^{-7} \text{ m} = 410.4 \text{ nm}$

Visible light with $\lambda = 410.4$ nm will excite an electron from the $n = 2$ to the $n = 6$ energy level.

56. Ionization from $n = 1$ corresponds to the transition $n_i = 1 \to n_f = \infty$ where $E_\infty = 0$.

$\Delta E = E_\infty - E_1 = -E_1 = 2.178 \times 10^{-18} \left(\dfrac{1}{1^2} \right) = 2.178 \times 10^{-18} \text{ J} = E_{photon}$

$\lambda = \dfrac{hc}{E} = \dfrac{6.6261 \times 10^{-34} \text{ J s} \times 2.9979 \times 10^8 \text{ m/s}}{2.178 \times 10^{-18} \text{ J}} = 9.120 \times 10^{-8} \text{ m} = 91.20 \text{ nm}$

To ionize from $n = 2$, $\Delta E = E_\infty - E_2 = -E_2 = 2.178 \times 10^{-18} \left(\dfrac{1}{2^2} \right) = 5.445 \times 10^{-19} \text{ J}$

$\lambda = \dfrac{6.6261 \times 10^{-34} \text{ J s} \times 2.9979 \times 10^8 \text{ m/s}}{5.445 \times 10^{-19} \text{ J}} = 3.648 \times 10^{-7} \text{ m} = 364.8 \text{ nm}$

57. $\Delta E = h\nu = 6.6261 \times 10^{-34} \text{ J s} \times 1.141 \times 10^{14} \text{ s}^{-1} = 7.560 \times 10^{-20} \text{ J}$

$\Delta E = -7.560 \times 10^{-20} \text{ J}$ since light is emitted.

$\Delta E = E_4 - E_n, \; -7.560 \times 10^{-20} \text{ J} = -2.178 \times 10^{-18} \text{ J} \left(\dfrac{1}{4^2} - \dfrac{1}{n^2} \right)$

$3.471 \times 10^{-2} = 6.250 \times 10^{-2} - \dfrac{1}{n^2}, \; \dfrac{1}{n^2} = 2.779 \times 10^{-2}, \; n^2 = 35.98, \; n = 6$

58. $|\Delta E| = E_{photon} = \dfrac{hc}{\lambda} = \dfrac{6.6261 \times 10^{-34}\,J\,s \times 2.9979 \times 10^8\,m/s}{397.2 \times 10^{-9}\,m} = 5.001 \times 10^{-19}\,J$

$\Delta E = -5.001 \times 10^{-19}\,J$ since we have an emission.

$-5.001 \times 10^{-19}\,J = E_2 - E_n = -2.178 \times 10^{-18}\,J \left(\dfrac{1}{2^2} - \dfrac{1}{n^2} \right)$

$0.2296 = \dfrac{1}{4} - \dfrac{1}{n^2}, \quad \dfrac{1}{n^2} = 0.0204, \quad n = 7$

Quantum Mechanics, Quantum Numbers, and Orbitals

59.   a. $\Delta(mv) = m\Delta v = 9.11 \times 10^{-31}\,kg \times 0.100\,m/s = \dfrac{9.11 \times 10^{-32}\,kg\,m}{s}$

$\Delta(mv) \bullet \Delta x \geq \dfrac{h}{4\pi}, \quad \Delta x = \dfrac{h}{4\pi\Delta(mv)} = \dfrac{6.626 \times 10^{-34}\,J\,s}{4 \times 3.142 \times (9.11 \times 10^{-32}\,kg\,m/s)} = 5.79 \times 10^{-4}\,m$

b. $\Delta x = \dfrac{h}{4\pi m\Delta v} = \dfrac{6.626 \times 10^{-34}\,J\,s}{4 \times 3.142 \times 0.145\,kg \times 0.100\,m/s} = 3.64 \times 10^{-33}\,m$

c.   The diameter of an H atom is roughly 1.0×10^{-8} cm. The uncertainty in position is much larger than the size of the atom.

d.   The uncertainty is insignificant compared to the size of a baseball.

60.   Units of $\Delta E \bullet \Delta t = J \times s$, the same as the units of Planck's constant.

Units of $\Delta(mv) \bullet \Delta x = kg \times \dfrac{m}{s} \times m = \dfrac{kg\,m^2}{s} = \dfrac{kg\,m^2}{s^2} \times s = J \times s$

61. $n = 1, 2, 3, \dots; \quad \ell = 0, 1, 2, \dots (n - 1); \quad m_\ell = -\ell \dots -2, -1, 0, 1, 2, \dots +\ell$

62.   1p: $n = 1, \ell = 1$ is not possible; 3f: $n = 3, \ell = 3$ is not possible; 2d: $n = 2, \ell = 2$ is not possible; In all three incorrect cases, $n = \ell$. The maximum value ℓ can have is $n - 1$, not n.

63.   b. ℓ must be smaller than n.   d.   For $\ell = 0$, $m_\ell = 0$ is the only allowed value.

64.   b.   For $\ell = 3$, m_ℓ can range from -3 to +3; thus +4 is not allowed.

c. n cannot equal zero.   d. ℓ cannot be a negative number.

65. ψ^2 gives the probability of finding the electron at that point.

66.   The diagrams of the orbitals in the text give only 90% probabilities of where the electron may reside. We can never be 100% certain of the location of the electrons due to Heisenburg's uncertainty principle.

Polyelectronic Atoms

67.   5p: three orbitals; $3d_{z^2}$: one orbital;   4d: five orbitals

$n = 5$: $\ell = 0$ (1 orbital), $\ell = 1$ (3 orbitals), $\ell = 2$ (5 orbitals), $\ell = 3$ (7 orbitals), $\ell = 4$ (9 orbitals)

Total for $n = 5$ is 25 orbitals.

$n = 4$: $\ell = 0$ (l), $\ell = 1$ (3), $\ell = 2$ (5), $\ell = 3$ (7);   Total for $n = 4$ is 16 orbitals.

68.   1p, 0 electrons ($\ell \neq 1$ when $n = 1$); $6d_{x^2-y^2}$, 2 electrons (specifies one atomic orbital);   4f, 14 electrons (7 orbitals have 4f designation); $7p_y$, 2 electrons (specifies one atomic orbital);   2s, 2 electrons (specifies one atomic orbital); $n = 3$, 18 electrons (3s, 3p and 3d orbitals are possible; there are one 3s orbital, three 3p orbitals and five 3d orbitals).

69.   a. $n = 4$: ℓ can be 0, 1, 2, or 3.   Thus we have s (2 e⁻), p (6 e⁻), d (10 e⁻) and f (14 e⁻) orbitals present.   Total number of electrons to fill these orbitals is 32.

b. $n = 5$, $m_\ell = +1$: For $n = 5$, $\ell = 0, 1, 2, 3, 4$. For $\ell = 1, 2, 3, 4$, all can have $m_\ell = +1$. Four distinct orbitals, thus 8 electrons.

c. $n = 5$, $m_s = +1/2$: For $n = 5$, $\ell = 0, 1, 2, 3, 4$. Number of orbitals = 1, 3, 5, 7, 9 for each value of ℓ, respectively. There are 25 orbitals with $n = 5$. They can hold 50 electrons and 25 of these electrons can have $m_s = +1/2$.

d. $n = 3$, $\ell = 2$: These quantum numbers define a set of 3d orbitals. There are 5 degenerate 3d orbitals which can hold a total of 10 electrons.

e. $n = 2$, $\ell = 1$: These define a set of 2p orbitals. There are 3 degenerate 2p orbitals which can hold a total of 6 electrons.

70.   a.   It is impossible for $n = 0$. Thus, no electrons can have this set of quantum numbers.

b.   The four quantum numbers completely specify a single electron.

c. $n = 3$: 3s, 3p and 3d orbitals all have $n = 3$. These orbitals can hold up to 18 electrons.

d. $n = 2$, $\ell = 2$: This combination is not possible ($\ell \neq 2$ for $n = 2$). Zero electrons in an atom can have these quantum numbers.

e. $n = 1$, $\ell = 0$, $m_\ell = 0$: These define a 1s orbital which can hold 2 electrons.

71.   Si: $1s^2 2s^2 2p^6 3s^2 3p^2$ or [Ne]$3s^2 3p^2$; Ga: $1s^2 2s^2 2p^6 3s^2 3p^6 4s^2 3d^{10} 4p^1$ or [Ar]$4s^2 3d^{10} 4p^1$

As: [Ar]$4s^2 3d^{10} 4p^3$; Ge: [Ar]$4s^2 3d^{10} 4p^2$; Al: [Ne]$3s^2 3p^1$; Cd: [Kr]$5s^2 4d^{10}$

S: [Ne]$3s^2 3p^4$; Se: [Ar]$4s^2 3d^{10} 4p^4$

72.   Cu: $[Ar]4s^23d^9$ (using periodic table), $[Ar]4s^13d^{10}$ (actual)

O: $1s^22s^22p^4$; La: $[Xe]6s^25d^1$; Y: $[Kr]5s^24d^1$; Ba: $[Xe]6s^2$

Tl: $[Xe]6s^24f^{14}5d^{10}6p^1$; Bi: $[Xe]6s^24f^{14}5d^{10}6p^3$

73.   The following are complete electron configurations. Noble gas shorthand notation could also be used.

Sc: $1s^22s^22p^63s^23p^64s^23d^1$; Fe: $1s^22s^22p^63s^23p^64s^23d^6$

P: $1s^22s^22p^63s^23p^3$; Cs: $1s^22s^22p^63s^23p^64s^23d^{10}4p^65s^24d^{10}5p^66s^1$

Eu: $1s^22s^22p^63s^23p^64s^23d^{10}4p^65s^24d^{10}5p^66s^24f^65d^1$*

Pt: $1s^22s^22p^63s^23p^64s^23d^{10}4p^65s^24d^{10}5p^66s^24f^{14}5d^8$*

Xe: $1s^22s^22p^63s^23p^64s^23d^{10}4p^65s^24d^{10}5p^6$; Br: $1s^22s^22p^63s^23p^64s^23d^{10}4p^5$

*Note: These electron configurations were predicted using only the periodic table.

Actual electron configurations are: Eu: $[Xe]6s^24f^7$ and Pt: $[Xe]6s^14f^{14}5d^9$

74.   Cl: $1s^22s^22p^63s^23p^5$ or $[Ne]3s^23p^5$ As: $1s^22s^22p^63s^23p^64s^23d^{10}4p^3$ or $[Ar]4s^23d^{10}4p^3$

Sr: $1s^22s^22p^63s^23p^64s^23d^{10}4p^65s^2$ or $[Kr]5s^2$ W: $[Xe]6s^24f^{14}5d^4$

Pb: $[Xe]6s^24f^{14}5d^{10}6p^2$ Cf: $[Rn]7s^25f^{10}$

Predicting electron configurations for lanthanide and actinide elements is difficult since they have 0, 1 or 2 electrons in d orbitals.

75.   Exceptions: Cr, Cu, Nb, Mo, Tc, Ru, Rh, Pd, Ag, Pt, Au; Tc, Ru, Rh, Pd and Pt do not correspond to the supposed extra stability of half-filled and filled subshells.

76.   No, in the solid and liquid state the electrons may interact (bonding occurs), and the experimental configuration will not be valid. That is, the experimental electron configurations are for isolated atoms; in condensed phases, the atoms are no longer isolated.

77.   a.   Both In and I have one unpaired 5p electron, but only the nonmetal I would be expected to form a covalent compound with the nonmetal F. One would predict an ionic compound to form between the metal In and the nonmetal F.

I: $[Kr]5s^24d^{10}5p^5$ $\underline{\uparrow\downarrow}$ $\underline{\uparrow\downarrow}$ $\underline{\uparrow}$
                                          5p

b.   From the periodic table, this will be element 120.   Element 120: $[Rn]7s^25f^{14}6d^{10}7p^68s^2$

c. Rn: $[Xe]6s^24f^{14}5d^{10}6p^6$; Note that the next discovered noble gas will also have 4f electrons (as well as 5f electrons).

d. This is chromium, which is an exception to the predicted filling order. Cr has 6 unpaired electrons and the next most is 5 unpaired electrons for Mn.

Cr: $[Ar]4s^13d^5$ ↑   ↑ ↑ ↑ ↑ ↑
              4s   3d

78. a. As: $1s^22s^22p^63s^23p^64s^23d^{10}4p^3$

b. Element 116 will be below Po in the periodic table: $[Rn]7s^25f^{14}6d^{10}7p^4$

c. Ta: $[Xe]6s^24f^{14}5d^3$ or Ir: $[Xe]6s^24f^{14}5d^7$

d. At: $[Xe]6s^24f^{14}5d^{10}6p^5$.   Note that element 117 (when it is discovered) will also have electrons in the 6p atomic orbitals (as well as electrons in the 7p atomic orbitals).

79. B : $1s^22s^22p^1$

	n	ℓ	m_ℓ	m_s
1s	1	0	0	+1/2
1s	1	0	0	-1/2
2s	2	0	0	+1/2
2s	2	0	0	-1/2
2p*	2	1	-1	+1/2

*This is only one of several possibilities for the 2p electron. The 2p electron in B could have m_ℓ = -1, 0 or +1, and m_s = +1/2 or -1/2, a total of six possibilities.

N : $1s^22s^22p^3$

	n	ℓ	m_ℓ	m_s
1s	1	0	0	+1/2
1s	1	0	0	-1/2
2s	2	0	0	+1/2
2s	2	0	0	-1/2
2p	2	1	-1	+1/2
2p	2	1	0	+1/2
2p	2	1	+1	+1/2

(Or all 2p electrons could have m_s = -1/2.)

80.   Ti : $[Ar]4s^23d^2$

	n	ℓ	m_ℓ	m_s
4s	4	0	0	+1/2
4s	4	0	0	-1/2
3d	3	2	-2	+1/2
3d	3	2	-1	+1/2

Only one of 10 possible combinations of m_ℓ and m_s for the first d electron. For the ground state, the second d electron should be in a different orbital with spin parallel; 4 possibilities.

81.   O: $1s^22s^22p_x^22p_y^2$ ($\uparrow\downarrow$ $\uparrow\downarrow$ _); There are no unpaired electrons in this oxygen atom. This configuration would be an excited state, and in going to the more stable ground state ($\uparrow\downarrow$ $\uparrow$ $\uparrow$), energy would be released.

82.   The number of unpaired electrons is in parentheses.

a. excited state of boron   (1)   b. ground state of neon   (0)

   B ground state: $1s^22s^22p^1$ (1)   Ne ground state: $1s^22s^22p^6$

c. exited state of fluorine   (3)   d. excited state of iron   (6)

   F ground state: $1s^22s^22p^5$ (l)   Fe ground state: $[Ar]4s^23d^6$ (4)

   $\underline{\uparrow\downarrow}$ $\underline{\uparrow\downarrow}$ $\underline{\uparrow}$ $\underline{\uparrow\downarrow}$ $\underline{\uparrow}$ $\underline{\uparrow}$ $\underline{\uparrow}$ $\underline{\uparrow}$
        2p   3d

83.   We get the number of unpaired electrons by examining the incompletely filled subshells. The paramagnetic substances have unpaired electrons, and the ones with no unpaired electrons are not paramagnetic (they are called diamagnetic).

   Li: $1s^22s^1$ $\underline{\uparrow}$; Paramagnetic with 1 unpaired electron.
               2s

   N: $1s^22s^22p^3$ $\underline{\uparrow}$ $\underline{\uparrow}$ $\underline{\uparrow}$; Paramagnetic with 3 unpaired electrons.
                 2p

   Ni: $[Ar]4s^23d^8$ $\underline{\uparrow\downarrow}$ $\underline{\uparrow\downarrow}$ $\underline{\uparrow\downarrow}$ $\underline{\uparrow}$ $\underline{\uparrow}$; Paramagnetic with 2 unpaired electrons.
                  3d

   Te: $[Kr]5s^24d^{10}5p^4$ $\underline{\uparrow\downarrow}$ $\underline{\uparrow}$ $\underline{\uparrow}$; Paramagnetic with 2 unpaired electrons.
                      5p

   Ba: $[Xe]6s^2$ $\underline{\uparrow\downarrow}$; Not paramagnetic since no unpaired electrons.
             6s

   Hg: $[Xe]6s^24f^{14}5d^{10}$ $\underline{\uparrow\downarrow}$ $\underline{\uparrow\downarrow}$ $\underline{\uparrow\downarrow}$ $\underline{\uparrow\downarrow}$ $\underline{\uparrow\downarrow}$; Not paramagnetic since no unpaired electrons.
                     5d

84.   We get the number of unpaired electrons by examining the incompletely filled subshells.

O: $[He]2s^22p^4$ $2p^4$:   ↑↓ ↑ ↑   two unpaired e^-

O^+: $[He]2s^22p^3$ $2p^3$:   ↑ ↑ ↑   three unpaired e^-

O^-: $[He]2s^22p^5$ $2p^5$:   ↑↓ ↑↓ ↑   one unpaired e^-

Os: $[Xe]6s^24f^{14}5d^6$ $5d^6$:   ↑↓ ↑ ↑ ↑ ↑   four unpaired e^-

Zr: $[Kr]5s^24d^2$ $4d^2$:   ↑ ↑ _ _ _   two unpaired e^-

S: $[Ne]3s^23p^4$ $3p^4$:   ↑↓ ↑ ↑   two unpaired e^-

F: $[He]2s^22p^5$ $2p^5$:   ↑↓ ↑↓ ↑   one unpaired e^-

Ar: $[Ne]3s^23p^6$ $3p^6$:   ↑↓ ↑↓ ↑↓   zero unpaired e^-

The Periodic Table and Periodic Properties

85.   Size (radii) decreases left to right across the periodic table, and size increases from top to bottom of the periodic table.

    a.   Be < Mg < Ca   b.   Xe < I < Te   c.   Ge < Ga < In

86.   a.   Be < Na < Rb   b.   Ne < Se < Sr   c.   O < P < Fe
                                                        (All follow the general radii trend.)

87.   The ionization energy trend is the opposite of the radii trend; ionization energy (IE), in general, increases left to right across the periodic table and decreases from top to bottom of the periodic table.

    a.   Ca < Mg < Be   b.   Te < I < Xe   c.   In < Ga < Ge

88.   a.   Rb < Na < Be   b.   Sr < Se < Ne   c.   Fe < P < O   (All follow the general IE trend.)

89.   a.   Li   b.   P

    c. O^+. This ion has the fewest electrons as compared to the other oxygen species present. O^+ has the smallest amount of electron-electron repulsions, which makes it the smallest ion with the largest ionization energy.

    d.   From the radii trend, Ar < Cl < S and Kr > Ar. Since variation in size down a family is greater than the variation across a period, we would predict Cl to be the smallest of the three.

    e.   Cu

90.   a. Ba   b. K

       c. O; In general, group 6A elements have a lower ionization energy than neighboring group 5A
           elements. This is an exception to the general ionization energy trend across the periodic table.

       d. S^{2-}; This ion has the most electrons as compared to the other sulfur species present. S^{2-} has the
           largest amount of electron-electron repulsions which leads to S^{2-} having the largest size and
           smallest ionization energy.

       e. Cs; This follows the general ionization energy trend.

91.   a. 106: $[Rn]7s^2 5f^{14} 6d^4$ b. W   c. SgO_3 or Sg_2O_3 and SgO_4^{2-} or $Sg_2O_7^{2-}$
                                                            (similar to Cr; Sg = 106)

92.   a. Uus will have 117 electrons. $[Rn]7s^2 5f^{14} 6d^{10} 7p^5$

       b. It will be in the halogen family and most similar to astatine, At.

       c. Uus should form -1 charged anions like the other halogens.

           $NaUus$, $Mg(Uus)_2$, $C(Uus)_4$, $O(Uus)_2$

       d. Assuming Uus is like the other halogens: $UusO^-$, $UusO_2^-$, $UusO_3^-$, $UusO_4^-$

93.   As: $[Ar]4s^2 3d^{10} 4p^3$; Se: $[Ar]4s^2 3d^{10} 4p^4$; The general ionization energy trend predicts that Se
       should have a higher ionization energy than As. Se is an exception to the general ionization energy
       trend. There are extra electron-electron repulsions in Se because two electrons are in the same 4p
       orbital, resulting in a lower ionization energy for Se than predicted.

94.   Al (-44), Si(-120), P (-74), S (-200.4), Cl (-348.7); Based on the increasing nuclear charge, we
       would expect the electron affinity (EA) values to become more exothermic as we go from left to
       right in the period. Phosphorus is out of line. The reaction for the EA of P is:

$$P(g) + e^- \rightarrow P^-(g)$$

$$[Ne]3s^2 3p^3 \quad\quad [Ne]3s^2 3p^4$$

       The additional electron in P^- will have to go into an orbital that already has one electron. There
       will be greater repulsions between the paired electrons in P^-, causing the EA of P to be less
       favorable than predicted based solely on attractions to the nucleus.

95.   a. More favorable EA: C and Br; The electron affinity trend is very erratic. Both N and Ar have
           positive EA values (unfavorable) due to their electron configurations (see text for detailed
           explanation).

       b. Higher IE: N and Ar (follows the IE trend)

       c. Larger size: C and Br (follows the radii trend)

96.   a.   More favorable EA: K and Cl; Mg has a positive EA value, and F has a more positive EA value than expected from its position relative to Cl.

       b.   Higher IE: Mg and F   c.   Larger radius: K and Cl

97.   The electron affinity trend is very erratic. In general, EA decreases down the periodic table, and the trend across the table is too erratic to be of much use.

       a.   Se < S; S is more exothermic.   b.   I < Br < F < Cl; Cl is most exothermic (F is an exception).

98.   a.   N < O < F; F is most exothermic.   b.   Al < P < Si; Si is most exothermic.

99.   Electron-electron repulsions are much greater in O^- than in S^- because the electron goes into a smaller 2p orbital vs. the larger 3p orbital in sulfur. This results in a more favorable (more exothermic) EA for sulfur.

100.   O; The electron-electron repulsions will be much more severe for $O^- + e^- \rightarrow O^{2-}$ than for $O + e^- \rightarrow O^-$.

101.   a.   The electron affinity of Mg^{2+} is ΔH for $Mg^{2+}(g) + e^- \rightarrow Mg^+(g)$. This is just the reverse of the second ionization energy, or $EA(Mg^{2+}) = -IE_2(Mg) = -1445$ kJ/mol (Table 7.5).

       b.   EA of Al^+ is ΔH for $Al^+(g) + e^- \rightarrow Al(g)$. $EA(Al^+) = -IE_1(Al) = -580$ kJ/mol (Table 7.5)

102.   a.   IE of Cl^- is ΔH for $Cl^-(g) \rightarrow Cl(g) + e^-$. $IE(Cl^-) = -EA(Cl) = 348.7$ kJ/mol (Table 7.7)

       b. $Cl(g) \rightarrow Cl^+(g) + e^-$ IE = 1255 kJ/mol (Table 7.5)

       c. $Cl^+(g) + e^- \rightarrow Cl(g)$ $\Delta H = -IE_1 = -1255$ kJ/mol $= EA(Cl^+)$

Alkali Metals

103.   It should be potassium peroxide, K_2O_2, since K^+ ions are much more stable than K^{2+} ions. The second ionization energy of K is very large as compared to the first ionization energy.

104.   a. Li_3N; lithium nitride   b. NaBr; sodium bromide   c. K_2S;   potassium sulfide

105. $\nu = \dfrac{c}{\lambda} = \dfrac{2.9979 \times 10^8 \text{ m/s}}{455.5 \times 10^{-9} \text{ m}} = 6.582 \times 10^{14} \text{ s}^{-1}$

       $E = h\nu = 6.6261 \times 10^{-34} \text{ J s} \times 6.582 \times 10^{14} \text{ s}^{-1} = 4.361 \times 10^{-19} \text{ J}$

106.   For 589.0 nm: $\nu = \dfrac{c}{\lambda} = \dfrac{2.9979 \times 10^8 \text{ m/s}}{589.0 \times 10^{-9} \text{ m}} = 5.090 \times 10^{14} \text{ s}^{-1}$

$E = h\nu = 6.6261 \times 10^{-34} \text{ J s} \times 5.090 \times 10^{14} \text{ s}^{-1} = 3.373 \times 10^{-19} \text{ J}$

For 589.6 nm: $\nu = c/\lambda = 5.085 \times 10^{14} \text{ s}^{-1}$; $E = h\nu = 3.369 \times 10^{-19} \text{ J}$

The energies in kJ/mol are:

$3.373 \times 10^{-19} \text{ J} \times \dfrac{1 \text{ kJ}}{1000 \text{ J}} \times \dfrac{6.0221 \times 10^{23}}{\text{mol}} = 203.1 \text{ kJ/mol}$

$3.369 \times 10^{-19} \text{ J} \times \dfrac{1 \text{ kJ}}{1000 \text{ J}} \times \dfrac{6.0221 \times 10^{23}}{\text{mol}} = 202.9 \text{ kJ/mol}$

107.   Yes; the ionization energy general trend is to decrease down a group, and the atomic radius trend is to increase down a group. The data in Table 7.8 confirm both of these general trends.

108.   It should be element #119 with the ground state electron configuration: $[\text{Rn}]7s^25f^{14}6d^{10}7p^68s^1$

109.   a. $4 \text{ Li}(s) + O_2(g) \rightarrow 2 \text{ Li}_2O(s)$ b. $2 \text{ K}(s) + S(s) \rightarrow K_2S(s)$

110.   a. $2 \text{ Cs}(s) + 2 \text{ H}_2O(l) \rightarrow 2 \text{ CsOH}(aq) + H_2(g)$ b. $2 \text{ Na}(s) + Cl_2(g) \rightarrow 2 \text{ NaCl}(s)$

Additional Exercises

111. $E = \dfrac{310. \text{ kJ}}{\text{mol}} \times \dfrac{1 \text{ mol}}{6.022 \times 10^{23}} = 5.15 \times 10^{-22} \text{ kJ} = 5.15 \times 10^{-19} \text{ J}$

$E = \dfrac{hc}{\lambda}, \quad \lambda = \dfrac{hc}{E} = \dfrac{6.626 \times 10^{-34} \text{ J s} \times 2.998 \times 10^8 \text{ m/s}}{5.15 \times 10^{-19}} = 3.86 \times 10^{-7} \text{ m} = 386 \text{ nm}$

112.   Energy to make water boil $= s \times m \times \Delta T = \dfrac{4.18 \text{ J}}{g \, ^\circ C} \times 50.0 \text{ g} \times 75.0 ^\circ C = 1.57 \times 10^4 \text{ J}$

$E_{\text{photon}} = \dfrac{hc}{\lambda} = \dfrac{6.626 \times 10^{-34} \text{ J s} \times 2.998 \times 10^8 \text{ m/s}}{9.75 \times 10^{-2} \text{ m}} = 2.04 \times 10^{-24} \text{ J}$

$1.57 \times 10^4 \text{ J} \times \dfrac{1 \text{ sec}}{750. \text{ J}} = 20.9 \text{ sec}$; $1.57 \times 10^4 \text{ J} \times \dfrac{1 \text{ photon}}{2.04 \times 10^{-24} \text{ J}} = 7.70 \times 10^{27} \text{ photons}$

113. $60 \times 10^6 \text{ km} \times \dfrac{1000 \text{ m}}{\text{km}} \times \dfrac{1 \text{ s}}{3.00 \times 10^8 \text{ m}} = 200 \text{ s}$ (about 3 minutes)

114. $\lambda = \dfrac{\text{hc}}{\text{E}} = \dfrac{6.626 \times 10^{-34} \text{ J s} \times 2.998 \times 10^8 \text{ m/s}}{3.90 \times 10^{-19} \text{ J}} = 5.09 \times 10^{-7} \text{ m} \times \dfrac{100 \text{ cm}}{\text{m}} = 5.09 \times 10^{-5} \text{ cm}$

From the spectrum, $\lambda = 5.09 \times 10^{-5}$ cm is green light.

115. $\Delta E = -R_H \left(\dfrac{1}{n_f^2} - \dfrac{1}{n_i^2} \right) = -2.178 \times 10^{-18} \text{ J} \left(\dfrac{1}{2^2} - \dfrac{1}{6^2} \right) = -4.840 \times 10^{-19} \text{ J}$

$\lambda = \dfrac{\text{hc}}{|\Delta E|} = \dfrac{6.6261 \times 10^{-34} \text{ J s} \times 2.9979 \times 10^8 \text{ m/s}}{4.840 \times 10^{-19} \text{ J}} = 4.104 \times 10^{-7} \text{ m} \times \dfrac{100 \text{ cm}}{\text{m}} = 4.104 \times 10^{-5} \text{ cm}$

From the spectrum, $\lambda = 4.104 \times 10^{-5}$ cm is violet light, so the $n = 6$ to $n = 2$ visible spectrum line is violet.

116.   a.   False;   It takes less energy to ionize an electron from $n = 3$ than from the ground state.

b.   True

c.   False;   The energy difference from $n = 3 \rightarrow n = 2$ is less than the energy difference from $n = 3 \rightarrow n = 1$, thus, the wavelength is larger for $n = 3 \rightarrow n = 2$ than for $n = 3 \rightarrow n = 1$.

d.   True

e.   False; $n = 2$ is the first excited state and $n = 3$ is the second excited state.

117.   a.   True for H only.   b.   True for all atoms.   c.   True for all atoms.

118. $n = 5$; $m_\ell = -4, -3, -2, -1, 0, 1, 2, 3, 4$; 18 electrons

119.   Cd: $1s^2 2s^2 2p^6 3s^2 3p^6 4s^2 3d^{10} 4p^6 5s^2 4d^{10}$

a. $\ell = 2$ are d orbitals. There are 20 electrons in d orbitals ($3d^{10}$ and $4d^{10}$).

b.   For $n = 4$, there are 4s, 4p, 4d and 4f orbitals. Cd only uses the 4s, 4p and 4d orbitals. Since all are filled with electrons, then Cd has $2 + 6 + 10 = 18$ electrons with $n = 4$.

c.   All p, d and f orbitals have one orbital which has $m_\ell = -1$. So each of the 2p, 3p, 4p, 3d and 4d orbitals (which are filled) have an orbital with $m_\ell = -1$; $5(2) = 10$ electrons with $m_\ell = -1$.

d.   Each orbital used has one of the two paired electrons with $m_s = -1/2$. Therefore, one-half of the total electrons in Cd have $m_s = -1/2$; $(48/2) = 24$ electrons with $m_s = -1/2$.

120.   He: $1s^2$; Ne: $1s^22s^22p^6$; Ar: $1s^22s^22p^63s^23p^6$; Each peak in the diagram corresponds to a subshell with different values of n. Corresponding subshells are closer to the nucleus for heavier elements because of the increased nuclear charge.

121.   The general ionization energy trend says that ionization energy increases going left to right across the periodic table. However, one of the exceptions to this trend occurs between groups 2A and 3A. Between these two groups, group 3A elements usually have a lower ionization energy than group 2A elements. Therefore, Al should have the lowest first ionization energy value, followed by Mg, with Si having the largest ionization energy. Looking at the values for the first ionization energy in the graph, the green plot is Al, the blue plot is Mg, and the red plot is Si.

Mg (the blue plot) is the element with the huge jump between I_2 and I_3. Mg has two valence electrons, so the third electron removed is an inner core electron. Inner core electrons are always much more difficult to remove compared to valence electrons since they are closer to the nucleus, on average, than the valence electrons.

122.   a.   The 4+ ion contains 20 electrons. Thus, the electrically neutral atom will contain 24 electrons. The atomic number is 24.

   b.   The ground state electron configuration of the ion must be: $1s^22s^22p^63s^23p^64s^03d^2$; There are 6 electrons in s orbitals.

   c.   12   d.   2

   e.   Because of the mass, this is an isotope of $^{50}_{24}\text{Cr}$. There are 26 neutrons in the nucleus.

   f. $1s^22s^22p^63s^23p^64s^13d^5$ is the ground state electron configuration for Cr. Cr is an exception to the normal filling order.

123.   Valence electrons are easier to remove than inner core electrons. The large difference in energy between I_2 and I_3 indicates that this element has two valence electrons. This element is most likely an alkaline earth metal since alkaline earth metal elements all have two valence electrons.

124.   Since all oxygen family elements have ns^2np^4 valence electron configurations, this nonmetal is from the oxygen family.

   a.   2 + 4 = 6 valence electrons

   b.   O, S, Se and Te are the nonmetals from the oxygen family (Po is a metal).

   c.   Since oxygen family nonmetals form -2 charged ions in ionic compounds, K_2X would be the predicted formula where X is the unknown nonmetal.

   d.   From the size trend, this element would have a smaller radius than barium.

   e.   From the ionization energy trend, this element would have a smaller ionization energy than fluorine.

125. a. $Na(g) \rightarrow Na^+(g) + e^-$ $I_1 = 495$ kJ
          $Cl(g) + e^- \rightarrow Cl^-(g)$ $EA = -348.7$ kJ

          _____

          $Na(g) + Cl(g) \rightarrow Na^+(g) + Cl^-(g)$ $\Delta H = 146$ kJ

      b. $Mg(g) \rightarrow Mg^+(g) + e^-$ $I_1 = 735$ kJ
          $F(g) + e^- \rightarrow F^-(g)$ $EA = -327.8$ kJ

          _____

          $Mg(g) + F(g) \rightarrow Mg^+(g) + F^-(g)$ $\Delta H = 407$ kJ

      c. $Mg^+(g) \rightarrow Mg^{2+}(g) + e^-$ $I_2 = 1445$ kJ
          $F(g) + e^- \rightarrow F^-(g)$ $EA = -327.8$ kJ

          _____

          $Mg^+(g) + F(g) \rightarrow Mg^{2+}(g) + F^-(g)$ $\Delta H = 1117$ kJ

      d. From parts b and c we get:

          $Mg(g) + F(g) \rightarrow Mg^+(g) + F^-(g)$ $\Delta H = 407$ kJ
          $Mg^+(g) + F(g) \rightarrow Mg^{2+}(g) + F^-(g)$ $\Delta H = 1117$ kJ

          _____

          $Mg(g) + 2\ F(g) \rightarrow Mg^{2+}(g) + 2\ F^-(g)$ $\Delta H = 1524$ kJ

126. a. $Se^{3+}(g) \rightarrow Se^{4+}(g) + e^-$ b. $S^-(g) + e^- \rightarrow S^{2-}(g)$

      c. $Fe^{3+}(g) + e^- \rightarrow Fe^{2+}(g)$ d. $Mg(g) \rightarrow Mg^+(g) + e^-$

Challenge Problems

127. $E_{photon} = \dfrac{hc}{\lambda} = \dfrac{6.6261 \times 10^{-34}\ \text{J s} \times 2.9979 \times 10^8\ \text{m/s}}{253.4 \times 10^{-9}\ \text{m}} = 7.839 \times 10^{-19}$ J; $\Delta E = -7.839 \times 10^{-19}$ J

The general energy equation for one-electron ions is $E_n = -2.178 \times 10^{-18}\text{J}\ (Z^2)/n^2$ where $Z =$ atomic number.

$$\Delta E = -2.178 \times 10^{-18}\ \text{J}\ (Z)^2 \left(\dfrac{1}{n_f^2} - \dfrac{1}{n_i^2} \right) ,\quad Z = 4 \text{ for } Be^{3+}$$

$$-7.839 \times 10^{-19}\ \text{J} = -2.178 \times 10^{-18}\ (4)^2 \left(\dfrac{1}{n_f^2} - \dfrac{1}{5^2} \right)$$

$$\dfrac{7.839 \times 10^{-19}}{2.178 \times 10^{-18} \times 16} + \dfrac{1}{25} = \dfrac{1}{n_f^2},\quad \dfrac{1}{n_f^2} = 0.06249,\ n_f = 4$$

This emission line corresponds to the $n = 5 \rightarrow n = 4$ electronic transition.

128. a. Since wavelength is inversely proportional to energy, the spectral line to the right of B (at a larger wavelength) represents the lowest possible energy transition; this is $n = 4$ to $n = 3$. The B line represents the next lowest energy transition, which is $n = 5$ to $n = 3$ and the A line corresponds to the $n = 6$ to $n = 3$ electronic transition.

   b. Since this spectrum is for a one-electron ion, then $E_n = -2.178 \times 10^{-18}$ J (Z^2/n^2). To determine ΔE and, in turn, the wavelength of spectral line A, we must determine Z, the atomic number of the one-electron species. Use spectral line B data to determine Z.

$$\Delta E_{5 \rightarrow 3} = -2.178 \times 10^{-18} \text{ J } (Z^2)\left(\frac{1}{3^2} - \frac{1}{5^2} \right) = -2.178 \times 10^{-18} \left(\frac{Z^2}{3^2} - \frac{Z^2}{5^2} \right)$$

$$\Delta E_{5 \rightarrow 3} = -2.178 \times 10^{-18} \left(\frac{16 Z^2}{9 \times 25} \right)$$

$$E = \frac{hc}{\lambda} = \frac{6.6261 \times 10^{-34} \text{ J s} \times 2.9979 \times 10^8 \text{ m/s}}{142.5 \times 10^{-9} \text{ m}} = 1.394 \times 10^{-18} \text{ J}$$

Since an emission occurs, $\Delta E_{5 \rightarrow 3} = -1.394 \times 10^{-18}$ J.

$$\Delta E = -1.394 \times 10^{-18} \text{ J} = -2.178 \times 10^{-18} \text{ J} \left(\frac{16 Z^2}{9 \times 25} \right), \ \ Z^2 = 9.001, \ \ Z = 3; \ \text{ The ion is Li}^{2+}.$$

Solving for the wavelength of line A:

$$\Delta E_{6 \rightarrow 3} = -2.178 \times 10^{-18} \ (3)^2 \left(\frac{1}{3^2} - \frac{1}{6^2} \right) = -1.634 \times 10^{-18} \text{ J}$$

$$\lambda = \frac{hc}{|\Delta E|} = \frac{6.6261 \times 10^{-34} \text{ J s} \times 2.9979 \times 10^8 \text{ m/s}}{1.634 \times 10^{-18} \text{ J}} = 1.216 \times 10^{-7} \text{ m} = 121.6 \text{ nm}$$

129. For $r = a_0$ and $\theta = 0°$ (Z = 1 for H):

$$\psi_{2p_z} = \frac{1}{4(2\pi)^{1/2}} \left(\frac{1}{5.29 \times 10^{-11}} \right)^{3/2} (1) \ e^{-1/2} \cos 0 = 1.57 \times 10^{14}; \ \psi^2 = 2.46 \times 10^{28}$$

For $r = a_0$ and $\theta = 90°$, $\psi_{2p_z} = 0$ since $\cos 90° = 0$; $\psi^2 = 0$

There is no probability of finding an electron in the $2p_z$ orbital with $\theta = 0°$. As expected, the xy plane, which corresponds to $\theta = 0°$, is a node for the $2p_z$ atomic orbital.

130. a. Each orbital could hold 3 electrons.

b. The first period corresponds to $n = 1$ which can only have 1s orbitals. The 1s orbital could hold 3 electrons, hence the first period would have three elements. The second period corresponds to $n = 2$, which has 2s and 2p orbitals. These four orbitals can each hold three electrons. A total of 12 elements would be in the second period.

c. 15   d. 21

131.   a. 1st period: $p = 1,\ q = 1,\ r = 0,\ \ s = \pm 1/2$ (2 elements)

           2nd period: $p = 2,\ q = 1,\ r = 0,\ \ s = \pm 1/2$ (2 elements)

           3rd period: $p = 3,\ q = 1,\ r = 0,\ \ s = \pm 1/2$ (2 elements)

                             $p = 3,\ q = 3,\ r = -2,\ s = \pm 1/2$ (2 elements)

                             $p = 3,\ q = 3,\ r = 0,\ \ s = \pm 1/2$ (2 elements)

                             $p = 3,\ q = 3,\ r = +2,\ s = \pm 1/2$ (2 elements)

           4th period: $p = 4$; q and r values are the same as with $p = 3$ (8 total elements)

1							2
3							4
5	6	7	8	9	10	11	12
13	14	15	16	17	18	19	20

b. Elements 2, 4, 12 and 20 all have filled shells and will be least reactive.

c. Draw similarities to the modern periodic table.

XY could be X^+Y^-, $X^{2+}Y^{2-}$ or $X^{3+}Y^{3-}$. Possible ions for each are:

    X^+ could be elements 1, 3, 5 or 13; Y^- could be 11 or 19.

    X^{2+} could be 6 or 14; Y^{2-} could be 10 or 18.

    X^{3+} could be 7 or 15; Y^{3-} could be 9 or 17.

Note: X^{4+} and Y^{4-} ions probably won't form.

XY_2 will be $X^{2+}(Y^-)_2$; See above for possible ions.

X_2Y will be $(X^+)_2Y^{2-}$; See above for possible ions.

XY_3 will be $X^{3+}(Y^-)_3$; See above for possible ions.

X_2Y_3 will be $(X^{3+})_2(Y^{2-})_3$; See above for possible ions.

d. $p = 4$, $q = 3$, $r = -2$, $s = \pm 1/2$ (2)

   $p = 4$, $q = 3$, $r = 0$, $s = \pm 1/2$ (2)

   $p = 4$, $q = 3$, $r = +2$, $s = \pm 1/2$ (2)

   A total of 6 electrons can have $p = 4$ and $q = 3$.

e. $p = 3$, $q = 0$, $r = 0$:   This is not allowed; q must be odd. Zero electrons can have these quantum numbers.

f. $p = 6$, $q = 1$, $r = 0$, $s = \pm 1/2$ (2)

   $p = 6$, $q = 3$, $r = -2, 0, +2$; $s = \pm 1/2$ (6)

   $p = 6$, $q = 5$, $r = -4, -2, 0, +2, +4$; $s = \pm 1/2$ (10)

   Eighteen electrons can have $p = 6$.

132.   Size also decreases going across a period. Sc & Ti and Y & Zr are adjacent elements. There are 14 elements (the lanthanides) between La and Hf, making Hf considerable smaller.

133.   a.   As we remove succeeding electrons, the electron being removed is closer to the nucleus, and there are fewer electrons left repelling it. The remaining electrons are more strongly attracted to the nucleus, and it takes more energy to remove these electrons.

   b.   Al : $1s^22s^22p^63s^23p^1$; For I_4, we begin by removing an electron with $n = 2$. For I_3, we remove an electron with $n = 3$. In going from $n = 3$ to $n = 2$, there is a big jump in ionization energy because the $n = 2$ electrons (inner core electrons) are much closer to the nucleus on average than $n = 3$ electrons (valence electrons). Since the $n = 2$ electrons are closer to the nucleus, they are held more tightly and require a much larger amount of energy to remove them compared to the $n = 3$ electrons.

   c. Al^{4+}; The electron affinity for Al^{4+} is ΔH for the reaction:

   $$Al^{4+}(g) + e^- \rightarrow Al^{3+}(g) \quad \Delta H = -I_4 = -11,600 \text{ kJ/mol}$$

   d.   The greater the number of electrons, the greater the size.

   Size trend: $Al^{4+} < Al^{3+} < Al^{2+} < Al^+ < Al$

134.   None of the noble gases and no subatomic particles had been discovered when Mendeleev published his periodic table. Thus, there was not an element out of place in terms of reactivity. There was no reason to predict an entire family of elements. Mendeleev ordered his table by mass; he had no way of knowing there were gaps in atomic numbers (they hadn't been discovered yet).

CHAPTER EIGHT

BONDING: GENERAL CONCEPTS

Questions

13.   a.   Electronegativity: The ability of an atom <u>in a molecule</u> to attract electrons to itself.

Electron affinity: The energy change for $M(g) + e^- \rightarrow M^-(g)$. EA deals with isolated atoms in the gas phase.

b.   Covalent bond: Sharing of electron pair(s); Polar covalent bond: Unequal sharing of electron pair(s).

c.   Ionic bond: Electrons are no longer shared, i.e., complete transfer of electron(s) from one atom to another to form ions.

14.   Isoelectronic: same number of electrons. Two variables, the number of protons and the number of electrons, determine the size of an ion. Keeping the number of electrons constant, we only have to consider the number of protons to predict trends in size.

15. H_2O and NH_3 have lone pair electrons on the central atoms. These lone pair electrons require more room than the bonding electrons, which tends to compress the angles between the bonding pairs. The bond angle for H_2O is the smallest since oxygen has two lone pairs on the central atom, and the bond angle is compressed more than in NH_3 where N has only one lone pair.

16.   Resonance occurs when more than one valid Lewis structure can be drawn for a particular molecule. A common characteristic in resonance structures is a multiple bond(s) that moves from one position to another. We say this multiple bond(s) is delocalized about the entire molecule, which helps us rationalize why the bonds in a molecule that exhibits resonance are all equivalent in length and strength.

17.   The two general requirements for a polar molecular are:

1.   polar bonds

2.   a structure such that the bond dipoles of the polar bonds do not cancel.

18.   Nonmetals form covalent compounds. Nonmetals have valence electrons in the s and p orbitals. Since there are 4 total s and p orbitals, there is room for only eight electrons (the octet rule).

19.   Of the compounds listed, P_2O_5 is the only compound containing only covalent bonds. $(NH_4)_2SO_4$, $Ca_3(PO_4)_2$, K_2O, and KCl are all compounds composed of ions so they exhibit ionic bonding. The ions in $(NH_4)_2SO_4$ are NH_4^+ and SO_4^{2-}. Covalent bonds exist between the N and H atoms in NH_4^+ and between the S and O atoms in SO_4^{2-}. Therefore, $(NH_4)_2SO_4$ contains both ionic and covalent bonds. The same is true for $Ca_3(PO_4)_2$. The bonding is ionic between the Ca^{2+} and PO_4^{3-} ions and covalent between the P and O atoms in PO_4^{3-}. Therefore, $(NH_4)_2SO_4$ and $Ca_3(PO_4)_2$ are the compounds with both ionic and covalent bonds.

20.   Ionic solids are held together by strong electrostatic forces which are omnidirectional.

    i.   For electrical conductivity, charged species must be free to move. In ionic solids the charged ions are held rigidly in place. Once the forces are disrupted (melting or dissolution), the ions can move about (conduct).

    ii.   Melting and boiling disrupts the attractions of the ions for each other. If the forces are strong, it will take a lot of energy (high temperature) to accomplish this.

    iii. If we try to bend a piece of material, the ions must slide across each other. For an ionic solid the following might happen:

           strong attraction   strong repulsion

       Just as the layers begin to slide, there will be very strong repulsions causing the solid to snap across a fairly clean plane.

    iv.   Polar molecules are attracted to ions and can break up the lattice.

These properties and their correlation to chemical forces will be discussed in detail in Chapters 10 and 11.

Exercises

Chemical Bonds and Electronegativity

21.   The general trend for electronegativity is:
      1)   increase as we go from left to right across a period and
      2)   decrease as we go down a group

Using these trends, the expected orders are:

    a.   C < N < O   b.   Se < S < Cl   c.   Sn < Ge < Si   d.   Tl < Ge < S

22.   a.   Rb < K < Na   b.   Ga < B < O   c.   Br < Cl < F   d.   S < O < F

23.   The most polar bond will have the greatest difference in electronegativity between the two atoms. From positions in the periodic table, we would predict:

a. Ge–F   b. P–Cl   c. S–F   d. Ti–Cl

24.   a. Sn–H   b. Tl–Br   c. Si–O   d. O–F

25.   The general trends in electronegativity used in Exercises 8.21 and 8.23 are only rules of thumb. In this exercise, we use experimental values of electronegativities and can begin to see several exceptions. The order of EN from Figure 8.3 is:

a. C (2.5) < N (3.0) < O (3.5)   same as predicted

b. Se (2.4) < S (2.5) < Cl (3.0)   same

c. Si = Ge = Sn (1.8)   different

d. Tl (1.8) = Ge (1.8) < S (2.5)   different

Most polar bonds using actual EN values:

a. Si–F and Ge–F have equal polarity (Ge–F predicted).

b. P–Cl (same as predicted)

c. S–F (same as predicted)   d. Ti–Cl (same as predicted)

26.   a. Rb (0.8) = K (0.8) < Na (0.9), different   b. Ga (1.6) < B (2.0) < O (3.5), same

c. Br (2.8) < Cl (3.0) < F (4.0), same   d. S (2.5) < O (3.5) < F (4.0), same

Most polar bonds using actual EN values.

a. C–H most polar (Sn–H predicted)

b. Al–Br most polar (Tl–Br predicted).   c. Si–O (same as predicted).

d. Each bond has the same polarity, but the bond dipoles point in opposite directions. Oxygen is the positive end in the O–F bond dipole, and oxygen is the negative end in the O–Cl bond dipole. (O–F predicted.)

27.   Electronegativity values increase from left to right across the periodic table. The order of electronegativities for the atoms from smallest to largest electronegativity will be H = P < C < N < O < F. The most polar bond will be F–H since it will have the largest difference in electronegativities, and the least polar bond will be P–H since it will have the smallest difference in electronegativities (ΔEN = 0). The order of the bonds in decreasing polarity will be F–H > O–H > N–H > C–H > P–H.

28.   Ionic character is proportional to the difference in electronegativity values between the two elements forming the bond. Using the trend in electronegativity, the order will be:

Br–Br < N–O < C–F < Ca–O < K–F
least   most
ionic character   ionic character

Note that Br–Br, N–O and C–F bonds are all covalent bonds since the elements are all nonmetals. The Ca–O and K–F bonds are ionic as is generally the case when a metal forms a bond with a nonmetal.

Ions and Ionic Compounds

29. Rb^+: $[Ar]4s^23d^{10}4p^6$; Ba^{2+}: $[Kr]5s^24d^{10}5p^6$; Se^{2-}: $[Ar]4s^23d^{10}4p^6$

I^-: $[Kr]5s^24d^{10}5p^6$

30.   a. Mg^{2+}: $1s^22s^22p^6$; K^+: $1s^22s^22p^63s^23p^6$; Al^{3+}: $1s^22s^22p^6$

b. N^{3-}, O^{2-} and F^-: $1s^22s^22p^6$; Te^{2-}: $[Kr]5s^24d^{10}5p^6$

31.   a. Sc^{3+} b. Te^{2-} c. Ce^{4+} and Ti^{4+} d. Ba^{2+}

All of these have the number of electrons of a noble gas.

32.   a. Cs_2S is composed of Cs^+ and S^{2-}. Cs^+ has the same electron configuration as Xe, and S^{2-} has the same configuration as Ar.

b. SrF_2; Sr^{2+} has the Kr electron configuration and F^- has the Ne configuration.

c. Ca_3N_2; Ca^{2+} has the Ar electron configuration and N^{3-} has the Ne configuration.

d. $AlBr_3$; Al^{3+} has the Ne electron configuration and Br^- has the Kr configuration.

33.   There are many possible ions with 10 electrons. Some are: N^{3-}, O^{2-}, F^-, Na^+, Mg^{2+} and Al^{3+}. In terms of size, the ion with the most protons will hold the electrons the tightest and will be the smallest. The largest ion will be the ion with the fewest protons. The size trend is:

Al^{3+} < Mg^{2+} < Na^+ < F^- < O^{2-} < N^{3-}
smallest   largest

34.   Some possibilities which contain 18 electrons are: P^{3-}, S^{2-}, Cl^-, K^+, Ca^{2+} and Sc^{3+}. The size trend is proportional to the number of protons in the nucleus. Size trend:

Sc^{3+} < Ca^{2+} < K^+ < Cl^- < S^{2-} < P^{3-}
smallest   largest

35. a. $Cu > Cu^+ > Cu^{2+}$ b. $Pt^{2+} > Pd^{2+} > Ni^{2+}$ c. $O^{2-} > O^- > O$

d. $La^{3+} > Eu^{3+} > Gd^{3+} > Yb^{3+}$ e. $Te^{2-} > I^- > Cs^+ > Ba^{2+} > La^{3+}$

For answer a, as electrons are removed from an atom, size decreases. Answers b and d follow the radii trend. For answer c, as electrons are added to an atom, size increases. Answer e follows the trend for an isoelectronic series, i.e., the smallest ion has the most protons.

36. a. $V > V^{2+} > V^{3+} > V^{5+}$ b. $Cs^+ > Rb^+ > K^+ > Na^+$ c. $Te^{2-} > I^- > Cs^+ > Ba^{2+}$

d. $P^{3-} > P^{2-} > P^- > P$ e. $Te^{2-} > Se^{2-} > S^{2-} > O^{2-}$

37. a. Li^+ and N^{3-} are the expected ions. The formula of the compound would be Li_3N (lithium nitride).

b. Ga^{3+} and O^{2-}; Ga_2O_3, gallium(III) oxide or gallium oxide

c. Rb^+ and Cl^-; RbCl, rubidium chloride   d. Ba^{2+} and S^{2-}; BaS, barium sulfide

38. a. Al^{3+} and Cl^-; $AlCl_3$, aluminum chloride   b. Na^+ and O^{2-}; Na_2O, sodium oxide

c. Sr^{2+} and F^-; SrF_2, strontium fluoride   d. Ca^{2+} and P^{3-}; Ca_3P_2, calcium phosphide

39. Lattice energy is proportional to Q_1Q_2/r where Q is the charge of the ions and r is the distance between the ions. In general, charge effects on lattice energy are much greater than size effects.

a. NaCl; Na^+ is smaller than K^+.   b. LiF; F^- is smaller than Cl^-.

c. MgO; O^{2-} has a greater charge than OH^-.   d. $Fe(OH)_3$; Fe^{3+} has a greater charge than Fe^{2+}.

e. Na_2O; O^{2-} has a greater charge than Cl^-.   f. MgO; The ions are smaller in MgO.

40. a. LiF; Li^+ is smaller than Cs^+.   b. NaBr; Br^- is smaller than I^-.

c. BaO; O^{2-} has a greater charge than Cl^-.   d. $CaSO_4$; Ca^{2+} has a greater charge than Na^+.

e. K_2O; O^{2-} has a greater charge than F^-.   f. Li_2O; The ions are smaller in Li_2O.

41.
$$Na(s) \rightarrow Na(g) \qquad \Delta H = 109 \text{ kJ (sublimation)}$$
$$Na(g) \rightarrow Na^+(g) + e^- \qquad \Delta H = 495 \text{ kJ (ionization energy)}$$
$$1/2\ Cl_2(g) \rightarrow Cl(g) \qquad \Delta H = 239/2 \text{ kJ (bond energy)}$$
$$Cl(g) + e^- \rightarrow Cl^-(g) \qquad \Delta H = -349 \text{ kJ (electron affinity)}$$
$$Na^+(g) + Cl^-(g) \rightarrow NaCl(s) \qquad \Delta H = -786 \text{ kJ (lattice energy)}$$

$$Na(s) + 1/2\ Cl_2(g) \rightarrow NaCl(s) \qquad \Delta H^{\circ}_f = -412 \text{ kJ/mol}$$

42.

$Mg(s) \rightarrow Mg(g)$	$\Delta H = 150. \text{ kJ}$	(sublimation)
$Mg(g) \rightarrow Mg^+(g) + e^-$	$\Delta H = 735 \text{ kJ}$	(IE_1)
$Mg^+(g) \rightarrow Mg^{2+}(g) + e^-$	$\Delta H = 1445 \text{ kJ}$	(IE_2)
$F_2(g) \rightarrow 2 F(g)$	$\Delta H = 154 \text{ kJ}$	(BE)
$2 F(g) + 2 e^- \rightarrow 2 F^-(g)$	$\Delta H = 2(-328) \text{ kJ}$	(EA)
$Mg^{2+}(g) + 2 F^-(g) \rightarrow MgF_2(s)$	$\Delta H = -3916 \text{ kJ}$	(LE)

$$Mg(s) + F_2(g) \rightarrow MgF_2(s) \qquad \Delta H_f^\circ = -2088 \text{ kJ/mol}$$

43.   From the data given, it takes less energy to produce $Mg^+(g) + O^-(g)$ than to produce $Mg^{2+}(g) + O^{2-}(g)$. However, the lattice energy for $Mg^{2+}O^{2-}$ will be much more exothermic than that for Mg^+O^- due to the greater charges in $Mg^{2+}O^{2-}$. The favorable lattice energy term dominates and $Mg^{2+}O^{2-}$ forms.

44.   Let us look at the complete cycle for Li_2S.

$2 Li(s) \rightarrow 2 Li(g)$	$2 \Delta H_{sub, Li} = 2(161) \text{ kJ}$
$2 Li(g) \rightarrow 2 Li^+(g) + 2 e^-$	$2 IE = 2(520.) \text{ kJ}$
$S(s) \rightarrow S(g)$	$\Delta H_{sub, S} = 277 \text{ kJ}$
$S(g) + e^- \rightarrow S^-(g)$	$EA_1 = -200. \text{ kJ}$
$S^-(g) + e^- \rightarrow S^{2-}(g)$	$EA_2 = ?$
$2 Li^+(g) + S^{2-}(g) \rightarrow Li_2S$	$LE = -2472 \text{ kJ}$

$$2 Li(s) + S(s) \rightarrow Li_2S(s) \qquad \Delta H_f^\circ = -500. \text{ kJ}$$

$\Delta H_f^\circ = 2 \Delta H_{sub, Li} + 2 IE + \Delta H_{sub, S} + EA_1 + EA_2 + LE$, $-500. = -1033 + EA_2$, $EA_2 = 533 \text{ kJ}$

For each salt: $\Delta H_f^\circ = 2 \Delta H_{sub, M} + 2 IE + 277 - 200. + LE + EA_2$

Na_2S: $-365 = 2(109) + 2(495) + 277 - 200. - 2203 + EA_2$, $EA_2 = 553 \text{ kJ}$

K_2S: $-381 = 2(90.) + 2(419) + 277 - 200. - 2052 + EA_2$, $EA_2 = 576 \text{ kJ}$

Rb_2S: $-361 = 2(82) + 2(409) + 277 - 200. - 1949 + EA_2$, $EA_2 = 529 \text{ kJ}$

Cs_2S: $-360. = 2(78) + 2(382) + 277 - 200. - 1850. + EA_2$, $EA_2 = 493 \text{ kJ}$

We get values from 493 to 576 kJ.

The mean value is: $\dfrac{533 + 553 + 576 + 529 + 493}{5} = 537 \text{ kJ}$

We can represent the results as $EA_2 = 540 \pm 50 \text{ kJ}$.

45. Ca^{2+} has a greater charge than Na^+, and Se^{2-} is smaller than Te^{2-}. The effect of charge on the lattice energy is greater than the effect of size. We expect the trend from most exothermic to least exothermic to be:

$$CaSe \; > \; CaTe \; > \; Na_2Se \; > \; Na_2Te$$
$$(-2862) \quad (-2721) \quad (-2130) \quad (-2095) \quad \text{This is what we observe.}$$

46. Lattice energy is proportional to the charge of the cation times the charge of the anion, Q_1Q_2.

Compound	Q_1Q_2	Lattice Energy
$FeCl_2$	$(+2)(-1) = -2$	-2631 kJ/mol
$FeCl_3$	$(+3)(-1) = -3$	-5359 kJ/mol
Fe_2O_3	$(+3)(-2) = -6$	-14,744 kJ/mol

Bond Energies

47. a.   H—H + Cl——Cl ⟶ 2 H—Cl

Bonds broken: Bonds formed:

1 H – H (432 kJ/mol)   2 H – Cl (427 kJ/mol)
1 Cl – Cl (239 kJ/mol)

$\Delta H = \Sigma D_{broken} - \Sigma D_{formed}$, $\Delta H = 432$ kJ + 239 kJ - 2(427) kJ = -183 kJ

b. 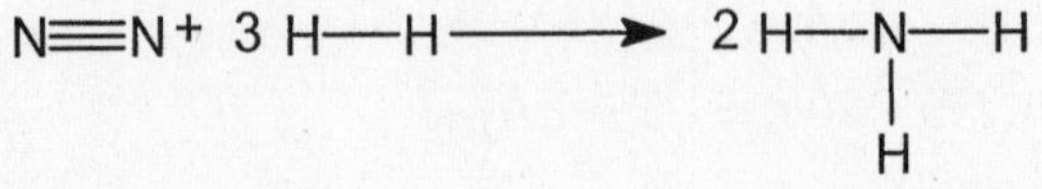   N≡N + 3 H—H ⟶ 2 H—N—H
                                               |
                                               H

Bonds broken: Bonds formed:

1 N ≡ N (941 kJ/mol)   6 N – H (391 kJ/mol)
3 H – H (432 kJ/mol)

$\Delta H = 941$ kJ + 3(432) kJ - 6(391) kJ = -109 kJ

48. Sometimes some of the bonds remain the same between reactants and products. To save time, only break and form bonds that are involved in the reaction.

a. 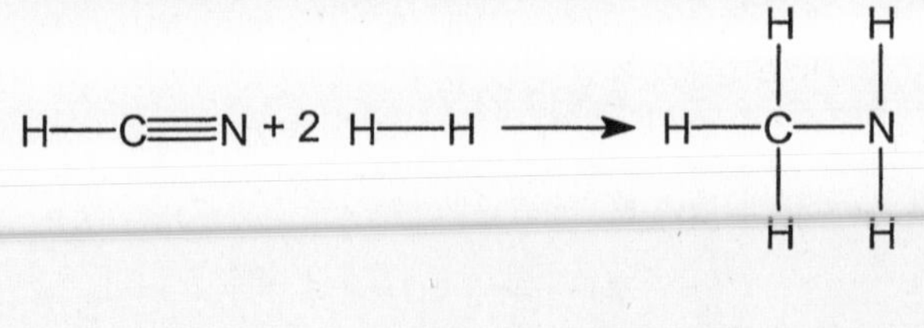   H—C≡N + 2 H—H ⟶ H—C—N with H, H on C and H, H on N

Bonds broken: Bonds formed:

$\quad$ 1 C $\equiv$ N (891 kJ/mol) $\quad$ 1 C – N (305 kJ/mol)
$\quad$ 2 H – H (432 kJ/mol) $\quad$ 2 C – H (413 kJ/mol)
$\qquad$ $\quad$ 2 N – H (391 kJ/mol)

$\Delta H = 891\ kJ + 2(432\ kJ) - [305\ kJ + 2(413\ kJ) + 2(391\ kJ)] = -158\ kJ$

b. + 2 F—F $\longrightarrow$ 4 H—F + N$\equiv$N

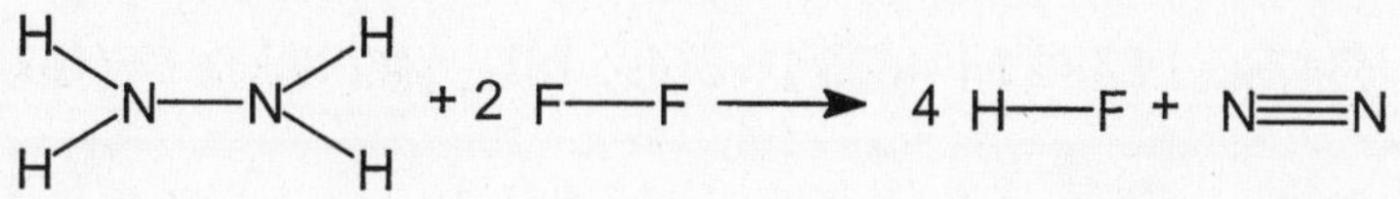

Bonds broken: Bonds formed:

$\quad$ 1 N – N (160. kJ/mol) $\quad$ 4 H – F (565 kJ/mol)
$\quad$ 4 N – H (391 kJ/mol) $\quad$ 1 N $\equiv$ N (941 kJ/mol)
$\quad$ 2 F – F (154 kJ/mol)

$\Delta H = 160.\ kJ + 4(391\ kJ) + 2(154\ kJ) - [4(565\ kJ) + 941\ kJ] = -1169\ kJ$

49.

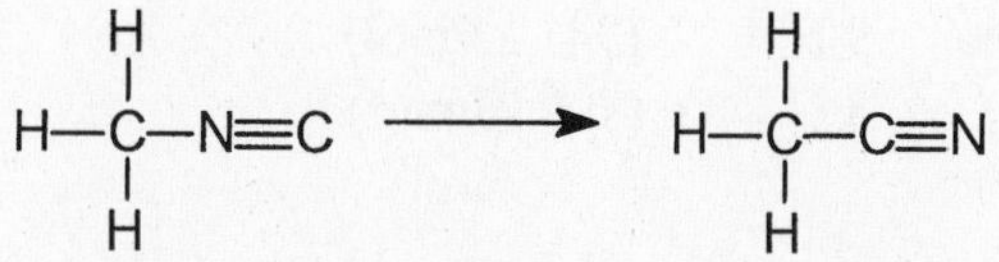

Bonds broken: 1 C – N (305 kJ/mol)$\qquad$ Bonds formed: 1 C – C (347 kJ/mol)

$\Delta H = \Sigma D_{broken} - \Sigma D_{formed},\ \ \Delta H = 305 - 347 = -42\ kJ$

Note: Sometimes some of the bonds remain the same between reactants and products. To save time, only break and form bonds that are involved in the reaction.

50.

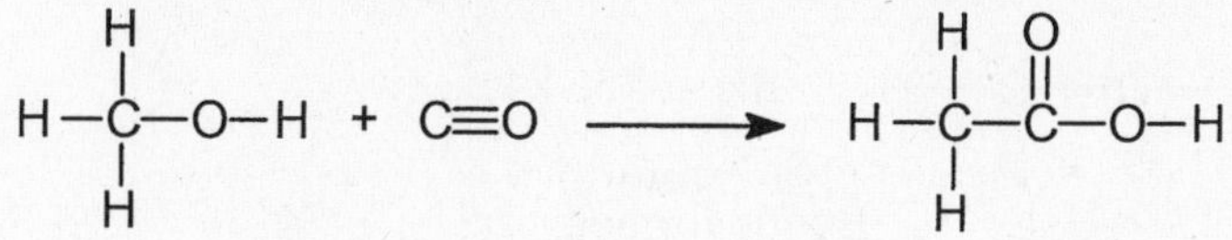

Bonds broken: Bonds formed:

$\quad$ 1 C $\equiv$ O (1072 kJ/mol) $\quad$ 1 C – C (347 kJ/mol)
$\quad$ 1 C – O (358 kJ/mol) $\quad$ 1 C = O (745 kJ/mol)
$\qquad$ $\quad$ 1 C – O (358 kJ/mol)

$\Delta H = 1072 + 358 - [347 + 745 + 358] = -20.\ kJ$

51.

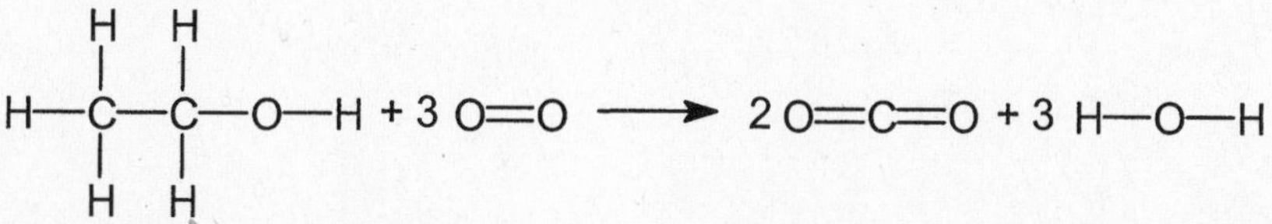

Bonds broken:	Bonds formed:
5 C – H (413 kJ/mol)	2 × 2 C = O (799 kJ/mol)
1 C – C (347 kJ/mol)	3 × 2 O – H (467 kJ/mol)
1 C – O (358 kJ/mol)	
1 O – H (467 kJ/mol)	
3 O = O (495 kJ/mol)	

$$\Delta H = 5(413 \text{ kJ}) + 347 \text{ kJ} + 358 \text{ kJ} + 467 \text{ kJ} + 3(495 \text{ kJ}) - [4(799 \text{ kJ}) + 6(467 \text{ kJ})]$$

$$= -1276 \text{ kJ}$$

52.

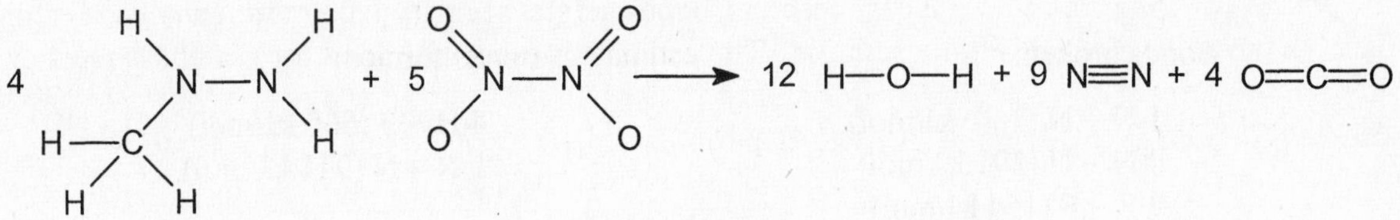

Bonds broken:	Bonds made:
9 N – N (160. kJ/mol)	24 O – H (467 kJ/mol)
4 N – C (305 kJ/mol)	9 N ≡ N (941 kJ/mol)
12 C – H (413 kJ/mol)	8 C = O (799 kJ/mol)
12 N – H (391 kJ/mol)	
10 N = O (607 kJ/mol)	
10 N – O (201 kJ/mol)	

$$\Delta H = 9(160.) + 4(305) + 12(413) + 12(391) + 10(607) + 10(201)$$

$$- [24(467) + 9(941) + 8(799)]$$

$$\Delta H = 20,388 \text{ kJ} - 26,069 \text{ kJ} = -5681 \text{ kJ}$$

53.   H—H + O=O $\longrightarrow$ H—O—O—H $\Delta H = -153$ kJ

Bonds broken:	Bonds formed:
1 H – H (432 kJ/mol)	2 O – H (467 kJ/mol)
1 O = O (495 kJ/mol)	1 O – O (D_{OO} kJ/mol)

$$\Delta H = -153 \text{ kJ} = 432 \text{ kJ} + 495 \text{ kJ} - [2(467 \text{ kJ}) + D_{OO}], \ D_{OO} = 146 \text{ kJ/mol}$$

54.

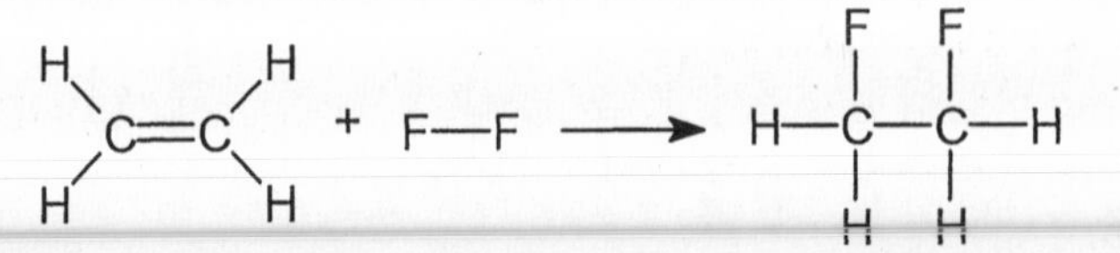

$\Delta H = -549$ kJ

Bonds broken: Bonds formed:

   1 C = C (614 kJ/mol)   1 C – C (347 kJ/mol)

   1 F – F (154 kJ/mol)   2 C – F (D_{CF})

ΔH = -549 kJ = 614 kJ + 154 kJ - [347 kJ + 2 D_{CF}], 2 D_{CF} = 970., D_{CF} = 485 kJ/mol

55.   a. $\Delta H° = 2\ \Delta H°_{f\ HCl}$ = 2 mol (-92 kJ/mol) = -184 kJ (= -183 kJ from bond energies)

     b. $\Delta H° = 2\ \Delta H°_{f\ NH_3}$ = 2 mol (-46 kJ/mol) = -92 kJ (= -109 kJ from bond energies)

Comparing the values for each reaction, bond energies seem to give a reasonably good estimate of the enthalpy change for a reaction. The estimate is especially good for gas phase reactions.

56. $CH_3OH(g) + CO(g) \rightarrow CH_3COOH(l)$

$\Delta H°$ = -484 kJ - [(-201 kJ) + (-110.5 kJ)] = -173 kJ

Using bond energies, ΔH = -20. kJ. For this reaction, bond energies give a much poorer estimate for ΔH as compared to the gas phase reactions in Exercise 8.47. The reason is that not all species are gases in Exercise 8.50. Bond energies do not account for the energy changes that occur when liquids and solids form instead of gases. These energy changes are due to intermolecular forces and will be discussed in Chapter 10.

57.   a.   Using SF_4 data: $SF_4(g) \rightarrow S(g) + 4\ F(g)$

        $\Delta H° = 4\ D_{SF}$ = 278.8 + 4 (79.0) - (-775) = 1370. kJ

        $D_{SF} = \dfrac{1370.\ kJ}{4\ mol\ SF\ bonds}$ = 342.5 kJ/mol

        Using SF_6 data: $SF_6(g) \rightarrow S(g) + 6\ F(g)$

        $\Delta H° = 6\ D_{SF}$ = 278.8 + 6 (79.0) - (-1209) = 1962 kJ

        $D_{SF} = \dfrac{1962\ kJ}{6}$ = 327.0 kJ/mol

     b.   The S – F bond energy in the table is 327 kJ/mol. The value in the table was based on the S – F bond in SF_6.

     c.   S(g) and F(g) are not the most stable forms of the elements at 25 °C. The most stable forms are $S_8(s)$ and $F_2(g)$; $\Delta H°_f$ = 0 for these two species.

58. $NH_3(g) \rightarrow N(g) + 3\ H(g);\quad \Delta H^\circ = 3\ D_{NH} = 472.7 + 3(216.0) - (-46.1) = 1166.8\ kJ$

$$D_{NH} = \frac{1166.8\ kJ}{3\ mol\ NH\ bonds} = 388.93\ kJ/mol \approx 389\ kJ/mol$$

$D_{calc} = 389\ kJ/mol$ as compared to 391 kJ/mol in the table. There is good agreement.

59. $N_2 + 3\ H_2 \rightarrow 2\ NH_3;\quad \Delta H = D_{N_2} + 3\ D_{H_2} - 6\ D_{NH};\quad \Delta H^\circ = 2(-46\ kJ) = -92\ kJ$

$-92\ kJ = 941\ kJ + 3(432\ kJ) - (6\ D_{N-H}),\ \ 6\ D_{N-H} = 2329\ kJ,\ \ D_{N-H} = 388.2\ kJ/mol$

Table in text: 391 kJ/mol; There is good agreement.

60.

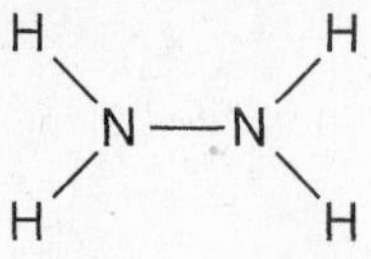

$2\ N(g)\ +\ 4\ H(g)\quad \Delta H = D_{N-N} + 4\ D_{N-H} = D_{N-N} + 4(388.9)$

$\Delta H^\circ = 2\,\Delta H^\circ_{f,N} + 4\,\Delta H^\circ_{f,H} - \Delta H^\circ_{f,N_2H_4} = 2(472.7\ kJ) + 4(216.0\ kJ) - 95.4\ kJ = 1714.0\ kJ$

$\Delta H^\circ = 1714.0\ kJ = D_{N-N} + 4(388.9)$

$D_{N-N} = 158.4\ kJ/mol$ (160. kJ/mol in Table 8.5)

Lewis Structures and Resonance

61. Drawing Lewis structures is mostly trial and error. However, the first two steps are always the same. These steps are 1) count the valence electrons available in the molecule, and 2) attach all atoms to each other with single bonds (called the skeletal structure). Unless noted otherwise, the atom listed first is assumed to be the atom in the middle (called the central atom) and all other atoms in the formula are attached to this atom. The most notable exceptions to the rule are formulas which begin with H, e.g., H_2O, H_2CO, etc. Hydrogen can never be a central atom since this would require H to have more than two electrons.

After counting valence electrons and drawing the skeletal structure, the rest is trial and error. We place the remaining electrons around the various atoms in an attempt to satisfy the octet rule (or duet rule for H). Keep in mind that practice makes perfect. After practicing you can (and will) become very adept at drawing Lewis structures.

a. HCN has $1 + 4 + 5 = 10$ valence electrons.

b. PH_3 has $5 + 3(1) = 8$ valence electrons.

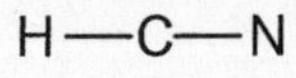

H—C—N   H—C≡N:

Skeletal   Lewis
structure   structure

Skeletal structures uses 4 e⁻ ; 6 e⁻ remain

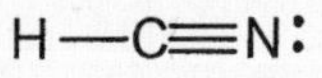

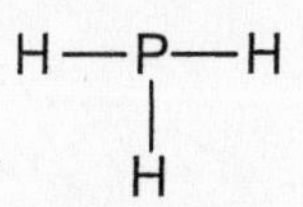

H—P—H   H—P̈—H
      | |
      H   H

Skeletal   Lewis
structure   structure

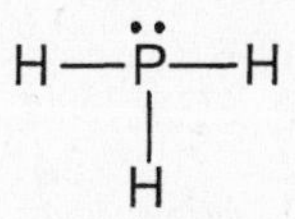

Skeletal structure uses 6 e⁻; 2 e⁻ remain

c. $CHCl_3$ has $4 + 1 + 3(7) = 26$ valence electrons.

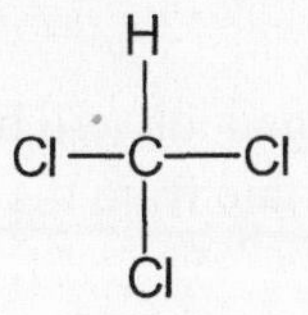

H   H
| |
Cl—C—Cl   :C̈l—C—C̈l:
| |
Cl   :C̈l:

Skeletal   Lewis
structure   structure

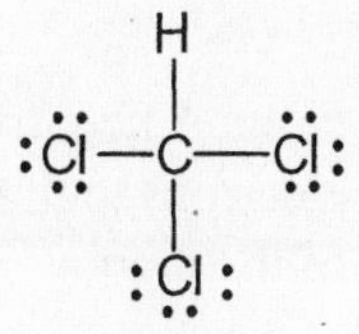

Skeletal structure uses 8 e⁻; 18 e⁻ remain

d. NH_4^+ has $5 + 4(1) - 1 = 8$ valence electrons.

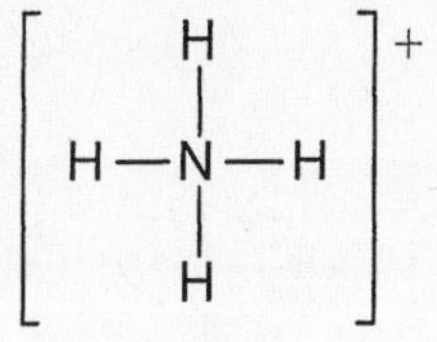

$$\left[\begin{array}{c} H \\ | \\ H—N—H \\ | \\ H \end{array} \right]^+$$

Note: Subtract valence electrons for positive charged ions.

Lewis
structure

e. H_2CO has $2(1) + 4 + 6 = 12$ valence electrons.

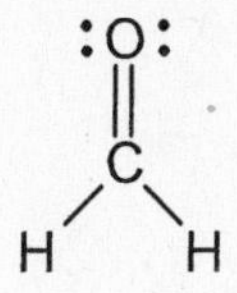

:O:
‖
C
/   \
H   H

f. SeF_2 has $6 + 2(7) = 20$ valence electrons.

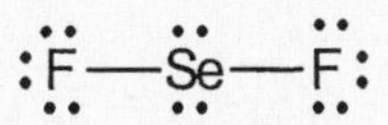

:F̈—S̈e—F̈:

g. CO_2 has $4 + 2(6) = 16$ valence electrons.

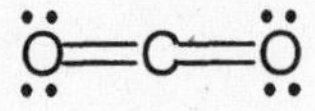

Ö=C=Ö

h. O_2 has $2(6) = 12$ valence electrons.

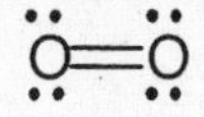

Ö=Ö

i. HBr has $1 + 7 = 8$ valence electrons.

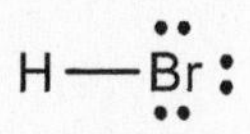

H—B̈r:

62.   a. $POCl_3$ has $5 + 6 + 3(7) = 32$ valence electrons.

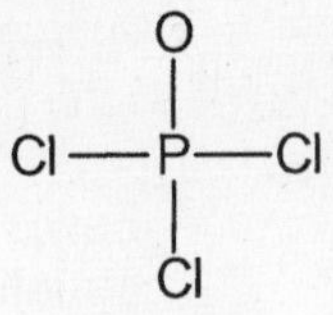

Skeletal   Lewis
structure   structure

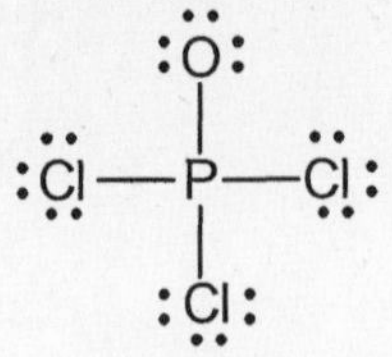

This structure uses all 32 e⁻ while satisfying the octet rule for all atoms. This is a valid Lewis structure.

SO_4^{2-} has $6 + 4(6) + 2 = 32$ valence electrons.

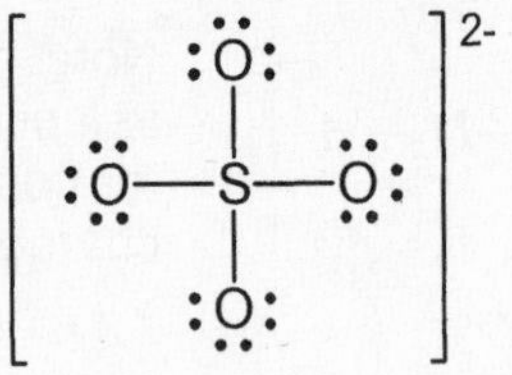

Note: A negatively charged ion will have additional electrons to those that come from the valence shells of the atoms.

XeO_4, $8 + 4(6) = 32$ e⁻ PO_4^{3-}, $5 + 4(6) + 3 = 32$ e⁻

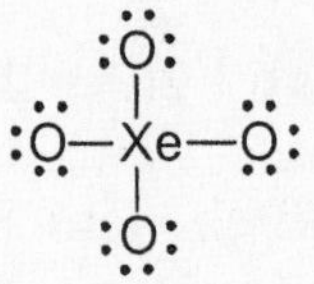

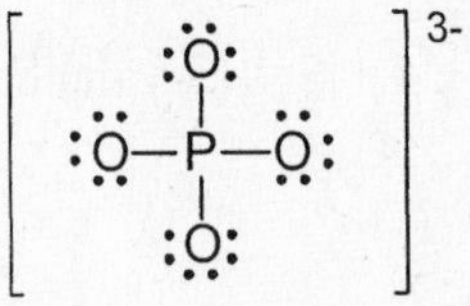

ClO_4^{-} has $7 + 4(6) + 1 = 32$ valence electrons.

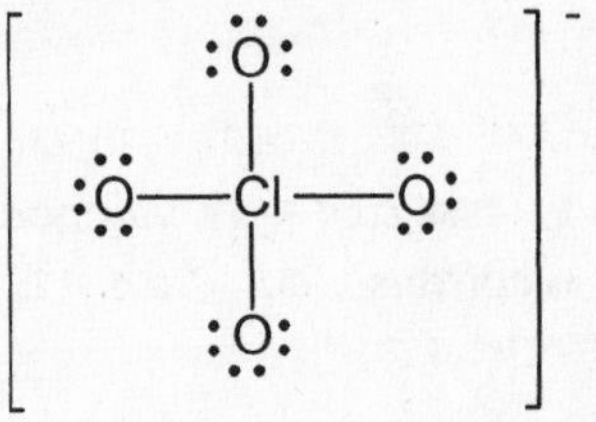

Note: All of these species have the same number of atoms and the same number of valence electrons. They also have the same Lewis structure.

b. NF_3 has $5 + 3(7) = 26$ valence electrons. SO_3^{2-}, $6 + 3(6) + 2 = 26$ e⁻

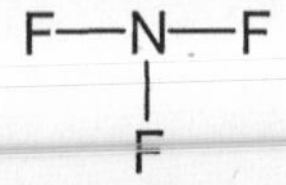

Skeletal   Lewis
structure   structure

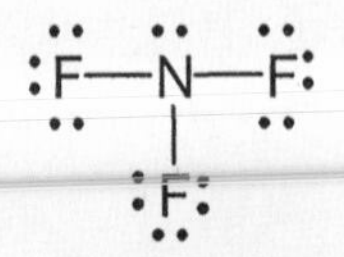

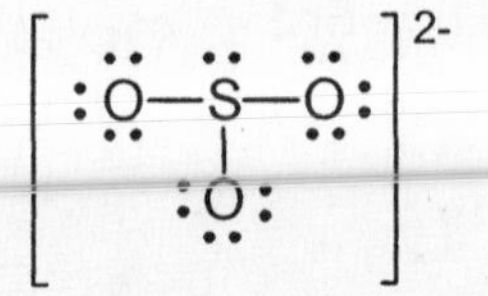

PO_3^{3-}, $5 + 3(6) + 3 = 26$ e⁻ ClO_3^-, $7 + 3(6) + 1 = 26$ e⁻

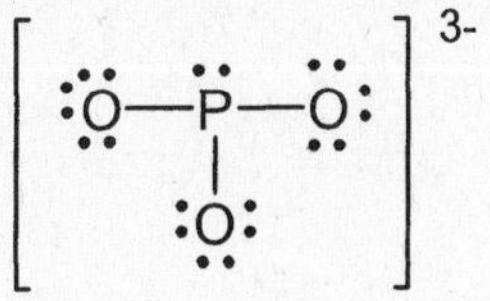

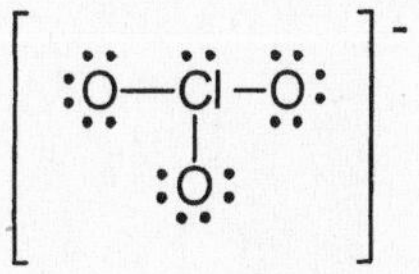

Note: Species with the same number of atoms and valence electrons have similar Lewis structures.

c. ClO_2^- has $7 + 2(6) + 1 = 20$ valence electrons.

Skeletal   Lewis
structure   structure

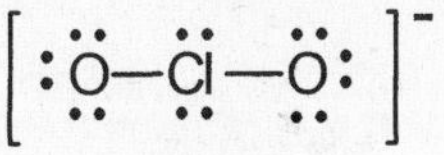

SCl_2, $6 + 2(7) = 20$ e⁻ PCl_2^-, $5 + 2(7) + 1 = 20$ e⁻

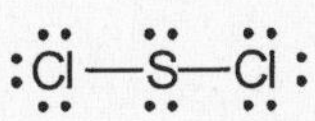

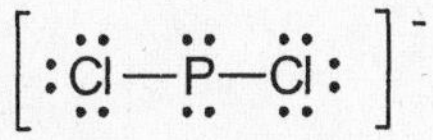

Note: Species with the same number of atoms and valence electrons have similar Lewis structures.

d.   Molecules/ions that have the same number of valence electrons and the same number of atoms will have similar Lewis structures.

63.   In each case in this problem, the octet rule cannot be satisfied for the central atom. BeH_2 and BH_3 have too few electrons around the central atom, and all the others have too many electrons around the central atom. Always try to satisfy the octet rule for every atom, but when it is impossible, the central atom is the species which will disobey the octet rule.

PF_5, $5 + 5(7) = 40$ e⁻ BeH_2, $2 + 2(1) = 4$ e⁻

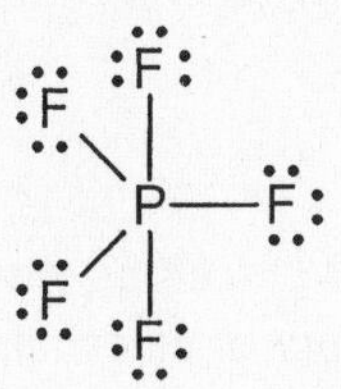

H – Be – H

BH_3, $3 + 3(1) = 6\ e^-$ Br_3^-, $3(7) + 1 = 22\ e^-$

SF_4, $6 + 4(7) = 34\ e^-$ XeF_4, $8 + 4(7) = 36\ e^-$

ClF_5, $7 + 5(7) = 42\ e^-$ SF_6, $6 + 6(7) = 48\ e^-$

64. ClF_3 has $7 + 3(7) = 28$ valence BrF_3 also has 28 valence
     electrons.   electrons.

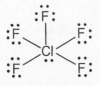   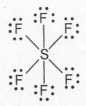

We expand the octet of the central Cl atom in ClF_3 and the central Br atom in BrF_3.

65. a. NO_2^- has $5 + 2(6) + 1 = 18$ valence electrons. The skeletal structure is: O – N – O

To get an octet about the nitrogen and only use $18\ e^-$, we must form a double bond to one of the oxygen atoms.

$$\left[\ \ddot{O}=N-\ddot{\underset{..}{O}}:\ \right]^- \longleftrightarrow \left[\ :\ddot{\underset{..}{O}}-N=\ddot{O}\ \right]^-$$

Since there is no reason to have the double bond to a particular oxygen atom, we can draw two resonance structures. Each Lewis structure uses the correct number of electrons and satisfies the octet rule, so each is a valid Lewis structure. Resonance structures occur when you have multiple bonds that can be in various positions. We say the actual structure is an average of these two resonance structures.

NO_3^- has $5 + 3(6) + 1 = 24$ valence electrons. We can draw three resonance structures for NO_3^-, with the double bond rotating among the three oxygen atoms.

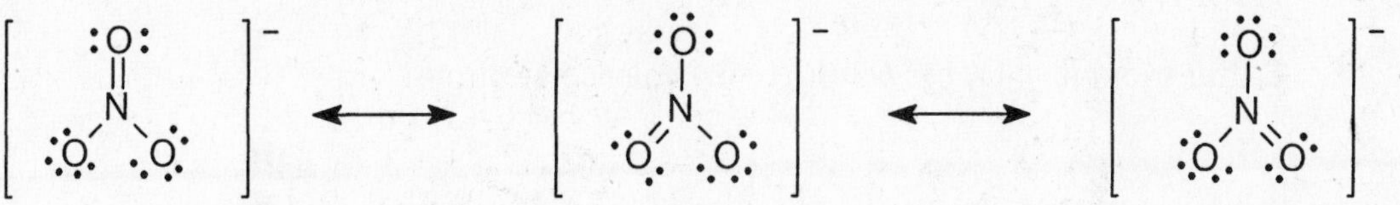

N_2O_4 has $2(5) + 4(6) = 34$ valence electrons. We can draw four resonance structures for N_2O_4.

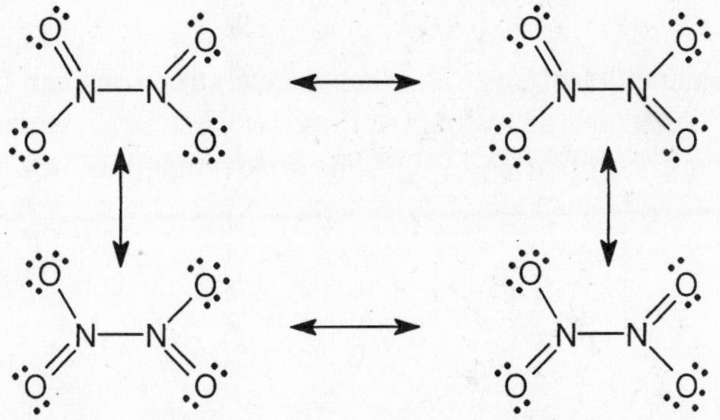

b. OCN^- has $6 + 4 + 5 + 1 = 16$ valence electrons. We can draw three resonance structures for OCN^-.

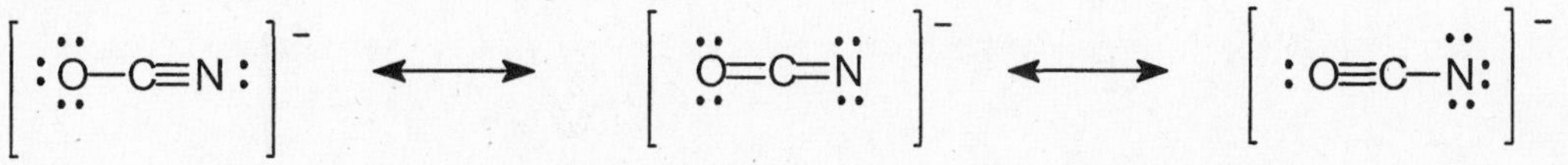

SCN^- has $6 + 4 + 5 + 1 = 16$ valence electrons. Three resonance structures can be drawn.

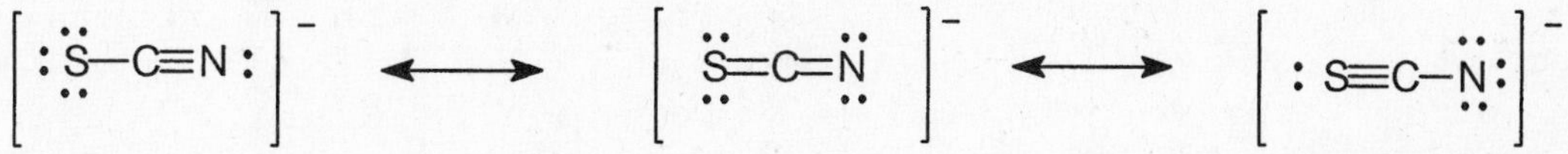

N_3^- has $3(5) + 1 = 16$ valence electrons. As with OCN^- and SCN^-, three different resonance structures can be drawn.

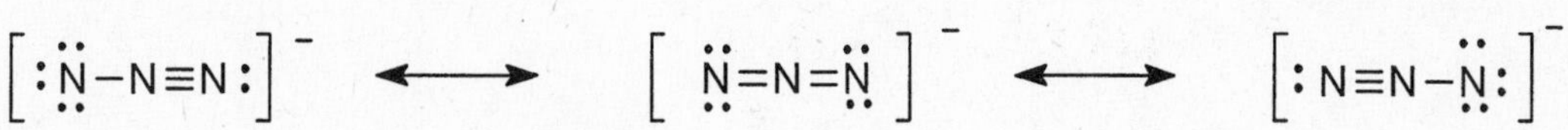

66.   Ozone: O_3 has $3(6) = 18$ valence electrons.

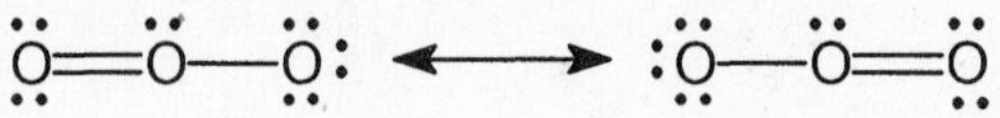

Sulfur dioxide: SO_2 has $6 + 2(6) = 18$ valence electrons.

Sulfur trioxide: SO_3 has $6 + 3(6) = 24$ valence electrons.

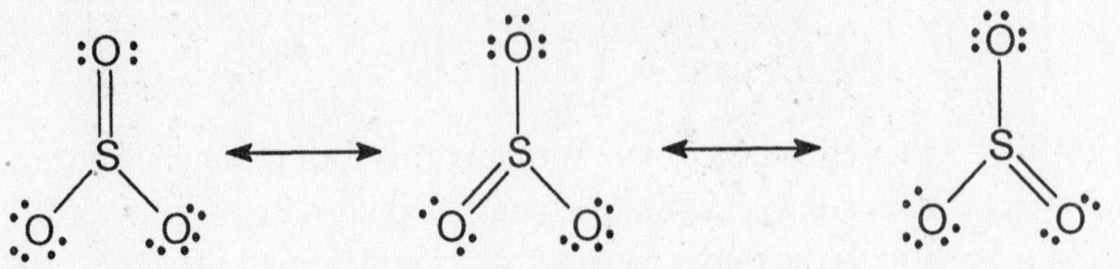

67.   Benzene has $6(4) + 6(1) = 30$ valence electrons. Two resonance structures can be drawn for benzene. The actual structure of benzene is an average of these two resonance structures, that is, all carbon-carbon bonds are equivalent with a bond length and bond strength somewhere between a single and a double bond.

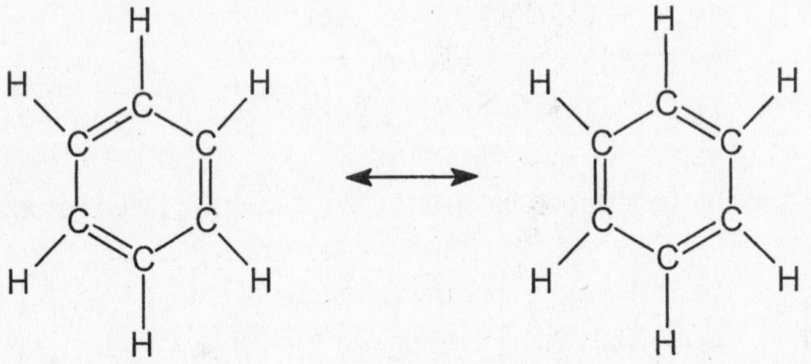

68.   Borazine $(B_3N_3H_6)$ has $3(3) + 3(5) + 6(1) = 30$ valence electrons. The possible resonance structures are similar to those of benzene in Exercise 8.69.

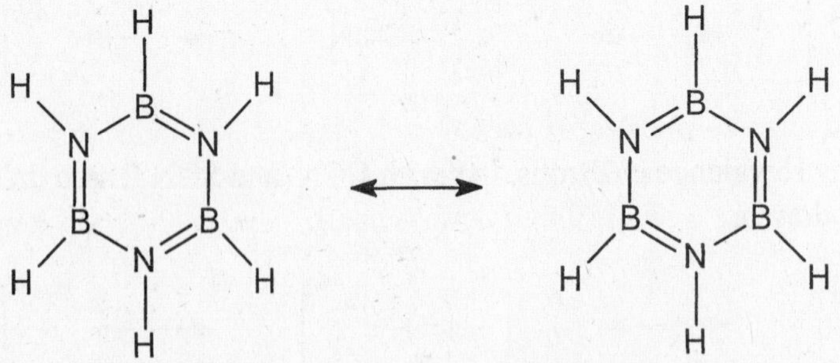

69.   We will use a hexagon to represent the six-member carbon ring, and we will omit the 4 hydrogen atoms and the three lone pairs of electrons on each chlorine. If no resonance existed, we could draw 4 different molecules:

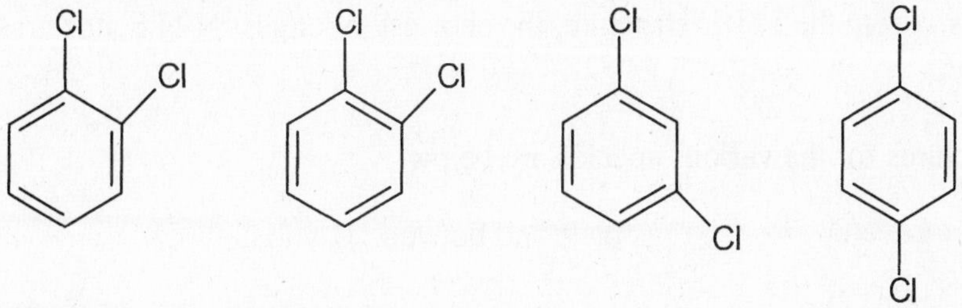

If the double bonds in the benzene ring exhibit resonance, then we can draw only three different dichlorobenzenes. The circle in the hexagon represents the delocalization of the three double bonds in the benzene ring (see Exercise 8.67).

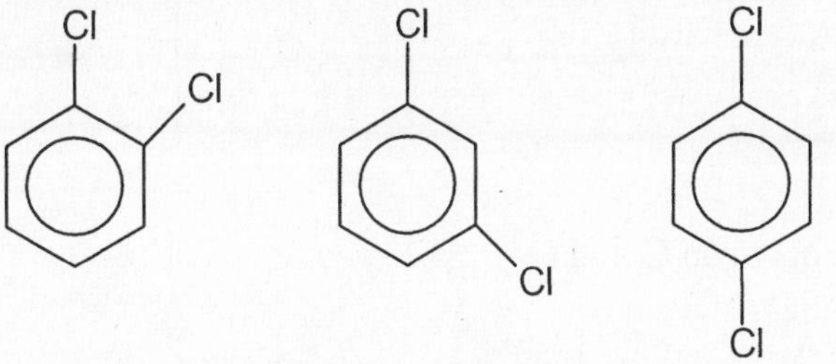

With resonance, all carbon-carbon bonds are equivalent. We can't distinguish between a single and double bond between adjacent carbons that have a chlorine attached. That only 3 isomers are observed provides evidence for the existence of resonance.

70. CO_3^{2-} has $4 + 3(6) + 2 = 24$ valence electrons.

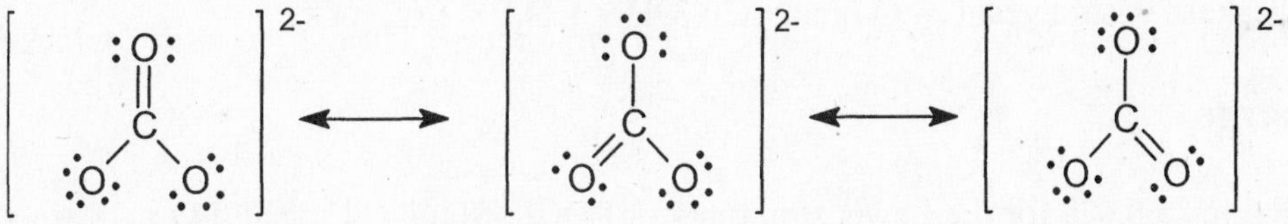

Three resonance structures can be drawn for CO_3^{2-}. The actual structure for CO_3^{2-} is an average of these three resonance structures. That is, the three C – O bond lengths are all equivalent, with a length somewhere between a single and a double bond. The actual bond length of 136 pm is consistent with this resonance view of CO_3^{2-}.

71. N_2 (10 e$^-$):   :N≡N: Triple bond between N and N.

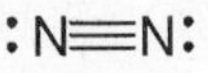

       N_2F_4 (38 e$^-$):   Single bond between N and N.

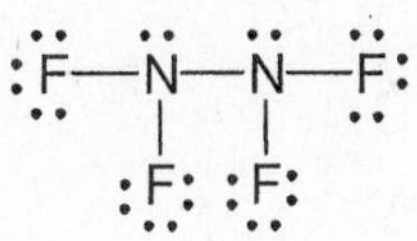

       N_2F_2 (24 e$^-$):   :F—N=N—F: Double bond between N and N.

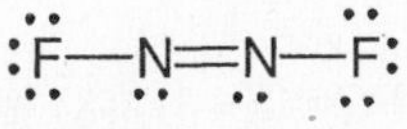

As the number of bonds increase between two atoms, bond strength increases and bond length decreases. From the Lewis structure, the shortest to longest N-N bonds are: $N_2 < N_2F_2 < N_2F_4$.

72.   The Lewis structures for the various species are below:

CO (10 e⁻):   :C≡O: Triple bond between C and O

CO_2 (16 e⁻):   Ö=C=Ö   Double bond between C and O

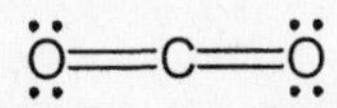

CO_3^{2-} (24 e⁻):

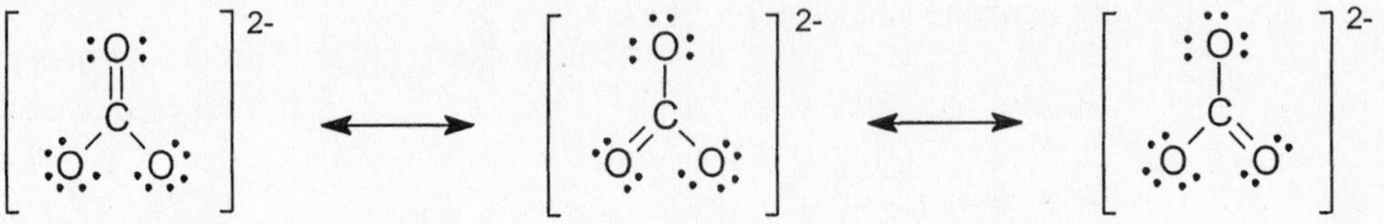

Average of 1 1/3 bond between C and O

CH_3OH (14 e⁻):   H–C–Ö–H   Single bond between C and O

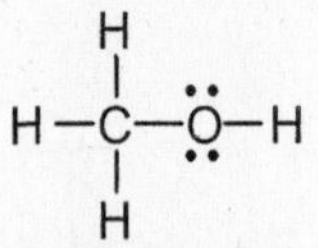

As the number of bonds increases between two atoms, bond length decreases and bond strength increases. With this in mind, then:

longest → shortest C – O bond: $CH_3OH > CO_3^{2-} > CO_2 > CO$

weakest → strongest C – O bond: $CH_3OH < CO_3^{2-} < CO_2 < CO$

Formal Charge

73.   See Exercise 8.62a for the Lewis structures of $POCl_3$, SO_4^{2-}, ClO_4^- and PO_4^{3-}. Formal charge = [number of valence electrons on free atom] - [number of lone pair electrons on atom - 1/2 (number of shared electrons of atom)].

a. $POCl_3$: P, FC = 5 - 1/2(8) = +1   b. SO_4^{2-}: S, FC = 6 - 1/2(8) = +2

c. ClO_4^-: Cl, FC = 7 - 1/2(8) = +3   d. PO_4^{3-}: P, FC = 5 - 1/2(8) = +1

e. SO_2Cl_2, 6 + 2(6) + 2(7) = 32 e⁻   f. XeO_4, 8 + 4(6) = 32 e⁻

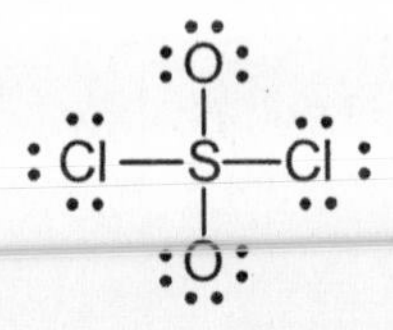

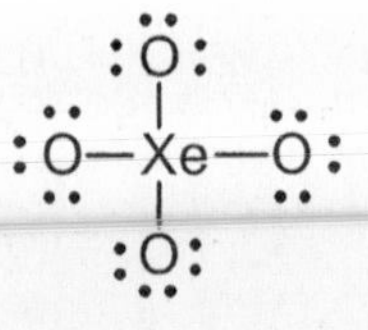

S, FC = 6 - 1/2(8) = +2   Xe, FC = 8 - 1/2(8) = +4

g. ClO_3^-, $7 + 3(6) + 1 = 26$ e$^-$ h. NO_4^{3-}, $5 + 4(6) + 3 = 32$ e$^-$

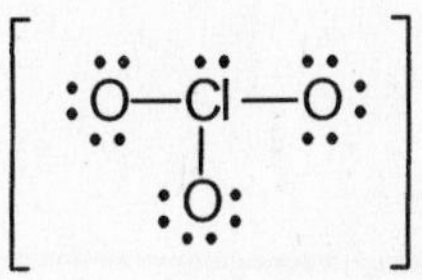

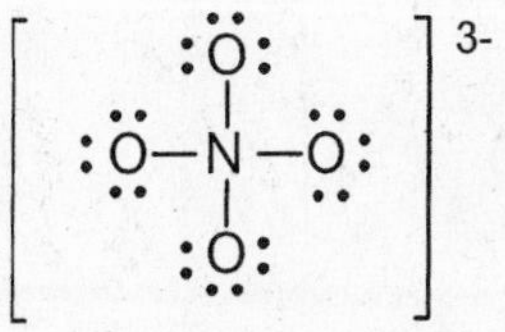

Cl, FC = $7 - 2 - 1/2(6) = +2$ N, FC = $5 - 1/2(8) = +1$

74.   For SO_4^{2-}, ClO_4^-, PO_4^{3-} and ClO_3^-, only one of the possible resonance structures is drawn.

a. Must have five bonds to P to minimize formal charge of P. The best choice is to form a double bond to O since this will give O a formal charge of zero and single bonds to Cl for the same reason.

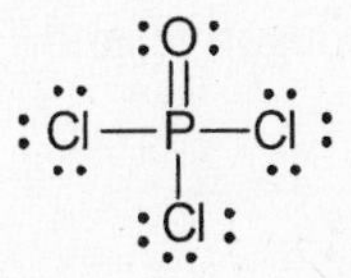

b. Must form six bonds to S to minimize formal charge of S.

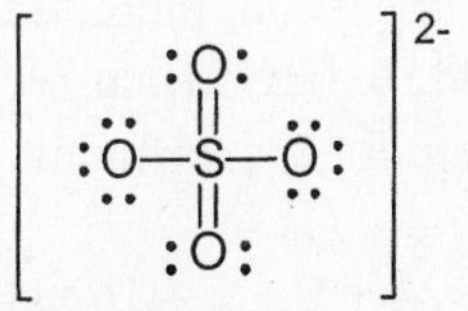

P, FC = 0   S, FC = 0

c. Must form seven bonds to Cl to minimize formal charge.

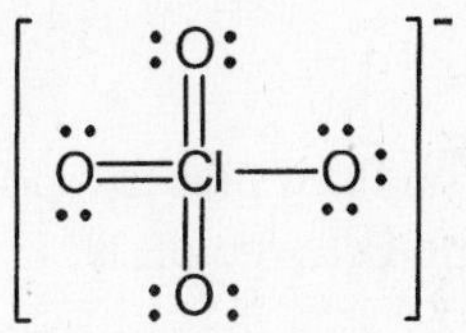

d. Must form five bonds to P to to minimize formal charge.

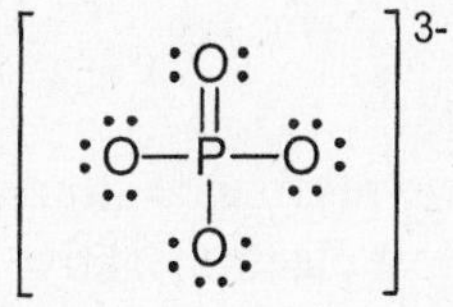

Cl, FC = 0   P, FC = 0

e.

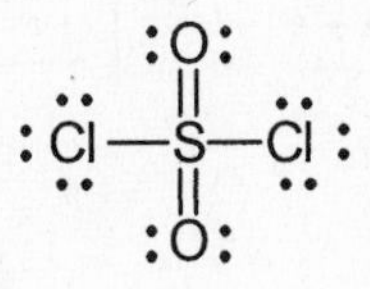

S, FC = 0
Cl, FC = 0
O, FC = 0

f.

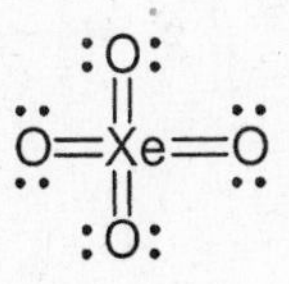

Xe, FC = 0

g.

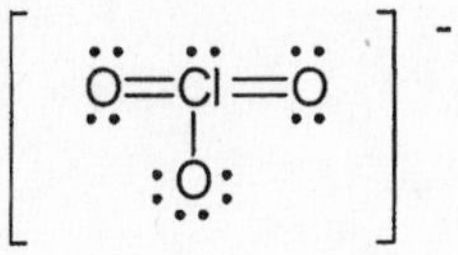

Cl, FC = 0

h.   We can't. The following structure has a zero formal charge for N:

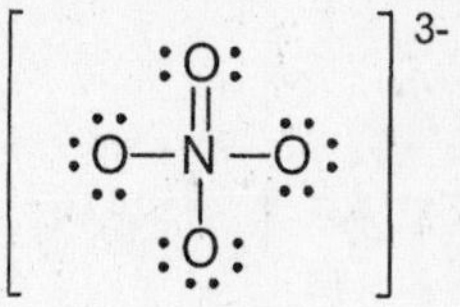

but N does not expand its octet. We wouldn't expect this resonance form to exist.

75. O_2F_2 has $2(6) + 2(7) = 26$ valence e^-.   The formal charge and oxidation number of each atom is below the Lewis structure of O_2F_2.

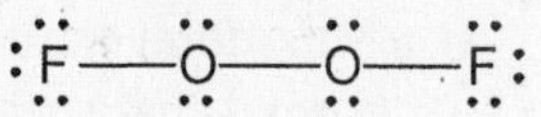

Formal Charge	0	0	0	0
Oxid. Number	-1	+1	+1	-1

Oxidation numbers are more useful when accounting for the reactivity of O_2F_2. We are forced to assign +1 as the oxidation number for oxygen. Oxygen is very electronegative, and +1 is not a stable oxidation state for this element.

76. OCN^- has $6 + 4 + 5 + 1 = 16$ valence electrons.

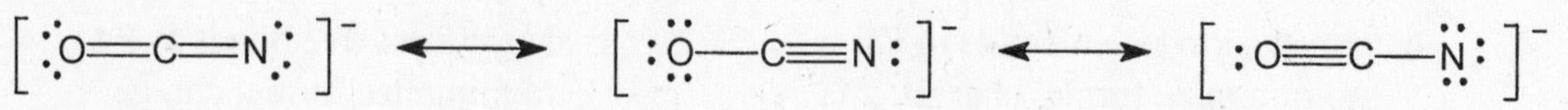

Formal
charge   0   0   -1   -1   0   0   +1   0   -2

Only the first two resonance structures should be important. The third places a positive formal charge on the most electronegative atom in the ion and a -2 formal charge on N.

CNO^-:

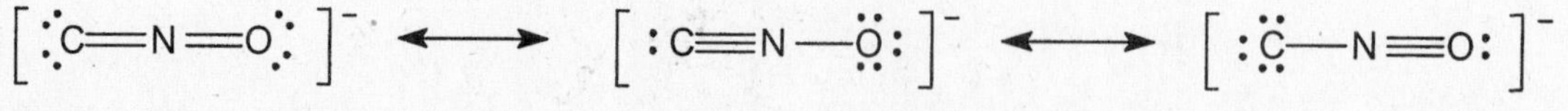

Formal
charge   -2   +1   0   -1   +1   -1   -3   +1   +1

All of the resonance structures for fulminate (CNO^-) involve greater formal charges than in cyanate (OCN^-), making fulminate more reactive (less stable).

Molecular Geometry and Polarity

77.   The first step always is to draw a valid Lewis structure when predicting molecular structure. When resonance is possible, only one of the possible resonance structures is necessary to predict the correct structure since all resonance structures give the same structure. The Lewis structures are in Exercises 8.61 and 8.65. The structures and bond angles for each follow.

8.61   a.   HCN: linear, 180°   b. PH_3: trigonal pyramid, < 109.5°

       c. $CHCl_3$: tetrahedral, 109.5°   d. NH_4^+: tetrahedral, 109.5°

       e. H_2CO: trigonal planar, 120°   f. SeF_2: V-shaped or bent, < 109.5°

       g. CO_2: linear, 180°   h and i. O_2 and HBr are both linear, but there is

                                                            no bond angle in either.

Note: PH_3 and SeF_2 both have lone pairs of electrons on the central atom which result in bond angles that are something less than predicted from a tetrahedral arrangement (109.5°). However, we cannot predict the exact number. For the solutions manual, we will insert a less than sign to indicate this phenomenon. For bond angles equal to 120°, the lone pair phenomenon isn't as significant as compared to smaller bond angles. For these molecules, e.g., NO_2^-, we will insert an approximate sign in front of the 120° to note that there may be a slight distortion from the VSEPR predicted bond angle.

8.65   a. NO_2^-: V-shaped, ≈ 120°; NO_3^-: trigonal planar, 120°;

            N_2O_4: trigonal planar, 120° about both N atoms

       b. OCN^-, SCN^- and N_3^- are all linear with 180° bond angles.

78.   See Exercises 8.62 and 8.66 for the Lewis structures.

8.62   a.   All are tetrahedral; 109.5°

       b.   All are trigonal pyramid; < 109.5°

       c.   All are V-shaped; < 109.5°

8.66 O_3 and SO_2 are V-shaped (or bent) with a bond angle ≈ 120°. SO_3 is trigonal planar with 120° bond angles.

79.   From the Lewis structures (see Exercises 8.63 and 8.64), Br_3^- would have a linear molecular structure, ClF_3 and BrF_3 would have a T-shaped molecular structure and SF_4 would have a see-saw molecular structure. For example, consider ClF_3 (28 valence electrons):

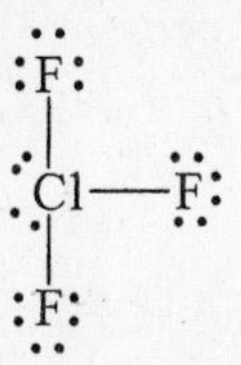

The central Cl atom is surrounded by 5 electron pairs, which requires a trigonal bipyramid geometry. Since there are 3 bonded atoms and 2 lone pairs of electrons about Cl, we describe the molecular structure of ClF_3 as T-shaped with predicted bond angles of about 90°. The actual bond angles would be slightly less than 90° due to the stronger repulsive effect of the lone pair electrons as compared to the bonding electrons.

80.   From the Lewis structures (see Exercise 8.63), XeF_4 would have a square planar molecular structure, and ClF_5 would have a square pyramid molecular structure.

81.   a. SeO_3, $6 + 3(6) = 24$ e⁻

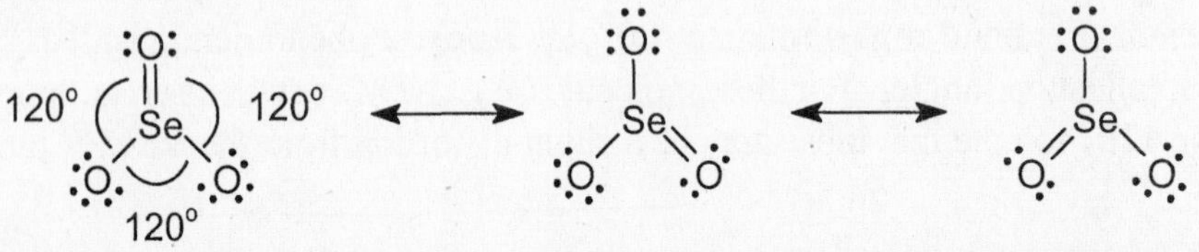

SeO_3 has a trigonal planar molecular structure with all bond angles equal to 120°. Note that any one of the resonance structures could be used to predict molecular structure and bond angles.

   b. SeO_2, $6 + 2(6) = 18$ e⁻

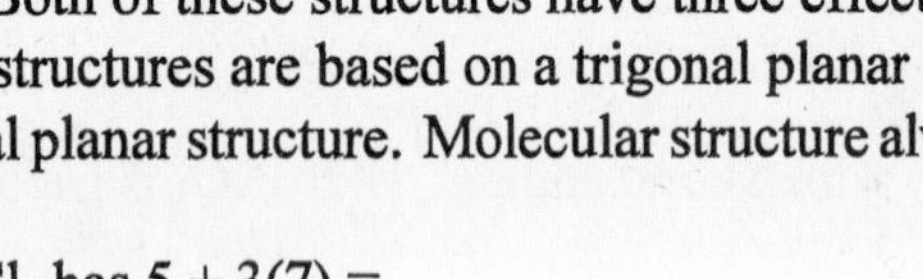

≈ 120°

SeO_2 has a V-shaped molecular structure. We would expect the bond angle to be approximately 120° as expected for trigonal planar geometry.

Note: Both of these structures have three effective pairs of electrons about the central atom. All of the structures are based on a trigonal planar geometry, but only SeO_3 is described as having a trigonal planar structure. Molecular structure always describes the relative positions of the atoms.

82.   a. PCl_3 has $5 + 3(7) =$ 26 valence electrons.   b. SCl_2 has $6 + 2(7) =$ 20 valence electrons

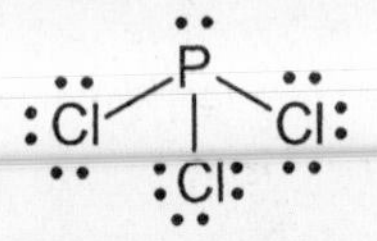

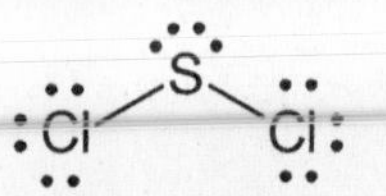

Trigonal pyramid; all angles are < 109.5°.   V-shaped; angle is < 109.5°.

c. SiF_4 has $4 + 4(7) = 32$ valence electrons.

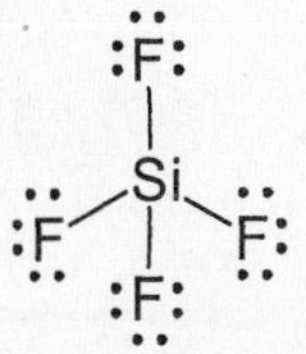

Tetrahedral; all angles are $109.5°$.

Note: There are 4 pairs of electrons about the central atom in each case in this exercise. All of the structures are based on a tetrahedral geometry, but only SiF_4 has a tetrahedral structure. We consider only the relative positions of the atoms when describing the molecular structure.

83.   a. $XeCl_2$ has $8 + 2(7) = 22$ valence electrons.

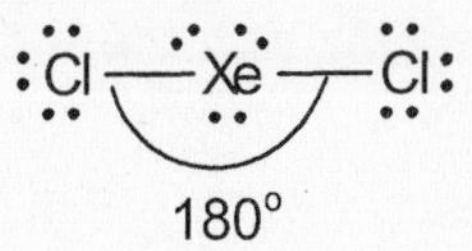

$180°$

There are 5 pairs of electrons about the central Xe atom. The structure will be based on a trigonal bipyramid geometry. The most stable arrangement of the atoms in $XeCl_2$ is a linear molecular structure with a $180°$ bond angle.

b. ICl_3 has $7 + 3(7) = 28$ valence electrons.

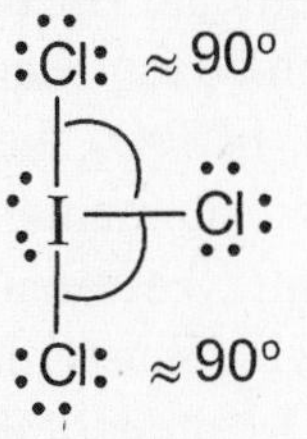

T-shaped; The ClICl angles are $\approx 90°$. Since the lone pairs will take up more space, the ClICl bond angles will probably be slightly less than $90°$.

c. TeF_4 has $6 + 4(7) = 34$ valence electrons.

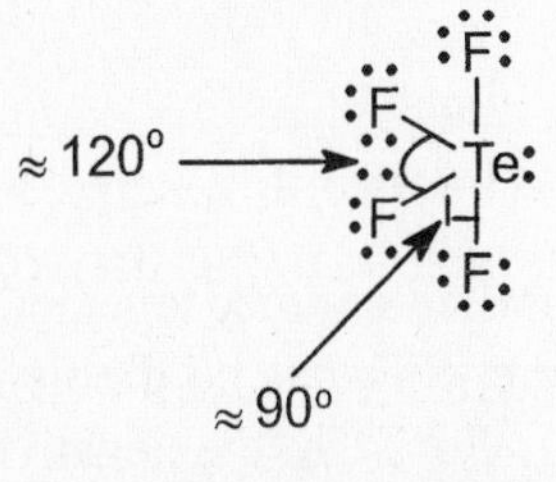

d. PCl_5 has $5 + 5(7) = 40$ valence electrons.

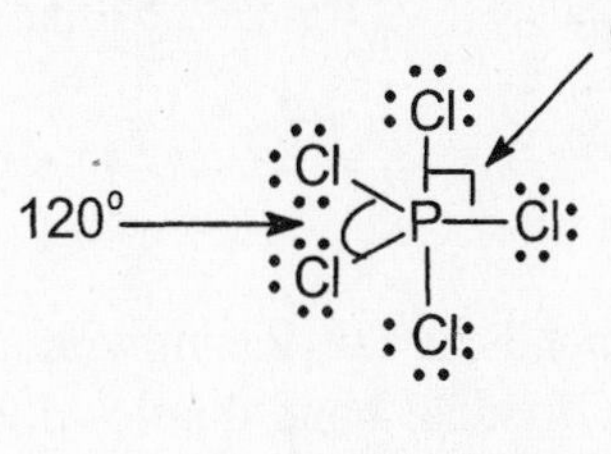

See-saw or teeter-totter or distorted tetrahedron

Trigonal bipyramid

All of the species in this exercise have 5 pairs of electrons around the central atom. All of the structures are based on a trigonal bipyramid geometry, but only in PCl_5 are all of the pairs bonding pairs. Thus, PCl_5 is the only one we describe as a trigonal bipyramid molecular structure. Still, we had to begin with the trigonal bipyramid geometry to get to the structures of the others.

84.   a. ICl_5, $7 + 5(7) = 42$ e⁻   b. $XeCl_4$, $8 + 4(7) = 36$ e⁻

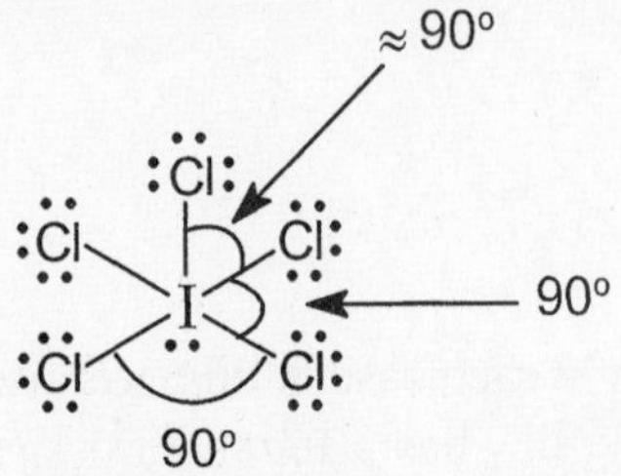

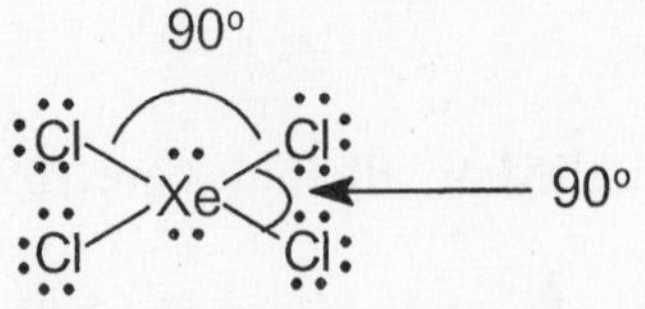

Square pyramid, $\approx 90°$ bond angles   Square planar, $90°$ bond angles

c. $SeCl_6$ has $6 + 6(7) = 48$ valence electrons.

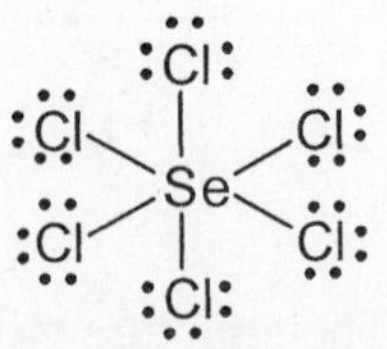

Octahedral, $90°$ bond angles

Note: All these species have 6 pairs of electrons around the central atom. All three structures are based on the octahedron, but only $SeCl_6$ has an octahedral molecular structure.

85. SeO_3 and SeO_2 both have polar bonds but only SeO_2 has a dipole moment. The three bond dipoles from the three polar $Se-O$ bonds in SeO_3 will all cancel when summed together. Hence, SeO_3 is nonpolar since the overall molecule has no resulting dipole moment. In SeO_2, the two $Se-O$ bond dipoles do not cancel when summed together, hence SeO_2 has a dipole moment (is polar). Since O is more electronegative than Se, the negative end of the dipole moment is between the two O atoms, and the positive end is around the Se atom. The arrow in the following illustration represents the overall dipole moment in SeO_2. Note that to predict polarity for SeO_2, either of the two resonance structures can be used.

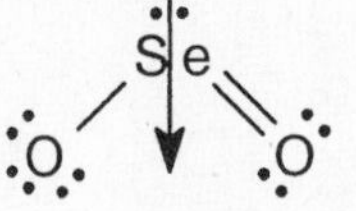

86.   All have polar bonds; in SiF_4 the individual bond dipoles cancel when summed together, and in PCl_3 and SCl_2 the individual bond dipoles do not cancel. Therefore, SiF_4 has no dipole moment (is nonpolar), and PCl_3 and SCl_2 have dipole moments (are polar). For PCl_3, the negative end of the dipole moment is between the more electronegative chlorine atoms and the positive end is around P. For SCl_2, the negative end is between the more electronegative Cl atoms, and the positive end of the dipole moment is around S.

87. All have polar bonds, but only TeF_4 and ICl_3 have dipole moments. The bond dipoles from the five P–Cl bonds in PCl_5 cancel each other when summed together, so PCl_5 has no dipole moment. The bond dipoles in $XeCl_2$ also cancel:

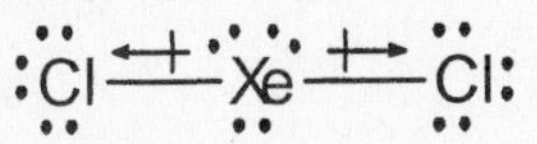

Since the bond dipoles from the two Xe – Cl bonds are equal in magnitude but point in opposite directions, they cancel each other and $XeCl_2$ has no dipole moment (is nonpolar). For TeF_4 and ICl_3, the arrangement of these molecules is such that the individual bond dipoles do not all cancel, so each has an overall dipole moment.

88. All have polar bonds, but only ICl_5 has an overall dipole moment. The six bond dipoles in $SeCl_6$ all cancel each other, so $SeCl_6$ has no dipole moment. The same is true for $XeCl_4$:

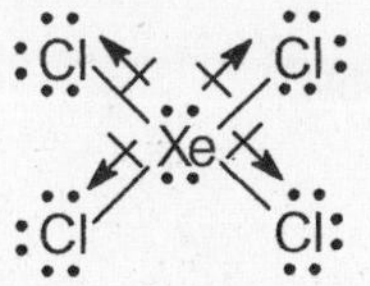

When the four bond dipoles are added together, they all cancel each other, and $XeCl_4$ has no overall dipole moment. ICl_5 has a structure where the individual bond dipoles do not all cancel, hence ICl_5 has a dipole moment.

89. Molecules which have an overall dipole moment are called polar molecules, and molecules which do not have an overall dipole moment are called nonpolar molecules.

a. OCl_2, $6 + 2(7) = 20$ e⁻

KrF_2, $8 + 2(7) = 22$ e⁻

V-shaped, polar; OCl_2 is polar because the two O – Cl bond dipoles don't cancel each other. The resultant dipole moment is shown in the drawing.

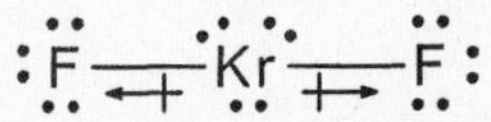

Linear, nonpolar; The molecule is nonpolar because the two Kr – F bond dipoles cancel each other.

BeH_2, $2 + 2(1) = 4$ e⁻

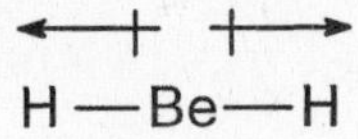

SO_2, $6 + 2(6) = 18$ e⁻

Linear, nonpolar; Be – H bond dipoles are equal and point in opposite directions. They cancel each other. BeH_2 is nonpolar.

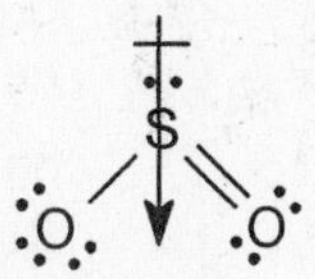

V-shaped, polar; The S – O bond dipoles do not cancel, so SO_2 is polar (has a dipole moment). Only one resonance structure is shown.

Note: All four species contain three atoms. They have different structures because the number of lone pairs of electrons around the central atom is different in each case.

b. SO_3, $6 + 3(6) = 24$ e⁻ NF_3, $5 + 3(7) = 26$ e⁻

Trigonal planar, nonpolar;   Trigonal pyramid, polar;
Bond dipoles cancel. Only one   Bond dipoles do not cancel.
resonance structure is shown.

IF_3 has $7 + 3(7) = 28$ valence electrons.

T-shaped, polar; bond dipoles do not cancel.

Note: Each molecule has the same number of atoms, but the structures are different because of differing numbers of lone pairs around each central atom.

c. CF_4, $4 + 4(7) = 32$ e⁻ SeF_4, $6 + 4(7) = 34$ e⁻

Tetrahedral, nonpolar;   See-saw, polar;
Bond dipoles cancel.   Bond dipoles do not cancel.

KrF_4, $8 + 4(7) = 36$ valence electrons

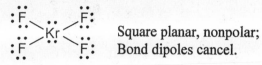

Square planar, nonpolar;
Bond dipoles cancel.

Again, each molecule has the same number of atoms, but a different structure because of differing numbers of lone pairs around the central atom.

d. IF_5, $7 + 5(7) = 42$ e$^-$ AsF_5, $5 + 5(7) = 40$ e$^-$

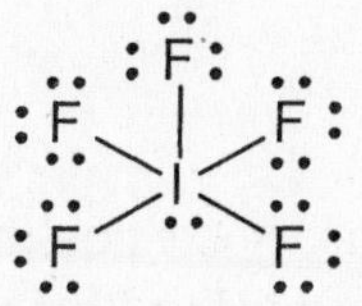

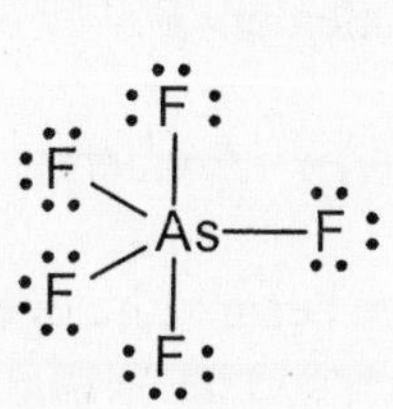

Square pyramid, polar;   Trigonal bipyramid, nonpolar;
Bond dipoles do not cancel.   Bond dipoles cancel.

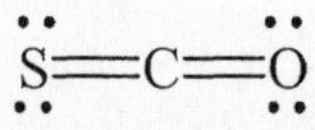

Yet again, the molecules have the same number of atoms, but different structures because of the
presence of differing numbers of lone pairs.

90.   a.   b.

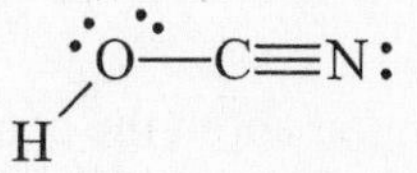

Polar; The bond dipoles do   Polar; The C – O bond is a more
not cancel.   polar bond than the C – S bond. So
                                      the two bond dipoles do not cancel
                                      each other.

c.   d.

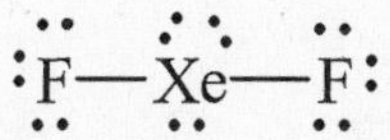

Nonpolar; The two Xe – F bond   Polar; All the bond dipoles are not
dipoles cancel each other.   equivalent, so they don't cancel each
                                      other.

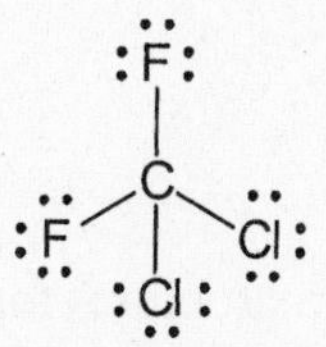

e.   f.

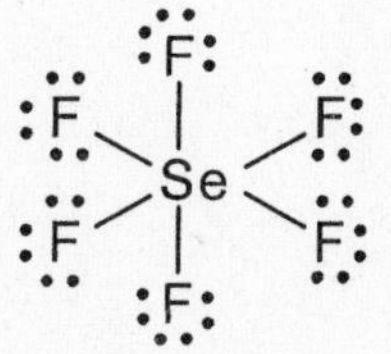

Nonpolar; The six Se – F bond dipoles   Polar; Bond dipoles are not equivalent
cancel each other.   so they don't cancel each other.

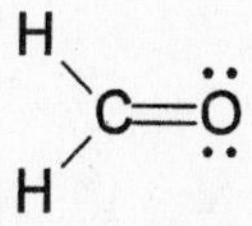

91. EO_3^- is the formula of the ion.   The Lewis structure has 26 valence electrons. Let x = number of valence electrons of element E.

       $26 = x + 3(6) + 1, \ x = 7$ valence electrons

       Element E is a halogen because halogens have 7 valence electrons. Some possible identities are F, Cl, Br and I. The EO_3^- ion has a trigonal pyramid molecular structure with bond angles < 109.5°.

92.   The formula is EF_2O^{2-} and the Lewis structure has 28 valence electrons.

       $28 = x + 2(7) + 6 + 2, \ x = 6$ valence electrons for element E

       Element E must belong to the group 6A elements since E has 6 valence electrons. E must also be a row 3 or heavier element since this ion has more than 8 electrons around the central E atom (row 2 elements never have more than 8 electrons around them). Some possible identities for E are S, Se and Te. The ion has a T-shaped molecular structure (see Exercise 8.79) with bond angles of ≈ 90°.

93.   All these molecules have polar bonds that are symmetrically arranged about the central atoms. In each molecule, the individual bond dipoles cancel to give no net overall dipole moment.

94. XeF_2Cl_2, $8 + 2(7) + 2(7) = 36 \ e^-$

       polar   nonpolar

       The two possible structures for XeF_2Cl_2 are above. In the first structure, the F atoms are 90° apart from each other and the Cl atoms are also 90° apart. The individual bond dipoles would not cancel in this molecule, so this molecule is polar. In the second possible structure, the F atoms are 180° apart as are the Cl atoms. Here, the bond dipoles are symmetrically arranged so they do cancel out each other, and this molecule is nonpolar. Therefore, measurement of the dipole moment would differentiate between the two compounds.

Additional Exercises

95.   a.   Radius: $N^+ < N < N^-$;   IE: $N^- < N < N^+$

            N^+ has the fewest electrons held by the 7 protons in the nucleus while N^- has the most electrons held by the 7 protons. The 7 protons in the nucleus will hold the electrons most tightly in N^+ and least tightly in N^-. Therefore, N^+ has the smallest radius with the largest ionization energy (IE) and N^- is the largest species with the smallest IE.

b. Radius: $Cl^+ < Cl < Se < Se^-$;   IE: $Se^- < Se < Cl < Cl^+$

The general trends tell us that Cl has a smaller radius than Se and a larger IE than Se. Cl^+, with fewer electron-electron repulsions than Cl, will be smaller than Cl and have a larger IE. Se^-, with more electron-electron repulsions than Se, will be larger than Se and have a smaller IE.

c. Radius: $Sr^{2+} < Rb^+ < Br^-$;   IE: $Br^- < Rb^+ < Sr^{2+}$

These ions are isoelectronic. The species with the most protons (Sr^{2+}) will hold the electrons most tightly and will have the smallest radius and largest IE. The ion with the fewest protons (Br^-) will hold the electrons least tightly and will have the largest radius and smallest IE.

96. a. $Na^+(g) + Cl^-(g) \rightarrow NaCl(s)$ b. $NH_4^+(g) + Br^-(g) \rightarrow NH_4Br(s)$

c. $Mg^{2+}(g) + S^{2-}(g) \rightarrow MgS(s)$ d. $O_2(g) \rightarrow 2\, O(g)$

97. a.

$HF(g) \rightarrow H(g) + F(g)$	$\Delta H = 565$ kJ
$H(g) \rightarrow H^+(g) + e^-$	$\Delta H = 1312$ kJ
$F(g) + e^- \rightarrow F^-(g)$	$\Delta H = -327.8$ kJ
$HF(g) \rightarrow H^+(g) + F^-(g)$	$\Delta H = 1549$ kJ

b.

$HCl(g) \rightarrow H(g) + Cl(g)$	$\Delta H = 427$ kJ
$H(g) \rightarrow H^+(g) + e^-$	$\Delta H = 1312$ kJ
$Cl(g) + e^- \rightarrow Cl^-(g)$	$\Delta H = -348.7$ kJ
$HCl(g) \rightarrow H^+(g) + Cl^-(g)$	$\Delta H = 1390.$ kJ

c.

$HI(g) \rightarrow H(g) + I(g)$	$\Delta H = 295$ kJ
$H(g) \rightarrow H^+(g) + e^-$	$\Delta H = 1312$ kJ
$I(g) + e^- \rightarrow I^-(g)$	$\Delta H = -295.2$ kJ
$HI(g) \rightarrow H^+(g) + I^-(g)$	$\Delta H = 1312$ kJ

d.

$H_2O(g) \rightarrow OH(g) + H(g)$	$\Delta H = 467$ kJ
$H(g) \rightarrow H^+(g) + e^-$	$\Delta H = 1312$ kJ
$OH(g) + e^- \rightarrow OH^-(g)$	$\Delta H = -180.$ kJ
$H_2O(g) \rightarrow H^+(g) + OH^-(g)$	$\Delta H = 1599$ kJ

98. CO_3^{2-} has $4 + 3(6) + 2 = 24$ valence electrons.

HCO_3^- has $1 + 4 + 3(6) + 1 = 24$ valence electrons.

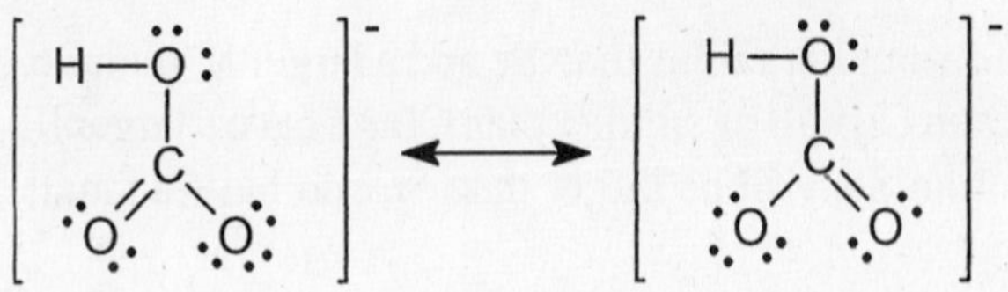

H_2CO_3 has $2(1) + 4 + 3(6) = 24$ valence electrons.

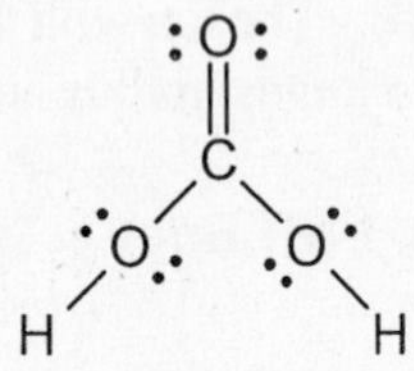

The Lewis structures for the reactants and products are:

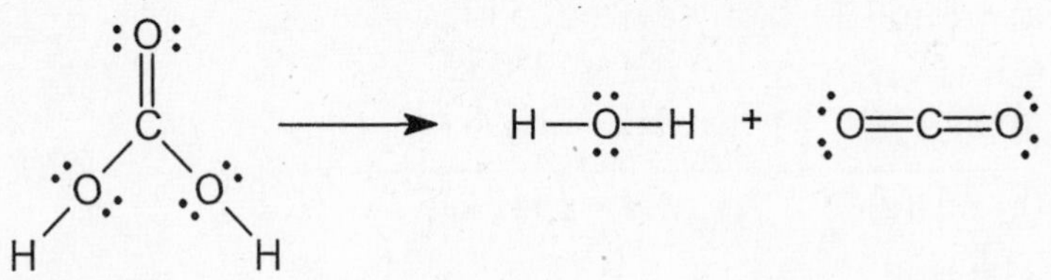

Bonds broken:	Bonds formed:
2 C – O (358 kJ/mol)	1 C = O (799 kJ/mol)
1 O – H (467 kJ/mol)	1 O – H (467 kJ/mol)

$\Delta H = 2(358) + 467 - [799 + 467] = -83$ kJ; The carbon-oxygen double bond is stronger than two carbon-oxygen single bonds, hence CO_2 and H_2O are more stable than H_2CO_3.

99.   The stable species are:

a.   NaBr: In $NaBr_2$, the sodium ion would have a +2 charge assuming each bromine has a -1 charge. Sodium doesn't form stable Na^{2+} compounds.

b. ClO_4^-: ClO_4 has 31 valence electrons so it is impossible to satisfy the octet rule for all atoms in ClO_4. The extra electron from the -1 charge in ClO_4^- allows for complete octets for all atoms.

c. XeO_4: We can't draw a Lewis structure that obeys the octet rule for SO_4 (30 electrons), unlike XeO_4 (32 electrons).

d. SeF_4: Both compounds require the central atom to expand its octet. O is too small and doesn't have low energy d orbitals to expand its octet (which is true for all row 2 elements).

100. a. NO_2, $5 + 2(6) = 17$ e^- N_2O_4, $2(5) + 4(6) = 34$ e^-

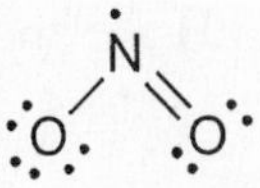

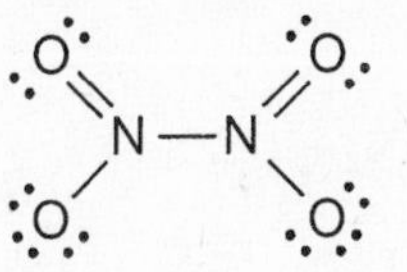

plus other resonance structures   plus other resonance structures

b. BF_3, $3 + 3(7) = 24$ e^- NH_3, $5 + 3(1) = 8$ e^-

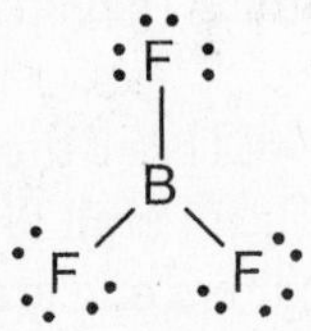

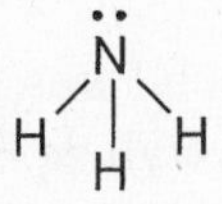

BF_3NH_3, $24 + 8 = 32$ e^-

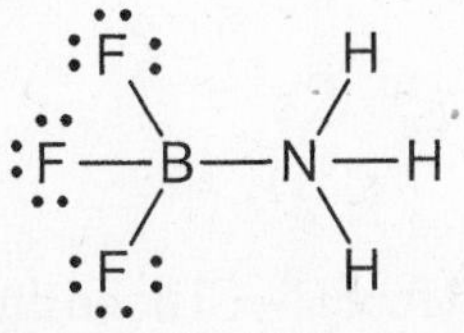

In reaction a, NO_2 has an odd number of electrons so it is impossible to satisfy the octet rule. By dimerizing to form N_2O_4, the odd electron on two NO_2 molecules can pair up, giving a species whose Lewis structure can satisfy the octet rule. In general, odd electron species are very reactive. In reaction b, BF_3 can be considered electron deficient. Boron has only six electrons around it. By forming BF_3NH_3, the boron atom satisfies the octet rule by accepting a lone pair of electrons from NH_3 to form a fourth bond.

101. The general structure of the trihalide ions is: $\left[\; :\ddot{X}-\ddot{X}-\ddot{X}: \; \right]^-$

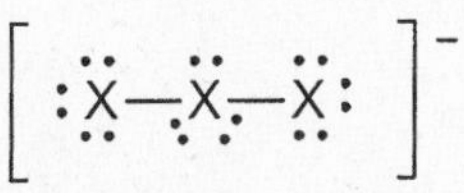

Bromine and iodine are large enough and have low energy, empty d-orbitals to accommodate the expanded octet. Fluorine is small, its valence shell contains only 2s and 2p orbitals (4 orbitals) and it does not expand its octet. The lowest energy d orbitals in F are 3d orbitals; they are too high in energy as compared to 2s and 2p to be used in bonding.

102. CS_2 has $4 + 2(6) = 16$ valence electrons. C_3S_2 has $3(4) + 2(6) = 24$ valence electrons.

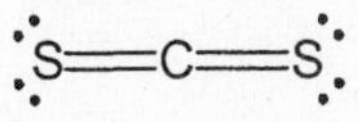

linear; 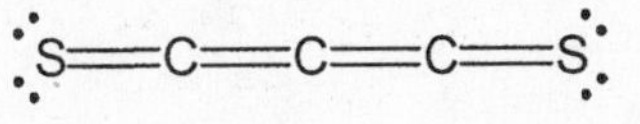   linear

103. Yes, each structure has the same number of effective pairs around the central atom. (A multiple bond is counted as a single group of electrons.)

104.   a.

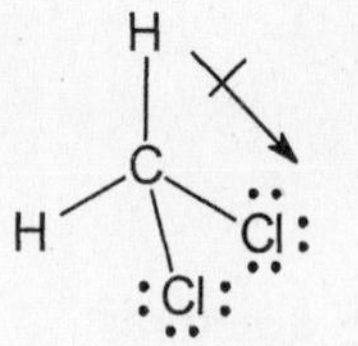

The C–H bonds are assumed nonpolar since the electronegativities of C and H are about equal.

δ+   δ-
C – Cl is the charge distribution for each C – Cl bond. In CH_2Cl_2, the two individual C–Cl bond dipoles add together to give an overall dipole moment for the molecule. The overall dipole will point from C (positive end) to the midpoint of the two Cl atoms (negative end).

In $CHCl_3$, the C – H bond is essentially nonpolar. The three C – Cl bond dipoles in $CHCl_3$ add together to give an overall dipole moment for the molecule. The overall dipole will have the negative end at the midpoint of the three chlorines and the positive end around the carbon.

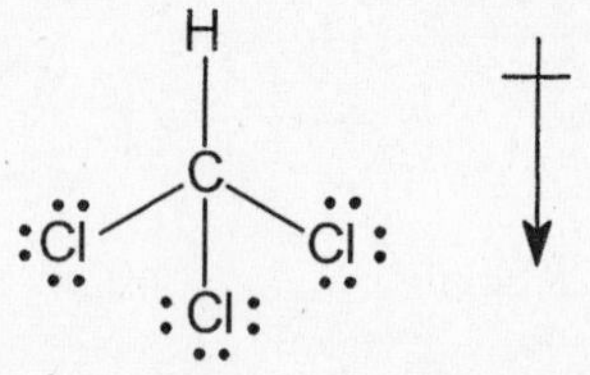

CCl_4 is nonpolar. CCl_4 is a tetrahedral molecule where all four C – Cl bond dipoles cancel when added together. Let's consider just the C and two of the Cl atoms. There will be a net dipole pointing in the direction of the middle of the two Cl atoms.

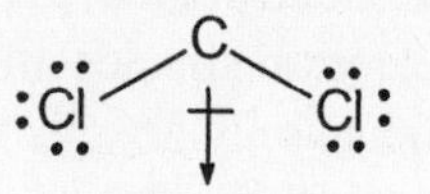

There will be an equal and opposite dipole arising from the other two Cl atoms. Combining:

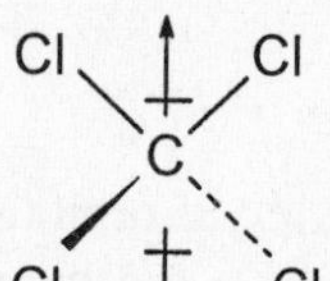

The two dipoles cancel and CCl_4 is nonpolar.

b. CO_2 is nonpolar. CO_2 is a linear molecule with two equivalence bond dipoles that cancel. N_2O is polar since the bond dipoles do not cancel.

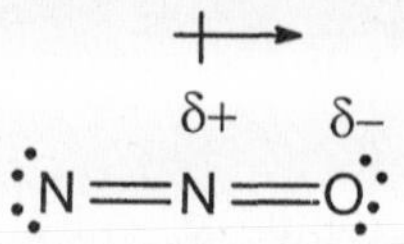

c. NH_3 is polar. The 3 N – H bond dipoles add together to give a net dipole in the direction of the lone pair. We would predict PH_3 to be nonpolar on the basis of electronegativitity, i.e., P – H bonds are nonpolar. However, the presence of the lone pair makes the PH_3 molecule slightly polar. The net dipole is in the direction of the lone pair and has a magnitude about one third that of the NH_3 dipole.

105. TeF_5^- has $6 + 5(7) + 1 = 42$ valence electrons.

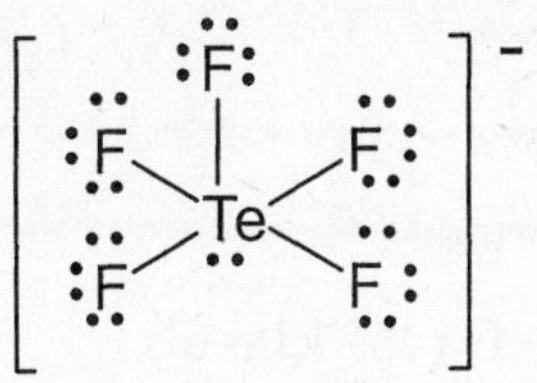

The lone pair of electrons around Te exerts a stronger repulsion than the bonding pairs, pushing the four square planar F's away from the lone pair and thus reducing the bond angles between the axial F atom and the square planar F atoms.

Challenge Problems

106.

	(IE - EA)	(IE - EA)/502	EN (text)	$2006/502 = 4.0$
F	2006 kJ/mol	4.0	4.0	
Cl	1604	3.2	3.0	
Br	1463	2.9	2.8	
I	1302	2.6	2.5	

The values calculated from IE and EA show the same trend (and agree fairly closely) to the values given in the text.

107. As the halogen atoms get larger, it becomes more difficult to fit three halogen atoms around the small nitrogen atom, and the NX_3 molecule becomes less stable.

108. a. I.

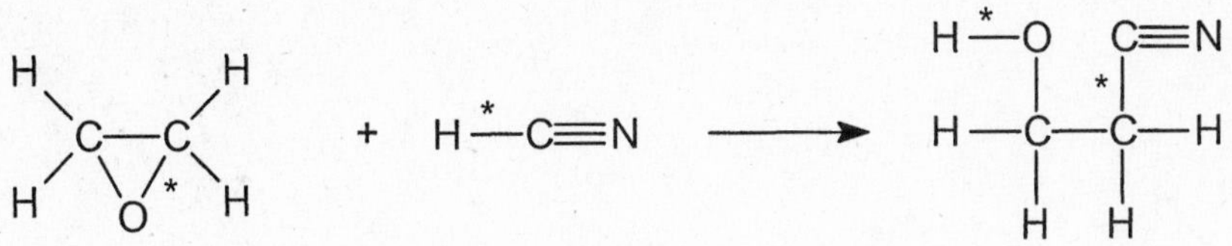

Bonds broken (*):   Bonds formed (*):

1 C – O (358 kJ)   1 O – H (467 kJ)
1 H – C (413 kJ)   1 C – C (347 kJ)

$\Delta H_I = 358 \text{ kJ} + 413 \text{ kJ} - [467 \text{ kJ} + 347 \text{ kJ}] = -43 \text{ kJ}$

II.

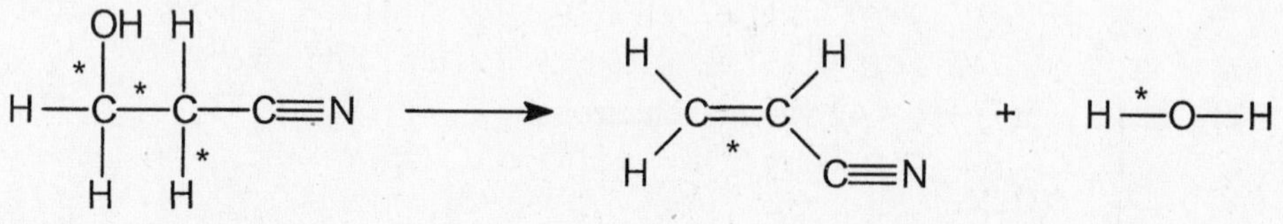

Bonds broken (*):   Bonds formed (*):

1 C – O (358 kJ/mol)   1 H – O (467 kJ/mol)
1 C – H (413 kJ/mol)   1 C = C (614 kJ/mol)
1 C – C (347 kJ/mol)

$\Delta H_{II} = 358 \text{ kJ} + 413 \text{ kJ} + 347 \text{ kJ} - [467 \text{ kJ} + 614 \text{ kJ}] = +37 \text{ kJ}$

$\Delta H_{overall} = \Delta H_I + \Delta H_{II} = -43 \text{ kJ} + 37 \text{ kJ} = -6 \text{ kJ}$

b.

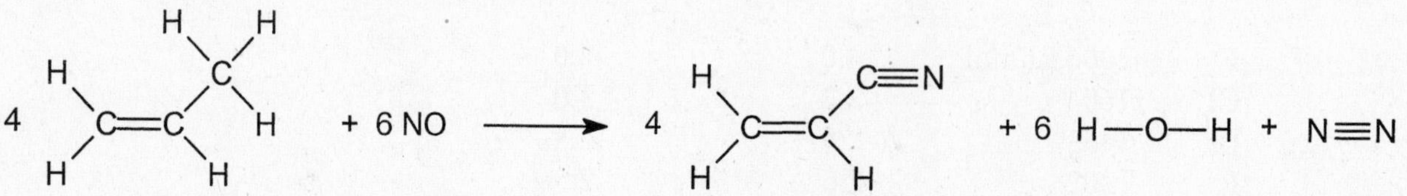

Bonds broken: Bonds formed:

4 × 3 C – H (413 kJ/mol)   4 C ≡ N (891 kJ/mol)
6 N = O (630. kJ/mol)   6 × 2 H – O (467 kJ/mol)
                                1 N ≡ N (941 kJ/mol)

$\Delta H = 12(413) + 6(630.) - [4(891) + 12(467) + 941] = -1373 \text{ kJ}$

c.

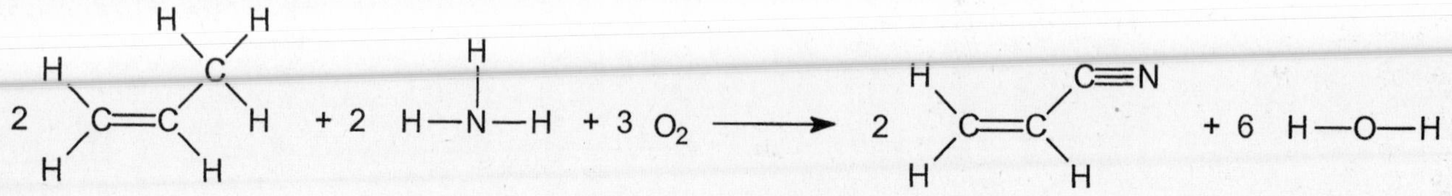

Bonds broken: Bonds formed:

2×3 C $-$ H (413 kJ/mol)   2 C $\equiv$ N (891 kJ/mol)
2×3 N $-$ H (391 kJ/mol) 6×2 O $-$ H (467 kJ/mol)
    3 O $=$ O (495 kJ/mol)

$\Delta H = 6(413) + 6(391) + 3(495) - [2(891) + 12(467)] = -1077$ kJ

    d.   Since both reactions are highly exothermic, the high temperature is not needed to provide energy. It must be necessary for some other reason. The reason is to increase the speed of the reaction. This is discussed in Chapter 12 on kinetics.

109.   a. i. $C_6H_6N_{12}O_{12} \rightarrow 6\ CO + 6\ N_2 + 3\ H_2O + 3/2\ O_2$

The NO_2 groups are assumed to have one N $-$ O single bond and one N $=$ O double bond and each carbon atom has one C $-$ H single bond. We must break and form all bonds.

Bonds broken: Bonds formed:

  3 C $-$ C (347 kJ/mol)   6 C $\equiv$ O (1072 kJ/mol)
  6 C $-$ H (413 kJ/mol)   6 N $\equiv$ N (941 kJ/mol)
12 C $-$ N (305 kJ/mol)   6 H $-$ O (467 kJ/mol)
  6 N $-$ N (160. kJ/mol)   3/2 O $=$ O (495 kJ/mol)
  6 N $-$ O (201 kJ/mol)
  6 N $=$ O (607 kJ/mol) $\Sigma D_{formed} = 15{,}623$ kJ

$\Sigma D_{broken} = 12{,}987$ kJ

$\Delta H = \Sigma D_{broken} - \Sigma D_{formed} = 12{,}987$ kJ $- 15{,}623$ kJ $= -2636$ kJ

   ii. $C_6H_6N_{12}O_{12} \rightarrow 3\ CO + 3\ CO_2 + 6\ N_2 + 3\ H_2O$

Note: The bonds broken will be the same for all three reactions.

Bonds formed:

  3 C $\equiv$ O (1072 kJ/mol)
  6 C $=$ O (799 kJ/mol)
  6 N $\equiv$ N (941 kJ/mol)
  6 H $-$ O (467 kJ/mol)

  $\Sigma D_{formed} = 16{,}458$ kJ

$\Delta H = 12{,}987$ kJ $- 16{,}458$ kJ $= -3471$ kJ

  iii. $C_6H_6N_{12}O_{12} \rightarrow 6\ CO_2 + 6\ N_2 + 3\ H_2$

Bonds formed:

12 C $=$ O (799 kJ/mol)
  6 N $\equiv$ N (941 kJ/mol)
  3 H $-$ H (432 kJ/mol)

  $\Sigma D_{formed} = 16{,}530.$ kJ

$\Delta H = 12{,}987$ kJ $- 16{,}530.$ kJ $= -3543$ kJ

b. Reaction iii yields the most energy per mole of CL-20 so it will yield the most energy per kg.

$$\frac{-3543 \text{ kJ}}{\text{mol}} \times \frac{1 \text{ mol}}{438.23 \text{ g}} \times \frac{1000 \text{ g}}{\text{kg}} = \text{-8085 kJ/kg}$$

110. If we can draw resonance forms for the anion after the loss of H⁺, we can argue that the extra stability of the anion causes the proton to be more readily lost, i.e., makes the compound a better acid.

a.

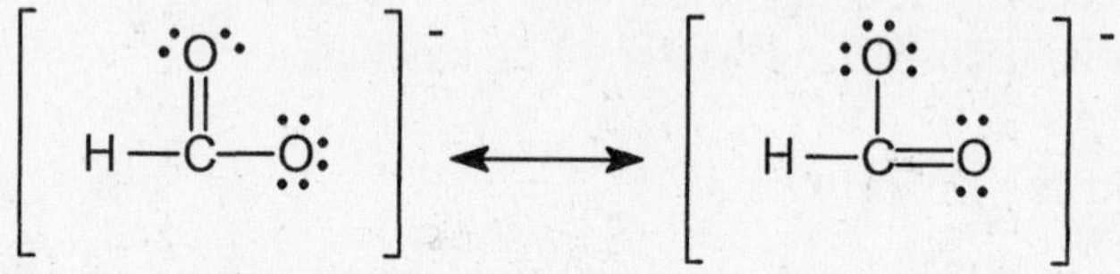

b.

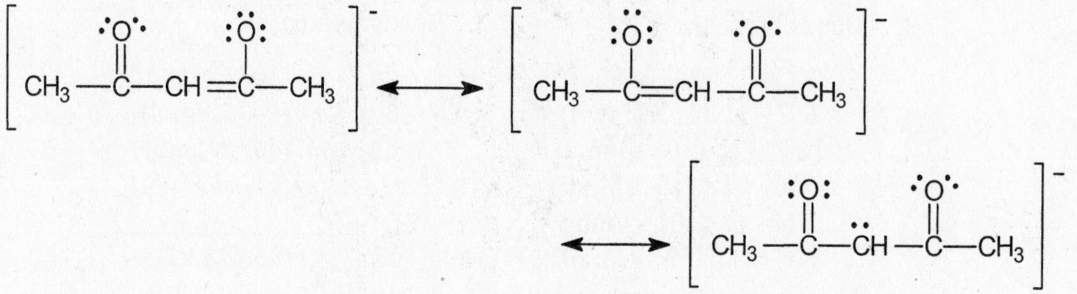

c.

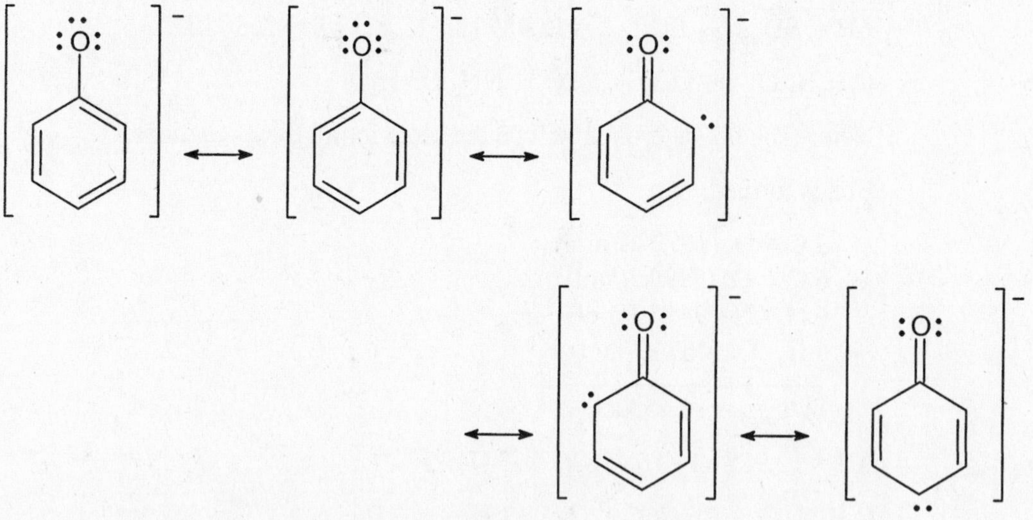

In all 3 cases, extra resonance forms can be drawn for the anion that are not possible when the H⁺ is present, which leads to enhanced stability.

111.   PAN ($H_3C_2NO_5$) has $3(1) + 2(4) + 5 + 5(6) = 46$ valence electrons.

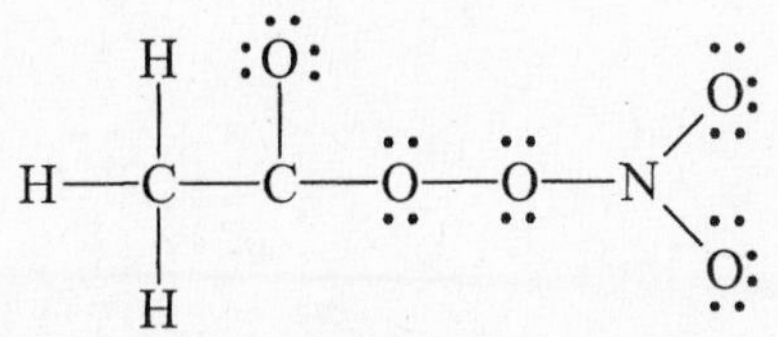

Skeletal structure with complete octets about oxygen atoms (46 electrons used).

This structure has used all 46 electrons, but there are only six electrons around one of the carbon atoms and the nitrogen atom. Two unshared pairs must become shared, that is we must form two double bonds.

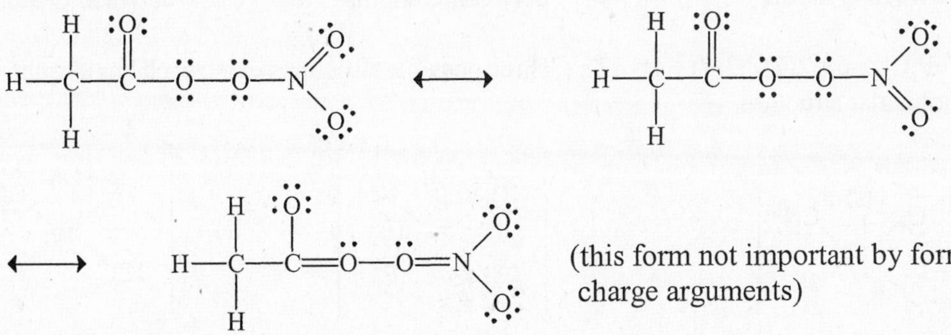

(this form not important by formal charge arguments)

112.

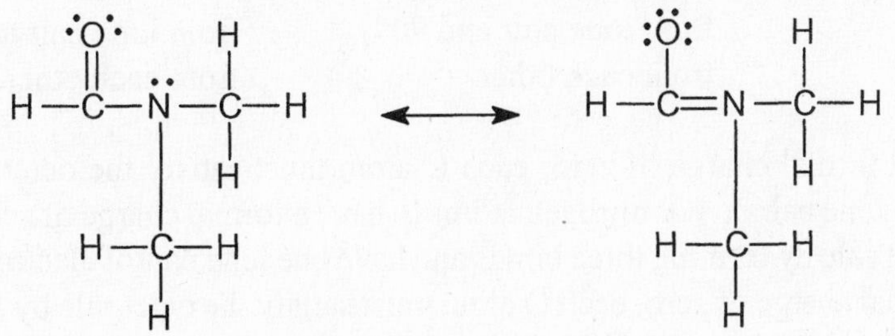

113.   a. $BrFI_2$, $7 + 7 + 2(7) = 28$ e⁻; Two possible structures exist; each has a T-shaped molecular structure.

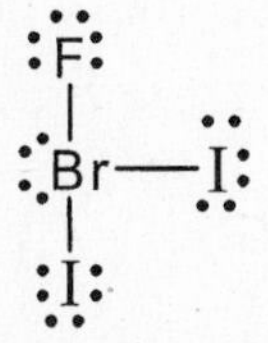

90° bond angles between I atoms 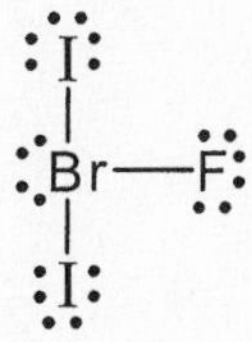   180° bond angles between I atoms

b. XeO_2F_2, $8 + 2(6) + 2(7) = 34$ e⁻; Three possible structures exist; each has a see-saw molecular structure.

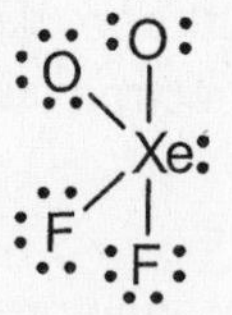

| 90° bond angle between O atoms | 180° bond angle between O atoms | 120° bond angle between O atoms |

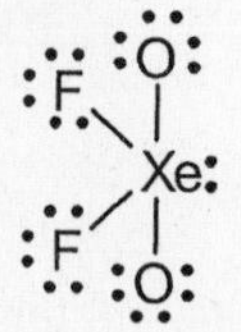

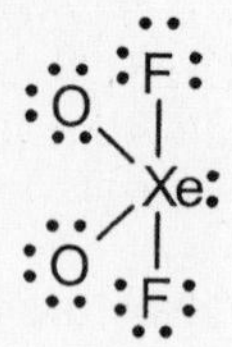

c. $TeF_2Cl_3^-$; $6 + 2(7) + 3(7) + 1 = 42$ e⁻; Three possible structures exist; each has a square pyramid molecular structure.

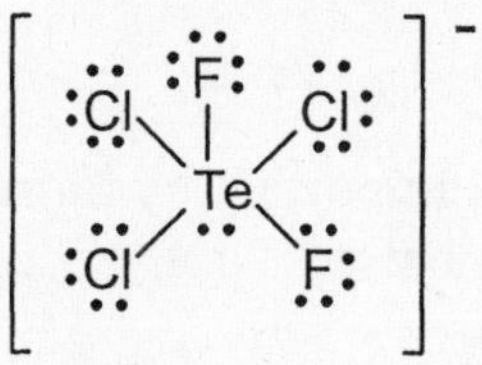

| One F is 180° from lone pair. | Both F atoms are 90° from lone pair and 90° from each other. | Both F atoms are 90° from lone pair and 180° from each other. |

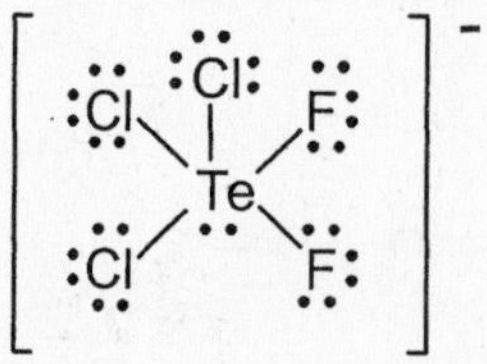

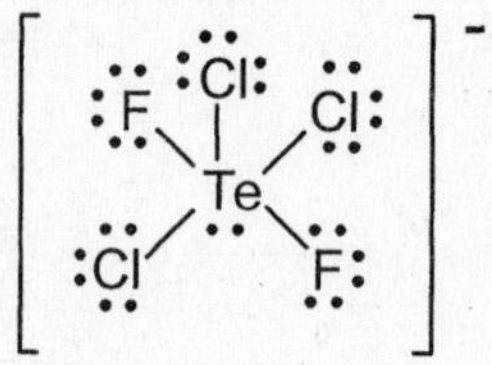

114.   For carbon atoms to have a formal charge of zero, each C atom must satisfy the octet rule by forming four bonds (with no lone pairs). For nitrogen atoms to have a formal charge of zero, each N atom must satisfy the octet rule by forming three bonds and have one lone pair of electrons. For oxygen atoms to have a formal charge of zero, each O atom must satisfy the octet rule by forming two bonds and have two lone pairs of electrons. With these bonding requirements in mind, the Lewis structure of histidine, where all atoms have a formal charge of zero, is:

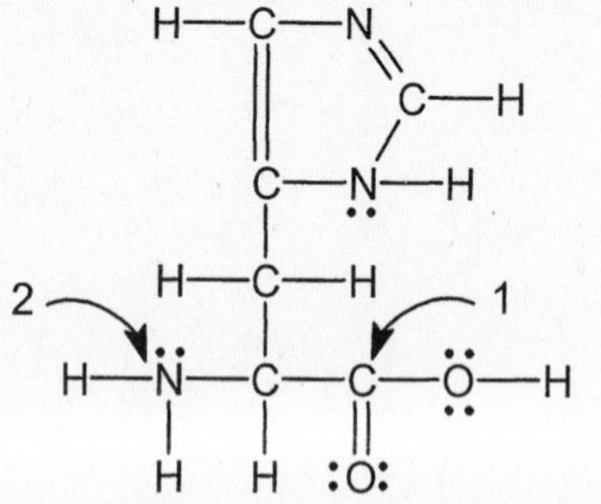

We would expect 120° bond angles about the carbon atom labeled 1 and ~109.5° bond angles about the nitrogen atom labeled 2. The nitrogen bond angles are slightly smaller than 109.5° due to the lone pair on nitrogen.

115.   The nitrogen-nitrogen bond length of 112 pm is between a double (120 pm) and a triple (110 pm) bond. The nitrogen-oxygen bond length of 119 pm is between a single (147 pm) and a double bond (115 pm). The last resonance structure doesn't appear to be as important as the other two since there is no evidence from bond lengths for a nitrogen-oxygen triple bond or a nitrogen-nitrogen single bond as in the third resonance form. We can adequately describe the structure of N_2O using the resonance forms:

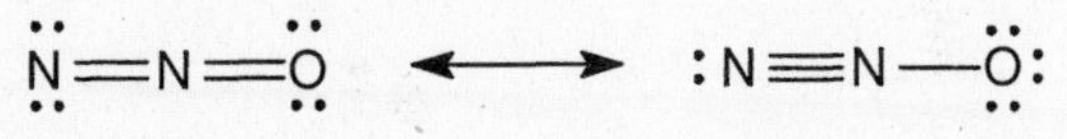

$$\ddot{N}=N=\ddot{O} \longleftrightarrow :N\equiv N-\ddot{O}:$$

Assigning formal charges for all 3 resonance forms:

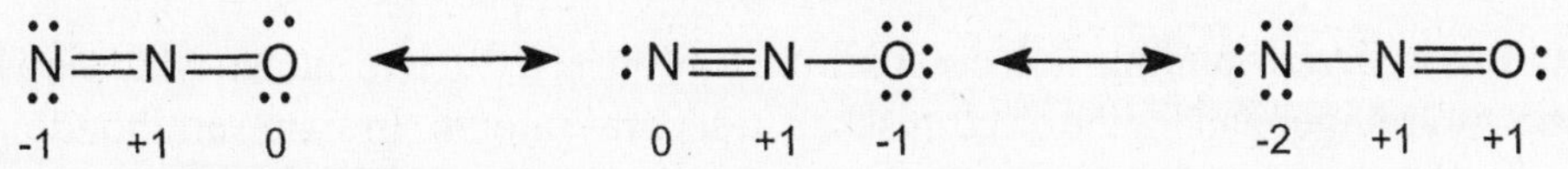

$$\ddot{N}=N=\ddot{O} \longleftrightarrow :N\equiv N-\ddot{O}: \longleftrightarrow :\ddot{N}-N\equiv O:$$
$$\;\;-1 \quad\; +1 \quad\;\; 0 \qquad\qquad\quad 0 \quad +1 \quad -1 \qquad\qquad -2 \quad +1 \quad +1$$

For:

$$\left(\ddot{N}=\right),\; FC = 5 - 4 - 1/2(4) = -1$$

$$\left(=N=\right),\; FC = 5 - 1/2(8) = +1,\; \text{Same for }\left(\equiv N-\right)\text{ and }\left(-N\equiv\right)$$

$$\left(:\ddot{N}-\right),\; FC = 5 - 6 - 1/2(2) = -2;\; \left(:N\equiv\right),\; FC = 5 - 2 - 1/2(6) = 0$$

$$\left(=\ddot{O}\right),\; FC = 6 - 4 - 1/2(4) = 0;\; \left(-\ddot{O}:\right),\; FC = 6 - 6 - 1/2(2) = -1$$

$$\left(\equiv O:\right),\; FC = 6 - 2 - 1/2(6) = +1$$

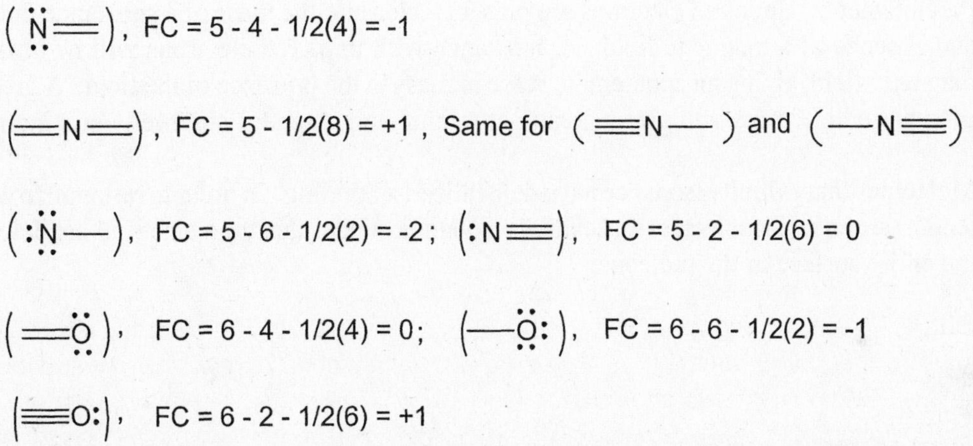

We should eliminate $N - N \equiv O$ since it has a formal charge of +1 on the most electronegative element (O). This is consistent with the observation that the $N - N$ bond is between a double and triple bond and that the $N - O$ bond is between a single and double bond.

CHAPTER NINE

COVALENT BONDING: ORBITALS

Questions

7. Bond energy is directly proportional to bond order. Bond length is inversely proportional to bond order. Bond energy and bond length can be measured.

8. The electrons in sigma bonding molecular orbitals are attracted to two nuclei, which is a lower, more stable energy arrangement for the electrons than in separate atoms. In sigma antibonding molecular orbitals, the electrons are mainly outside the space between the nuclei, which is a higher, less stable energy arrangement than in the separated atoms.

9. Paramagnetic: Unpaired electrons are present. Measure the mass of a substance in the presence and absence of a magnetic field. A substance with unpaired electrons will be attracted by the magnetic field, giving an apparent increase in mass in the presence of the field. A greater number of unpaired electrons will give a greater attraction and a greater observed mass increase.

10. Molecules that exhibit resonance have delocalized π bonding. In order to rationalize why the bond lengths are equal in molecules that exhibit resonance, we say that the π electrons are delocalized over the entire surface of the molecule.

Exercises

The Localized Electron Model and Hybrid Orbitals

11. H_2O has $2(1) + 6 = 8$ valence electrons.

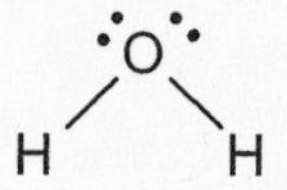

H₂O has a tetrahedral arrangement of the electron pairs about the O atom that requires sp^3 hybridization. Two of the four sp^3 hybrid orbitals are used to form bonds to the two hydrogen atoms and the other two sp^3 hybrid orbitals hold the two lone pairs of oxygen. The two $O - H$ bonds are formed from overlap of the sp^3 hybrid orbitals on oxygen with the $1s$ atomic orbitals on the hydrogen atoms.

12. CCl_4 has $4 + 4(7) = 32$ valence electrons.

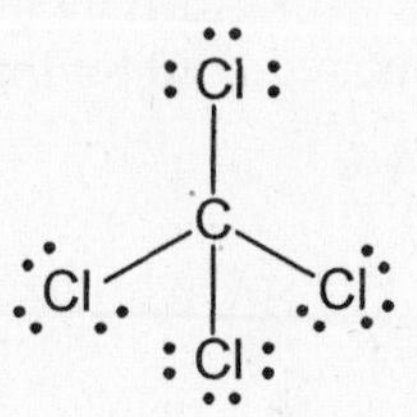

CCl$_4$ has a tetrahedral arrangement of the electron pairs about the carbon atom which requires sp^3 hybridization. The four sp^3 hybrid orbitals on carbon are used to form the four bonds to chlorine. The chlorine atoms also have a tetrahedral arrangement of electron pairs and we will assume that they are also sp^3 hybridized. The C – Cl sigma bonds are all formed from overlap of sp^3 hybrid orbitals on carbon with sp^3 hybrid orbitals on each chlorine atom.

13. H_2CO has $2(1) + 4 + 6 = 12$ valence electrons.

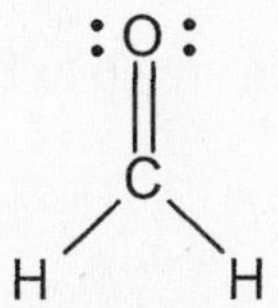

The central carbon atom has a trigonal planar arrangement of the electron pairs which requires sp^2 hybridization. The two C – H sigma bonds are formed from overlap of the sp^2 hybrid orbitals on carbon with the hydrogen 1s atomic orbitals. The double bond between carbon and oxygen consists of one σ and one π bond. The oxygen atom, like the carbon atom, also has a trigonal planar arrangement of the electrons which requires sp^2 hybridization. The σ bond in the double bond is formed from overlap of a carbon sp^2 hybrid orbital with an oxygen sp^2 hybrid orbital. The π bond in the double bond is formed from overlap of the unhybridized p atomic orbitals. Carbon and oxygen each have one unhybridized p atomic orbital which are parallel to each other. When two parallel p atomic orbitals overlap, a π bond results.

14. C_2H_2 has $2(4) + 2(1) = 10$ valence electrons.

Each carbon atom in C_2H_2 is sp hybridized since each carbon atom is surrounded by two effective pairs of electrons, i.e., each carbon atom has a linear arrangement of electrons. Since each carbon atom is sp hybridized, each carbon atom has two unhybridized p atomic orbitals. The two C – H sigma bonds are formed from overlap of carbon sp hybrid orbitals with hydrogen 1s atomic orbitals. The triple bond is composed of one σ bond and two π bonds. The sigma bond between the carbon atoms is formed from overlap of sp hybrid orbitals on each carbon atom. The two π bonds of the triple bond are formed from parallel overlap of the two unhybridized p atomic orbitals on each carbon.

15.   See Exercises 8.61 and 8.65 for the Lewis structures. To predict the hybridization, first determine the arrangement of electron pairs about each central atom using the VSEPR model; then utilize the information in Figure 9.24 of the text to deduce the hybridization required for that arrangement of electron pairs.

    8.61   a.   HCN; C is sp hybridized.   b. PH_3; P is sp^3 hybridized.

          c. $CHCl_3$;   C is sp^3 hybridized.   d. NH_4^+; N is sp^3 hybridized.

          e. H_2CO; C is sp^2 hybridized.   f. SeF_2; Se is sp^3 hybridized.

          g. CO_2; C is sp hybridized.   h. O_2; Each O atom is sp^2 hybridized.

          i.   HBr; Br is sp^3 hybridized.

    8.65   a.   The central N atom is sp^2 hybridized in NO_2^- and NO_3^-. In N_2O_4, both central N atoms are sp^2 hybridized.

          b.   In OCN^- and SCN^-, the central carbon atoms in each ion are sp hybridized and in N_3^-, the central N atom is also sp hybridized.

16.   See Exercises 8.62 and 8.66 for the Lewis structures.

    8.62   a.   All the central atoms are sp^3 hybridized.

          b.   All the central atoms are sp^3 hybridized.

          c.   All the central atoms are sp^3 hybridized.

    8.66   In O_3 and in SO_2, the central atoms are sp^2 hybridized and in SO_3, the central sulfur atom is also sp^2 hybridized.

17.   See Exercise 8.63 for the Lewis structures.

    PF_5:   P is dsp^3 hybridized. BeH_2:   Be is sp hybridized.

    BH_3:   B is sp^2 hybridized. Br_3^-:   Br is dsp^3 hybridized.

    SF_4:   S is dsp^3 hybridized. XeF_4:   Xe is d^2sp^3 hybridized.

    ClF_5:   Cl is d^2sp^3 hybridized. SF_6:   S is d^2sp^3 hybridized.

18.   In ClF_3, the central Cl atom is dsp^3 hybridized and in BrF_3, the central Br atom is also dsp^3 hybridized. See Exercise 8.64 for the Lewis structures.

19.   The molecules in Exercise 8.81 all have a trigonal planar arrangement of electron pairs about the central atom so all have central atoms with sp^2 hybridization. The molecules in Exercise 8.82 all have a tetrahedral arrangement of electron pairs about the central atom so all have central atoms with sp^3 hybridization. See Exercises 8.81 and 8.82 for the Lewis structures.

20.   The molecules in Exercise 8.83 all have central atoms with dsp^3 hybridization since all are based on the trigonal bipyramid arrangement of electron pairs. The molecules in Exercise 8.84 all have central atoms with d^{2}sp^3 hybridization since all are based on the octahedral arrangement of electron pairs. See Exercises 8.83 and 8.84 for the Lewis structures.

21.   a.

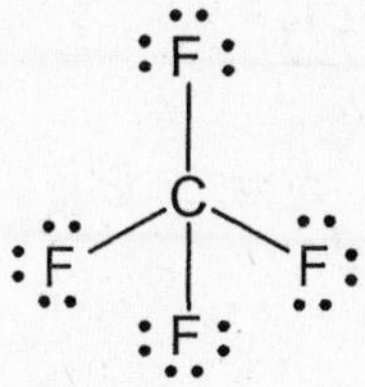

      b.

tetrahedral	sp^3	trigonal pyramid	sp^3
109.5°	nonpolar	< 109.5°	polar

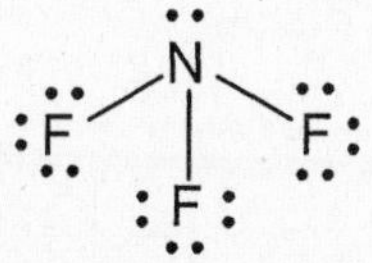

The angles in NF$_3$ should be slightly less than 109.5° because the lone pair requires more space than the bonding pairs.

c.

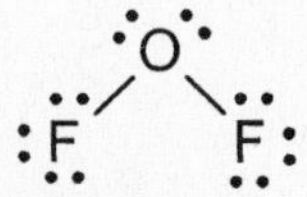

      d.

V-shaped	sp^3	trigonal planar	sp^2
< 109.5°	polar	120°	nonpolar

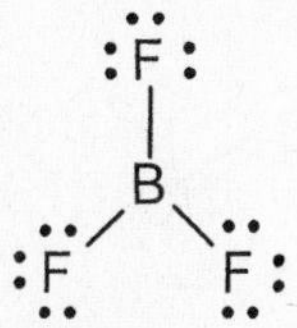

e.

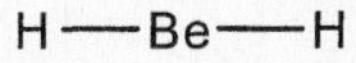

H —— Be —— H

      f.

linear	sp	see-saw	
180°	nonpolar	a. ≈ 120°, b. ≈ 90°	
		dsp^3	polar

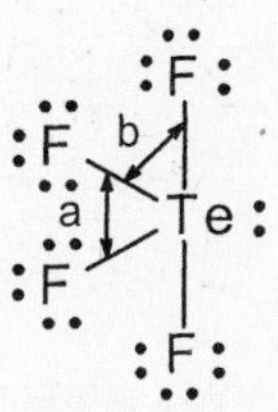

g.

      h.

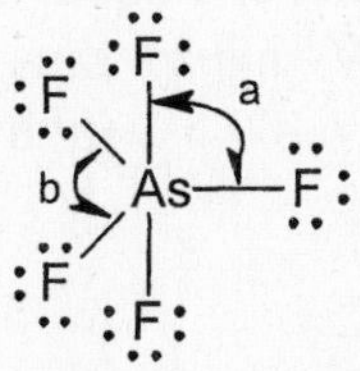

trigonal bipyramid	dsp^3	linear	dsp^3
a. 90°, b. 120°	nonpolar	180°	nonpolar

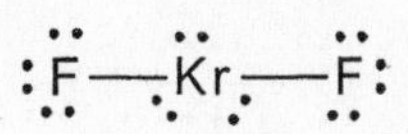

i.

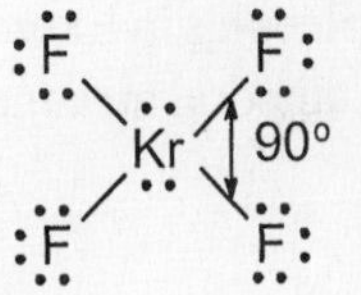

square planar d^2sp^3
90°   nonpolar

j.

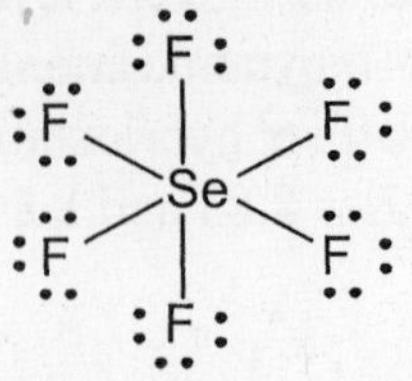

octahedral d^2sp^3
90°   nonpolar

k.

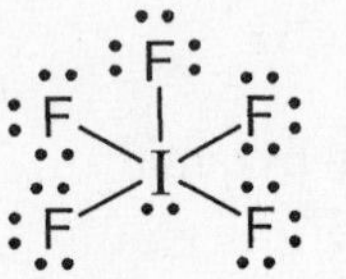

square pyramid d^2sp^3
≈ 90°   polar

l.

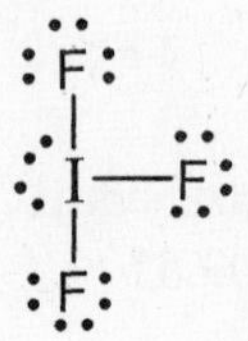

T-shaped dsp^3
≈ 90°   polar

22.   a.

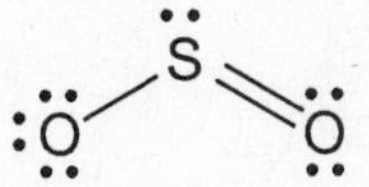

V-shaped
≈120°
sp^2

Only one resonance form is shown. Resonance does not change the position of the atoms. We can predict the geometry and hybridization from any one of the resonance structures.

b.

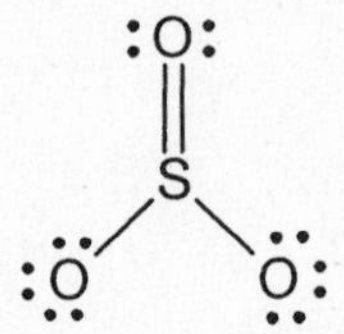

plus two other resonance structures
trigonal planar   120°
sp^2

c.

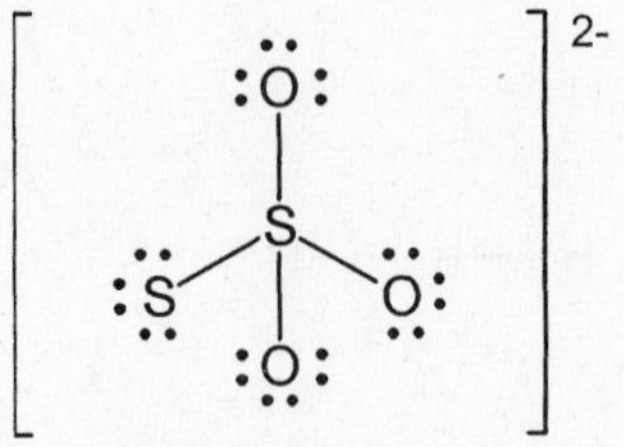

tetrahedral   109.5°
sp^3

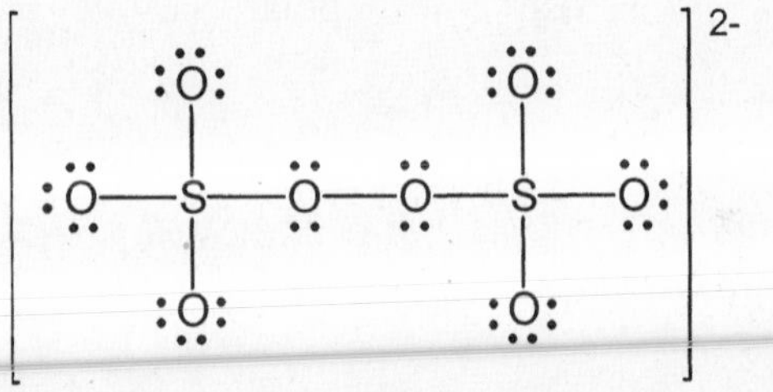

Tetrahedral geometry about each S, 109.5°, sp^3 hybrids; V-shaped arrangement about peroxide O's, ≈ 109.5°, sp^3 hybrids

d.

e.

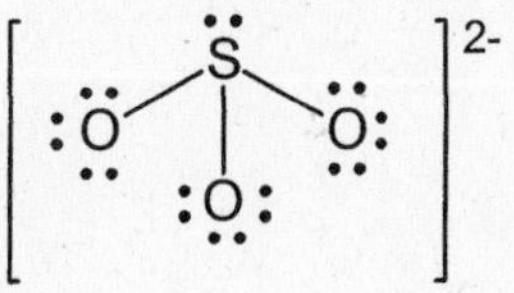

trigonal pyramid
< 109.5°
sp³

f.   g.

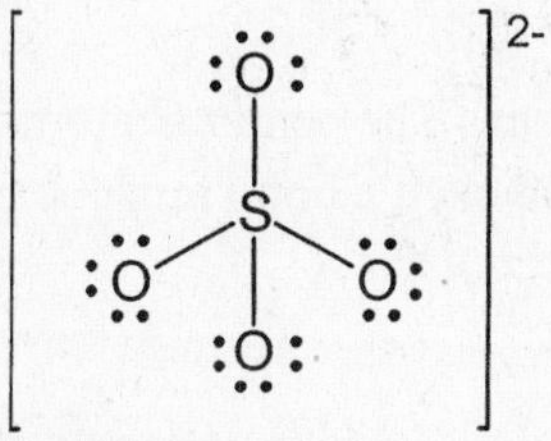

tetrahedral   109.5°   V-shaped   < 109.5°
sp³   sp³

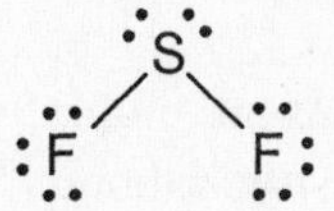

h.   i.

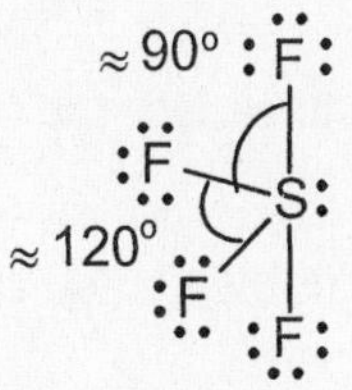

see-saw   ≈ 90°, ≈ 120°   octahedral   90°
dsp³   d²sp³

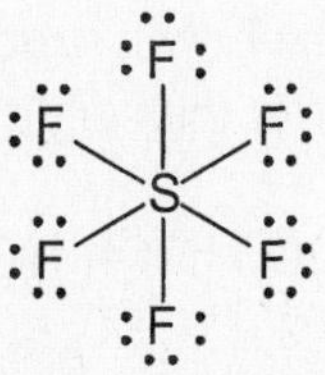

j.

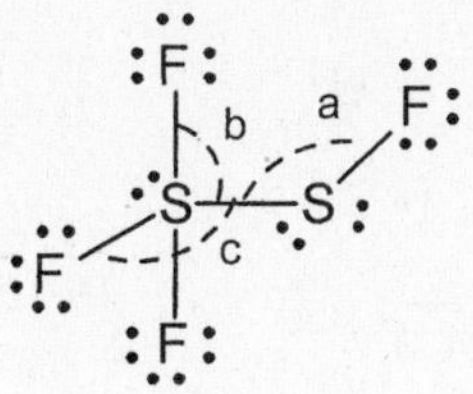

a)   ≈ 109.5°   b) ≈ 90°   c) ≈ 120°

See-saw about S atom with one lone pair (dsp³);
bent about S atom with two lone pairs (sp³)

k.

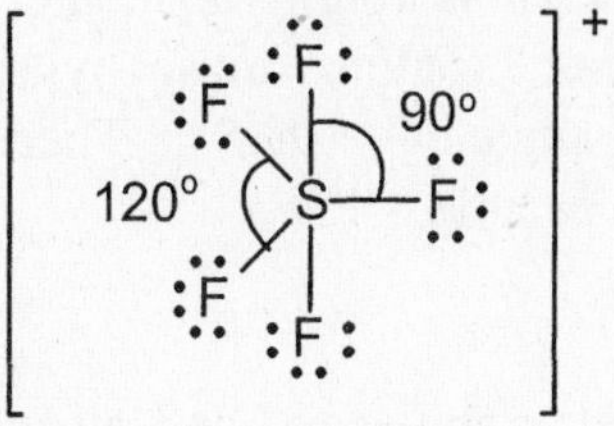

trigonal bipyramid
90° and 120°, dsp³

23.

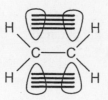

For the p-orbitals to properly line up to form the π bond, all six atoms are forced into the same plane. If the atoms were not in the same plane, the π bond could not form since the p-orbitals would no longer be parallel to each other.

24.   No, the CH_2 planes are mutually perpendicular to each other. The center C atom is sp hybridized and is involved in two π-bonds. The p-orbitals used to form each π bond must be perpendicular to each other. This forces the two CH_2 planes to be perpendicular.

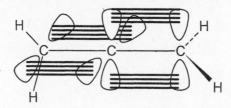

25.   To complete the Lewis structures, just add lone pairs of electrons to satisfy the octet rule for the atoms with fewer than eight electrons.

Biacetyl ($C_4H_6O_2$) has 4(4) + 6(1) + 2(6) = 34 valence electrons.

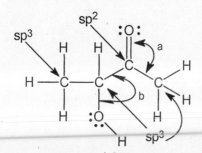

All CCO angles are 120°. The six atoms are not in the same plane because of free rotation about the carbon- carbon single (sigma) bonds. There are 11 σ and 2 π bonds in biacetyl.

Acetoin ($C_4H_8O_2$) has 4(4) + 8(1) + 2(6) = 36 valence electrons.

The carbon with the doubly-bonded O is sp² hybridized. The other 3 C atoms are sp³ hybridized. Angle a = 120° and angle b = 109.5°. There are 13 σ and 1 π bonds in acetoin.

Note: All single bonds are σ bonds, all double bonds are one σ and one π bond, and all triple bonds are one σ and two π bonds.

26.   Acrylonitrile: C_3H_3N has $3(4) + 3(1) + 5 = 20$ valence electrons.

a.   120°
b.   120°
c.   180°

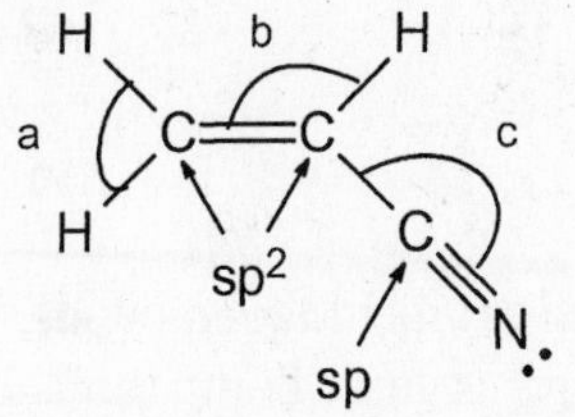

6 σ and 3 π bonds

All atoms of acrylonitrile lie in the same plane. The π bond in the double bond dictates that the C and H atoms are all in the same plane, and the triple bond dictates that N is in the same plane with the other atoms.

Methyl methacrylate ($C_5H_8O_2$) has $5(4) + 8(1) + 2(6) = 40$ valence electrons.

d.   120°
e.   120°
f.   ≈ 109.5°

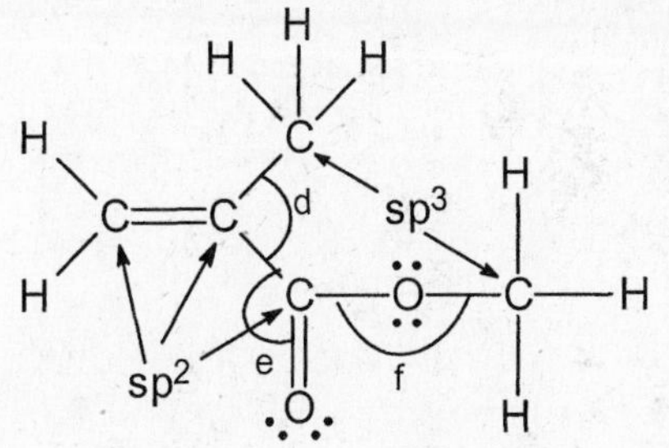

14 σ and 2 π bonds

27.   To complete the Lewis structure, just add lone pairs of electrons to satisfy the octet rule for the atoms that have fewer than eight electrons.

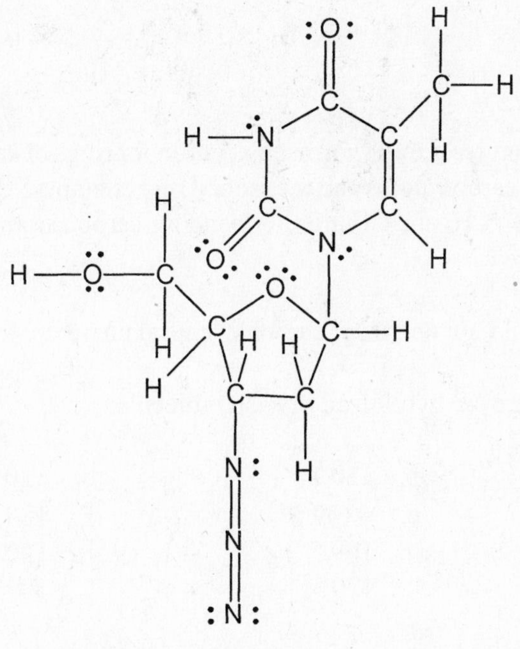

a. 6   b. 4   c. The center N in $-N=N=N$ group

d. 33 σ   e. 5 π bonds   f. 180°

g. < 109.5°   h. sp³

28.   a.   Piperine and capsaicin are molecules classified as organic compounds, i.e., compounds based on
            carbon. The majority of Lewis structures for organic compounds have all atoms with zero
            formal charge. Therefore, carbon atoms in organic compounds will usually form four bonds,
            nitrogen atoms will form three bonds and complete the octet with one lone pair of electrons, and
            oxygen atoms will form two bonds and complete the octet with two lone pairs of electrons.
            Using these guidelines, the Lewis structures are:

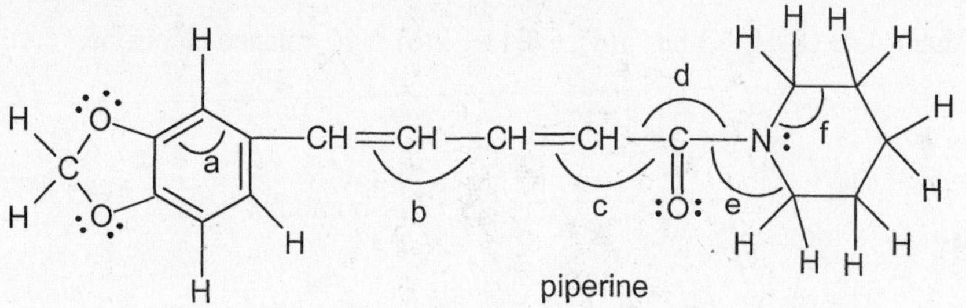

piperine

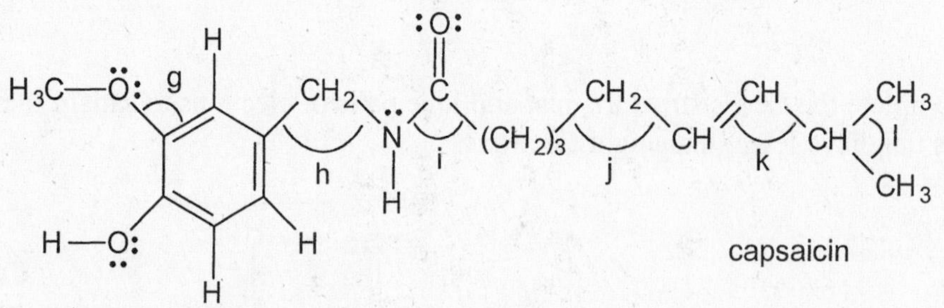

capsaicin

       Note: The ring structures are all shorthand notation for rings of carbon atoms. In piperine, the
       first ring contains 6 carbon atoms and the second ring contains 5 carbon atoms (plus nitrogen).
       Also notice that CH₃, CH₂ and CH are shorthand for carbon atoms singly bonded to hydrogen
       atoms.

       b.   piperine: 0 sp, 11 sp² and 6 sp³ carbons; capsaicin: 0 sp, 9 sp² and 9 sp³ carbons

       c.   The nitrogens are sp³ hybridized in each molecule.

       d.   a.   120°   b. 120°   c. 120°
            d.   120°   e. ≈109.5°   f. 109.5°
            g.   120°   h. 109.5°   i. 120°
            j.   109.5°   k. 120°   l. 109.5°

29.

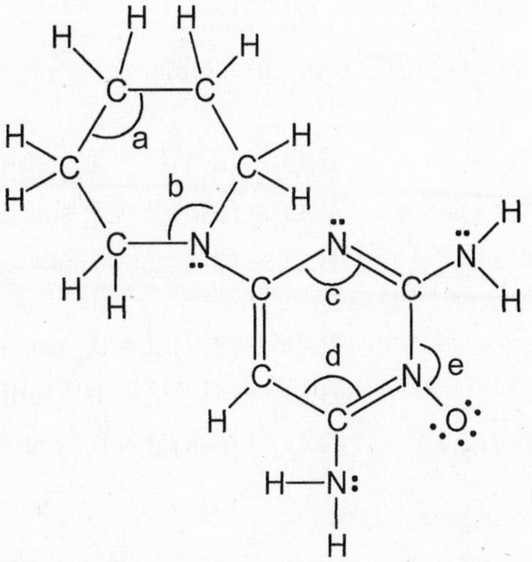

a. The two nitrogens in the ring with double bonds are sp^2 hybridized. The other three nitrogens are sp^3 hybridized.

b. The five carbon atoms in the ring with one nitrogen are all sp^3 hybridized. The four carbon atoms in the other ring with double bonds are all sp^2 hybridized.

c. Angles a and b: $\approx 109.5°$; angles c, d, and e: $\approx 120°$

d. 31 sigma bonds

e. 3 pi bonds (Each double bond consists of one sigma and one pi bond.)

30. CO, $4 + 6 = 10$ e$^-$; CO_2, $4 + 2(6) = 16$ e$^-$; C_3O_2, $3(4) + 2(6) = 24$ e$^-$

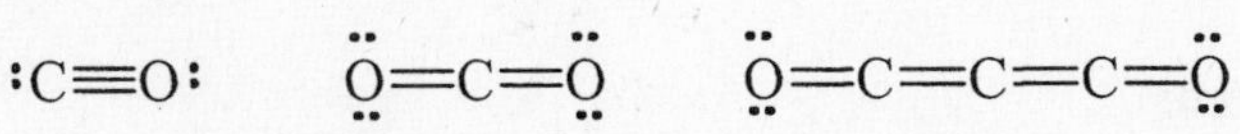

$$:C\equiv O: \qquad \ddot{O}=C=\ddot{O} \qquad \ddot{O}=C=C=C=\ddot{O}$$

There is no molecular structure for the diatomic CO molecule. The carbon in CO is sp hybridized. CO_2 is a linear molecule, and the central carbon atom is sp hybridized. C_3O_2 is a linear molecule with all of the central carbon atoms exhibiting sp hybridization.

The Molecular Orbital Model

31. If we calculate a non-zero bond order for a molecule, then we predict that it can exist (is stable).

a. H_2^+: $(\sigma_{1s})^1$ B.O. $= (1-0)/2 = 1/2$, stable

   H_2: $(\sigma_{1s})^2$ B.O. $= (2-0)/2 = 1$, stable

   H_2^-: $(\sigma_{1s})^2(\sigma_{1s}*)^1$ B.O. $= (2-1)/2 = 1/2$, stable

   H_2^{2-}: $(\sigma_{1s})^2(\sigma_{1s}*)^2$ B.O. $= (2-2)/2 = 0$, not stable

b. He_2^{2+}: $(\sigma_{1s})^2$ B.O. = (2-0)/2 = 1, stable

    He_2^+: $(\sigma_{1s})^2(\sigma_{1s}*)^1$ B.O. = (2-1)/2 = 1/2, stable

    He_2: $(\sigma_{1s})^2(\sigma_{1s}*)^2$ B.O. = (2-2)/2 = 0, not stable

32. a. N_2^{2-}: $(\sigma_{2s})^2(\sigma_{2s}*)^2(\pi_{2p})^4(\sigma_{2p})^2(\pi_{2p}*)^2$ B.O. = (8-4)/2 = 2, stable

      O_2^{2-}: $(\sigma_{2s})^2(\sigma_{2s}*)^2(\sigma_{2p})^2(\pi_{2p})^4(\pi_{2p}*)^4$ B.O. = (8-6)/2 = 1, stable

      F_2^{2-}: $(\sigma_{2s})^2(\sigma_{2s}*)^2(\sigma_{2p})^2(\pi_{2p})^4(\pi_{2p}*)^4(\sigma_{2p}*)^2$ B.O. = (8-8)/2 = 0, not stable

  b. Be_2: $(\sigma_{2s})^2(\sigma_{2s}*)^2$ B.O. = (2-2)/2 = 0, not stable

      B_2: $(\sigma_{2s})^2(\sigma_{2s}*)^2(\pi_{2p})^2$ B.O. = (4-2)/2 = 1, stable

      Ne_2: $(\sigma_{2s})^2(\sigma_{2s}*)^2(\sigma_{2p})^2(\pi_{2p})^4(\pi_{2p}*)^4(\sigma_{2p}*)^2$ B.O. = (8-8)/2 = 0, not stable

33. The electron configurations are:

  a. Li_2: $(\sigma_{2s})^2$ B.O. = (2-0)/2 = 1, diamagnetic (0 unpaired e⁻)

  b. C_2: $(\sigma_{2s})^2(\sigma_{2s}*)^2(\pi_{2p})^4$ B.O. = (6-2)/2 = 2, diamagnetic (0 unpaired e⁻)

  c. S_2: $(\sigma_{3s})^2(\sigma_{3s}*)^2(\sigma_{3p})^2(\pi_{3p})^4(\pi_{3p}*)^2$ B.O. = (8-4)/2 = 2, paramagnetic (2 unpaired e⁻)

34. C_2^{2-} has 10 valence electrons. The Lewis structure predicts sp hybridization for each carbon with two unhybridized p orbitals on each carbon.

$$\left[:C\equiv C:\right]^{2-}$$ sp hybrid orbitals form the σ bond, and the two unhybridized
                          p atomic orbitals from each carbon form the two π bonds.

MO: $(\sigma_{2s})^2(\sigma_{2s}*)^2(\pi_{2p})^4(\sigma_{2p})^2$, B.O. = (8 - 2)/2 = 3

Both give the same picture, a triple bond composed of one σ and two π-bonds. Both predict the ion will be diamagnetic. Lewis structures deal well with diamagnetic (all electrons paired) species. The Lewis model cannot really predict magnetic properties.

35. The electron configurations are:

    O_2^+: $(\sigma_{2s})^2(\sigma_{2s}*)^2(\sigma_{2p})^2(\pi_{2p})^4(\pi_{2p}*)^1$

    O_2: $(\sigma_{2s})^2(\sigma_{2s}*)^2(\sigma_{2p})^2(\pi_{2p})^4(\pi_{2p}*)^2$

    O_2^-: $(\sigma_{2s})^2(\sigma_{2s}*)^2(\sigma_{2p})^2(\pi_{2p})^4(\pi_{2p}*)^3$

    O_2^{2-}: $(\sigma_{2s})^2(\sigma_{2s}*)^2(\sigma_{2p})^2(\pi_{2p})^4(\pi_{2p}*)^4$

	O_2^+	O_2	O_2^-	O_2^{2-}
Bond order	2.5	2	1.5	1
# of unpaired electrons	1	2	1	0

Bond energy: $O_2^{2-} < O_2^- < O_2 < O_2^+$; Bond length: $O_2^+ < O_2 < O_2^- < O_2^{2-}$

Bond energy is directly proportional to bond order, and bond length is inversely proportional to bond order.

36.   The electron configurations are:

F_2^+: $(\sigma_{2s})^2(\sigma_{2s}*)^2(\sigma_{2p})^2(\pi_{2p})^4(\pi_{2p}*)^3$ B.O. = (8-5)/2 = 1.5; 1 unpaired e⁻

F_2: $(\sigma_{2s})^2(\sigma_{2s}*)^2(\sigma_{2p})^2(\pi_{2p})^4(\pi_{2p}*)^4$ B.O. = (8-6)/2 = 1;   0 unpaired e⁻

F_2^-: $(\sigma_{2s})^2(\sigma_{2s}*)^2(\sigma_{2p})^2(\pi_{2p})^4(\pi_{2p}*)^4(\sigma_{2p}*)^1$ B.O. = (8-7)/2 = 0.5; 1 unpaired e⁻

From the calculated bond orders, the order of bond lengths should be: $F_2^+ < F_2 < F_2^-$

37.   The electron configurations are (assuming the same orbital order as that for N_2):

a. CO: $(\sigma_{2s})^2(\sigma_{2s}*)^2(\pi_{2p})^4(\sigma_{2p})^2$ B.O. = (8-2)/2 = 3, diamagnetic

b. CO⁺: $(\sigma_{2s})^2(\sigma_{2s}*)^2(\pi_{2p})^4(\sigma_{2p})^1$ B.O. = (7-2)/2 = 2.5, paramagnetic

c. CO²⁺: $(\sigma_{2s})^2(\sigma_{2s}*)^2(\pi_{2p})^4$ B.O. = (6-2)/2 = 2, diamagnetic

Since bond order is directly proportional to bond energy and inversely proportional to bond length, then:

shortest → longest bond length: CO < CO⁺ < CO²⁺

smallest → largest bond energy: CO²⁺ < CO⁺ < CO

38.   The electron configurations are (assuming the same orbital order as that for N_2):

a. NO⁺: $(\sigma_{2s})^2(\sigma_{2s}*)^2(\pi_{2p})^4(\sigma_{2p})^2$ B.O. = (8-2)/2 = 3, diamagnetic

b. NO: $(\sigma_{2s})^2(\sigma_{2s}*)^2(\pi_{2p})^4(\sigma_{2p})^2(\pi_{2p}*)^1$ B.O. = (8-3)/2 = 2.5, paramagnetic

c. NO⁻: $(\sigma_{2s})^2(\sigma_{2s}*)^2(\pi_{2p})^4(\sigma_{2p})^2(\pi_{2p}*)^2$ B.O. = (8-4)/2 = 2, paramagnetic

shortest → longest bond length: NO⁺ < NO < NO⁻

smallest → largest bond energy: NO⁻ < NO < NO⁺

39. H_2: $(\sigma_{1s})^2$

B_2: $(\sigma_{2s})^2(\sigma_{2s}*)^2(\pi_{2p})^2$

N_2: $(\sigma_{2s})^2(\sigma_{2s}*)^2(\pi_{2p})^4(\sigma_{2p})^2$

OF: $(\sigma_{2s})^2(\sigma_{2s}*)^2(\sigma_{2p})^2(\pi_{2p})^4(\pi_{2p}*)^3$

The bond strength will weaken if the electron removed comes from a bonding orbital. Of the molecules listed, H_2, B_2, and N_2 would be expected to have their bond strength weaken as an electron is removed. OF has the electron removed from an antibonding orbital, so its bond strength increases.

40.   CN: $(\sigma_{2s})^2(\sigma_{2s}*)^2(\pi_{2p})^4(\sigma_{2p})^1$

NO: $(\sigma_{2s})^2(\sigma_{2s}*)^2(\pi_{2p})^4(\pi_{2p})^2(\pi_{2p}*)^1$

O_2^{2+}: $(\sigma_{2s})^2(\sigma_{2s}*)^2(\sigma_{2p})^2(\pi_{2p})^4$

N_2^{2+}: $(\sigma_{2s})^2(\sigma_{2s}*)^2(\pi_{2p})^4$

If the added electron goes into a bonding orbital, the bond order would increase, making the species more stable and more likely to form. Between CN and NO, CN would most likely form CN⁻ since the bond order increases (unlike NO⁻ where the added electron goes into an antibonding orbital). Between O_2^{2+} and N_2^{2+}, N_2^+ would most likely form since the bond order increases (unlike O_2^+).

41.   The two types of overlap that result in bond formation for p orbitals are side to side overlap (π bond) and head to head overlap (σ bond).

42.

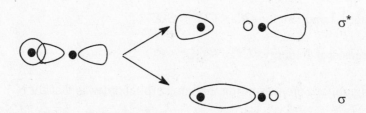

These molecular orbitals are sigma MOs since the electron density is cylindrically symmetric about the internuclear axis.

43.   a.   The electron density would be closer to F on the average. The F atom is more electronegative than the H atom, and the 2p orbital of F is lower in energy than the 1s orbital of H.

b.   The bonding MO would have more fluorine 2p character since it is closer in energy to the fluorine 2p atomic orbital.

c.   The antibonding MO would place more electron density closer to H and would have a greater contribution from the higher energy hydrogen 1s atomic orbital.

44.   a.   The antibonding MO will have more hydrogen 1s character because the hydrogen 1s atomic orbital is closer in energy to the antibonding MO.

b.   No, the overall overlap is zero. The p_x orbital does not have proper symmetry to overlap with a 1s orbital. The $2p_x$ and $2p_y$ orbitals are called nonbonding orbitals.

c.

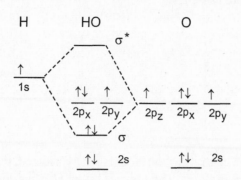

d.   Bond order $= \dfrac{2 - 0}{2} = 1$; Note: The 2s, $2p_x$, and $2p_y$ electrons have no effect on the bond order.

e.   To form OH^+, a nonbonding electron is removed from OH. Since the number of bonding electrons and antibonding electrons are unchanged, the bond order is still equal to one.

45. O_3 and NO_2^- are isoelectronic, so we only need consider one of them since the same bonding ideas apply to both. The Lewis structures for O_3 are:

For each of the two resonance forms, the central O atom is sp^2 hybridized with one unhybridized p atomic orbital. The sp^2 hybrid orbitals are used to form the two sigma bonds to the central atom. The localized electron view of the π bond utilizes unhybridized p atomic orbitals. The π bond resonates between the two positions in the Lewis structures:

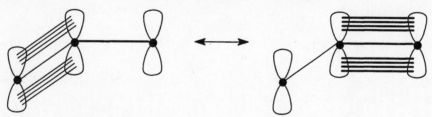

In the MO picture of the π bond, all three unhybridized p-orbitals overlap at the same time, resulting in π electrons that are delocalized over the entire surface of the molecule. This is represented as:

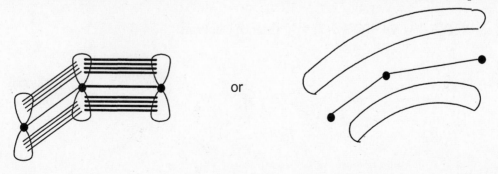

or

46.   The Lewis structures for CO_3^{2-} are (24 e$^-$):

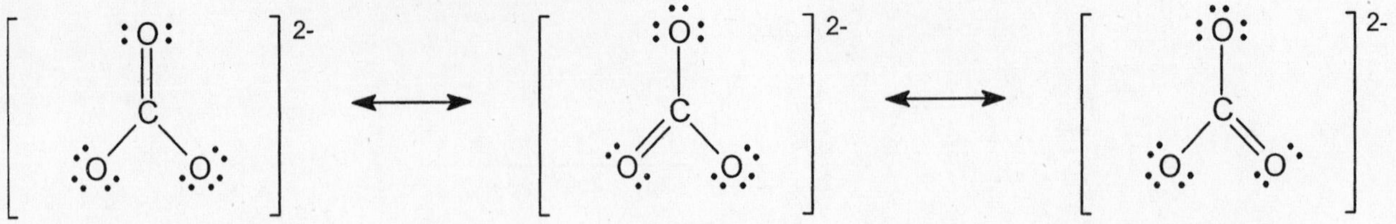

In the localized electron view, the central carbon atom is sp^2 hybridized; the sp^2 hybrid orbitals are used to form the three sigma bonds in CO_3^{2-}. The central C atom also has one unhybridized p atomic orbital which overlaps with another p atomic orbital from one of the oxygen atoms to form the π bond in each resonance structure. This localized π bond moves (resonates) from one position to another. In the molecular orbital model for CO_3^{2-}, all four atoms in CO_3^{2-} have a p atomic orbital which is perpendicular to the plane of the ion. All four of these p orbitals overlap at the same time to form a delocalized π bonding system where the π electrons can roam over the entire surface of the ion. The π molecular orbital system for CO_3^{2-} is analogous to that for NO_3^- which is shown in Figure 9.49 of the text.

Additional Exercises

47.   a. XeO_3, $8 + 3(6) = 26$ e$^-$ b. XeO_4, $8 + 4(6) = 32$ e$^-$

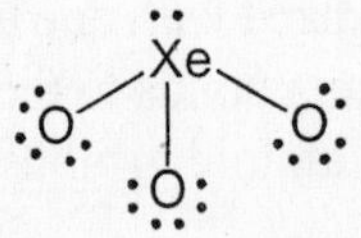

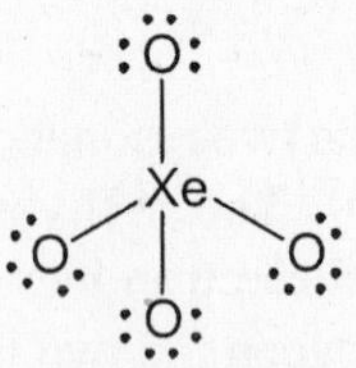

trigonal pyramid; sp^3 tetrahedral; sp^3

c. $XeOF_4$, $8 + 6 + 4(7) = 42$ e$^-$ d. $XeOF_2$, $8 + 6 + 2(7) = 28$ e$^-$

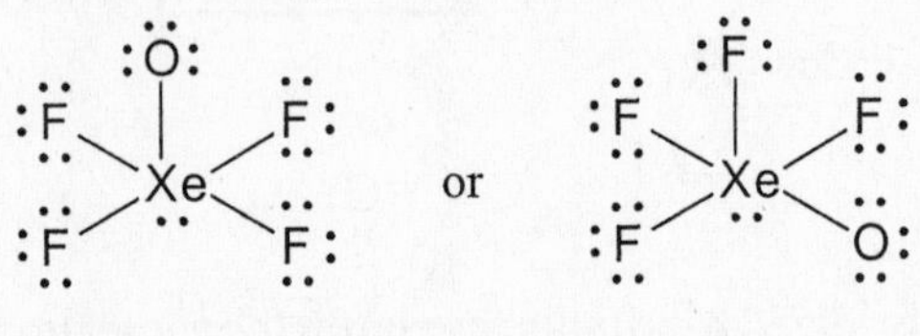

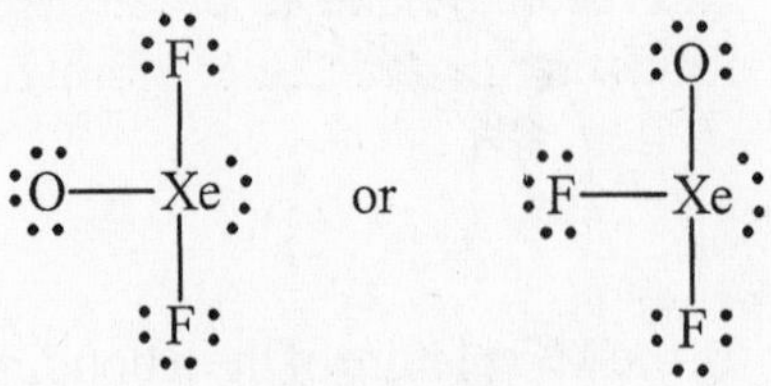

square pyramid; d^2sp^3 T-shaped; dsp^3

e. XeO_3F_2 has $8 + 3(6) + 2(7) = 40$ valence electrons.

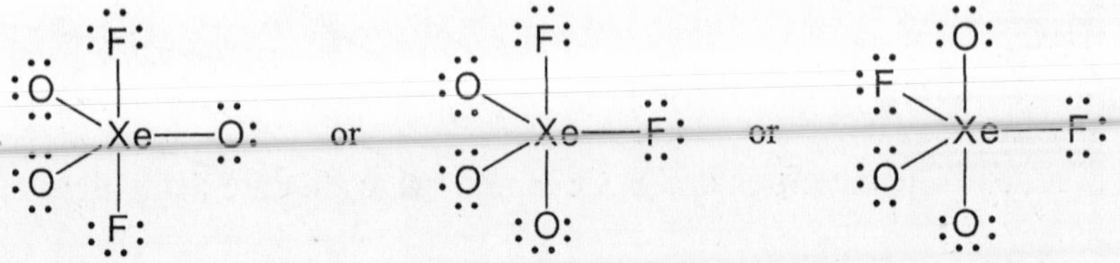

trigonal
bipyramid;
dsp^3

48. $FClO_2 + F^- \rightarrow F_2ClO_2^-$ $F_3ClO + F^- \rightarrow F_4ClO^-$

$F_2ClO_2^-$, $2(7) + 7 + 2(6) + 1 = 34$ e$^-$ F_4ClO^-, $4(7) + 7 + 6 + 1 = 42$ e$^-$

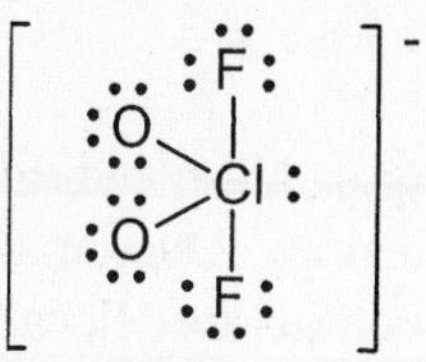

       see-saw, dsp^3 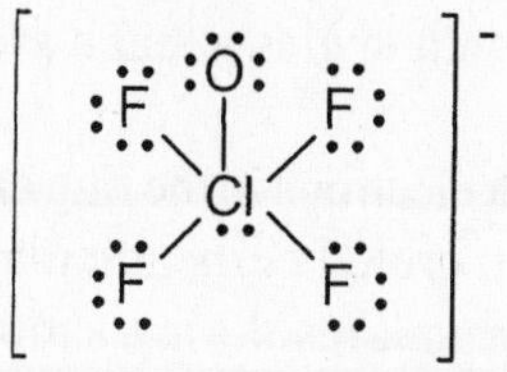   square pyramid, d^{2}sp^3

Note: Similar to Exercises 9.51 c, d and e, $F_2ClO_2^-$ has two additional Lewis structures that are possible, and F_4ClO^- has one additional Lewis structure that is possible. The predicted hybridization is unaffected.

$F_3ClO \rightarrow F^- + F_2ClO^+$ $F_3ClO_2 \rightarrow F^- + F_2ClO_2^+$

F_2ClO^+, $2(7) + 7 + 6 - 1 = 26$ e$^-$ $F_2ClO_2^+$, $2(7) + 7 + 2(6) - 1 = 32$ e$^-$

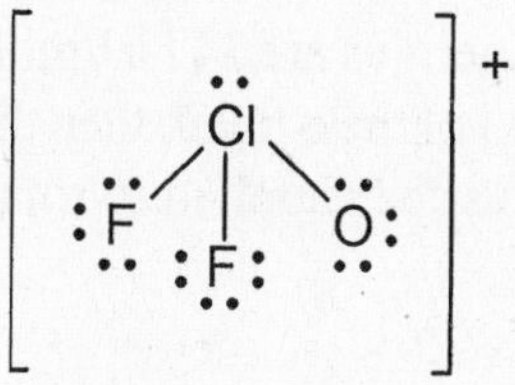

       trigonal pyramid, sp^3 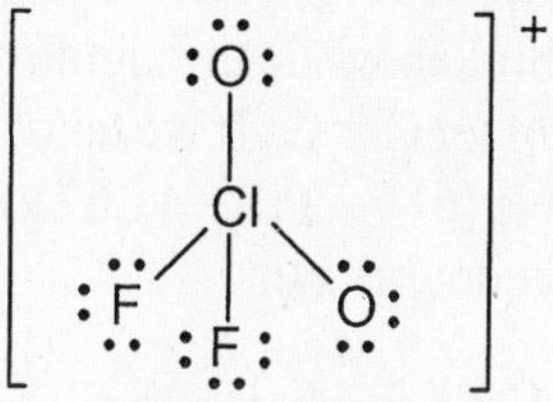   tetrahedral, sp^3

49.   For carbon, nitrogen, and oxygen atoms to have formal charge values of zero, each C atom will form four bonds to other atoms and have no lone pairs of electrons, each N atom will form three bonds to other atoms and have one lone pair of electrons, and each O atom will form two bonds to other atoms and have two lone pairs of electrons. Following these bonding requirements gives the following two resonance structures for vitamin B_6:

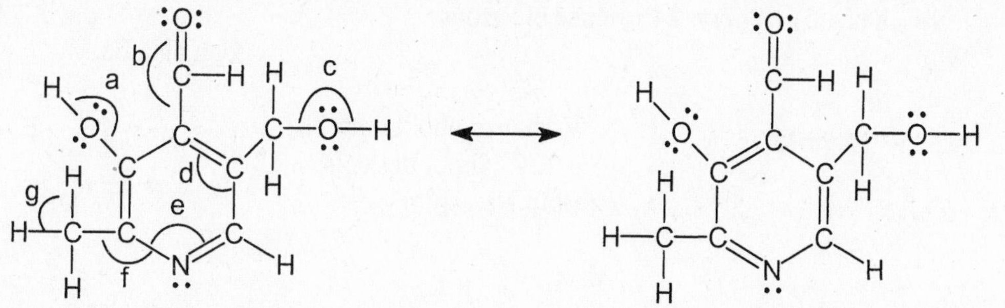

a.   21 σ bonds; 4 π bonds (The electrons in the 3 π bonds in the ring are delocalized.)

b.   angles a, c, and g: ≈ 109.5°; angles b, d, e and f: ≈ 120°

c. 6 sp² carbons; the 5 carbon atoms in the ring are sp² hybridized, as is the carbon with the double bond to oxygen.

d. 4 sp³ atoms; the 2 carbons which are not sp² hybridized are sp³ hybridized, and the oxygens marked with angles a and c are sp³ hybridized.

e. Yes, the π electrons in the ring are delocalized. The atoms in the ring are all sp² hybridized. This leaves a p orbital perpendicular to the plane of the ring from each atom. Overlap of all six of these p orbitals results in a π molecular orbital system where the electrons are delocalized above and below the plane of the ring (similar to benzene in Figure 9.48 of the text).

50.

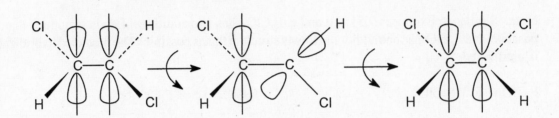

In order to rotate about the double bond, the molecule must go through an intermediate stage where the π bond is broken while the sigma bond remains intact. Bond energies are 347 kJ/mol for C – C and 614 kJ/mol for C=C. If we take the single bond as the strength of the σ bond, then the strength of the π bond is (614 - 347 =) 267 kJ/mol. Thus, 267 kJ/mol must be supplied to rotate about a carbon-carbon double bond.

51.   a. $COCl_2$ has $4 + 6 + 2(7) = 24$ valence electrons.

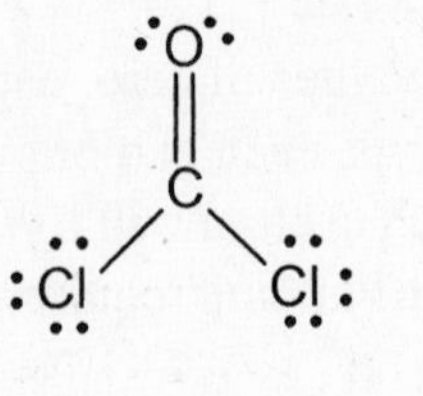

trigonal planar
polar
120°
sp²

b. N_2F_2 has $2(5) + 2(7) = 24$ valence electrons.

Can also be:

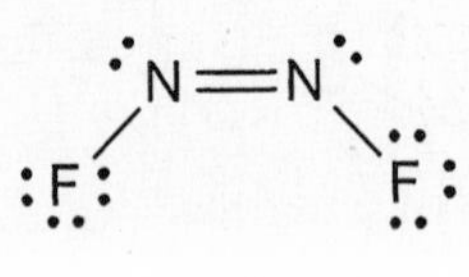

V-shaped about both Ns;
≈ 120° about both Ns;
Both Ns: sp²

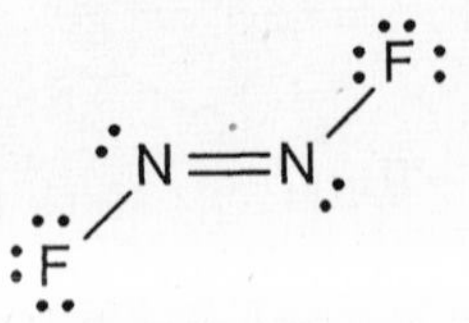

polar

nonpolar

These are distinctly different molecules.

c.   COS has $4 + 6 + 6 = 16$ valence electrons.

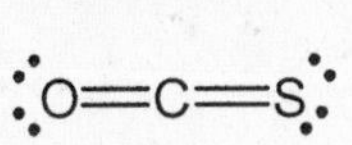

linear, polar, 180°, sp

d. ICl_3 has $7 + 3(7) = 28$ valence electrons.

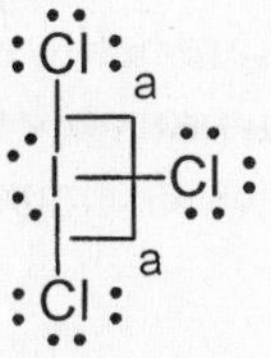

T-shaped
polar
a. ≈ 90°
dsp^3

52.   a.   Yes, both have 4 sets of electrons about the P. We would predict a tetrahedral structure for both. See part d for Lewis structures.

b.   The hybridization is sp^3 for each P since both structures are tetrahedral.

c.   P has to use one of its d orbitals to form the π bond since the p orbitals are all used to form the hybrid orbitals.

d.   Formal charge = number of valence electrons of an atom - [(number of lone pair electrons) + 1/2 (number of shared electrons)]. The formal charges calculated for the O and P atoms are next to the atoms in the following Lewis structures.

In both structures, the formal charges of the Cl atoms are all zeros. The structure with the $P = O$ bond is favored on the basis of formal charge since it has a zero formal charge for all atoms.

53.   a.   The Lewis structures for NNO and NON are:

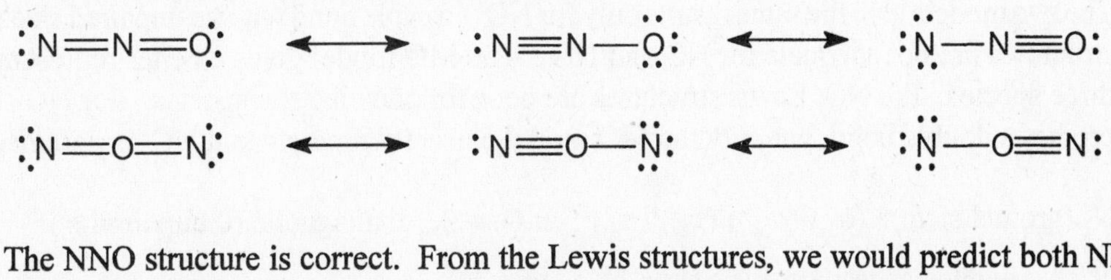

The NNO structure is correct. From the Lewis structures, we would predict both NNO and NON to be linear. However, we would predict NNO to be polar and NON to be nonpolar. Since experiments show N_2O to be polar, NNO is the correct structure.

b. Formal charge = number of valence electrons of atoms - [(number of lone pair electrons) + 1/2 (number of shared electrons)].

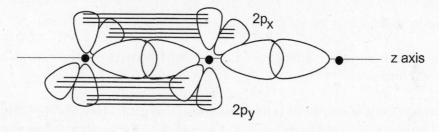

$$:N=N=O: \longleftrightarrow \quad :N\equiv N-O: \longleftrightarrow \quad :N-N\equiv O:$$

$$\quad -1 \quad +1 \quad 0 \qquad\qquad 0 \quad +1 \quad -1 \qquad\qquad -2 \quad +1 \quad +1$$

The formal charges for the atoms in the various resonance structures are below each atom. The central N is sp hybridized in all of the resonance structures. We can probably ignore the third resonance structure on the basis of the relatively large formal charges as compared to the first two resonance structures.

c. The sp hybrid orbitals on the center N overlap with atomic orbitals (or hybrid orbitals) on the other two atoms to form the two sigma bonds. The remaining two unhybridized p orbitals on the center N overlap with two p orbitals on the peripheral N to form the two π bonds.

54. Lewis structures:

NO$^+$: $\left[:N\equiv O: \right]^+$ NO$^-$: $\left[\ddot{N}=\ddot{O} \right]^-$

NO: $\ddot{N}=\ddot{O} \longleftrightarrow \ddot{N}=\ddot{O} \longleftrightarrow \dot{N}=\ddot{O}$

M.O. model:

NO$^+$: $(\sigma_{2s})^2(\sigma_{2s}*)^2(\pi_{2p})^4(\sigma_{2p})^2$, B.O. = 3, 0 unpaired e$^-$ (diamagnetic)

NO: $(\sigma_{2s})^2(\sigma_{2s}*)^2(\pi_{2p})^4(\sigma_{2p})^2(\pi_{2p}*)^1$, B.O. = 2.5, 1 unpaired e$^-$ (paramagnetic)

NO$^-$: $(\sigma_{2s})^2(\sigma_{2s}*)^2(\pi_{2p})^4(\sigma_{2p})^2(\pi_{2p}*)^2$ B.O. = 2, 2 unpaired e$^-$ (paramagnetic)

The two models give the same results only for NO$^+$ (a triple bond with no unpaired electrons). Lewis structures are not adequate for NO and NO$^-$. The MO model gives a better representation for all three species. For NO, Lewis structures are poor for odd electron species. For NO$^-$, both models predict a double bond, but only the MO model correctly predicts that NO$^-$ is paramagnetic.

55. N$_2$ (ground state): $(\sigma_{2s})^2(\sigma_{2s}*)^2(\pi_{2p})^4(\sigma_{2p})^2$, B.O. = 3, diamagnetic (0 unpaired e$^-$)

N$_2$ (1st excited state): $(\sigma_{2s})^2(\sigma_{2s}*)^2(\pi_{2p})^4(\sigma_{2p})^1(\pi_{2p}*)^1$

B.O. = (7-3)/2 = 2, paramagnetic (2 unpaired e$^-$)

The first excited state of N$_2$ should have a weaker bond and should be paramagnetic.

56.   Considering only the twelve valence electrons in O_2, the MO models would be:

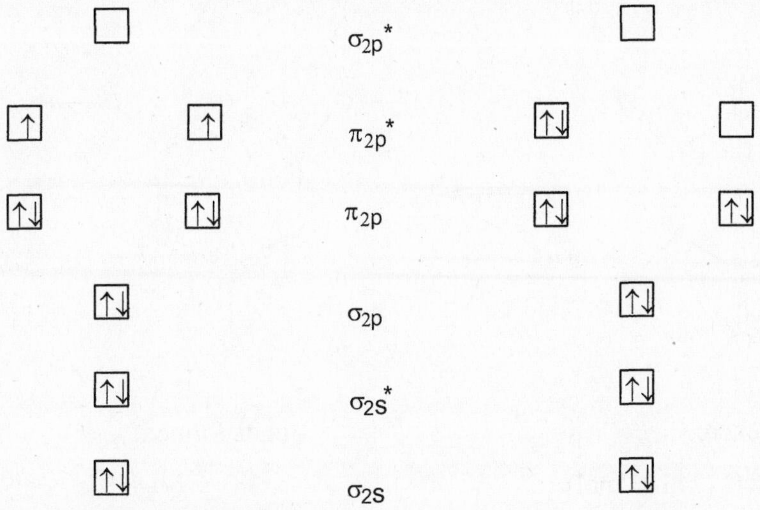

O_2 ground state			Arrangement (Lewis)	
☐		$\sigma_{2p}{}^{*}$	☐	
↑	↑	$\pi_{2p}{}^{*}$	↑↓	☐
↑↓	↑↓	π_{2p}	↑↓	↑↓
↑↓		σ_{2p}	↑↓	
↑↓		$\sigma_{2s}{}^{*}$	↑↓	
↑↓		σ_{2s}	↑↓	

O_2 ground state

Arrangement of electrons consistent with the Lewis structure (double bond and no unpaired electrons).

It takes energy to pair electrons in the same orbital. Thus, the structure with no unpaired electrons is at a higher energy; it is an excited state.

57. F_2: $(\sigma_{2s})^2(\sigma_{2s}{}^{*})^2(\sigma_{2p})^2(\pi_{2p})^4(\pi_{2p}{}^{*})^4$; F_2 should have a lower ionization energy than F. The electron removed from F_2 is in a $\pi_{2p}{}^{*}$ antibonding molecular orbital that is higher in energy than the 2p atomic orbitals from which the electron in atomic fluorine is removed. Since the electron removed from F_2 is higher in energy than the electron removed from F, it should be easier to remove an electron from F_2 than from F.

58.   Side to side overlap of these d-orbitals would produce a π molecular orbital. There would be no probability of finding an electron on the axis joining the two nuclei, which is characteristic of π MOs.

Challenge Problems

59.   a.   No, some atoms are in different places. Thus, these are not resonance structures; they are different compounds.

   b.   For the first Lewis structure, all nitrogens are sp^3 hybridized and all carbons are sp^2 hybridized. In the second Lewis structure, all nitrogens and carbons are sp^2 hybridized.

c.   For the reaction:

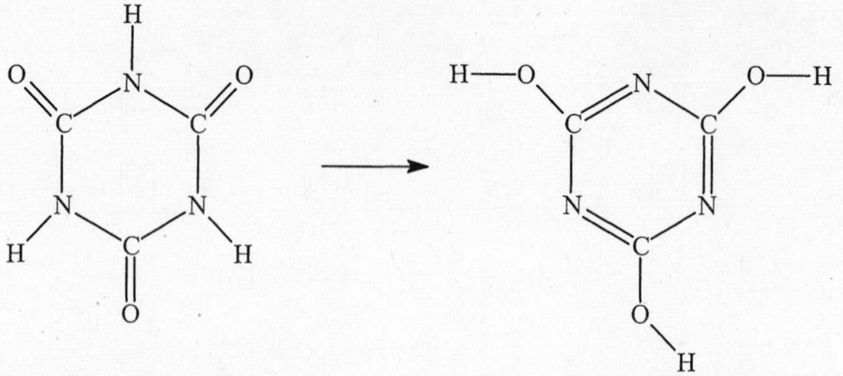

Bonds broken:

3 C = O (745 kJ/mol)

3 C – N (305 kJ/mol)

3 N – H (391 kJ/mol)

Bonds formed:

3 C = N (615 kJ/mol)

3 C – O (358 kJ/mol)

3 O – H (467 kJ/mol)

$\Delta H = 3(745) + 3(305) + 3(391) - [3(615) + 3(358) + 3(467)]$

$\Delta H = 4323 \text{ kJ} - 4320 \text{ kJ} = 3 \text{ kJ}$

The bonds are slightly stronger in the first structure with the carbon-oxygen double bonds since ΔH for the reaction is positive. However, the value of ΔH is so small that the best conclusion is that the bond strengths are comparable in the two structures.

60.   The complete Lewis structure follows. All but two of the carbon atoms are sp^3 hybridized. The two carbon atoms which contain the double bond are sp^2 hybridized (see *).

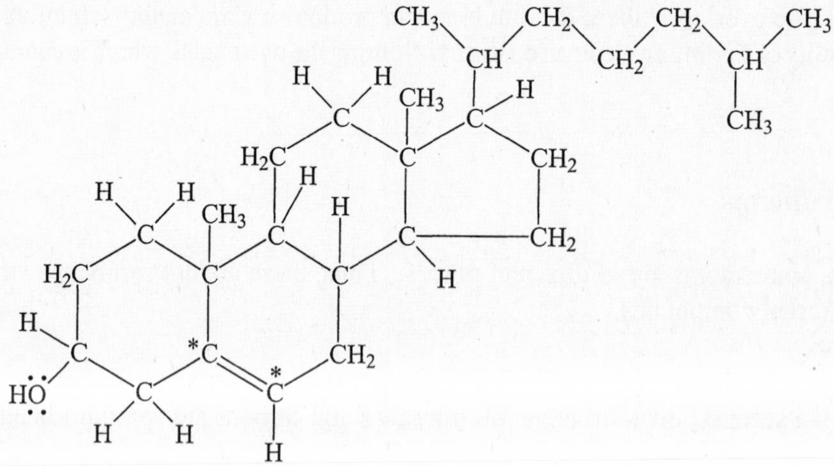

No; most of the carbons are not in the same plane since a majority of carbon atoms exhibit a tetrahedral structure.

61.   a. NCN^{2-} has $5 + 4 + 5 + 2 = 16$ valence electrons.

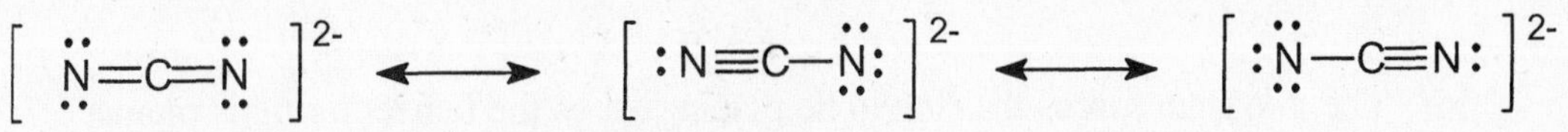

H_2NCN has $2(1) + 5 + 4 + 5 = 16$ valence electrons.

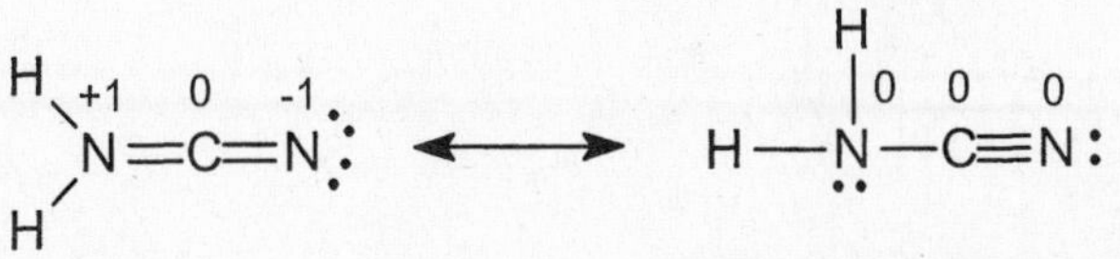

favored by formal charge

$NCNC(NH_2)_2$ has $5 + 4 + 5 + 4 + 2(5) + 4(1) = 32$ valence electrons.

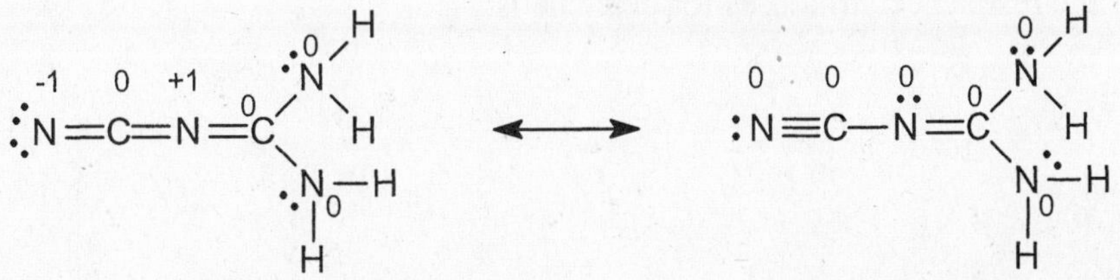

favored by formal charge

Melamine $(C_3N_6H_6)$ has $3(4) + 6(5) + 6(1) = 48$ valence electrons.

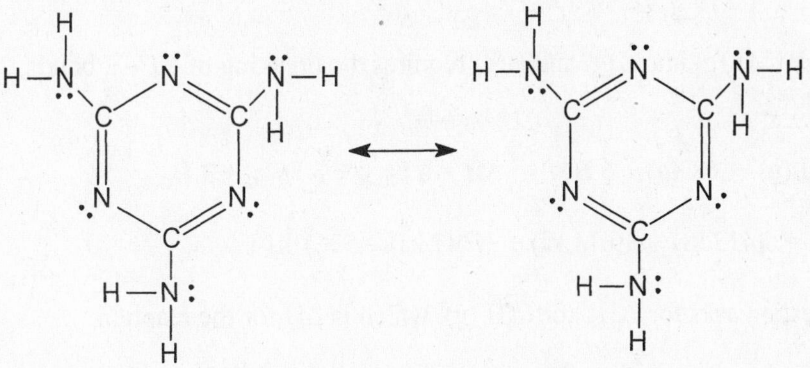

   b. NCN^{2-}:   C is sp hybridized. Depending on the resonance form, N can be sp, sp^2, or sp^3 hybridized. For the remaining compounds, we will give hybrids for the favored resonance structures as predicted from formal charge considerations.

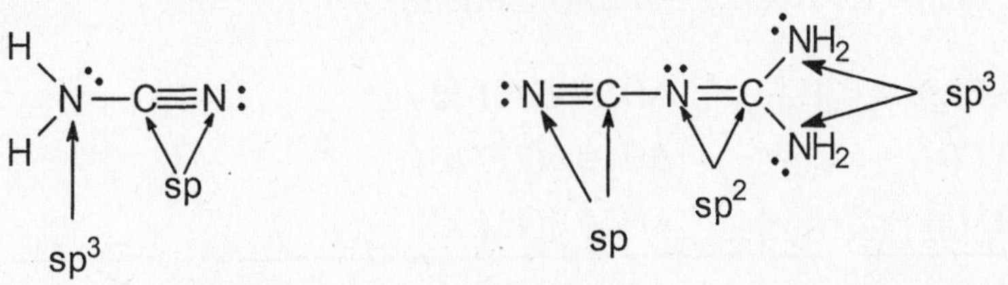

Melamine: N in NH_2 groups are all sp^3 hybridized. Atoms in ring are all sp^2 hybridized.

c. NCN^{2-}: 2 σ and 2 π bonds; H$_2$NCN: 4 σ and 2 π bonds; dicyandiamide: 9 σ and 3 π bonds; melamine: 15 σ and 3 π bonds

d. The π-system forces the ring to be planar just as the benzene ring is planar.

e. The structure:

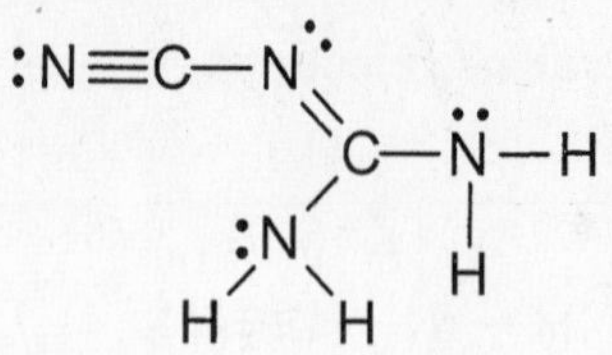

is the most important since it has three different CN bonds. This structure is also favored on the basis of formal charge.

62.   One of the resonance structures for benzene is:

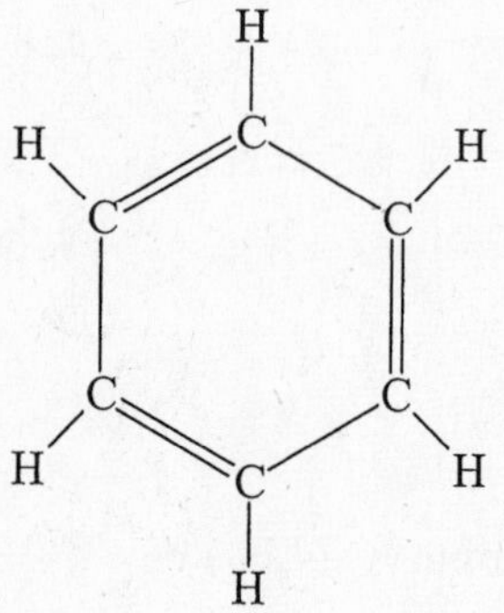

To break C$_6$H$_6$(g) into C(g) and H(g) requires the breaking of 6 C–H bonds, 3 C=C bonds and 3 C–C bonds:

C$_6$H$_6$(g) → 6 C(g) + 6 H(g)   ΔH = 6 D$_{C-H}$ + 3 D$_{C=C}$ + 3 D$_{C-C}$

ΔH = 6(413 kJ) + 3(614 kJ) + 3(347 kJ) = 5361 kJ

The question asks for ΔH$_f^\circ$ for C$_6$H$_6$(g), which is ΔH for the reaction:

6 C(s) + 3 H$_2$(g) → C$_6$H$_6$(g)   ΔH = ΔH$^\circ_{f\, C_6H_6(g)}$

To calculate ΔH for this reaction, we will use Hess's law along with the ΔH$_f^\circ$ value for C(g) and the bond energy value for H$_2$ (D$_{H_2}$ = 432 kJ/mol).

6 C(g) + 6 H(g) → C$_6$H$_6$(g)   ΔH$_1$ = -5361 kJ
6 C(s) → 6 C(g)   ΔH$_2$ = 6(717 kJ)
3 H$_2$(g) → 6 H(g)   ΔH$_3$ = 3(432 kJ)
_____
6 C(s) + 3 H$_2$(g) → C$_6$H$_6$(g)   ΔH = ΔH$_1$ + ΔH$_2$ + ΔH$_3$ = 237 kJ; ΔH$^\circ_{f\, C_6H_6(g)}$ = 237 kJ/mol

The experimental ΔH_f° for $C_6H_6(g)$ is more stable (lower in energy) by 154 kJ as compared to ΔH_f° calculated from bond energies (83 - 237 = -154 kJ). This extra stability is related to benzene's ability to exhibit resonance. Two equivalent Lewis structures can be drawn for benzene. The π bonding system implied by each Lewis structure consists of three localized π bonds. This is not correct as all C–C bonds in benzene are equivalent. We say the π electrons in benzene are delocalized over the entire surface of C_6H_6 (see Section 9.5 of the text). The large discrepancy between ΔH_f° values is due to the delocalized π electrons, whose effect was not accounted for in the calculated ΔH_f° value. The extra stability associated with benzene can be called resonance stabilization. In general, molecules that exhibit resonance are usually more stable than predicted using bond energies.

63.   a. $E = \dfrac{hc}{\lambda} = \dfrac{(6.626 \times 10^{-34}\ \text{J s}) (2.998 \times 10^8\ \text{m/s})}{25 \times 10^{-9}\ \text{m}} = 7.9 \times 10^{-18}\ \text{J}$

   $7.9 \times 10^{-18}\ \text{J} \times \dfrac{6.022 \times 10^{23}}{\text{mol}} \times \dfrac{1\ \text{kJ}}{1000\ \text{J}} = 4800\ \text{kJ/mol}$

   Using ΔH values from the various reactions, 25 nm light has sufficient energy to ionize N_2 and N and to break the triple bond. Thus, N_2, N_2^+, N, and N^+ will all be present, assuming excess N_2.

   b.   To produce atomic nitrogen but no ions, the range of energies of the light must be from 941 kJ/mol to just below 1402 kJ/mol.

   $\dfrac{941\ \text{kJ}}{\text{mol}} \times \dfrac{1\ \text{mol}}{6.022 \times 10^{23}} \times \dfrac{1000\ \text{J}}{\text{kJ}} = 1.56 \times 10^{-18}\ \text{J/photon}$

   $\lambda = \dfrac{hc}{E} = \dfrac{(6.626 \times 10^{-34}\ \text{J s}) (2.998 \times 10^8\ \text{m/s})}{1.56 \times 10^{-18}\ \text{J}} = 1.27 \times 10^{-7}\ \text{m} = 127\ \text{nm}$

   $\dfrac{1402\ \text{kJ}}{\text{mol}} \times \dfrac{1\ \text{mol}}{6.0221 \times 10^{23}} \times \dfrac{1000\ \text{J}}{\text{kJ}} = 2.328 \times 10^{-18}\ \text{J/photon}$

   $\lambda = \dfrac{hc}{E} = \dfrac{(6.6261 \times 10^{-34}\ \text{J s}) (2.9979 \times 10^8\ \text{m/s})}{2.328 \times 10^{-18}\ \text{J}} = 8.533 \times 10^{-8}\ \text{m} = 85.33\ \text{nm}$

   Light with wavelengths in the range of 85.33 nm $< \lambda \leq$ 127 nm will produce N but no ions.

   c. N_2: $(\sigma_{2s})^2(\sigma_{2s}^*)^2(\pi_{2p})^4(\sigma_{2p})^2$; The electron removed from N_2 is in the σ_{2p} molecular orbital which is lower in energy than the 2p atomic orbital from which the electron in atomic nitrogen is removed. Since the electron removed from N_2 is lower in energy than the electron in N, the ionization energy of N_2 is greater than for N.

64.   The π bonds between two S atoms and between C and S atoms are not as strong. The orbitals do not overlap with each other as well as the smaller atomic orbitals of C and O overlap.

65.   O=N–Cl: The bond order of the NO bond in NOCl is 2 (a double bond).

       NO: From molecular orbital theory, the bond order of this NO bond is 2.5.

       Both reactions apparently involve only the breaking of the N–Cl bond. However, in the reaction
       ONCl → NO + Cl, some energy is released in forming the stronger NO bond, lowering the value of
       ΔH. Therefore, the apparent N–Cl bond energy is artificially low for this reaction. The first reaction
       involves only the breaking of the N–Cl bond.

66.   The molecular orbitals for BeH_2 are formed from the two hydrogen 1s orbitals and the 2s and one
       of the 2p orbitals from beryllium. One of the sigma bonding orbitals forms from overlap of the
       hydrogen 1s orbitals with a 2s orbital from beryllium. Assuming the z-axis is the internuclear axis in
       the linear BeH_2 molecule, then the $2p_z$ orbital from beryllium has proper symmetry to overlap with
       the 1s orbitals from hydrogen; the $2p_x$ and $2p_y$ orbitals are nonbonding orbitals since they don't have
       proper symmetry necessary to overlap with 1s orbitals. The type of bond formed from the $2p_z$ and
       1s orbitals is a sigma bond since the orbitals overlap head to head. The MO diagram for BeH_2 is:

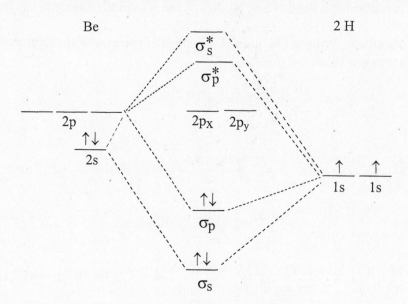

       Bond Order = (4 - 0)/2 = 2; The MO diagram predicts BeH_2 to be a stable species and also predicts
       that BeH_2 is diamagnetic. Note: The σ_s MO is a mixture of the two hydrogen 1s orbitals with the
       2s orbital from beryllium and the σ_p MO is a mixture of the two hydrogen 1s orbitals with the $2p_z$
       orbital from beryllium. The MOs are not localized between any two atoms; instead, they extend over
       the entire surface of the three atoms.

67.   a.   The CO bond is polar with the negative end around the more electronegative oxygen atom. We
             would expect metal cations to be attracted to and bond to the oxygen end of CO on the basis of
             electronegativity.

b. $:C \equiv O:$ FC (carbon) = 4 - 2 - 1/2(6) = -1

FC (oxygen) = 6 - 2 - 1/2(6) = +1

From formal charge, we would expect metal cations to bond to the carbon (with the negative formal charge).

c.   In molecular orbital theory, only orbitals with proper symmetry overlap to form bonding orbitals. The metals that form bonds to CO are usually transition metals, all of which have outer electrons in the d orbitals. The only molecular orbitals of CO that have proper symmetry to overlap with d orbitals are the $\pi_{2p}*$ orbitals, whose shape is similar to the d orbitals (see Figure 9.34). Since the antibonding molecular orbitals have more carbon character (carbon is less electronegative than oxygen), one would expect the bond to form through carbon.

CHAPTER TEN

LIQUIDS AND SOLIDS

Questions

12.   Dipole forces are the forces that act between polar molecules. The electrostatic attraction between the positive end of one polar molecule and the negative end of another is the dipole force. Dipole forces are generally weaker than hydrogen bonding. Both of these forces are due to dipole moments in molecules. Hydrogen bonding is given a separate name from dipole forces because hydrogen bonding is a particularly strong dipole force.

     London dispersion forces can be referred to as accidental-induced dipole forces. As the size of the molecule increases, the strength of the London dispersion forces increases. This is because, as the electron cloud about a molecule gets larger, it is easier for the electrons to be drawn away from the nucleus. The molecule is said to be more polarizable.

13.   London dispersion (LD) < dipole-dipole < H bonding < metallic bonding, covalent network, ionic.

     Yes, there is considerable overlap. Consider some of the examples in Exercise 10.92. Benzene (only LD forces) has a higher boiling point than acetone (dipole-dipole forces). Also, there is even more overlap among the stronger forces (metallic, covalent, and ionic).

14.   As the strengths of intermolecular forces increase: surface tension, viscosity, melting point and boiling point increase, while vapor pressure decreases.

15.   a. Polarizability of an atom refers to the ease of distorting the electron cloud. It can also refer to distorting the electron clouds in molecules or ions. Polarity refers to the presence of a permanent dipole moment in a molecule.

     b. London dispersion (LD) forces are present in all substances. LD forces can be referred to as accidental dipole-induced dipole forces. Dipole-dipole forces involve the attraction of molecules with permanent dipoles for each other.

     c. inter: between; intra: within; For example, in Br_2 the covalent bond is an intramolecular force holding the two Br atoms together in the molecule. The much weaker London dispersion forces are the intermolecular forces of attraction which hold different molecules of Br_2 together in the liquid phase.

16.   Liquids and solids both have characteristic volume and are not very compressible. Liquids and gases flow and assume the shape of their container.

17.   Atoms have an approximately spherical shape (on the average). It is impossible to pack spheres together without some empty space among the spheres.

18.   Critical temperature: The temperature above which a liquid cannot exist, i.e., the gas cannot be liquified by increased pressure.

       Critical pressure: The pressure that must be applied to a substance at its critical temperature to produce a liquid.

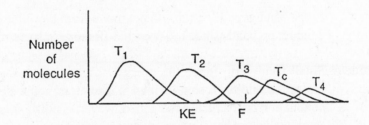

       The kinetic energy distribution changes as one raises the temperature ($T_4 > T_c > T_3 > T_2 > T_1$). At the critical temperature, T_c, all molecules have kinetic energies greater than the intermolecular forces, F, and a liquid can't form. Note: The distributions above are not to scale.

19.   As the intermolecular forces increase, the critical temperature increases.

20.   Evaporation takes place when some molecules at the surface of a liquid have enough energy to break the intermolecular forces holding them in the liquid phase. When a liquid evaporates, the molecules that escape have high kinetic energies. The average kinetic energy of the remaining molecules is lower, thus, the temperature of the liquid is lower.

21.   a.   Crystalline solid: Regular, repeating structure

             Amorphous solid: Irregular arrangement of atoms or molecules

       b.   Ionic solid: Made up of ions held together by ionic bonding.

             Molecular solid: Made up of discrete covalently bonded molecules held together in the solid phase by weaker forces (LD, dipole or hydrogen bonds).

       c.   Molecular solid: Discrete, individual molecules

             Network solid: No discrete molecules; A network solid is one large molecule. The intermolecular forces are the covalent bonds between atoms.

    d. Metallic solid: Completely delocalized electrons, conductor of electricity (ions in a sea of electrons)

       Network solid: Localized electrons; Insulator or semiconductor

22.   A crystalline solid will because a regular, repeating arrangement is necessary to produce planes of atoms that will diffract the X-rays in regular patterns. An amorphous solid does not have a regular repeating arrangement and will produce a complicated diffraction pattern.

23.   Conductor: The energy difference between the filled and unfilled molecular orbitals is minimal. We call this energy difference the band gap. Since the band gap is minimal, electrons can easily move into the conduction bands (the unfilled molecular orbitals).

       Insulator: Large band gap; Electrons do not move from the filled molecular orbitals to the conduction bands since the energy difference is large.

       Semiconductor: Small band gap; Since the energy difference between the filled and unfilled molecular orbitals is smaller than in insulators, some electrons can jump into the conduction bands. The band gap, however, is not as small as with conductors, so semiconductors have intermediate conductivity.

    a. As the temperature is increased, more electrons in the filled molecular orbitals have sufficient kinetic energy to jump into the conduction bands (the unfilled molecular orbitals).

    b. A photon of light is absorbed by an electron which then has sufficient energy to jump into the conduction bands.

    c. An impurity either adds electrons at an energy near that of the conduction bands (n-type) or creates holes (unfilled energy levels) at energies in the previously filled molecular orbitals (p-type).

24.   In conductors, electrical conductivity is inversely proportional to temperature. Increases in temperature increase the motions of the atoms, which gives rise to increased resistance (decreased conductivity). In a semiconductor, electrical conductivity is directly proportional to temperature. An increase in temperature provides more electrons with enough kinetic energy to jump from the filled molecular orbitals to the conduction bands, increasing conductivity.

25.   To produce an n-type semiconductor, dope Ge with a substance that has more than 4 valence electrons, e.g., a group 5A element. Phosphorus or arsenic are two substances which will produce n-type semiconductors when they are doped into germanium. To produce a p-type semiconductor, dope Ge with a substance that has fewer than 4 valence electrons, e.g., a group 3A element. Gallium or indium are two substances which will produce p-type semiconductors when they are doped into germanium.

26.   An alloy is a substance that contains a mixture of elements and has metallic properties. In a substitutional alloy, some of the host metal atoms are replaced by other metal atoms of similar size, e.g., brass, pewter, plumber's solder. An interstitial alloy is formed when some of the interstices (holes) in the closest packed metal structure are occupied by smaller atoms, e.g., carbon steels.

27.   a.   Condensation: vapor → liquid   b.   Evaporation: liquid → vapor

      c.   Sublimation: solid → vapor

      d.   A supercooled liquid is a liquid which is at a temperature below its freezing point.

28.   Equilibrium: There is no change in composition; the vapor pressure is constant.

      Dynamic: Two processes, vapor → liquid and liquid → vapor, are both occurring but with equal
                     rates so the composition of the vapor is constant.

29.   a.   As the intermolecular forces increase, the rate of evaporation decreases.

      b.   As temperature increases, the rate of evaporation increases.

      c.   As surface area increases, the rate of evaporation increases.

30.   A volatile liquid is one that evaporates relatively easily. Volatile liquids have large vapor pressures
      because the intermolecular forces that prevent evaporation are relatively weak.

31. $C_2H_5OH(l) \rightarrow C_2H_5OH(g)$ is an endothermic process. Heat is absorbed when liquid ethanol
      vaporizes; the internal heat from the body provides this heat which results in the cooling of the body.

32.   Sublimation will occur allowing water to escape as $H_2O(g)$.

33.   The phase change, $H_2O(g) \rightarrow H_2O(l)$, releases heat that can cause additional damage. Also steam
      can be at a temperature greater than 100°C.

34.   Fusion refers to a solid converting to a liquid, and vaporization refers to a liquid converting to a gas.
      Only a fraction of the hydrogen bonds are broken in going from the solid phase to the liquid phase.
      Most of the hydrogen bonds are still present in the liquid phase and must be broken during the liquid
      to gas phase transition. Thus, the enthalpy of vaporization is much larger than the enthalpy of fusion
      since more intermolecular forces are broken during the vaporization process.

Exercises

Intermolecular Forces and Physical Properties

35.   Ionic compounds have ionic forces. Covalent compounds all have London Dispersion (LD) forces,
      while polar covalent compounds have dipole forces and/or hydrogen bonding forces. For H bonding
      forces, the covalent compound must have either a N-H, O-H or F-H bond in the molecule.

      Ar   HCl   HF
      a.   LD only   b.   dipole, LD   c.   H bonding, LD

      d.   ionic   e.   LD only (CH_4 is a nonpolar covalent compound.)
      $CaCl_2$

CO   *ionic*

f. dipole, LD   g. ionic

36. a. ionic

b. LD mostly; C – F bonds are polar, but polymers like teflon are so large the LD forces are the predominant intermolecular forces.

c. LD   d. dipole, LD   e. H bonding, LD

f. dipole, LD   g. LD

37. a. OCS; OCS is polar and has dipole-dipole forces in addition to London dispersion (LD) forces. All polar molecules have dipole forces. CO_2 is nonpolar and only has LD forces. To predict polarity, draw the Lewis structure and deduce whether the individual bond dipoles cancel.

b. SeO_2; Both SeO_2 and SO_2 are polar compounds, so they both have dipole forces as well as LD forces. However, SeO_2 is a larger molecule, so it would have stronger LD forces.

c. $H_2NCH_2CH_2NH_2$; More extensive hydrogen bonding is possible.   *weakest*

d. H_2CO; H_2CO is polar while CH_3CH_3 is nonpolar. H_2CO has dipole forces in addition to LD forces.

e. CH_3OH; CH_3OH can form relatively strong H bonding interactions, unlike H_2CO.

38. a. Neopentane is more compact than n-pentane. There is less surface area contact among neopentane molecules. This leads to weaker LD forces and a lower boiling point.

b. Ethanol is capable of H bonding; dimethyl ether is not.

c. HF is capable of H bonding; HCl is not.

d. LiCl is ionic, and HCl is a molecular solid with only dipole forces and LD forces. Ionic forces are much stronger than the forces for molecular solids.

e. n-pentane is a larger molecule so has stronger LD forces.

f. Dimethyl ether is polar so has dipole forces in addition to LD forces, unlike n-propane which has only LD forces.

39. See Question 10.14 to review the dependence of some physical properties on the strength of the intermolecular forces.

a. HCl; HCl is polar while Ar and F_2 are nonpolar. HCl has dipole forces unlike Ar and F_2.

b. NaCl; Ionic forces are much stronger than molecular forces.

c. I_2; All are nonpolar, so the largest molecule (I_2) will have the strongest LD forces and the lowest vapor pressure.

d. N_2; Nonpolar and smallest, so has the weakest intermolecular forces.

e. CH_4; Smallest, nonpolar molecule so has the weakest LD forces.

f. HF; HF can form relatively strong H bonding interactions unlike the others.

g. $CH_3CH_2CH_2OH$; H bonding, unlike the others, so has strongest intermolecular forces.

40. a. CBr_4; Largest of these nonpolar molecules so has strongest LD forces.

b. Cl_2; Ionic forces in LiF are much stronger than the covalent forces in Cl_2 and HBr. HBr has dipole forces that the nonpolar Cl_2 does not exhibit; so Cl_2 has the weakest intermolecular forces.

c. CH_3CH_2OH; Can form H bonding interactions unlike the others.

d. H_2O_2; $H-O-O-H$ structure produces stronger H bonding interactions than HF, so has greatest viscosity.

e. H_2CO; H_2CO is polar so has dipole forces, unlike the other nonpolar covalent compounds.

f. I_2; I_2 has only LD forces while CsBr and CaO have much stronger ionic forces. I_2 has weakest intermolecular forces so has smallest ΔH_{fusion}.

Properties of Liquids

41. The attraction of H_2O for glass is stronger than the H_2O-H_2O attraction. The miniscus is concave to increase the area of contact between glass and H_2O. The $Hg-Hg$ attraction is greater than the $Hg-glass$ attraction. The miniscus is convex to minimize the $Hg-glass$ contact.

42. A molecule at the surface of a waterdrop is subject to attractions only by molecules below it and to each side. The effect of this uneven pull on the surface molecules tends to draw them into the body of the liquid and causes the droplet to assume the shape that has the minimum surface area, a sphere.

43. The structure of H_2O_2 is $H-O-O-H$, which produces greater hydrogen bonding than water. Long chains of hydrogen bonded H_2O_2 molecules then get tangled together.

44. CO_2 is a gas at room temperature. As mp and bp increase, the strength of the intermolecular forces also increases. Therefore, the strength of forces is $CO_2 < CS_2 < CSe_2$. From a structural standpoint this is expected. All three are linear, nonpolar molecules. Thus, only London dispersion forces are present. Since the molecules increase in size from $CO_2 < CS_2 < CSe_2$, the strength of the intermolecular forces will increase in the same order.

Structures and Properties of Solids

45. $n\lambda = 2d \sin \theta$, $d = \dfrac{n\lambda}{2 \sin \theta} = \dfrac{1 \times 1.54 \text{ Å}}{2 \times \sin 14.22°} = 3.13 \text{ Å} = 3.13 \times 10^{-10} \text{ m} = 313 \text{ pm}$

46. $n\lambda = 2d \sin \theta$, $d = \dfrac{n\lambda}{2 \sin \theta} = \dfrac{1 \times 2.63 \text{ Å}}{2 \times \sin 15.55°} = 4.91 \text{ Å} = 4.91 \times 10^{-10} \text{ m} = 491 \text{ pm}$

$\sin \theta = \dfrac{n\lambda}{2d} = \dfrac{2 \times 2.63 \text{ Å}}{2 \times 4.91 \text{ Å}} = 0.536,$ $\theta = 32.4°$

47.   A cubic closest packed structure has a face-centered cubic unit cell. In a face-centered cubic unit, there are:

$$8 \text{ corners} \times \frac{1/8 \text{ atom}}{\text{corner}} + 6 \text{ faces} \times \frac{1/2 \text{ atom}}{\text{face}} = 4 \text{ atoms}$$

The atoms in a face-centered cubic unit cell touch along the face diagonal of the cubic unit cell. Using the Pythagorean formula where l = length of the face diagonal and r = radius of the atom:

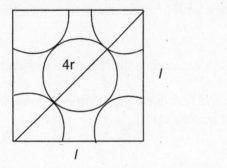

$l^2 + l^2 = (4r)^2$

$2 l^2 = 16 r^2$

$l = r \sqrt{8}$

$l = r \sqrt{8} = 197 \times 10^{-12} \text{ m} \times \sqrt{8} = 5.57 \times 10^{-10} \text{ m} = 5.57 \times 10^{-8} \text{ cm}$

Volume of a unit cell = $l^3 = (5.57 \times 10^{-8} \text{ cm})^3 = 1.73 \times 10^{-22} \text{ cm}^3$

Mass of a unit cell = $4 \text{ Ca atoms} \times \dfrac{1 \text{ mol Ca}}{6.022 \times 10^{23} \text{ atoms}} \times \dfrac{40.08 \text{ g Ca}}{\text{mol Ca}} = 2.662 \times 10^{-22} \text{ g Ca}$

$\text{density} = \dfrac{\text{mass}}{\text{volume}} = \dfrac{2.662 \times 10^{-22} \text{ g}}{1.73 \times 10^{-22} \text{ cm}^3} = 1.54 \text{ g/cm}^3$

48.   There are 4 Ni atoms in each unit cell: For a unit cell:

$$\text{density} = \dfrac{\text{mass}}{\text{volume}} = 6.84 \text{ g/cm}^3 = \dfrac{4 \text{ Ni atoms} \times \dfrac{1 \text{ mol Ni}}{6.022 \times 10^{23} \text{ atoms}} \times \dfrac{58.69 \text{ g Ni}}{\text{mol Ni}}}{l^3}$$

Solving: $l = 3.85 \times 10^{-8} \text{ cm} = \text{cube edge length}$

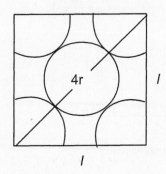

For a face centered cube:

$$(4r)^2 = l^2 + l^2 = 2\,l^2$$

$$r\sqrt{8} = l,\ r = l/\sqrt{8}$$

$$r = 3.85 \times 10^{-8}\ \text{cm}/\sqrt{8}$$

$$r = 1.36 \times 10^{-8}\ \text{cm} = 136\ \text{pm}$$

49.   The volume of a unit cell is:

$$V = l^3 = (383.3 \times 10^{-10}\ \text{cm})^3 = 5.631 \times 10^{-23}\ \text{cm}^3$$

There are 4 Ir atoms in the unit cell, as is the case for all face-centered cubic unit cells. The mass of atoms in a unit cell is:

$$\text{mass} = 4\ \text{Ir atoms} \times \frac{1\ \text{mol Ir}}{6.022 \times 10^{23}\ \text{atoms}} \times \frac{192.2\ \text{g Ir}}{\text{mol Ir}} = 1.277 \times 10^{-21}\ \text{g}$$

$$\text{density} = \frac{\text{mass}}{\text{volume}} = \frac{1.277 \times 10^{-21}\ \text{g}}{5.631 \times 10^{-23}\ \text{cm}^3} = 22.68\ \text{g/cm}^3$$

50.   A face-centered cubic unit cell contains 4 atoms. For a unit cell:

$$\text{mass of X} = \text{volume} \times \text{density} = (4.09 \times 10^{-8}\ \text{cm})^3 \times 10.5\ \text{g/cm}^3 = 7.18 \times 10^{-22}\ \text{g}$$

$$\text{mol X} = 4\ \text{atoms X} \times \frac{1\ \text{mol X}}{6.022 \times 10^{23}\ \text{atoms}} = 6.642 \times 10^{-24}\ \text{mol X}$$

$$\text{Molar mass} = \frac{7.18 \times 10^{-22}\ \text{g X}}{6.642 \times 10^{-24}\ \text{mol X}} = 108\ \text{g/mol};\ \text{The metal is silver (Ag).}$$

51.   For a body-centered unit cell: $8\ \text{corners} \times \dfrac{1/8\ \text{Ti}}{\text{corner}} + \text{Ti at body center} = 2\ \text{Ti atoms}$

All body-centered unit cells have 2 atoms per unit cell. For a unit cell:

$$\text{density} = 4.50\ \text{g/cm}^3 = \frac{2\ \text{atoms Ti} \times \dfrac{1\ \text{mol Ti}}{6.022 \times 10^{23}\ \text{atoms}} \times \dfrac{47.88\ \text{g Ti}}{\text{mol Ti}}}{l^3},\ l = \text{cube edge length}$$

Solving: $l = \text{edge length of unit cell} = 3.28 \times 10^{-8}\ \text{cm} = 328\ \text{pm}$

Assume Ti atoms just touch along the body diagonal of the cube, so body diagonal = 4 × radius of atoms = 4r.

The triangle we need to solve is:

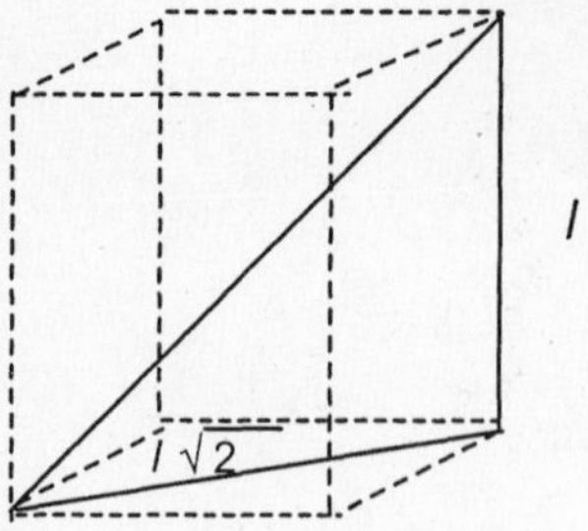

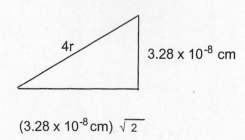

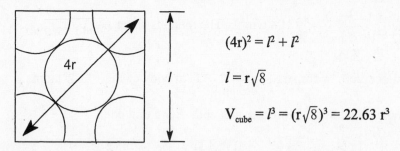

$(4r)^2 = (3.28 \times 10^{-8} \text{ cm})^2 + [(3.28 \times 10^{-8} \text{ cm}) \sqrt{2}\,]^2$, $r = 1.42 \times 10^{-8}$ cm = 142 pm

For a body-centered unit cell (bcc), the radius of the atom is related to the cube edge length by $4r = l\sqrt{3}$ or $l = 4r/\sqrt{3}$.

52.   From Exercise 10.51:

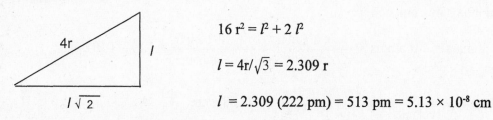

$16\, r^2 = l^2 + 2\, l^2$

$l = 4r/\sqrt{3} = 2.309\, r$

$l = 2.309\, (222 \text{ pm}) = 513 \text{ pm} = 5.13 \times 10^{-8}$ cm

In a bcc, there are 2 atoms/unit cell. For a unit cell:

$$\text{density} = \frac{\text{mass}}{\text{volume}} = \frac{2 \text{ atoms Ba} \times \dfrac{1 \text{ mol Ba}}{6.022 \times 10^{23} \text{ atoms}} \times \dfrac{137.3 \text{ g Ba}}{\text{mol Ba}}}{(5.13 \times 10^{-8} \text{ cm})^3} = \frac{3.38 \text{ g}}{\text{cm}^3}$$

53.   In a face-centered unit cell (ccp structure), the atoms touch along the face diagonal:

$(4r)^2 = l^2 + l^2$

$l = r\sqrt{8}$

$V_{cube} = l^3 = (r\sqrt{8})^3 = 22.63\, r^3$

There are four atoms in a face-centered cubic cell (see Exercise 10.47). Each atom has a volume of $4/3\ \pi r^3$.

$V_{atoms} = 4 \times \dfrac{4}{3}\ \pi r^3 = 16.76\, r^3$

So, $\dfrac{V_{atom}}{V_{cube}} = \dfrac{16.76\, r^3}{22.63\, r^3} = 0.7406$ or 74.06% of the volume of each unit cell is occupied by atoms.

In a simple cubic unit cell, the atoms touch along the cube edge (l):

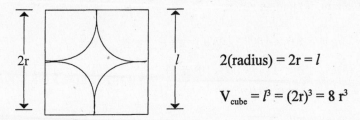

$2(\text{radius}) = 2r = l$

$V_{cube} = l^3 = (2r)^3 = 8\,r^3$

There is one atom per simple cubic cell (8 corner atoms × 1/8 atom per corner = 1 atom/unit cell). Each atom has an assumed volume of 4/3 πr^3 = volume of a sphere.

$$V_{atom} = \frac{4}{3}\,\pi r^3 = 4.189\,r^3$$

So, $\dfrac{V_{atom}}{V_{cube}} = \dfrac{4.189\,r^3}{8\,r^3} = 0.5236$ or 52.36% of the volume of each unit cell is occupied by atoms.

A cubic closest packed structure packs the atoms much more efficiently than a simple cubic structure.

54.   From Exercise 10.51, a body-centered unit cell contains 2 atoms, and the length of a cube edge (l) is related to the radius of the atom (r) by the equation $l = 4r/\sqrt{3}$.

Volume of unit cell = $l^3 = (4\,r/\sqrt{3})^3 = 12.32\,r^3$

Volume of atoms in unit cell = $2 \times \dfrac{4}{3}\pi\,r^3 = 8.378\,r^3$

So, $\dfrac{V_{atom}}{V_{cube}} = \dfrac{8.378\,r^3}{12.32\,r^3} = 0.6800 = 68.00\%$ occupied

To determine the radius of the Fe atoms, we need to determine the cube edge length (l).

Volume of unit cell = $\left(2\ \text{Fe atoms} \times \dfrac{1\ \text{mol Fe}}{6.022 \times 10^{23}\ \text{atoms}} \times \dfrac{55.85\ \text{g Fe}}{\text{mol Fe}} \right) \times \dfrac{1\ \text{cm}^3}{7.86\ \text{g}}$

$= 2.36 \times 10^{-23}\ \text{cm}^3$

Volume = $l^3 = 2.36 \times 10^{-23}\ \text{cm}^3,\ \ l = 2.87 \times 10^{-8}\ \text{cm}$

$l = 4r/\sqrt{3},\ \ r = l\sqrt{3}/4 = 2.87 \times 10^{-8}\ \text{cm} \times \sqrt{3}/4 = 1.24 \times 10^{-8}\ \text{cm}$

55.   In has fewer valence electrons than Se, thus, Se doped with In would be a p-type semiconductor.

56.   To make a p-type semiconductor we need to dope the material with atoms that have fewer valence electrons. The average number of valence electrons is four when 50-50 mixtures of group 3A and group 5A elements are considered. We could dope with more of the Group 3A element or with atoms of Zn or Cd. Cadmium is the most common impurity used to produce p-type GaAs

semiconductors. To make an n-type GaAs semiconductor, dope with an excess group 5A element or dope with a Group 6A element such as sulfur.

57. $E_{gap} = 2.5 \text{ eV} \times 1.6 \times 10^{-19} \text{ J/eV} = 4.0 \times 10^{-19} \text{ J}$; We want $E_{gap} = E_{light}$, so:

$$E_{light} = \frac{hc}{\lambda}, \ \lambda = \frac{hc}{E} = \frac{(6.63 \times 10^{-34} \text{ J s}) (3.00 \times 10^8 \text{ m/s})}{4.0 \times 10^{-19} \text{ J}} = 5.0 \times 10^{-7} \text{ m} = 5.0 \times 10^2 \text{ nm}$$

58. $$E = \frac{hc}{\lambda} = \frac{(6.626 \times 10^{-34} \text{ J s}) (2.998 \times 10^8 \text{ m/s})}{730. \times 10^{-9} \text{ m}} = 2.72 \times 10^{-19} \text{ J} = \text{energy of band gap}$$

59. a. $8 \text{ corners} \times \dfrac{1/8 \text{ Cl}}{\text{corner}} + 6 \text{ faces} \times \dfrac{1/2 \text{ Cl}}{\text{face}} = 4 \text{ Cl ions}$

$12 \text{ edges} \times \dfrac{1/4 \text{ Na}}{\text{edge}} + 1 \text{ Na at body center} = 4 \text{ Na ions};$ NaCl is the formula.

b. 1 Cs ion at body center; $8 \text{ corners} \times \dfrac{1/8 \text{ Cl}}{\text{corner}} = 1 \text{ Cl ion};$ CsCl is the formula.

c. There are 4 Zn ions inside the cube.

$8 \text{ corners} \times \dfrac{1/8 \text{ S}}{\text{corner}} + 6 \text{ faces} \times \dfrac{1/2 \text{ S}}{\text{face}} = 4 \text{ S ions};$ ZnS is the formula.

d. $8 \text{ corners} \times \dfrac{1/8 \text{ Ti}}{\text{corner}} + 1 \text{ Ti at body center} = 2 \text{ Ti ions}$

$4 \text{ faces} \times \dfrac{1/2 \text{ O}}{\text{face}} + 2 \text{ O inside cube} = 4 \text{ O ions};$ TiO_2 is the formula.

60. Both As ions are inside the unit cell. $8 \text{ corners} \times \dfrac{1/8 \text{ Ni}}{\text{corner}} + 4 \text{ edges} \times \dfrac{1/4 \text{ Ni}}{\text{edge}} = 2 \text{ Ni ions}$

The unit cell contains 2 ions of Ni and 2 ions of As which gives a formula of NiAs.

61. There is one octahedral hole per closest packed anion in a closest packed structure. If half of the octahedral holes are filled, there is a 2:1 ratio of fluoride ions to cobalt ions in the crystal. The formula is CoF_2.

62. There are 2 tetrahedral holes per closest packed anion. Let f = fraction of tetrahedral holes filled by the cations.

Na_2O: cation to anion ratio $= \dfrac{2}{1} = \dfrac{2f}{1}$, f = 1; All of the tetrahedral holes are filled by Na^+ cations.

CdS: cation to anion ratio $= \dfrac{1}{1} = \dfrac{2f}{1}$, f $= \dfrac{1}{2}$; $\dfrac{1}{2}$ of the tetrahedral holes are filled by Cd^{2+} cations.

ZrI_4: cation to anion ratio $= \dfrac{1}{4} = \dfrac{2f}{1}$, f $= \dfrac{1}{8}$; $\dfrac{1}{8}$ of the tetrahedral holes are filled by Zr^{4+} cations.

63.   8 F⁻ ions at corners × $\dfrac{1/8\,F^-}{\text{corner}}$ = 1 F⁻ ion per unit cell; Since there is one cubic hole per cubic unit

cell, there is a 2:1 ratio of F⁻ ions to metal ions in the crystal. The formula is MF_2 where M^{2+} is the metal ion.

64.   Mn ions at 8 corners: 8(1/8) = 1 Mn ion; F ions at 12 edges: 12(1/4) = 3 F ions

Formula is MnF_3. Assuming fluoride is -1 charged, the charge on Mn is +3.

65.   Since magnesium oxide has the same structure as NaCl, each unit cell contains 4 Mg^{2+} ions and 4 O^{2-} ions. The mass of a unit cell is:

$$4\text{ MgO formula units}\left(\frac{1\text{ mol MgO}}{6.022\times10^{23}\text{ formula units}}\right)\left(\frac{40.31\text{ g MgO}}{1\text{ mol MgO}}\right)=2.678\times10^{-22}\text{ g MgO}$$

$$\text{Volume of unit cell}=2.678\times10^{-22}\text{ g MgO}\left(\frac{1\text{ cm}^3}{3.58\text{ g}}\right)=7.48\times10^{-23}\text{ cm}^3$$

Volume of unit cell = l^3, l = cube edge length; $l = (7.48 \times 10^{-23}\text{ cm}^3)^{1/3} = 4.21 \times 10^{-8}$ cm = 421 pm

From the NaCl structure in Figure 10.35 of the text, Mg^{2+} and O^{2-} ions should touch along the cube edge, l:

$$l = 2\,r_{Mg^{2+}} + 2\,r_{O^{2-}} = 2\,(65\text{ pm}) + 2\,(140.\text{ pm}) = 410.\text{ pm}$$

The two values agree within 3%. In the actual crystals, the Mg^{2+} and O^{2-} ions may not touch, which is assumed in calculating the 410. pm value.

66.   CsCl is a simple cubic array of Cl⁻ ions with Cs⁺ in the middle of each unit cell. There is one Cs⁺ and one Cl⁻ ion in each unit cell. Cs⁺ and Cl⁻ touch along the body diagonal.

body diagonal = $2r_{Cs^+} + 2r_{Cl^-} = \sqrt{3}\,l$, l = length of cube edge

In each unit cell:

$$\text{mass} = 1\text{ CsCl formula unit}\left(\frac{1\text{ mol CsCl}}{6.022\times10^{23}\text{ formula units}}\right)\left(\frac{168.4\text{ g CsCl}}{\text{mol CsCl}}\right)=2.796\times10^{-22}\text{ g}$$

$$\text{volume} = l^3 = 2.796\times10^{-22}\text{ g CsCl}\times\frac{1\text{ cm}^3}{3.97\text{ g CsCl}}=7.04\times10^{-23}\text{ cm}^3$$

$l^3 = 7.04 \times 10^{-23}$ cm³, $l = 4.13 \times 10^{-8}$ cm = 413 pm = length of cube edge

$2r_{Cs^+} + 2r_{Cl^-} = \sqrt{3}\,l = \sqrt{3}(413\text{ pm}) = 715$ pm

The distance between ion centers = $r_{Cs^+} + r_{Cl^-}$ = 715 pm/2 = 358 pm

From ionic radius: r_{Cs^+} = 169 pm and r_{Cl^-} = 181 pm; r_{Cs^+} + r_{Cl^-} = 169 + 181 = 350. pm

The actual distance is 8 pm (2.3%) greater than that calculated from values of ionic radii.

67.   a. CO_2: molecular   b. SiO_2: network   c. Si: atomic, network

       d. CH_4: molecular   e. Ru: atomic, metallic   f. I_2: molecular

       g. KBr: ionic   h. H_2O: molecular   i. NaOH: ionic

       j. U: atomic, metallic   k. $CaCO_3$: ionic   l. PH_3: molecular

68.   a. diamond: atomic, network   b. PH_3: molecular   c. H_2: molecular

       d. Mg: atomic, metallic   e. KCl: ionic   f. quartz: network

       g. NH_4NO_3: ionic   h. SF_2: molecular   i. Ar: atomic, group 8A

       j. Cu: atomic, metallic   k. $C_6H_{12}O_6$: molecular

69.   a. The unit cell consists of Ni at the cube corners and Ti at the body center, or Ti at the cube corners and Ni at the body center.

       b. 8 × 1/8 = 1 atom from corners + 1 atom at body center; Empirical formula = NiTi

       c. Both have a coordination number of 8 (both are surrounded by 8 atoms).

70.   8 corners × $\dfrac{1/8\ Xe}{corner}$ + 1 Xe inside cell = 2 Xe; 8 edges × $\dfrac{1/4\ F}{edge}$ + 2 F inside cell = 4 F

Empirical formula is XeF_2. This is also the molecular formula.

71.   Structure 1   Structure 2

       8 corners × $\dfrac{1/8\ Ca}{corner}$ = 1 Ca atom   8 corners × $\dfrac{1/8\ Ti}{corner}$ = 1 Ti atom

       6 faces × $\dfrac{1/2\ O}{face}$ = 3 O atoms   12 edges × $\dfrac{1/4\ O}{edge}$ = 3 O atoms

       1 Ti at body center. Formula = $CaTiO_3$ 1 Ca at body center. Formula = $CaTiO_3$

In the extended lattice of both structures, each Ti atom is surrounded by six O atoms.

72.   There are four sulfur ions per unit cell since the sulfur ions are cubic closest packed (face-centered cubic unit cell). Since there are one octahedral hole and two tetrahedral holes per closest packed ion, each unit cell has 4 octahedral holes and 8 tetrahedral holes. This gives 4 (1/2) = 2 Al ions per unit cell and 8 (1/8) = 1 Zn ion per unit cell. The formula of the mineral is Al_2ZnS_4.

73.   a.   Y: 1 Y in center; Ba: 2 Ba in center

Cu: 8 corners $\times \dfrac{1/8 \text{ Cu}}{\text{corner}}$ = 1 Cu, 8 edges $\times \dfrac{1/4 \text{ Cu}}{\text{edge}}$ = 2 Cu, total = 3 Cu atoms

O: 20 edges $\times \dfrac{1/4 \text{ O}}{\text{edge}}$ = 5 oxygen, 8 faces $\times \dfrac{1/2 \text{ O}}{\text{face}}$ = 4 oxygen, total = 9 O atoms

Formula: $YBa_2Cu_3O_9$

b.   The structure of this superconductor material follows the second perovskite structure described in Exercise 10.71. The $YBa_2Cu_3O_9$ structure is three of these cubic perovskite unit cells stacked on top of each other. The oxygen atoms are in the same places, Cu takes the place of Ti, two of the calcium atoms are replaced by two barium atoms, and one Ca is replaced by Y.

c.   Y, Ba, and Cu are the same. Some oxygen atoms are missing.

12 edges $\times \dfrac{1/4 \text{ O}}{\text{edge}}$ = 3 O, 8 faces $\times \dfrac{1/2 \text{ O}}{\text{face}}$ = 4 O, total = 7 O atoms

Superconductor formula is $YBa_2Cu_3O_7$.

74.   a.   Structure (a):

Ba: 2 Ba inside unit cell; Tl: 8 corners $\times \dfrac{1/8 \text{ Tl}}{\text{corner}}$ = 1 Tl; Cu: 4 edges $\times \dfrac{1/4 \text{ Cu}}{\text{edge}}$ = 1 Cu

O: 6 faces $\times \dfrac{1/2 \text{ O}}{\text{face}}$ + 8 edges $\times \dfrac{1/4 \text{ O}}{\text{edge}}$ = 5 O; Formula = $TlBa_2CuO_5$

Structure (b):

Tl and Ba are the same as in structure (a).

Ca: 1 Ca inside unit cell; Cu: 8 edges $\times \dfrac{1/4 \text{ Cu}}{\text{edge}}$ = 2 Cu

O: 10 faces $\times \dfrac{1/2 \text{ O}}{\text{face}}$ + 8 edges $\times \dfrac{1/4 \text{ O}}{\text{edge}}$ = 7 O; Formula = $TlBa_2CaCu_2O_7$

Structure (c):

Tl and Ba are the same, and two Ca are located inside the unit cell.

$$Cu:\ 12\ edges \times \frac{1/4\ Cu}{edge} = 3\ Cu;\quad O:\ 14\ faces \times \frac{1/2\ O}{face} + 8\ edges \times \frac{1/4\ O}{edge} = 9\ O$$

Formula: $TlBa_2Ca_2Cu_3O_9$

Structure (d):   Following similar calculations, formula = $TlBa_2Ca_3Cu_4O_{11}$

b. Structure (a) has one planar sheet of Cu and O atoms, and the number increases by one for each of the remaining structures. The order of superconductivity temperature from lowest to highest temperature is: (a) < (b) < (c) < (d).

c. $TlBa_2CuO_5$: $3 + 2(2) + x + 5(-2) = 0$, $x = +3$
Only Cu^{3+} is present in each formula unit.

$TlBa_2CaCu_2O_7$: $3 + 2(2) + 2 + 2(x) + 7(-2) = 0$, $x = +5/2$
Each formula unit contains 1 Cu^{2+} and 1 Cu^{3+}.

$TlBa_2Ca_2Cu_3O_9$: $3 + 2(2) + 2(2) + 3(x) + 9(-2) = 0$, $x = +7/3$
Each formula unit contains 2 Cu^{2+} and 1 Cu^{3+}.

$TlBa_2Ca_3Cu_4O_{11}$: $3 + 2(2) + 3(2) + 4(x) + 11(-2) = 0$, $x = +9/4$
Each formula unit contains 3 Cu^{2+} and 1 Cu^{3+}.

d. This superconductor material achieves variable copper oxidation states by varying the numbers of Ca, Cu and O in each unit cell. The mixtures of copper oxidation states are discussed above. The superconductor material in Exercise 10.73 achieves variable copper oxidation states by omitting oxygen at various sites in the lattice.

Phase Changes and Phase Diagrams

75. If we graph $\ln P_{vap}$ vs $1/T$, the slope of the resulting straight line will be $-\Delta H_{vap}/R$.

P_{vap}	$\ln P_{vap}$	T (Li)	$1/T$	T (Mg)	$1/T$
1 torr	0	1023 K	$9.775 \times 10^{-4}\ K^{-1}$	893 K	$11.2 \times 10^{-4}\ K^{-1}$
10.	2.3	1163	8.598×10^{-4}	1013	9.872×10^{-4}
100.	4.61	1353	7.391×10^{-4}	1173	8.525×10^{-4}
400.	5.99	1513	6.609×10^{-4}	1313	7.616×10^{-4}
760.	6.63	1583	6.317×10^{-4}	1383	7.231×10^{-4}

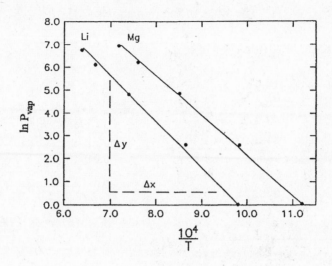

For Li:

We get the slope by taking two points (x, y) that are on the line we draw. For a line:

$$\text{slope} = \frac{\Delta y}{\Delta x} = \frac{y_2 - y_1}{x_2 - x_1}$$

or we can determine the straight line equation using a computer or calculator. The general straight line equation is $y = mx + b$ where m = slope and b = y-intercept.

The equation of the Li line is: $\ln P_{vap} = -1.90 \times 10^4(1/T) + 18.6$, slope $= -1.90 \times 10^4$ K

Slope $= -\Delta H_{vap}/R$, $\Delta H_{vap} = $ -slope $\times$ R $= 1.90 \times 10^4$ K $\times$ 8.3145 J/K•mol

$\Delta H_{vap} = 1.58 \times 10^5$ J/mol $= 158$ kJ/mol

For Mg:

The equation of the line is: $\ln P_{vap} = -1.67 \times 10^4(1/T) + 18.7$, slope $= -1.67 \times 10^4$ K

$\Delta H_{vap} = $ -slope $\times$ R $= 1.67 \times 10^4$ K $\times$ 8.3145 J/K•mol, $\Delta H_{vap} = 1.39 \times 10^5$ J/mol $= 139$ kJ/mol

The bonding is stronger in Li since ΔH_{vap} is larger for Li.

76.   Again we graph $\ln P_{vap}$ vs $1/T$. The slope of the line equals $-\Delta H_{vap}/R$.

T(K)	$10^3/T$ (K^{-1})	P_{vap} (torr)	$\ln P_{vap}$
273	3.66	14.4	2.67
283	3.53	26.6	3.28
293	3.41	47.9	3.87
303	3.30	81.3	4.40
313	3.19	133	4.89
323	3.10	208	5.34
353	2.83	670.	6.51

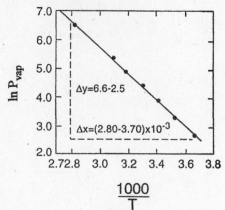

$$\text{slope} = \frac{6.6 - 2.5}{(2.80 \times 10^{-3} - 3.70 \times 10^{-3}) \text{ K}^{-1}} = -4600 \text{ K}$$

$$-4600 \text{ K} = \frac{-\Delta H_{vap}}{R} = \frac{-\Delta H_{vap}}{8.3145 \text{ J/K} \bullet \text{mol}}$$

$\Delta H_{vap} = 38$ kJ/mol

To determine the normal boiling point, we can use the following formula:

$$\ln\left(\frac{P_1}{P_2}\right) = \frac{\Delta H_{vap}}{R}\left(\frac{1}{T_2} - \frac{1}{T_1}\right)$$

At the normal boiling point, the vapor pressure equals 1.00 atm or 760. torr. At 273 K, the vapor pressure is 14.4. torr (from data in the problem).

$$\ln\left(\frac{14.4}{760.}\right) = \frac{38,000 \text{ J/mol}}{8.3145 \text{ J/K} \bullet \text{mol}}\left(\frac{1}{T_2} - \frac{1}{273 \text{ K}}\right), \; -3.97 = 4.6 \times 10^3 \, (1/T_2 - 3.66 \times 10^{-3})$$

$- 8.6 \times 10^{-4} + 3.66 \times 10^{-3} = 1/T_2 = 2.80 \times 10^{-3}$, $T_2 = 357$ K = normal boiling point

77.   At $100.°C$ (373 K), the vapor pressure of H_2O is 1.00 atm = 760. torr.
For water, $\Delta H_{vap} = 40.7$ kJ/mol.

$$\ln\left(\frac{P_1}{P_2}\right) = \frac{\Delta H_{vap}}{R}\left(\frac{1}{T_2} - \frac{1}{T_1}\right) \text{ or } \ln\left(\frac{P_2}{P_1}\right) = \frac{\Delta H_{vap}}{R}\left(\frac{1}{T_1} - \frac{1}{T_2}\right)$$

$$\ln\left(\frac{520. \text{ torr}}{760. \text{ torr}}\right) = \frac{40.7 \times 10^3 \text{ J/mol}}{8.3145 \text{ J/K} \bullet \text{mol}}\left(\frac{1}{373 \text{ K}} - \frac{1}{T_2}\right), \; -7.75 \times 10^{-5} = \left(\frac{1}{373} - \frac{1}{T_2}\right)$$

$-7.75 \times 10^{-5} = 2.68 \times 10^{-3} - \frac{1}{T_2}$, $\frac{1}{T_2} = 2.76 \times 10^{-3}$, $T_2 = \frac{1}{2.76 \times 10^{-3}} = 362$ K or $89°C$

78. $\ln\left(\frac{P_2}{1.00}\right) = \frac{40.7 \times 10^3 \text{ J/mol}}{8.3145 \text{ J/K}\bullet\text{mol}}\left(\frac{1}{373 \text{ K}} - \frac{1}{623 \text{ K}}\right)$, $\ln P_2 = 5.27$, $P_2 = e^{5.27} = 194$ atm

79. $\ln\left(\frac{P_1}{P_2}\right) = \frac{\Delta H_{vap}}{R}\left(\frac{1}{T_2} - \frac{1}{T_1}\right)$; $P_1 = 760.$ torr, $T_1 = 630.$ K; $P_2 = ?$, $T_2 = 298$ K

$$\ln\left(\frac{760.}{P_2}\right) = \frac{59.1 \times 10^3 \text{ J/mol}}{8.3145 \text{ J/K}\bullet\text{mol}}\left(\frac{1}{298 \text{ K}} - \frac{1}{630. \text{ K}}\right) = 12.6$$

$760./P_2 = e^{12.6}$, $P_2 = 760./(2.97 \times 10^5) = 2.56 \times 10^{-3}$ torr

80. $\ln\left(\frac{P_1}{P_2}\right) = \frac{\Delta H_{vap}}{R}\left(\frac{1}{T_2} - \frac{1}{T_1}\right)$

$P_1 = 760.$ torr, $T_1 = 56.5°C + 273.2 = 329.7$ K; $P_2 = 630.$ torr, $T_2 = ?$

$$\ln\left(\frac{760.}{630.}\right) = \frac{32.0 \times 10^3 \text{ J/mol}}{8.3145 \text{ J/K}\bullet\text{mol}}\left(\frac{1}{T_2} - \frac{1}{329.7}\right),\ 0.188 = 3.85 \times 10^3\left(\frac{1}{T_2} - 3.033 \times 10^{-3}\right)$$

$\frac{1}{T_2} - 3.033 \times 10^{-3} = 4.88 \times 10^{-5}$, $\frac{1}{T_2} = 3.082 \times 10^{-3}$, $T_2 = 324.5 \text{ K} = 51.3°C$

$$\ln\left(\frac{630. \text{ torr}}{P_2}\right) = \frac{32.0 \times 10^3 \text{ J/mol}}{8.3145 \text{ J/K}\bullet\text{mol}}\left(\frac{1}{298.2} - \frac{1}{324.5}\right),\ \ln 630. - \ln P_2 = 1.05$$

$\ln P_2 = 5.40$, $P_2 = e^{5.40} = 221$ torr

81.

Temp(°C)
80
60
40
20
0
-20
-40
-60
Time
1
2
3
4
5
Slope 5 > Slope 3 > Slope 1
Time 4 = 4 × Time 2

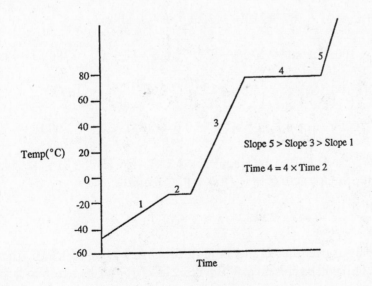

82. $X(g, 100.°C) \rightarrow X(g, 75°C)$, $\Delta T = -25°C$

$q_1 = s_{gas} \times m \times \Delta T = \dfrac{1.0 \text{ J}}{\text{g °C}} \times 250. \text{ g} \times (-25°C) = -6300 \text{ J} = -6.3 \text{ kJ}$

$X(g, 75°C) \rightarrow X(l, 75°C)$, $q_2 = 250. \text{ g} \times \dfrac{1 \text{ mol}}{75.0 \text{ g}} \times \dfrac{-20. \text{ kJ}}{\text{mol}} = -67 \text{ kJ}$

$X(l, 75°C) \rightarrow X(l, -15°C)$, $q_3 = \dfrac{2.5 \text{ J}}{\text{g °C}} \times 250. \text{ g} \times (-90.°C) = -56,000 \text{ J} = -56 \text{ kJ}$

$X(l, -15°C) \rightarrow X(s, -15°C)$, $q_4 = 250. \text{ g} \times \dfrac{1 \text{ mol}}{75.0 \text{ g}} \times \dfrac{-5.0 \text{ kJ}}{\text{mol}} = -17 \text{ kJ}$

$X(s, -15°C) \rightarrow X(s, -50.°C)$, $q_5 = \dfrac{3.0 \text{ J}}{\text{g °C}} \times 250. \text{ g} \times (-35°C) = -26,000 \text{ J} = -26 \text{ kJ}$

$q_{total} = q_1 + q_2 + q_3 + q_4 + q_5 = -6.3 - 67 - 56 - 17 - 26 = -172 \text{ kJ}$

83. $H_2O(s, -20.°C) \rightarrow H_2O(s, 0°C)$, $\Delta T = 20.°C$

$q_1 = s_{ice} \times m \times \Delta T = \dfrac{2.1 \text{ J}}{\text{g °C}} \times 5.00 \times 10^2 \text{ g} \times 20.°C = 2.1 \times 10^4 \text{ J} = 21 \text{ kJ}$

$H_2O(s, 0°C) \rightarrow H_2O(l, 0°C)$, $q_2 = 5.00 \times 10^2 \text{ g H}_2\text{O} \times \dfrac{1 \text{ mol}}{18.02 \text{ g}} \times \dfrac{6.02 \text{ kJ}}{\text{mol}} = 167 \text{ kJ}$

$H_2O(l, 0°C) \rightarrow H_2O(l, 100°C)$, $q_3 = \dfrac{4.2 \text{ J}}{\text{g °C}} \times 5.00 \times 10^2 \text{ g} \times 100.°C = 2.1 \times 10^5 \text{ J} = 210 \text{ kJ}$

$H_2O(l, 100°C) \rightarrow H_2O(g, 100°C)$, $q_4 = 5.00 \times 10^2 \text{ g} \times \dfrac{1 \text{ mol}}{18.02 \text{ g}} \times \dfrac{40.7 \text{ kJ}}{\text{mol}} = 1130 \text{ kJ}$

$H_2O(g, 100°C) \rightarrow H_2O(g, 250°C)$, $q_5 = \dfrac{2.0 \text{ J}}{\text{g °C}} \times 5.00 \times 10^2 \text{ g} \times 150.°C = 1.5 \times 10^5 \text{ J} = 150 \text{ kJ}$

$q_{total} = q_1 + q_2 + q_3 + q_4 + q_5 = 21 + 167 + 210 + 1130 + 150 = 1680 \text{ kJ}$

84. $H_2O(g, 125°C) \rightarrow H_2O(g, 100.°C)$, $q_1 = 2.0 \text{ J/g} \bullet °C \times 75.0 \text{ g} \times (-25°C) = -3800 \text{ J} = -3.8 \text{ kJ}$

$H_2O(g, 100.°C) \rightarrow H_2O(l, 100.°C)$, $q_2 = 75.0 \text{ g} \times \dfrac{1 \text{ mol}}{18.02 \text{ g}} \times \dfrac{-40.7 \text{ kJ}}{\text{mol}} = -169 \text{ kJ}$

$H_2O(l, 100.°C) \rightarrow H_2O(l, 0°C)$, $q_3 = 4.2 \text{ J/g} \bullet °C \times 75.0 \text{ g} \times (-100.°C) = -32,000 \text{ J} = -32 \text{ kJ}$

To convert $H_2O(g)$ at 125°C to $H_2O(l)$ at 0°C requires (-3.8 kJ - 169 kJ - 32 kJ =) -205 kJ of heat removed. To convert from $H_2O(l)$ at 0°C to $H_2O(s)$ at 0°C requires:

$q_4 = 75.0 \text{ g} \times \dfrac{1 \text{ mol}}{18.02 \text{ g}} \times \dfrac{-6.02 \text{ kJ}}{\text{mol}} = -25 \text{ kJ}$

This amount of energy puts us over the -215 kJ limit (-205 kJ - 25 kJ = -230. kJ). Therefore, a mixture of $H_2O(s)$ and $H_2O(l)$ will be present at 0°C when 215 kJ of heat are removed from the gas sample.

85.   Total mass H_2O = 18 cubes $\times \dfrac{30.0\ g}{cube}$ = 540. g;   540. g $H_2O \times \dfrac{1\ mol\ H_2O}{18.02\ g}$ = 30.0 mol H_2O

Heat removed to produce ice at -5.0°C:

$$\dfrac{4.18\ J}{g\ °C} \times 540.\ g \times 22.0\ °C + \dfrac{6.02 \times 10^3\ J}{mol} \times 30.0\ mol + \dfrac{2.08\ J}{g\ °C} \times 540.\ g \times 5.0\ °C$$

$$= 4.97 \times 10^4\ J + 1.81 \times 10^5\ J + 5.6 \times 10^3\ J = 2.36 \times 10^5\ J$$

$2.36 \times 10^5\ J \times \dfrac{1\ g\ CF_2Cl_2}{158\ J} = 1.49 \times 10^3\ g\ CF_2Cl_2$ must be vaporized.

86.   Heat released = 0.250 g Na $\times \dfrac{1\ mol}{22.99\ g} \times \dfrac{368\ kJ}{2\ mol}$ = 2.00 kJ

To melt 50.0 g of ice requires: 50.0 g ice $\times \dfrac{1\ mol\ H_2O}{18.02\ g} \times \dfrac{6.02\ kJ}{mol}$ = 16.7 kJ

The reaction doesn't release enough heat to melt all of the ice. The temperature will remain at 0°C.

87.   A: solid   B: liquid   C: vapor

      D: solid + vapor   E: solid + liquid + vapor

      F: liquid + vapor G: liquid + vapor   H: vapor

      triple point: E   critical point: G

normal freezing point: temperature at which solid - liquid line is at 1.0 atm (see plot below).

normal boiling point: temperature at which liquid - vapor line is at 1.0 atm (see plot below).

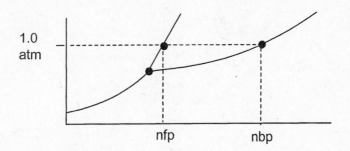

Since the solid-liquid line has a positive slope, the solid phase is denser than the liquid phase.

88.   a. 3

      b. Triple point at 95.31°C: rhombic, monoclinic, gas
          Triple point at 115.18°C: monoclinic, liquid, gas
          Triple point at 153°C: rhombic, monoclinic, liquid

c. From the phase diagram, the monoclinic solid phase is stable at T = 100°C and P = 1 atm.

d. Normal melting point = 115.21°C; normal boiling point = 444.6°C; The normal melting and boiling points occur at P = 1.0 atm.

e. Rhombic is the densest phase since the rhombic-monoclinic equilibrium line has a positive slope and since the solid-liquid lines also have positive slopes.

f. No; P = 1.0 × 10⁻⁵ atm is at a pressure somewhere between the 95.31°C and 115.18°C triple points. At this pressure, the rhombic and gas phases are never in equilibrium with each other, so rhombic sulfur cannot sublime at P = 1.0 × 10⁻⁵ atm. However, monoclinic sulfur can sublime at this pressure.

g. From the phase diagram, we would start off with gaseous sulfur. At 100°C and ~1 × 10⁻⁵ atm, S(g) would convert to the solid monoclinic form of sulfur. Finally at 100°C and some large pressure less than 1420 atm, S(s, monoclinic) would convert to the solid rhombic form of sulfur. Summarizing, the phase changes are S(g) → S(monoclinic) → S(rhombic).

89. a. two

b. Higher pressure triple point: graphite, diamond and liquid; Lower pressure triple point: graphite, liquid and vapor

c. It is converted to diamond (the more dense solid form).

d. Diamond is more dense, which is why graphite can be converted to diamond by applying pressure.

90. The following sketch of the Br₂ phase diagram is not to scale. Since the triple point of Br₂ is at a temperature below the freezing point of Br₂, the slope of the solid-liquid line is positive.

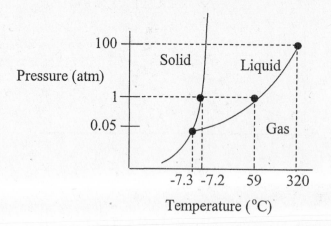

The positive slopes of all the lines indicate that Br₂(s) is more dense than Br₂(l) which is more dense than Br₂(g). At room temperature (~22°C) and 1 atm, Br₂(l) is the stable phase. Br₂(l) cannot exist at a temperature below the triple point temperature of -7.3°C and at a temperature above the critical point temperature of 320°C. The phase changes that occur as temperature is increased at 0.10 atm are solid → liquid → gas.

Additional Exercises

91. $C_{25}H_{52}$ has the stronger intermolecular forces because it has the higher boiling point. Even though $C_{25}H_{52}$ is nonpolar, it is so large that its London dispersion forces are much stronger than the sum of the London dispersion and hydrogen bonding interactions found in H_2O.

92. Benzene   Naphthalene

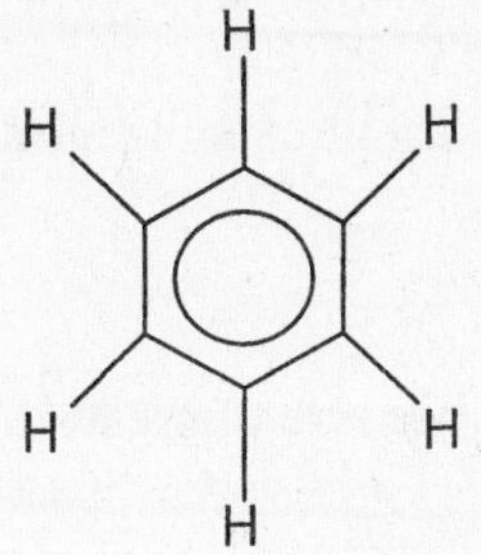

LD forces only   LD forces only

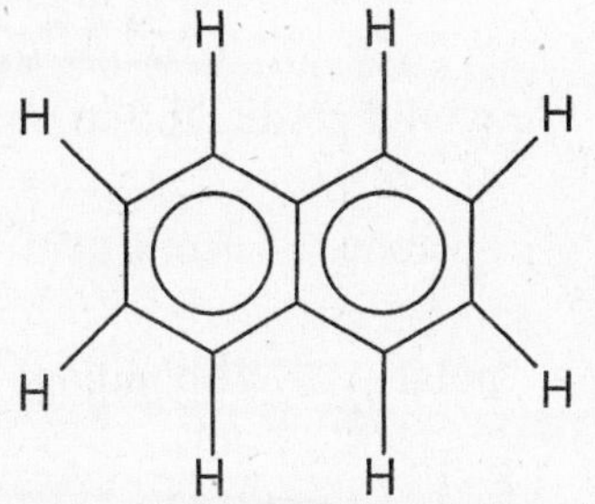

Note: London dispersion forces in molecules like benzene and naphthalene are fairly large. The molecules are flat, and there is efficient surface area contact among molecules. Large surface area contact leads to stronger London dispersion forces.

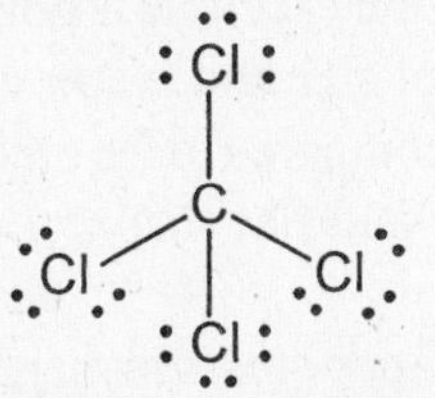

Carbon tetrachloride (CCl_4) has polar bonds but is a nonpolar molecule. CCl_4 only has LD forces.

In terms of size and shape: $CCl_4 < C_6H_6 < C_{10}H_8$

The strengths of the LD forces are proportional to size and are related to shape. Although CCl_4 is fairly large, its overall spherical shape gives rise to relatively weak LD forces as compared to flat molecules like benzene and naphthalene. The physical properties given in the problem are consistent with the order listed above. Each of the physical properties will increase with an increase in intermolecular forces.

Acetone   Acetic Acid

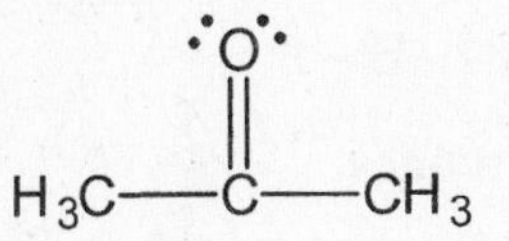

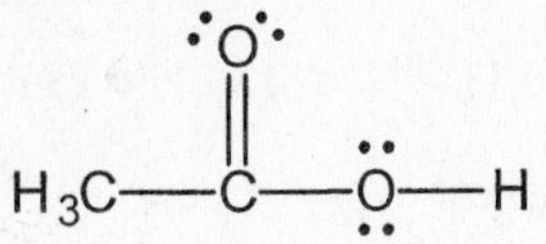

LD, dipole   LD, dipole, H bonding

Benzoic Acid

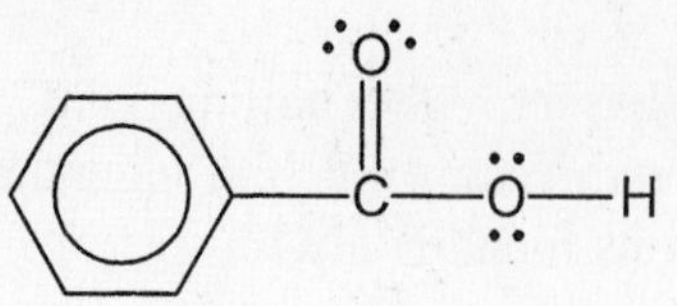

LD, dipole, H bonding

We would predict the strength of intermolecular forces for the last three molecules to be:

acetone < acetic acid < benzoic acid

polar   H bonding   H bonding, but large LD forces because of greater size and shape.

This ordering is consistent with the values given for bp, mp, and ΔH_{vap}.

The overall order of the strengths of intermolecular forces based on physical properties are:

acetone < CCl_4 < C_6H_6 < acetic acid < naphthalene < benzoic acid

The order seems reasonable except for acetone and naphthalene. Since acetone is polar, we would not expect it to boil at the lowest temperature. However, in terms of size and shape, acetone is the smallest molecule, and the LD forces in acetone must be very small compared to the other molecules. Naphthalene must have very strong LD forces because of its size and flat shape.

93. At any temperature, the plot tells us that substance A has a higher vapor pressure than substance B, with substance C having the lowest vapor pressure. Therefore, the substance with the weakest intermolecular forces is A, and the substance with the strongest intermolecular forces is C.

NH_3 can form hydrogen bonding interactions while the others cannot. Substance C is NH_3. The other two are nonpolar compounds with only London dispersion forces. Since CH_4 is smaller than SiH_4, CH_4 will have weaker LD forces and is substance A. Therefore, substance B is SiH_4.

94. As the electronegativity of the atoms covalently bonded to H increases, the strength of the hydrogen bonding interaction increases.

$$N \cdots H - N < N \cdots H - O < O \cdots H - O < O \cdots H - F < F \cdots H - F$$

weakest   strongest

95. $n\lambda = 2d \sin\theta, \ \lambda = \dfrac{2d \sin\theta}{n} = \dfrac{2 \times 201 \text{ pm} \times \sin 34.68°}{1}, \ \lambda = 229 \text{ pm} = 2.29 \times 10^{-10} \text{ m} = 0.229 \text{ nm}$

96. If a face-centered cubic structure, then 4 atoms/unit cell and from Exercise 10.47:

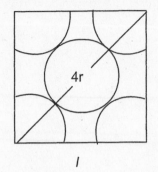

$$2\,l^2 = 16\,r^2$$

$$l = r\sqrt{8} = 144 \text{ pm } \sqrt{8} = 407 \text{ pm}$$

$$l = 407 \times 10^{-12} \text{ m} = 407 \times 10^{-10} \text{ cm}$$

$$\text{density} = \frac{4 \text{ atoms Au} \times \dfrac{1 \text{ mol Au}}{6.022 \times 10^{23} \text{ atoms}} \times \dfrac{197.0 \text{ g Au}}{\text{mol Au}}}{(4.07 \times 10^{-8} \text{ cm})^3} = 19.4 \text{ g/cm}^3$$

If a body-centered cubic structure, then 2 atoms/unit cell and from Exercise 10.51:

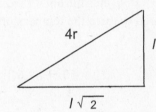

$$16\,r^2 = l^2 + 2\,l^2$$

$$l = 4r/\sqrt{3} = 333 \text{ pm} = 333 \times 10^{-12} \text{ m}$$

$$l = 333 \times 10^{-10} \text{ cm} = 3.33 \times 10^{-8} \text{ cm}$$

$$\text{density} = \frac{2 \text{ atoms Au} \times \dfrac{1 \text{ mol Au}}{6.022 \times 10^{23} \text{ atoms}} \times \dfrac{197.0 \text{ g Au}}{\text{mol Au}}}{(3.33 \times 10^{-8} \text{ cm})^3} = 17.7 \text{ g/cm}^3$$

The measured density is consistent with a face-centered cubic unit cell.

97.   If TiO_2 conducts electricity as a liquid, then it is an ionic solid; if not, then TiO_2 is a network solid.

98.   One B atom and one N atom together have the same number of electrons as two C atoms. The description of physical properties sounds a lot like the properties of graphite and diamond, the two solid forms of carbon. The two forms of BN have structures similar to graphite and diamond.

99. B_2H_6:   This compound contains only nonmetals so it is probably a molecular solid with covalent bonding. The low boiling point confirms this.

        SiO_2:   This is the empirical formula for quartz, which is a network solid.

        CsI: This is a metal bonded to a nonmetal, which generally form ionic solids. The electrical conductivity in aqueous solution confirms this.

        W: Tungsten is a metallic solid as the conductivity data confirms.

100.   In order to set up an equation, we need to know what phase exists at the final temperature. To heat 20.0 g of ice from -10.0°C to 0.0°C requires:

$$q = \frac{2.08\,J}{g\,°C} \times 20.0\,g \times 10.0\,°C = 416\,J$$

To convert ice to water at 0.0°C requires:

$$q = 20.0\,g \times \frac{1\,mol}{18.02} \times \frac{6.02\,kJ}{mol} = 6.68\,kJ = 6680\,J$$

To chill 100.0 g of water from 80.0°C to 0.0° requires:

$$q = \frac{4.18\,J}{g\,°C} \times 100.0\,g \times 80.0\,°C = 33,400\,J \text{ of heat removed}$$

From the heat values above, the liquid phase exists once the final temperature is reached (a lot more heat is lost when the 100.0 g of water is cooled to 0.0°C than the heat required to convert the ice into water). To calculate the final temperature, we will equate the heat gain by the ice to the heat loss by the water. We will keep all quantities positive in order to avoid sign errors. The heat gain by the ice will be the 416 J required to convert the ice to 0.0°C plus the 6680 J required to convert the ice at 0.0°C into water at 0.0°C plus the heat required to raise the temperature from 0.0°C to the final temperature.

$$\text{heat gain by ice} = 416\,J + 6680\,J + \frac{4.18\,J}{g\,°C} \times 20.0\,g \times (T_f - 0.0°C) = 7.10 \times 10^3 + 83.6\,T_f$$

$$\text{heat loss by water} = \frac{4.18\,J}{g\,°C} \times 100.0\,g \times (80.0°C - T_f) = 3.34 \times 10^4 - 418\,T_f$$

Solving for the final temperature:

$$7.10 \times 10^3 + 83.6\,T_f = 3.34 \times 10^4 - 418\,T_f, \quad 502\,T_f = 2.63 \times 10^4, \quad T_f = 52.4°C$$

101. $1.00\,lb \times \dfrac{454\,g}{lb} = 454\,g\ H_2O$; A change of 1.00°F is equal to a change of 5/9°C.

The amount of heat in J in 1 Btu is: $\dfrac{4.18\,J}{g\,°C} \times 454\,g \times \dfrac{5}{9}°C = 1.05 \times 10^3\,J$ or $1.05\,kJ$

It takes 40.7 kJ to vaporize 1 mol H_2O (ΔH_{vap}). Combining these:

$$\frac{1.00 \times 10^4\,Btu}{hr} \times \frac{1.05\,kJ}{Btu} \times \frac{1\,mol\,H_2O}{40.7\,kJ} = 258\,mol/hr$$

or: $\dfrac{258\,mol}{hr} \times \dfrac{18.02\,g\,H_2O}{mol} = 4650\,g/hr = 4.65\,kg/hr$

102.   The critical temperature is the temperature above which the vapor cannot be liquefied no matter what pressure is applied. Since N_2 has a critical temperature below room temperature (~22°C), it cannot be liquefied at room temperature. NH_3, with a critical temperature above room temperature, can be liquefied at room temperature.

Challenge Problems

103.   A single hydrogen bond in H_2O has a strength of 21 kJ/mol. Each H_2O molecule forms two H bonds. Thus, it should take 42 kJ/mol of energy to break all of the H bonds in water. Consider the phase transitions:

$$\text{solid} \overset{6.0\,kJ}{\to} \text{liquid} \overset{40.7\,kJ}{\to} \text{vapor} \qquad \Delta H_{sub} = \Delta H_{fus} + \Delta H_{vap}$$

It takes a total of 46.7 kJ/mol to convert solid H_2O to vapor (ΔH_{sub}). This would be the amount of energy necessary to disrupt all of the intermolecular forces in ice. Thus, $(42 \div 46.7) \times 100 = 90\%$ of the attraction in ice can be attributed to H bonding.

104.   Both molecules are capable of H bonding. However, in oil of wintergreen the hydrogen bonding is <u>intra</u>molecular.

In methyl-4-hydroxybenzoate, the H bonding is <u>inter</u>molecular, resulting in stronger intermolecular forces and a higher melting point.

105. $NaCl$, $MgCl_2$, NaF, MgF_2 AlF_3 all have very high melting points indicative of strong intermolecular forces. They are all ionic solids. $SiCl_4$, SiF_4, F_2, Cl_2, PF_5 and SF_6 are nonpolar covalent molecules. Only LD forces are present. PCl_3 and SCl_2 are polar molecules. LD forces and dipole forces are present. In these 8 molecular substances, the intermolecular forces are weak and the melting points low. $AlCl_3$ doesn't seem to fit in as well. From the melting point, there are much stronger forces present than in the nonmetal halides, but they aren't as strong as we would expect for an ionic solid. $AlCl_3$ illustrates a gradual transition from ionic to covalent bonding, from an ionic solid to discrete molecules.

106.   a.   The NaCl unit cell has a face centered cubic arrangement of the anions with cations in the octahedral holes. There are 4 NaCl formula units per unit cell and, since there is a 1:1 ratio of cations to anions in MnO, there would be 4 MnO formula units per unit cell, assuming an NaCl type structure. The CsCl unit cell has a simple cubic structure of anions with the cations in the cubic holes. There is one CsCl formula unit per unit cell, so there would be one MnO formula unit per unit cell if a CsCl structure is observed.

$$\frac{\text{molecules MnO}}{\text{unit cell}} = (4.47 \times 10^{-8}\,cm)^3 \times \frac{5.28\,g\,MnO}{cm^3} \times \frac{1\,mol\,MnO}{70.94\,g\,MnO}$$

$$\times \frac{6.022 \times 10^{23}\,\text{molecules MnO}}{mol\,MnO_4} = 4.00\,\text{molecules MnO}$$

From the calculation, MnO crystallizes in the NaCl type structure.

b.   From the NaCl structure and assuming the ions touch each other, then ℓ = cube edge length $= 2r_{Mn^{2+}} + 2r_{O^{2-}}$.

$\ell = 4.47 \times 10^{-8}\,cm = 2r_{Mn^{2+}} + 2(1.40 \times 10^{-8}\,cm)$, $r_{Mn^{2+}} = 8.35 \times 10^{-8}\,cm = \mathbf{84\,pm}$

107.   Out of 100.00 g: $28.31 \text{ g O} \times \dfrac{1 \text{ mol}}{16.00 \text{ g}} = 1.769 \text{ mol O}$; $71.69 \text{ g Ti} \times \dfrac{1 \text{ mol}}{47.88 \text{ g}} = 1.497 \text{ mol Ti}$

$\dfrac{1.769}{1.497} = 1.182$; $\dfrac{1.497}{1.769} = 0.8462$; The formula is $TiO_{1.182}$ or $Ti_{0.8462}O$.

For $Ti_{0.8462}O$, let $x = Ti^{2+}$ per mol O^{2-} and $y = Ti^{3+}$ per mol O^{2-}. Setting up two equations and solving:

$x + y = 0.8462$ (mass balance) and $2x + 3y = 2$ (charge balance); $2x + 3(0.8462 - x) = 2$

$x = 0.539 \text{ mol } Ti^{2+}/\text{mol } O^{2-}$ and $y = 0.307 \text{ mol } Ti^{3+}/\text{mol } O^{2-}$

$\dfrac{0.539}{0.8462} \times 100 = 63.7\%$ of the titanium ions is Ti^{2+} and 36.3% is Ti^{3+} (a 1.75:1 ion ratio).

108.   First we need to get the empirical formula of spinel. Assume 100.0 g of spinel.

$37.9 \text{ g Al} \times \dfrac{1 \text{ mol Al}}{26.98 \text{ g Al}} = 1.40 \text{ mol Al}$

$17.1 \text{ g Mg} \times \dfrac{1 \text{ mol Mg}}{24.31 \text{ g Mg}} = 0.703 \text{ mol Mg}$

The mole ratios are 2:1:4.

Empirical Formula = Al_2MgO_4

$45.0 \text{ g O} \times \dfrac{1 \text{ mol O}}{16.00 \text{ g O}} = 2.81 \text{ mol O}$

Assume each unit cell contains an integral value (n) of Al_2MgO_4 formula units. Each Al_2MgO_4 formula unit has a mass of: $24.31 + 2(26.98) + 4(16.00) = 142.27$ g/mol

$\text{density} = \dfrac{n \text{ formula units} \times \dfrac{1 \text{ mol}}{6.022 \times 10^{23} \text{ formula units}} \times \dfrac{142.27 \text{ g}}{\text{mol}}}{(8.09 \times 10^{-8} \text{ cm})^3} = \dfrac{3.57 \text{ g}}{\text{cm}^3}$, Solving: n = 8.00

Each unit cell has 8 formula units of Al_2MgO_4 or 16 Al, 8 Mg and 32 O atoms.

109. $\dfrac{\text{density}_{Mn}}{\text{density}_{Cu}} = \dfrac{\text{mass}_{Mn} \times \text{volume}_{Cu}}{\text{volume}_{Mn} \times \text{mass}_{Cu}} = \dfrac{\text{mass}_{Mn}}{\text{mass}_{Cu}} \times \dfrac{\text{volume}_{Cu}}{\text{volume}_{Mn}}$

The type of cubic cell formed is not important; only that Cu and Mn crystallize in the same type of cubic unit cell is important. Each cubic unit cell has a specific relationship between the cube edge length, l, and the radius, r. In all cases $l \propto r$. Therefore, $V \propto l^3 \propto r^3$. For the mass ratio, we can use the molar masses of Mn and Cu since each unit cell must contain the same number of Mn and Cu atoms. Solving:

$\dfrac{\text{density}_{Mn}}{\text{density}_{Cu}} = \dfrac{\text{mass}_{Mn}}{\text{mass}_{Cu}} \times \dfrac{\text{volume}_{Cu}}{\text{volume}_{Mn}} = \dfrac{54.94 \text{ g/mol}}{63.55 \text{ g/mol}} \times \dfrac{(r_{Cu})^3}{(1.056 \, r_{Cu})^3}$

$$\frac{density_{Mn}}{density_{Cu}} = 0.8645 \times \left(\frac{1}{1.056}\right)^3 = 0.7341$$

$$density_{Mn} = 0.7341 \times density_{Cu} = 0.7341 \times 8.96 \text{ g/cm}^3 = 6.58 \text{ g/cm}^3$$

110.   a.   The arrangement of the layers are:

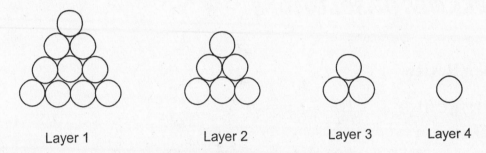

Layer 1   Layer 2   Layer 3   Layer 4

A total of 20 cannon balls will be needed.

   b.   The layering alternates abcabc which is cubic closest packing.

   c.   tetrahedron

111.

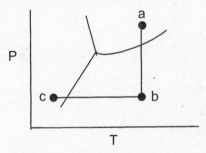

As P is lowered, we go from a to b on the phase diagram. The water boils. The boiling of water is endothermic and the water is cooled (b →c), forming some ice. If the pump is left on, the ice will sublime until none is left. This is the basis of freeze drying.

112. $w = -P\Delta V$;   Assuming a constant P of 1.00 atm.

$$V_{373} = \frac{nRT}{P} = \frac{1.00 \,(0.08206)\,(373)}{1.00} = 30.6 \text{ L for one mol of water vapor}$$

Since the density of $H_2O(l)$ is 1.00 g/cm^3, 1.00 mol of $H_2O(l)$ occupies 18.0 cm^3 or 0.0180 L.

$w = -1.00 \text{ atm } (30.6 \text{ L} - 0.0180 \text{ L}) = -30.6 \text{ L atm}$

$w = -30.6 \text{ L atm} \times 101.3 \text{ J/L·atm} = -3.10 \times 10^3 \text{ J} = -3.10 \text{ kJ}$

$\Delta E = q + w = 40.7 \text{ kJ} - 3.10 \text{ kJ} = 37.6 \text{ kJ}$

$\frac{37.6}{40.7} \times 100 = 92.4\%$ of the energy goes to increase the internal energy of the water.

The remainder of the energy (7.6%) goes to do work against the atmosphere.

CHAPTER ELEVEN

PROPERTIES OF SOLUTIONS

Solution Review

9. $125 \text{ g } C_{12}H_{22}O_{11} \times \dfrac{1 \text{ mol}}{342.30 \text{ g}} = 0.365 \text{ mol}; \quad M = \dfrac{0.365 \text{ mol}}{1.00 \text{ L}} = \dfrac{0.365 \text{ mol sucrose}}{L}$

10. $0.250 \text{ L} \times \dfrac{0.100 \text{ mol}}{L} \times \dfrac{134.00 \text{ g}}{\text{mol}} = 3.35 \text{ g } Na_2C_2O_4$

11. $25.00 \times 10^{-3} \text{ L} \times \dfrac{0.308 \text{ mol}}{L} = 7.70 \times 10^{-3} \text{ mol}; \quad \dfrac{7.70 \times 10^{-3} \text{ mol}}{0.500 \text{ L}} = \dfrac{1.54 \times 10^{-2} \text{ mol } NiCl_2}{L}$

   $NiCl_2(s) \rightarrow Ni^{2+}(aq) + 2 \, Cl^-(aq); \quad M_{Ni^{2+}} = \dfrac{1.54 \times 10^{-2} \text{ mol}}{L}; \quad M_{Cl^-} = \dfrac{3.08 \times 10^{-2} \text{ mol}}{L}$

12. a. $HNO_3(l) \rightarrow H^+(aq) + NO_3^-(aq)$ b. $Na_2SO_4(s) \rightarrow 2 \, Na^+(aq) + SO_4^{2-}(aq)$

    c. $Al(NO_3)_3(s) \rightarrow Al^{3+}(aq) + 3 \, NO_3^-(aq)$ d. $SrBr_2(s) \rightarrow Sr^{2+}(aq) + 2 \, Br^-(aq)$

    e. $KClO_4(s) \rightarrow K^+(aq) + ClO_4^-(aq)$ f. $NH_4Br(s) \rightarrow NH_4^+(aq) + Br^-(aq)$

    g. $NH_4NO_3(s) \rightarrow NH_4^+(aq) + NO_3^-(aq)$ h. $CuSO_4(s) \rightarrow Cu^{2+}(aq) + SO_4^{2-}(aq)$

    i. $NaOH(s) \rightarrow Na^+(aq) + OH^-(aq)$

Questions

13. $\text{Molarity} = \dfrac{\text{moles solute}}{\text{L solution}}; \quad \text{Molality} = \dfrac{\text{moles solute}}{\text{kg solvent}}$

    Volume is temperature dependent and mass is not. Therefore, molarity is temperature dependent and molality is temperature independent. In determining ΔT_f and ΔT_b, we are interested in how some temperature depends on composition. Thus, we don't want our expression of composition to also depend on temperature.

14.   The nature of the intermolecular forces. Polar solutes and ionic solutes dissolve in polar solvents, and nonpolar solutes dissolve in nonpolar solvents.

15.   hydrophobic: water hating; hydrophilic: water loving

16.   As the temperature increases, the gas molecules will have a greater average kinetic energy. A greater fraction of the gas molecules in solution will have kinetic energy greater than the attractive forces between the gas molecules and the solvent molecules. More gas molecules will escape to the vapor phase and the solubility of the gas will decrease.

17.   Since the solute is volatile, then both the water and solute will transfer back and forth between the two beakers. The volume in each beaker will become constant when the concentrations of solute in the beakers are equal to each other. Since the solute is less volatile than water, one would expect there to be a larger net transfer of water molecules into the right beaker than the net transfer of solute molecules into the left beaker. This results in a larger solution volume in the right beaker when equilibrium is reached, i.e., when the solute concentration is identical in each beaker.

18.   Solutions of A and B have vapor pressures less than ideal (see Figure 11.13 of the text), so this plot shows negative deviations from Rault's law. Negative deviations occur when the intermolecular forces are stronger in solution than in pure solvent and solute. This results in an exothermic enthalpy of solution. The only statement that is false is e. A substance boils when the vapor pressure equals the external pressure. Since $X_b = 0.6$ has a lower vapor pressure at the temperature of the plot than either pure A or pure B, then one would expect this solution to require the highest temperature in order for the vapor pressure to reach the external pressure. Therefore, the solution with $X_B = 0.6$ will have a higher boiling point than either pure A or pure B. (Note that since $P°_B > P°_A$, then B is more volatile than A.).

19.   No, the solution is not ideal. For an ideal solution, the strength of intermolecular forces in solution is the same as in pure solute and pure solvent. This results in $\Delta H_{soln} = 0$ for an ideal solution. ΔH_{soln} for methanol/water is not zero. Since $\Delta H_{soln} < 0$, this solution shows a negative deviation from Raoult's law.

20.   Ion pairing can occur, resulting in fewer particles than expected. This results in smaller freezing point depressions and smaller boiling point elevations ($\Delta T = km$). Ion pairing will increase as the concentration of electrolyte increases.

21.   With addition of salt or sugar, the osmotic pressure inside the fruit cells (and bacteria) is less than outside the cell. Water will leave the cells which will dehydrate bacteria present, causing them to die.

22.   A strong electrolyte dissociates completely into ions in solution. A weak electrolyte dissociates only partially into ions in solution. Colligative properties depend on the total number of particles in solution. By measuring a property such as freezing point depression, boiling point elevation or osmotic pressure, we can calculate the van't Hoff factor (i) to see if an electrolyte is strong or weak.

23.   Both solutions and colloids have suspended particles in some medium. The major difference between the two is the size of the particles. A colloid is a suspension of relatively large particles as compared to a solution. Because of this, colloids will scatter light while solutions will not. The scattering of light by a colloidal suspension is called the Tyndall effect.

24.   The micelles form so the ionic ends of the detergent molecules, the SO_4^- ends, are exposed to the polar water molecules on the outside, while the nonpolar hydrocarbon chains from the detergent molecules are hidden from the water by pointing toward the inside of the micelle. Dirt, which is basically nonpolar, is stabilized in the nonpolar interior of the micelle and is washed away.

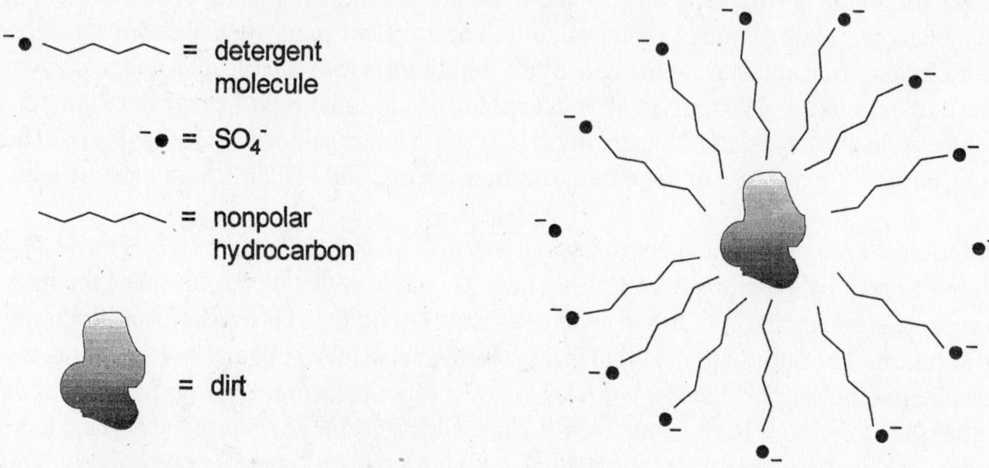

Exercises

Concentration of Solutions

25. $1.00 \text{ L} \times \dfrac{1000 \text{ mL}}{\text{L}} \times \dfrac{1.00 \text{ g}}{\text{mL}} = 1.00 \times 10^3 \text{ g solution}$

mass % NaCl $= \dfrac{25 \text{ g NaCl}}{1.00 \times 10^3 \text{ g solution}} \times 100 = 2.5\%$

molarity $= \dfrac{25 \text{ g NaCl}}{1.00 \text{ L solution}} \times \dfrac{1 \text{ mol NaCl}}{58.44 \text{ g NaCl}} = 0.43 \ M$

1.00×10^3 g solution contains 25 g NaCl and 975 g $H_2O \approx 980$ g H_2O.

molality $= \dfrac{0.43 \text{ mol NaCl}}{0.98 \text{ kg solvent}} = 0.44 \text{ molal}$

0.43 mol NaCl; $980 \text{ g} \times \dfrac{1 \text{ mol}}{18.0 \text{ g}} = 54 \text{ mol } H_2O$; $\chi_{\text{NaCl}} = \dfrac{0.43}{0.43 + 54} = 7.9 \times 10^{-3}$

26. $\text{molality} = \dfrac{40.0 \text{ g EG}}{60.0 \text{ g H}_2\text{O}} \times \dfrac{1000 \text{ g}}{\text{kg}} \times \dfrac{1 \text{ mol EG}}{62.07 \text{ g}} = 10.7 \text{ mol/kg}$ where EG = ethylene glycol

$(\text{C}_2\text{H}_6\text{O}_2)$

$\text{molarity} = \dfrac{40.0 \text{ g EG}}{100.0 \text{ g solution}} \times \dfrac{1.05 \text{ g}}{\text{cm}^3} \times \dfrac{1000 \text{ cm}^3}{\text{L}} \times \dfrac{1 \text{ mol}}{62.07 \text{ g}} = 6.77 \text{ mol/L}$

$40.0 \text{ g EG} \times \dfrac{1 \text{ mol}}{62.07 \text{ g}} = 0.644 \text{ mol EG}$; $60.0 \text{ g H}_2\text{O} \times \dfrac{1 \text{ mol}}{18.02 \text{ g}} = 3.33 \text{ mol H}_2\text{O}$

$\chi_{EG} = \dfrac{0.644}{3.33 + 0.644} = 0.162 = \text{mole fraction ethylene glycol}$

27.   Hydrochloric acid:

$\text{molarity} = \dfrac{38 \text{ g HCl}}{100. \text{ g soln}} \times \dfrac{1.19 \text{ g soln}}{\text{cm}^3 \text{ soln}} \times \dfrac{1000 \text{ cm}^3}{\text{L}} \times \dfrac{1 \text{ mol HCl}}{36.46 \text{ g}} = 12 \text{ mol/L}$

$\text{molality} = \dfrac{38 \text{ g HCl}}{62 \text{ g solvent}} \times \dfrac{1000 \text{ g}}{\text{kg}} \times \dfrac{1 \text{ mol HCl}}{36.46 \text{ g}} = 17 \text{ mol/kg}$

$38 \text{ g HCl} \times \dfrac{1 \text{ mol}}{36.46 \text{ g}} = 1.0 \text{ mol HCl}$; $62 \text{ g H}_2\text{O} \times \dfrac{1 \text{ mol}}{18.02 \text{ g}} = 3.4 \text{ mol H}_2\text{O}$

$\text{mole fraction of HCl} = \chi_{HCl} = \dfrac{1.0}{3.4 + 1.0} = 0.23$

Nitric acid:

$\dfrac{70. \text{ g HNO}_3}{100. \text{ g soln}} \times \dfrac{1.42 \text{ g soln}}{\text{cm}^3 \text{ soln}} \times \dfrac{1000 \text{ cm}^3}{\text{L}} \times \dfrac{1 \text{ mol HNO}_3}{63.02 \text{ g}} = 16 \text{ mol/L}$

$\dfrac{70. \text{ g HNO}_3}{30. \text{ g solvent}} \times \dfrac{1000 \text{ g}}{\text{kg}} \times \dfrac{1 \text{ mol HNO}_3}{63.02 \text{ g}} = 37 \text{ mol/kg}$

$70. \text{ g HNO}_3 \times \dfrac{1 \text{ mol}}{63.02 \text{ g}} = 1.1 \text{ mol HNO}_3$; $30. \text{ g H}_2\text{O} \times \dfrac{1 \text{ mol}}{18.02 \text{ g}} = 1.7 \text{ mol H}_2\text{O}$

$\chi_{HNO_3} = \dfrac{1.1}{1.7 + 1.1} = 0.39$

Sulfuric acid:

$\dfrac{95 \text{ g H}_2\text{SO}_4}{100. \text{ g soln}} \times \dfrac{1.84 \text{ g soln}}{\text{cm}^3 \text{ soln}} \times \dfrac{1000 \text{ cm}^3}{\text{L}} \times \dfrac{1 \text{ mol H}_2\text{SO}_4}{98.09 \text{ g H}_2\text{SO}_4} = 18 \text{ mol/L}$

$\dfrac{95 \text{ g H}_2\text{SO}_4}{5 \text{ g H}_2\text{O}} \times \dfrac{1000 \text{ g}}{\text{kg}} \times \dfrac{1 \text{ mol}}{98.09 \text{ g}} = 194 \text{ mol/kg} \approx 200 \text{ mol/kg}$

$$95 \text{ g H}_2\text{SO}_4 \times \frac{1 \text{ mol}}{98.09 \text{ g}} = 0.97 \text{ mol H}_2\text{SO}_4; \quad 5 \text{ g H}_2\text{O} \times \frac{1 \text{ mol}}{18.02 \text{ g}} = 0.3 \text{ mol H}_2\text{O}$$

$$\chi_{\text{H}_2\text{SO}_4} = \frac{0.97}{0.97 + 0.3} = 0.76$$

Acetic Acid:

$$\frac{99 \text{ g HC}_2\text{H}_3\text{O}_2}{100. \text{ g soln}} \times \frac{1.05 \text{ g soln}}{\text{cm}^3 \text{ soln}} \times \frac{1000 \text{ cm}^3}{\text{L}} \times \frac{1 \text{ mol}}{60.05 \text{ g}} = 17 \text{ mol/L}$$

$$\frac{99 \text{ g HC}_2\text{H}_3\text{O}_2}{1 \text{ g H}_2\text{O}} \times \frac{1000 \text{ g}}{\text{kg}} \times \frac{1 \text{ mol}}{60.05 \text{ g}} = 1600 \text{ mol/kg} \approx 2000 \text{ mol/kg}$$

$$99 \text{ g HC}_2\text{H}_3\text{O}_2 \times \frac{1 \text{ mol}}{60.05 \text{ g}} = 1.6 \text{ mol HC}_2\text{H}_3\text{O}_2; \quad 1 \text{ g H}_2\text{O} \times \frac{1 \text{ mol}}{18.02 \text{ g}} = 0.06 \text{ mol H}_2\text{O}$$

$$\chi_{\text{HC}_2\text{H}_3\text{O}_2} = \frac{1.6}{1.6 + 0.06} = 0.96$$

Ammonia:

$$\frac{28 \text{ g NH}_3}{100. \text{ g soln}} \times \frac{0.90 \text{ g}}{\text{cm}^3} \times \frac{1000 \text{ cm}^3}{\text{L}} \times \frac{1 \text{ mol}}{17.03 \text{ g}} = 15 \text{ mol/L}$$

$$\frac{28 \text{ g NH}_3}{72 \text{ g H}_2\text{O}} \times \frac{1000 \text{ g}}{\text{kg}} \times \frac{1 \text{ mol}}{17.03 \text{ g}} = 23 \text{ mol/kg}$$

$$28 \text{ g NH}_3 \times \frac{1 \text{ mol}}{17.03 \text{ g}} = 1.6 \text{ mol NH}_3; \quad 72 \text{ g H}_2\text{O} \times \frac{1 \text{ mol}}{18.02 \text{ g}} = 4.0 \text{ mol H}_2\text{O}$$

$$\chi_{\text{NH}_3} = \frac{1.6}{4.0 + 1.6} = 0.29$$

28.   a.   If we use 100. mL (100. g) of H_2O, we need:

$$0.100 \text{ kg H}_2\text{O} \times \frac{2.0 \text{ mol KCl}}{\text{kg}} \times \frac{74.55 \text{ g}}{\text{mol KCl}} = 14.9 \text{ g} = 15 \text{ g KCl}$$

Dissolve 15 g KCl in 100. mL H_2O to prepare a 2.0 m KCl solution. This will give us slightly more than 100 mL, but this will be the easiest way to make the solution. Since we don't know the density of the solution, we can't calculate the molarity and use a volumetric flask to make exactly 100 mL of solution.

b.   If we took 15 g NaOH and 85 g H_2O, the volume would probably be less than 100 mL. To make sure we have enough solution, let's use 100. mL H_2O (100. g). Let x = mass of NaCl.

$$\text{mass \%} = 15 = \frac{x}{100. + x} \times 100, \quad 1500 + 15 \, x = 100. \, x, \quad x = 17.6 \text{ g} \approx 18 \text{ g}$$

Dissolve 18 g NaOH in 100. mL H_2O to make a 15% NaOH solution by mass.

c. In a fashion similar to part b, let's use 100. mL CH_3OH. Let x = mass of NaOH.

$$100. \text{ mL } CH_3OH \times \frac{0.79 \text{ g}}{\text{mL}} = 79 \text{ g } CH_3OH$$

$$\text{mass } \% = 25 = \frac{x}{79 + x} \times 100, \quad 25(79) + 25\,x = 100.\,x, \quad x = 26.3 \text{ g} \approx 26 \text{ g}$$

Dissolve 26 g NaOH in 100. mL CH_3OH.

d. To make sure we have enough solution, let's use 100. mL (100. g) of H_2O. Let x = mol $C_6H_{12}O_6$.

$$100. \text{ g } H_2O \times \frac{1 \text{ mol } H_2O}{18.02 \text{ g}} = 5.55 \text{ mol } H_2O$$

$$\chi_{C_6H_{12}O_6} = 0.10 = \frac{x}{x + 5.55}; \quad 0.10\,x + 0.56 = x, \quad x = 0.62 \text{ mol } C_6H_{12}O_6$$

$$0.62 \text{ mol } C_6H_{12}O_6 \times \frac{180.16 \text{ g}}{\text{mol}} = 110 \text{ g } C_6H_{12}O_6$$

Dissolve 110 g $C_6H_{12}O_6$ in 100. mL of H_2O to prepare a solution with $\chi_{C_6H_{12}O_6} = 0.10$.

29. $25 \text{ mL } C_5H_{12} \times \dfrac{0.63 \text{ g}}{\text{mL}} = 16 \text{ g } C_5H_{12}; \quad 25 \text{ mL} \times \dfrac{0.63}{\text{mL}} \times \dfrac{1 \text{ mol}}{72.15 \text{ g}} = 0.22 \text{ mol } C_5H_{12}$

$45 \text{ mL } C_6H_{14} \times \dfrac{0.66 \text{ g}}{\text{mL}} = 30. \text{ g } C_6H_{14}; \quad 45 \text{ mL} \times \dfrac{0.66 \text{ g}}{\text{mL}} \times \dfrac{1 \text{ mol}}{86.17 \text{ g}} = 0.34 \text{ mol } C_6H_{14}$

$$\text{mass } \% \text{ pentane} = \frac{\text{mass pentane}}{\text{total mass}} \times 100 = \frac{16 \text{ g}}{16 \text{ g} + 30. \text{ g}} \times 100 = 35\%$$

$$\chi_{\text{pentane}} = \frac{\text{mol pentane}}{\text{total mol}} = \frac{0.22 \text{ mol}}{0.22 \text{ mol} + 0.34 \text{ mol}} = 0.39$$

$$\text{molality} = \frac{\text{mol pentane}}{\text{kg hexane}} = \frac{0.22 \text{ mol}}{0.030 \text{ kg}} = 7.3 \text{ mol/kg}$$

$$\text{molarity} = \frac{\text{mol pentane}}{\text{L solution}} = \frac{0.22 \text{ mol}}{25 \text{ mL} + 45 \text{ mL}} \times \frac{1000 \text{ mL}}{1 \text{ L}} = 3.1 \text{ mol/L}$$

30.   If there are 100.0 mL of wine:

$$12.5 \text{ mL } C_2H_5OH \times \frac{0.789 \text{ g}}{\text{mL}} = 9.86 \text{ g } C_2H_5OH \text{ and } 87.5 \text{ mL } H_2O \times \frac{1.00 \text{ g}}{\text{mL}} = 87.5 \text{ g } H_2O$$

$$\text{mass } \% \text{ ethanol} = \frac{9.86}{87.5 + 9.86} \times 100 = 10.1\% \text{ by mass}$$

$$\text{molality} = \frac{9.86 \text{ g } C_2H_5OH}{0.0875 \text{ kg } H_2O} \times \frac{1 \text{ mol}}{46.07 \text{ g}} = 2.45 \text{ mol/kg}$$

31.   If we have 1.00 L of solution:

$$1.37 \text{ mol citric acid} \times \frac{192.12 \text{ g}}{\text{mol}} = 263 \text{ g citric acid } (H_3C_6H_5O_7)$$

$$1.00 \times 10^3 \text{ mL solution} \times \frac{1.10 \text{ g}}{\text{mL}} = 1.10 \times 10^3 \text{ g solution}$$

$$\text{mass \% of citric acid} = \frac{263 \text{ g}}{1.10 \times 10^3 \text{ g}} \times 100 = 23.9\%$$

In 1.00 L of solution, we have 263 g citric acid and $(1.10 \times 10^3 - 263) = 840$ g of H_2O.

$$\text{molality} = \frac{1.37 \text{ mol citric acid}}{0.84 \text{ kg } H_2O} = 1.6 \text{ mol/kg}$$

$$840 \text{ g } H_2O \times \frac{1 \text{ mol}}{18.02 \text{ g}} = 47 \text{ mol } H_2O; \quad \chi_{\text{citric acid}} = \frac{1.37}{47 + 1.37} = 0.028$$

Since citric acid is a triprotic acid, the number of protons citric acid can provide is three times the molarity. Therefore, normality = 3 × molarity:

$$\text{normality} = 3 \times 1.37 \, M = 4.11 \, N$$

32. $$\frac{1.00 \text{ mol acetone}}{1.00 \text{ kg ethanol}} = 1.00 \text{ molal}; \quad 1.00 \times 10^3 \text{ g } C_2H_5OH \times \frac{1 \text{ mol}}{46.07 \text{ g}} = 21.7 \text{ mol } C_2H_5OH$$

$$\chi_{\text{acetone}} = \frac{1.00}{1.00 + 21.7} = 0.0441$$

$$1 \text{ mol } CH_3COCH_3 \times \frac{58.08 \text{ g } CH_3COCH_3}{\text{mol } CH_3COCH_3} \times \frac{1 \text{ mL}}{0.788 \text{ g}} = 73.7 \text{ mL } CH_3COCH_3$$

$$1.00 \times 10^3 \text{ g ethanol} \times \frac{1 \text{ mL}}{0.789 \text{ g}} = 1270 \text{ mL}; \quad \text{Total volume} = 1270 + 73.7 = 1340 \text{ mL}$$

$$\text{molarity} = \frac{1.00 \text{ mol}}{1.34 \text{ L}} = 0.746 \, M$$

Energetics of Solutions and Solubility

33.   Using Hess's law:

$NaI(s) \rightarrow Na^+(g) + I^-(g)$	$\Delta H = -\Delta H_{LE} = -(-686 \text{ kJ/mol})$
$Na^+(g) + I^-(g) \rightarrow Na^+(aq) + I^-(aq)$	$\Delta H = \Delta H_{\text{hyd}} = -694 \text{ kJ/mol}$
$NaI(s) \rightarrow Na^+(aq) + I^-(aq)$	$\Delta H_{\text{soln}} = -8 \text{ kJ/mol}$

ΔH_{soln} refers to the heat released or gained when a solute dissolves in a solvent. Here, an ionic compound dissolves in water.

34.   a.
$$CaCl_2(s) \rightarrow Ca^{2+}(g) + 2\ Cl^-(g) \qquad \Delta H = -\Delta H_{LE} = -(-2247\ kJ)$$
$$Ca^{2+}(g) + 2\ Cl^-(g) \rightarrow Ca^{2+}(aq) + 2\ Cl^-(aq) \qquad \Delta H = \Delta H_{hyd}$$

$$\overline{CaCl_2(s) \rightarrow Ca^{2+}(aq) + 2\ Cl^-(aq) \qquad \Delta H_{soln} = -46\ kJ}$$

$-46\ kJ = 2247\ kJ + \Delta H_{hyd}$, $\Delta H_{hyd} = -2293\ kJ$

$$CaI_2(s) \rightarrow Ca^{2+}(g) + 2\ I^-(g) \qquad \Delta H = -\Delta H_{LE} = -(-2059\ kJ)$$
$$Ca^{2+}(g) + 2\ I^-(g) \rightarrow Ca^{2+}(aq) + 2\ I^-(aq) \qquad \Delta H = \Delta H_{hyd}$$

$$\overline{CaI_2(s) \rightarrow Ca^{2+}(aq) + 2\ I^-(aq) \qquad \Delta H_{soln} = -104\ kJ}$$

$-104\ kJ = 2059\ kJ + \Delta H_{hyd}$, $\Delta H_{hyd} = -2163\ kJ$

   b.   The enthalpy of hydration for $CaCl_2$ is more exothermic than for CaI_2. Any differences must be due to differences in hydration between Cl^- and I^-. Thus, the chloride ion is more strongly hydrated as compared to the iodide ion.

35.   Both $Al(OH)_3$ and $NaOH$ are ionic compounds. Since the lattice energy is proportional to the charge of the ions, the lattice energy of aluminum hydroxide is greater than that of sodium hydroxide. The attraction of water molecules for Al^{3+} and OH^- cannot overcome the larger lattice energy and $Al(OH)_3$ is insoluble. For $NaOH$, the favorable hydration energy is large enough to overcome the smaller lattice energy and $NaOH$ is soluble.

36.   The dissolving of an ionic solute in water can be thought of as taking place in two steps. The first step, called the lattice energy term, refers to breaking apart the ionic compound into gaseous ions. This step, as indicated in the problem requires a lot of energy and is unfavorable. The second step, called the hydration energy term, refers to the energy released when the separated gaseous ions are stabilized as water molecules surround the ions. Since the interactions between water molecules and ions are strong, a lot of energy is released when ions are hydrated. Thus, the dissolution process for ionic compounds can be thought of as consisting of an unfavorable and a favorable energy term. These two processes basically cancel each other out; so when ionic solids dissolve in water, the heat released or gained is minimal, and the temperature change is minimal.

37.   Water is a polar solvent and dissolves polar solutes and ionic solutes. Carbon tetrachloride (CCl_4) is a nonpolar solvent and dissolves nonpolar solutes (like dissolves like).

   a. CCl_4; CO_2 is a nonpolar molecule.   b.   Water; NH_4NO_3 is an ionic solid.

   c.   Water; CH_3COCH_3 is polar molecule. d.   Water; $HC_2H_3O_2$ (acetic acid) is a polar molecule.

   e. CCl_4; $CH_3CH_2CH_2CH_2CH_3$ is a nonpolar molecule.

38.   a. water   b. water   c. hexane   d. water

39.   Water is a polar molecule capable of hydrogen bonding. Polar molecules, especially molecules capable of hydrogen bonding, and ions are all attracted to water. For covalent compounds, as polarity increases, the attraction to water increases. For ionic compounds, as the charge of the ions increases and/or the size of the ions decreases, the attraction to water increases.

a. CH_3CH_2OH; CH_3CH_2OH is polar while $CH_3CH_2CH_3$ is nonpolar.

b. $CHCl_3$; $CHCl_3$ is polar while CCl_4 is nonpolar.

c. CH_3CH_2OH; CH_3CH_2OH is much more polar than $CH_3(CH_2)_{14}CH_2OH$.

40.   For ionic compounds, as the charge of the ions increases and/or the size of the ions decreases, the attraction to water (hydration) increases.

a. Mg^{2+}; smaller size, higher charge   b. Be^{2+}; smaller

c. Fe^{3+}; smaller size, higher charge   d. F^-; smaller

e. Cl^-; smaller   f. SO_4^{2-}; higher charge

41.   As the length of the hydrocarbon chain increases, the solubility decreases. The –OH end of the alcohols can hydrogen bond with water. The hydrocarbon chain, however, is basically nonpolar and interacts poorly with water. As the hydrocarbon chain gets longer, a greater portion of the molecule cannot interact with the water molecules and the solubility decreases, i.e., the effect of the –OH group decreases as the alcohols get larger.

42.   Benzoic acid is capable of hydrogen bonding, but a significant part of benzoic acid is the nonpolar benzene ring which is composed of only carbon and hydrogen. In benzene, a hydrogen bonded dimer forms:

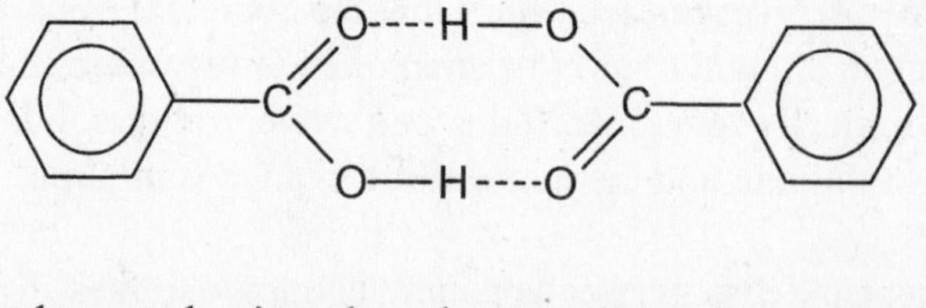

The dimer is relatively nonpolar since the polar part of benzoic acid is hidden in the dimer formation. Thus, benzoic acid is more soluble in benzene than in water due to the dimer formation.

Benzoic acid would be more soluble in 0.1 M NaOH because of the reaction:

$$C_6H_5CO_2H + OH^- \rightarrow C_6H_5CO_2^- + H_2O$$

By removing the proton from benzoic acid, an anion forms, and like all anions, the species becomes more soluble in water.

43. $C = kP$, $\dfrac{8.21 \times 10^{-4} \text{ mol}}{L} = k \times 0.790 \text{ atm}$, $k = 1.04 \times 10^{-3} \text{ mol/L}\bullet\text{atm}$

$C = kP$, $C = \dfrac{1.04 \times 10^{-3} \text{ mol}}{L \text{ atm}} \times 1.10 \text{ atm} = 1.14 \times 10^{-3} \text{ mol/L}$

44.   750. mL grape juice $\times \dfrac{12 \text{ mL } C_2H_5OH}{100. \text{ mL juice}} \times \dfrac{0.79 \text{ g } C_2H_5OH}{mL} \times \dfrac{1 \text{ mol } C_2H_5OH}{46.07 \text{ g}} \times \dfrac{2 \text{ mol } CO_2}{2 \text{ mol } C_2H_5OH}$

$= 1.54 \text{ mol } CO_2$ (carry extra significant figure)

1.54 mol CO_2 = total mol CO_2 = mol $CO_2(g)$ + mol $CO_2(aq)$ = $n_g + n_{aq}$

$$P_{CO_2} = \frac{n_g RT}{V} = \frac{n_g\left(\dfrac{0.08206 \text{ L atm}}{\text{mol K}}\right)(298 \text{ K})}{75 \times 10^{-3} \text{ L}} = 326 \, n_g;\ P_{CO_2} = \frac{C}{k} = \frac{\dfrac{n_{aq}}{0.750 \text{ L}}}{\dfrac{3.1 \times 10^{-2} \text{ mol}}{\text{L atm}}} = 43.0 \, n_{aq}$$

$P_{CO_2} = 326 \, n_g = 43.0 \, n_{aq}$ and from above $n_{aq} = 1.54 - n_g$; Solving:

326 n_g = 43.0(1.54 - n_g), 369 n_g = 66.2, n_g = 0.18 mol

P_{CO_2} = 326(0.18) = 59 atm in gas phase

$C = kP_{CO_2} = \dfrac{3.1 \times 10^{-2} \text{ mol}}{L \text{ atm}} \times 59 \text{ atm}$, $C = 1.8 \text{ mol } CO_2/L$ in wine

Vapor Pressures of Solutions

45. $P_{H_2O} = \chi_{H_2O} P^\circ_{H_2O}$; $\chi_{H_2O} = \dfrac{\text{mol } H_2O \text{ in solution}}{\text{total mol in solution}}$

50.0 g $C_6H_{12}O_6 \times \dfrac{1 \text{ mol } C_6H_{12}O_6}{180.16 \text{ g } C_6H_{12}O_6} = 0.278 \text{ mol glucose}$

600.0 g $H_2O \times \dfrac{1 \text{ mol}}{18.02 \text{ g}} = 33.30 \text{ mol } H_2O$; Total mol = 0.278 + 33.30 = 33.58 mol

$\chi_{H_2O} = \dfrac{33.30}{33.58} = 0.9917$; $P_{H_2O} = \chi_{H_2O} P^\circ_{H_2O} = 0.9917 \times 23.8 \text{ torr} = 23.6 \text{ torr}$

46. $P_{C_2H_5OH} = \chi_{C_2H_5OH} P^\circ_{C_2H_5OH}$; $\chi_{C_2H_5OH} = \dfrac{\text{mol } C_2H_5OH \text{ solution}}{\text{total mol in solution}}$

53.6 g $C_3H_8O_3 \times \dfrac{1 \text{ mol } C_3H_8O_3}{92.09 \text{ g}} = 0.582 \text{ mol } C_3H_8O_3$

$$133.7 \text{ g C}_2\text{H}_5\text{OH} \times \frac{1 \text{ mol C}_2\text{H}_5\text{OH}}{46.07 \text{ g}} = 2.90 \text{ mol C}_2\text{H}_5\text{OH}; \quad \text{total mol} = 0.582 + 2.90 = 3.48 \text{ mol}$$

$$113 \text{ torr} = \frac{2.90 \text{ mol}}{3.48 \text{ mol}} \times P^\circ_{\text{C}_2\text{H}_5\text{OH}}, \quad P^\circ_{\text{C}_2\text{H}_5\text{OH}} = 136 \text{ torr}$$

47. $P_B = \chi_B P^\circ_B$, $\chi_B = P_B/P^\circ_B = 0.900 \text{ atm}/0.930 \text{ atm} = 0.968$

$$0.968 = \frac{\text{mol benzene}}{\text{total mol}}; \quad \text{mol benzene} = 78.11 \text{ g C}_6\text{H}_6 \times \frac{1 \text{ mol}}{78.11 \text{ g}} = 1.000 \text{ mol}$$

Let x = mol solute, then: $\chi_B = 0.968 = \dfrac{1.000 \text{ mol}}{1.000 + x}$, $\quad 0.968 + 0.968 x = 1.000$, $\quad x = 0.033 \text{ mol}$

$$\text{molar mass} = \frac{10.0 \text{ g}}{0.033 \text{ mol}} = 303 \text{ g/mol} \approx 3.0 \times 10^2 \text{ g/mol}$$

48. $19.6 \text{ torr} = \chi_{\text{H}_2\text{O}} (23.8 \text{ torr})$, $\chi_{\text{H}_2\text{O}} = 0.824$; $\chi_{\text{solute}} = 1.000 - 0.824 = 0.176$

0.176 is the mol fraction of all the solute particles present. Since NaCl dissolves to produce two ions in solution (Na^+ and Cl^-), 0.176 is the mole fraction of Na^+ and Cl^- ions present. The mole fraction of NaCl is 1/2 (0.176) = 0.0880 = χ_{NaCl}.

At 45°C, $P_{\text{H}_2\text{O}} = 0.824 (71.9 \text{ torr}) = 59.2 \text{ torr}$

49. a. $25 \text{ mL C}_5\text{H}_{12} \times \dfrac{0.63 \text{ g}}{\text{mL}} \times \dfrac{1 \text{ mol}}{72.15 \text{ g}} = 0.22 \text{ mol C}_5\text{H}_{12}$

$$45 \text{ mL C}_6\text{H}_{14} \times \frac{0.66 \text{ g}}{\text{mL}} \times \frac{1 \text{ mol}}{86.17 \text{ g}} = 0.34 \text{ mol C}_6\text{H}_{14}; \quad \text{total mol} = 0.22 + 0.34 = 0.56 \text{ mol}$$

$$\chi^L_{\text{pen}} = \frac{\text{mol pentane in solution}}{\text{total mol in solution}} = \frac{0.22 \text{ mol}}{0.56 \text{ mol}} = 0.39, \quad \chi^L_{\text{hex}} = 1.00 - 0.39 = 0.61$$

$$P_{\text{pen}} = \chi^L_{\text{pen}} P^\circ_{\text{pen}} = 0.39(511 \text{ torr}) = 2.0 \times 10^2 \text{ torr}; \quad P_{\text{hex}} = 0.61(150. \text{ torr}) = 92 \text{ torr}$$

$$P_{\text{total}} = P_{\text{pen}} + P_{\text{hex}} = 2.0 \times 10^2 + 92 = 292 \text{ torr} = 290 \text{ torr}$$

b. From Chapter 5 on gases, the partial pressure of a gas is proportional to the number of moles of gas present. For the vapor phase:

$$\chi^V_{\text{pen}} = \frac{\text{mol pentane in vapor}}{\text{total mol vapor}} = \frac{P_{\text{pen}}}{P_{\text{total}}} = \frac{2.0 \times 10^2 \text{ torr}}{290 \text{ torr}} = 0.69$$

Note: In the Solutions Guide, we added V or L to the mole fraction symbol to emphasize which value we are solving. If the L or V is omitted, then the liquid phase is assumed.

50. $P_{total} = P_{CH_2Cl_2} + P_{CH_2Br_2}$; $P = \chi^L P°$; $\chi^L_{CH_2Cl_2} = \dfrac{0.0300 \text{ mol } CH_2Cl_2}{0.0800 \text{ mol total}} = 0.375$

$P_{total} = 0.375 \, (133 \text{ torr}) + (1.000 - 0.375) \, (11.4 \text{ torr}) = 49.9 + 7.13 = 57.0 \text{ torr}$

In the vapor: $\chi^V_{CH_2Cl_2} = \dfrac{P_{CH_2Cl_2}}{P_{total}} = \dfrac{49.9 \text{ torr}}{57.0 \text{ torr}} = 0.875$; $\chi^V_{CH_2Br_2} = 1.000 - 0.875 = 0.125$

51. $P_{total} = P_{meth} + P_{prop}$, $174 \text{ torr} = \chi^L_{meth} \, (303 \text{ torr}) + \chi^L_{prop} \, (44.6 \text{ torr})$; $\chi^L_{prop} = 1.000 - \chi^L_{meth}$

$174 = 303 \, \chi^L_{meth} + (1.000 - \chi^L_{meth}) \, 44.6$, $\dfrac{129}{258} = \chi^L_{meth} = 0.500$; $\chi^L_{prop} = 1.000 - 0.500 = 0.500$

52. $P_{tol} = \chi^L_{tol} P°_{tol}$; $P_{ben} = \chi^L_{ben} P°_{ben}$; For the vapor, $\chi^V_A = P_A / P_{total}$. Since the mole fractions of benzene and toluene are equal in the vapor phase, then $P_{tol} = P_{ben}$.

$\chi^L_{tol} P°_{tol} = \chi^L_{ben} P°_{ben} = (1.00 - \chi^L_{tol}) \, P°_{ben}$, $\chi^L_{tol} (28 \text{ torr}) = (1.00 - \chi^L_{tol}) \, 95 \text{ torr}$

$123 \, \chi^L_{tol} = 95$, $\chi^L_{tol} = 0.77$; $\chi^L_{ben} = 1.00 - 0.77 = 0.23$

53.   Compared to H_2O, solution d (methanol/water) will have the highest vapor pressure because methanol is more volatile than water. Both solution b (glucose/water) and solution c (NaCl/water) will have a lower vapor pressure than water by Raoult's law. NaCl dissolves to give Na^+ ions and Cl^- ions; glucose is a nonelectrolyte. Since there are more solute particles in solution c, the vapor pressure of solution c will be the lowest.

54.   Solution d (methanol/water); Methanol is more volatile than water, which will increase the total vapor pressure to a value greater than the vapor pressure of pure water at this temperature.

55. $50.0 \text{ g } CH_3COCH_3 \times \dfrac{1 \text{ mol}}{58.08 \text{ g}} = 0.861 \text{ mol acetone}$

$50.0 \text{ g } CH_3OH \times \dfrac{1 \text{ mol}}{32.04 \text{ g}} = 1.56 \text{ mol methanol}$

$\chi^L_{acetone} = \dfrac{0.861}{0.861 + 1.56} = 0.356$; $\chi^L_{methanol} = 1.000 - \chi^L_{acetone} = 0.644$

$P_{total} = P_{methanol} + P_{acetone} = 0.644 (143 \text{ torr}) + 0.356 (271 \text{ torr}) = 92.1 \text{ torr} + 96.5 \text{ torr} = 188.6 \text{ torr}$

Since partial pressures are proportional to the moles of gas present, then in the vapor phase:

$\chi^V_{acetone} = \dfrac{P_{acetone}}{P_{total}} = \dfrac{96.5 \text{ torr}}{188.6 \text{ torr}} = 0.512$; $\chi^V_{methanol} = 1.000 - 0.512 = 0.488$

The actual vapor pressure of the solution (161 torr) is less than the calculated pressure assuming ideal behavior (188.6 torr). Therefore, the solution exhibits negative deviations from Raoult's law. This occurs when the solute-solvent interactions are stronger than in pure solute and pure solvent.

56.   a.   An ideal solution would have a vapor pressure at any mole fraction of H_2O between that of pure propanol and pure water (between 74.0 torr and 71.9 torr). The vapor pressures of the solutions are not between these limits, so water and propanol do not make ideal solutions.

       b.   From the data, the vapor pressures of the various solutions are greater than in the ideal solution (positive deviation from Raoult's law). This occurs when the intermolecular forces in solution are weaker than the intermolecular forces in pure solvent and pure solute. This gives rise to endothermic (positive) ΔH_{soln} values.

       c.   The interactions between propanol and water molecules are weaker than between the pure substances since this solution exhibits a positive deviation from Raoult's law.

       d.   At $\chi_{H_2O} = 0.54$, the vapor pressure is highest as compared to the other solutions. Since a solution boils when the vapor pressure of the solution equals the external pressure, the $\chi_{H_2O} = 0.54$ solution should have the lowest normal boiling point; this solution will have a vapor pressure equal to 1 atm at a lower temperature than the other solutions.

Colligative Properties

57. $$\text{molality} = m = \frac{\text{mol solute}}{\text{kg solvent}} = \frac{4.9 \text{ g } C_{12}H_{22}O_{11}}{175 \text{ g } H_2O} \times \frac{1000 \text{ g}}{\text{kg}} \times \frac{1 \text{ mol } C_{12}H_{22}O_{11}}{342.30 \text{ g } C_{12}H_{22}O_{11}} = 0.082 \text{ molal}$$

$$\Delta T_b = K_b m = \frac{0.51°C}{\text{molal}} \times 0.082 \text{ molal} = 0.042°C$$

The boiling point is raised from 100.000°C to 100.042°C. We assumed P = 1 atm and ample significant figures in the boiling point of pure water.

58. $$\Delta T_b = 77.85°C - 76.50°C = 1.35°C; \quad m = \frac{\Delta T_b}{K_b} = \frac{1.35°C}{5.03°C \text{ kg/mol}} = 0.268 \text{ mol/kg}$$

$$\text{mol biomolecule} = 0.0150 \text{ kg solvent} \times \frac{0.268 \text{ mol hydrocarbon}}{\text{kg solvent}} = 4.02 \times 10^{-3} \text{ mol}$$

From the problem, 2.00 g biomolecule was used that must contain 4.02×10^{-3} mol biomolecule. The molar mass of the biomolecule is:

$$\frac{2.00 \text{ g}}{4.02 \times 10^{-3} \text{ mol}} = 498 \text{ g/mol}$$

59. $$\Delta T_f = K_f m, \quad \Delta T_f = 1.50°C = \frac{1.86°C}{\text{molal}} \times m, \quad m = 0.806 \text{ mol/kg}$$

$$0.200 \text{ kg } H_2O \times \frac{0.806 \text{ mol } C_3H_8O_3}{\text{kg } H_2O} \times \frac{92.09 \text{ g } C_3H_8O_3}{\text{mol } C_3H_8O_3} = 14.8 \text{ g } C_3H_8O_3$$

60. $\Delta T_f = 25.50°C - 24.59°C = 0.91°C = K_f m,$ $m = \dfrac{0.91°C}{9.1°C/molal} = 0.10 \text{ mol/kg}$

mass $H_2O = 0.0100 \text{ kg t-butanol} \left(\dfrac{0.10 \text{ mol } H_2O}{\text{kg t-butanol}} \right) \left(\dfrac{18.02 \text{ g } H_2O}{\text{mol } H_2O} \right) = 0.018 \text{ g } H_2O$

61.   molality $= m = \dfrac{50.0 \text{ g } C_2H_6O_2}{50.0 \text{ g } H_2O} \times \dfrac{1000 \text{ g}}{\text{kg}} \times \dfrac{1 \text{ mol}}{62.07 \text{ g}} = 16.1 \text{ mol/kg}$

$\Delta T_f = K_f m = 1.86°C/molal \times 16.1 \text{ molal} = 29.9°C;$ $T_f = 0.0°C - 29.9°C = -29.9°C$

$\Delta T_b = K_b m = 0.51°C/molal \times 16.1 \text{ molal} = 8.2°C;$ $T_b = 100.0°C + 8.2°C = 108.2°C$

62. $m = \dfrac{\Delta T_f}{K_f} = \dfrac{30.0°C}{1.86°C \text{ kg/mol}} = 16.1 \text{ mol } C_2H_6O_2/kg$

Since the density of water is 1.00 g/cm^3, the moles of $C_2H_6O_2$ needed are:

$15.0 \text{ L } H_2O \times \dfrac{1.00 \text{ kg } H_2O}{\text{L } H_2O} \times \dfrac{16.1 \text{ mol } C_2H_6O_2}{\text{kg } H_2O} = 242 \text{ mol } C_2H_6O_2$

Volume $C_2H_6O_2 = 242 \text{ mol } C_2H_6O_2 \times \dfrac{62.07 \text{ g}}{\text{mol } C_2H_6O_2} \times \dfrac{1 \text{ cm}^3}{1.11 \text{ g}} = 13,500 \text{ cm}^3 = 13.5 \text{ L}$

$\Delta T_b = K_b m = \dfrac{0.51°C}{molal} \times 16.1 \text{ molal} = 8.2°C;$ $T_b = 100.0°C + 8.2°C = 108.2°C$

63. $\Delta T_f = K_f m,$ $m = \dfrac{\Delta T_f}{K_f} = \dfrac{0.240°C}{4.70°C \text{ kg/mol}} = \dfrac{5.11 \times 10^{-2} \text{ mol biomolecule}}{\text{kg solvent}}$

The mol of biomolecule present is:

$0.0150 \text{ kg solvent} \times \dfrac{5.11 \times 10^{-2} \text{ mol biomolecule}}{\text{kg solvent}} = 7.67 \times 10^{-4} \text{ mol biomolecule}$

From the problem, 0.350 g biomolecule were used that must contain 7.67×10^{-4} mol biomolecule. The molar mass of the biomolecule is:

molar mass $= \dfrac{0.350 \text{ g}}{7.67 \times 10^{-4} \text{ mol}} = 456 \text{ g/mol}$

64.   empirical formula mass $\approx 7(12) + 4(1) + 16 = 104 \text{ g/mol}$

$\Delta T_f = K_f m,$ $m = \dfrac{\Delta T_f}{K_f} = \dfrac{22.3°C}{40.°C/molal} = 0.56 \text{ molal}$

mol anthraquinone $= 0.0114 \text{ kg solvent} \times \dfrac{0.56 \text{ mol anthraquinone}}{\text{kg solvent}} = 6.4 \times 10^{-3} \text{ mol}$

$$\text{molar mass} = \frac{1.32 \text{ g}}{6.4 \times 10^{-3} \text{ mol}} = 210 \text{ g/mol}$$

$$\frac{\text{molar mass}}{\text{empirical formula mass}} = \frac{210}{104} = 2.0; \quad \text{molecular formula} = C_{14}H_8O_2$$

65.   a. $M = \dfrac{1.0 \text{ g protein}}{L} \times \dfrac{1 \text{ mol}}{9.0 \times 10^4 \text{ g}} = 1.1 \times 10^{-5} \text{ mol/L}; \quad \pi = MRT$

At 298 K: $\pi = \dfrac{1.1 \times 10^{-5} \text{ mol}}{L} \times \dfrac{0.08206 \text{ L atm}}{\text{mol K}} \times 298 \text{ K} \times \dfrac{760 \text{ torr}}{\text{atm}}, \quad \pi = 0.20 \text{ torr}$

Since d = 1.0 g/cm³, 1.0 L solution has a mass of 1.0 kg. Since only 1.0 g of protein is present per liter of solution, 1.0 kg of H_2O is present and molality equals molarity.

$$\Delta T_f = K_f m = \frac{1.86 °C}{\text{molal}} \times 1.1 \times 10^{-5} \text{ molal} = 2.0 \times 10^{-5} °C$$

b. Osmotic pressure is better for determining the molar mass of large molecules. A temperature change of 10^{-5} °C is very difficult to measure. A change in height of a column of mercury by 0.2 mm (0.2 torr) is not as hard to measure precisely.

66. $M = \dfrac{\pi}{RT} = \dfrac{0.745 \text{ torr} \times \dfrac{1 \text{ atm}}{760 \text{ torr}}}{\dfrac{0.08206 \text{ L atm}}{\text{mol K}} \times 300. \text{ K}} = 3.98 \times 10^{-5} \text{ mol/L}$

$$1.00 \text{ L} \times \frac{3.98 \times 10^{-5} \text{ mol}}{L} = 3.98 \times 10^{-5} \text{ mol catalase}$$

$$\text{molar mass} = \frac{10.00 \text{ g}}{3.98 \times 10^{-5} \text{ mol}} = 2.51 \times 10^5 \text{ g/mol}$$

67. $\pi = MRT, \quad M = \dfrac{\pi}{RT} = \dfrac{8.00 \text{ atm}}{0.08206 \text{ L atm/mol} \cdot \text{K} \times 298 \text{ K}} = 0.327 \text{ mol/L}$

68. $M = \dfrac{\pi}{RT} = \dfrac{15 \text{ atm}}{0.08206 \times 295 \text{ K}} = 0.62 \text{ } M \text{ solute particles}$

This represents the total molarity of the solute particles. NaCl is a soluble ionic compound that breaks up into two ions, Na^+ and Cl^-. Therefore, the concentration of NaCl needed is $0.62/2 = 0.31$ M.

$$1.0 \text{ L} \times \frac{0.31 \text{ mol NaCl}}{L} \times \frac{58.44 \text{ g NaCl}}{\text{mol NaCl}} = 18.1 \approx 18 \text{ g NaCl}$$

Dissolve 18 g of NaCl in some water and dilute to 1.0 L in a volumetric flask. To get 0.31 ± 0.01 mol/L, we need $18.1 \text{ g} \pm 0.6 \text{ g}$ NaCl.

Properties of Electrolyte Solutions

69. $Na_3PO_4(s) \rightarrow 3\ Na^+(aq) + PO_4^{3-}(aq)$, i = 4.0; $CaBr_2(s) \rightarrow Ca^{2+}(aq) + 2\ Br^-(aq)$, i = 3.0

$KCl(s) \rightarrow K^+(aq) + Cl^-(aq)$, i = 2.0.

The effective particle concentrations of the solutions are:

4.0(0.010 molal) = 0.040 molal for Na_3PO_4 solution; 3.0(0.020 molal) = 0.060 molal for $CaBr_2$ solution; 2.0(0.020 molal) = 0.040 molal for KCl solution; slightly greater than 0.020 molal for HF solution since HF only partially dissociates in water (it is a weak acid).

a.   The 0.010 m Na_3PO_4 solution and the 0.020 m KCl solution both have effective particle concentrations of 0.040 m (assuming complete dissociation), so both of these solutions should have the same boiling point as the 0.040 m $C_6H_{12}O_6$ solution (a nonelectrolyte).

b. $P = \chi P°$; As the solute concentration decreases, the solvent's vapor pressure increases since χ increases. Therefore, the 0.020 m HF solution will have the highest vapor pressure since it has the smallest effective particle concentration.

c. $\Delta T = K_f m$; The 0.020 m $CaBr_2$ solution has the largest effective particle concentration so it will have the largest freezing point depression (largest ΔT).

70.   The solutions of $C_{12}H_{22}O_{11}$, NaCl and $CaCl_2$ will all have lower freezing points, higher boiling points and higher osmotic pressures than pure water. The solution with the largest particle concentration will have the lowest freezing point, the highest boiling point and the highest osmotic pressure. The $CaCl_2$ solution will have the largest effective particle concentration because it produces three ions per mol of compound.

a.   pure water   b. $CaCl_2$ solution   c. $CaCl_2$ solution

d.   pure water   e. $CaCl_2$ solution

71.   a. $MgCl_2(s) \rightarrow Mg^{2+}(aq) + 2\ Cl^-(aq)$, i = 3.0 mol ions/mol solute

$\Delta T_f = iK_f m = 3.0 \times 1.86°C/molal \times 0.050\ molal = 0.28°C$; $T_f = -0.28°C$ (Assuming water freezes at 0.00°C.)

$\Delta T_b = iK_b m = 3.0 \times 0.51°C/molal \times 0.050\ molal = 0.077°C$; $T_b = 100.077°C$ (Assuming water boils at 100.000°C.)

b. $FeCl_3(s) \rightarrow Fe^{3+}(aq) + 3\ Cl^-(aq)$, i = 4.0 mol ions/mol solute

$\Delta T_f = iK_f m = 4.0 \times 1.86°C/molal \times 0.050\ molal = 0.37°C$; $T_f = -0.37°C$

$\Delta T_b = iK_b m = 4.0 \times 0.51°C/molal \times 0.050\ molal = 0.10°C$; $T_b = 100.10°C$

72. $NaCl(s) \rightarrow Na^+(aq) + Cl^-(aq)$, $i = 2.0$

$$\pi = iMRT = 2.0 \times \frac{0.10 \text{ mol}}{L} \times \frac{0.08206 \text{ L atm}}{\text{mol K}} \times 293 \text{ K} = 4.8 \text{ atm}$$

A pressure greater than 4.8 atm should be applied to insure purification by reverse osmosis.

73. $\Delta T_f = iK_f m$, $i = \dfrac{\Delta T_f}{K_f m} = \dfrac{0.110°C}{1.86° \text{ C/molal} \times 0.0225 \text{ molal}} = 2.63$ for 0.0225 m $CaCl_2$

$i = \dfrac{0.440}{1.86 \times 0.0910} = 2.60$ for 0.0910 m $CaCl_2$; $i = \dfrac{1.330}{1.86 \times 0.278} = 2.57$ for 0.278 m $CaCl_2$

$i_{ave} = (2.63 + 2.60 + 2.57)/3 = 2.60$

Note that i is less than the ideal value of 3.0 for $CaCl_2$. This is due to ion pairing in solution. Also note that as molality increases, i decreases. More ion pairing occurs as concentration increases.

74.   a. $MgCl_2$, i(observed) = 2.7

         $\Delta T_f = iK_f m = 2.7 \times 1.86°C/\text{molal} \times 0.050 \text{ molal} = 0.25°C$; $T_f = -0.25°C$

         $\Delta T_b = iK_b m = 2.7 \times 0.51°C/\text{molal} \times 0.050 \text{ molal} = 0.069°C$; $T_b = 100.069°C$

       b. $FeCl_3$, i(observed) = 3.4

         $\Delta T_f = iK_f m = 3.4 \times 1.86 °C/\text{molal} \times 0.050 \text{ molal} = 0.32°C$; $T_f = -0.32°C$

         $\Delta T_b = iK_b m = 3.4 \times 0.51°C/\text{molal} \times 0.050 \text{ molal} = 0.087°C$; $T_b = 100.087°C$

75. $\pi = iMRT = 3.0 \times 0.50 \text{ mol/L} \times 0.08206 \text{ L atm/K}\bullet\text{mol} \times 298 \text{ K} = 37 \text{ atm}$

Because of ion pairing in solution, we would expect i to be less than 3.0, which results in fewer solute particles in solution, which results in a lower osmotic pressure than calculated above.

76.   a. $T_C = 5(T_F - 32)/9 = 5(-29 - 32)/9 = -34°C$

         Assuming the solubility of $CaCl_2$ is temperature independent, the molality of a saturated $CaCl_2$ solution is:

$$\frac{74.5 \text{ g } CaCl_2}{100.0 \text{ g } H_2O} \times \frac{1000 \text{ g}}{kg} \times \frac{1 \text{ mol } CaCl_2}{110.98 \text{ g } CaCl_2} = \frac{6.71 \text{ mol } CaCl_2}{kg \, H_2O}$$

         $\Delta T_f = iK_f m = 3.00 \times 1.86°C \text{ kg/mol} \times 6.71 \text{ mol/kg} = 37.4°C$

         Assuming i = 3.00, a saturated solution of $CaCl_2$ can lower the freezing point of water to -37.4°C. Assuming these conditions, a saturated $CaCl_2$ solution should melt ice at -34°C (-29°F).

    b. From Exercise 11.73, $i_{ave} = 2.60$; $\Delta T_f = iK_f m = 2.60 \times 1.86 \times 6.71 = 32.4°C$; $T_f = -32.4°C$

        Assuming $i = 2.60$, a saturated $CaCl_2$ solution will not melt ice at $-34°C(-29°F)$.

Additional Exercises

77.   a. $NH_4NO_3(s) \rightarrow NH_4^+(aq) + NO_3^-(aq)$ $\Delta H_{soln} = ?$

        Heat gain by dissolution process = heat loss by solution; We will keep all quantities positive in order to avoid sign errors. Since the temperature of the water decreased, the dissolution of NH_4NO_3 is endothermic (ΔH is positive). Mass of solution $= 1.60 + 75.0 = 76.6$ g.

        heat loss by solution $= \dfrac{4.18\,J}{g\,°C} \times 76.6\,g \times (25.00°C - 23.34°C) = 532\,J$

        $\Delta H_{soln} = \dfrac{532\,J}{1.60\,g\,NH_4NO_3} \times \dfrac{80.05\,g\,NH_4NO_3}{mol\,NH_4NO_3} = 2.66 \times 10^4\,J/mol = 26.6\,kJ/mol$

    b. We will use Hess's law to solve for the lattice energy. The lattice energy equation is:

        $NH_4^+(g) + NO_3^-(g) \rightarrow NH_4NO_3(s)$ ΔH = lattice energy

        $NH_4^+(g) + NO_3^-(g) \rightarrow NH_4^+(aq) + NO_3^-(aq)$ $\Delta H = \Delta H_{hyd} = -630.\,kJ/mol$
        $NH_4^+(aq) + NO_3^-(aq) \rightarrow NH_4NO_3(s)$ $\Delta H = -\Delta H_{soln} = -26.6\,kJ/mol$

          ———

        $NH_4^+(g) + NO_3^-(g) \rightarrow NH_4NO_3(s)$ $\Delta H = \Delta H_{hyd} - \Delta H_{soln} = -657\,kJ/mol$

78.   The main intermolecular forces are:

        hexane (C_6H_{14}): London dispersion; chloroform ($CHCl_3$): dipole-dipole, London dispersion

        methanol (CH_3OH): H bonding; H_2O: H bonding (two places)

    There is a gradual change in the nature of the intermolecular forces (weaker to stronger). Each preceding solvent is miscible in its predecessor because there is not a great change in the strengths of the intermolecular forces from one solvent to the next.

79.   a. Water boils when the vapor pressure equals the pressure above the water. In an open pan $P_{atm} \approx 1.0$ atm. In a pressure cooker, $P_{inside} > 1.0$ atm, and water boils at a higher temperature. The higher the cooking temperature, the faster the cooking time.

    b. Salt dissolves in water forming a solution with a melting point lower than that of pure water ($\Delta T_f = K_f m$). This happens in water on the surface of ice. If it is not too cold, the ice melts. This won't work if the ambient temperature is lower than the depressed freezing point of the salt solution.

c. When water freezes from a solution, it freezes as pure water, leaving behind a more concentrated salt solution.   Therefore, the melt of frozen sea ice is pure water.

d. On the CO_2 phase diagram in chapter 10, the triple point is above 1 atm, so $CO_2(g)$ is the stable phase at 1 atm and room temperature. $CO_2(l)$ can't exist at normal atmospheric pressures. Therefore, dry ice sublimes instead of boils. In a fire extinguisher, P > 1 atm and $CO_2(l)$ can exist. When CO_2 is released from the fire extinguisher, $CO_2(g)$ forms as predicted from the phase diagram.

e. Adding a solute to a solvent increases the boiling point and decreases the freezing point of the solvent. Thus, the solvent is a liquid over a wider range of temperatures when a solute is dissolved.

80. $CO_2(g) + OH^-(aq) \rightarrow HCO_3^-(aq)$; No, the reaction of CO_2 with OH^- greatly increases the solubility of CO_2 in water by forming the soluble bicarbonate anion.

81.   Since partial pressures are proportional to the moles of gas present, then $\chi_{CS_2}^V = P_{CS_2}/P_{tot}$.

$$P_{CS_2} = \chi_{CS_2}^V P_{tot} = 0.855 \,(263 \text{ torr}) = 225 \text{ torr}$$

$$P_{CS_2} = \chi_{CS_2}^L P_{CS_2}^\circ, \quad \chi_{CS_2}^L = \frac{P_{CS_2}}{P_{CS_2}^\circ} = \frac{225 \text{ torr}}{375 \text{ torr}} = 0.600$$

82. $$\pi = MRT = \frac{0.1 \text{ mol}}{L} \times \frac{0.08206 \text{ L atm}}{\text{mol K}} \times 298 \text{ K} = 2.45 \text{ atm} \approx 2 \text{ atm}$$

$$\pi = 2 \text{ atm} \times \frac{760 \text{ mm Hg}}{\text{atm}} \approx 2000 \text{ mm} \approx 2 \text{ m}$$

The osmotic pressure would support a mercury column of ≈ 2 m. The height of a fluid column in a tree will be higher because Hg is more dense than the fluid in a tree. If we assume the fluid in a tree is mostly H_2O, the fluid has a density of 1.0 g/cm^3. The density of Hg is 13.6 g/cm^3.

Height of fluid $\approx 2 \text{ m} \times 13.6 \approx 30 \text{ m}$

83.   Out of 100.00 g, there are:

$$31.57 \text{ g C} \times \frac{1 \text{ mol C}}{12.01 \text{ g}} = 2.629 \text{ mol C}; \quad \frac{2.629}{2.629} = 1.000$$

$$5.30 \text{ g H} \times \frac{1 \text{ mol H}}{1.008 \text{ g}} = 5.26 \text{ mol H}; \quad \frac{5.26}{2.629} = 2.00$$

$$63.13 \text{ g O} \times \frac{1 \text{ mol O}}{16.00 \text{ g}} = 3.946 \text{ mol O}; \quad \frac{3.946}{2.629} = 1.501$$

empirical formula: $C_2H_4O_3$; Use the freezing point data to determine the molar mass.

$$m = \frac{\Delta T_f}{K_f} = \frac{5.20°C}{1.86°C/molal} = 2.80 \text{ molal}$$

$$\text{mol solute} = 0.0250 \text{ kg} \times \frac{2.80 \text{ mol solute}}{kg} = 0.0700 \text{ mol solute}$$

$$\text{molar mass} = \frac{10.56 \text{ g}}{0.0700 \text{ mol}} = 151 \text{ g/mol}$$

The empirical formula mass of $C_2H_4O_3$ = 76.05 g/mol. Since the molar mass is about twice the empirical mass, the molecular formula is $C_4H_8O_6$, which has a molar mass of 152.10 g/mol.

Note: We use the experimental molar mass to determine the molecular formula. Knowing this, we calculate the molar mass precisely from the molecular formula using the periodic table.

84.   a.   As discussed in Figure 11.18 of the text, the water would migrate from right to left. Initially, the level of liquid in the right arm would go down and the level in the left arm would go up. At some point, the rate of solvent transfer would be the same in both directions and the levels of the liquids in the two arms would stabilize. The height difference between the two arms is a measure of the osmotic pressure of the NaCl solution.

      b.   Initially, H_2O molecules will have a net migration into the NaCl side. However, Na^+ and Cl^- ions can now migrate into the H_2O side. Because solute and solvent transfer are both possible, the levels of the liquids will be equal once the rate of solute and solvent transfer is equal in both directions. At this point, the concentration of Na^+ and Cl^- ions will be equal in both chambers and the levels of liquid will be equal.

85.   If ideal, NaCl dissociates completely and i = 2.00. $\Delta T_f = iK_f m$; Assuming water freezes at 0.00°C:

      1.28°C = 2 × 1.86°C kg/mol × m, m = 0.344 mol NaCl/kg H_2O

      Assume an amount of solution which contains 1.00 kg of water (solvent).

      $0.344 \text{ mol NaCl} \times \frac{58.44 \text{ g}}{mol} = 20.1 \text{ g NaCl};$ $\text{mass \% NaCl} = \frac{20.1 \text{ g}}{1.00 \times 10^3 \text{ g} + 20.1 \text{ g}} \times 100 = 1.97\%$

Challenge Problems

86.   a. $\pi = iMRT$, $iM = \frac{\pi}{RT} = \frac{7.83 \text{ atm}}{0.08206 \text{ L atm/K} \bullet \text{mol} \times 298 \text{ K}} = 0.320 \text{ mol/L}$

      Assuming 1.000 L of solution:

      total mol solute particles = mol Na^+ + mol Cl^- + mol NaCl = 0.320 mol

$$\text{mass solution} = 1000.\ \text{mL} \times \frac{1.071\ \text{g}}{\text{mL}} = 1071\ \text{g solution}$$

$$\text{mass NaCl in solution} = 0.0100 \times 1071\ \text{g} = 10.7\ \text{g NaCl}$$

$$\text{mol NaCl added to solution} = 10.7\ \text{g} \times \frac{1\ \text{mol}}{58.44\ \text{g}} = 0.183\ \text{mol NaCl}$$

Some of this NaCl dissociates into Na^+ and Cl^- (two mol ions per mol NaCl) and some remains undissociated. Let x = mol undissociated NaCl = mol ion pairs.

$$\text{mol solute particles} = 0.320\ \text{mol} = 2(0.183 - x) + x$$

$$0.320 = 0.366 - x, \ \ x = 0.046\ \text{mol ion pairs}$$

$$\text{fraction of ion pairs} = \frac{0.046}{0.183} = 0.25, \ \ \text{or 25\%}$$

b. $\Delta T = K_f m$ where $K_f = 1.86\ °C\ \text{kg/mol}$; From part a, 1.000 L of solution contains 0.320 mol of solute particles. To calculate the molality of the solution, we need the kg of solvent present in 1.000 L solution.

$$\text{mass of 1.000 L solution} = 1071\ \text{g}; \ \ \text{mass of NaCl} = 10.7\ \text{g}$$

$$\text{mass of solvent in 1.000 L solution} = 1071\ \text{g} - 10.7\ \text{g} = 1060.\ \text{g}$$

$$\Delta T = 1.86\ °C\ \text{kg/mol} \times \frac{0.320\ \text{mol}}{1.060\ \text{kg}} = 0.562\,°C$$

Assuming water freezes at 0.000 °C, then $T_f = -0.562\,°C$.

87. $\chi_{pen}^{V} = 0.15 = \dfrac{P_{pen}}{P_{total}}$; $\ P_{pen} = \chi_{pen}^{L} P_{pen}^{\circ} = \chi_{pen}^{L}(511\ \text{torr})$; $\ P_{total} = P_{pen} + P_{hex} = \chi_{pen}^{L}(511) + \chi_{hex}^{L}(150.)$

Since $\chi_{hex}^{L} = 1.000 - \chi_{pen}^{L}$, then: $\ P_{total} = \chi_{pen}^{L}(511) + (1.000 - \chi_{pen}^{L})(150.) = 150. + 361\,\chi_{pen}^{L}$

$$\chi_{pen}^{V} = \frac{P_{pen}}{P_{total}}, \ \ 0.15 = \frac{\chi_{pen}^{L}(511)}{150. + 361\,\chi_{pen}^{L}}, \ \ 0.15\,(150. + 361\,\chi_{pen}^{L}) = 511\,\chi_{pen}^{L}$$

$$23 + 54\,\chi_{pen}^{L} = 511\,\chi_{pen}^{L}, \ \ \chi_{pen}^{L} = \frac{23}{457} = 0.050$$

88. a. $m = \dfrac{\Delta T_f}{K_f} = \dfrac{1.32\,°C}{5.12\,°C\ \text{kg/mol}} = 0.258\ \text{mol/kg}$

$$\text{mol unknown} = 0.01560\ \text{kg} \times \frac{0.258\ \text{mol unknown}}{\text{kg}} = 4.02 \times 10^{-3}\ \text{mol}$$

$$\text{molar mass of unknown} = \frac{1.22\ \text{g}}{4.02 \times 10^{-3}\ \text{mol}} = 303\ \text{g/mol}$$

Uncertainty in temperature $= \dfrac{0.04}{1.32} \times 100 = 3\%$;   A 3% uncertainty in 303 g/mol = 9 g/mol.

So, molar mass = 303 ± 9 g/mol.

b. No, codeine could not be eliminated since its molar mass is in the possible range including the uncertainty.

c. We would really like the uncertainty to be ± 1 g/mol. We need the freezing point depression to be about 10 times what it was in this problem. Two possibilities are:

1. make the solution ten times more concentrated (may be a solubility problem) or
2. use a solvent with a larger K_f value, e.g., camphor

89. $\Delta T_f = 5.51 - 2.81 = 2.70°C$; $m = \dfrac{\Delta T_f}{K_f} = \dfrac{2.7°C}{5.12°C/molal} = 0.527$ molal

Let x = mass of naphthalene (molar mass: 128.2 g/mol). Then $1.60 - x$ = mass of anthracene (molar mass = 178.2 g/mol).

$\dfrac{x}{128.2}$ = moles naphthalene and $\dfrac{1.60 - x}{178.2}$ = moles anthracene

$\dfrac{0.527 \text{ mol solute}}{\text{kg solvent}} = \dfrac{\dfrac{x}{128.2} + \dfrac{1.60 - x}{178.2}}{0.0200 \text{ kg solvent}}$, $1.05 \times 10^{-2} = \dfrac{178.2\,x + 1.60\,(128.2) - 128.2\,x}{128.2\,(178.2)}$

$50.0\,x + 205 = 240.$, $50.0\,x = 35$, $x = 0.70$ g naphthalene

So mixture is: $\dfrac{0.70 \text{ g}}{1.60 \text{ g}} \times 100 = 44\%$ naphthalene by mass and 56% anthracene by mass

90. $iM = \dfrac{\pi}{RT} = \dfrac{0.3950 \text{ atm}}{0.08206 \text{ L atm/mol} \cdot \text{K } (298.2 \text{ K})} = 0.01614$ mol/L = total ion concentration

$0.01614 \text{ mol/L} = M_{Mg^{2+}} + M_{Na^+} + M_{Cl^-}$; $M_{Cl^-} = 2\,M_{Mg^{2+}} + M_{Na^+}$ (charge balance)

Combining: $0.01614 = 3\,M_{Mg^{2+}} + 2\,M_{Na^+}$

Let x = mass $MgCl_2$ and y = mass NaCl, then $x + y = 0.5000$ g.

$M_{Mg^{2+}} = \dfrac{x}{95.21}$ and $M_{Na^+} = \dfrac{y}{58.44}$ (Since V = 1.000 L.)

Total ion concentration $= \dfrac{3\,x}{95.21} + \dfrac{2\,y}{58.44} = 0.01614$ mol/L;   Rearranging: $3\,x + 3.258\,y = 1.537$

Solving by simultaneous equations:

$$3x \ + \ 3.258y \ = 1.537$$
$$-3(x \ + \quad \ y) \ = -3(0.5000)$$

$$0.258y \ = \ 0.037, \ y = 0.14 \text{ g NaCl}$$

mass $MgCl_2$ = 0.5000 g - 0.14 g = 0.36 g;   mass % $MgCl_2$ = $\dfrac{0.36 \text{ g}}{0.5000 \text{ g}}$ × 100 = 72%

91. $HCO_2H \rightarrow H^+ + HCO_2^-$; Only 4.2% of HCO_2H ionizes. The amount of H^+ or HCO_2^- produced is:

0.042 × 0.10 M = 0.0042 M

The amount of HCO_2H remaining in solution after ionization is:

0.10 M - 0.0042 M = 0.10 M

The total molarity of species present = $M_{HCO_2H} + M_{H^+} + M_{HCO_2^-}$ = 0.10 + 0.0042 + 0.0042 = 0.11 M.

Assuming 0.11 M = 0.11 molal and assuming ample significant figures in the freezing point and boiling point of water at P = 1 atm:

$\Delta T = K_f m$ = 1.86°C/molal × 0.11 molal = 0.20°C;   freezing point = -0.20°C

$\Delta T = K_b m$ = 0.51°C/molal × 0.11 molal = 0.056°C;   boiling point = 100.056°C

92.   a.   The average values for each ion are:

       300. mg Na^+;   15.7 mg K^+;   5.45 mg Ca^{2+};   388 mg Cl^-;   246 mg lactate, $C_3H_5O_3^-$

Note: Since we can precisely weigh to ± 0.1 mg on an analytical balance, we'll carry extra significant figures and calculate results to ± 0.1 mg.

The only source of lactate is $NaC_3H_5O_3$.

246 mg lactate × $\dfrac{112.06 \text{ mg } NaC_3H_5O_3}{89.07 \text{ mg } C_3H_5O_3^-}$ = 309.5 mg sodium lactate

The only source of Ca^{2+} is $CaCl_2 \cdot 2H_2O$.

5.45 mg Ca^{2+} × $\dfrac{147.0 \text{ mg } CaCl_2 \cdot 2H_2O}{40.08 \text{ mg } Ca^{2+}}$ = 19.99 or 20.0 mg $CaCl_2 \cdot 2H_2O$

The only source of K^+ is KCl.

15.7 mg K^+ × $\dfrac{74.55 \text{ mg KCl}}{39.10 \text{ mg } K^+}$ = 29.9 mg KCl

From what we have used already, let's calculate the mass of Na^+ and Cl^- added.

309.5 mg sodium lactate = 246.0 mg lactate + 63.5 mg Na^+

Thus, we need to add an additional 236.5 mg Na^+ to get the desired 300. mg.

$$236.5 \text{ mg } Na^+ \times \frac{58.44 \text{ mg NaCl}}{22.99 \text{ mg } Na^+} = 601.2 \text{ mg NaCl}$$

Let's check the mass of Cl^- added:

$$20.0 \text{ mg } CaCl_2 \cdot 2H_2O \times \frac{70.90 \text{ mg } Cl^-}{147.0 \text{ mg } CaCl_2 \cdot 2H_2O} = 9.6 \text{ mg } Cl^-$$

$$
\begin{aligned}
20.0 \text{ mg } CaCl_2 \cdot 2H_2O &= 9.6 \text{ mg } Cl^- \\
29.9 \text{ mg KCl} - 15.7 \text{ mg } K^+ &= 14.2 \text{ mg } Cl^- \\
601.2 \text{ mg NaCl} - 236.5 \text{ mg } Na^+ &= 364.7 \text{ mg } Cl^- \\
\hline
\text{Total } Cl^- &= 388.5 \text{ mg } Cl^-
\end{aligned}
$$

This is the quantity of Cl^- we want (the average amount of Cl^-).

An analytical balance can weigh to the nearest 0.1 mg. We would use 309.5 mg sodium lactate, 20.0 mg $CaCl_2 \cdot 2H_2O$, 29.9 mg KCl and 601.2 mg NaCl.

b. To get the range of osmotic pressure, we need to calculate the molar concentration of each ion at its minimum and maximum values. At minimum concentrations, we have:

$$\frac{285 \text{ mg } Na^+}{100. \text{ mL}} \times \frac{1 \text{ mmol}}{22.99 \text{ mg}} = 0.124 \ M; \quad \frac{14.1 \text{ mg } K^+}{100. \text{ mL}} \times \frac{1 \text{ mmol}}{39.10 \text{ mg}} = 0.00361 \ M$$

$$\frac{4.9 \text{ mg } Ca^{2+}}{100. \text{ mL}} \times \frac{1 \text{ mmol}}{40.08 \text{ mg}} = 0.0012 \ M; \quad \frac{368 \text{ mg } Cl^-}{100. \text{ mL}} \times \frac{1 \text{ mmol}}{35.45 \text{ mg}} = 0.104 \ M$$

$$\frac{231 \text{ mg } C_3H_5O_3^-}{100. \text{ mL}} \times \frac{1 \text{ mmol}}{89.07 \text{ mg}} = 0.0259 \ M$$

Total = 0.124 + 0.00361 + 0.0012 + 0.104 + 0.0259 = 0.259 M

$$\pi = MRT = \frac{0.259 \text{ mol}}{L} \times \frac{0.08206 \text{ L atm}}{\text{mol K}} \times 310. \text{ K} = 6.59 \text{ atm}$$

Similarly at maximum concentrations, the concentration of each ion is:

Na^+: 0.137 M; K^+: 0.00442 M; Ca^{2+}: 0.0015 M; Cl^-: 0.115 M; $C_3H_5O_3^-$: 0.0293 M

The total concentration of all ions is the sum, 0.287 M.

$$\pi = \frac{0.287 \text{ mol}}{L} \times \frac{0.08206 \text{ L atm}}{\text{mol K}} \times 310. \text{ K} = 7.30 \text{ atm}$$

Osmotic pressure ranges from 6.59 atm to 7.30 atm.

93.   a. Assuming $MgCO_3(s)$ does not dissociate, the solute concentration in water is:

$$\frac{560 \; \mu g \; MgCO_3(s)}{mL} = \frac{560 \; mg}{L} = \frac{560 \times 10^{-3} \; g}{L} \times \frac{1 \; mol \; MgCO_3}{84.32 \; g} = 6.6 \times 10^{-3} \; mol \; MgCO_3/L$$

An applied pressure of 8.0 atm will purify water up to a solute concentration of:

$$M = \frac{\pi}{RT} = \frac{8.0 \; atm}{0.08206 \; L \; atm/mol \cdot K \times 300. \; K} = \frac{0.32 \; mol}{L}$$

When the concentration of $MgCO_3(s)$ reaches 0.32 mol/L, the reverse osmosis unit can no longer purify the water. Let V = volume (L) of water remaining after purifying 45 L of H_2O. When V + 45 L of water have been processed, the moles of solute particles will equal:

   $6.6 \times 10^{-3} \; mol/L \times (45 \; L + V) = 0.32 \; mol/L \times V$

Solving: $0.30 = (0.32 - 0.0066) \times V,$ V = 0.96 L

The minimum total volume of water that must be processed is 45 L + 0.96 L = 46 L.

Note: If $MgCO_3$ does dissociate into Mg^{2+} and CO_3^{2-} ions, the solute concentration will increase to $1.3 \times 10^{-2} \; M$ and at least 47 L of water must be processed.

   b. No; A reverse osmosis system that applies 8.0 atm can only purify water with a solute concentration less than 0.32 mol/L. Salt water has a solute concentration of 2(0.60 M) = 1.20 M ions. The solute concentration of salt water is much too high for this reverse osmosis unit to work.

CHAPTER TWELVE

CHEMICAL KINETICS

Questions

9.   The rate of a chemical reaction varies with time. Consider the general reaction:

$$A \rightarrow Products \text{ where rate} = \frac{-\Delta[A]}{\Delta t}$$

If we graph [A] vs. t, it would roughly look like the dark line in the following plot.

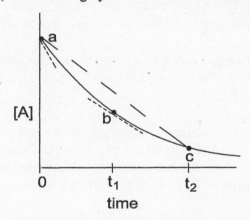

An instantaneous rate is the slope of a tangent line to the graph of [A] vs. t. We can determine the instantaneous rate at any time during the reaction. On the plot, tangent lines at $t = 0$ and $t = t_1$ are drawn. The slope of these tangent lines would be the instantaneous rates at $t \approx 0$ and $t = t_1$. We call the instantaneous rate at $t \approx 0$ the initial rate. The average rate is measured over a period of time. For example, the slope of the line connecting points a and c is the average rate of the reaction over the entire length of time 0 to t_2 (average rate = $\Delta[A]/\Delta t$). An average rate is determined over some time period while an instantaneous rate is determined at one specific time. The rate which is largest is generally the initial rate. At $t \approx 0$, the slope of the tangent line is greatest, which means the rate is largest at $t \approx 0$.

10.   a.   An elementary step (reaction) is one in which the rate law can be written from the molecularity, i.e., from the coefficients in the balanced equation.

291

b.   The mechanism of a reaction is the series of proposed elementary reactions that may occur to give the overall reaction. The sum of all the steps in the mechanism gives the balanced chemical reaction.

c.   The rate-determining step is the slowest elementary reaction in any given mechanism.

11.   a.   The greater the frequency of collisions, the greater the opportunities for molecules to react, and, hence, the greater the rate.

b.   Chemical reactions involve the making and breaking of chemical bonds. The kinetic energy of the collisions can be used to break bonds. So, as the kinetic energy of the collisions increases, the rate increases.

c.   For a reaction to occur, it is the reactive portion of each molecule that must be involved in a collision. Only some of all the possible collisions have the correct orientation to convert reactants to products.

12.   In a unimolecular reaction, a single reactant molecule decomposes to products. In a bimolecular reaction, two molecules collide to give products. The probability of the simultaneous collision of three molecules with enough energy and orientation is very small, making termolecular steps very unlikely.

13.   A catalyst increases the rate of a reaction by providing reactants with an alternate pathway (mechanism) to convert to products. This alternate pathway has a lower activation energy, thus increasing the rate of the reaction.

A heterogeneous catalyst is in a different phase than the reactants. The catalyst is usually a solid, although a catalyst in a liquid phase can act as a heterogeneous catalyst for some gas phase reactions. Since the catalyzed reaction has a different mechanism than the uncatalyzed reaction, the catalyzed reaction most likely will have a different rate law.

14.   Some energy must be added to get the reaction started, that is, to overcome the activation energy barrier. Chemically what happens is:

$$\text{Energy} + H_2 \rightarrow 2\,H$$

The hydrogen atoms initiate a chain reaction that proceeds very rapidly. Collisions of H_2 and O_2 molecules at room temperature do not have sufficient kinetic energy to form hydrogen atoms and initiate the reaction.

15.   a.   Activation energy and ΔE are independent of each other. Activation energy depends on the path reactants to take to convert to products. The overall energy change, ΔE, depends only on the initial and final energy states of the reactants and products. ΔE is path independent.

b.   The rate law can be determined only from experiment, not from the overall balanced reaction.

c.   Most reactions occur by a series of steps. The rate of the reaction is determined by the rate of the slowest step in the mechanism.

16. All of these choices would affect the rate of the reaction, but only b and c affect the rate by affecting the value of the rate constant k. The value of the rate constant is dependent on temperature. The value of the rate constant also depends on the activation energy. A catalyst will change the value of k because the activation energy changes. Increasing the concentration (partial pressure) of either H_2 or NO does not affect the value of k, but it does increase the rate of the reaction because both concentrations appear in the rate law.

Exercises

Reaction Rates

17. The coefficients in the balanced reaction relate the rate of disappearance of reactants to the rate of production of products. From the balanced reaction, the rate of production of P_4 will be 1/4 the rate of disappearance of PH_3, and the rate of production of H_2 will be 6/4 the rate of disappearance of PH_3. By convention, all rates are given as positive values.

$$\text{Rate} = -\frac{\Delta[PH_3]}{\Delta t} = \frac{(0.0048 \text{ mol}/2.0 \text{ L})}{s} = 2.4 \times 10^{-3} \text{ mol/L} \cdot s$$

$$\frac{\Delta[P_4]}{\Delta t} = -\frac{1}{4}\frac{\Delta[PH_3]}{\Delta t} = 2.4 \times 10^{-3}/4 = 6.0 \times 10^{-4} \text{ mol/L} \cdot s$$

$$\frac{\Delta[H_2]}{\Delta t} = -\frac{6}{4}\frac{\Delta[PH_3]}{\Delta t} = 6(2.4 \times 10^{-3})/4 = 3.6 \times 10^{-3} \text{ mol/L} \cdot s$$

18. $\dfrac{\Delta[H_2]}{\Delta t} = 3\dfrac{\Delta[N_2]}{\Delta t}$ and $\dfrac{\Delta[NH_3]}{\Delta t} = -2\dfrac{\Delta[N_2]}{\Delta t}$; So, $-\dfrac{1}{3}\dfrac{\Delta[H_2]}{\Delta t} = \dfrac{1}{2}\dfrac{\Delta[NH_3]}{\Delta t}$

or $\dfrac{\Delta[NH_3]}{\Delta t} = -\dfrac{2}{3}\dfrac{\Delta[H_2]}{\Delta t}$

Ammonia is produced at a rate equal to 2/3 of the rate of consumption of hydrogen.

19. a. The units for rate are always mol/L·s.   b. Rate = k; k must have units of mol/L·s.

   c. Rate = k[A], $\dfrac{\text{mol}}{\text{L s}} = k\left(\dfrac{\text{mol}}{\text{L}}\right)$ d. Rate = k[A]^2, $\dfrac{\text{mol}}{\text{L s}} = k\left(\dfrac{\text{mol}}{\text{L}}\right)^2$

   k must have units of s^{-1}.   k must have units of L/mol·s.

   e. $L^2/\text{mol}^2 \cdot s$

20. Rate = k[Cl]$^{1/2}$[CHCl$_3$], $\dfrac{\text{mol}}{\text{L s}} = k\left(\dfrac{\text{mol}}{\text{L}}\right)^{1/2}\left(\dfrac{\text{mol}}{\text{L}}\right)$, k must have units of $L^{1/2}/\text{mol}^{1/2} \cdot s$.

Rate Laws from Experimental Data: Initial Rates Method

21.   a.   In the first two experiments, [NO] is held constant and $[Cl_2]$ is doubled. The rate also doubled.
            Thus, the reaction is first order with respect to Cl_2. Or mathematically: Rate = $k[NO]^x[Cl_2]^y$

$$\frac{0.36}{0.18} = \frac{k(0.10)^x(0.20)^y}{k(0.10)^x(0.10)^y} = \frac{(0.20)^y}{(0.10)^y}, \ 2.0 = 2.0^y, \ y = 1$$

We can get the dependence on NO from the second and third experiments. Here, as the NO
concentration doubles (Cl_2 concentration is constant), the rate increases by a factor of four.
Thus, the reaction is second order with respect to NO. Or mathematically:

$$\frac{1.45}{0.36} = \frac{k(0.20)^x(0.20)}{k(0.10)^x(0.20)} = \frac{(0.20)^x}{(0.10)^x}, \ 4.0 = 2.0^x, \ x = 2; \ \text{So, Rate} = k[NO]^2[Cl_2]$$

Try to examine experiments where only one concentration changes at a time. The more
variables that change, the harder it is to determine the orders. Also, these types of problems can
usually be solved by inspection. In general, we will solve using a mathematical approach, but
keep in mind you probably can solve for the orders by simple inspection of the data.

   b.   The rate constant k can be determined from the experiments. From experiment 1:

$$\frac{0.18 \text{ mol}}{\text{L min}} = k\left(\frac{0.10 \text{ mol}}{\text{L}}\right)^2\left(\frac{0.10 \text{ mol}}{\text{L}}\right), \ k = 180 \text{ L}^2/\text{mol}^2\bullet\text{min}$$

From the other experiments:

$$k = 180 \text{ L}^2/\text{mol}^2\bullet\text{min (2nd exp.)}; \ k = 180 \text{ L}^2/\text{mol}^2\bullet\text{min (3rd exp.)}$$

The average rate constant is $k_{mean} = 1.8 \times 10^2 \text{ L}^2/\text{mol}^2\bullet\text{min}$.

22.   a.   Rate = $k[I^-]^x[S_2O_8^{2-}]^y$; $\ \frac{12.5 \times 10^{-6}}{6.25 \times 10^{-6}} = \frac{k(0.080)^x(0.040)^y}{k(0.040)^x(0.040)^y}, \ 2.00 = 2.0^x, \ x = 1$

$$\frac{12.5 \times 10^{-6}}{6.25 \times 10^{-6}} = \frac{k(0.080)(0.040)^y}{k(0.080)(0.020)^y}, \ 2.00 = 2.0^y, \ y = 1; \ \text{Rate} = k[I^-][S_2O_8^{2-}]$$

   b.   For the first experiment:

$$\frac{12.5 \times 10^{-6} \text{ mol}}{\text{L s}} = k\left(\frac{0.080 \text{ mol}}{\text{L}}\right)\left(\frac{0.040 \text{ mol}}{\text{L}}\right), \ k = 3.9 \times 10^{-3} \text{ L/mol}\bullet\text{s}$$

Each of the other experiments also gives $k = 3.9 \times 10^{-3}$ L/mol•s, so $k_{mean} = 3.9 \times 10^{-3}$ L/mol•s.

23.   a.   Rate = k[NOCl]^n; Using experiments two and three:

$$\frac{2.66 \times 10^4}{6.64 \times 10^3} = \frac{k(2.0 \times 10^{16})^n}{k(1.0 \times 10^{16})^n}, \quad 4.01 = 2.0^n, \quad n = 2; \quad Rate = k[NOCl]^2$$

       b. $$\frac{5.98 \times 10^4 \text{ molecules}}{cm^3 \text{ s}} = k\left(\frac{3.0 \times 10^{16} \text{ molecules}}{cm^3}\right)^2, \quad k = 6.6 \times 10^{-29} \text{ cm}^3/\text{molecules}\bullet\text{s}$$

       The other three experiments give (6.7, 6.6 and 6.6) $\times$ 10^{-29} cm^3/molecules•s, respectively.

       The mean value for k is 6.6 $\times$ 10^{-29} cm^3/molecules•s.

       c. $$\frac{6.6 \times 10^{-29} \text{ cm}^3}{\text{molecules s}} \times \frac{1\text{ L}}{1000\text{ cm}^3} \times \frac{6.022 \times 10^{23} \text{molecules}}{\text{mol}} = \frac{4.0 \times 10^{-8}\text{ L}}{\text{mol s}}$$

24.   Rate = k[N$_2$O$_5$]^x; The rate laws for the first two experiments are:

       $2.26 \times 10^{-3} = k(0.190)^x$ and $8.90 \times 10^{-4} = k(0.0750)^x$

       Dividing: $2.54 = \dfrac{(0.190)^x}{(0.0750)^x} = (2.53)^x, \quad x = 1; \quad Rate = k[N_2O_5]$

       $k = \dfrac{Rate}{[N_2O_5]} = \dfrac{8.90 \times 10^{-4}\text{ mol/L}\bullet\text{s}}{0.0750\text{ mol/L}} = 1.19 \times 10^{-2}\text{ s}^{-1}; \quad k_{mean} = 1.19 \times 10^{-2}\text{ s}^{-1}$

25.   a.   Rate = k[Hb]^x[CO]^y; Comparing the first two experiments, [CO] is unchanged, [Hb] doubles, and the rate doubles. Therefore, the reaction is first order in Hb. Comparing the second and third experiments, [Hb] is unchanged, [CO] triples. and the rate triples. Therefore, y = 1 and the reaction is first order in CO.

       b.   Rate = k[Hb][CO]

       c.   From the first experiment:

             0.619 μmol/L•s = k (2.21 μmol/L)(1.00 μmol/L), k = 0.280 L/μmol•s

       The second and third experiments give similar k values, so k$_{mean}$ = 0.280 L/μmol•s.

       d.   Rate = k[Hb][CO] = $\dfrac{0.280\text{ L}}{\mu\text{mol s}} \times \dfrac{3.36\ \mu\text{mol}}{L} \times \dfrac{2.40\ \mu\text{mol}}{L} = 2.26\ \mu$mol/L•s

26.   a.   Rate = k[ClO$_2$]^x[OH$^-$]^y; From the first two experiments:

             $2.30 \times 10^{-1} = k(0.100)^x(0.100)^y$ and $5.75 \times 10^{-2} = k(0.0500)^x(0.100)^y$

Dividing the two rate laws: $4.00 = \dfrac{(0.100)^x}{(0.0500)^x} = 2.00^x$, $x = 2$

Comparing the second and third experiments:

$2.30 \times 10^{-1} = k(0.100)(0.100)^y$ and $1.15 \times 10^{-1} = k(0.100)(0.0500)^y$

Dividing: $2.00 = \dfrac{(0.100)^y}{(0.0500)^y} = 2.00^y$, $y = 1$

The rate law is: Rate $= k[ClO_2]^2[OH^-]$

2.30×10^{-1} mol/L•s $= k(0.100\ \text{mol/L})^2(0.100\ \text{mol/L})$, $k = 2.30 \times 10^2\ L^2/mol^2$•s $= k_{mean}$

b. Rate $= k[ClO_2]^2[OH^-] = \dfrac{2.30 \times 10^2\ L^2}{mol^2\ s} \times \left(\dfrac{0.175\ mol}{L}\right)^2 \times \dfrac{0.0844\ mol}{L} = 0.594$ mol/L•s

Integrated Rate Laws

27. The first assumption to make is that the reaction is first order. For a first-order reaction, a graph of ln $[H_2O_2]$ vs time will yield a straight line. If this plot is not linear, then the reaction is not first order and we make another assumption. The data and plot for the first-order assumption is below.

Time (s)	$[H_2O_2]$ (mol/L)	ln $[H_2O_2]$
0	1.00	0.000
120.	0.91	-0.094
300.	0.78	-0.25
600.	0.59	-0.53
1200.	0.37	-0.99
1800.	0.22	-1.51
2400.	0.13	-2.04
3000.	0.082	-2.50
3600.	0.050	-3.00

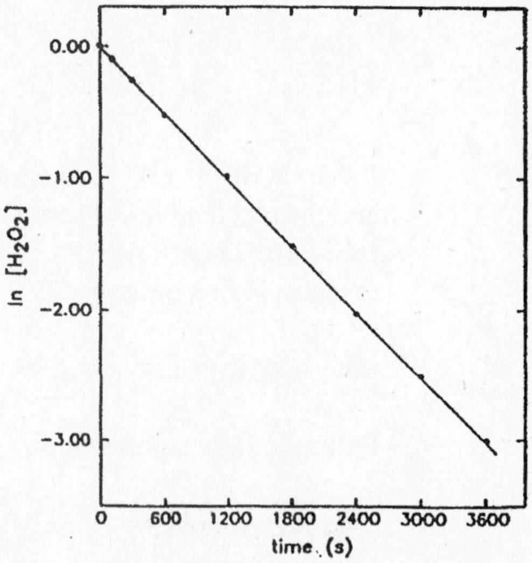

Note: We carried extra significant figures in some of the ln values in order to reduce round-off error. For the plots, we will do this most of the time when the ln function is involved.

The plot of ln $[H_2O_2]$ vs. time is linear. Thus, the reaction is first order. The rate law and integrated rate law are: Rate $= k[H_2O_2]$ and ln $[H_2O_2] = -kt + $ ln $[H_2O_2]_o$.

We determine the rate constant k by determining the slope of the ln $[H_2O_2]$ vs time plot (slope $= -k$).

Using two points on the curve gives:

$$\text{slope} = -k = \frac{\Delta y}{\Delta x} = \frac{0 - (3.00)}{0 - 3600.} = -8.3 \times 10^{-4}\ s^{-1},\ \ k = 8.3 \times 10^{-4}\ s^{-1}$$

To determine $[H_2O_2]$ at 4000. s, use the integrated rate law where at $t = 0$, $[H_2O_2]_0 = 1.00\ M$.

$$\ln [H_2O_2] = -kt + \ln [H_2O_2]_0 \ \ \text{or}\ \ \ln\left(\frac{[H_2O_2]}{[H_2O_2]_0}\right) = -kt$$

$$\ln\left(\frac{[H_2O_2]}{1.00}\right) = -8.3 \times 10^{-4}\ s^{-1} \times 4000.\ s,\ \ \ln [H_2O_2] = -3.3,\ \ [H_2O_2] = e^{-3.3} = 0.037\ M$$

28.   a.   Since the ln[A] vs time plot was linear, the reaction is first order in A. The slope of the ln[A] vs. time plot equals -k. Therefore, the rate law, the integrated rate law and the rate constant value are:

   $$\text{Rate} = k[A];\ \ \ln[A] = -kt + \ln[A]_0;\ \ k = 2.97 \times 10^{-2}\ min^{-1}$$

   b.   The half-life expression for a first-order rate law is:

   $$t_{1/2} = \frac{\ln 2}{k} = \frac{0.6931}{k},\ \ t_{1/2} = \frac{0.6931}{2.97 \times 10^{-2}\ min^{-1}} = 23.3\ min$$

   c. $2.50 \times 10^{-3}\ M$ is 1/8 of the original amount of A present initially, so the reaction is 87.5% complete. When a first-order reaction is 87.5% complete (or 12.5% remains), the reaction has gone through 3 half-lives:

   $$100\% \xrightarrow[t_{1/2}]{} 50.0\% \xrightarrow[t_{1/2}]{} 25\% \xrightarrow[t_{1/2}]{} 12.5\%;\ \ t = 3 \times t_{1/2} = 3 \times 23.3\ min = 69.9\ min$$

   Or we can use the integrated rate law:

   $$\ln\left(\frac{[A]}{[A]_0}\right) = -kt,\ \ \ln\left(\frac{2.50 \times 10^{-3}\ M}{2.00 \times 10^{-2}\ M}\right) = -(2.97 \times 10^{-2}\ min^{-1})\, t,\ \ t = \frac{\ln (0.125)}{-2.97 \times 10^{-2}\ min^{-1}}$$

   $$= 70.0\ min$$

29.   Assume the reaction is first order and see if the plot of ln $[NO_2]$ vs. time is linear. If this isn't linear, try the second-order plot of $1/[NO_2]$ vs. time. The data and plots follow.

Time (s)	$[NO_2]$ (M)	ln $[NO_2]$	$1/[NO_2]$ (M^{-1})
0	0.500	-0.693	2.00
1.20×10^3	0.444	-0.812	2.25
3.00×10^3	0.381	-0.965	2.62
4.50×10^3	0.340	-1.079	2.94
9.00×10^3	0.250	-1.386	4.00
1.80×10^4	0.174	-1.749	5.75

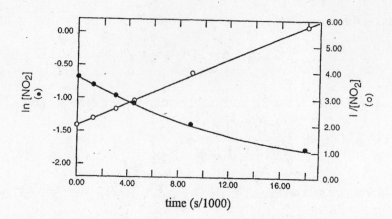

The plot of $1/[NO_2]$ vs. time is linear. The reaction is second order in NO_2. The rate law and integrated rate law are: Rate $= k[NO_2]^2$ and $\dfrac{1}{[NO_2]} = kt + \dfrac{1}{[NO_2]_o}$.

The slope of the plot $1/[NO_2]$ vs. t gives the value of k. Using a couple of points on the plot:

$$\text{slope} = k = \frac{\Delta y}{\Delta x} = \frac{(5.75 - 2.00)\,M^{-1}}{(1.80 \times 10^4 - 0)\,\text{s}} = 2.08 \times 10^{-4}\ \text{L/mol·s}$$

To determine $[NO_2]$ at 2.70×10^4 s, use the integrated rate law where $1/[NO_2]_o = 1/0.500\,M = 2.00\,M^{-1}$.

$$\frac{1}{[NO_2]} = kt + \frac{1}{[NO_2]_o},\quad \frac{1}{[NO_2]} = \frac{2.08 \times 10^{-4}\,\text{L}}{\text{mol s}} \times 2.70 \times 10^4\ \text{s} + 2.00\,M^{-1}$$

$$\frac{1}{[NO_2]} = 7.62,\quad [NO_2] = 0.131\,M$$

30.   a.   Since the $1/[A]$ vs. time plot was linear, the reaction is second order in A. The slope of the $1/[A]$ vs. time plot equals the rate constant k. Therefore, the rate law, the integrated rate law and the rate constant value are:

$$\text{Rate} = k[A]^2;\quad \frac{1}{[A]} = kt + \frac{1}{[A]_o};\quad k = 3.60 \times 10^{-2}\ \text{L/mol·s}$$

b.   The half-life expression for a second-order reaction is: $t_{1/2} = \dfrac{1}{k[A]_o}$

For this reaction: $t_{1/2} = \dfrac{1}{3.60 \times 10^{-2}\ \text{L/mol·s} \times 2.80 \times 10^{-3}\ \text{mol/L}} = 9.92 \times 10^3$ s

Note: We could have used the integrated rate law to solve for $t_{1/2}$ where $[A] = (2.80 \times 10^{-3}\,/2)$ mol/L.

c. Since the half-life for a second-order reaction depends on concentration, we must use the integrated rate law to solve.

$$\frac{1}{[A]} = kt + \frac{1}{[A]_o}, \quad \frac{1}{7.00 \times 10^{-4} \, M} = \frac{3.60 \times 10^{-2} \, L}{mol \bullet s} \times t + \frac{1}{2.80 \times 10^{-3} \, M}$$

$$1.43 \times 10^3 - 357 = 3.60 \times 10^{-2} \, t, \quad t = 2.98 \times 10^4 \, s$$

31. a. Since the $[C_2H_5OH]$ vs. time plot was linear, the reaction is zero order in C_2H_5OH. The slope of the $[C_2H_5OH]$ vs. time plot equals -k. Therefore, the rate law, the integrated rate law and the rate constant value are: Rate $= k[C_2H_5OH]^0 = k$; $[C_2H_5OH] = -kt + [C_2H_5OH]_o$; $k = 4.00 \times 10^{-5}$ mol/L$\bullet$s.

b. The half-life expression for a zero-order reaction is: $t_{1/2} = [A]_o/2k$.

$$t_{1/2} = \frac{[C_2H_5OH]_o}{2k} = \frac{1.25 \times 10^{-2} \, mol/L}{2 \times 4.00 \times 10^{-5} \, mol/L \bullet s} = 156 \, s$$

Note: we could have used the integrated rate law to solve for $t_{1/2}$ where $[C_2H_5OH] = (1.25 \times 10^{-2}/2)$ mol/L.

c. $[C_2H_5OH] = -kt + [C_2H_5OH]_o$, 0 mol/L $= -(4.00 \times 10^{-5} \, mol/L \bullet s) \, t + 1.25 \times 10^{-2}$ mol/L

$$t = \frac{1.25 \times 10^{-2} \, mol/L}{4.00 \times 10^{-5} \, mol/L \bullet s} = 313 \, s$$

32. From the data, the pressure of C_2H_5OH decreases at a constant rate of 13 torr for every 100. s. Since the rate of disappearance of C_2H_5OH is not dependent on concentration, the reaction is zero order in C_2H_5OH.

$$k = \frac{13 \, torr}{100. \, s} \times \frac{1 \, atm}{760 \, torr} = 1.7 \times 10^{-4} \, atm/s$$

The rate law and integrated rate law are:

$$Rate = k = 1.7 \times 10^{-4} \, atm/s; \quad P_{C_2H_5OH} = -kt + 250. \, torr\left(\frac{1 \, atm}{760 \, torr}\right) = -kt + 0.329 \, atm$$

At 900. s: $P_{C_2H_5OH} = -1.7 \times 10^{-4} \, atm/s \times 900. \, s + 0.329 \, atm = 0.176 \, atm = 0.18 \, atm = 130 \, torr$

33.   The first assumption to make is that the reaction is first order. For a first-order reaction, a graph of ln $[C_4H_6]$ vs. t should yield a straight line. If this isn't linear, then try the second-order plot of $1/[C_4H_6]$ vs. t. The data and the plots follow.

Time	195	604	1246	2180	6210 s
$[C_4H_6]$	1.6×10^{-2}	1.5×10^{-2}	1.3×10^{-2}	1.1×10^{-2}	0.68×10^{-2} M
ln $[C_4H_6]$	-4.14	-4.20	-4.34	-4.51	-4.99
$1/[C_4H_6]$	62.5	66.7	76.9	90.9	147 M^{-1}

Note: To reduce round-off error, we carried extra sig. figs. in the data points.

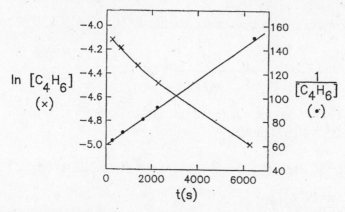

The natural log plot is not linear, so the reaction is not first order. Since the second-order plot of $1/[C_4H_6]$ vs. t is linear, we can conclude that the reaction is second order in butadiene. The rate law is:

Rate = $k[C_4H_6]^2$

For a second order reaction, the integrated rate law is: $\dfrac{1}{[C_4H_6]} = kt + \dfrac{1}{[C_4H_6]_o}$

The slope of the straight line equals the value of the rate constant. Using the points on the line at 1000. and 6000. s:

k = slope = $\dfrac{144 \text{ L/mol} - 73 \text{ L/mol}}{6000.\text{ s} - 1000.\text{ s}}$ = 1.4×10^{-2} L/mol•s

34.   a. First, assume the reaction to be first order with respect to O. A graph of ln [O] vs. t should be linear if the reaction is first order.

t(s)	[O] (atoms/cm^3)	ln[O]
0	5.0×10^9	22.33
$10. \times 10^{-3}$	1.9×10^9	21.37
$20. \times 10^{-3}$	6.8×10^8	20.34
$30. \times 10^{-3}$	2.5×10^8	19.34

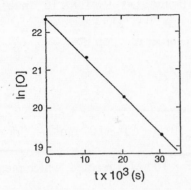

Since the graph is linear, we can conclude that the reaction is first order with respect to O.

b.   The overall rate law is: Rate = k[NO$_2$][O]

Since NO$_2$ was in excess, its concentration is constant. So for this experiment, the rate law is:
Rate = k'[O] where k' = k[NO$_2$] . In a typical first-order plot, the slope equals -k. For this
experiment, the slope equals -k' = -k[NO$_2$]. From the graph:

$$\text{slope} = \frac{19.34 - 22.33}{(30. \times 10^{-3} - 0)\,\text{s}} = -1.0 \times 10^2\,\text{s}^{-1}, \ \ k' = -\text{slope} = 1.0 \times 10^2\,\text{s}^{-1}$$

To determine k, the actual rate constant:

$$k' = k[NO_2], \ \ 1.0 \times 10^2\,\text{s}^{-1} = k(1.0 \times 10^{13}\,\text{molecules/cm}^3), \ \ k = 1.0 \times 10^{-11}\,\text{cm}^3/\text{molecules}\bullet\text{s}$$

35.   Since the 1/[A] vs. time plot is linear with a positive slope, the reaction is second order with respect
to A. The y-intercept in the plot will equal 1/[A]$_o$. Extending the plot, the y-intercept will be about
10, so 1/10 = 0.1 M = [A]$_o$.

36.   The slope of the 1/[A] vs time plot in Exercise 12.35 with equal k.

$$\text{slope} = k = \frac{(60 - 20)\,\text{L/mol}}{(5 - 1)\,\text{s}} = 10\,\text{L/mol}\bullet\text{s}$$

a. $\dfrac{1}{[A]} = kt + \dfrac{1}{[A]_o} = \dfrac{10\,\text{L}}{\text{mol}\,\text{s}} \times 9\,\text{s} + \dfrac{1}{0.1\,\text{M}} = 100, \ \ [A] = 0.01\,M$

b.   For a second-order reaction, the half-life does depend on concentration: $t_{1/2} = \dfrac{1}{k[A]_0}$.

First half-life: $t_{1/2} = \dfrac{1}{\dfrac{10\,\text{L}}{\text{mol}\,\text{s}} \times \dfrac{0.1\,\text{mol}}{\text{L}}} = 1\,\text{s}$

Second half-life ([A]$_o$ is now 0.05 M): $t_{1/2} = 1/(10 \times 0.05) = 2\,\text{s}$

Third half-life ([A]$_o$ is now 0.025 M): $t_{1/2} = 1/(10 \times 0.025) = 4\,\text{s}$

37.   a. $[A] = - kt + [A]_o$, $[A] = -(5.0 \times 10^{-2} \text{ mol/L} \cdot \text{s}) \, t + 1.0 \times 10^{-3} \text{ mol/L}$

       b.   The half-life expression for a zero-order reaction is: $t_{1/2} = \dfrac{[A]_o}{2 \, k}$

            $t_{1/2} = \dfrac{1.0 \times 10^{-3} \text{ mol/L}}{2 \times 5.0 \times 10^{-2} \text{ mol/L} \cdot \text{s}} = 1.0 \times 10^{-2} \text{ s}$

       c. $[A] = -5.0 \times 10^{-2} \text{ mol/L} \cdot \text{s} \times 5.0 \times 10^{-3} \text{ s} + 1.0 \times 10^{-3} \text{ mol/L} = 7.5 \times 10^{-4} \text{ mol/L}$

            Since 7.5×10^{-4} M A remains, 2.5×10^{-4} M A reacted, which means that 2.5×10^{-4} M B has been produced.

38. $\ln\left(\dfrac{[A]}{[A]_o}\right) = -kt; \quad k = \dfrac{\ln 2}{t_{1/2}} = \dfrac{0.6931}{14.3 \text{ d}} = 4.85 \times 10^{-2} \text{ d}^{-1}$

       If $[A]_o = 100.0$, then after 95.0% completion, $[A] = 5.0$.

       $\ln\left(\dfrac{5.0}{100.0}\right) = -4.85 \times 10^{-2} \text{ d}^{-1} \times t, \ \ t = 62 \text{ days}$

39.   a.   If the reaction is 38.5% complete, then 38.5% of the original concentration is consumed, leaving 61.5%.

            $[A] = 61.5\% \text{ of } [A]_o \text{ or } [A] = 0.615 \, [A]_o; \ \ \ln\left(\dfrac{[A]}{[A]_o}\right) = -kt, \ \ \ln\left(\dfrac{0.615 \, [A]_o}{[A]_o}\right) = -k(480. \text{ s})$

            $\ln(0.615) = -k(480. \text{ s}), \ -0.486 = -k(480. \text{ s}), \ k = 1.01 \times 10^{-3} \text{ s}^{-1}$

       b. $t_{1/2} = (\ln 2)/k = 0.6931/1.01 \times 10^{-3} \text{ s}^{-1} = 686 \text{ s}$

       c.   25% complete: $[A] = 0.75 \, [A]_o; \ \ln(0.75) = -1.01 \times 10^{-3} \, (t), \ \ t = 280 \text{ s}$

            75% complete: $[A] = 0.25 \, [A]_o; \ \ln(0.25) = -1.01 \times 10^{-3} \, (t), \ \ t = 1.4 \times 10^3 \text{ s}$

            Or, we know it takes $2 \times t_{1/2}$ for reaction to be 75% complete. $t = 2 \times 686 \text{ s} = 1370 \text{ s}$

            95% complete: $[A] = 0.05 \, [A]_o; \ \ln(0.05) = -1.01 \times 10^{-3} \, (t), \ \ t = 3 \times 10^3 \text{ s}$

40.   For a first-order reaction, the integrated rate law is: $\ln([A]/[A]_o) = -kt$. Solving for k:

       $\ln\left(\dfrac{0.250 \text{ mol/L}}{1.00 \text{ mol/L}}\right) = -k \times 120. \text{ s}, \ \ k = 0.0116 \text{ s}^{-1}$

       $\ln\left(\dfrac{0.350 \text{ mol/L}}{2.00 \text{ mol/L}}\right) = - 0.0116 \text{ s}^{-1} \times t, \ \ t = 150. \text{ s}$

41.   For a second-order reaction: $t_{1/2} = \dfrac{1}{k[A]_o}$ or $k = \dfrac{1}{t_{1/2}[A]_o}$

$k = \dfrac{1}{143\text{ s}(0.060\text{ mol/L})} = 0.12\text{ L/mol}\bullet\text{s}$

42.   a.   The integrated rate law for a second-order reaction is: $1/[A] = kt + 1/[A]_o$, and the half-life
       expression is: $t_{1/2} = 1/k[A]_o$. We could use either to solve for $t_{1/2}$. Using the integrated rate law:

$\dfrac{1}{(0.900/2)\text{ mol/L}} = k \times 2.00\text{ s} + \dfrac{1}{0.900\text{ mol/L}}$, $k = \dfrac{1.11\text{ L/mol}}{2.00\text{ s}} = 0.555\text{ L/mol}\bullet\text{s}$

       b. $\dfrac{1}{0.100\text{ mol/L}} = 0.555\text{ L/mol}\bullet\text{s} \times t + \dfrac{1}{0.900\text{ mol/L}}$, $t = \dfrac{8.9\text{ L/mol}}{0.555\text{ L/mol}\bullet\text{s}} = 16\text{ s}$

43.   Successive half-lives increase in time for a second-order reaction. Therefore, assume the reaction
       is second order in A.

$t_{1/2} = \dfrac{1}{k[A]_o}$, $k = \dfrac{1}{t_{1/2}[A]_o} = \dfrac{1}{10.0\text{ min }(0.10\ M)} = 1.0\text{ L/mol}\bullet\text{min}$

       a. $\dfrac{1}{[A]} = kt + \dfrac{1}{[A]_o} = \dfrac{1.0\text{ L}}{\text{mol min}} \times 80.0\text{ min} + \dfrac{1}{0.10\ M} = 90.\ M^{-1}$, $[A] = 1.1 \times 10^{-2}\ M$

       b.   30.0 min = 2 half-lives, so 25% of original A is remaining.

       $[A] = 0.25(0.10\ M) = 0.025\ M$

44.   Since $[B]_o >> [A]_o$, the B concentration is essentially constant during this experiment, so rate $= k'[A]$
       where $k' = k[B]^2$. For this experiment, the reaction is a pseudo-first-order reaction in A.

       a. $\ln\left(\dfrac{[A]}{[A]_o}\right) = -k't$, $\ln\left(\dfrac{3.8 \times 10^{-3}\ M}{1.0 \times 10^{-2}\ M}\right) = -k' \times 8.0\text{ s}$, $k' = 0.12\text{ s}^{-1}$

       For the reaction: $k' = k[B]^2$, $k = 0.12\text{ s}^{-1}/(3.0\text{ mol/L})^2 = 1.3 \times 10^{-2}\text{ L}^2/\text{mol}^2\bullet\text{s}$

       b. $t_{1/2} = \dfrac{\ln 2}{k'} = \dfrac{0.693}{0.12\text{ s}^{-1}} = 5.8\text{ s}$

       c. $\ln\left(\dfrac{[A]}{1.0 \times 10^{-2}\ M}\right) = -0.12\text{ s}^{-1} \times 13.0\text{ s}$, $\dfrac{[A]}{1.0 \times 10^{-2}} = e^{-0.12(13.0)} = 0.21$, $[A] = 2.1 \times 10^{-3}\ M$

       d. $[A]_{\text{reacted}} = 0.010\ M - 0.0021\ M = 0.008\ M$; $[C]_{\text{reacted}} = 0.008\ M \times \dfrac{2\text{ mol C}}{1\text{ mol A}} = 0.016\ M \approx 0.02\ M$

       $[C]_{\text{remaining}} = 2.0\ M - 0.02\ M = 2.0\ M$; As expected, the concentration of C basically remains
       constant during this experiment since $[C]_o >> [A]_o$.

Reaction Mechanisms

45.   For elementary reactions, the rate law can be written using the coefficients in the balanced equation to determine orders.

    a. Rate = $k[CH_3NC]$ b. Rate = $k[O_3][NO]$

    c. Rate = $k[O_3]$ d. Rate = $k[O_3][O]$

46.   The observed rate law for this reaction is: Rate = $k[NO]^2[H_2]$. For a mechanism to be plausible, the sum of all the steps must give the overall balanced equation (true for all the proposed mechanisms in this problem), and the rate law derived from the mechanism must agree with the observed mechanism. In each mechanism (I - III), the first elementary step is the rate-determining step (the slow step), so the derived rate law for each mechanism will be the rate of the first step. The derived rate laws follow:

Mechanism I: Rate = $k[H_2]^2[NO]^2$

Mechanism II: Rate = $k[H_2][NO]$

Mechanism III: Rate = $k[H_2][NO]^2$

Only in Mechanism III does the derived rate law agree with the observed rate law. Thus, only Mechanism III is a plausible mechanism for this reaction.

47.   A mechanism consists of a series of elementary reactions where the rate law for each step can be determined using the coefficients in the balanced equations. For a plausible mechanism, the rate law derived from a mechanism must agree with the rate law determined from experiment. To derive the rate law from the mechanism, the rate of the reaction is assumed to equal the rate of the slowest step in the mechanism.

Since step 1 is the rate-determining step, the rate law for this mechanism is: Rate = $k[C_4H_9Br]$. To get the overall reaction, we sum all the individual steps of the mechanism.

Summing all steps gives:

$$C_4H_9Br \rightarrow C_4H_9^+ + Br^-$$
$$C_4H_9^+ + H_2O \rightarrow C_4H_9OH_2^+$$
$$C_4H_9OH_2^+ + H_2O \rightarrow C_4H_9OH + H_3O^+$$

$$\overline{C_4H_9Br + 2\ H_2O \rightarrow C_4H_9OH + Br^- + H_3O^+}$$

Intermediates in a mechanism are species that are neither reactants nor products, but that are formed and consumed during the reaction sequence. The intermediates for this mechanism are $C_4H_9^+$ and $C_4H_9OH_2^+$.

48.   Since the rate of the slowest elementary step equals the rate of a reaction, then:

Rate = rate of step 1 = $k[NO_2]^2$

The sum of all steps in a plausible mechanism must give the overall balanced reaction. Summing all steps gives:

$$NO_2 + NO_2 \rightarrow NO_3 + NO$$
$$NO_3 + CO \rightarrow NO_2 + CO_2$$

$$\overline{NO_2 + CO \rightarrow NO + CO_2}$$

Temperature Dependence of Rate Constants and the Collision Model

49.   In the following plot, R = reactants, P = products, E_a = activation energy and RC = reaction coordinate which is the same as reaction progress. Note for this reaction that ΔE is positive since the products are at a higher energy than the reactants.

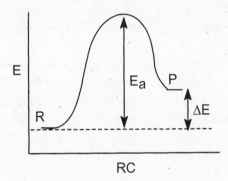

50.   When ΔE is positive, the products are at a higher energy relative to reactants and, when ΔE is negative, the products are at a lower energy relative to reactants.

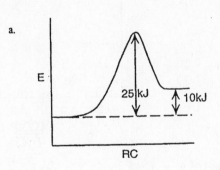

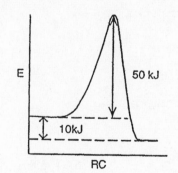

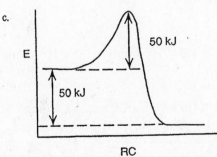

51.

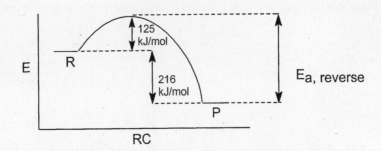

The activation energy for the reverse reaction is:

$$E_{a, \text{ reverse}} = 216 \text{ kJ/mol} + 125 \text{ kJ/mol} = 341 \text{ kJ/mol}$$

52.   When ΔE is negative, then $E_{a, \text{ reverse}} > E_{a, \text{ forward}}$ (see energy profile in Exercise 12.51). When ΔE is positive (the products have higher energy than the reactants as represented in the energy profile for Exercise 12.49), then $E_{a, \text{ forward}} > E_{a, \text{ reverse}}$. Therefore, this reaction has a positive ΔE value.

53.   The Arrhenius equation is: $k = A \exp(-E_a/RT)$ or in logarithmic form, $\ln k = -E_a/RT + \ln A$. Hence, a graph of $\ln k$ vs. $1/T$ should yield a straight line with a slope equal to $-E_a/R$ since the logarithmic form of the Arrhenius equation is in the form of a straight line equation, $y = mx + b$. Note: We carried extra significant figures in the following $\ln k$ values in order to reduce round off error.

T (K)	1/T (K^{-1})	k (s^{-1})	ln k
338	2.96×10^{-3}	4.9×10^{-3}	-5.32
318	3.14×10^{-3}	5.0×10^{-4}	-7.60
298	3.36×10^{-3}	3.5×10^{-5}	-10.26

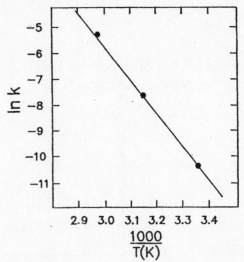

$$\text{Slope} = \frac{-10.76 - (-5.85)}{(3.40 \times 10^{-3} - 3.00 \times 10^{-3}) \text{ K}^{-1}} = -1.2 \times 10^4 \text{ K} = -E_a/R$$

$$E_a = -\text{slope} \times R = 1.2 \times 10^4 \text{ K} \times \frac{8.3145 \text{ J}}{\text{K mol}}, \quad E_a = 1.0 \times 10^5 \text{ J/mol} = 1.0 \times 10^2 \text{ kJ/mol}$$

54.   From the Arrhenius equation in logarithmic form ($\ln k = -E_a/RT + \ln A$), a graph of $\ln k$ vs. $1/T$ should yield a straight line with a slope equal to $-E_a/R$ and a y-intercept equal to $\ln A$.

a.   slope = $-E_a/R$, $E_a = 1.10 \times 10^4 \text{ K} \times \dfrac{8.3145 \text{ J}}{\text{K mol}} = 9.15 \times 10^4 \text{ J/mol} = 91.5 \text{ kJ/mol}$

b. The units for A are the same as the units for $k(s^{-1})$.

y-intercept = ln A, $A = e^{33.5} = 3.54 \times 10^{14}\ s^{-1}$

c. $\ln k = -E_a/RT + \ln A$ or $k = A\ exp(-E_a/RT)$

$$k = 3.54 \times 10^{14}\ s^{-1} \times exp\left(\frac{-9.15 \times 10^4\ J/mol}{8.3145\ J/K \bullet mol \times 298\ K}\right) = 3.24 \times 10^{-2}\ s^{-1}$$

55. $k = A\ exp(-E_a/RT)$ or $\ln k = \dfrac{-E_a}{RT} + \ln A$ (the Arrhenius equation)

For two conditions: $\ln\left(\dfrac{k_2}{k_1}\right) = \dfrac{E_a}{R}\left(\dfrac{1}{T_1} - \dfrac{1}{T_2}\right)$ (Assuming A is temperature independent.)

Let $k_1 = 2.0 \times 10^3\ s^{-1}$, $T_1 = 298\ K$; $k_2 = ?$, $T_2 = 348\ K$; $E_a = 15.0 \times 10^3\ J/mol$

$$\ln\left(\frac{k_2}{2.0 \times 10^3\ s^{-1}}\right) = \frac{15.0 \times 10^3\ J/mol}{8.3145\ J/mol \bullet K}\left(\frac{1}{298\ K} - \frac{1}{348\ K}\right) = 0.87$$

$$\ln\left(\frac{k_2}{2.0 \times 10^3}\right) = 0.87,\ \frac{k_2}{2.0 \times 10^3} = e^{0.87} = 2.4,\ k_2 = 2.4(2.0 \times 10^3) = 4.8 \times 10^3\ s^{-1}$$

56.   For two conditions: $\ln\left(\dfrac{k_2}{k_1}\right) = \dfrac{E_a}{R}\left(\dfrac{1}{T_1} - \dfrac{1}{T_2}\right)$ (Assuming A factor is T independent.)

$$\ln\left(\frac{8.1 \times 10^{-2}\ s^{-1}}{4.6 \times 10^{-2}\ s^{-1}}\right) = \frac{E_a}{8.3145\ J/mol \bullet K}\left(\frac{1}{273\ K} - \frac{1}{293\ K}\right)$$

$$0.57 = \frac{E_a}{8.3145}(2.5 \times 10^{-4}),\ E_a = 1.9 \times 10^4\ J/mol = 19\ kJ/mol$$

57. $\ln\left(\dfrac{k_2}{k_1}\right) = \dfrac{E_a}{R}\left(\dfrac{1}{T_1} - \dfrac{1}{T_2}\right)$; $\dfrac{k_2}{k_1} = 7.00$, $T_1 = 295\ K$, $E_a = 54.0 \times 10^3\ J/mol$

$$\ln(7.00) = \frac{54.0 \times 10^3\ J/mol}{8.3145\ J/mol \bullet K}\left(\frac{1}{295\ K} - \frac{1}{T_2}\right),\ \frac{1}{295} - \frac{1}{T_2} = 3.00 \times 10^{-4}$$

$$\frac{1}{T_2} = 3.09 \times 10^{-3},\ T_2 = 324\ K = 51\,^{\circ}C$$

58. $\ln\left(\dfrac{k_2}{k_1}\right) = \dfrac{E_a}{R}\left(\dfrac{1}{T_1} - \dfrac{1}{T_2}\right)$; Since the rate doubles, then $k_2 = 2\,k_1$.

$\ln(2.00) = \dfrac{E_a}{8.3145\ \text{J/mol·K}}\left(\dfrac{1}{298\ \text{K}} - \dfrac{1}{308\ \text{K}}\right)$, $E_a = 5.3 \times 10^4$ J/mol = 53 kJ/mol

59. $H_3O^+(aq) + OH^-(aq) \rightarrow 2\,H_2O(l)$ should have the faster rate. H_3O^+ and OH^- will be electrostatically attracted to each other; Ce^{4+} and Hg_2^{2+} will repel each other (so E_a is much larger).

60.   Carbon cannot form the fifth bond necessary for the transition state because of the small atomic size of carbon and because carbon doesn't have low energy d orbitals available to expand the octet.

Catalysts

61.   a.   NO is the catalyst. NO is present in the first step of the mechanism on the reactant side, but it is not a reactant since it is regenerated in the second step.

b. NO_2 is an intermediate. Intermediates also never appear in the overall balanced equation. In a mechanism, intermediates always appear first on the product side while catalysts always appear first on the reactant side.

c. $k = A\exp(-E_a/RT)$; $\dfrac{k_{cat}}{k_{un}} = \dfrac{A\exp[-E_a(cat)/RT]}{A\exp[-E_a(un)/RT]} = \exp\left(\dfrac{E_a(un) - E_a(cat)}{RT}\right)$

$\dfrac{k_{cat}}{k_{un}} = \exp\left(\dfrac{2100\ \text{J/mol}}{8.3145\ \text{J/mol·K} \times 298\ \text{K}}\right) = e^{0.85} = 2.3$

The catalyzed reaction is 2.3 times faster than the uncatalyzed reaction at 25°C.

62.   The mechanism for the chlorine catalyzed destruction of ozone is:

$O_3 + Cl \rightarrow O_2 + ClO$ (slow)
$ClO + O \rightarrow O_2 + Cl$ (fast)

_____

$O_3 + O \rightarrow 2\,O_2$

Since the chlorine atom-catalyzed reaction has a lower activation energy, then the Cl-catalyzed rate is faster. Hence, Cl is a more effective catalyst. Using the activation energy, we can estimate the efficiency with which Cl atoms destroy ozone as compared to NO molecules (see Exercise 12.61c).

At 25°C: $\dfrac{k_{Cl}}{k_{NO}} = \exp\left(\dfrac{-E_a(Cl)}{RT} + \dfrac{E_a(NO)}{RT}\right) = \exp\left(\dfrac{(-2100 + 11,900)\ \text{J/mol}}{(8.3145 \times 298)\ \text{J/mol}}\right) = e^{3.96} = 52$

At 25°C, the Cl-catalyzed reaction is roughly 52 times faster than the NO-catalyzed reaction, assuming the frequency factor A is the same for each reaction.

63.   The reaction at the surface of the catalyst is assumed to follow the steps:

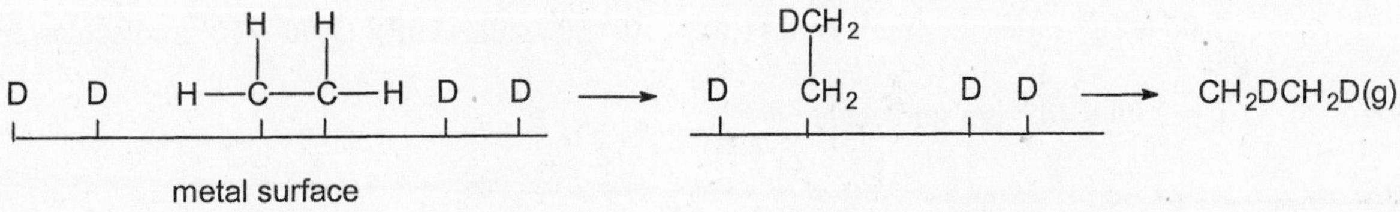

metal surface

Thus, CH_2D–CH_2D should be the product. If the mechanism is possible, then the reaction must be:

$$C_2H_4 + D_2 \rightarrow CH_2DCH_2D$$

If we got this product, then we could conclude that this is a possible mechanism. If we got some other product, e.g., CH_3CHD_2, then we would conclude that the mechanism is wrong. Even though this mechanism correctly predicts the products of the reaction, we cannot say conclusively that this is the correct mechanism; we might be able to conceive of other mechanisms that would give the same products as our proposed one.

64.   a.   W since it has a lower activation energy than the Os catalyst.

       b. $k_w = A_w \exp[-E_a(W)/RT]$; $k_{uncat} = A_{uncat} \exp[-E_a(uncat)/RT]$; Assume $A_w = A_{uncat}$

$$\frac{k_w}{k_{uncat}} = \exp\left(\frac{-E_a(W)}{RT} + \frac{E_a(uncat)}{RT} \right)$$

$$\frac{k_w}{k_{uncat}} = \exp\left(\frac{-163,000 \text{ J/mol} + 335,000 \text{ J/mol}}{8.3145 \text{ J/mol·K} \times 298 \text{ K}} \right) = 1.41 \times 10^{30}$$

       The W-catalyzed reaction is approximately 10^{30} times faster than the uncatalyzed reaction.

       c.   Since $[H_2]$ is in the denominator of the rate law, then H_2 decreases the rate of the reaction. For the decomposition to occur, NH_3 molecules must be adsorbed on the surface of the catalyst. If H_2 is also adsorbed on the catalyst surface, then there are fewer sites for NH_3 molecules to be adsorbed and the rate decreases.

Additional Exercises

65.   Rate $= k[NO]^x[O_2]^y$; comparing the first two experiments, $[O_2]$ is unchanged, $[NO]$ is tripled, and the rate increases by a factor of nine. Therefore, the reaction is second order in NO ($3^2 = 9$). The order of O_2 is more difficult to determine. Comparing the second and third experiments;

$$\frac{3.13 \times 10^{17}}{1.80 \times 10^{17}} = \frac{k(2.50 \times 10^{18})^2(2.50 \times 10^{18})^y}{k(3.00 \times 10^{18})^2(1.00 \times 10^{18})^y}, \ 1.74 = 0.694 \,(2.50)^y, \ 2.51 = 2.50^y, \ y = 1$$

Rate = $k[NO]^2[O_2]$; From experiment 1:

2.00×10^{16} molecules/cm^3•s = k $(1.00 \times 10^{18}$ molecules/cm$^3)^2$ $(1.00 \times 10^{18}$ molecules/cm$^3)$

$k = 2.00 \times 10^{-38}$ cm^6/molecules^2•s = k_{mean}

$$\text{Rate} = \frac{2.00 \times 10^{-38}\ \text{cm}^6}{\text{molecules}^2 \cdot \text{s}} \times \left(\frac{6.21 \times 10^{18}\ \text{molecules}}{\text{cm}^3}\right)^2 \times \frac{7.36 \times 10^{18}\ \text{molecules}}{\text{cm}^3}$$

$$= 5.68 \times 10^{18}\ \text{molecules/cm}^3 \cdot \text{s}$$

66.   The pressure of a gas is directly proportional to concentration. Therefore, we can use the pressure data to solve the problem since Rate = $-\Delta[SO_2Cl_2]/\Delta t \propto -\Delta P_{SO_2Cl_2}/\Delta t$.

Assuming a first order equation, the data and plot follow.

Time (hour)	0.00	1.00	2.00	4.00	8.00	16.00
$P_{SO_2Cl_2}$ (atm)	4.93	4.26	3.52	2.53	1.30	0.34
ln $P_{SO_2Cl_2}$	1.595	1.449	1.258	0.928	0.262	-1.08

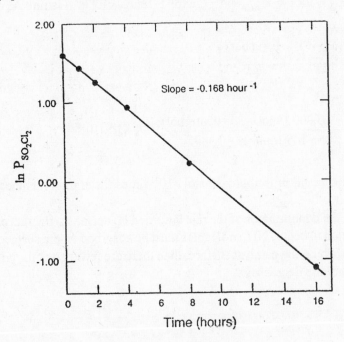

Since the ln $P_{SO_2Cl_2}$ vs. time plot is linear, the reaction is first order in SO_2Cl_2.

a.   Slope of ln(P) vs. t plot is -0.168 hour^{-1} = -k, k = 0.168 hour^{-1} = 4.67×10^{-5} s^{-1}

Since concentration units don't appear in first-order rate constants, this value of k determined from the pressure data will be the same as if concentration data in molarity units were used.

b. $t_{1/2} = \dfrac{\ln 2}{k} = \dfrac{0.6931}{k} = \dfrac{0.6931}{0.168\ h^{-1}} = 4.13$ hour

c. $\ln\left(\dfrac{P_{SO_2Cl_2}}{P_o}\right) = -kt = -0.168\ h^{-1}\ (20.0\ hr) = -3.36,\quad \left(\dfrac{P_{SO_2Cl_2}}{P_o}\right) = e^{-3.36} = 3.47 \times 10^{-2}$

Fraction left = 0.0347 = 3.47%

67.   From 338 K data, a plot of $\ln[N_2O_5]$ vs. t is linear and the slope = -4.86×10^{-3} (plot not included). This tells us the reaction is first order in N_2O_5 with k = 4.86×10^{-3} at 338 K.

From 318 K data, the slope of $\ln[N_2O_5]$ vs t plot is equal to -4.98×10^{-4}, so k = 4.98×10^{-4} at 318 K. We now have two values of k at two temperatures, so we can solve for E_a.

$$\ln\left(\frac{k_2}{k_1}\right) = \frac{E_a}{R}\left(\frac{1}{T_1} - \frac{1}{T_2}\right),\quad \ln\left(\frac{4.86 \times 10^{-3}}{4.98 \times 10^{-4}}\right) = \frac{E_a}{8.3145\ J/K \bullet mol}\left(\frac{1}{318\ K} - \frac{1}{338\ K}\right)$$

$E_a = 1.0 \times 10^5$ J/mol = 1.0×10^2 kJ/mol

68.   The Arrhenius equation is: k = A exp ($-E_a/RT$) or in logarithmic form, ln k = $-E_a/RT$ + ln A. Hence, a graph of ln k vs. 1/T should yield a straight line with a slope equal to $-E_a/R$ since the logarithmic form of the Arrhenius equation is in the form of a straight line equation, y = mx + b. Note: We carried one extra significant figure in the following ln k values in order to reduce round off error.

T (K)	1/T (K^{-1})	k (L/mol$\bullet$s)	ln k
195	5.13×10^{-3}	1.08×10^9	20.80
230.	4.35×10^{-3}	2.95×10^9	21.81
260.	3.85×10^{-3}	5.42×10^9	22.41
298	3.36×10^{-3}	12.0×10^9	23.21
369	2.71×10^{-3}	35.5×10^9	24.29

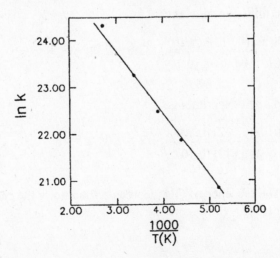

From "eyeballing" the line on the graph:

$$\text{slope} = \frac{20.95 - 23.65}{(5.00 \times 10^{-3} - 3.00 \times 10^{-3})\ K^{-1}} = \frac{-2.70}{2.00 \times 10^{-3}} = -1.35 \times 10^3\ K = \frac{-E_a}{R}$$

$$E_a = 1.35 \times 10^3\ K \times \frac{8.3145\ J}{K\ mol} = 1.12 \times 10^4\ J/mol = 11.2\ kJ/mol$$

From a graphing calculator: slope $= -1.43 \times 10^3$ K and $E_a = 11.9$ kJ/mol

69.   At high [S], the enzyme is completely saturated with substrate. Once the enzyme is completely saturated, the rate of decomposition of ES can no longer increase, and the overall rate remains constant.

70. $k = A \exp(-E_a/RT)$; $\dfrac{k_{cat}}{k_{uncat}} = \dfrac{A_{cat} \exp(-E_{a,cat}/RT)}{A_{uncat} \exp(-E_{a,uncat}/RT)} = \exp\left(\dfrac{-E_{a,cat} + E_{a,uncat}}{RT}\right)$

$$2.50 \times 10^3 = \frac{k_{cat}}{k_{uncat}} = \exp\left(\frac{-E_{a,cat} + 5.00 \times 10^4\ J/mol}{8.3145\ J/K \bullet mol \times 310.\ K}\right)$$

$$\ln(2.50 \times 10^3)\ \times 2.58\ \times 10^3\ J/mol = -E_{a,cat}\ + 5.00 \times 10^4\ J/mol$$

$$E_{a,cat} = 5.00 \times 10^4\ J/mol - 2.02 \times 10^4\ J/mol = 2.98 \times 10^4\ J/mol = 29.8\ kJ/mol$$

Challenge Problems

71.   Rate $= k[I^-]^x[OCl^-]^y[OH^-]^z$; Comparing the first and second experiments:

$$\frac{18.7 \times 10^{-3}}{9.4 \times 10^{-3}} = \frac{k(0.0026)^x (0.012)^y (0.10)^z}{k(0.0013)^x (0.012)^y (0.10)^z},\ 2.0 = 2.0^x,\ x = 1$$

Comparing the first and third experiments:

$$\frac{9.4 \times 10^{-3}}{4.7 \times 10^{-3}} = \frac{k(0.0013) (0.012)^y (0.10)^z}{k(0.0013) (0.0060)^y (0.10)^z},\ 2.0 = 2.0^y,\ y = 1$$

Comparing the first and sixth experiments:

$$\frac{4.7 \times 10^{-3}}{9.4 \times 10^{-3}} = \frac{k(0.0013) (0.012) (0.20)^z}{k(0.0013) (0.012) (0.10)^z},\ 1/2 = 2.0^z,\ z = -1$$

Rate $= \dfrac{k[I^-][OCl^-]}{[OH^-]}$; The presence of OH^- decreases the rate of the reaction.

For the first experiment:

$$\frac{9.4 \times 10^{-3} \text{ mol}}{\text{L s}} = k \frac{(0.0013 \text{ mol/L}) (0.012 \text{ mol/L})}{(0.10 \text{ mol/L})}, \quad k = 60.3 \text{ s}^{-1} = 60. \text{ s}^{-1}$$

For all experiments, $k_{mean} = 60. \text{ s}^{-1}$.

72.   For second order kinetics: $\dfrac{1}{[A]} - \dfrac{1}{[A]_o} = kt$ and $t_{1/2} = \dfrac{1}{k[A]_o}$

a. $\dfrac{1}{[A]} = (0.250 \text{ L/mol} \bullet \text{s})t + \dfrac{1}{[A]_o}, \quad \dfrac{1}{[A]} = 0.250 \times 180. \text{ s} + \dfrac{1}{1.00 \times 10^{-2} M}$

$\dfrac{1}{[A]} = 145 \ M^{-1}, \ [A] = 6.90 \times 10^{-3} \ M$

Amount of A that reacted = 0.0100 - 0.00690 = 0.0031 M

$[A_2] = \dfrac{1}{2}(3.1 \times 10^{-3} \ M) = 1.6 \times 10^{-3} \ M$

b.   After 3.00 minutes (180. s): $[A] = 3.00 \ [B], \ 6.90 \times 10^{-3} \ M = 3.00 \ [B], \ [B] = 2.30 \times 10^{-3} \ M$

$\dfrac{1}{[B]} = k_2 t + \dfrac{1}{[B]_o}, \quad \dfrac{1}{2.30 \times 10^{-3} M} = k_2(180. \text{ s}) + \dfrac{1}{2.50 \times 10^{-2} M}, \quad k_2 = 2.19 \text{ L/mol} \bullet \text{s}$

c. $t_{1/2} = \dfrac{1}{k[A]_o} = \dfrac{1}{0.250 \text{ L/mol} \bullet \text{s} \times 1.00 \times 10^{-2} \text{ mol/L}} = 4.00 \times 10^2 \text{ s}$

73.   a.   We check for first-order dependence by graphing ln [concentration] vs. time for each set of data. The rate dependence on NO is determined from the first set of data since the ozone concentration is relatively large compared to the NO concentration, so $[O_3]$ is effectively constant.

Time (ms)	[NO] (molecules/cm^3)	ln [NO]
0	6.0×10^8	20.21
100.	5.0×10^8	20.03
500.	2.4×10^8	19.30
700.	1.7×10^8	18.95
1000.	9.9×10^7	18.41

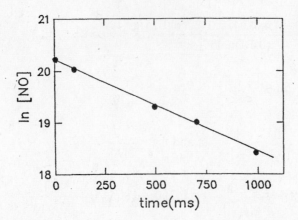

Since ln [NO] vs. t is linear, the reaction is first order with respect to NO.

We follow the same procedure for ozone using the second set of data. The data and plot are:

Time (ms)	$[O_3]$ (molecules/cm^3)	ln $[O_3]$
0	1.0×10^{10}	23.03
50.	8.4×10^9	22.85
100.	7.0×10^9	22.67
200.	4.9×10^9	22.31
300.	3.4×10^9	21.95

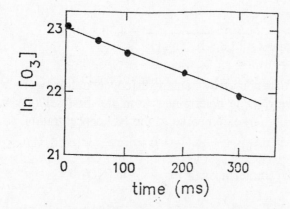

The plot of ln $[O_3]$ vs. t is linear. Hence, the reaction is first order with respect to ozone.

b. Rate = k[NO][O_3] is the overall rate law.

c. For NO experiment, Rate = k′[NO] and k′ = -(slope from graph of ln [NO] vs. t).

$$k' = \text{-slope} = -\frac{18.41 - 20.21}{(1000. - 0) \times 10^{-3} \text{ s}} = 1.8 \text{ s}^{-1}$$

For ozone experiment, Rate = $k''[O_3]$ and k'' = -(slope from ln $[O_3]$ vs. t).

$$k'' = \text{-slope} = -\frac{(21.95 - 23.03)}{(300. - 0) \times 10^{-3}\ s} = 3.6\ s^{-1}$$

d. From NO experiment, Rate = $k[NO][O_3] = k'[NO]$ where $k' = k[O_3]$.

$k' = 1.8\ s^{-1} = k(1.0 \times 10^{14}\ \text{molecules/cm}^3),\ k = 1.8 \times 10^{-14}\ \text{cm}^3/\text{molecules}\bullet s$

We can check this from the ozone data. Rate = $k''[O_3] = k[NO][O_3]$ where $k'' = k[NO]$.

$k'' = 3.6\ s^{-1} = k(2.0 \times 10^{14}\ \text{molecules/cm}^3),\ k = 1.8 \times 10^{-14}\ \text{cm}^3/\text{molecules}\bullet s$

Both values of k agree.

74. On the energy profile to the right, R = reactants, P = products, E_a = activation energy, ΔE = overall energy change for the reaction and I = intermediate.

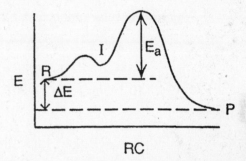

a - d. See plot to the right.

e. This is a two-step reaction since an intermediate plateau appears between the reactant and the products. This plateau represents the energy of the intermediate. The general reaction mechanism for this reaction is:

$$\begin{array}{l} R \rightarrow I \\ \underline{I \rightarrow P} \\ R \rightarrow P \end{array}$$

In a mechanism, the rate of the slowest step determines the rate of the reaction. The activation energy for the slowest step will be the largest energy barrier that the reaction must overcome. Since the second hump in the diagram is at the highest energy, the second step has the largest activation energy and will be the rate-determining step (the slow step).

75. a. If the interval between flashes is 16.3 sec, then the rate is:

1 flash/16.3 s = $6.13 \times 10^{-2}\ s^{-1}$ = k

Interval	k	T
16.3 s	$6.13 \times 10^{-2}\ s^{-1}$	21.0°C (294.2 K)
13.0 s	$7.69 \times 10^{-2}\ s^{-1}$	27.8°C (301.0 K)

$\ln\left(\dfrac{k_2}{k_1}\right) = \dfrac{E_a}{R}\left(\dfrac{1}{T_1} - \dfrac{1}{T_2}\right)$; Solving: $E_a = 2.5 \times 10^4$ J/mol = 25 kJ/mol

b. $\ln\left(\dfrac{k}{6.13 \times 10^{-2}}\right) = \dfrac{2.5 \times 10^4 \text{ J/mol}}{8.3145 \text{ J/K} \bullet \text{mol}}\left(\dfrac{1}{294.2 \text{ K}} - \dfrac{1}{303.2 \text{ K}}\right) = 0.30$

$k = e^{0.30} \times 6.13 \times 10^{-2} = 8.3 \times 10^{-2} \text{ s}^{-1}$; Interval $= 1/k = 12$ seconds

c.

T	Interval	54-2(Intervals)
21.0 °C	16.3 s	21 °C
27.8 °C	13.0 s	28 °C
30.0 °C	12 s	30. °C

This rule of thumb gives excellent agreement to two significant figures.

76.   We need the value of k at 500. K. $\ln\left(\dfrac{k_2}{k_1}\right) = \dfrac{E_a}{R}\left(\dfrac{1}{T_1} - \dfrac{1}{T_2}\right)$

$\ln\left(\dfrac{k_2}{2.3 \times 10^{-12} \text{ L/mol}\bullet\text{s}}\right) = \dfrac{1.11 \times 10^5 \text{ J/mol}}{8.3145 \text{ J/K}\bullet\text{mol}}\left(\dfrac{1}{273 \text{ K}} - \dfrac{1}{500. \text{ K}}\right) = 22.2$

$\dfrac{k_2}{2.3 \times 10^{-12}} = e^{22.2}$, $k_2 = 1.0 \times 10^{-2}$ L/mol$\bullet$s

Since the decomposition reaction is an elementary reaction, then the rate law can be written using the coefficients in the balanced equation. For this reaction: Rate $= k[NO_2]^2$. To solve for the time, we must use the integrated rate law for second-order kinetics. The major problem now is converting units so they match. Rearranging the ideal gas law gives n/V = P/RT. Substituting P/RT for concentration units in the second-order integrated rate law equation:

$$\dfrac{1}{[NO_2]} = kt + \dfrac{1}{[NO_2]_o}, \quad \dfrac{1}{P/RT} = kt + \dfrac{1}{P_o/RT}, \quad \dfrac{RT}{P} - \dfrac{RT}{P_o} = kt, \quad t = \dfrac{RT}{k}\left(\dfrac{P_o - P}{P \times P_o}\right)$$

$$t = \dfrac{(0.08206 \text{ L atm/K}\bullet\text{mol}) (500. \text{ K})}{1.0 \times 10^{-2} \text{ L/mol}\bullet\text{s}} \times \left(\dfrac{2.5 \text{ atm} - 1.5 \text{ atm}}{1.5 \text{ atm} \times 2.5 \text{ atm}}\right) = 1.1 \times 10^3 \text{ s}$$

CHAPTER THIRTEEN

CHEMICAL EQUILIBRIUM

Questions

9.   a.   The rates of the forward and reverse reactions are equal at equilibrium.

          b.   There is no net change in the composition (as long as temperature is constant).

10.   False. Equilibrium and rates of reaction (kinetics) are independent of each other. A reaction with a large equilibrium constant value may be a fast reaction or a slow reaction. The same is true for a reaction with a small equilibrium constant value. Kinetics is discussed in detail in Chapter 12 of the text.

11.   The equilibrium constant is a number that tells us the relative concentrations (pressures) of reactants and products at equilibrium. An equilibrium position is a set of concentrations that satisfy the equilibrium constant expression. More than one equilibrium position can satisfy the same equilibrium constant expression.

       Table 13.1 of the text illustrates this nicely. Each of the three experiments in Table 13.1 have different equilibrium positions; that is, each experiment has different equilibrium concentrations. However, when these equilibrium concentrations are inserted into the equilibrium constant expression, each experiment gives the same value for K. The equilibrium position depends on the initial concentrations one starts with. Since there are an infinite number of initial conditions, there are an infinite number of equilibrium positions. However, each of these infinite equilibrium positions will always give the same value for the equilibrium constant (assuming temperature is constant).

12. $2\ NOCl(g) \rightleftharpoons 2\ NO(g) + Cl_2(g)$ $K = 1.6 \times 10^{-5}$

       The expression for K is the product concentrations divided by the reactant concentrations. When K has a value much less than one, the product concentrations are relatively small and the reactant concentrations are relatively large.

       $2\ NO(g) \rightleftharpoons N_2(g) + O_2(g)$ $K = 1 \times 10^{31}$

       When K has a value much greater than one, the product concentrations are relatively large and the reactant concentrations are relatively small. In both cases, however, the rate of the forward reaction equals the rate of the reverse reaction at equilibrium (this is a definition of equilibrium).

13.   For the gas phase reaction $a A + b B \rightleftharpoons c C + d D$

the equilibrium constant expression is: $K = \dfrac{[C]^c [D]^d}{[A]^a [B]^b}$

and the reaction quotient has the same form: $Q = \dfrac{[C]^c [D]^d}{[A]^a [B]^b}$

The difference is that in the expression for K we use equilibrium concentrations, i.e., [A], [B], [C], and [D] are all in equilibrium with each other. Any set of concentrations can be plugged into the reaction quotient expression. Typically, we plug initial concentrations into the Q expression and then compare the value of Q to K to see how far we are from equilibrium. If Q = K, the reaction is at equilibrium with these concentrations. If Q ≠ K, then the reaction will have to shift either to products or to reactants to reach equilibrium.

14.   A change in volume will change the partial pressure of all reactants and products by the same factor. The shift in equilibrium depends on the number of gaseous particles on each side. An increase in volume will shift the equilibrium to the side with the greater number of particles in the gas phase. A decrease in volume will favor the side with lesser gas phase particles. If there are the same number of gas phase particles on each side of the reaction, a change in volume will not shift the equilibrium.

When we change the pressure by adding an unreactive gas, we do not change the partial pressures (or concentrations) of any of the substances in equilibrium with each other since the volume of the container did not change. If the partial pressures (and concentrations) are unchanged, the reaction is still at equilibrium.

Exercises

Characteristics of Chemical Equilibrium

15.   No, equilibrium is a dynamic process. Both reactions:

$$H_2O + CO \rightarrow H_2 + CO_2 \text{ and } H_2 + CO_2 \rightarrow H_2O + CO$$

are occurring, but at equal rates. Thus, ^{14}C atoms will be distributed between CO and CO_2.

16.   No, it doesn't matter which direction the equilibrium position is reached. Both experiments will give the same equilibrium position since both experiments started with stoichiometric amounts of reactants or products.

17. $H_2O(g) + CO(g) \rightleftharpoons H_2(g) + CO_2(g)$ $K = \dfrac{[H_2][CO_2]}{[H_2O][CO]} = 2.0$

K is a unitless number since there is an equal number of moles of product gases as compared to moles of reactant gases in the balanced equation. Therefore, we can use units of molecules per liter instead of moles per liter to determine K.

By trial and error, if 3 molecules of CO react, then 3 molecules of H_2O must react, and 3 molecules each of H_2 and CO_2 are formed. We would have $6 - 3 = 3$ molecules CO, $8 - 3 = 5$ molecules H_2O, $0 + 3 = 3$ molecules H_2, and $0 + 3 = 3$ molecules CO_2 present. This will be an equilibrium mixture if $K = 2.0$:

$$K = \frac{\left(\dfrac{3 \text{ molecules } H_2}{L}\right)\left(\dfrac{3 \text{ molecules } CO_2}{L}\right)}{\left(\dfrac{5 \text{ molecules } H_2O}{L}\right)\left(\dfrac{3 \text{ molecules } CO}{L}\right)} = \frac{3}{5}$$

Since this mixture does not give a value of $K = 2.0$, this is not an equilibrium mixture. Let's try 4 molecules of CO reacting to reach equilibrium.

molecules CO remaining $= 6 - 4 = 2$ molecules CO;
molecules H_2O remaining $= 8 - 4 = 4$ molecules H_2O;
molecules H_2 present $= 0 + 4 = 4$ molecules H_2;
molecules CO_2 present $= 0 + 4 = 4$ molecules CO_2

$$K = \frac{\left(\dfrac{4 \text{ molecules } H_2}{L}\right)\left(\dfrac{4 \text{ molecules } CO_2}{L}\right)}{\left(\dfrac{4 \text{ molecules } H_2O}{L}\right)\left(\dfrac{2 \text{ molecules } CO}{L}\right)} = 2.0$$

Since $K = 2.0$ for this reaction mixture, we are at equilibrium.

18.   When equilibrium is reached, there is no net change in the amount of reactants and products present since the rates of the forward and reverse reactions are equal to each other. The first diagram has 4 A_2B molecules, 2 A_2 molecules and 1 B_2 molecule present. The second diagram has 2 A_2B molecules, 4 A_2 molecules, and 2 B_2 molecules. Therefore, the first diagram cannot represent equilibrium since there was a net change in reactants and products. Is the second diagram the equilibrium mixture? That depends on whether there is a net change between reactants and products when going from the second diagram to the third diagram. The third diagram contains the same number and type of molecules as the second diagram, so the second diagram is the first illustration that represents equilibrium.

The reaction container initially contained only A_2B. From the first diagram, 2 A_2 molecules and 1 B_2 molecule are present (along with 4 A_2B molecules). From the balanced reaction, these 2 A_2 molecules and 1 B_2 molecule were formed when 2 A_2B molecules decomposed. Therefore, the initial number of A_2B molecules present equals $4 + 2 = 6$ molecules A_2B.

The Equilibrium Constant

19.   a. $K = \dfrac{[NO]^2}{[N_2][O_2]}$ b. $K = \dfrac{[NO_2]^2}{[N_2O_4]}$

c. $K = \dfrac{[SiCl_4][H_2]^2}{[SiH_4][Cl_2]^2}$ d. $K = \dfrac{[PCl_3]^2[Br_2]^3}{[PBr_3]^2[Cl_2]^3}$

20. a. $K_p = \dfrac{P_{NO}^2}{P_{N_2} \times P_{O_2}}$ b. $K_p = \dfrac{P_{NO_2}^2}{P_{N_2O_4}}$

c. $K_p = \dfrac{P_{SiCl_4} \times P_{H_2}^2}{P_{SiH_4} \times P_{Cl_2}^2}$ d. $K_p = \dfrac{P_{PCl_3}^2 \times P_{Br_2}^3}{P_{PBr_3}^2 \times P_{Cl_2}^3}$

21. $K = 1.3 \times 10^{-2} = \dfrac{[NH_3]^2}{[N_2][H_2]^3}$ for $N_2(g) + 3\,H_2(g) \rightleftharpoons 2\,NH_3(g)$

When a reaction is reversed, then $K_{new} = 1/K_{original}$. When a reaction is multiplied through by a value of n, then $K_{new} = (K_{original})^n$.

a. $1/2\,N_2(g) + 3/2\,H_2(g) \rightleftharpoons NH_3(g)$ $K' = \dfrac{[NH_3]}{[N_2]^{1/2}[H_2]^{3/2}} = K^{1/2} = (1.3 \times 10^{-2})^{1/2} = 0.11$

b. $2\,NH_3(g) \rightleftharpoons N_2(g) + 3\,H_2(g)$ $K'' = \dfrac{[N_2][H_2]^3}{[NH_3]^2} = \dfrac{1}{K} = \dfrac{1}{1.3 \times 10^{-2}} = 77$

c. $NH_3(g) \rightleftharpoons 1/2\,N_2(g) + 3/2\,H_2(g)$ $K''' = \dfrac{[N_2]^{1/2}[H_2]^{3/2}}{[NH_3]} = \left(\dfrac{1}{K}\right)^{1/2} = \left(\dfrac{1}{1.3 \times 10^{-2}}\right)^{1/2} = 8.8$

d. $2\,N_2(g) + 6\,H_2(g) \rightleftharpoons 4\,NH_3(g)$ $K = \dfrac{[NH_3]^4}{[N_2]^2[H_2]^6} = (K)^2 = (1.3 \times 10^{-2})^2 = 1.7 \times 10^{-4}$

22. $H_2(g) + Br_2(g) \rightleftharpoons 2\,HBr(g)$ $K_p = \dfrac{P_{HBr}^2}{(P_{H_2})(P_{Br_2})} = 3.5 \times 10^4$

a. $HBr \rightleftharpoons 1/2\,H_2 + 1/2\,Br_2$ $K_p' = \dfrac{(P_{H_2})^{1/2}(P_{Br_2})^{1/2}}{P_{HBr}} = \left(\dfrac{1}{K_p}\right)^{1/2} = \left(\dfrac{1}{3.5 \times 10^4}\right)^{1/2} = 5.3 \times 10^{-3}$

b. $2\,HBr \rightleftharpoons H_2 + Br_2$ $K_p'' = \dfrac{(P_{H_2})(P_{Br_2})}{P_{HBr}^2} = \dfrac{1}{K_p} = \dfrac{1}{3.5 \times 10^4} = 2.9 \times 10^{-5}$

c. $1/2\,H_2 + 1/2\,Br_2 \rightleftharpoons HBr$ $K_p''' = \dfrac{P_{HBr}}{(P_{H_2})^{1/2}(P_{Br_2})^{1/2}} = (K_p)^{1/2} = 190$

23. $K = \dfrac{[NO]^2}{[N_2][O_2]} = \dfrac{(4.7 \times 10^{-4})^2}{(0.041)(0.0078)} = 6.9 \times 10^{-4}$

24. $K = \dfrac{[NCl_3]^2}{[N_2][Cl_2]^3} = \dfrac{(0.19)^2}{(1.4 \times 10^{-3})(4.3 \times 10^{-4})^3} = 3.2 \times 10^{11}$

25. $[NO] = \dfrac{4.5 \times 10^{-3}\ mol}{3.0\ L} = 1.5 \times 10^{-3}\ M;\quad [Cl_2] = \dfrac{2.4\ mol}{3.0\ L} = 0.80\ M$

       $[NOCl] = \dfrac{1.0\ mol}{3.0\ L} = 0.33\ M;\quad K = \dfrac{[NO]^2[Cl_2]}{[NOCl]^2} = \dfrac{(1.5 \times 10^{-3})^2(0.80)}{(0.33)^2} = 1.7 \times 10^{-5}$

26. $[N_2O] = \dfrac{2.00 \times 10^{-2}\ mol}{2.00\ L};\quad [N_2] = \dfrac{2.80 \times 10^{-4}\ mol}{2.00\ L};\quad [O_2] = \dfrac{2.50 \times 10^{-5}\ mol}{2.00\ L}$

       $K = \dfrac{[N_2O]^2}{[N_2]^2[O_2]} = \dfrac{\left(\dfrac{2.00 \times 10^{-2}}{2.00}\right)^2}{\left(\dfrac{2.80 \times 10^{-4}}{2.00}\right)^2\left(\dfrac{2.50 \times 10^{-5}}{2.00}\right)} = \dfrac{(1.00 \times 10^{-2})^2}{(1.40 \times 10^{-4})^2(1.25 \times 10^{-5})} = 4.08 \times 10^8$

27. $K_p = \dfrac{P_{NO}^2 \times P_{O_2}}{P_{NO_2}^2} = \dfrac{(6.5 \times 10^{-5})^2(4.5 \times 10^{-5})}{(0.55)^2} = 6.3 \times 10^{-13}$

28. $K_p = \dfrac{P_{NH_3}^2}{P_{N_2} \times P_{H_2}^3} = \dfrac{(3.1 \times 10^{-2})^2}{(0.85)(3.1 \times 10^{-3})^3} = 3.8 \times 10^4$

29. $K_p = K(RT)^{\Delta n}$ where Δn = sum of gaseous product coefficients - sum of gaseous reactant coefficients. For this reaction, $\Delta n = 1 - 2 = -1$.

       $K_p = \dfrac{3.7 \times 10^9}{(0.08206 \times 298)} = 1.5 \times 10^8$

30. $K_p = K(RT)^{\Delta n},\ K = \dfrac{K_p}{(RT)^{\Delta n}};\ \Delta n = 2 - 3 = -1;\ K = 0.25 \times (0.08206 \times 1100) = 23$

31.   Solids and liquids do not appear in the equilibrium expression. Only gases and dissolved solutes appear in the equilibrium expression.

       a. $K = \dfrac{1}{[O_2]^5};\ K_p = \dfrac{1}{P_{O_2}^5}$
       b. $K = [N_2O][H_2O]^2;\ K_p = P_{N_2O} \times P_{H_2O}^2$

c. $K = \dfrac{1}{[CO_2]}$; $K_p = \dfrac{1}{P_{CO_2}}$ d. $K = \dfrac{[SO_2]^8}{[O_2]^8}$; $K_p = \dfrac{P_{SO_2}^8}{P_{O_2}^8}$

32. $K_p = K(RT)^{\Delta n}$ where Δn equals the difference in the sum of the coefficients between gaseous products and gaseous reactants (Δn = mol gaseous products - mol gaseous reactants).   When Δn = 0, then $K_p = K$. In Exercise 13.31, only reaction d has $\Delta n = 0$ so only reaction d has $K_p = K$.

33.   Since solids do not appear in the equilibrium constant expression, $K = 1/[O_2]^3$.

$$[O_2] = \dfrac{1.0 \times 10^{-3}\ mol}{2.0\ L};\quad K = \dfrac{1}{[O_2]^3} = \dfrac{1}{\left(\dfrac{1.0 \times 10^{-3}}{2.0}\right)^3} = \dfrac{1}{(5.0 \times 10^{-4})^3} = 8.0 \times 10^9$$

34. $K_p = \dfrac{P_{H_2}^4}{P_{H_2O}^4}$; $P_{tot} = P_{H_2O} + P_{H_2}$,   36.3 torr = 15.0 torr + P_{H_2}, P_{H_2} = 21.3 torr

Since 1 atm = 760 torr: $K_p = \dfrac{(21.3/760)^4}{(15.0/760)^4} = 4.07$

Equilibrium Calculations

35. $2\ NO(g) \leftrightharpoons N_2(g) + O_2(g)$ $K = \dfrac{[N_2][O_2]}{[NO]^2} = 2.4 \times 10^3$

Use the reaction quotient Q to determine which way the reaction shifts to reach equilibrium. For the reaction quotient, initial concentrations given in a problem are used to calculate the value for Q. If Q < K, then the reaction shifts right to reach equilibrium. If Q > K, then the reaction shifts left to reach equilibrium. If Q = K, then the reaction does not shift in either direction since the reaction is at equilibrium.

a. $[N_2] = \dfrac{2.0\ mol}{1.0\ L} = 2.0\ M$; $[O_2] = \dfrac{2.6\ mol}{1.0\ L} = 2.6\ M$; $[NO] = \dfrac{0.024\ mol}{1.0\ L} = 0.024\ M$

$$Q = \dfrac{[N_2]_o[O_2]_o}{[NO]_o^2} = \dfrac{(2.0)(2.6)}{(0.024)^2} = 9.0 \times 10^3$$

Q > K so the reaction shifts left to produce more reactants in order to reach equilibrium.

b. $[N_2] = \dfrac{0.62\ mol}{2.0\ L} = 0.31\ M$; $[O_2] = \dfrac{4.0\ mol}{2.0\ L} = 2.0\ M$; $[NO] = \dfrac{0.032\ mol}{2.0\ L} = 0.016\ M$

$$Q = \dfrac{(0.31)(2.0)}{(0.016)^2} = 2.4 \times 10^3 = K;\ \ at\ equilibrium$$

c. $[N_2] = \dfrac{2.4\ \text{mol}}{3.0\ \text{L}} = 0.80\ M$; $[O_2] = \dfrac{1.7\ \text{mol}}{3.0\ \text{L}} = 0.57\ M$; $[NO] = \dfrac{0.060\ \text{mol}}{3.0\ \text{L}} = 0.020\ M$

$$Q = \dfrac{(0.80)(0.57)}{(0.020)^2} = 1.1 \times 10^3 < K; \text{ Reaction shifts right to reach equilibrium.}$$

36.   As in Exercise 13.35, determine Q for each reaction, and compare this value to K_p (2.4×10^3) to determine which direction the reaction shifts to reach equilibrium. Note that, for this reaction, $K = K_p$ since $\Delta n = 0$.

a. $Q = \dfrac{P_{N_2} \times P_{O_2}}{P_{NO}^2} = \dfrac{(0.11)(2.0)}{(0.010)^2} = 2.2 \times 10^3$

$Q < K_p$ so the reaction shifts right to reach equilibrium.

b. $Q = \dfrac{(0.36)(0.67)}{(0.0078)^2} = 4.0 \times 10^3 > K_p$

Reaction shifts left to reach equilibrium.

c. $Q = \dfrac{(0.51)(0.18)}{(0.0062)^2} = 2.4 \times 10^3 = K_p$; at equilbrium

37. $CaCO_3(s) \rightleftharpoons CaO(s) + CO_2(g)$ $K_p = P_{CO_2} = 1.04$

a. $Q = P_{CO_2}$; We only need the partial pressure of CO_2 to determine Q since solids do not appear in equilibrium expressions (or Q expressions). At this temperature all CO_2 will be in the gas phase. $Q = 2.55$ so $Q > K_p$; Reaction will shift to the left to reach equilibrium; the mass of CaO will decrease.

b. $Q = 1.04 = K_p$ so the reaction is at equilibrium; mass of CaO will not change.

c. $Q = 1.04 = K_p$ so the reaction is at equilibrium; mass of CaO will not change.

d. $Q = 0.211 < K_p$; The reaction will shift to the right to reach equilibrium; the mass of CaO will increase.

38. $CH_3CO_2H + C_2H_5OH \rightleftharpoons CH_3CO_2C_2H_5 + H_2O$ $K = \dfrac{[CH_3CO_2C_2H_5][H_2O]}{[CH_3CO_2H][C_2H_5OH]} = 2.2$

a. $Q = \dfrac{(0.22)(0.10)}{(0.010)(0.010)} = 220 > K$;   Reaction will shift left to reach equilibrium so the concentration of water will decrease.

b. $Q = \dfrac{(0.22)\,(0.0020)}{(0.0020)\,(0.10)} = 2.2 = K;$ Reaction is at equilibrium, so the concentration of water will remain the same.

c. $Q = \dfrac{(0.88)\,(0.12)}{(0.044)\,(6.0)} = 0.40 < K;$ Since Q < K, the concentration of water will increase since the reaction shifts right to reach equilibrium.

d. $Q = \dfrac{(4.4)\,(4.4)}{(0.88)\,(10.0)} = 2.2 = K;$ At equilibrium, so the water concentration is unchanged.

e. $K = 2.2 = \dfrac{(2.0)\,[H_2O]}{(0.10)\,(5.0)},\ [H_2O] = 0.55\ M$

f. Water is a product of the reaction, but it is not the solvent. Thus, the concentration of water must be included in the equilibrium expression since it is a solute in the reaction. Only when water is the solvent do we not include it in the equilibrium expression.

39. $K = \dfrac{[HF]^2}{[H_2]\,[F_2]} = 2.1 \times 10^3;\ 2.1 \times 10^3 = \dfrac{[HF]^2}{(0.0021)\,(0.0021)},\ [HF] = 0.096\ M$

40. $K_P = \dfrac{P_{NOBr}^2}{P_{NO}^2 \times P_{Br_2}},\ 109 = \dfrac{(0.0768)^2}{P_{NO}^2 \times 0.0159},\ P_{NO} = 0.0583\ atm$

41. $SO_2(g) + NO_2(g) \rightleftharpoons SO_3(g) + NO(g)\quad K = \dfrac{[SO_3]\,[NO]}{[SO_2]\,[NO_2]}$

To determine K, we must calculate the equilibrium concentrations. The initial concentrations are:

$$[SO_3]_o = [NO]_o = 0;\quad [SO_2]_o = [NO_2]_o = \dfrac{2.00\ mol}{1.00\ L} = 2.00\ M$$

Next, we determine the change required to reach equilibrium. At equilibrium, [NO] = 1.30 mol/1.00 L = 1.30 M. Since there was zero NO present initially, 1.30 M of SO_2 and 1.30 M NO_2 must have reacted to produce 1.30 M NO as well as 1.30 M SO_3, all required by the balanced reaction. The equilibrium concentration for each substance is the sum of the initial concentration plus the change in concentration necessary to reach equilibrium. The equilibrium concentrations are:

$$[SO_3] = [NO] = 0 + 1.30\ M = 1.30\ M;\quad [SO_2] = [NO_2] = 2.00\ M - 1.30\ M = 0.70\ M$$

We now use these equilibrium concentrations to calculate K:

$$K = \dfrac{[SO_3]\,[NO]}{[SO_2]\,[NO_2]} = \dfrac{(1.30)\,(1.30)}{(0.70)\,(0.70)} = 3.4$$

42. $S_8(g) \rightleftharpoons 4 S_2(g)$ $K_P = \dfrac{P_{S_2}^4}{P_{S_8}}$

Initially: $P_{S_8} = 1.00$ atm and $P_{S_2} = 0$ atm

Change: Since 0.25 atm of S_8 remain at equilibrium, then 1.00 atm - 0.25 atm = 0.75 atm of S_8 must have reacted in order to reach equilibrium. Since there is a 4:1 mol ratio between S_2 and S_8 (from the balanced reaction), then 4(0.75 atm) = 3.0 atm of S_2 must have been produced when the reaction went to equilibrium (moles and pressure are directly related at constant T and V).

Equilibrium: $P_{S_8} = 0.25$ atm, $P_{S_2} = 0 + 3.0$ atm = 3.0 atm; Solving for K_P:

$$K_P = \frac{(3.0)^4}{0.25} = 3.2 \times 10^2$$

43.   When solving equilibrium problems, a common method to summarize all the information in the problem is to set up a table. We call this table the ICE table since it summarizes initial concentrations, changes that must occur to reach equilibrium and equilibrium concentrations (the sum of the initial and change columns). For the change column, we will generally use the variable x, which will be defined as the amount of reactant (or product) that must react to reach equilibrium. In this problem, the reaction must shift right to reach equilibrium since there are no products present initially. Therefore, x is defined as the amount of reactant SO_3 that reacts to reach equilibrium, and we use the coefficients in the balanced equation to relate the net change in SO_3 to the net change in SO_2 and O_2. The general ICE table for this problem is:

	$2 SO_3(g)$	$\rightleftharpoons$	$2 SO_2(g)$	+	$O_2(g)$	$K = \dfrac{[SO_2]^2[O_2]}{[SO_3]^2}$
Initial	12.0 mol/3.0 L		0		0	
	Let x mol/L of SO_3 react to reach equilibrium					
Change	-x	$\rightarrow$	+x		+x/2	
Equil.	4.0 - x		x		x/2	

From the problem, we are told that the equilibrium SO_2 concentration is 3.0 mol/3.0 L = 1.0 M ($[SO_2]_e = 1.0\ M$). From the ICE table, $[SO_2]_e = x$ so $x = 1.0$. Solving for the other equilibrium concentrations: $[SO_3]_e = 4.0 - x = 4.0 - 1.0 = 3.0\ M$; $[O_2] = x/2 = 1.0/2 = 0.50\ M$.

$$K = \frac{[SO_2]^2[O_2]}{[SO_3]^2} = \frac{(1.0)^2 (0.50)}{(3.0)^2} = 0.056$$

Alternate Method: Fractions in the change column can be avoided (if you want) by defining x differently. If we were to let $2x$ mol/L of SO_3 react to reach equilibrium, then the ICE table is:

	$2 SO_3(g)$	$\rightleftharpoons$	$2 SO_2(g)$	+	$O_2(g)$	$K = \dfrac{[SO_2]^2[O_2]}{[SO_3]^2}$
Initial	4.0 M		0		0	
	Let $2x$ mol/L of SO_3 react to reach equilibrium					
Change	-$2x$	$\rightarrow$	+$2x$		+x	
Equil.	4.0 - $2x$		$2x$		x	

Solving: $2x = [SO_2]_e = 1.0\ M$, $x = 0.50\ M$; $[SO_3]_e = 4.0 - 2(0.50) = 3.0\ M$; $[O_2]_e = x = 0.50\ M$

These are exactly the same equilibrium concentrations as solved for previously, thus K will be the same (as it must be). The moral of the story is to define x in a manner that is most comfortable for you. Your final answer is independent of how you define x initially.

44.
$$2\ NH_3(g) \rightleftharpoons N_2(g) + 3\ H_2(g) \qquad K = \frac{[N_2][H_2]^3}{[NH_3]^2}$$

Initial	4.0 mol/2.0 L	0	0

Let $2x$ mol/L of NH_3 react to reach equilibrium

Change	$-2x$	$\rightarrow$	$+x$	$+3x$
Equil.	$2.0 - 2x$		x	$3x$

From the problem: $[NH_3]_e = 2.0$ mol/2.0 L $= 1.0\ M = 2.0 - 2x$, $x = 0.50\ M$

$[N_2] = x = 0.50\ M$; $[H_2] = 3x = 3(0.50\ M) = 1.5\ M$

$$K = \frac{[N_2][H_2]^3}{[NH_3]^2} = \frac{(0.50)(1.5)^3}{(1.0)^2} = 1.7$$

45. $Q = 1.00$, which is less than K. Reaction shifts to the right to reach equilibrium. Summarizing the equilibrium problem in a table:

$$SO_2(g) + NO_2(g) \rightleftharpoons SO_3(g) + NO(g) \qquad K = 3.75$$

Initial	0.800 M	0.800 M	0.800 M	0.800 M

x mol/L of SO_2 reacts to reach equilibrium

Change	$-x$	$-x$	$\rightarrow$	$+x$	$+x$
Equil.	$0.800 - x$	$0.800 - x$		$0.800 + x$	$0.800 + x$

Plug the equilibrium concentrations into the equilibrium constant expression:

$$K = \frac{[SO_3][NO]}{[SO_2][NO_2]} = 3.75 = \frac{(0.800 + x)^2}{(0.800 - x)^2};\ \text{Take the square root of both sides and solve for } x:$$

$\dfrac{0.800 + x}{0.800 - x} = 1.94$, $0.800 + x = 1.55 - 1.94\ x$, $2.94\ x = 0.75$, $x = 0.26\ M$

The equilibrium concentrations are:

$[SO_3] = [NO] = 0.800 + x = 0.800 + 0.26 = 1.06\ M$; $[SO_2] = [NO_2] = 0.800 - x = 0.54\ M$

46. $Q = 1.00$, which is less than K. Reaction shifts right to reach equilibrium.

$$H_2(g) \quad + \quad I_2(g) \quad \rightleftharpoons \quad 2\ HI(g) \quad K = \frac{[HI]^2}{[H_2][I_2]} = 100.$$

Initial	1.00 M	1.00 M	1.00 M
	x mol/L of H_2 reacts to reach equilibrium		
Change	$-x$	$-x$ $\rightarrow$	$+2x$
Equil.	1.00 - x	1.00 - x	1.00 + 2x

$K = 100. = \dfrac{(1.00 + 2x)^2}{(1.00 - x)^2}$; Taking the square root of both sides:

$10.0 = \dfrac{1.00 + 2x}{1.00 - x}$, $10.0 - 10.0\ x = 1.00 + 2x$, $12.0\ x = 9.0$, $x = 0.75\ M$

$[H_2] = [I_2] = 1.00 - 0.75 = 0.25\ M$; $[HI] = 1.00 + 2(0.75) = 2.50\ M$

47.   Since only reactants are present initially, the reaction must proceed to the right to reach equilibrium. Summarizing the problem in a table:

$$N_2(g) \quad + \quad O_2(g) \quad \rightleftharpoons \quad 2\ NO(g) \quad K_p = 0.050$$

Initial	0.80 atm	0.20 atm	0
	x atm of N_2 reacts to reach equilibrium		
Change	$-x$	$-x$ $\rightarrow$	$+2x$
Equil.	0.80 - x	0.20 - x	2x

$K_p = 0.050 = \dfrac{P_{NO}^2}{P_{N_2} \times P_{O_2}} = \dfrac{(2x)^2}{(0.80 - x)(0.20 - x)}$, $0.050(0.16 - 1.00\ x + x^2) = 4\ x^2$

$4\ x^2 = 8.0 \times 10^{-3} - 0.050\ x + 0.050\ x^2$, $3.95\ x^2 + 0.050\ x - 8.0 \times 10^{-3} = 0$

Solving using the quadratic formula (see Appendix 1.4 of the text):

$x = \dfrac{-b \pm (b^2 - 4ac)^{1/2}}{2a} = \dfrac{-0.050 \pm [(0.050)^2 - 4(3.95)(-8.0 \times 10^{-3})]^{1/2}}{2(3.95)}$

$x = 3.9 \times 10^{-2}$ atm or $x = -5.2 \times 10^{-2}$ atm; Only $x = 3.9 \times 10^{-2}$ atm makes sense (x cannot be negative), so the equilibrium NO partial pressure is:

$P_{NO} = 2x = 2(3.9 \times 10^{-2}$ atm$) = 7.8 \times 10^{-2}$ atm

48. $H_2O(g) + Cl_2O(g) \rightleftharpoons 2\ HOCl(g) \qquad K = 0.090 = \dfrac{[HOCl]^2}{[H_2O][Cl_2O]}$

a.   The initial concentrations of H_2O and Cl_2O are:

$\dfrac{1.0\ g\ H_2O}{1.0\ L} \times \dfrac{1\ mol}{18.02\ g} = 5.5 \times 10^{-2}$ mol/L; $\dfrac{2.0\ g\ Cl_2O}{1.0\ L} \times \dfrac{1\ mol}{86.90\ g} = 2.3 \times 10^{-2}$ mol/L

$$H_2O(g) \quad + \quad Cl_2O(g) \quad \rightleftharpoons \quad 2\ HOCl(g)$$

Initial	$5.5 \times 10^{-2}\ M$	$2.3 \times 10^{-2}\ M$	0

x mol/L of H_2O reacts to reach equilibrium

Change	$-x$	$-x$	$\rightarrow$	$+2x$
Equil.	$5.5 \times 10^{-2} - x$	$2.3 \times 10^{-2} - x$		$2x$

$$K = 0.090 = \frac{(2x)^2}{(5.5 \times 10^{-2} - x)(2.3 \times 10^{-2} - x)}, \quad 1.14 \times 10^{-4} - 7.02 \times 10^{-3}\,x + 0.090\,x^2 = 4\,x^2$$

$3.91\,x^2 + 7.02 \times 10^{-3}\,x - 1.14 \times 10^{-4} = 0$ (We carried extra significant figures.)

Solving using the quadratic formula:

$$x = \frac{-7.02 \times 10^{-3} \pm (4.93 \times 10^{-5} + 1.78 \times 10^{-3})^{1/2}}{7.82} = 4.6 \times 10^{-3}\ \text{or}\ -6.4 \times 10^{-3}$$

A negative answer makes no physical sense; we can't have less than nothing.
So $x = 4.6 \times 10^{-3}\ M$.

$[HOCl] = 2x = 9.2 \times 10^{-3}\ M$; $[Cl_2O] = 2.3 \times 10^{-2} - x = 0.023 - 0.0046 = 1.8 \times 10^{-2}\ M$

$[H_2O] = 5.5 \times 10^{-2} - x = 0.055 - 0.0046 = 5.0 \times 10^{-2}\ M$

b. $$\qquad\qquad H_2O(g) \quad + \quad Cl_2O(g) \quad \rightleftharpoons \quad 2\ HOCl(g)$$

Initial	0	0	1.0 mol/2.0 L = 0.50 M

$2x$ mol/L of HOCl reacts to reach equilibrium

Change	$+x$	$+x$	$\leftarrow$	$-2x$
Equil.	x	x		$0.50 - 2x$

$$K = 0.090 = \frac{[HOCl]^2}{[H_2O]\,[Cl_2O]} = \frac{(0.50 - 2x)^2}{x^2}$$

The expression is a perfect square, so we can take the square root of each side:

$$0.30 = \frac{0.50 - 2x}{x}, \quad 0.30\,x = 0.50 - 2x, \quad 2.30\,x = 0.50$$

$x = 0.217$ (We carried extra significant figures.)

$x = [H_2O] = [Cl_2O] = 0.217 = 0.22\ M$; $[HOCl] = 0.50 - 2x = 0.50 - 0.434 = 0.07\ M$

49. $$\qquad\qquad 2\ SO_2(g) \quad + \quad O_2(g) \quad \rightleftharpoons \quad 2\ SO_3(g) \quad K_p = 0.25$$

Initial	0.50 atm	0.50 atm	0

$2x$ atm of SO_2 reacts to reach equilibrium

Change	$-2x$	$-x$	$\rightarrow$	$+2x$
Equil.	$0.50 - 2x$	$0.50 - x$		$2x$

$$K_p = 0.25 = \frac{P_{SO_3}^2}{P_{SO_2}^2 \times P_{O_2}} = \frac{(2x)^2}{(0.50 - 2x)^2(0.50 - x)}$$

This will give a cubic equation. Graphing calculators can be used to solve this expression. If you don't have a graphing calculator, an alternative method for solving a cubic equation is to use the method of successive approximations (see Appendix 1.4 of the text). The first step is to guess a value for x. Since the value of K is small (K < 1), then not much of the forward reaction will occur to reach equilibrium. This tells us that x is small. Lets guess that $x = 0.050$ atm. Now we take this estimated value for x and substitute it into the equation everywhere that x appears except for one. For equilibrium problems, we will substitute the estimated value for x into the denominator, then solve for the numerator value of x. We continue this process until the estimated value of x and the calculated value of x converge on the same number. This is the same answer we would get if we were to solve the cubic equation exactly. Applying the method of successive approximations and carrying extra significant figures:

$$\frac{4x^2}{[0.50 - 2(0.050)]^2 [0.50 - (0.050)]} = \frac{4x^2}{(0.40)^2(0.45)} = 0.25, \; x = 0.067$$

$$\frac{4x^2}{[0.50 - 2(0.067)]^2 [0.50 - (0.067)]} = \frac{4x^2}{(0.366)^2(0.433)} = 0.25, \; x = 0.060$$

$$\frac{4x^2}{(0.38)^2(0.44)} = 0.25, \; x = 0.063; \quad \frac{4x^2}{(0.374)^2(0.437)} = 0.25, \; x = 0.062$$

The next trial gives the same value for $x = 0.062$ atm. We are done except for determining the equilibrium concentrations. They are:

$$P_{SO_2} = 0.50 - 2x = 0.50 - 2(0.062) = 0.376 = 0.38 \text{ atm}$$

$$P_{O_2} = 0.50 - x = 0.438 = 0.44 \text{ atm}; \quad P_{SO_3} = 2x = 0.124 = 0.12 \text{ atm}$$

50.   a.   The reaction must proceed to products to reach equilibrium. Summarizing the problem in a table where x atm of N_2O_4 reacts to reach equilibrium:

$$N_2O_4(g) \quad \rightleftharpoons \quad 2 NO_2(g) \quad K_p = 0.25$$

Initial	4.5 atm	0
Change	-x	+2x
Equil.	4.5 - x	2x

$$K_p = \frac{P_{NO_2}^2}{P_{N_2O_4}} = \frac{(2x)^2}{4.5 - x} = 0.25, \; 4x^2 = 1.125 - 0.25\,x, \; 4x^2 + 0.25\,x - 1.125 = 0$$

We carried extra significant figures in this expression (as will be typical when we solve an expression using the quadratic formula). Solving using the quadratic formula (Appendix 1.4 of text):

$$x = \frac{-0.25 \pm [(0.25)^2 - 4(4)(-1.125)]^{1/2}}{2(4)} = \frac{-0.25 \pm 4.25}{8}, \quad x = 0.50 \text{ (Other value is negative.)}$$

$P_{NO_2} = 2x = 1.0$ atm; $P_{N_2O_4} = 4.5 - x = 4.0$ atm

b. The reaction must shift to reactants (shifts left) to reach equilibrium.

$$N_2O_4(g) \quad \rightleftharpoons \quad 2\,NO_2(g)$$

Initial   0   9.0 atm
Change   +x   ←   -2x
Equil.   x   9.0 - 2x

$K_p = \dfrac{(9.0 - 2x)^2}{x} = 0.25$, $4x^2 - 36.25\,x + 81 = 0$ (carrying extra sig. figs.)

Solving using quadratic formula: $x = \dfrac{-(-36.25) \pm [(-36.25)^2 - 4(4)(81)]^{1/2}}{2(4)}, \quad x = 4.0$ atm

The other value, 5.1, is impossible. $P_{N_2O_4} = x = 4.0$ atm; $P_{NO_2} = 9.0 - 2x = 1.0$ atm

c. No, we get the same equilibrium position starting with either pure N_2O_4 or pure NO_2 in stoichiometric amounts.

51.   a. The reaction must proceed to products to reach equilibrium since only reactants are present initially. Summarizing the problem in a table:

	2 NOCl(g)	$\rightleftharpoons$	2 NO(g)	+	Cl$_2$(g)	K = 1.6 × 10⁻⁵
Initial	$\dfrac{2.0 \text{ mol}}{2.0 \text{ L}}$		0		0	

2x mol/L of NOCl reacts to reach equilibrium

Change	-2x	$\rightarrow$	+2x	+x
Equil.	1.0 - 2x		2x	x

$$K = 1.6 \times 10^{-5} = \frac{[NO]^2[Cl_2]}{[NOCl]^2} = \frac{(2x)^2\,(x)}{(1.0 - 2x)^2}$$

If we assume that $1.0 - 2x \approx 1.0$ (from the size of K, we know that not much reaction will occur so x is small), then:

$$1.6 \times 10^{-5} = \frac{4x^3}{1.0^2}, \quad x = 1.6 \times 10^{-2}; \text{ Now we must check the assumption.}$$

$1.0 - 2x = 1.0 - 2(0.016) = 0.97 = 1.0$ (to proper significant figures)

Our error is about 3%, i.e., $2x$ is 3.2% of $1.0\,M$. Generally, if the error we introduce by making simplifying assumptions is less than 5%, we go no further and the assumption is said to be valid. We call this the 5% rule. Solving for the equilibrium concentrations:

$[NO] = 2x = 0.032\ M;\ [Cl_2] = x = 0.016\ M;\ [NOCl] = 1.0 - 2x = 0.97\ M \approx 1.0\ M$

Note: If we were to solve this cubic equation exactly (a long and tedious process), we would get $x = 0.016$. This is the exact same answer we determined by making a simplifying assumption. We saved time and energy. Whenever K is a very small value, always make the assumption that x is small. If the assumption introduces an error of less than 5%, then the answer you calculated making the assumption will be considered the correct answer.

b.
$$2\ NOCl(g) \rightleftharpoons 2\ NO(g) + Cl_2(g)$$

Initial	1.0 M	1.0 M	0

2x mol/L of NOCl reacts to reach equilibrium

Change	-2x	$\rightarrow$	+2x	+x
Equil.	1.0 - 2x		1.0 + 2x	x

$$1.6 \times 10^{-5} = \frac{(1.0 + 2x)^2(x)}{(1.0 - 2x)^2} \approx \frac{(1.0)^2(x)}{(1.0)^2} \quad \text{(assuming } 2x \ll 1.0)$$

$x = 1.6 \times 10^{-5}$; Assumptions are great ($2x$ is 3.2×10^{-3}% of 1.0).

$[Cl_2] = 1.6 \times 10^{-5}\ M$ and $[NOCl] = [NO] = 1.0\ M$

c.
$$2\ NOCl(g) \rightleftharpoons 2\ NO(g) + Cl_2(g)$$

Initial	2.0 M	0	1.0 M

2x mol/L of NOCl reacts to reach equilibrium

Change	-2x	$\rightarrow$	+2x	+x
Equil.	2.0 - 2x		2x	1.0 + x

$$1.6 \times 10^{-5} = \frac{(2x)^2(1.0 + x)}{(2.0 - 2x)^2} \approx \frac{4x^2}{4.0} \quad \text{(assuming } x \ll 1.0)$$

Solving: $x = 4.0 \times 10^{-3}$; Assumptions good (x is 0.4% of 1.0 and $2x$ is 0.4% of 2.0).

$[Cl_2] = 1.0 + x = 1.0\ M;\ [NO] = 2(4.0 \times 10^{-3}) = 8.0 \times 10^{-3};\ [NOCl] = 2.0\ M$

52.
$$N_2O_4(g) \rightleftharpoons 2\ NO_2(g) \qquad K = \frac{[NO_2]^2}{[N_2O_4]} = 4.0 \times 10^{-7}$$

Initial	1.0 mol/10.0 L		0

x mol/L of N_2O_4 reacts to reach equilibrium

Change	-x	$\rightarrow$	+2x
Equil.	0.10 - x		2x

$$K = \frac{[NO_2]^2}{[N_2O_4]} = \frac{(2x)^2}{0.10 - x} = 4.0 \times 10^{-7};\quad \text{Since K has a small value, assume that } x \text{ is small}$$
$$\text{compared to 0.10 so that } 0.10 - x \approx 0.10.\ \text{Solving:}$$

$$4.0 \times 10^{-7} \approx \frac{4x^2}{0.10}, \ 4x^2 = 4.0 \times 10^{-8}, \ x = 1.0 \times 10^{-4} \ M$$

Checking the assumption by the 5% rule: $\dfrac{x}{0.10} \times 100 = \dfrac{1.0 \times 10^{-4}}{0.10} \times 100 = 0.10\%$

Since this number is less than 5%, we will say that the assumption is valid.

$[N_2O_4] = 0.10 - 1.0 \times 10^{-4} = 0.10 \ M; \ [NO_2] = 2x = 2(1.0 \times 10^{-4}) = 2.0 \times 10^{-4} \ M$

53.
$$2 \ CO_2(g) \quad \rightleftharpoons \quad 2 \ CO(g) \quad + \quad O_2(g) \qquad K = \frac{[CO]^2 \ [O_2]}{[CO_2]^2} = 2.0 \times 10^{-6}$$

Initial	2.0 mol/5.0 L	0	0
	$2x$ mol/L of CO_2 reacts to reach equilibrium		
Change	$-2x$ $\rightarrow$	$+2x$	$+x$
Equil.	$0.40 - 2x$	$2x$	x

$$K = 2.0 \times 10^{-6} = \frac{[CO]^2[O_2]}{[CO_2]^2} = \frac{(2x)^2(x)}{(0.40 - 2x)^2}; \ \text{Assuming } 2x << 0.40:$$

$$2.0 \times 10^{-6} \approx \frac{4x^3}{(0.40)^2}, \ 2.0 \times 10^{-6} = \frac{4x^3}{0.16}, \ x = 4.3 \times 10^{-3} \ M$$

Checking assumption: $\dfrac{2(4.3 \times 10^{-3})}{0.40} \times 100 = 2.2\%$; Assumption valid by the 5% rule.

$[CO_2] = 0.40 - 2x = 0.40 - 2(4.3 \times 10^{-3}) = 0.39 \ M$

$[CO] = 2x = 2(4.3 \times 10^{-3}) = 8.6 \times 10^{-3} \ M; \ [O_2] = x = 4.3 \times 10^{-3} \ M$

54.
$$COCl_2(g) \quad \rightleftharpoons \quad CO(g) \quad + \quad Cl_2(g) \qquad K_p = \frac{P_{CO} \times P_{Cl_2}}{P_{COCl_2}} = 6.8 \times 10^{-9}$$

Initial	1.0 atm	0	0
	x atm of $COCl_2$ reacts to reach equilibrium		
Change	$-x$ $\rightarrow$	$+x$	$+x$
Equil.	$1.0 - x$	x	x

$$6.8 \times 10^{-9} = \frac{P_{CO} \times P_{Cl_2}}{P_{COCl_2}} = \frac{x^2}{1.0 - x} \approx \frac{x^2}{1.0} \quad \text{(Assuming } 1.0 - x \approx 1.0.)$$

$x = 8.2 \times 10^{-5}$ atm; Assumption good (x is $8.2 \times 10^{-3}\%$ of 1.0).

$P_{COCl_2} = 1.0 - x = 1.0 - 8.2 \times 10^{-5} = 1.0$ atm; $P_{CO} = P_{Cl_2} = x = 8.2 \times 10^{-5}$ atm

55. This is a typical equilibrium problem except that the reaction contains a solid. Whenever solids and liquids are present, we basically ignore them in the equilibrium problem.

$$NH_4OCONH_2(s) \rightleftharpoons 2 NH_3(g) + CO_2(g) \quad K_p = 2.9 \times 10^{-3}$$

Initial   -   0   0

               Some NH_4OCONH_2 decomposes to produce $2x$ atm of NH_3 and x atm of CO_2.

Change   - $\rightarrow$ $+2x$ $+x$

Equil.   - $2x$ x

$$K_p = 2.9 \times 10^{-3} = P_{NH_3}^2 \times P_{CO_2} = (2x)^2(x) = 4x^3$$

$$x = \left(\frac{2.9 \times 10^{-3}}{4} \right)^{1/3} = 9.0 \times 10^{-2} \text{ atm}; \quad P_{NH_3} = 2x = 0.18 \text{ atm}; \quad P_{CO_2} = x = 9.0 \times 10^{-2} \text{ atm}$$

$$P_{total} = P_{NH_3} + P_{CO_2} = 0.18 \text{ atm} + 0.090 \text{ atm} = 0.27 \text{ atm}$$

56. a. $$2 AsH_3(g) \rightleftharpoons 2 As(s) + 3 H_2(g)$$

Initial   392.0 torr   -   0

Equil.   392.0 - $2x$ - $3x$

Using Dalton's Law of Partial Pressure:

$$P_{total} = 488.0 \text{ torr} = P_{AsH_3} + P_{H_2} = 392.0 - 2x + 3x, \quad x = 96.0 \text{ torr}$$

$$P_{H_2} = 3x = 3(96.0) = 288 \text{ torr} \times \frac{1 \text{ atm}}{760 \text{ torr}} = 0.379 \text{ atm}$$

b. $$P_{AsH_3} = 392.0 - 2(96.0) = 200.0 \text{ torr} \times \frac{1 \text{ atm}}{760 \text{ torr}} = 0.2632 \text{ atm}$$

$$K_p = \frac{(P_{H_2})^3}{(P_{AsH_3})^2} = \frac{(0.379)^3}{(0.2632)^2} = 0.786$$

Le Chatelier's Principle

57. a. No effect; Adding more of a pure solid or pure liquid has no effect on the equilibrium position.

b. Shifts left; HF(g) will be removed by reaction with the glass. As HF(g) is removed, the reaction will shift left to produce more HF(g).

c. Shifts right; As $H_2O(g)$ is removed, the reaction will shift right to produce more $H_2O(g)$.

58.   When the volume of a reaction container is increased, the reaction itself will want to increase its own volume by shifting to the side of the reaction which contains the most molecules of gas. When the molecules of gas are equal on both sides of the reaction, then the reaction will remain at equilibrium no matter what happens to the volume of the container.

   a.   Reaction shifts left (to reactants) since the reactants contain 4 molecules of gas compared to 2 molecules of gas on the product side.

   b.   Reaction shifts right (to products) since there are more product molecules of gas (2) than reactant molecules (1).

   c.   No change since there are equal reactant and product molecules of gas.

   d.   Reaction shifts right.

   e.   Reaction shifts right to produce more $CO_2(g)$. One can ignore the solids and only concentrate on the gases since gases occupy a relatively large volume compared to solids. We make the same assumption when liquids are present (only worry about the gas molecules).

59.   a.   right   b.   right   c.   no effect; He(g) is neither a reactant nor a product.

   d.   left; Since the reaction is exothermic, heat is a product:

$$CO(g) + H_2O(g) \rightarrow H_2(g) + CO_2(g) + Heat$$

   Increasing T will add heat. The equilibrium shifts to the left to use up the added heat.

   e.   no effect; Since there are equal numbers of gas molecules on both sides of the reaction, a change in volume has no effect on the equilibrium position.

60.   a.   The moles of SO_3 will increase since the reaction will shift left to use up the added $O_2(g)$.

   b.   Increase; Since there are fewer reactant gas molecules than product gas molecules, the reaction shifts left with a decrease in volume.

   c.   No effect; The partial pressures of sulfur trioxide, sulfur dioxide, and oxygen are unchanged.

   d.   Increase; Heat $+ 2 SO_3 \rightleftharpoons 2 SO_2 + O_2$; Decreasing T will remove heat, shifting this endothermic reaction to the left.

   e.   Decrease

61.   a.   left   b.   right   c.   left

   d.   no effect (reactant and product concentrations are unchanged)

e. no effect; Since there are equal numbers of product and reactant gas molecules, a change in volume has no effect on the equilibrium position.

f. right; A decrease in temperature will shift the equilibrium to the right since heat is a product in this reaction (as is true in all exothermic reactions).

62. a. shift to left

b. shift to right; Since the reaction is endothermic (heat is a reactant), an increase in temperature will shift the equilibrium to the right.

c. no effect   d. shift to right

e. shift to right; Since there are more gaseous product molecules than gaseous reactant molecules, the equilibrium will shift right with an increase in volume.

63. In an exothermic reaction, heat is a product. To maximize product yield, one would want as low a temperature as possible since high temperatures would shift the reaction left (away from products). Since temperature changes also change the value of K, then at low temperatures the value of K will be largest, which maximizes yield of products.

64. As temperature increases, the value of K decreases. This is consistent with an exothermic reaction. In an exothermic reaction, heat is a product and an increase in temperature shifts the equilibrium to the reactant side (as well as lowering the value of K).

Additional Exercises

65.
$$O(g) + NO(g) \rightleftharpoons NO_2(g) \qquad K = 1/6.8 \times 10^{-49} = 1.5 \times 10^{48}$$
$$NO_2(g) + O_2(g) \rightleftharpoons NO(g) + O_3(g) \qquad K = 1/5.8 \times 10^{-34} = 1.7 \times 10^{33}$$

$$O_2(g) + O(g) \rightleftharpoons O_3(g) \qquad K = (1.5 \times 10^{48})(1.7 \times 10^{33}) = 2.6 \times 10^{81}$$

66. a. $N_2(g) + O_2(g) \rightleftharpoons 2\,NO(g) \quad K_p = 1 \times 10^{-31} = \dfrac{P_{NO}^2}{P_{N_2} \times P_{O_2}} = \dfrac{P_{NO}^2}{(0.8)(0.2)}$

$$P_{NO} = 1 \times 10^{-16}\ atm; \quad n_{NO} = \dfrac{PV}{RT} = \dfrac{(1 \times 10^{-16}\ atm)\,(1.0 \times 10^{-3}\ L)}{\left(\dfrac{0.08206\ L\ atm}{mol\,K}\right)(298\ K)} = 4 \times 10^{-21}\ mol\ NO$$

$$\dfrac{4 \times 10^{-21}\ mol\ NO}{cm^3} \times \dfrac{6.02 \times 10^{23}\ molecules}{mol\ NO} = \dfrac{2 \times 10^3\ molecules\ NO}{cm^3}$$

b. There is more NO in the atmosphere than we would expect from the value of K. The answer must lie in the rates of the reaction. At 25°C the rates of both reactions:

$$N_2 + O_2 \rightarrow 2\,NO \quad \text{and} \quad 2\,NO \rightarrow N_2 + O_2$$

are so slow that they are essentially zero. Very strong bonds must be broken; the activation energy is very high. Nitric oxide is produced in high energy or high temperature environments. In nature, some NO is produced by lightning, and the primary manmade source is from automobiles. The production of NO is endothermic ($\Delta H = +90$ kJ/mol). At high temperatures, K will increase, and the rates of the reaction will also increase, resulting in a higher production of NO. Once the NO gets into a more normal temperature environment, it doesn't go back to N_2 and O_2 because of the slow rate.

67. $$3\,H_2(g) \quad + \quad N_2(g) \quad \rightleftharpoons \quad 2\,NH_3\,(g)$$

Initial	$[H_2]_o$	$[N_2]_o$	0
	x mol/L of N_2 reacts to reach equilibrium		
Change	$-3x$	$-x \quad \rightarrow$	$+2x$
Equil	$[H_2]_o - 3x$	$[N_2]_o - x$	$2x$

From the problem:

$[NH_3]_e = 4.0\ M = 2x$, $x = 2.0\ M$; $[H_2]_e = 5.0\ M = [H_2]_o - 3x$; $[N_2]_e = 8.0\ M = [N_2]_o - x$

$5.0\ M = [H_2]_o - 3(2.0\ M)$, $[H_2]_o = 11.0\ M$; $8.0\ M = [N_2]_o - 2.0\ M$, $[N_2]_o = 10.0\ M$

68. $$FeSCN^{2+}(aq) \quad \rightleftharpoons \quad Fe^{3+}(aq) \quad + \quad SCN^-(aq) \qquad K = 9.1 \times 10^{-4}$$

Initial	$2.0\ M$	0	0
	x mol/L of $FeSCN^{2+}$ reacts to reach equilibrium		
Change	$-x \quad \rightarrow$	$+x$	$+x$
Equil.	$2.0 - x$	x	x

$$9.1 \times 10^{-4} = \frac{[Fe^{3+}][SCN^-]}{[FeSCN^{2+}]} = \frac{x^2}{2.0 - x} \approx \frac{x^2}{2.0} \qquad (\text{Assuming } 2.0 - x \approx 2.0.)$$

$x = 4.3 \times 10^{-2}\ M$; Assumption good by the 5% rule (x is 2.2% of 2.0).

$[FeSCN^{2+}] = 2.0 - x = 2.0 - 4.3 \times 10^{-2} = 2.0\ M$; $[Fe^{3+}] = [SCN^-] = x = 4.3 \times 10^{-2}\ M$

69. a. $\displaystyle P_{PCl_5} = \frac{nRT}{V} = \frac{\dfrac{2.450\ \text{g PCl}_5}{208.22\ \text{g/mol}} \times \dfrac{0.08206\ \text{L atm}}{\text{mol K}} \times 600.\ \text{K}}{0.500\ \text{L}} = 1.16\ \text{atm}$

b. $$PCl_5(g) \quad \rightleftharpoons \quad PCl_3(g) \quad + \quad Cl_2(g) \qquad K_p = \frac{P_{PCl_3} \times P_{Cl_2}}{P_{PCl_5}} = 11.5$$

Initial	1.16 atm	0	0
	x atm of PCl_5 reacts to reach equilibrium		
Change	$-x \quad \rightarrow$	$+x$	$+x$
Equil.	$1.16 - x$	x	x

$$K_p = \frac{x^2}{1.16 - x} = 11.5, \quad x^2 + 11.5\,x - 13.3 = 0; \quad \text{Using the quadratic formula: } x = 1.06 \text{ atm}$$

$$P_{PCl_5} = 1.16 - 1.06 = 0.10 \text{ atm}$$

c. $P_{PCl_3} = P_{Cl_2} = 1.06$ atm; $P_{PCl_5} = 0.10$ atm

$$P_{tot} = P_{PCl_5} + P_{PCl_3} + P_{Cl_2} = 0.10 + 1.06 + 1.06 = 2.22 \text{ atm}$$

d.   Percent dissociation $= \dfrac{x}{1.16} \times 100 = \dfrac{1.06}{1.16} \times 100 = 91.4\%$

70. $SO_2Cl_2(g) \rightleftharpoons Cl_2(g) + SO_2(g)$

Initial	P_0	0	0	P_0 = initial pressure of SO_2Cl_2
Change	$-x$	$\rightarrow$ $+x$	$+x$	
Equil.	$P_0 - x$	x	x	

$$P_{total} = 0.900 \text{ atm} = P_0 - x + x + x = P_0 + x$$

$$\frac{x}{P_0} \times 100 = 12.5, \quad P_0 = 8.00\,x$$

Solving: $0.900 = P_0 + x = 9.00\,x, \quad x = 0.100$ atm

$x = 0.100$ atm $= P_{Cl_2} = P_{SO_2}$; $P_0 - x = 0.800 - 0.100 = 0.700$ atm $= P_{SO_2Cl_2}$

$$K_p = \frac{P_{Cl_2} \times P_{SO_2}}{P_{SO_2Cl_2}} = \frac{(0.100)^2}{0.700} = 1.43 \times 10^{-2} \text{ atm}$$

71. $K = \dfrac{[HF]^2}{[H_2][F_2]} = \dfrac{(0.400)^2}{(0.0500)\,(0.0100)} = 320.$;   0.200 mol F_2/5.00 L = 0.0400 $M\,F_2$ added

From LeChatelier's principle, added F_2 causes the reaction to shift right to reestablish equilibrium.

                $H_2(g) + F_2(g) \rightleftharpoons 2\,HF(g)$

Initial	0.0500 M	0.0500 M	0.400 M
	x mol/L of F_2 reacts to reach equilibrium		
Change	$-x$	$-x$	$\rightarrow$ $+2x$
Equil.	0.0500 $-x$	0.0500 $-x$	0.400 $+ 2x$

$K = 320. = \dfrac{(0.400 + 2x)^2}{(0.0500 - x)^2}$;   Taking the square root of the equation:

$17.9 = \dfrac{0.400 + 2x}{0.0500 - x}$, $0.895 - 17.9\,x = 0.400 + 2x$, $19.9\,x = 0.495$, $x = 0.0249$ mol/L

$[HF] = 0.400 + 2(0.0249) = 0.450\ M$; $[H_2] = [F_2] = 0.0500 - 0.0249 = 0.0251\ M$

72.   a.   Doubling the volume will decrease all concentrations by a factor of one-half.

$$Q = \frac{\frac{1}{2}[FeSCN^{2+}]_{eq}}{\left(\frac{1}{2}[Fe^{3+}]_{eq}\right)\left(\frac{1}{2}[SCN^-]_{eq}\right)} = 2K, \quad Q > K$$

The reaction will shift to the left to reestablish equilibrium.

b.   Adding Ag^+ will remove SCN^- through the formation of AgSCN(s). The reaction will shift to the left to produce more SCN^-.

c.   Removing Fe^{3+} as $Fe(OH)_3$(s) will shift the reaction to the left to produce more Fe^{3+}.

d.   Reaction will shift to the right as Fe^{3+} is added.

73. $H^+ + OH^- \rightarrow H_2O$; Sodium hydroxide (NaOH) will react with the H^+ on the product side of the reaction. This effectively removes H^+ from the equilibrium, which will shift the reaction to the right to produce more H^+ and CrO_4^{2-}. Since more CrO_4^{2-} is produced, the solution turns yellow.

74. $N_2(g) + 3 H_2(g) \rightleftharpoons 2 NH_3(g) + heat$

a.   This reaction is exothermic, so an increase in temperature will decrease the value of K (see Table 13.3 of text.) This has the effect of lowering the amount of NH_3(g) produced at equilibrium. The temperature increase, therefore, must be for kinetics reasons. As temperature increases, the reaction reaches equilibrium much faster. At low temperatures, this reaction is very slow, too slow to be of any use.

b.   As NH_3(g) is removed, the reaction shifts right to produce more NH_3(g).

c.   A catalyst has no effect on the equilibrium position. The purpose of a catalyst is to speed up a reaction so it reaches equilibrium more quickly.

d.   When the pressure of reactants and products is high, the reaction shifts to the side that has fewer gas molecules. Since the product side contains 2 molecules of gas as compared to 4 molecules of gas on the reactant side, then the reaction shifts right to products at high pressures of reactants and products.

Challenge Problems

75.   There is a little trick we can use to solve this problem in order to avoid solving a cubic equation. Since K for this reaction is very small, the dominant reaction is the reverse reaction. We will let the products react to completion by the reverse reaction, then we will solve the forward equilibrium problem to determine the equilibrium concentrations. Summarizing these steps to solve in a table:

$$2\,NOCl(g) \rightleftharpoons 2\,NO(g) + Cl_2(g) \qquad K = 1.6 \times 10^{-5}$$

Before	0	2.0 M	1.0 M	

Let 1.0 mol/L Cl_2 react completely.   (K is small, reactants dominate.)

Change	+2.0	$\leftarrow$	-2.0	-1.0	React completely
After	2.0		0	0	New initial conditions

2x mol/L of NOCl reacts to reach equilibrium

Change	-2x	$\rightarrow$	+2x	+x
Equil.	2.0 - 2x		2x	x

$$K = 1.6 \times 10^{-5} = \frac{(2x)^2\,(x)}{(2.0 - 2x)^2} \approx \frac{4x^3}{2.0^2} \qquad \text{(assuming 2.0 - 2x} \approx 2.0)$$

$x^3 = 1.6 \times 10^{-5}$, $x = 2.5 \times 10^{-2}$ Assumption good by the 5% rule (2x is 2.5% of 2.0).

$[NOCl] = 2.0 - 0.050 = 1.95\,M = 2.0\,M$; $[NO] = 0.050\,M$; $[Cl_2] = 0.025\,M$

Note: If we do not break this problem into two parts (a stoichiometric part and an equilibrium part), we are faced with solving a cubic equation. The set-up would be:

$$2\,NOCl \rightleftharpoons 2\,NO + Cl_2$$

Initial	0	2.0 M	1.0 M
Change	+2y	$\leftarrow$ -2y	-y
Equil.	2y	2.0 - 2y	1.0 - y

$1.6 \times 10^{-5} = \dfrac{(2.0 - 2y)^2\,(1.0 - y)}{(2y)^2}$;   If we say that y is small to simplify the problem, then:

$1.6 \times 10^{-5} = \dfrac{2.0^2}{4y^2}$;   We get $y = 250$.   This is impossible!

To solve this equation, we cannot make any simplifying assumptions; we have to solve a cubic equation. If you don't have a graphing calculator, this is difficult. Alternatively, we can use some chemical common sense and solve the problem as illustrated above.

76.   a. $$2\,NO(g) + Br_2(g) \rightleftharpoons 2\,NOBr(g)$$

Initial	98.4 torr	41.3 torr	0

2x torr of NO reacts to reach equilibrium

Change	-2x	-x	$\rightarrow$	+2x
Equil.	98.4 - 2x	41.3 - x		2x

$P_{total} = P_{NO} + P_{Br_2} + P_{NOBr} = (98.4 - 2x) + (41.3 - x) + 2x = 139.7 - x$

$P_{total} = 110.5 = 139.7 - x$, $x = 29.2$ torr; $P_{NO} = 98.4 - 2(29.2) = 40.0$ torr $= 0.0526$ atm

$P_{Br_2} = 41.3 - 29.2 = 12.1 \text{ torr} = 0.0159 \text{ atm}; \quad P_{NOBr} = 2(29.2) = 58.4 \text{ torr} = 0.0768 \text{ atm}$

$$K_p = \frac{P_{NOBr}^2}{P_{NO}^2 \times P_{Br_2}} = \frac{(0.0768)^2}{(0.0526)^2 (0.0159)} = 134$$

b.
$$2\,NO(g) \quad + \quad Br_2(g) \quad \rightleftharpoons \quad 2\,NOBr(g)$$

Initial	0.30 atm	0.30 atm	0

$2x$ atm of NO reacts to reach equilibrium

Change	-2x	-x	$\rightarrow$	+2x
Equil.	0.30 - 2x	0.30 - x		2x

This would yield a cubic equation. For those students without a graphing calculator, a strategy to solve this is to first notice that K_p is pretty large. Since K_p is large, let us approach equilibrium in two steps; assume the reaction goes to completion then solve the back equilibrium problem.

$$2\,NO \quad + \quad Br_2 \quad \rightleftharpoons \quad 2\,NOBr$$

Before	0.30 atm	0.30 atm	0

Let 0.30 atm NO react completely.

Change	-0.30	-0.15	$\rightarrow$	+0.30	React completely
After	0	0.15		0.30	New initial conditions

$2y$ atm of NOBr reacts to reach equilibrium

Change	+2y	+y	$\leftarrow$	-2y
Equil.	2y	0.15 + y		0.30 - 2y

$$\frac{(0.30 - 2y)^2}{(2y)^2 (0.15 + y)} = 134, \quad \frac{(0.30 - 2y)^2}{(0.15 + y)} = 134 \times 4\,y^2 = 536\,y^2$$

If $y \ll 0.15$: $\dfrac{(0.30)^2}{0.15} \approx 536\,y^2$ and $y = 0.034$; Assumptions are poor (y is 23% of 0.15).

Use 0.034 as approximation for y and solve by successive approximations:

$$\frac{(0.30 - 0.068)^2}{0.15 + 0.034} = 536\,y^2, \quad y = 0.023; \quad \frac{(0.30 - 0.046)^2}{0.15 + 0.023} = 536\,y^2, \quad y = 0.026$$

$$\frac{(0.30 - 0.052)^2}{0.15 + 0.026} = 536\,y^2, \quad y = 0.026$$

So: $P_{NO} = 2y = 0.052 \text{ atm}; \quad P_{Br_2} = 0.15 + y = 0.18 \text{ atm}; \quad P_{NOBr} = 0.30 - 2y = 0.25 \text{ atm}$

77. $N_2(g)$ + $3 H_2(g)$ $\rightleftharpoons$ $2 NH_3(g)$

Initial   0   0 P_0 P_0 = initial pressure of NH_3 in atm
                 $2x$ atm of NH_3 reacts to reach equilibrium
Change $+x$ $+3x$ $\leftarrow$ $-2x$
Equil. x $3x$ $P_0 - 2x$

From problem, $P_0 - 2x = \dfrac{P_0}{2.00}$, so $P_0 = 4.00\, x$

$$K_p = \frac{(4.00\, x - 2x)^2}{(x)(3x)^3} = \frac{(2.00\, x)^2}{(x)(3x)^3} = \frac{4.00\, x^2}{27 x^4} = \frac{4.00}{27 x^2} = 5.3 \times 10^5, \ x = 5.3 \times 10^{-4}\ \text{atm}$$

$P_0 = 4.00\, x = 4.00 \times (5.3 \times 10^{-4})\ \text{atm} = 2.1 \times 10^{-3}\ \text{atm}$

78. $P_4(g) \rightleftharpoons 2 P_2(g)$ $K_p = 0.100 = \dfrac{P_{P_2}^2}{P_{P_4}}$; $P_{P_4} + P_{P_2} = P_{total} = 1.00\ \text{atm}$, $P_{P_4} = 1.00 - P_{P_2}$

Let $y = P_{P_2}$ at equilibrium, then $\dfrac{y^2}{1.00 - y} = 0.100$

Solving: $y = 0.270\ \text{atm} = P_{P_2}$; $P_{P_4} = 1.00 - 0.270 = 0.73\ \text{atm}$

To solve for the fraction dissociated, we need the initial pressure of P_4 (mol $\propto$ pressure).

                       $P_4(g)$ $\rightleftharpoons$ $2 P_2(g)$

Initial P_0 0 P_0 = initial pressure of P_4 in atm.
                 x atm of P_4 reacts to reach equilibrium
Change $-x$ $\rightarrow$ $+2x$
Equil. $P_0 - x$ $2x$

$P_{total} = P_0 - x + 2x = 1.00\ \text{atm} = P_0 + x$

Solving: $0.270\ \text{atm} = P_{P_2} = 2x$, $x = 0.135\ \text{atm}$; $P_0 = 1.00 - 0.135 = 0.87\ \text{atm}$

Fraction dissociation = $\dfrac{x}{P_0} = \dfrac{0.135}{0.87} = 0.16$ or 16% of P_4 is dissociated to reach equilibrium.

79. $N_2O_4(g) \rightleftharpoons 2 NO_2(g)$ $K_p = \dfrac{P_{NO_2}^2}{P_{N_2O_4}} = \dfrac{(1.20)^2}{0.34} = 4.2$

Doubling the volume decreases each partial pressure by a factor of 2 ($P = nRT/V$).

$P_{NO_2} = 0.600\ \text{atm}$ and $P_{N_2O_4} = 0.17\ \text{atm}$ are the new partial pressures.

$Q = \dfrac{(0.600)^2}{0.17} = 2.1$, so $Q < K$;   Equilibrium will shift to the right.

$$N_2O_4(g) \quad \rightleftharpoons \quad 2\,NO_2(g)$$

Initial   0.17 atm   0.600 atm

x atm of N_2O_4 reacts to reach equilibrium

Change $-x$ $\rightarrow$ $+2x$

Equil. $0.17 - x$ $0.600 + 2x$

$$K_p = 4.2 = \frac{(0.600 + 2x)^2}{(0.17 - x)}, \quad 4x^2 + 6.6\,x - 0.354 = 0 \text{ (carrying extra sig. figs.)}$$

Solving using the quadratic formula: $x = 0.052$

$$P_{NO_2} = 0.600 + 2(0.052) = 0.704 \text{ atm}; \quad P_{N_2O_4} = 0.17 - 0.052 = 0.12 \text{ atm}$$

80.   a.

$$2\,NaHCO_3(s) \quad \rightleftharpoons \quad Na_2CO_3(s) \quad + \quad CO_2(g) \quad + \quad H_2O(g)$$

Initial   -   -   0   0

$NaHCO_3(s)$ decomposes to form x atm each of $CO_2(g)$ and $H_2O(g)$ at equilibrium.

Change   - $\rightarrow$ - $+x$ $+x$

Equil.   -   - x x

$$0.25 = K_p = P_{CO_2} \times P_{H_2O}, \quad 0.25 = x^2, \quad x = P_{CO_2} = P_{H_2O} = 0.50 \text{ atm}$$

b. $$n_{CO_2} = \frac{PV}{RT} = \frac{(0.50 \text{ atm})\,(1.00 \text{ L})}{(0.08206 \text{ L atm/mol} \cdot \text{K})\,(398 \text{ K})} = 1.5 \times 10^{-2} \text{ mol } CO_2$$

Mass of Na_2CO_3 produced:

$$1.5 \times 10^{-2} \text{ mol } CO_2 \times \frac{1 \text{ mol } Na_2CO_3}{\text{mol } CO_2} \times \frac{105.99 \text{ g } Na_2CO_3}{\text{mol } Na_2CO_3} = 1.6 \text{ g } Na_2CO_3$$

Mass of $NaHCO_3$ reacted:

$$1.5 \times 10^{-2} \text{ mol } CO_2 \times \frac{2 \text{ mol } NaHCO_3}{1 \text{ mol } CO_2} \times \frac{84.01 \text{ g } NaHCO_3}{\text{mol}} = 2.5 \text{ g } NaHCO_3$$

Mass of $NaHCO_3$ remaining = 10.0 - 2.5 = 7.5 g

c. $$10.0 \text{ g } NaHCO_3 \times \frac{1 \text{ mol } NaHCO_3}{84.01 \text{ g } NaHCO_3} \times \frac{1 \text{ mol } CO_2}{2 \text{ mol } NaHCO_3} = 5.95 \times 10^{-2} \text{ mol } CO_2$$

When all of the $NaHCO_3$ has been just consumed, we will have 5.95×10^{-2} mol CO_2 gas at a pressure of 0.50 atm (from a).

$$V = \frac{nRT}{P} = \frac{(5.95 \times 10^{-2} \text{ mol})\,(0.08206 \text{ L atm/mol} \cdot \text{K})\,(398 \text{ K})}{0.50 \text{ atm}} = 3.9 \text{ L}$$

81. $SO_3(g)$ $\rightleftharpoons$ $SO_2(g)$ + $1/2\ O_2(g)$

Initial	P_0		0	0	P_0 = initial pressure of SO_3
Change	$-x$	$\rightarrow$	$+x$	$+x/2$
Equil.	$P_0 - x$		x	$x/2$

The average molar mass of the mixture is:

$$\text{average molar mass} = \frac{dRT}{P} = \frac{(1.60\ \text{g/L})\ (0.08206\ \text{L atm mol}^{-1}\ \text{K}^{-1})\ (873\ \text{K})}{1.80\ \text{atm}} = 63.7\ \text{g/mol}$$

The average molar mass is determined by:

$$\text{average molar mass} = \frac{n_{SO_3}\ (80.07\ \text{g/mol}) + n_{SO_2}\ (64.07\ \text{g/mol}) + n_{O_2}\ (32.00\ \text{g/mol})}{n_{\text{total}}}$$

Since χ_A = mol fraction of component A = $n_A/n_{\text{total}} = P_A/P_{\text{total}}$, then:

$$63.7\ \text{g/mol} = \frac{P_{SO_3}\ (80.07) + P_{SO_2}\ (64.07) + P_{O_2}\ (32.00)}{P_{\text{total}}}$$

$P_{\text{total}} = P_0 - x + x + x/2 = P_0 + x/2 = 1.80\ \text{atm},\ P_0 = 1.80 - x/2$

$$63.7 = \frac{(P_0 - x)\ (80.07) + x(64.07) + \frac{x}{2}(32.00)}{1.80}$$

$$63.7 = \frac{(1.80 - 3/2x)\ (80.07) + x(64.07) + \frac{x}{2}(32.00)}{1.80}$$

$115 = 144 - 120.1\ x + 64.07\ x + 16.00\ x,\ 40.0\ x = 29,\ x = 0.73\ \text{atm}$

$P_{SO_3} = P_0 - x = 1.80 - 3/2\ x = 0.71\ \text{atm};\ P_{SO_2} = 0.73\ \text{atm};\ P_{O_2} = x/2 = 0.37\ \text{atm}$

$$K_p = \frac{P_{SO_2} \times P_{O_2}^{1/2}}{P_{SO_3}} = \frac{(0.73)\ (0.37)^{1/2}}{(0.71)} = 0.63$$

82.   The first reaction produces equal amounts of SO_3 and SO_2. Using the second reaction, calculate the SO_3, SO_2 and O_2 partial pressures at equilibrium.

                        $SO_3(g)$ $\rightleftharpoons$ $SO_2(g)$ + $1/2\ O_2(g)$

Initial	P_0		P_0	0	P_0 = initial pressure of SO_3 and SO_2
Change	$-x$	$\rightarrow$	$+x$	$+x/2$	after first reaction occurs.
Equil.	$P_0 - x$		$P_0 + x$	$x/2$

$P_{total} = P_0 - x + P_0 + x + x/2 = 2\,P_0 + x/2 = 0.836$ atm

$P_{O_2} = x/2 = 0.0275$ atm, $x = 0.0550$ atm

$2\,P_0 + x/2 = 0.836$ atm; $2\,P_0 = 0.836 - 0.0275 = 0.809$ atm, $P_0 = 0.405$ atm

$P_{SO_3} = P_0 - x = 0.405 - 0.0550 = 0.350$ atm; $P_{SO_2} = P_0 + x = 0.405 + 0.0550 = 0.460$ atm

For $2\,FeSO_4(s) \rightleftharpoons Fe_2O_3(s) + SO_3(g) + SO_2(g)$:

$\quad K_p = P_{SO_2} \times P_{SO_3} = (0.460)(0.350) = 0.161$

For $SO_3(g) \rightleftharpoons SO_2(g) + 1/2\,O_2(g)$:

$$K_p = \frac{P_{SO_2} \times P_{O_2}^{1/2}}{P_{SO_3}} = \frac{(0.460)\,(0.0275)^{1/2}}{0.350} = 0.218$$

CHAPTER FOURTEEN

ACIDS AND BASES

Questions

16. The Arrhenius definitions are: acids produce H^+ in water and bases produce OH^- in water. The difference between strong and weak acids and bases is the amount of H^+ and OH^- produced.

    a. A strong acid is 100% dissociated in water.
    b. A strong base is 100% dissociated in water.
    c. A weak acid is much less than 100% dissociated in water (typically 1-10% dissociated).
    d. A weak base is one that has only a small percentage of molecules react with water to form OH^-.

17.   a. Arrhenius acid: produce H^+ in water
    b. Brønsted-Lowry acid: proton (H^+) donor
    c. Lewis acid: electron pair acceptor

    The Lewis definition is most general. The Lewis definition can apply to all Arrhenius and Brønsted-Lowry acids; H^+ has an empty 1s orbital and forms bonds to all bases by accepting a pair of electrons from the base. In addition, the Lewis definition incorporates other reactions not typically considered acid-base reactions, e.g., $BF_3(g) + NH_3(g) \rightarrow F_3B-NH_3(s)$. NH_3 is something we usually consider a base and it is a base in this reaction using the Lewis definition; NH_3 donates a pair of electrons to form the N–B bond.

18. When a strong acid (HX) is added to water, the reaction $HX + H_2O \rightarrow H_3O^+ + X^-$ basically goes to completion. All strong acids in water are completely converted into H_3O^+ and X^-. Thus, no acid stronger than H_3O^+ will remain undissociated in water. Similarly, when a strong base (B) is added to water, the reaction $B + H_2O \rightarrow BH^+ + OH^-$ basically goes to completion. All bases stronger than OH^- are completely converted into OH^- and BH^+. Even though there are acids and bases stronger than H_3O^+ and OH^-, in water these acids and bases are completely converted into H_3O^+ and OH^-.

19. $H_2O \rightleftharpoons H^+ + OH^-$ $K_w = [H^+][OH^-] = 1.0 \times 10^{-14}$

    Neutral solution: $[H^+] = [OH^-]$; $[H^+] = 1.0 \times 10^{-7} M$; $pH = -\log(1.0 \times 10^{-7}) = 7.00$

20. 10.78 (4 S.F.); 6.78 (3 S.F.); 0.78 (2 S.F.); A pH value is a logarithm. The numbers to the left of the decimal place identy the power of ten to which $[H^+]$ is expressed in scientific notation, e.g., 10^{-11}, 10^{-7}, 10^{-1}. The number of decimal places in a pH value identifies the number of significant figures in $[H^+]$. In all three pH values, the $[H^+]$ should be expressed only to two significant figures since these pH values have only two decimal places.

21.   Neutrally charged organic compounds containing at least one nitrogen atom generally behave as weak bases. The nitrogen atom has an unshared pair of electrons around it. This lone pair of electrons is used to form a bond to H^+.

22.   a.   The weaker the X – H bond, the stronger the acid.
        b.   As the electronegativity of neighboring atoms increases, the strength of the acid increases.
        c.   As the number of oxygen atoms increases, the strength of the acid increases.

23.   In general, the weaker the acid, the stronger the conjugate base and vice versa.

        a.   Since acid strength increases as the X – H bond strength decreases, conjugate base strength will increase as the strength of the X – H bond increases.
        b.   Since acid strength increases as the electronegativity of neighboring atoms increases, conjugate base strength will decrease as the electronegativity of neighboring atoms increases.
        c.   Since acid strength increases as the number of oxygen atoms increases, conjugate base strength decreases as the number of oxygen atoms increase.

24.   A Lewis acid must have an empty orbital to accept an electron pair, and a Lewis base must have an unshared pair of electrons.

25.   a. H_2O and $CH_3CO_2^-$

        b.   An acid-base reaction can be thought of as a competition between two opposing bases. Since this equilibrium lies far to the left ($K_a < 1$), $CH_3CO_2^-$ is a stronger base than H_2O.

        c.   The acetate ion is a better base than water and produces basic solutions in water. When we put acetate ion into solution as the only major basic species, the reaction is:

$$CH_3CO_2^- + H_2O \rightleftharpoons CH_3CO_2H + OH^-$$

        Now the competition is between $CH_3CO_2^-$ and OH^- for the proton. Hydroxide ion is the strongest base possible in water. The equilibrium above lies far to the left, resulting in a K_b value less than one. Those species we specifically call weak bases ($10^{-14} < K_b < 1$) lie between H_2O and OH^- in base strength. Weak bases are stronger bases than water but are weaker bases than OH^-.

26.   The NH_4^+ ion is a weak acid because it lies between H_2O and H_3O^+ (H^+) in terms of acid strength. Weak acids are better acids than water, thus their aqueous solutions are acidic. They are weak acids because they are not as strong as H_3O^+ (H^+). Weak acids only partially dissociate in water and have K_a values between 10^{-14} and 1.

27.   When an acid dissociates, ions are produced. The conductivity of the solution is a measure of the number of ions. In addition, the colligative properties, which are discussed in Chapter 11, depend on the number of particles present. So measurements of osmotic pressure, vapor pressure lowering, freezing point depression or boiling point elevation will also allow us to determine the extent of ionization.

28. $K_a \times K_b = K_w$, $-\log (K_a \times K_b) = -\log K_w$

$-\log K_a - \log K_b = -\log K_w$, $pK_a + pK_b = pK_w = 14.00$ (at $25\,°C$)

Exercises

Nature of Acids and Bases

29.   a. $HClO_4(aq) + H_2O(l) \rightarrow H_3O^+(aq) + ClO_4^-(aq)$. Only the forward reaction is indicated since $HClO_4$ is a strong acid and is basically 100% dissociated in water. For acids, the dissociation reaction is commonly written without water as a reactant. The common abbreviation for this reaction is: $HClO_4(aq) \rightarrow H^+(aq) + ClO_4^-(aq)$. This reaction is also called the K_a reaction as the equilibrium constant for this reaction is called K_a.

        b.   Propanoic acid is a weak acid, so it is only partially dissociated in water. The dissociation reaction is: $CH_3CH_2CO_2H(aq) + H_2O(l) \rightleftharpoons H_3O^+(aq) + CH_3CH_2CO_2^-(aq)$ or $CH_3CH_2CO_2H(aq) \rightleftharpoons H^+(aq) + CH_3CH_2CO_2^-(aq)$.

        c. NH_4^+ is a weak acid. Similar to propanoic acid, the dissociation reaction is:

             $NH_4^+(aq) + H_2O(l) \rightleftharpoons H_3O^+(aq) + NH_3(aq)$ or $NH_4^+(aq) \rightleftharpoons H^+(aq) + NH_3(aq)$

30.   The dissociation reaction (the K_a reaction) of an acid in water commonly omits water as a reactant. We will follow this practice. All dissociation reactions produce H^+ and the conjugate base of the acid that is dissociated.

        a. $HC_2H_3O_2(aq) \rightleftharpoons H^+(aq) + C_2H_3O_2^-(aq)$ $K_a = \dfrac{[H^+][C_2H_3O_2^-]}{[HC_2H_3O_2]}$

        b. $Co(H_2O)_6^{3+}(aq) \rightleftharpoons H^+(aq) + Co(H_2O)_5(OH)^{2+}(aq)$ $K_a = \dfrac{[H^+][Co(H_2O)_5(OH)^{2+}]}{[Co(H_2O)_6^{3+}]}$

        c. $CH_3NH_3^+(aq) \rightleftharpoons H^+(aq) + CH_3NH_2(aq)$ $K_a = \dfrac{[H^+][CH_3NH_2]}{[CH_3NH_3^+]}$

31.   An acid is a proton (H^+) donor and a base is a proton acceptor. A conjugate acid-base pair differs by only a proton (H^+).

	Acid	Base	Conjugate Base of Acid	Conjugate Acid of Base
a.	HF	H_2O	F^-	H_3O^+
b.	H_2SO_4	H_2O	HSO_4^-	H_3O^+
c.	HSO_4^-	H_2O	SO_4^{2-}	H_3O^+

32.

	Acid	Base	Conjugate Base of Acid	Conjugate Acid of Base
a.	$Al(H_2O)_6^{3+}$	H_2O	$Al(H_2O)_5(OH)^{2+}$	H_3O^+
b.	$HONH_3^+$	H_2O	$HONH_2$	H_3O^+
c.	$HOCl$	$C_6H_5NH_2$	OCl^-	$C_6H_5NH_3^+$

33.   Strong acids have a $K_a \gg 1$ and weak acids have $K_a < 1$. Table 14.2 in the text lists some K_a values for weak acids. K_a values for strong acids are hard to determine so they are not listed in the text. However, there are only a few common strong acids so if you memorize the strong acids, then all other acids will be weak acids. The strong acids to memorize are HCl, HBr, HI, HNO_3, $HClO_4$ and H_2SO_4.

   a. $HClO_4$ is a strong acid.
   b. $HOCl$ is a weak acid ($K_a = 3.5 \times 10^{-8}$).
   c. H_2SO_4 is a strong acid.
   d. H_2SO_3 is a weak diprotic acid since the K_{a1} and K_{a2} values are less than one.

34.   The beaker on the left represents a strong acid in solution; the acid, HA, is 100% dissociated into the H^+ and A^- ions. The beaker on the right represents a weak acid in solution; only a little bit of the acid, HB, dissociates into ions, so the acid exists mostly as undissociated HB molecules in water.

   a. HNO_2: weak acid beaker
   b. HNO_3: strong acid beaker
   c. HCl: strong acid beaker
   d. HF: weak acid beaker
   e. $HC_2H_3O_2$: weak acid beaker

35.   The K_a value is directly related to acid strength. As K_a increases, acid strength increases. For water, use K_w when comparing the acid strength of water to other species. The K_a values are:

        HNO_3: strong acid ($K_a \gg 1$); $HOCl$: $K_a = 3.5 \times 10^{-8}$

        NH_4^+: $K_a = 5.6 \times 10^{-10}$; H_2O: $K_a = K_w = 1.0 \times 10^{-14}$

        From the K_a values, the ordering is: $HNO_3 > HOCl > NH_4^+ > H_2O$.

36.   Except for water, these are the conjugate bases of the acids in the previous exercise. In general, the weaker the acid, the stronger the conjugate base. NO_3^- is the conjugate base of a strong acid; it is a terrible base (worse than water). The ordering is: $NH_3 > OCl^- > H_2O > NO_3^-$

37.   a. HCl is a strong acid and water is a very weak acid with $K_a = K_w = 1.0 \times 10^{-14}$. HCl is a much stronger acid than H_2O.

        b. H_2O, $K_a = K_w = 1.0 \times 10^{-14}$; HNO_2, $K_a = 4.0 \times 10^{-4}$; HNO_2 is a stronger acid than H_2O since K_a for $HNO_2 > K_a$ for H_2O.

c. HOC_6H_5, $K_a = 1.6 \times 10^{-10}$; HCN, $K_a = 6.2 \times 10^{-10}$; HCN is a stronger acid than HOC_6H_5 since K_a for HCN > K_a for HOC_6H_5.

38. a. H_2O; The conjugate bases of strong acids are terrible bases ($K_b < 10^{-14}$).

b. NO_2^-; The conjugate bases of weak acids are weak bases ($10^{-14} < K_b < 1$).

c. $OC_6H_5^-$; For a conjugate acid-base pair, $K_a \times K_b = K_w$. From this relationship, the stronger the acid the weaker the conjugate base (K_b decreases as K_a increases). Since HCN is a stronger acid than HOC_6H_5 (K_a for HCN > K_a for HOC_6H_5), $OC_6H_5^-$ will be a stronger base than CN^-.

Autoionization of Water and the pH Scale

39. At 25°C, the relationship: $[H^+][OH^-] = K_w = 1.0 \times 10^{-14}$ always holds for aqueous solutions. When $[H^+]$ is greater than 1.0×10^{-7} M, the solution is acidic; when $[H^+]$ is less than 1.0×10^{-7} M, the solution is basic; when $[H^+] = 1.0 \times 10^{-7}$ M, the solution is neutral. In terms of $[OH^-]$, an acidic solution has $[OH^-] < 1.0 \times 10^{-7}$ M, a basic solution has $[OH^-] > 1.0 \times 10^{-7}$ M, and a neutral solution has $[OH^-] = 1.0 \times 10^{-7}$ M.

a. $[OH^-] = \dfrac{K_w}{[H^+]} = \dfrac{1.0 \times 10^{-14}}{1.0 \times 10^{-7}} = 1.0 \times 10^{-7}$ M; The solution is neutral.

b. $[OH^-] = \dfrac{1.0 \times 10^{-14}}{6.7 \times 10^{-4}} = 1.5 \times 10^{-11}$ M; The solution is acidic.

c. $[OH^-] = \dfrac{1.0 \times 10^{-14}}{1.9 \times 10^{-11}} = 5.3 \times 10^{-4}$ M; The solution is basic.

d. $[OH^-] = \dfrac{1.0 \times 10^{-14}}{2.3} = 4.3 \times 10^{-15}$ M; The solution is acidic.

40. a. $[H^+] = \dfrac{K_w}{[OH^-]} = \dfrac{1.0 \times 10^{-14}}{3.6} = 2.8 \times 10^{-15}$ M; basic

b. $[H^+] = \dfrac{1.0 \times 10^{-14}}{9.7 \times 10^{-9}} = 1.0 \times 10^{-6}$ M; acidic

c. $[H^+] = \dfrac{1.0 \times 10^{-14}}{2.2 \times 10^{-3}} = 4.5 \times 10^{-12}$ M; basic

d. $[H^+] = \dfrac{1.0 \times 10^{-14}}{1.0 \times 10^{-7}} = 1.0 \times 10^{-7}$ M; neutral

41. a. Since the value of the equilibrium constant increases as the temperature increases, the reaction is endothermic. In endothermic reactions, heat is a reactant so an increase in temperature (heat) shifts the reaction to produce more products and increases K in the process.

b. $H_2O(l) \rightleftharpoons H^+(aq) + OH^-(aq)$ $K_w = 5.47 \times 10^{-14} = [H^+][OH^-]$ at 50°C

In pure water $[H^+] = [OH^-]$, so $5.47 \times 10^{-14} = [H^+]^2$, $[H^+] = 2.34 \times 10^{-7} M = [OH^-]$

42. a. $H_2O(l) \rightleftharpoons H^+(aq) + OH^-(aq)$ $K_w = 2.92 \times 10^{-14} = [H^+][OH^-]$

In pure water $[H^+] = [OH^-]$, so $2.92 \times 10^{-14} = [H^+]^2$, $[H^+] = 1.71 \times 10^{-7} M = [OH^-]$

b. $pH = -\log[H^+] = -\log(1.71 \times 10^{-7}) = 6.767$

c. $[H^+] = K_w/[OH^-] = 2.92 \times 10^{-14}/0.10 = 2.9 \times 10^{-13} M$; $pH = -\log(2.9 \times 10^{-13}) = 12.54$

43. $pH = -\log[H^+]$; $pOH = -\log[OH^-]$; At 25°C, $pH + pOH = 14.00$; For Exercise 13.39:

a. $pH = -\log[H^+] = -\log(1.0 \times 10^{-7}) = 7.00$; $pOH = 14.00 - pH = 14.00 - 7.00 = 7.00$

b. $pH = -\log(6.7 \times 10^{-4}) = 3.17$; $pOH = 14.00 - 3.17 = 10.83$

c. $pH = -\log(1.9 \times 10^{-11}) = 10.72$; $pOH = 14.00 - 10.72 = 3.28$

d. $pH = -\log(2.3) = -0.36$; $pOH = 14.00 - (-0.36) = 14.36$

Note that pH is less than zero when $[H^+]$ is greater than 1.0 M (an extremely acidic solution). For Exercise 13.40:

a. $pOH = -\log[OH^-] = -\log(3.6) = -0.56$; $pH = 14.00 - pOH = 14.00 - (-0.56) = 14.56$

b. $pOH = -\log(9.7 \times 10^{-9}) = 8.01$; $pH = 14.00 - 8.01 = 5.99$

c. $pOH = -\log(2.2 \times 10^{-3}) = 2.66$; $pH = 14.00 - 2.66 = 11.34$

d. $pOH = -\log(1.0 \times 10^{-7}) = 7.00$; $pH = 14.00 - 7.00 = 7.00$

Note that pH is greater than 14.00 when $[OH^-]$ is greater than 1.0 M (an extremely basic solution).

44. a. $[H^+] = 10^{-pH}$, $[H^+] = 10^{-7.40} = 4.0 \times 10^{-8} M$

$pOH = 14.00 - pH = 14.00 - 7.40 = 6.60$; $[OH^-] = 10^{-pOH} = 10^{-6.60} = 2.5 \times 10^{-7} M$

or $[OH^-] = \dfrac{K_w}{[H^+]} = \dfrac{1.0 \times 10^{-14}}{4.0 \times 10^{-8}} = 2.5 \times 10^{-7} M$; This solution is basic since pH > 7.00.

b. $[H^+] = 10^{-15.3} = 5 \times 10^{-16} M$; $pOH = 14.00 - 15.3 = -1.3$; $[OH^-] = 10^{-(-1.3)} = 20 M$; basic

c. $[H^+] = 10^{-(-1.0)} = 10 M$; $pOH = 14.0 - (-1.0) = 15.0$; $[OH^-] = 10^{-15.0} = 1 \times 10^{-15} M$; acidic

d. $[H^+] = 10^{-3.20} = 6.3 \times 10^{-4} M$; $pOH = 14.00 - 3.20 = 10.80$; $[OH^-] = 10^{-10.80} = 1.6 \times 10^{-11} M$; acidic

e. $[OH^-] = 10^{-5.0} = 1 \times 10^{-5}$ M; pH = 14.0 - pOH = 14.0 - 5.0 = 9.0; $[H^+] = 10^{-9.0} = 1 \times 10^{-9}$ M; basic

f. $[OH^-] = 10^{-9.60} = 2.5 \times 10^{-10}$ M; pH = 14.00 - 9.60 = 4.40; $[H^+] = 10^{-4.40} = 4.0 \times 10^{-5}$ M; acidic

45.   pOH = 14.00 - pH = 14.00 - 6.77 = 7.23; $[H^+] = 10^{-pH} = 10^{-6.77} = 1.7 \times 10^{-7}$ M

$$[OH^-] = \frac{K_w}{[H^+]} = \frac{1.0 \times 10^{-14}}{1.7 \times 10^{-7}} = 5.9 \times 10^{-8} \ M \text{ or } [OH^-] = 10^{-pOH} = 10^{-7.23} = 5.9 \times 10^{-8} \ M$$

The sample of milk is slightly acidic since the pH is less than 7.00 at 25°C.

46.   pH = 14.00 - pOH = 14.00 - 5.74 = 8.26; $[H^+] = 10^{-pH} = 10^{-8.26} = 5.5 \times 10^{-9}$ M

$$[OH^-] = \frac{K_w}{[H^+]} = \frac{1.0 \times 10^{-14}}{5.5 \times 10^{-9}} = 1.8 \times 10^{-6} \ M \text{ or } [OH^-] = 10^{-pOH} = 10^{-5.74} = 1.8 \times 10^{-6} \ M$$

The solution of baking soda is basic since the pH is greater than 7.00 at 25°C.

Solutions of Acids

47.   All the acids in this problem are strong acids that are always assumed to completely dissociate in water. The general dissociation reaction for a strong acid is: $HA(aq) \rightarrow H^+(aq) + A^-(aq)$ where A^- is the conjugate base of the strong acid HA. For 0.250 M solutions of these strong acids, 0.250 M H^+ and 0.250 M A^- are present when the acids completely dissociate. The amount of H^+ donated from water will be insignificant in this problem since H_2O is a very weak acid.

a.   Major species present after dissociation = H^+, ClO_4^- and H_2O;

pH = -log $[H^+]$ = -log (0.250) = 0.602

b.   Major species = H^+, NO_3^- and H_2O; pH = 0.602

48.   Strong acids are assumed to completely dissociate in water: $HCl(aq) \rightarrow H^+(aq) + Cl^-(aq)$

a.   A 0.10 M HCl solution gives 0.10 M H^+ and 0.10 M Cl^- since HCl completely dissociates. The amount of H^+ from H_2O will be insignificant.   pH = -log $[H^+]$ = -log (0.10) = 1.00

b.   5.0 M H^+ is produced when 5.0 M HCl completely dissociates. The amount of H^+ from H_2O will be insignificant. pH = -log (5.0) = -0.70 (Negative pH values just indicate very concentrated acid solutions).

c. 1.0×10^{-11} M H^+ is produced when 1.0×10^{-11} M HCl completely dissociates. If you take the negative log of 1.0×10^{-11} this gives pH = 11.00. This is impossible! We dissolved an acid in water and got a basic pH. What we must consider in this problem is that water by itself donates 1.0×10^{-7} M H^+. We can normally ignore the small amount of H^+ from H_2O except when we have a very dilute solution of an acid (as is the case here). Therefore, the pH is that of neutral water (pH = 7.00) since the amount of HCl present is insignificant.

49. Both are strong acids.

$0.0500 \text{ L} \times 0.050 \text{ mol/L} = 2.5 \times 10^{-3} \text{ mol HCl} = 2.5 \times 10^{-3} \text{ mol H}^+ + 2.5 \times 10^{-3} \text{ mol Cl}^-$

$0.1500 \text{ L} \times 0.10 \text{ mol/L} = 1.5 \times 10^{-2} \text{ mol HNO}_3 = 1.5 \times 10^{-2} \text{ mol H}^+ + 1.5 \times 10^{-2} \text{ mol NO}_3^-$

$[\text{H}^+] = \dfrac{(2.5 \times 10^{-3} + 1.5 \times 10^{-2}) \text{ mol}}{0.2000 \text{ L}} = 0.088 \ M; \ [\text{OH}^-] = \dfrac{\text{K}_w}{[\text{H}^+]} = 1.1 \times 10^{-13} \ M$

$[\text{Cl}^-] = \dfrac{2.5 \times 10^{-3} \text{ mol}}{0.2000 \text{ L}} = 0.013 \ M; \ [\text{NO}_3^-] = \dfrac{1.5 \times 10^{-2} \text{ mol}}{0.2000 \text{ L}} = 0.075 \ M$

50. $90.0 \times 10^{-3} \text{ L} \times \dfrac{5.00 \text{ mol}}{\text{L}} = 0.450 \text{ mol H}^+ \text{ from HCl}$

$30.0 \times 10^{-3} \text{ L} \times \dfrac{8.00 \text{ mol}}{\text{L}} = 0.240 \text{ mol H}^+ \text{ from HNO}_3$

$[\text{H}^+] = \dfrac{0.450 \text{ mol} + 0.240 \text{ mol}}{1.00 \text{ L}} = 0.690 \ M; \ \text{pH} = -\log(0.690) = 0.161$

$\text{pOH} = 14.000 - 0.161 = 13.839; \ [\text{OH}^-] = 10^{-13.839} = 1.45 \times 10^{-14} \ M$

51. $[\text{H}^+] = 10^{-\text{pH}} = 10^{-2.50} = 3.2 \times 10^{-3} \ M.$ Since HCl is a strong acid, a $3.2 \times 10^{-3} \ M$ HCl solution will produce $3.2 \times 10^{-3} \ M \ \text{H}^+$ giving a pH = 2.50.

52. $[\text{H}^+] = 10^{-\text{pH}} = 10^{-5.10} = 7.9 \times 10^{-6} \ M.$ Since HNO$_3$ is a strong acid, a $7.9 \times 10^{-6} \ M$ HNO$_3$ solution is necessary to produce a pH = 5.10 solution.

53. a. HNO$_2$ ($\text{K}_a = 4.0 \times 10^{-4}$) and H$_2$O ($\text{K}_a = \text{K}_w = 1.0 \times 10^{-14}$) are the major species. HNO$_2$ is a much stronger acid than H$_2$O so it is the major source of H$^+$. However, HNO$_2$ is a weak acid ($\text{K}_a < 1$) so it only partially dissociates in water. We must solve an equilibrium problem to determine [H$^+$]. In the Solutions Guide, we will summarize the initial, change and equilibrium concentrations into one table called the ICE table. Solving the weak acid problem:

$$\text{HNO}_2 \quad \rightleftharpoons \quad \text{H}^+ \quad + \quad \text{NO}_2^-$$

	HNO$_2$	H$^+$	NO$_2^-$
Initial	0.250 M	~0	0

x mol/L HNO$_2$ dissociates to reach equilibrium

	HNO$_2$	H$^+$	NO$_2^-$
Change	-x $\rightarrow$	+x	+x
Equil.	0.250 -x	x	x

$\text{K}_a = \dfrac{[\text{H}^+][\text{NO}_2^-]}{[\text{HNO}_2]} = 4.0 \times 10^{-4} = \dfrac{x^2}{0.250 - x};$ If we assume $x \ll 0.250$, then:

$4.0 \times 10^{-4} \approx \dfrac{x^2}{0.250}, \ x = \sqrt{4.0 \times 10^{-4}(0.250)} = 0.010 \ M$

We must check the assumption: $\dfrac{x}{0.250} \times 100 = \dfrac{0.010}{0.250} \times 100 = 4.0\%$

All the assumptions are good. The H^+ contribution from water (10^{-7} M) is negligible, and x is small compared to 0.250 (percent error = 4.0%). If the percent error is less than 5% for an assumption, we will consider it a valid assumption (called the 5% rule). Finishing the problem: $x = 0.010$ $M = [H^+]$; pH = -log(0.010) = 2.00

b. CH_3CO_2H ($K_a = 1.8 \times 10^{-5}$) and H_2O ($K_a = K_w = 1.0 \times 10^{-14}$) are the major species. CH_3CO_2H is the major source of H^+. Solving the weak acid problem:

$$CH_3CO_2H \; \rightleftharpoons \; H^+ \; + \; CH_3CO_2^-$$

Initial	0.250 M	~0	0

x mol/L CH_3CO_2H dissociates to reach equilibrium

Change	-x	$\rightarrow$ +x	+x
Equil.	0.250 - x	x	x

$$K_a = \frac{[H^+][CH_3CO_2^-]}{[CH_3CO_2H]} = 1.8 \times 10^{-5} = \frac{x^2}{0.250 - x} \approx \frac{x^2}{0.250} \quad \text{(assuming } x \ll 0.250\text{)}$$

$x = 2.1 \times 10^{-3}$ M; Checking assumption: $\dfrac{2.1 \times 10^{-3}}{0.250} \times 100 = 0.84\%$. Assumptions good.

$[H^+] = x = 2.1 \times 10^{-3}$ M; pH = -log (2.1×10^{-3}) = 2.68

54.   a. HOC_6H_5 ($K_a = 1.6 \times 10^{-10}$) and H_2O ($K_a = K_w = 1.0 \times 10^{-14}$) are the major species. The major equilibrium is the dissociation of HOC_6H_5. Solving the weak acid problem:

$$HOC_6H_5 \; \rightleftharpoons \; H^+ \; + \; OC_6H_5^-$$

Initial	0.250 M	~0	0

x mol/L HOC_6H_5 dissociates to reach equilibrium

Change	-x	$\rightarrow$ +x	+x
Equil.	0.250 - x	x	x

$$K_a = 1.6 \times 10^{-10} = \frac{[H^+][OC_6H_5^-]}{[HOC_6H_5]} = \frac{x^2}{0.250 - x} \approx \frac{x^2}{0.250} \quad \text{(assuming } x \ll 0.250\text{)}$$

$x = [H^+] = 6.3 \times 10^{-6}$ M; Checking assumption: x is 2.5×10^{-3}% of 0.250, so assumption is valid by the 5% rule.

pH = -log(6.3×10^{-6}) = 5.20

b.   HCN ($K_a = 6.2 \times 10^{-10}$) and H_2O are the major species. HCN is the major source of H^+.

$$HCN \rightleftharpoons H^+ + CN^-$$

Initial	0.250 M	~0	0
	x mol/L HCN dissociates to reach equilibrium		
Change	$-x$ $\rightarrow$	$+x$	$+x$
Equil.	0.250 - x	x	x

$$K_a = 6.2 \times 10^{-10} = \frac{[H^+][CN^-]}{[HCN]} = \frac{x^2}{0.250 - x} \approx \frac{x^2}{0.250} \quad \text{(assuming } x \ll 0.250)$$

$x = [H^+] = 1.2 \times 10^{-5}\ M$; Checking assumption: x is 4.8×10^{-3}% of 0.250

Assumptions good. pH = -log (1.2×10^{-5}) = 4.92

55.   20.0 mL glacial acetic acid $\times \dfrac{1.05\ g}{mL} \times \dfrac{1\ mol}{60.05\ g} = 0.350$ mol $HC_2H_3O_2$

Initial concentration of $HC_2H_3O_2 = \dfrac{0.350\ mol}{0.2500\ L} = 1.40\ M$

$$HC_2H_3O_2 \rightleftharpoons H^+ + C_2H_3O_2^- \qquad K_a = 1.8 \times 10^{-5}$$

Initial	1.40 M	~0	0
	x mol/L $HC_2H_3O_2$ dissociates to reach equilibrium		
Change	$-x$ $\rightarrow$	$+x$	$+x$
Equil.	1.40 - x	x	x

$$K_a = 1.8 \times 10^{-5} = \frac{[H^+][C_2H_3O_2^-]}{[HC_2H_3O_2]} = \frac{x^2}{1.40 - x} \approx \frac{x^2}{1.40}$$

$x = [H^+] = 5.0 \times 10^{-3}\ M$; pH = 2.30   Assumptions good (x is 0.36% of 1.40).

56. $HC_3H_5O_2$ ($K_a = 1.3 \times 10^{-5}$) and H_2O ($K_a = K_w = 1.0 \times 10^{-14}$) are the major species present. $HC_3H_5O_2$ will be the dominant producer of H^+ since $HC_3H_5O_2$ is a stronger acid than H_2O. Solving the weak acid problem:

$$HC_3H_5O_2 \rightleftharpoons H^+ + C_3H_5O_2^-$$

Initial	0.100 M	~0	0
	x mol/L $HC_3H_5O_2$ dissociates to reach equilibrium		
Change	$-x$ $\rightarrow$	$+x$	$+x$
Equil.	0.100 - x	x	x

$$K_a = 1.3 \times 10^{-5} = \frac{[H^+][C_3H_5O_2^-]}{[HC_3H_5O_2]} = \frac{x^2}{0.100 - x} \approx \frac{x^2}{0.100}$$

$x = [H^+] = 1.1 \times 10^{-3}\ M$; $pH = -\log (1.1 \times 10^{-3}) = 2.96$

Assumption follows the 5% rule (x is 1.1% of 0.100).

$[H^+] = [C_3H_5O_2^-] = 1.1 \times 10^{-3}\ M$; $[OH^-] = K_w/[H^+] = 9.1 \times 10^{-12}\ M$

$[HC_3H_5O_2] = 0.100 - 1.1 \times 10^{-3} = 0.099\ M$

Percent dissociation $= \dfrac{[H^+]}{[HC_3H_5O_2]_o} \times 100 = \dfrac{1.1 \times 10^{-3}}{0.100} \times 100 = 1.1\%$

57.   This is a weak acid in water. Solving the weak acid problem:

$$HF \quad\rightleftharpoons\quad H^+ \quad + \quad F^- \qquad K_a = 7.2 \times 10^{-4}$$

Initial	0.020 M	~0	0

x mol/L HF dissociates to reach equilibrium

Change	$-x$	$\rightarrow$ $+x$	$+x$
Equil.	0.020 - x	x	x

$K_a = 7.2 \times 10^{-4} = \dfrac{[H^+][F^-]}{[HF]} = \dfrac{x^2}{0.020 - x} \approx \dfrac{x^2}{0.020}$ (assuming $x \ll 0.020$)

$x = [H^+] = 3.8 \times 10^{-3}\ M$; Check assumptions: $\dfrac{x}{0.020} \times 100 = \dfrac{3.8 \times 10^{-3}}{0.020} \times 100 = 19\%$

The assumption $x \ll 0.020$ is not good (x is more than 5% of 0.020). We must solve $x^2/(0.020 - x) = 7.2 \times 10^{-4}$ exactly by using either the quadratic formula or by the method of successive approximations (see Appendix 1.4 of text). Using successive approximations, we let 0.016 M be a new approximation for [HF]. That is, in the denominator, try $x = 0.0038$ (the value of x we calculated making the normal assumption), so $0.020 - 0.0038 = 0.016$, then solve for a new value of x in the numerator.

$\dfrac{x^2}{0.020 - x} \approx \dfrac{x^2}{0.016} = 7.2 \times 10^{-4}$, $x = 3.4 \times 10^{-3}$

We use this new value of x to further refine our estimate of [HF], i.e., $0.020 - x = 0.020 - 0.0034 = 0.0166$ (carry extra significant figure).

$\dfrac{x^2}{0.020 - x} \approx \dfrac{x^2}{0.0166} = 7.2 \times 10^{-4}$, $x = 3.5 \times 10^{-3}$

We repeat until we get an answer that repeats itself. This would be the same answer we would get solving exactly using the quadratic equation. In this case it is: $x = 3.5 \times 10^{-3}$

So: $[H^+] = [F^-] = x = 3.5 \times 10^{-3}\ M$; $[OH^-] = K_w/[H^+] = 2.9 \times 10^{-12}\ M$

$[HF] = 0.020 - x = 0.020 - 0.0035 = 0.017\ M$; pH = 2.46

Note: When the 5% assumption fails, use whichever method you are most comfortable with to solve exactly. The method of successive approximations is probably fastest when the percent error is less than ~25% (unless you have a calculator that can solve quadratic equations).

58.   Major species: HIO_3, H_2O; Major source of H^+: HIO_3 (a weak acid, $K_a = 0.17$)

$$HIO_3 \rightleftharpoons H^+ + IO_3^-$$

Initial	0.20 M	~0	0

x mol/L HIO_3 dissociates to reach equilibrium

Change	$-x$ $\rightarrow$	$+x$	$+x$
Equil.	0.20 - x	x	x

$$K_a = 0.17 = \frac{[H^+][IO_3^-]}{[HIO_3]} = \frac{x^2}{0.20 - x} \approx \frac{x^2}{0.20}, \quad x = 0.18; \quad \text{Check assumption.}$$

Assumption is horrible (x is 90% of 0.20). When the assumption is this poor, it is generally quickest to solve exactly using the quadratic formula (see Appendix 1.4 in text). The method of successive approximations will require many trials to finally converge on the answer. For this problem, 5 trials were required. Using the quadratic formula and carrying extra significant figures:

$$0.17 = \frac{x^2}{0.20 - x}, \quad x^2 = 0.17(0.20 - x), \quad x^2 + 0.17x - 0.034 = 0$$

$$x = \frac{-0.17 \pm [(0.17)^2 - 4(1)(-0.034)]^{1/2}}{2(1)} = \frac{-0.17 \pm 0.406}{2}, \quad x = 0.12 \text{ or } -0.29$$

Only $x = 0.12$ makes sense. $x = 0.12\ M = [H^+]$; pH = -log (0.12) = 0.92

59.   Major species: $HC_2H_2ClO_2$ ($K_a = 1.35 \times 10^{-3}$) and H_2O; Major source of H^+: $HC_2H_2ClO_2$

$$HC_2H_2ClO_2 \rightleftharpoons H^+ + C_2H_2ClO_2^-$$

Initial	0.10 M	~0	0

x mol/L $HC_2H_2ClO_2$ dissociates to reach equilibrium

Change	$-x$ $\rightarrow$	$+x$	$+x$
Equil.	0.10 - x	x	x

$$K_a = 1.35 \times 10^{-3} = \frac{x^2}{0.10 - x} \approx \frac{x^2}{0.10}, \quad x = 1.2 \times 10^{-2}\ M$$

Checking the assumptions finds that x is 12% of 0.10 which fails the 5% rule. We must solve $1.35 \times 10^{-3} = x^2/(0.10 - x)$ exactly using either the method of successive approximations or the quadratic equation. Using either method gives $x = [H^+] = 1.1 \times 10^{-2}\ M$. pH = -log [H^+] = -log $(1.1 \times 10^{-2}) = 1.96$.

60. $HClO_2$ $\rightleftharpoons$ H^+ + ClO_2^- $K_a = 1.2 \times 10^{-2}$

Initial	0.22 M		~0	0

x mol/L $HClO_2$ dissociates to reach equilibrium

Change	-x	$\rightarrow$	+x	+x
Equil.	0.22 - x		x	x

$$K_a = 1.2 \times 10^{-2} = \frac{[H^+][ClO_2^-]}{[HClO_2]} = \frac{x^2}{0.22 - x} \approx \frac{x^2}{0.22}, \ x = 5.1 \times 10^{-2}$$

The assumption that x is small is not good (x is 23% of 0.22). Using the method of successive approximations and carrying extra significant figures:

$$\frac{x^2}{0.22 - 0.051} = \frac{x^2}{0.169} = 1.2 \times 10^{-2}, \ x = 4.5 \times 10^{-2}$$

$$\frac{x^2}{0.175} = 1.2 \times 10^{-2}, \ x = 4.6 \times 10^{-2}; \ x = 4.6 \times 10^{-2} \text{ repeats}$$

$[H^+] = [ClO_2^-] = x = 4.6 \times 10^{-2} \ M;$ % dissociation $= \dfrac{4.6 \times 10^{-2}}{0.22} \times 100 = 21\%$

61.   a.   HCl is a strong acid. It will produce 0.10 M H^+. HOCl is a weak acid. Let's consider the equilibrium:

                             HOCl $\rightleftharpoons$ H^+ + OCl^- $K_a = 3.5 \times 10^{-8}$

Initial	0.10 M		0.10 M	0

x mol/L HOCl dissociates to reach equilibrium

Change	-x	$\rightarrow$	+x	+x
Equil.	0.10 - x		0.10 + x	x

$$K_a = 3.5 \times 10^{-8} = \frac{[H^+][OCl^-]}{[HOCl]} = \frac{(0.10 + x)(x)}{0.10 - x} \approx x, \ x = 3.5 \times 10^{-8} \ M$$

Assumptions are great (x is 3.5×10^{-5}% of 0.10). We are really assuming that HCl is the only important source of H^+, which it is. The $[H^+]$ contribution from HOCl, x, is negligible. Therefore, $[H^+] = 0.10 \ M$; pH = 1.00

       b. HNO_3 is a strong acid, giving an initial concentration of H^+ equal to 0.050 M. Consider the equilibrium:

                             $HC_2H_3O_2$ $\rightleftharpoons$ H^+ + $C_2H_3O_2^-$ $K_a = 1.8 \times 10^{-5}$

Initial	0.50 M		0.050 M	0

x mol/L $HC_2H_3O_2$ dissociates to reach equilibrium

Change	-x	$\rightarrow$	+x	+x
Equil.	0.50 - x		0.050 + x	x

$$K_a = 1.8 \times 10^{-5} = \frac{[H^+][C_2H_3O_2^-]}{[HC_2H_3O_2]} = \frac{(0.050 + x)x}{(0.50 - x)} \approx \frac{0.050\,x}{0.50}$$

$x = 1.8 \times 10^{-4}$; Assumptions are good (well within the 5% rule).

$[H^+] = 0.050 + x = 0.050\,M$ and pH = 1.30

62.   HF and HOC_6H_5 are both weak acids with K_a values of 7.2×10^{-4} and 1.6×10^{-10}, respectively. Since the K_a value for HF is much greater than the K_a value for HOC_6H_5, HF will be the dominant producer of H^+ (we can ignore the amount of H^+ produced from HOC_6H_5 since it will be insignificant).

	HF	$\rightleftharpoons$	H^+	+	F^-
Initial	1.0 M		~0		0
	x mol/L HF dissociates to reach equilibrium				
Change	-x	$\rightarrow$	+x		+x
Equil.	1.0 - x		x		x

$$K_a = 7.2 \times 10^{-4} = \frac{[H^+][F^-]}{[HF]} = \frac{x^2}{1.0 - x} \approx \frac{x^2}{1.0}$$

$x = [H^+] = 2.7 \times 10^{-2}\,M$; pH = -log$(2.7 \times 10^{-2})$ = 1.57   Assumptions good.

Solving for $[OC_6H_5^-]$ using $HOC_6H_5 \rightleftharpoons H^+ + OC_6H_5^-$ equilibrium:

$$K_a = 1.6 \times 10^{-10} = \frac{[H^+][OC_6H_5^-]}{[HOC_6H_5]} = \frac{(2.7 \times 10^{-2})[OC_6H_5^-]}{1.0}, \quad [OC_6H_5^-] = 5.9 \times 10^{-9}\,M$$

Note that this answer indicates that only $5.9 \times 10^{-9}\,M\ HOC_6H_5$ dissociates, which indicates that HF is truly the only significant producer of H^+ in this solution.

63.   In all parts of this problem, acetic acid $(HC_2H_3O_2)$ is the best weak acid present. We must solve a weak acid problem.

a.

	$HC_2H_3O_2$	$\rightleftharpoons$	H^+	+	$C_2H_3O_2^-$
Initial	0.50 M		~0		0
	x mol/L $HC_2H_3O_2$ dissociates to reach equilibrium				
Change	-x	$\rightarrow$	+x		+x
Equil.	0.50 - x		x		x

$$K_a = 1.8 \times 10^{-5} = \frac{[H^+][C_2H_3O_2^-]}{[HC_2H_3O_2]} = \frac{x^2}{0.50 - x} \approx \frac{x^2}{0.50}$$

$x = [H^+] = [C_2H_3O_2^-] = 3.0 \times 10^{-3}\,M$ Assumptions good.

$$\text{Percent dissociation} = \frac{[H^+]}{[HC_2H_3O_2]_0} \times 100 = \frac{3.0 \times 10^{-3}}{0.50} \times 100 = 0.60\%$$

b.   The setups for solutions b and c are similar to solution a except the final equation is slightly different, reflecting the new concentration of $HC_2H_3O_2$.

$$K_a = 1.8 \times 10^{-5} = \frac{x^2}{0.050 - x} \approx \frac{x^2}{0.050}$$

$x = [H^+] = [C_2H_3O_2^-] = 9.5 \times 10^{-4}\ M$ Assumptions good.

$$\% \text{ dissociation} = \frac{9.5 \times 10^{-4}}{0.050} \times 100 = 1.9\%$$

c. $$K_a = 1.8 \times 10^{-5} = \frac{x^2}{0.0050 - x} \approx \frac{x^2}{0.0050}$$

$x = [H^+] = [C_2H_3O_2^-] = 3.0 \times 10^{-4}\ M;$ Check assumptions.

Assumption that x is negligible is borderline (6.0% error). We should solve exactly. Using the method of successive approximations (see Appendix 1.4 of text):

$$1.8 \times 10^{-5} = \frac{x^2}{0.0050 - 3.0 \times 10^{-4}} = \frac{x^2}{0.0047}, \quad x = 2.9 \times 10^{-4}$$

Next trial also gives $x = 2.9 \times 10^{-4}$.

$$\% \text{ dissociation} = \frac{2.9 \times 10^{-4}}{5.0 \times 10^{-3}} \times 100 = 5.8\%$$

d.   As we dilute a solution, all concentrations decrease. Dilution will shift the equilibrium to the side with the greater number of particles. For example, suppose we double the volume of an equilibrium mixture of a weak acid by adding water, then:

$$Q = \frac{\left(\dfrac{[H^+]_{eq}}{2}\right)\left(\dfrac{[X^-]_{eq}}{2}\right)}{\left(\dfrac{[HX]_{eq}}{2}\right)} = \frac{1}{2}K_a$$

$Q < K_a$, so the equilibrium shifts to the right or towards a greater percent dissociation.

e. $[H^+]$ depends on the initial concentration of weak acid and on how much weak acid dissociates. For solutions a-c the initial concentration of acid decreases more rapidly than the percent dissociation increases. Thus, $[H^+]$ decreases.

64.   a. HNO_3 is a strong acid; it is assumed 100% dissociated in solution.

b.

$$HNO_2 \rightleftharpoons H^+ + NO_2^- \qquad K_a = 4.0 \times 10^{-4}$$

Initial	0.20 M	~0	0

x mol/L HNO_2 dissociates to reach equilibrium

Change	-x	$\rightarrow$	+x	+x
Equil.	0.20 - x		x	x

$$K_a = 4.0 \times 10^{-4} = \frac{[H^+][NO_2^-]}{[HNO_2]} = \frac{x^2}{0.20 - x} \approx \frac{x^2}{0.20}$$

$x = [H^+] = [NO_2^-] = 8.9 \times 10^{-3}\ M$; Assumptions good.

$$\% \text{ dissociation} = \frac{[H^+]}{[HNO_2]_o} \times 100 = \frac{8.9 \times 10^{-3}}{0.20} \times 100 = 4.5\%$$

c.

$$HOC_6H_5 \rightleftharpoons H^+ + OC_6H_5^- \qquad K_a = 1.6 \times 10^{-10}$$

Initial	0.20 M	~0	0

x mol/L HOC_6H_5 dissociates to reach equilibrium

Change	-x	$\rightarrow$	+x	+x
Equil.	0.20 - x		x	x

$$K_a = 1.6 \times 10^{-10} = \frac{[H^+][OC_6H_5^-]}{[HOC_6H_5]} = \frac{x^2}{0.20 - x} \approx \frac{x^2}{0.20}$$

$x = [H^+] = [OC_6H_5^-] = 5.7 \times 10^{-6}\ M$;   Assumptions good.

$$\% \text{ dissociation} = \frac{5.7 \times 10^{-6}}{0.20} \times 100 = 2.9 \times 10^{-3}\%$$

d. For the same initial concentration, the percent dissociation increases as the strength of the acid increases (as K_a increases).

65.   Let HX symbolize the weak acid. Setup the problem like a typical weak acid equilibrium problem.

$$HX \rightleftharpoons H^+ + X^-$$

Initial	0.15 M	~0	0

x mol/L HX dissociates to reach equilibrium

Change	-x	$\rightarrow$	+x	+x
Equil.	0.15 - x		x	x

If the acid is 3.0% dissociated, then $x = [H^+]$ is 3.0% of 0.15: $x = 0.030 \times (0.15\ M) = 4.5 \times 10^{-3}\ M$. Now that we know the value of x, we can solve for K_a.

$$K_a = \frac{[H^+][X^-]}{[HX]} = \frac{x^2}{0.15 - x} = \frac{(4.5 \times 10^{-3})^2}{0.15 - 4.5 \times 10^{-3}} = 1.4 \times 10^{-4}$$

66.
$$HF \rightleftharpoons H^+ + F^-$$

Initial   0.100 M ~0   0
               x mol/L HF dissociates to reach equilibrium
Change   -x $\rightarrow$ +x +x
Equil.   0.100 - x x x

$$K_a = \frac{[H^+][F^-]}{[HF]} = \frac{x^2}{0.100 - x}; \quad x = [H^+] = [F^-] = 0.081 \times (0.100\ M) = 8.1 \times 10^{-3}\ M$$

$$[HF] = 0.100 - 8.1 \times 10^{-3} = 0.092\ M; \quad K_a = \frac{(8.1 \times 10^{-3})^2}{0.092} = 7.1 \times 10^{-4}$$

67.   Setup the problem using the K_a equilibrium reaction for HOCN.

$$HOCN \rightleftharpoons H^+ + OCN^-$$

Initial   0.0100 M ~0   0
               x mol/L HOBr dissociates to reach equilibrium
Change   -x $\rightarrow$ +x +x
Equil.   0.0100 - x x x

$$K_a = \frac{[H^+][OCN^-]}{[HOCN]} = \frac{x^2}{0.0100 - x}; \quad \text{Since pH} = 2.77, \text{ then: } x = [H^+] = 10^{-pH} = 10^{-2.77} = 1.7 \times 10^{-3}\ M$$

$$K_a = \frac{(1.7 \times 10^{-3})^2}{0.0100 - 1.7 \times 10^{-3}} = 3.5 \times 10^{-4}$$

68. $HClO_4$ is a strong acid with $[H^+] = 0.040\ M$. This equals the $[H^+]$ in the trichloroacetic acid solution. Now setup the problem using the K_a equilibrium reaction for CCl_3CO_2H.

$$CCl_3CO_2H \rightleftharpoons H^+ + CCl_3CO_2^-$$

Initial   0.050 M ~0   0
Equil.   0.050 - x x x

$$K_a = \frac{[H^+][CCl_3CO_2^-]}{[CCl_3CO_2H]} = \frac{x^2}{0.050 - x}; \quad x = [H^+] = 4.0 \times 10^{-2}\ M$$

$$K_a = \frac{(4.0 \times 10^{-2})^2}{0.050 - (4.0 \times 10^{-2})} = 0.16$$

69.   Major species: HCOOH and H_2O; Major source of H^+: HCOOH

$$HCOOH \rightleftharpoons H^+ + HCOO^-$$

Initial   C   ~0   0   where C = $[HCOOH]_o$
               x mol/L HCOOH dissociates to reach equilibrium
Change   -x $\rightarrow$ +x +x
Equil.   C - x x x

$$K_a = 1.8 \times 10^{-4} = \frac{[H^+][HCOO^-]}{[HCOOH]} = \frac{x^2}{C - x} \text{ where } x = [H^+]$$

$$1.8 \times 10^{-4} = \frac{[H^+]^2}{C - [H^+]}; \text{ Since pH} = 2.70, \text{ then: } [H^+] = 10^{-2.70} = 2.0 \times 10^{-3} M$$

$$1.8 \times 10^{-4} = \frac{(2.0 \times 10^{-3})^2}{C - (2.0 \times 10^{-3})}, \quad C - (2.0 \times 10^{-3}) = \frac{4.0 \times 10^{-6}}{1.8 \times 10^{-4}}, \quad C = 2.4 \times 10^{-2} M$$

A 0.024 M formic acid solution will have pH = 2.70.

70. $[HA]_o = \dfrac{1.0 \text{ mol}}{2.0 \text{ L}} = 0.50 \text{ mol/L}$; Solve using the K_a equilibrium reaction.

	HA	$\rightleftharpoons$	H^+	+	A^-
Initial	0.50 M		~0		0
Equil.	0.50 - x		x		x

$$K_a = \frac{[H^+][A^-]}{[HA]} = \frac{x^2}{0.50 - x}; \text{ In this problem, } [HA] = 0.45 \; M \text{ so:}$$

$$[HA] = 0.45 \; M = 0.50 \; M - x, \quad x = 0.05 \; M; \quad K_a = \frac{(0.05)^2}{0.45} = 6 \times 10^{-3}$$

Solutions of Bases

71.   a. $NH_3(aq) + H_2O(l) \rightleftharpoons NH_4^+(aq) + OH^-(aq)$ $K_b = \dfrac{[NH_4^+][OH^-]}{[NH_3]}$

        b. $C_5H_5N(aq) + H_2O(l) \rightleftharpoons C_5H_5NH^+(aq) + OH^-(aq)$ $K_b = \dfrac{[C_5H_5NH^+][OH^-]}{[C_5H_5N]}$

72.   a. $C_6H_5NH_2(aq) + H_2O(l) \rightleftharpoons C_6H_5NH_3^+(aq) + OH^-(aq)$ $K_b = \dfrac{[C_6H_5NH_3^+][OH^-]}{[C_6H_5NH_2]}$

        b. $(CH_3)_2NH(aq) + H_2O(l) \rightleftharpoons (CH_3)_2NH_2^+(aq) + OH^-(aq)$ $K_b = \dfrac{[(CH_3)_2NH_2^+][OH^-]}{[(CH)_3)_2NH]}$

73. NO_3^-: $K_b \ll K_w$ since HNO_3 is a strong acid. All conjugate bases of strong acids have no base strength. H_2O: $K_b = K_w = 1.0 \times 10^{-14}$; NH_3: $K_b = 1.8 \times 10^{-5}$; C_5H_5N: $K_b = 1.7 \times 10^{-9}$

        $NH_3 > C_5H_5N > H_2O > NO_3^-$ (As K_b increases, base strength increases.)

74.   Excluding water, these are the conjugate acids of the bases in the previous exercise. In general, the stronger the base, the weaker the conjugate acid. Note: Even though NH_4^+ and $C_5H_5NH^+$ are conjugate acids of weak bases, they are still weak acids with K_a values between K_w and 1. Prove this to yourself by calculating the K_a values for NH_4^+ and $C_5H_5NH^+$ ($K_a = K_w/K_b$).

        $HNO_3 > C_5H_5NH^+ > NH_4^+ > H_2O$

75.   a. NH_3 b. NH_3 c. OH^- d. CH_3NH_2

The base with the largest K_b value is the strongest base. OH^- is the strongest base possible in water.

76.   a. HNO_3 b. NH_4^+ c. NH_4^+

The acid with the largest K_a value is the strongest acid. To calculate K_a values for NH_4^+ and $CH_3NH_3^+$, use $K_a = K_w/K_b$ where K_b refers to the bases NH_3 or CH_3NH_2.

77. $NaOH(aq) \rightarrow Na^+(aq) + OH^-(aq)$; NaOH is a strong base which completely dissociates into Na^+ and OH^-. The initial concentration of NaOH will equal the concentration of OH^- donated by NaOH.

a. $[OH^-] = 0.10\ M$; $pOH = -\log[OH^-] = -\log(0.10) = 1.00$

pH = 14.00 - pOH = 14.00 - 1.00 = 13.00

Note that H_2O is also present, but the amount of OH^- produced by H_2O will be insignificant compared to the $0.10\ M\ OH^-$ produced from the NaOH.

b.   The $[OH^-]$ concentration donated by the NaOH is $1.0 \times 10^{-10}\ M$. Water by itself donates $1.0 \times 10^{-7}\ M$. In this problem, water is the major OH^- contributor and $[OH^-] = 1.0 \times 10^{-7}\ M$.

pOH = $-\log(1.0 \times 10^{-7}) = 7.00$; pH = 14.00 - 7.00 = 7.00

c. $[OH^-] = 2.0\ M$; pOH = $-\log(2.0) = -0.30$; pH = 14.00 - (-0.30) = 14.30

78.   a. $Ca(OH)_2 \rightarrow Ca^{2+} + 2\ OH^-$; $Ca(OH)_2$ is a strong base and dissociates completely.

$[OH^-] = 2(0.00040) = 8.0 \times 10^{-4}\ M$; pOH = $-\log[OH^-] = 3.10$; pH = 14.00 - pOH = 10.90

b. $\dfrac{25\ g\ KOH}{L} \times \dfrac{1\ mol\ KOH}{56.11\ g\ KOH} = 0.45\ mol\ KOH/L$

KOH is a strong base, so $[OH^-] = 0.45\ M$; pOH = $-\log(0.45) = 0.35$; pH = 13.65

c. $\dfrac{150.0\ g\ NaOH}{L} \times \dfrac{1\ mol}{40.00\ g} = 3.750\ M$; NaOH is a strong base, so $[OH^-] = 3.750\ M$.

pOH = $-\log(3.750) = -0.5740$ and pH = 14.0000 - (-0.5740) = 14.5740

Although we are justified in calculating the answer to four decimal places, in reality pH values are generally measured to two decimal places, sometimes three.

79.   a. Major species: K^+, OH^-, H_2O (KOH is a strong base.)

$[OH^-] = 0.015\ M$, pOH = $-\log(0.015) = 1.82$; pH = 14.00 - pOH = 12.18

b.   Major species: Ba^{2+}, OH^-, H_2O; $Ba(OH)_2(aq) \rightarrow Ba^{2+}(aq) + 2\ OH^-(aq)$; Since each mol of the strong base $Ba(OH)_2$ dissolves in water to produce two mol OH^-, then $[OH^-] = 2(0.015\ M) = 0.030\ M$.

pOH = $-\log(0.030) = 1.52$; pH = 14.00 - 1.52 = 12.48

80.   a. Major species: Na^+, Li^+, OH^-, H_2O (NaOH and LiOH are both strong bases.)

        $[OH^-] = 0.050 + 0.050 = 0.100\ M$; pOH = 1.000; pH = 13.000

        b. Major species: Ba^{2+}, Rb^+, OH^-, H_2O; Both $Ba(OH)_2$ and RbOH are strong bases and $Ba(OH)_2$ donates 2 mol OH^- per mol $Ba(OH)_2$.

        $[OH^-] = 2(0.0010) + 0.020 = 0.022\ M$; pOH = -log (0.022) = 1.66; pH = 12.34

81.   pH = 10.50; pOH = 14.00 - 10.50 = 3.50; $[OH^-] = 10^{-3.50} = 3.2 \times 10^{-4}\ M$

        $KOH(aq) \rightleftharpoons K^+(aq) + OH^-(aq)$; Since KOH is a strong base, a $3.2 \times 10^{-4}\ M$ KOH solution will produce a pH = 10.50 solution.

82.   pH = 10.50; pOH = 14.00 - 10.50 = 3.50; $[OH^-] = 10^{-3.50} = 3.2 \times 10^{-4}\ M$

        $Sr(OH)_2(aq) \rightarrow Sr^{2+}(aq) + 2\ OH^-(aq)$; $Sr(OH)_2$ donates two mol OH^- per mol $Sr(OH)_2$.

        $$[Sr(OH)_2] = 3.2 \times 10^{-4}\ M\ OH^- \times \left(\frac{1\ M\ Sr(OH)_2}{2\ M\ OH^-} \right) = 1.6 \times 10^{-4}\ M\ Sr(OH)_2$$

        A $1.6 \times 10^{-4}\ M$ $Sr(OH)_2$ solution will produce a pH = 10.50 solution.

83. NH_3 is a weak base with $K_b = 1.8 \times 10^{-5}$. The major species present will be NH_3 and H_2O ($K_b = K_w = 1.0 \times 10^{-14}$). Since NH_3 has a much larger K_b value compared to H_2O, NH_3 is the stronger base present and will be the major producer of OH^-. To determine the amount of OH^- produced from NH_3, we must perform an equilibrium calculation.

        $$NH_3(aq)\ +\ H_2O(l)\ \rightleftharpoons\ NH_4^+(aq)\ +\ OH^-(aq)$$

Initial	0.150 M	0	~0
	x mol/L NH_3 reacts with H_2O to reach equilibrium		
Change	$-x$	$\rightarrow$ $+x$	$+x$
Equil.	0.150 - x	x	x

        $$K_b = 1.8 \times 10^{-5} = \frac{[NH_4^+][OH^-]}{[NH_3]} = \frac{x^2}{0.150 - x} \approx \frac{x^2}{0.150}\quad \text{(assuming } x \ll 0.150)$$

        $x = [OH^-] = 1.6 \times 10^{-3}\ M$; Check assumptions: x is 1.1% of 0.150 so the assumption $0.150 - x \approx 0.150$ is valid by the 5% rule. Also, the contribution of OH^- from water will be insignificant (which will usually be the case). Finishing the problem, pOH = -log $[OH^-]$ = -log ($1.6 \times 10^{-3}\ M$) = 2.80; pH = 14.00 - pOH = 14.00 - 2.80 = 11.20.

84.   Major species: H_2NNH_2 ($K_b = 3.0 \times 10^{-6}$) and H_2O ($K_b = K_w = 1.0 \times 10^{-14}$); The weak base H_2NNH_2 will dominate OH^- production. We must perform a weak base equilibrium calculation.

$$H_2NNH_2 + H_2O \rightleftharpoons H_2NNH_3^+ + OH^- \qquad K_b = 3.0 \times 10^{-6}$$

Initial	2.0 M	0	~0

x mol/L H_2NNH_2 reacts with H_2O to reach equilibrium

Change	$-x$ $\rightarrow$	$+x$	$+x$
Equil.	2.0 - x	x	x

$$K_b = 3.0 \times 10^{-6} = \frac{[H_2NNH_3^+][OH^-]}{[H_2NNH_2]} = \frac{x^2}{2.0 - x} \approx \frac{x^2}{2.0} \quad \text{(assuming } x \ll 2.0)$$

$x = [OH^-] = 2.4 \times 10^{-3} \, M$; pOH = 2.62; pH = 11.38   Assumptions good (x is 0.12% of 2.0).

$[H_2NNH_3^+] = 2.4 \times 10^{-3} \, M$; $[H_2NNH_2] = 2.0 \, M$; $[H^+] = 10^{-11.38} = 4.2 \times 10^{-12} \, M$

85. These are solutions of weak bases in water. We must solve the equilibrium weak base problem.

a.
$$(C_2H_5)_3N + H_2O \rightleftharpoons (C_2H_5)_3NH^+ + OH^- \qquad K_b = 4.0 \times 10^{-4}$$

Initial	0.20 M	0	~0

x mol/L of $(C_2H_5)_3N$ reacts with H_2O to reach equilibrium

Change	$-x$ $\rightarrow$	$+x$	$+x$
Equil.	0.20 - x	x	x

$$K_b = 4.0 \times 10^{-4} = \frac{[(C_2H_5)_3NH^+][OH^-]}{[(C_2H_5)_3N]} = \frac{x^2}{0.20 - x} \approx \frac{x^2}{0.20}, \quad x = [OH^-] = 8.9 \times 10^{-3} \, M$$

Assumptions good (x is 4.5% of 0.20). $[OH^-] = 8.9 \times 10^{-3} \, M$

$$[H^+] = \frac{K_w}{[OH^-]} = \frac{1.0 \times 10^{-14}}{8.9 \times 10^{-3}} = 1.1 \times 10^{-12} \, M; \quad \text{pH} = 11.96$$

b.
$$HONH_2 + H_2O \rightleftharpoons HONH_3^+ + OH^- \qquad K_b = 1.1 \times 10^{-8}$$

Initial	0.20 M	0	~0
Equil.	0.20 - x	x	x

$$K_b = 1.1 \times 10^{-8} = \frac{x^2}{0.20 - x} \approx \frac{x^2}{0.20}, \quad x = [OH^-] = 4.7 \times 10^{-5} \, M; \quad \text{Assumptions good.}$$

$[H^+] = 2.1 \times 10^{-10} \, M$; pH = 9.68

86.   These are solutions of weak bases in water.

a. $C_6H_5NH_2 + H_2O \rightleftharpoons C_6H_5NH_3^+ + OH^-$ $K_b = 3.8 \times 10^{-10}$

Initial	0.20 M	0	~0

 x mol/L of $C_6H_5NH_2$ reacts with H_2O to reach equilibrium

Change	$-x$	$\rightarrow$	$+x$	$+x$
Equil.	$0.20 - x$		x	x

$$3.8 \times 10^{-10} = \frac{x^2}{0.20 - x} \approx \frac{x^2}{0.20}, \quad x = [OH^-] = 8.7 \times 10^{-6} \, M; \text{ Assumptions good.}$$

$[H^+] = K_w/[OH^-] = 1.1 \times 10^{-9} \, M$; pH = 8.96

b. $C_5H_5N + H_2O \rightleftharpoons C_5H_5NH^+ + OH^-$ $K_b = 1.7 \times 10^{-9}$

Initial	0.20 M	0	~0
Equil.	$0.20 - x$	x	x

$$K_b = 1.7 \times 10^{-9} = \frac{x^2}{0.20 - x} \approx \frac{x^2}{0.20}, \quad x = 1.8 \times 10^{-5} \, M; \text{ Assumptions good.}$$

$[OH^-] = 1.8 \times 10^{-5} \, M$; $[H^+] = 5.6 \times 10^{-10} \, M$; pH = 9.25

87.   This is a solution of a weak base in water. We must solve the weak base equilibrium problem.

                $C_2H_5NH_2 + H_2O \rightleftharpoons C_2H_5NH_3^+ + OH^-$ $K_b = 5.6 \times 10^{-4}$

Initial	0.20 M	0	~0

 x mol/L $C_2H_5NH_2$ reacts with H_2O to reach equilibrium

Change	$-x$	$\rightarrow$	$+x$	$+x$
Equil.	$0.20 - x$		x	x

$$K_b = \frac{[C_2H_5NH_3^+][OH^-]}{[C_2H_5NH_2]} = \frac{x^2}{0.20 - x} \approx \frac{x^2}{0.20} \quad \text{(assuming } x \ll 0.20)$$

$x = 1.1 \times 10^{-2}$; Checking assumption: $\dfrac{1.1 \times 10^{-2}}{0.20} \times 100 = 5.5\%$

Assumption fails the 5% rule. We must solve exactly using either the quadratic equation or the method of successive approximations (see Appendix 1.4 of the text). Using successive approximations and carrying extra significant figures:

$$\frac{x^2}{0.20 - 0.011} = \frac{x^2}{0.189} = 5.6 \times 10^{-4}, \quad x = 1.0 \times 10^{-2} \, M \quad \text{(consistent answer)}$$

$$x = [OH^-] = 1.0 \times 10^{-2} \, M; \quad [H^+] = \frac{K_w}{[OH^-]} = \frac{1.0 \times 10^{-14}}{1.0 \times 10^{-2}} = 1.0 \times 10^{-12} \, M; \quad \text{pH} = 12.00$$

88.
$$(C_2H_5)_2NH + H_2O \rightleftharpoons (C_2H_5)_2NH_2^+ + OH^- \quad K_b = 1.3 \times 10^{-3}$$

Initial   0.050 M 0   ~0

x mol/L $(C_2H_5)_2NH$ reacts with H_2O to reach equilibrium

Change $-x$ $\rightarrow$ $+x$ $+x$

Equil.   0.050 - x x x

$$K_b = 1.3 \times 10^{-3} = \frac{[(C_2H_5)_2NH_2^+][OH^-]}{[(C_2H_5)_2NH]} = \frac{x^2}{0.050 - x} \approx \frac{x^2}{0.050} \quad \text{(assuming } x << 0.050)$$

$x = 8.1 \times 10^{-3}$;   Assumption is bad (x is 16% of 0.20).

Using successive approximations:

$$1.3 \times 10^{-3} = \frac{x^2}{0.050 - 0.0081}, \quad x = 7.4 \times 10^{-3}$$

$$1.3 \times 10^{-3} = \frac{x^2}{0.050 - 0.0074}, \quad x = 7.4 \times 10^{-3} \text{ (consistent answer)}$$

$[OH^-] = x = 7.4 \times 10^{-3} \, M$; $[H^+] = K_w/[OH^-] = 1.4 \times 10^{-12} \, M$;   pH = 11.85

89.   To solve for percent ionization, just solve the weak base equilibrium problem.

a. $NH_3 + H_2O \rightleftharpoons NH_4^+ + OH^- \quad K_b = 1.8 \times 10^{-5}$

Initial   0.10 M 0   ~0

Equil.   0.10 - x x x

$$K_b = 1.8 \times 10^{-5} = \frac{x^2}{0.10 - x} \approx \frac{x^2}{0.10}, \quad x = [OH^-] = 1.3 \times 10^{-3} \, M; \quad \text{Assumptions good.}$$

$$\text{Percent ionization} = \frac{[OH^-]}{[NH_3]_o} \times 100 = \frac{1.3 \times 10^{-3} \, M}{0.10 \, M} \times 100 = 1.3\%$$

b. $NH_3 + H_2O \rightleftharpoons NH_4^+ + OH^-$

Initial   0.010 M 0   ~0

Equil.   0.010 - x x x

$$1.8 \times 10^{-5} = \frac{x^2}{0.010 - x} \approx \frac{x^2}{0.010}, \quad x = [OH^-] = 4.2 \times 10^{-4} \, M; \quad \text{Assumptions good.}$$

$$\text{Percent ionization} = \frac{4.2 \times 10^{-4}}{0.010} \times 100 = 4.2\%$$

Note: For the same base, the percent ionization increases as the initial concentration of base decreases.

90.   a. $HONH_2 + H_2O$ $\rightleftharpoons$ $HONH_3^+ + OH^-$ $K_b = 1.1 \times 10^{-8}$

Initial   0.10 M 0   ~0
Equil.   0.10 - x x x

$1.1 \times 10^{-8} = \dfrac{x^2}{0.10 - x} \approx \dfrac{x^2}{0.10}$, $x = [OH^-] = 3.3 \times 10^{-5}$ M; Assumptions good.

Percent ionization $= \dfrac{[OH^-]}{[HONH_2]_o} \times 100 = \dfrac{3.3 \times 10^{-5}}{0.10} \times 100 = 0.033\%$

b. $CH_3NH_2 + H_2O$ $\rightleftharpoons$ $CH_3NH_3^+ + OH^-$ $K_b = 4.38 \times 10^{-4}$

Initial   0.10 M 0   ~0
Equil.   0.10 - x x x

$4.38 \times 10^{-4} = \dfrac{x^2}{0.10 - x} \approx \dfrac{x^2}{0.10}$, $x = 6.6 \times 10^{-3}$; Assumption fails the 5% rule (x is 6.6% of 0.10).

Using successive approximations and carrying extra significant figures:

$$\dfrac{x^2}{0.10 - 0.0066} = \dfrac{x^2}{0.093} = 4.38 \times 10^{-4}, \ x = 6.4 \times 10^{-3} \quad \text{(consistent answer)}$$

Percent ionization $= \dfrac{6.4 \times 10^{-3}}{0.10} \times 100 = 6.4\%$

91.   Let cod = codeine, $C_{18}H_{21}NO_3$; using the K_b reaction to solve:

cod + H_2O $\rightleftharpoons$ $codH^+$ + OH^-

Initial 1.7×10^{-3} M 0   ~0
              x mol/L codeine reacts with H_2O to reach equilibrium
Change $-x$ $\rightarrow$ $+x$ $+x$
Equil. $1.7 \times 10^{-3} - x$ x x

$K_b = \dfrac{x^2}{1.7 \times 10^{-3} - x}$; Since pH = 9.59, then pOH = 14.00 - 9.59 = 4.41.

$[OH^-] = x = 10^{-4.41} = 3.9 \times 10^{-5}$ M; $K_b = \dfrac{(3.9 \times 10^{-5})^2}{1.7 \times 10^{-3} - 3.9 \times 10^{-5}} = 9.2 \times 10^{-7}$

92.   Using the K_b reaction to solve where PY = pyrrolidine, C_4H_8NH:

PY + H_2O $\rightleftharpoons$ PYH^+ + OH^-

Initial 1.00×10^{-3} M 0   ~0
Equil. $1.00 \times 10^{-3} - x$ x x

$$K_b = \frac{x^2}{1.00 \times 10^{-3} - x}; \quad \text{Since pH} = 10.82: \ \text{pOH} = 3.18 \ \text{and} \ [\text{OH}^-] = x = 10^{-3.18} = 6.6 \times 10^{-4} \ M$$

$$K_b = \frac{(6.6 \times 10^{-4})^2}{1.00 \times 10^{-3} - 6.6 \times 10^{-4}} = 1.3 \times 10^{-3}$$

Polyprotic Acids

93. $H_2SO_3(aq) \rightleftharpoons HSO_3^-(aq) + H^+(aq)$ $K_{a_1} = \dfrac{[HSO_3^-][H^+]}{[H_2SO_3]}$

        $HSO_3^-(aq) \rightleftharpoons SO_3^{2-}(aq) + H^+(aq)$ $K_{a_2} = \dfrac{[SO_3^{2-}][H^+]}{[HSO_3^-]}$

94. $H_3C_6H_5O_7(aq) \rightleftharpoons H_2C_6H_5O_7^-(aq) + H^+(aq)$ $K_{a_1} = \dfrac{[H_2C_6H_5O_7^-][H^+]}{[H_3C_6H_5O_7]}$

        $H_2C_6H_5O_7^-(aq) \rightleftharpoons HC_6H_5O_7^{2-}(aq) + H^+(aq)$ $K_{a_2} = \dfrac{[HC_6H_5O_7^{2-}][H^+]}{[H_2C_6H_5O_7^-]}$

        $HC_6H_5O_7^{2-}(aq) \rightleftharpoons C_6H_5O_7^{3-}(aq) + H^+(aq)$ $K_{a_3} = \dfrac{[C_6H_5O_7^{3-}][H^+]}{[HC_6H_5O_7^{2-}]}$

95.   In both these polyprotic acid problems, the dominate equilibrium is the K_{a_1} reaction. The amount of H^+ produced from the subsequent K_a reactions will be minimal since they are all have much smaller K_a values.

     a. $H_3PO_4 \quad \rightleftharpoons \quad H^+ \quad + \quad H_2PO_4^-$ $K_{a_1} = 7.5 \times 10^{-3}$

Initial	0.10 M	~0	0

            x mol/L H_3PO_4 dissociates to reach equilibrium

Change	-x	$\rightarrow$ +x	+x
Equil.	0.10 - x	x	x

$$K_{a_1} = 7.5 \times 10^{-3} = \frac{[H^+][H_2PO_4^-]}{[H_3PO_4]} = \frac{x^2}{0.10 - x} \approx \frac{x^2}{0.10}, \ x = 2.7 \times 10^{-2}$$

Assumption is bad (x is 27% of 0.10). Using successive approximations:

$$\frac{x^2}{0.10 - 0.027} = 7.5 \times 10^{-3}, \ x = 2.3 \times 10^{-2}; \quad \frac{x^2}{0.10 - 0.023} = 7.5 \times 10^{-3}, \ x = 2.4 \times 10^{-2}$$
$$\text{(consistent answer)}$$

$$x = [H^+] = 2.4 \times 10^{-2} \ M; \ \text{pH} = -\log(2.4 \times 10^{-2}) = 1.62$$

b. $H_2CO_3 \rightleftharpoons H^+ + HCO_3^-$ $K_{a_1} = 4.3 \times 10^{-7}$

Initial	0.10 M	~0	0
Equil.	0.10 - x	x	x

$$K_{a_1} = 4.3 \times 10^{-7} = \frac{[H^+][HCO_3^-]}{[H_2CO_3]} = \frac{x^2}{(0.10 - x)} \approx \frac{x^2}{0.10}$$

$x = [H^+] = 2.1 \times 10^{-4} M$; pH = 3.68; Assumptions good.

96.   The reactions are:

$$H_3AsO_4 \rightleftharpoons H^+ + H_2AsO_4^- \quad K_{a_1} = 5 \times 10^{-3}$$

$$H_2AsO_4^- \rightleftharpoons H^+ + HAsO_4^{2-} \quad K_{a_2} = 8 \times 10^{-8}$$

$$HAsO_4^{2-} \rightleftharpoons H^+ + AsO_4^{3-} \quad K_{a_3} = 6 \times 10^{-10}$$

We will deal with the reactions in order of importance, beginning with the largest K_a, K_{a_1}.

$H_3AsO_4 \rightleftharpoons H^+ + H_2AsO_4^-$ $K_{a_1} = 5 \times 10^{-3} = \dfrac{[H^+][H_2AsO_4^-]}{[H_3AsO_4]}$

Initial	0.20 M	~0	0
Equil.	0.20 - x	x	x

$5 \times 10^{-3} = \dfrac{x^2}{0.20 - x}$, $x = 2.9 \times 10^{-2} = 3 \times 10^{-2} M$ (By using successive approximations or the quadratic formula.)

$[H^+] = [H_2AsO_4^-] = 3 \times 10^{-2} M$; $[H_3AsO_4] = 0.20 - 0.03 = 0.17 M$

Since $K_{a_2} = \dfrac{[H^+][HAsO_4^{2-}]}{[H_2AsO_4^-]} = 8 \times 10^{-8}$ is much smaller than the K_{a_1} value, very little of

$H_2AsO_4^-$ (and $HAsO_4^{2-}$) dissociates compared to H_3AsO_4. Therefore, $[H^+]$ and $[H_2AsO_4^-]$ will not change significantly by the K_{a_2} reaction. Using the previously calculated concentrations of H^+ and $H_2AsO_4^-$ to calculate the concentration of $HAsO_4^{2-}$:

$$8 \times 10^{-8} = \frac{(3 \times 10^{-2})[HAsO_4^{2-}]}{3 \times 10^{-2}}, \quad [HAsO_4^{2-}] = 8 \times 10^{-8} M$$

Assumption that the K_{a2} reaction does not change $[H^+]$ and $[H_2AsO_4^-]$ is good. We repeat the process using K_{a_3} to get $[AsO_4^{3-}]$.

$$K_{a_3} = 6 \times 10^{-10} = \frac{[H^+][AsO_4^{3-}]}{[HAsO_4^{2-}]} = \frac{(3 \times 10^{-2})[AsO_4^{3-}]}{(8 \times 10^{-8})}$$

$[AsO_4^{3-}] = 1.6 \times 10^{-15} \approx 2 \times 10^{-15}\ M$ Assumption good.

So in 0.20 M analytical concentration of H_3AsO_4:

$[H_3AsO_4] = 0.17\ M$; $[H^+] = [H_2AsO_4^-] = 3 \times 10^{-2}\ M$

$[HAsO_4^{2-}] = 8 \times 10^{-8}\ M$; $[AsO_4^{3-}] = 2 \times 10^{-15}\ M$

$[OH^-] = K_w/[H^+] = 3 \times 10^{-13}\ M$

97.   The dominant H^+ producer is the strong acid H_2SO_4. A 2.0 M H_2SO_4 solution produces 2.0 M HSO_4^- and 2.0 M H^+. However, HSO_4^- is a weak acid which could also add H^+ to the solution.

	HSO_4^-	$\rightleftharpoons$	H^+	+	SO_4^{2-}
Initial	2.0 M		2.0 M		0
	x mol/L HSO_4^- dissociates to reach equilibrium				
Change	$-x$	$\rightarrow$	$+x$		$+x$
Equil.	2.0 $- x$		2.0 $+ x$		x

$$K_{a_2} = 1.2 \times 10^{-2} = \frac{[H^+][SO_4^{2-}]}{[HSO_4^-]} = \frac{(2.0 + x)(x)}{2.0 - x} \approx \frac{2.0\,(x)}{2.0},\ x = 1.2 \times 10^{-2}$$

Since x is 0.60% of 2.0, the assumption is valid by the 5% rule. The amount of additional H^+ from HSO_4^- is 1.2×10^{-2}. The total amount of H^+ present is:

$[H^+] = 2.0 + 1.2 \times 10^{-2} = 2.0\ M$; pH $= -\log(2.0) = -0.30$

Note: In this problem, H^+ from HSO_4^- could have been ignored. However, this is not usually the case, especially in more dilute solutions of H_2SO_4.

98.   For H_2SO_4, the first dissociation occurs to completion. The hydrogen sulfate ion, HSO_4^-, is a weak acid with $K_{a_2} = 1.2 \times 10^{-2}$. We will consider this equilibrium for additional H^+ production:

	HSO_4^-	$\rightleftharpoons$	H^+	+	SO_4^{2-}
Initial	0.0050 M		0.0050 M		0
	x mol/L HSO_4^- dissociates to reach equilibrium				
Change	$-x$	$\rightarrow$	$+x$		$+x$
Equil.	0.0050 $- x$		0.0050 $+ x$		x

$$K_{a_2} = 0.012 = \frac{(0.0050 + x)(x)}{(0.0050 - x)} \approx x,\ x = 0.012;\ \text{Assumption is horrible (240\% error).}$$

Using the quadratic formula:

$$6.0 \times 10^{-5} - 0.012\,x = x^2 + 0.0050\,x, \quad x^2 + 0.017\,x - 6.0 \times 10^{-5} = 0$$

$$x = \frac{-0.017 \pm (2.9 \times 10^{-4} + 2.4 \times 10^{-4})^{1/2}}{2} = \frac{-0.017 \pm 0.023}{2}, \quad x = 3.0 \times 10^{-3}\,M$$

$[H^+] = 0.0050 + x = 0.0050 + 0.0030 = 0.0080\,M;\ \ pH = 2.10$

Note: We had to consider both H_2SO_4 and HSO_4^- for H^+ production in this problem.

Acid-Base Properties of Salts

99. One difficult aspect of acid-base chemistry is recognizing what types of species are present in
     solution, i.e., whether a species is a strong acid, strong base, weak acid, weak base or a neutral
     species. Below are some ideas and generalizations to keep in mind that will help in recognizing types
     of species present.

   a. Memorize the following strong acids: HCl, HBr, HI, HNO_3, $HClO_4$ and H_2SO_4
   b. Memorize the following strong bases: LiOH, NaOH, KOH, RbOH, $Ca(OH)_2$, $Sr(OH)_2$ and
       $Ba(OH)_2$
   c. All weak acids have a K_a value less than 1 but greater than K_w. Some weak acids are in Table
       14.2 of the text. All weak bases have a K_b value less than 1 but greater than K_w. Some weak
       bases are in Table 14.3 of the text.
   d. All conjugate bases of weak acids are weak bases, i.e., all have a K_b value less than 1 but
       greater than K_w. Some examples of these are the conjugate bases of the weak acids in Table
       14.2 of the text.
   e. All conjugate acids of weak bases are weak acids, i.e., all have a K_a value less than 1 but greater
       than K_w. Some examples of these are the conjugate acids of the weak bases in Table 14.3 of
       the text.
   f. Alkali metal ions (Li^+, Na^+, K^+, Rb^+, Cs^+) and heavier alkaline earth metal ions (Ca^{2+}, Sr^{2+}, Ba^{2+})
       have no acidic or basic properties in water.
   g. All conjugate bases of strong acids (Cl^-, Br^-, I^-, NO_3^-, ClO_4^-, HSO_4^-) have no basic properties in
       water ($K_b \ll K_w$) and only HSO_4^- has any acidic properties in water.

   Lets apply these ideas to this problem to see what type of species are present. The letters in
   parenthesis is/are the generalization(s) above which identifies the species.

   KOH: strong base (b)
   KCl: neutral; K^+ and Cl^- have no acidic/basic properties (f and g).
   KCN: CN^- is a weak base, $K_b = 1.0 \times 10^{-14}/6.2 \times 10^{-10} = 1.6 \times 10^{-5}$ (c and d). Ignore K^+(f).
   NH_4Cl: NH_4^+ is a weak acid, $K_a = 5.6 \times 10^{-10}$ (c and e). Ignore Cl^-(g).
   HCl: strong acid (a)

   The most acidic solution will be the strong acid followed by the weak acid. The most basic solution
   will be the strong base followed by the weak base. The KCl solution will be between the acidic and
   basic solutions at pH = 7.00.

   Most acidic → most basic; HCl > NH_4Cl > KCl > KCN > KOH

100.   See Exercise 14.99 for some generalizations on acid-base properties of salts. The letters in parenthesis is/are the generalization(s) listed in Exercise 14.99 which identifies the species.

        $CaBr_2$:   neutral; Ca^{2+} and Br^- have no acidic/basic properties (f and g).

        KNO_2: NO_2^- is a weak base, $K_b = 1.0 \times 10^{-14}/4.0 \times 10^{-4} = 2.5 \times 10^{-11}$ (c and d).
                                      Ignore K^+ (f).

        $HClO_4$:   strong acid (a)

        HNO_2:   weak acid, $K_a = 4.0 \times 10^{-4}$ (c)

        $HONH_3ClO_4$: $HONH_3^+$ is a weak acid, $K_a = 1.0 \times 10^{-14}/1.1 \times 10^{-8} = 9.1 \times 10^{-7}$ (c and e).
                                        Ignore ClO_4^- (g).

Using the information above (identity and K_a or K_b values), the ordering is:

        most acidic → most basic: $HClO_4 > HNO_2 > HONH_3ClO_4 > CaBr_2 > KNO_2$

101.   From the K_a values, acetic acid is a stronger acid than hypochlorous acid. Conversely, the conjugate base of acetic acid, $C_2H_3O_2^-$, will be a weaker base than the conjugate base of hypochlorous acid, OCl^-. Thus, the hypochlorite ion, OCl^-, is a stronger base than the acetate ion, $C_2H_3O_2^-$. In general, the stronger the acid, the weaker the conjugate base. This statement comes from the relationship $K_w = K_a \times K_b$, which holds for all conjugate acid-base pairs.

102.   Since NH_3 is a weaker base (smaller K_b value) than CH_3NH_2, the conjugate acid of NH_3 will be a stronger acid than the conjugate acid of CH_3NH_2. Thus, NH_4^+ is a stronger acid than $CH_3NH_3^+$.

103. $NaN_3 \rightarrow Na^+ + N_3^-$; Azide, N_3^-, is a weak base since it is the conjugate base of a weak acid. All conjugate bases of weak acids are weak bases ($K_w < K_b < 1$). Ignore Na^+.

$$N_3^- \ + \ H_2O \ \rightleftharpoons \ HN_3 \ + \ OH^- \qquad K_b = \frac{K_w}{K_a} = \frac{1.0 \times 10^{-14}}{1.9 \times 10^{-5}} = 5.3 \times 10^{-10}$$

Initial	0.010 M	0	~0

                x mol/L of N_3^- reacts with H_2O to reach equilibrium

Change	-x	→ +x	+x
Equil.	0.010 - x	x	x

$$K_b = \frac{[HN_3][OH^-]}{[N_3^-]} = 5.3 \times 10^{-10} = \frac{x^2}{0.010 - x} \approx \frac{x^2}{0.010} \quad \text{(assuming } x \ll 0.010\text{)}$$

$$x = [OH^-] = 2.3 \times 10^{-6}\ M; \ [H^+] = \frac{1.0 \times 10^{-14}}{2.3 \times 10^{-6}} = 4.3 \times 10^{-9}\ M \quad \text{Assumptions good.}$$

$[HN_3] = [OH^-] = 2.3 \times 10^{-6}\ M;\ [Na^+] = 0.010\ M;\ [N_3^-] = 0.010 - 2.3 \times 10^{-6} = 0.010\ M$

104. $C_2H_5NH_3Cl \rightarrow C_2H_5NH_3^+ + Cl^-$; $C_2H_5NH_3^+$ is the conjugate acid of the weak base $C_2H_5NH_2$ ($K_b = 5.6 \times 10^{-4}$). As is true for all conjugate acids of weak bases, $C_2H_5NH_3^+$ is a weak acid. Cl^- has no basic (or acidic) properties. Ignore Cl^-. Solving the weak acid problem:

$$C_2H_5NH_3^+ \quad \rightleftharpoons \quad C_2H_5NH_2 \ + \ H^+ \quad K_a = K_w/5.6 \times 10^{-4} = 1.8 \times 10^{-11}$$

Initial   0.25 M 0   ~0

           x mol/L $C_2H_5NH_3^+$ dissociates to reach equilibrium

Change $-x$ $\rightarrow$ $+x$ $+x$

Equil.   0.25 $- x$ x x

$$K_a = 1.8 \times 10^{-11} = \frac{[C_2H_5NH_2][H^+]}{[C_2H_5NH_3^+]} = \frac{x^2}{0.25 - x} \approx \frac{x^2}{0.25} \quad \text{(assuming } x << 0.25)$$

$x = [H^+] = 2.1 \times 10^{-6} \ M; \ \text{pH} = 5.68; \ \text{Assumptions good.}$

$[C_2H_5NH_2] = [H^+] = 2.1 \times 10^{-6} \ M; \ [C_2H_5NH_3^+] = 0.25 \ M; \ [Cl^-] = 0.25 \ M$

$[OH^-] = K_w/[H^+] = 4.8 \times 10^{-9} \ M$

105.   a. $CH_3NH_3Cl \rightarrow CH_3NH_3^+ + Cl^-$: $CH_3NH_3^+$ is a weak acid. Cl^- is the conjugate base of a strong acid. Cl^- has no basic (or acidic) properties.

$$CH_3NH_3^+ \rightleftharpoons CH_3NH_2 + H^+ \quad K_a = \frac{[CH_3NH_2][H^+]}{[CH_3NH_3^+]} = \frac{K_w}{K_b} = \frac{1.00 \times 10^{-14}}{4.38 \times 10^{-4}} = 2.28 \times 10^{-11}$$

$$CH_3NH_3^+ \quad \rightleftharpoons \quad CH_3NH_2 \ + \ H^+$$

Initial   0.10 M 0   ~0

           x mol/L $CH_3NH_3^+$ dissociates to reach equilibrium

Change $-x$ $\rightarrow$ $+x$ $+x$

Equil.   0.10 $- x$ x x

$$K_a = 2.28 \times 10^{-11} = \frac{x^2}{0.10 - x} \approx \frac{x^2}{0.10} \quad \text{(assuming } x << 0.10)$$

$x = \ [H^+] = 1.5 \times 10^{-6} \ M; \ \text{pH} = 5.82 \qquad \text{Assumptions good.}$

   b. $NaCN \rightarrow Na^+ + CN^-$: CN^- is a weak base. Na^+ has no acidic (or basic) properties.

$$CN^- \ + \ H_2O \quad \rightleftharpoons \quad HCN \ + \ OH^- \quad K_b = \frac{K_w}{K_a} = \frac{1.0 \times 10^{-14}}{6.2 \times 10^{-10}} = 1.6 \times 10^{-5}$$

Initial   0.050 M 0   ~0

           x mol/L CN^- reacts with H_2O to reach equilibrium

Change $-x$ $\rightarrow$ $+x$ $+x$

Equil.   0.050 $- x$ x x

$$K_b = 1.6 \times 10^{-5} = \frac{[HCN][OH^-]}{[CN^-]} = \frac{x^2}{0.050 - x} \approx \frac{x^2}{0.050}$$

$x = [OH^-] = 8.9 \times 10^{-4} \ M; \ \text{pOH} = 3.05; \ \text{pH} = 10.95 \qquad \text{Assumptions good.}$

106.   a. $KNO_2 \rightarrow K^+ + NO_2^-$: NO_2^- is a weak base. Ignore K^+.

$$NO_2^- + H_2O \rightleftharpoons HNO_2 + OH^- \qquad K_b = \frac{K_w}{K_a} = \frac{1.0 \times 10^{-14}}{4.0 \times 10^{-4}} = 2.5 \times 10^{-11}$$

Initial   0.12 M 0   ~0
Equil.   0.12 - x x x

$$K_b = 2.5 \times 10^{-11} = \frac{[OH^-][HNO_2]}{[NO_2^-]} = \frac{x^2}{0.12 - x} \approx \frac{x^2}{0.12}$$

$x = [OH^-] = 1.7 \times 10^{-6}\ M$; pOH = 5.77; pH = 8.23   Assumptions good.

   b. $NaOCl \rightarrow Na^+ + OCl^-$: OCl^- is a weak base. Ignore Na^+.

$$OCl^- + H_2O \rightleftharpoons HOCl + OH^- \qquad K_b = \frac{K_w}{K_a} = \frac{1.0 \times 10^{-14}}{3.5 \times 10^{-8}} = 2.9 \times 10^{-7}$$

Initial   0.45 M 0   ~0
Equil.   0.45 - x x x

$$K_b = 2.9 \times 10^{-7} = \frac{[HOCl][OH^-]}{[OCl^-]} = \frac{x^2}{0.45 - x} \approx \frac{x^2}{0.45}$$

$x = [OH^-] = 3.6 \times 10^{-4}\ M$; pOH = 3.44; pH = 10.56   Assumptions good.

   c. $NH_4ClO_4 \rightarrow NH_4^+ + ClO_4^-$: NH_4^+ is a weak acid. ClO_4^- is the conjugate base of a strong acid. ClO_4^- has no basic (or acidic) properties.

$$NH_4^+ \rightleftharpoons NH_3 + H^+ \qquad K_a = \frac{K_w}{K_b} = \frac{1.0 \times 10^{-14}}{1.8 \times 10^{-5}} = 5.6 \times 10^{-10}$$

Initial   0.40 M 0   ~0
Equil.   0.40 - x x x

$$K_a = 5.6 \times 10^{-10} = \frac{[NH_3][H^+]}{[NH_4^+]} = \frac{x^2}{0.40 - x} \approx \frac{x^2}{0.40}$$

$x = [H^+] = 1.5 \times 10^{-5}\ M$; pH = 4.82;   Assumptions good.

107.   All these salts contain Na^+, which has no acidic/basic properties, and a conjugate base of a weak acid (except for NaCl where Cl^- is a neutral species.). All conjugate bases of weak acids are weak bases since K_b for these species is between K_w and 1. To identify the species, we will use the data given to determine the K_b value for the weak conjugate base. From the K_b value and data in Table 14.2 of the text, we can identify the conjugate base present by calculating the K_a value for the weak acid. We will use A^- as an abbreviation for the weak conjugate base.

$$A^- + H_2O \rightleftharpoons HA + OH^-$$

Initial	0.100 mol/1.00 L		0	~0

x mol/L A^- reacts with H_2O to reach equilibrium

Change $-x$ $\rightarrow$ $+x$ $+x$
Equil. $0.100 - x$ x x

$$K_b = \frac{[HA][OH^-]}{[A^-]} = \frac{x^2}{0.100 - x}; \quad \text{From the problem, pH = 8.07:}$$

$$pOH = 14.00 - 8.07 = 5.93; \quad [OH^-] = x = 10^{-5.93} = 1.2 \times 10^{-6}\ M$$

$$K_b = \frac{(1.2 \times 10^{-6})^2}{0.100 - 1.2 \times 10^{-6}} = 1.4 \times 10^{-11} = K_b \text{ value for the conjugate base of a weak acid.}$$

The K_a value for the weak acid equals K_w/K_b: $K_a = \dfrac{1.0 \times 10^{-14}}{1.4 \times 10^{-11}} = 7.1 \times 10^{-4}$

From Table 14.2 of the text, this K_a value is closest to HF. Therefore, the unknown salt is NaF.

108. $BHCl \rightarrow BH^+ + Cl^-$; Cl^- is the conjugate base of the strong acid HCl, so Cl^- has no acidic/basic properties. BH^+ is a weak acid since it is the conjugate acid of a weak base, B. Determining the K_a value for BH^+:

$$BH^+ \rightleftharpoons B + H^+$$

Initial	0.10 M		0	~0

x mol/L BH^+ dissociates to reach equilibrium

Change $-x$ $\rightarrow$ $+x$ $+x$
Equil. $0.10 - x$ x x

$$K_a = \frac{[B][H^+]}{[BH^+]} = \frac{x^2}{0.10 - x}; \quad \text{From the problem, pH = 5.82:}$$

$$[H^+] = x = 10^{-5.82} = 1.5 \times 10^{-6}\ M; \quad K_a = \frac{(1.5 \times 10^{-6})^2}{0.10 - 1.5 \times 10^{-6}} = 2.3 \times 10^{-11}$$

K_b for the base, $B = K_w \backslash K_a = 1.0 \times 10^{-14}/2.3 \times 10^{-11} = 4.3 \times 10^{-4}$.

From Table 14.3 of the text, this K_b value is closest to CH_3NH_2 so the unknown salt is CH_3NH_3Cl.

109. Major species present: $Al(H_2O)_6^{3+}$ ($K_a = 1.4 \times 10^{-5}$), NO_3^- (neutral) and H_2O ($K_w = 1.0 \times 10^{-14}$); $Al(H_2O)_6^{3+}$ is a stronger acid than water so it will be the dominant H^+ producer.

$$Al(H_2O)_6^{3+} \quad \rightleftharpoons \quad Al(H_2O)_5(OH)^{2+} \quad + \quad H^+$$

Initial   0.050 M 0   ~0

x mol/L $Al(H_2O)_6^{3+}$ dissociates to reach equilibrium

Change $-x$ $\rightarrow$ $+x$ $+x$

Equil.   0.050 $- x$ x x

$$K_a = 1.4 \times 10^{-5} = \frac{[Al(H_2O)_5(OH)^{2+}][H^+]}{[Al(H_2O)_6^{3+}]} = \frac{x^2}{0.050 - x} \approx \frac{x^2}{0.050}$$

$x = 8.4 \times 10^{-4} \, M = [H^+]$; pH $= -\log(8.4 \times 10^{-4}) = 3.08$; Assumptions good.

110.   Major species: $Co(H_2O)_6^{3+}$ ($K_a = 1.0 \times 10^{-5}$), Cl^- (neutral) and H_2O ($K_w = 1.0 \times 10^{-14}$); $Co(H_2O)_6^{3+}$ will determine the pH since it is a stronger acid than water. Solving the weak acid problem in the usual manner:

$$Co(H_2O)_6^{3+} \quad \rightleftharpoons \quad Co(H_2O)_5(OH)^{2+} \quad + \quad H^+ \quad K_a = 1.0 \times 10^{-5}$$

Initial   0.10 M 0   ~0

Equil.   0.10 $- x$ x x

$$K_a = 1.0 \times 10^{-5} = \frac{x^2}{0.10 - x} \approx \frac{x^2}{0.10}, \quad x = [H^+] = 1.0 \times 10^{-3} \, M$$

pH $= -\log(1.0 \times 10^{-3}) = 3.00$; Assumptions good.

111.   Reference Table 14.6 of the text and the solution to Exercise 14.99 for some generalizations on acid-base properties of salts.

   a. $NaNO_3 \rightarrow Na^+ + NO_3^-$ neutral; Neither species has any acidic/basic properties.

   b. $NaNO_2 \rightarrow Na^+ + NO_2^-$ basic; NO_2^- is a weak base and Na^+ has no effect on pH.

$$NO_2^- + H_2O \rightleftharpoons HNO_2 + OH^- \quad K_b = \frac{K_w}{K_{a, HNO_2}} = \frac{1.0 \times 10^{-14}}{4.0 \times 10^{-4}} = 2.5 \times 10^{-11}$$

   c. $C_5H_5NHClO_4 \rightarrow C_5H_5NH^+ + ClO_4^-$ acidic; $C_5H_5NH^+$ is a weak acid and ClO_4^- has no effect on pH.

$$C_5H_5NH^+ \rightleftharpoons H^+ + C_5H_5N \quad K_a = \frac{K_w}{K_{b, C_5H_5N}} = \frac{1.0 \times 10^{-14}}{1.7 \times 10^{-9}} = 5.9 \times 10^{-6}$$

   d. $NH_4NO_2 \rightarrow NH_4^+ + NO_2^-$ acidic; NH_4^+ is a weak acid ($K_a = 5.6 \times 10^{-10}$) and NO_2^- is a weak base ($K_b = 2.5 \times 10^{-11}$). Since $K_{a, NH_4^+} > K_{b, NO_2^-}$, then the solution is acidic.

$$NH_4^+ \rightleftharpoons H^+ + NH_3 \quad K_a = 5.6 \times 10^{-10}; \quad NO_2^- + H_2O \rightleftharpoons HNO_2 + OH^- \quad K_b = 2.5 \times 10^{-11}$$

e. $KOCl \rightarrow K^+ + OCl^-$ basic; OCl^- is a weak base and K^+ has no effect on pH.

$$OCl^- + H_2O \rightleftharpoons HOCl + OH^- \quad K_b = \frac{K_w}{K_{a,\,HOCl}} = \frac{1.0 \times 10^{-14}}{3.5 \times 10^{-8}} = 2.9 \times 10^{-7}$$

f. $NH_4OCl \rightarrow NH_4^+ + OCl^-$ basic; NH_4^+ is a weak acid and OCl^- is a weak base. Since $K_{b,\,OCl^-} > K_{a,\,NH_4^+}$, then the solution is basic.

$$NH_4^+ \rightleftharpoons NH_3 + H^+ \quad K_a = 5.6 \times 10^{-10}; \quad OCl^- + H_2O \rightleftharpoons HOCl + OH^- \quad K_b = 2.9 \times 10^{-7}$$

112. a. $KCl \rightarrow K^+ + Cl^-$ neutral; K^+ and Cl^- have no effect on pH.

b. $NH_4C_2H_3O_2 \rightarrow NH_4^+ + C_2H_3O_2^-$ neutral; NH_4^+ is a weak acid and $C_2H_3O_2^-$ is a weak base. Since $K_{a,\,NH_4^+} = K_{b,\,C_2H_3O_2^-}$, pH = 7.00.

$$NH_4^+ \rightleftharpoons NH_3 + H^+ \quad K_a = \frac{K_w}{K_{b,\,NH_3}} = \frac{1.0 \times 10^{-14}}{1.8 \times 10^{-5}} = 5.6 \times 10^{-10}$$

$$C_2H_3O_2^- + H_2O \rightleftharpoons HC_2H_3O_2 + OH^- \quad K_b = \frac{K_w}{K_{a,\,HC_2H_3O_2}} = \frac{1.0 \times 10^{-14}}{1.8 \times 10^{-5}} = 5.6 \times 10^{-10}$$

c. $CH_3NH_3Cl \rightarrow CH_3NH_3^+ + Cl^-$ acidic; $CH_3NH_3^+$ is a weak acid and Cl^- has no effect on pH.

$$CH_3NH_3^+ \rightleftharpoons H^+ + CH_3NH_2 \quad K_a = \frac{K_w}{K_{b,\,CH_3NH_2}} = \frac{1.00 \times 10^{-14}}{4.38 \times 10^{-4}} = 2.28 \times 10^{-11}$$

d. $KF \rightarrow K^+ + F^-$ basic; F^- is a weak base and K^+ has no effect on pH.

$$F^- + H_2O \rightleftharpoons HF + OH^- \quad K_b = \frac{K_w}{K_{a,\,HF}} = \frac{1.0 \times 10^{-14}}{7.2 \times 10^{-4}} = 1.4 \times 10^{-11}$$

e. $NH_4F \rightarrow NH_4^+ + F^-$ acidic; NH_4^+ is a weak acid and F^- is a weak base. Since $K_{a,\,NH_4^+} > K_{b,\,F^-}$, then the solution is acidic.

$$NH_4^+ \rightleftharpoons H^+ + NH_3 \quad K_a = 5.6 \times 10^{-10}; \quad F^- + H_2O \rightleftharpoons HF + OH^- \quad K_b = 1.4 \times 10^{-11}$$

f. $CH_3NH_3CN \rightarrow CH_3NH_3^+ + CN^-$ basic; $CH_3NH_3^+$ is a weak acid and CN^- is a weak base. Since $K_{b,\,CN^-} > K_{a,\,CH_3NH_3^+}$, the solution is basic.

$$CH_3NH_3^+ \rightleftharpoons H^+ + CH_3NH_2 \quad K_a = 2.28 \times 10^{-11}$$

$$CN^- + H_2O \rightleftharpoons HCN + OH^- \quad K_b = \frac{K_w}{K_{a,\,HCN}} = \frac{1.0 \times 10^{-14}}{6.2 \times 10^{-10}} = 1.6 \times 10^{-5}$$

Relationships Between Structure and Strengths of Acids and Bases

113.   a. $HIO_3 < HBrO_3$; As the electronegativity of the central atom increases, acid strength increases.

      b. $HNO_2 < HNO_3$; As the number of oxygen atoms attached to the central nitrogen atom increases, acid strength increases.

      c. $HOI < HOCl$; Same reasoning as in a.

      d. $H_3PO_3 < H_3PO_4$; Same reasoning as in b.

114.   a. $BrO_3^- < IO_3^-$; These are the conjugate bases of the acids in Exercise 14.113a. Since $HBrO_3$ is the stronger acid, the conjugate base of $HBrO_3$ (BrO_3^-) will be the weaker base. IO_3^- will be the stronger base since HIO_3 is the weaker acid.

      b. $NO_3^- < NO_2^-$; These are the conjugate bases of the acids in Exercise 14.113b. Conjugate base strength is inversely related to acid strength.

      c. $OCl^- < OI^-$. These are the conjugate bases of the acids in Exercise 14.113c.

115.   a. $H_2O < H_2S < H_2Se$; As the strength of the H – X bond decreases, acid strength increases.

      b. $CH_3CO_2H < FCH_2CO_2H < F_2CHCO_2H < F_3CCO_2H$; As the electronegativity of neighboring atoms increases, acid strength increases.

      c. $NH_4^+ < HONH_3^+$; Same reason as in b.

      d. $NH_4^+ < PH_4^+$; Same reason as in a.

116.   In general, the stronger the acid, the weaker the conjugate base.

      a. $SeH^- < SH^- < OH^-$; These are the conjugate bases of the acids in Exercise 14.115a. The ordering of the base strength is the opposite of the acids.

      b. $PH_3 < NH_3$ (See Exercise 14.115d.)

      c. $HONH_2 < NH_3$ (See Exercise 14.115c.)

117.   In general, metal oxides form basic solutions in water and nonmetal oxides form acidic solutions in water.

      a.   basic; $CaO(s) + H_2O(l) \rightarrow Ca(OH)_2(aq)$, $Ca(OH)_2$ is a strong base.

      b.   acidic; $SO_2(g) + H_2O(l) \rightarrow H_2SO_3(aq)$, H_2SO_3 is a weak diprotic acid.

      c.   acidic; $Cl_2O(g) + H_2O(l) \rightarrow 2\ HOCl(aq)$, HOCl is a weak acid.

118   a. basic; $Li_2O(s) + H_2O(l) \rightarrow 2\ LiOH(aq)$, LiOH is a strong base.

       b. acidic; $CO_2(g) + H_2O(l) \rightarrow H_2CO_3(aq)$, H_2CO_3 is a weak diprotic acid.

       c. basic; $SrO(s) + H_2O(l) \rightarrow Sr(OH)_2(aq)$, $Sr(OH)_2$ is a strong base.

Lewis Acids and Bases

119.   A Lewis base is an electron pair donor, and a Lewis acid is an electron pair acceptor.

        a. $B(OH)_3$, acid; H_2O, base   b. Ag^+, acid; NH_3, base   c. BF_3, acid; F^-, base

120.   a. Fe^{3+}, acid; H_2O, base   b. H_2O, acid; CN^-, base   c. HgI_2, acid; I^-, base

121. $Al(OH)_3(s) + 3\ H^+(aq) \rightarrow Al^{3+}(aq) + 3\ H_2O(l)$ (Brønsted-Lowry base, H^+ acceptor)

        $Al(OH)_3(s) + OH^-(aq) \rightarrow Al(OH)_4^-(aq)$ (Lewis acid, electron pair acceptor)

122. $Zn(OH)_2(s) + 2\ H^+(aq) \rightarrow Zn^{2+}(aq) + 2\ H_2O(l)$ (Brønsted-Lowry base)

        $Zn(OH)_2(s) + 2\ OH^-(aq) \rightarrow Zn(OH)_4^{2-}(aq)$ (Lewis acid)

123. Fe^{3+} should be the stronger Lewis acid. Fe^{3+} is smaller and has a greater positive charge. Because of this, Fe^{3+} will be more strongly attracted to lone pairs of electrons as compared to Fe^{2+}.

124.   The Lewis structures for the reactants and products are:

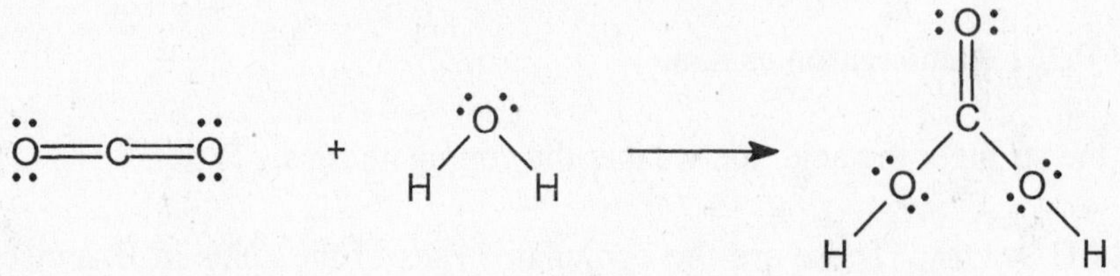

        In this reaction, H_2O donates a pair of electrons to carbon in CO_2, which is followed by a proton shift to form H_2CO_3. H_2O is the Lewis base, and CO_2 is the Lewis acid.

Additional Exercises

125.   At pH = 2.000, $[H^+] = 10^{-2.000} = 1.00 \times 10^{-2}\ M$; At pH = 4.000, $[H^+] = 10^{-4.000} = 1.00 \times 10^{-4}\ M$

        mol H^+ present = $0.0100\ L \times \dfrac{0.0100\ mol\ H^+}{L} = 1.00 \times 10^{-4}\ mol\ H^+$

        Let V = total volume of solution at pH = 4.000: $1.00 \times 10^{-4}\ mol/L = \dfrac{1.00 \times 10^{-4}\ mol\ H^+}{V}$, V = 1.00 L

        Volume of water added = 1.00 L - 0.0100 L = 0.99 L = 990 mL

126. $2.48 \text{ g TlOH} \times \dfrac{1 \text{ mol}}{221.4 \text{ g}} = 1.12 \times 10^{-2} \text{ mol}$;   TlOH is a strong base, so $[\text{OH}^-] = 1.12 \times 10^{-2} \, M$.

$\text{pOH} = -\log[\text{OH}^-] = 1.951$; $\text{pH} = 14.000 - \text{pOH} = 12.049$

127.   The light bulb is bright because a strong electrolyte is present, i.e., a solute is present that dissolves to produce a lot of ions in solution. The pH meter value of 4.6 indicates that a weak acid is present. (If a strong acid were present, the pH would be close to zero.) Of the possible substances, only HCl (strong acid), NaOH (strong base) and NH_4Cl are strong electrolytes. Of these three substances, only NH_4Cl contains a weak acid (the HCl solution would have a pH close to zero and the NaOH solution would have a pH close to 14.0). NH_4Cl dissociates into NH_4^+ and Cl^- ions when dissolved in water. Cl^- is the conjugate base of a strong acid, so it has no basic (or acidic properties) in water. NH_4^+, however, is the conjugate acid of the weak base NH_3, so NH_4^+ is a weak acid and would produce a solution with a pH = 4.6 when the concentration is ~1 M.

128.

	HBz	$\rightleftharpoons$	H^+	+	Bz^-	$HBz = C_6H_5CO_2H$
Initial	C		~0		0	$C = [HBz]_o = $ concentration of HBz that dissolves to give saturated solution.

x mol/L HBz dissociates to reach equilibrium

Change	$-x$	$\rightarrow$	$+x$	$+x$	
Equil.	$C - x$		x	x	

$K_a = \dfrac{[H^+][Bz^-]}{[HBz]} = 6.4 \times 10^{-5} = \dfrac{x^2}{C - x}$, where $x = [H^+]$

$6.4 \times 10^{-5} = \dfrac{[H^+]^2}{C - [H^+]}$; pH = 2.80; $[H^+] = 10^{-2.80} = 1.6 \times 10^{-3} \, M$

$C - 1.6 \times 10^{-3} = \dfrac{(1.6 \times 10^{-3})^2}{6.4 \times 10^{-5}} = 4.0 \times 10^{-2}$, $C = 4.0 \times 10^{-2} + 1.6 \times 10^{-3} = 4.2 \times 10^{-2} \, M$

The molar solubility of $C_6H_5CO_2H$ is 4.2×10^{-2} mol/L.

129.   For $H_2C_6H_6O_6$. $K_{a_1} = 7.9 \times 10^{-5}$ and $K_{a_2} = 1.6 \times 10^{-12}$. Since $K_{a_1} >> K_{a_2}$, then the amount of H^+ produced by the K_{a_2} reaction will be negligible.

$[H_2C_6H_6O_6]_o = \dfrac{0.500 \text{ g} \times \dfrac{1 \text{ mol } H_2C_6H_6O_6}{176.12 \text{ g}}}{0.2000 \text{ L}} = 0.0142 \, M$

	$H_2C_6H_6O_6(aq)$	$\rightleftharpoons$	$HC_6H_6O_6^-(aq)$	+	$H^+(aq)$	$K_{a_1} = 7.9 \times 10^{-5}$
Initial	0.0142 M		0		~0	
Equil.	0.0142 - x		x		x	

$K_{a_1} = 7.9 \times 10^{-5} = \dfrac{x^2}{0.0142 - x} \approx \dfrac{x^2}{0.0142}$, $x = 1.1 \times 10^{-3}$; Assumption fails the 5% rule.

Solving by the method of successive approximations:

$7.9 \times 10^{-5} = \dfrac{x^2}{0.0142 - 1.1 \times 10^{-3}}$, $x = 1.0 \times 10^{-3}\ M$ (consistent answer)

Since H^+ produced by the K_{a_2} reaction will be negligible, $[H^+] = 1.0 \times 10^{-3}$ and pH = 3.00.

130. $[H^+]_o = 1.0 \times 10^{-2} + 1.0 \times 10^{-2} = 2.0 \times 10^{-2}\ M$ from strong acids HCl and H_2SO_4.

HSO_4^- is a good weak acid ($K_a = 0.012$). However, HCN is a poor weak acid ($K_a = 6.2 \times 10^{-10}$) and can be ignored. Calculating the H^+ contribution from HSO_4^-:

	HSO_4^-	$\rightleftharpoons$	H^+	+	SO_4^{2-}	$K_a = 0.012$
Initial	0.010 M		0.020 M		0	
Equil.	0.010 - x		0.020 + x		x	

$K_a = \dfrac{x(0.020 + x)}{(0.010 - x)} = 0.012 \approx \dfrac{x(0.020)}{(0.010)}$, $x = 0.0060$; Assumption poor (60% error).

Using the quadratic formula: $x^2 + 0.032\,x - 1.2 \times 10^{-4} = 0$, $x = 3.4 \times 10^{-3}\ M$

$[H^+] = 0.020 + x = 0.020 + 3.4 \times 10^{-3} = 0.023\ M$; pH = 1.64

131. For this problem we will abbreviate $CH_2{=}CHCO_2H$ as Hacr and $CH_2{=}CHCO_2^-$ as acr$^-$.

a. Solving the weak acid problem:

	Hacr	$\rightleftharpoons$	H^+	+	acr$^-$	$K_a = 5.6 \times 10^{-5}$
Initial	0.10 M		~0		0	
Equil.	0.10 - x		x		x	

$\dfrac{x^2}{0.10 - x} = 5.6 \times 10^{-5} \approx \dfrac{x^2}{0.10}$, $x = [H^+] = 2.4 \times 10^{-3}\ M$; pH = 2.62; Assumptions good.

b. % dissociation $= \dfrac{[H^+]}{[\text{Hacr}]_o} \times 100 = \dfrac{2.4 \times 10^{-3}}{0.10} \times 100 = 2.4\%$

c. acr$^-$ is a weak base and the major source of OH^- in this solution.

	acr$^-$	+	H_2O	$\rightleftharpoons$	Hacr	+	OH^-	$K_b = \dfrac{K_w}{K_a} = \dfrac{1.0 \times 10^{-14}}{5.6 \times 10^{-5}}$
Initial	0.050 M				0		~0	$K_b = 1.8 \times 10^{-10}$
Equil.	0.050 - x				x		x	

$$K_b = \frac{[\text{Hacr}][\text{OH}^-]}{[\text{acr}^-]} = 1.8 \times 10^{-10} = \frac{x^2}{0.050 - x} \approx \frac{x^2}{0.050}$$

$x = [\text{OH}^-] = 3.0 \times 10^{-6} \, M$; pOH = 5.52; pH = 8.48   Assumptions good.

132.   From the pH, $C_7H_4ClO_2^-$ is a weak base. Use the weak base data to determine K_b for $C_7H_4ClO_2^-$ (which we will abbreviate as CB^-).

$$CB^- \quad + \quad H_2O \quad \rightleftharpoons \quad HCB \quad + \quad OH^-$$

Initial	0.20 M	0	~0
Equil.	0.20 - x	x	x

Since pH = 8.65, pOH = 5.35 and $[\text{OH}^-] = 10^{-5.35} = 4.5 \times 10^{-6} \, M = x$.

$$K_b = \frac{[\text{HCB}][\text{OH}^-]}{[\text{CB}^-]} = \frac{x^2}{0.20 - x} = \frac{(4.5 \times 10^{-6})^2}{0.20 - 4.5 \times 10^{-6}} = 1.0 \times 10^{-10}$$

Since CB^- is a weak base, HCB, chlorobenzoic acid, is a weak acid. Solving the weak acid problem:

$$HCB \quad \rightleftharpoons \quad H^+ \quad + \quad CB^-$$

Initial	0.20 M	~0	0
Equil.	0.20 - x	x	x

$$K_a = \frac{K_w}{K_b} = \frac{1.0 \times 10^{-14}}{1.0 \times 10^{-10}} = 1.0 \times 10^{-4} = \frac{x^2}{0.20 - x} \approx \frac{x^2}{0.20}$$

$x = [\text{H}^+] = 4.5 \times 10^{-3} \, M$; pH = 2.35   Assumptions good.

133.   a. $Fe(H_2O)_6^{3+} + H_2O \quad \rightleftharpoons \quad Fe(H_2O)_5(OH)^{2+} \quad + \quad H_3O^+$

Initial	0.10 M	0	~0
Equil.	0.10 - x	x	x

$$K_a = \frac{[Fe(H_2O)_5(OH)^{2+}][H_3O^+]}{[Fe(H_2O)_6^{3+}]} = 6.0 \times 10^{-3} = \frac{x^2}{0.10 - x} \approx \frac{x^2}{0.10}$$

$x = 2.4 \times 10^{-2}$; Assumption is poor (x is 24% of 0.10). Using successive approximations:

$$\frac{x^2}{0.10 - 0.024} = 6.0 \times 10^{-3}, \ x = 0.021$$

$$\frac{x^2}{0.10 - 0.021} = 6.0 \times 10^{-3}, \ x = 0.022; \quad \frac{x^2}{0.10 - 0.022} = 6.0 \times 10^{-3}, \ x = 0.022$$

$x = [H^+] = 0.022 \ M; \ pH = 1.66$

   b. Because of the lower charge, Fe^{2+}(aq) will not be as strong an acid as Fe^{3+}(aq). A solution of iron(II) nitrate will be less acidic (have a higher pH) than a solution with the same concentration of iron(III) nitrate.

134.   See generalizations in Exercise 14.99.

   a. HI: strong acid; HF: weak acid ($K_a = 7.2 \times 10^{-4}$)

       NaF: F^- is the conjugate base of the weak acid HF so F^- is a weak base. The K_b value for $F^- = K_w/K_{a, HF} = 1.4 \times 10^{-11}$. Na^+ has no acidic or basic properties.

       NaI: neutral (pH = 7.0); Na^+ and I^- have no acidic/basic properties.

                To place in order of increasing pH, we place the compounds from most acidic (lowest pH) to most basic (highest pH). Increasing pH: HI < HF < NaI < NaF.

   b. NH_4Br: NH_4^+ is a weak acid ($K_a = 5.6 \times 10^{-10}$) and Br^- is a neutral species.
       HBr: strong acid
       KBr: neutral; K^+ and Br^- have no acidic/basic properties
       NH_3:   weak base, $K_b = 1.8 \times 10^{-5}$

       Increasing pH: HBr < NH_4Br < KBr < NH_3
                       most   most
                       acidic   basic

   c. $C_6H_5NH_3NO_3$: $C_6H_5NH_3^+$ is a weak acid ($K_w/K_{b, C_6H_5NH_2} = 1.0 \times 10^{-14}/3.8 \times 10^{-10} = 2.6 \times 10^{-5}$) and NO_3^- is a neutral species.
       $NaNO_3$:   neutral; Na^+ and NO_3^- have no acidic/basic properties.
       NaOH: strong base
       HOC_6H_5:   weak acid ($K_a = 1.6 \times 10^{-10}$)
       KOC_6H_5: $OC_6H_5^-$ is a weak base ($K_b = K_w/K_{a, HOC_6H_5} = 6.3 \times 10^{-5}$) and K^+ is a neutral species.
       $C_6H_5NH_2$:   weak base ($K_b = 3.8 \times 10^{-10}$)
       HNO_3:   strong acid

       This is a little more difficult than the previous parts of this problem because two weak acids and two weak bases are present. Between the weak acids, $C_6H_5NH_3^+$ is a stronger weak acid than HOC_6H_5 since the K_a value for $C_6H_5NH_3^+$ is larger than the K_a value for HOC_6H_5. Between the two weak bases, since the K_b value for $OC_6H_5^-$ is larger than the K_b value for $C_6H_5NH_2$, then $OC_6H_5^-$ is a stronger weak base than $C_6H_5NH_2$.

Increasing pH: $HNO_3 < C_6H_5NH_3NO_3 < HOC_6H_5 < NaNO_3 < C_6H_5NH_2 < KOC_6H_5 < NaOH$
most   most
acidic   basic

135.   Solution is acidic from $HSO_4^- \rightleftharpoons H^+ + SO_4^{2-}$. Solving the weak acid problem:

$$HSO_4^- \quad\rightleftharpoons\quad H^+ \quad + \quad SO_4^{2-} \qquad K_a = 1.2 \times 10^{-2}$$

Initial   0.10 M ~0   0
Equil.   0.10 - x x x

$$1.2 \times 10^{-2} = \frac{[H^+][SO_4^{2-}]}{[HSO_4^-]} = \frac{x^2}{0.10 - x} \approx \frac{x^2}{0.10}, \ x = 0.035$$

Assumption is not good (x is 35% of 0.10). Using successive approximations:

$$\frac{x^2}{0.10 - x} \approx \frac{x^2}{0.10 - 0.035} = 1.2 \times 10^{-2}, \ x = 0.028$$

$$\frac{x^2}{0.10 - 0.028} = 1.2 \times 10^{-2}, \ x = 0.029; \quad \frac{x^2}{0.10 - 0.029} = 1.2 \times 10^{-2}, \ x = 0.029$$

$x = [H^+] = 0.029 \ M;$ pH = 1.54

136.   The relevant reactions are:

$$H_2CO_3 \rightleftharpoons H^+ + HCO_3^- \quad K_{a_1} = 4.3 \times 10^{-7}; \ HCO_3^- \rightleftharpoons H^+ + CO_3^{2-} \quad K_{a_2} = 5.6 \times 10^{-11}$$

Initially, we deal only with the first reaction (since $K_{a_1} \gg K_{a_2}$) and then let these results control values of concentrations in the second reaction.

$$H_2CO_3 \quad\rightleftharpoons\quad H^+ \quad + \quad HCO_3^-$$

Initial   0.010 M ~0   0
Equil.   0.010 - x x x

$$K_{a_1} = 4.3 \times 10^{-7} = \frac{[H^+][HCO_3^-]}{[H_2CO_3]} = \frac{x^2}{0.010 - x} \approx \frac{x^2}{0.010}$$

$x = 6.6 \times 10^{-5} \ M = [H^+] = [HCO_3^-]$ Assumptions good.

$$HCO_3^- \quad\rightleftharpoons\quad H^+ \quad + \quad CO_3^{2-}$$

Initial $6.6 \times 10^{-5} \ M$ $6.6 \times 10^{-5} \ M$ 0
Equil. $6.6 \times 10^{-5} - y$ $6.6 \times 10^{-5} + y$ y

If y is small, then $[H^+] = [HCO_3^-]$ and $K_{a_2} = 5.6 \times 10^{-11} = \dfrac{[H^+][CO_3^{2-}]}{[HCO_3^-]} \approx y$

$y = [CO_3^{2-}] = 5.6 \times 10^{-11} M$ Assumptions good.

The amount of H^+ from the second dissociation is $5.6 \times 10^{-11} M$ or:

$$\frac{5.6 \times 10^{-11}}{6.6 \times 10^{-5}} \times 100 = 8.5 \times 10^{-5}\,\%$$

This result justifies our treating the equilibria separately. If the second dissociation contributed a significant amount of H^+, then we would have to treat both equilibria simultaneously. The reaction that occurs when acid is added to a solution of HCO_3^- is:

$$HCO_3^-(aq) + H^+(aq) \rightarrow H_2CO_3(aq) \rightarrow H_2O(l) + CO_2(g)$$

The bubbles are $CO_2(g)$ and are formed by the breakdown of unstable H_2CO_3 molecules. We should write $H_2O(l) + CO_2(aq)$ or $CO_2(aq)$ for what we call carbonic acid. It is for convenience, however, that we write $H_2CO_3(aq)$.

137.   a.   In the lungs, there is a lot of O_2 and the equilibrium favors $Hb(O_2)_4$. In the cells, there is a deficiency of O_2, and the equilibrium favors HbH_4^{4+}.

b. CO_2 is a weak acid, $CO_2 + H_2O \rightleftharpoons HCO_3^- + H^+$. Removing CO_2 essentially decreases H^+. $Hb(O_2)_4$ is then favored and O_2 is not released by hemoglobin in the cells. Breathing into a paper bag increases CO_2 in the blood, thus increasing H^+, which shifts the reaction left.

c. CO_2 builds up in the blood and it becomes too acidic, driving the equilibrium to the left. Hemoglobin can't bind O_2 as strongly in the lungs. Bicarbonate ion acts as a base in water and neutralizes the excess acidity.

138.   a. $NH_3 + H_3O^+ \rightleftharpoons NH_4^+ + H_2O$

$$K_{eq} = \frac{[NH_4^+]}{[NH_3][H^+]} = \frac{1}{K_a \text{ for } NH_4^+} = \frac{K_b \text{ for } NH_3}{K_w} = \frac{1.8 \times 10^{-5}}{1.0 \times 10^{-14}} = 1.8 \times 10^9$$

b. $NO_2^- + H_3O^+ \rightleftharpoons HNO_2 + H_2O$ $K_{eq} = \dfrac{[HNO_2]}{[NO_2^-][H^+]} = \dfrac{1}{K_a \text{ for } HNO_2} = \dfrac{1}{4.0 \times 10^{-4}} = 2.5 \times 10^3$

c. $NH_4^+ + OH^- \rightleftharpoons NH_3 + H_2O$ $K_{eq} = \dfrac{1}{K_b \text{ for } NH_3} = \dfrac{1}{1.8 \times 10^{-5}} = 5.6 \times 10^4$

d. $HNO_2 + OH^- \rightleftharpoons H_2O + NO_2^-$

$$K_{eq} = \frac{[NO_2^-]}{[HNO_2][OH^-]} \times \frac{[H^+]}{[H^+]} = \frac{K_a \text{ for } HNO_2}{K_w} = \frac{4.0 \times 10^{-4}}{1.0 \times 10^{-14}} = 4.0 \times 10^{10}$$

139.   a. H_2SO_3 b. $HClO_3$ c. H_3PO_3

NaOH and KOH are soluble ionic compounds composed of Na^+ and K^+ cations and OH^- anions. All soluble ionic compounds dissolve to form the ions from which they are formed. In oxyacids, the compounds are all covalent compounds in which electrons are shared to form bonds (unlike ionic compounds). When these compounds are dissolved in water, the covalent bond between oxygen and hydrogen breaks to form H^+ ions.

Challenge Problems

140.   The pH of this solution is not 8.00 because water will donate a significant amount of H^+ from the autoionization of water. The pertinent reactions are:

$$H_2O \rightleftharpoons H^+ + OH^- \quad K_w = [H^+][OH^-] = 1.0 \times 10^{-14}$$

$$HCl \rightarrow H^+ + Cl^- \quad K_a \text{ is very large, so we assume that only the forward reaction occurs.}$$

In any solution, the overall net positive charge must equal the overall net negative charge (called the charge balance). For this problem:

[positive charge] = [negative charge], so $[H^+] = [OH^-] + [Cl^-]$

From K_w, $[OH^-] = K_w/[H^+]$, and from $1.0 \times 10^{-8}\ M$ HCl, $[Cl^-] = 1.0 \times 10^{-8}\ M$. Substituting into the charge balance equation:

$$[H^+] = \frac{1.0 \times 10^{-14}}{[H^+]} + 1.0 \times 10^{-8}, \quad [H^+]^2 - 1.0 \times 10^{-8}[H^+] - 1.0 \times 10^{-14} = 0$$

Using the quadratic formula to solve:

$$[H^+] = \frac{-(-1.0 \times 10^{-8}) \pm [(-1.0 \times 10^{-8})^2 - 4(1)(-1.0 \times 10^{-14})]^{1/2}}{2(1)}, \quad [H^+] = 1.1 \times 10^{-7}\ M$$

$$pH = -\log(1.1 \times 10^{-7}) = 6.96$$

141.   Since this is a very dilute solution of NaOH, we must worry about the amount of OH^- donated from the autoionization of water.

$$NaOH \rightarrow Na^+ + OH^-$$

$$H_2O \rightleftharpoons H^+ + OH^- \quad K_w = [H^+][OH^-] = 1.0 \times 10^{-14}$$

This solution, like all solutions, must be charge balanced, that is [positive charge] = [negative charge]. For this problem, the charge balance equation is:

$$[Na^+] + [H^+] = [OH^-], \text{ where } [Na^+] = 1.0 \times 10^{-7}\ M \text{ and } [H^+] = \frac{K_w}{[OH^-]}$$

Substituting into the charge balance equation:

$$1.0 \times 10^{-7} + \frac{1.0 \times 10^{-14}}{[OH^-]} = [OH^-], \quad [OH^-]^2 - 1.0 \times 10^{-7}\,[OH^-] - 1.0 \times 10^{-14} = 0$$

Using the quadratic formula to solve:

$$[OH^-] = \frac{-(-1.0 \times 10^{-7}) \pm [(-1.0 \times 10^{-7})^2 - 4(1)(-1.0 \times 10^{-14})]^{1/2}}{2(1)}$$

$[OH^-] = 1.6 \times 10^{-7}\,M$; pOH = -log (1.6×10^{-7}) = 6.80; pH = 7.20

142.   HA $\rightleftharpoons$ H^+ + A^- $K_a = 1.00 \times 10^{-6}$

Initial	C	~0	0	C = $[HA]_o$; For pH = 4.000,
Equil.	C - 1.00×10^{-4}	1.00×10^{-4}	1.00×10^{-4}	$x = [H^+] = 1.00 \times 10^{-4}\,M$

$$K_a = \frac{(1.00 \times 10^{-4})^2}{C - 1.00 \times 10^{-4}} = 1.00 \times 10^{-6}; \quad \text{Solving: } C = 0.0101\,M$$

The solution initially contains $50.0 \times 10^{-3}\,L \times 0.0101\,mol/L = 5.05 \times 10^{-4}\,mol\,HA$. We then dilute to a total volume, V, in liters. The resulting pH = 5.000, so $[H^+] = 1.00 \times 10^{-5}$. In the typical weak acid problem, $x = [H^+]$, so:

                        HA $\rightleftharpoons$ H^+ + A^-

Initial	5.05×10^{-4} mol/V	~0	0
Equil.	5.05×10^{-4}/V - 1.00×10^{-5}	1.00×10^{-5}	1.00×10^{-5}

$$K_a = \frac{(1.00 \times 10^{-5})^2}{5.05 \times 10^{-4}/V - 1.00 \times 10^{-5}} = 1.00 \times 10^{-6}, \quad 1.00 \times 10^{-4} = 5.05 \times 10^{-4}/V - 1.00 \times 10^{-5}$$

V = 4.59 L; 50.0 mL are present initially, so we need to add 4540 mL of water.

143.   HBrO $\rightleftharpoons$ H^+ + BrO^- $K_a = 2 \times 10^{-9}$

Initial	$1.0 \times 10^{-6}\,M$	~0	0
	x mol/L HBrO dissociates to reach equilibrium		
Change	-x	$\rightarrow$ +x	+x
Equil.	$1.0 \times 10^{-6} - x$	x	x

$$K_a = 2 \times 10^{-9} = \frac{x^2}{1.0 \times 10^{-6} - x} \approx \frac{x^2}{1.0 \times 10^{-6}}; \quad x = [H^+] = 4 \times 10^{-8}\,M; \quad pH = 7.4$$

Let's check the assumptions. This answer is impossible! We can't add a small amount of an acid to a neutral solution and get a basic solution. The highest possible pH for an acid in water is 7.0. In the correct solution, we would have to take into account the autoionization of water.

144.   Major species present are H_2O, $C_5H_5NH^+$ ($K_a = K_w/K_b(C_5H_5N) = 1.0 \times 10^{-14}/1.7 \times 10^{-9} =$ 5.9×10^{-6}) and F^- ($K_b = K_w/K_a(HF) = 1.0 \times 10^{-14}/7.2 \times 10^{-4} = 1.4 \times 10^{-11}$). The reaction to consider is the best acid present ($C_5H_5NH^+$) reacting with the best base present (F^-). Solving for the equilibrium concentrations:

$$C_5H_5NH^+(aq) \quad + \quad F^-(aq) \quad \rightleftharpoons \quad C_5H_5N(aq) + HF(aq)$$

Initial	0.200 M	0.200 M	0	0
Change	$-x$	$-x$ $\rightarrow$	$+x$	$+x$
Equil.	0.200 - x	0.200 - x	x	x

$$K = K_{a, C_5H_5NH^+} \times \frac{1}{K_{a, HF}} = 5.9 \times 10^{-6} (1/7.2 \times 10^{-4}) = 8.2 \times 10^{-3}$$

$$K = \frac{[C_5H_5N][HF]}{[C_5H_5NH^+][F^-]} = 8.2 \times 10^{-3} = \frac{x^2}{(0.200 - x)^2}; \text{ Taking the square root of both sides:}$$

$$0.091 = \frac{x}{0.200 - x} \quad x = 0.018 - 0.091\,x, \ x = 0.016\ M$$

From the setup to the problem, $x = [C_5H_5N] = [HF] = 0.016\ M$ and $0.200 - x = 0.200 - 0.016 = 0.184$ $M = [C_5H_5NH^+] = [F^-]$. To solve for the $[H^+]$, we can use either the K_a equilibrium for $C_5H_5NH^+$ or the K_a equilibrium for HF. Using $C_5H_5NH^+$ data:

$$K_{a, C_5H_5NH^+} = 5.9 \times 10^{-6} = \frac{[C_5H_5N][H^+]}{[C_5H_5NH^+]} = \frac{(0.016)[H^+]}{(0.184)}, \ [H^+] = 6.8 \times 10^{-5}\ M$$

pH = $-\log(6.8 \times 10^{-5}) = 4.17$

As one would expect, since the K_a for the weak acid is larger than the K_b for the weak base, then a solution of this salt should be acidic.

145.   Since NH_3 is so concentrated, we need to calculate the OH^- contribution from the weak base NH_3.

$$NH_3 \ + \ H_2O \quad \rightleftharpoons \quad NH_4^+ \ + \ OH^- \qquad K_b = 1.8 \times 10^{-5}$$

Initial	15.0 M	0	0.0100 M (Assume no volume change.)
Equil.	15.0 - x	x	0.0100 + x

$$K_b = 1.8 \times 10^{-5} = \frac{x(0.0100 + x)}{15.0 - x} \approx \frac{x(0.0100)}{15.0}, \ x = 0.027; \text{ Assumption is horrible } (x \text{ is } 270\%$$ of 0.0100).

Using the quadratic formula:

$$1.8 \times 10^{-5}(15.0 - x) = 0.0100\,x + x^2, \ x^2 + 0.0100\,x - 2.7 \times 10^{-4} = 0$$

$$x = 1.2 \times 10^{-2}, \ [OH^-] = 1.2 \times 10^{-2} + 0.0100 = 0.022\ M$$

146.   Molar mass $= \dfrac{dRT}{P} = \dfrac{\dfrac{5.11\text{ g}}{\text{L}} \times \dfrac{0.08206\text{ L atm}}{\text{mol K}} \times 298\text{ K}}{1.00\text{ atm}} = 125$ g/mol

$[HA]_o = \dfrac{1.50\text{ g} \times \dfrac{1\text{ mol}}{125\text{ g}}}{0.100\text{ L}} = 0.120\ M$; pH = 1.80, $[H^+] = 10^{-1.80} = 1.6 \times 10^{-2}\ M$

$$HA \rightleftharpoons H^+ + A^-$$
Equil. $0.120 - x$ x x $x = [H^+] = 1.6 \times 10^{-2}\ M$

$K_a = \dfrac{[H^+][A^-]}{[HA]} = \dfrac{(1.6 \times 10^{-2})^2}{0.120 - 0.016} = 2.5 \times 10^{-3}$

147. PO_4^{3-} is the conjugate base of HPO_4^{2-}. The K_a value for HPO_4^{2-} is $K_{a_3} = 4.8 \times 10^{-13}$.

$PO_4^{3-}(aq) + H_2O(l) \rightleftharpoons HPO_4^{2-}(aq) + OH^-(aq)$ $K_b = \dfrac{K_w}{K_{a_3}} = \dfrac{1.0 \times 10^{-14}}{4.8 \times 10^{-13}} = 0.021$

HPO_4^{2-} is the conjugate base of $H_2PO_4^-$ $(K_{a_2} = 6.2 \times 10^{-8})$.

$HPO_4^{2-} + H_2O \rightleftharpoons H_2PO_4^- + OH^-$ $K_b = \dfrac{K_w}{K_{a_2}} = \dfrac{1.0 \times 10^{-14}}{6.2 \times 10^{-8}} = 1.6 \times 10^{-7}$

$H_2PO_4^-$ is the conjugate base of H_3PO_4 $(K_{a_1} = 7.5 \times 10^{-3})$.

$H_2PO_4^- + H_2O \rightleftharpoons H_3PO_4 + OH^-$ $K_b = \dfrac{K_w}{K_{a_1}} = \dfrac{1.0 \times 10^{-14}}{7.5 \times 10^{-3}} = 1.3 \times 10^{-12}$

From the K_b values, PO_4^{3-} is the strongest base. This is expected since PO_4^{3-} is the conjugate base of the weakest acid (HPO_4^{2-}).

148.   a. $HCO_3^- + HCO_3^- \rightleftharpoons H_2CO_3 + CO_3^{2-}$

$K_{eq} = \dfrac{[H_2CO_3][CO_3^{2-}]}{[HCO_3^-][HCO_3^-]} \times \dfrac{[H^+]}{[H^+]} = \dfrac{K_{a_2}}{K_{a_1}} = \dfrac{5.6 \times 10^{-11}}{4.3 \times 10^{-7}} = 1.3 \times 10^{-4}$

   b. $[H_2CO_3] = [CO_3^{2-}]$ since the reaction in part a is the principle equilibrium reaction.

   c. $H_2CO_3 \rightleftharpoons 2\,H^+ + CO_3^{2-}$ $K_{eq} = \dfrac{[H^+]^2[CO_3^{2-}]}{[H_2CO_3]} = K_{a_1} \times K_{a_2}$

Since, $[H_2CO_3] = [CO_3^{2-}]$ from part b, then $[H^+]^2 = K_{a_1} \times K_{a_2}$.

$[H^+] = (K_{a_1} \times K_{a_2})^{1/2}$ or pH $= \dfrac{pK_{a_1} + pK_{a_2}}{2}$

   d. $[H^+] = [(4.3 \times 10^{-7}) \times (5.6 \times 10^{-11})]^{1/2}$, $[H^+] = 4.9 \times 10^{-9}\ M$; pH = 8.31

149.   Molality = m = $\dfrac{0.100 \text{ g} \times \dfrac{1 \text{ mol}}{100.0 \text{ g}}}{0.5000 \text{ kg}}$ = 2.00×10^{-3} mol/kg ≈ 2.00×10^{-3} mol/L   (dilute solution)

$\Delta T_f = iK_f m$, $0.0056°C = i(1.86°C/\text{molal})(2.00 \times 10^{-3} \text{ molal})$, i = 1.5

If i = 1.0, % dissociation = 0% and if i = 2.0, % dissociation = 100%. Since i = 1.5, the weak acid is 50.% dissociated.

$$HA \rightleftharpoons H^+ + A^- \qquad K_a = \dfrac{[H^+][A^-]}{[HA]}$$

Since the weak acid is 50.% dissociated, then:

$[H^+] = [A^-] = [HA]_o \times 0.50 = 2.00 \times 10^{-3}\,M \times 0.50 = 1.0 \times 10^{-3}\,M$

$[HA] = [HA]_o$ - amount HA reacted = $2.00 \times 10^{-3}\,M - 1.0 \times 10^{-3}\,M = 1.0 \times 10^{-3}\,M$

$K_a = \dfrac{[H^+][A^-]}{[HA]} = \dfrac{(1.0 \times 10^{-3})(1.0 \times 10^{-3})}{1.0 \times 10^{-3}} = 1.0 \times 10^{-3}$

150.   a.   Assuming no ion association between SO_4^{2-}(aq) and Fe^{3+}(aq), then i = 5 for $Fe_2(SO_4)_3$.

   $\pi = iMRT = 5(0.0500 \text{ mol/L})(0.08206 \text{ L atm K}^{-1} \text{ mol}^{-1})(298 \text{ K}) = 6.11$ atm

   b. $Fe_2(SO_4)_3$(aq) → 2 Fe^{3+}(aq) + 3 SO_4^{2-}(aq)

   Under ideal circumstances, 2/5 of π calculated above results from Fe^{3+} and 3/5 results from SO_4^{2-}. The contribution to π from SO_4^{2-} is 3/5 × 6.11 atm = 3.67 atm. Since SO_4^{2-} is assumed unchanged in solution, the SO_4^{2-} contribution in the actual solution will also be 3.67 atm. The contribution to the actual π from the $Fe(H_2O)_6^{3+}$ dissociation reaction is 6.73 - 3.67 = 3.06 atm.

   The initial concentration of $Fe(H_2O)_6^{2+}$ is 2(0.0500) = 0.100 M. The set-up for the weak acid problem is:

$$Fe(H_2O)_6^{3+} \rightleftharpoons H^+ + Fe(OH)(H_2O)_5^{2+} \qquad K_a = \dfrac{[H^+][Fe(OH)(H_2O)_5^{2+}]}{[Fe(H_2O)_6^{3+}]}$$

Initial   0.100 M   ~0   0
       x mol/L of $Fe(H_2O)_6^{3+}$ reacts to reach equilibrium
Equil.   0.100 - x x x

$\pi = iMRT$; Total ion concentration = iM = $\dfrac{\pi}{RT}$ = $\dfrac{3.06 \text{ atm}}{0.08206 \text{ L atm K}^{-1} \text{ mol}^{-1}(298 \text{ K})}$ = 0.125 M

0.125 M = 0.100 - x + x + x = 0.100 + x, x = 0.025 M

$K_a = \dfrac{[H^+][Fe(OH)(H_2O)_5^{2+}]}{[Fe(H_2O)_6^{3+}]} = \dfrac{x^2}{0.100 - x} = \dfrac{(0.025)^2}{(0.100 - 0.025)} = \dfrac{(0.025)^2}{0.075}$, $K_a = 8.3 \times 10^{-3}$

CHAPTER FIFTEEN

APPLICATIONS OF AQUEOUS EQUILIBRIA

Questions

13. A common ion is an ion that appears in an equilibrium reaction but came from a source other than that reaction. Addition of a common ion (H^+ or NO_2^-) to the reaction $HNO_2 \rightleftharpoons H^+ + NO_2^-$ will drive the equilibrium to the left as predicted by Le Chatelier's principle.

14. A buffered solution must contain both a weak acid and a weak base. Most buffered solutions are prepared using a weak acid plus the conjugate base of the weak acid (which is a weak base). Buffered solutions are useful for controlling the pH of a solution since they resist pH change. The capacity of a buffer is a measure of how much strong acid or strong base the buffer can neutralize. All the buffers listed have the same pH ($= pK_a = 4.74$) since they all have a 1:1 concentration ratio between the weak acid and the conjugate base. The 1.0 M buffer has the greatest capacity; the 0.01 M buffer the least capacity. In general, the larger the concentrations of weak acid and conjugate base, the greater the buffer capacity, i.e., the greater the ability to neutralize added strong acid or strong base.

15. No, as long as there are both a weak acid and a weak base present, the solution will be buffered. If the concentrations are the same, the buffer will have the same capacity towards added H^+ and added OH^-. In addition, buffers with equal concentrations of weak acid and conjugate base have pH $= pK_a$.

16. Between the starting point of the titration and the equivalence point, we are dealing with a buffer solution. The Henderson-Hasselbalch equation can be used to determine pH:

$$pH = pK_a + \log \frac{[\text{Base}]}{[\text{Acid}]}$$

At the halfway point to equivalence, enough OH^- has been added by the strong base to convert exactly one-half of the weak acid present initially into its conjugate base. Therefore, [conjugate base] = [weak acid] so pH = pK_a + log 1 = pK_a.

The K_a value can be determined at any point in a titration, from the initial point to the equivalence point. In Chapter 14, we calculated the K_a from a solution of only the weak acid. In the buffer region, we can calculate the ratio of the base form to the acid form and use the Henderson-Hasselbalch equation to determine the K_a value. The equivalence point data can be used to calculate the K_b value for the conjugate base which is related to K_a by the equation $K_a = K_w/K_b$.

17.   No, since there are three colored forms, there must be two proton transfer reactions. Thus, there must be at least two acidic protons in the acid (orange) form of thymol blue.

18.   Equivalence point: when enough titrant has been added to react exactly with the substance in the solution being titrated. End point: indicator changes color. We want the indicator to tell us when we have reached the equivalence point. We can detect the end point visually and assume it is the equivalence point for doing stoichiometric calculations. They don't have to be as close as 0.01 pH units since, at the equivalence point, the pH is changing very rapidly with added titrant. The range over which an indicator changes color only needs to be close to the pH of the equivalence point.

19.   The two forms of an indicator are different colors. The HIn form has one color and the In⁻ form has another color. To see only one color, that form must be in an approximately ten fold excess or greater over the other form. When the ratio of the two forms is less than 10, both colors are present. To go from $[HIn]/[In^-] = 10$ to $[HIn]/[In^-] = 0.1$ requires a change of 2 pH units (a 100-fold decrease in $[H^+]$) as the indicator changes from the HIn color to the In⁻ color.

20.   If the number of ions in the two salts is the same, the K_{sp}'s can be compared, i.e., 1:1 electrolytes (1 cation:1 anion) can be compared to each other; 2:1 electrolytes can be compared to each other, etc. If the number of ions is the same, the salt with the largest K_{sp} value has the largest molar solubility.

Exercises

Buffers

21.   When strong acid or strong base is added to a bicarbonate/carbonate mixture, the strong acid/base is neutralized. The reaction goes to completion, resulting in the strong acid/base being replaced with a weak acid/base, which results in a new buffer solution. The reactions are:

$$H^+(aq) + CO_3^{2-}(aq) \rightarrow HCO_3^-(aq); \quad OH^- + HCO_3^-(aq) \rightarrow CO_3^{2-}(aq) + H_2O(l)$$

22. $NH_3(aq) + H^+(aq) \rightarrow NH_4^+(aq); \quad NH_4^+(aq) + OH^-(aq) \rightarrow NH_3(aq) + H_2O(l)$

23.   a.   This is a weak acid problem. Let $HC_3H_5O_2 = HOPr$ and $C_3H_5O_2^- = OPr^-$.

$$HOPr \quad \rightleftharpoons \quad H^+ \quad + \quad OPr^- \qquad K_a = 1.3 \times 10^{-5}$$

	HOPr		H⁺		OPr⁻
Initial	0.100 M		~0		0

x mol/L HOPr dissociates to reach equilibrium

Change	-x	→	+x		+x
Equil.	0.100 - x		x		x

$$K_a = 1.3 \times 10^{-5} = \frac{[H^+][OPr^-]}{[HOPr]} = \frac{x^2}{0.100 - x} \approx \frac{x^2}{0.100}$$

$x = [H^+] = 1.1 \times 10^{-3}\ M;\ pH = 2.96$ Assumptions good by the 5% rule.

b. This is a weak base problem (Na^+ has no acidic/basic properties).

$$OPr^- \ + \ H_2O \ \rightleftharpoons \ HOPr \ + \ OH^- \qquad K_b = \frac{K_w}{K_a} = 7.7 \times 10^{-10}$$

Initial	0.100 M	0	~0

x mol/L OPr^- reacts with H_2O to reach equilibrium

Change	$-x$	$\rightarrow$	$+x$	$+x$
Equil.	$0.100 - x$		x	x

$$K_b = 7.7 \times 10^{-10} = \frac{[HOPr][OH^-]}{[OPr^-]} = \frac{x^2}{0.100 - x} \approx \frac{x^2}{0.100}$$

$x = [OH^-] = 8.8 \times 10^{-6} \ M$; pOH = 5.06; pH = 8.94   Assumptions good.

c. pure H_2O, $[H^+] = [OH^-] = 1.0 \times 10^{-7} \ M$; pH = 7.00

d. This solution contains a weak acid and its conjugate base. This is a buffer solution. We will solve for the pH through the weak acid equilibrium reaction.

$$HOPr \qquad \rightleftharpoons \qquad H^+ \ + \ OPr^- \qquad K_a = 1.3 \times 10^{-5}$$

Initial	0.100 M	~0	0.100 M

x mol/L HOPr dissociates to reach equilibrium

Change	$-x$	$\rightarrow$	$+x$	$+x$
Equil.	$0.100 - x$		x	$0.100 + x$

$$1.3 \times 10^{-5} = \frac{(0.100 + x)(x)}{0.100 - x} \approx \frac{(0.100)(x)}{0.100} = x = [H^+]$$

$[H^+] = 1.3 \times 10^{-5} \ M$; pH = 4.89   Assumptions good.

Alternatively, we can use the Henderson-Hasselbalch equation to calculate the pH of buffer solutions.

$$pH = pK_a + \log \frac{[Base]}{[Acid]} = pK_a + \log \frac{(0.100)}{(0.100)} = pK_a = -\log(1.3 \times 10^{-5}) = 4.89$$

The Henderson-Hasselbalch equation will be valid when an assumption of the type $0.1 + x \approx 0.1$ that we just made in this problem is valid. From a practical standpoint, this will almost always be true for useful buffer solutions. If the assumption is not valid, the solution will have such a low buffering capacity that it will be of no use to control the pH. Note: The Henderson-Hasselbalch equation can only be used to solve for the pH of buffer solutions.

24.   a.   Weak base problem:

$$HONH_2 \; + \; H_2O \; \rightleftharpoons \; HONH_3^+ \; + \; OH^- \qquad K_b = 1.1 \times 10^{-8}$$

Initial	0.100 M	0	~0
	x mol/L $HONH_2$ reacts with H_2O to reach equilibrium		
Change	$-x$ $\rightarrow$	$+x$	$+x$
Equil.	0.100 - x	x	x

$$K_b = 1.1 \times 10^{-8} = \frac{x^2}{0.100 - x} \approx \frac{x^2}{0.100}$$

$x = [OH^-] = 3.3 \times 10^{-5} \, M$; pOH = 4.48; pH = 9.52 Assumptions good.

b.   Weak acid problem (Cl⁻ has no acidic/basic properties):

$$HONH_3^+ \; \rightleftharpoons \; HONH_2 \; + \; H^+$$

Initial	0.100 M	0	~0
	x mol/L $HONH_3^+$ dissociates to reach equilibrium		
Change	$-x$ $\rightarrow$	$+x$	$+x$
Equil.	0.100 - x	x	x

$$K_a = \frac{K_w}{K_b} = 9.1 \times 10^{-7} = \frac{[HONH_2][H^+]}{[HONH_3^+]} = \frac{x^2}{0.100 - x} \approx \frac{x^2}{0.100}$$

$x = [H^+] = 3.0 \times 10^{-4} \, M$; pH = 3.52   Assumptions good.

c.   Pure H_2O, pH = 7.00

d.   Buffer solution where $pK_a = -\log(9.1 \times 10^{-7}) = 6.04$. Using the Henderson-Hasselbalch equation:

$$pH = pK_a + \log \frac{[Base]}{[Acid]} = 6.04 + \log \frac{[HONH_2]}{[HONH_3^+]} = 6.04 + \log \frac{(0.100)}{(0.100)} = 6.04$$

25.   0.100 $M \, HC_3H_5O_2$: percent dissociation $= \dfrac{[H^+]}{[HC_3H_5O_2]_o} \times 100 = \dfrac{1.1 \times 10^{-3} \, M}{0.100 \, M} \times 100 = 1.1\%$

0.100 $M \, HC_3H_5O_2 + 0.100 \, M \, NaC_3H_5O_2$: % dissociation $= \dfrac{1.3 \times 10^{-5}}{0.100} \times 100 = 1.3 \times 10^{-2}\%$

The percent dissociation of the acid decreases from 1.1% to 1.3×10^{-2} % when $C_3H_5O_2^-$ is present. This is known as the common ion effect. The presence of the conjugate base of the weak acid inhibits the acid dissociation reaction.

26.   0.100 M HONH$_2$: percent ionization = $\dfrac{[\text{OH}^-]}{[\text{HONH}_2]_o} \times 100 = \dfrac{3.3 \times 10^{-5}\,M}{0.100\,M} \times 100 = 3.3 \times 10^{-2}\,\%$

   0.100 M HONH$_2$ + 0.100 M HONH$_3{}^+$: % ionization = $\dfrac{1.1 \times 10^{-8}}{0.100} \times 100 = 1.1 \times 10^{-5}\,\%$

The percent ionization decreases by a factor of 3000. The presence of the conjugate acid of the weak base inhibits the weak base reaction with water. This is known as the common ion effect.

27.   a.   We have a weak acid (HOPr = HC$_3$H$_5$O$_2$) and a strong acid (HCl) present. The amount of H$^+$ donated by the weak acid will be negligible as compared to the 0.020 M H$^+$ from the strong acid. To prove it let's consider the weak acid equilibrium reaction:

$$\text{HOPr} \;\rightleftharpoons\; \text{H}^+ \;+\; \text{OPr}^- \qquad K_a = 1.3 \times 10^{-5}$$

Initial	0.100 M	0.020 M	0
	x mol/L HOPr dissociates to reach equilibrium		
Change	$-x$ $\rightarrow$	$+x$	$+x$
Equil.	0.100 - x	0.020 + x	x

$K_a = 1.3 \times 10^{-5} = \dfrac{(0.020 + x)(x)}{0.100 - x} \approx \dfrac{(0.020)(x)}{0.100}, \quad x = 6.5 \times 10^{-5}\,M$

[H$^+$] = 0.020 + x = 0.020 M; pH = 1.70   Assumptions good ($x = 6.5 \times 10^{-5}$ which is $\ll$ 0.020).

   b.   Added H$^+$ reacts completely with the best base present, OPr$^-$. Since all species present are in the same volume of solution, we can use molarity units to do the stoichiometry part of the problem (instead of moles). The stoichiometry problem is:

$$\text{OPr}^- \;+\; \text{H}^+ \;\rightarrow\; \text{HOPr}$$

Before	0.100 M	0.020 M	0	
Change	-0.020	-0.020 $\rightarrow$	+0.020	Reacts completely
After	0.080	0	0.020 M	

After reaction, a weak acid, HOPr , and its conjugate base, OPr$^-$, are present. This is a buffer solution. Using the Henderson-Hasselbalch equation where pK$_a$ = -log (1.3 $\times$ 10^{-5}) = 4.89:

$\text{pH} = \text{pK}_a + \log \dfrac{[\text{Base}]}{[\text{Acid}]} = 4.89 + \log \dfrac{(0.080)}{(0.020)} = 5.49 \qquad$ Assumptions good.

   c.   This is a strong acid problem. [H$^+$] = 0.020 M; pH = 1.70

   d.   Added H$^+$ reacts completely with the best base present, OPr$^-$.

$$\text{OPr}^- \;+\; \text{H}^+ \;\rightarrow\; \text{HOPr}$$

Before	0.100 M	0.020 M	0.100 M	
Change	-0.020	-0.020 $\rightarrow$	+0.020	Reacts completely
After	0.080	0	0.120	

A buffer solution results (weak acid + conjugate base). Using the Henderson-Hasselbalch equation:

$$pH = pK_a + \log \frac{[Base]}{[Acid]} = 4.89 + \log \frac{(0.080)}{(0.120)} = 4.71$$

28.   a.   Added H^+ reacts completely with $HONH_2$ (the best base present) to form $HONH_3^+$.

	$HONH_2$	+	H^+	$\rightarrow$	$HONH_3^+$	
Before	0.100 M		0.020 M		0	
Change	-0.020		-0.020	$\rightarrow$	+0.020	Reacts completely
After	0.080		0		0.020	

After this reaction, a buffer solution exists, i.e., a weak acid ($HONH_3^+$) and its conjugate base ($HONH_2$) are present at the same time. Using the Henderson-Hasselbalch equation to solve for the pH where $pK_a = - \log (K_w /K_b) = 6.04$:

$$pH = pK_a + \log \frac{[base]}{[acid]} = 6.04 + \log \frac{(0.080)}{(0.020)} = 6.04 + 0.60 = 6.64$$

b.   We have a weak acid and a strong acid present at the same time. The H^+ contribution from the weak acid, $HONH_3^+$, will be negligible. So, we have to consider only the H^+ from HCl. $[H^+] = 0.020$ M; pH = 1.70

c.   This is a strong acid in water. $[H^+] = 0.020$ M; pH = 1.70

d.   Major species: H_2O, Cl^-, $HONH_2$, $HONH_3^+$, H^+

H$^+$ will react completely with $HONH_2$, the best base present.

	$HONH_2$	+	H^+	$\rightarrow$	$HONH_3^+$	
Before	0.100 M		0.020 M		0.100 M	
Change	-0.020		-0.020	$\rightarrow$	+0.020	Reacts completely
After	0.080		0		0.120	

A buffer solution results after reaction. Using the Henderson-Hasselbalch equation:

$$pH = 6.04 + \log \frac{[HONH_2]}{[HONH_3^+]} = 6.04 + \log \frac{(0.080)}{(0.120)} = 6.04 - 0.18 = 5.86$$

29.   a.   OH$^-$ will react completely with the best acid present, HOPr.

	HOPr	+	OH$^-$	$\rightarrow$	OPr$^-$	+ H_2O	
Before	0.100 M		0.020 M		0		
Change	-0.020		-0.020	$\rightarrow$	+0.020		Reacts completely
After	0.080		0		0.020		

A buffer solution results after the reaction. Using the Henderson-Hasselbalch equation:

$$\text{pH} = \text{p}K_a + \log \frac{[\text{Base}]}{[\text{Acid}]} = 4.89 + \log \frac{(0.020)}{(0.080)} = 4.29$$

b. We have a weak base and a strong base present at the same time. The amount of OH^- added by the weak base will be negligible compared to the 0.020 M OH^- from the strong base. To prove it, let's consider the weak base equilibrium:

$$OPr^- \ + \ H_2O \ \rightleftharpoons \ HOPr \ + \ OH^- \qquad K_b = 7.7 \times 10^{-10}$$

Initial	0.100 M		0	0.020 M

x mol/L OPr^- reacts with H_2O to reach equilibrium

Change	-x	$\rightarrow$	+x	+x
Equil.	0.100 - x		x	0.020 + x

$$K_b = 7.7 \times 10^{-10} = \frac{x\,(0.020 + x)}{0.100 - x} \approx \frac{x(0.020)}{0.100}, \quad x = 3.9 \times 10^{-9} \ M$$

$[OH^-] = 0.020 + x = 0.020 \ M$; pOH = 1.70; pH = 12.30   Assumptions good.

c. This is a strong base in water. $[OH^-] = 0.020 \ M$; pOH = 1.70; pH = 12.30

d. OH^- will react completely with HOPr, the best acid present.

$$HOPr \ + \ OH^- \ \rightarrow \ OPr^- \ + \ H_2O$$

Before	0.100 M	0.020 M		0.100 M	
Change	-0.020	-0.020	$\rightarrow$	+0.020	Reacts completely
After	0.080	0		0.120	

Using the Henderson-Hasselbalch equation to solve for the pH of the resulting buffer solution:

$$\text{pH} = \text{p}K_a + \log \frac{[\text{Base}]}{[\text{Acid}]} = 4.89 + \log \frac{(0.120)}{(0.080)} = 5.07$$

30. a. We have a weak base and a strong base present at the same time. The OH^- contribution from the weak base, $HONH_2$, will be negligible. Consider only the added strong base as the primary source of OH^-.

[OH^-] = 0.020 M; pOH = 1.70; pH = 12.30

b. Added strong base will react to completion with the best acid present, $HONH_3^+$.

$$OH^- \ + \ HONH_3^+ \ \rightarrow \ HONH_2 \ + \ H_2O$$

Before	0.020 M	0.100 M		0	
Change	-0.020	-0.020	$\rightarrow$	+0.020	Reacts completely
After	0	0.080		0.020	

The resulting solution is a buffer (a weak acid and its conjugate base). Using the Henderson-Hasselbalch equation:

$$pH = 6.04 + \log\left(\frac{0.020}{0.080}\right) = 6.04 - 0.60 = 5.44$$

c. This is a strong base in water. $[OH^-] = 0.020 \; M$; $pOH = 1.70$; $pH = 12.30$

d. Major species: H_2O, Cl^-, Na^+, $HONH_2$, $HONH_3^+$, OH^-

Again, the added strong base reacts completely with the best acid present, $HONH_3^+$.

	$HONH_3^+$	$+$	OH^-	$\rightarrow$	$HONH_2$	$+$	H_2O	
Before	$0.100 \; M$		$0.020 \; M$		$0.100 \; M$			
Change	-0.020		-0.020	$\rightarrow$	$+0.020$			Reacts completely
After	0.080		0		0.120			

A buffer solution results. Using the Henderson-Hasselbalch equation:

$$pH = 6.04 + \log \frac{[HONH_2]}{[HONH_3^+]} = 6.04 + \log\left(\frac{0.120}{0.080}\right) = 6.04 + 0.18 = 6.22$$

31.   Consider all of the results to Exercises 15.23, 15.27, and 15.29:

Solution	Initial pH	after added acid	after added base
a	2.96	1.70	4.29
b	8.94	5.49	12.30
c	7.00	1.70	12.30
d	4.89	4.71	5.07

The solution in Exercise 15.23d is a buffer; it contains both a weak acid ($HC_3H_5O_2$) and a weak base ($C_3H_5O_2^-$). Solution d shows the greatest resistance to changes in pH when either strong acid or strong base is added, which is the primary property of buffers.

32.   Consider all of the results to Exercises 15.24, 15.28, and 15.30.

Solution	Initial pH	after added acid	after added base
a	9.52	6.64	12.30
b	3.52	1.70	5.44
c	7.00	1.70	12.30
d	6.04	5.86	6.22

The solution in Exercise 15.24d is a buffer; it shows the greatest resistance to a change in pH when strong acid or base is added. The solution in Exercise 15.24d contains a weak acid ($HONH_3^+$) and a weak base ($HONH_2$), which constitutes a buffer solution.

33.   Major species: HNO_2, NO_2^- and Na^+. Na^+ has no acidic or basic properties. The appropriate equilibrium reaction to use is the K_a reaction of HNO_2 which contains both HNO_2 and NO_2^-. Solving the equilibrium problem (called a buffer problem):

$$HNO_2 \rightleftharpoons NO_2^- + H^+$$

Initial	1.00 M	1.00 M	~0

x mol/L HNO_2 dissociates to reach equilibrium

Change	$-x$	$\rightarrow$	$+x$	$+x$
Equil.	$1.00 - x$		$1.00 + x$	x

$$K_a = 4.0 \times 10^{-4} = \frac{[NO_2^-][H^+]}{[HNO_2]} = \frac{(1.00 + x)(x)}{(1.00 - x)} \approx \frac{1.00(x)}{1.00} \quad \text{(assuming } x \ll 1.00)$$

$x = 4.0 \times 10^{-4} \, M = [H^+]$;   Assumptions good ($x$ is 4.0×10^{-2}% of 1.00).

$pH = -\log (4.0 \times 10^{-4}) = 3.40$

Note: We would get the same answer using the Henderson-Hasselbalch equation. Use whichever method you prefer.

34.   Major species: HF, F^- and K^+ (no acidic/basic properties). The appropriate equilibrium reaction to use is the K_a reaction of HF which contains both HF and F^-.

$$HF \rightleftharpoons F^- + H^+$$

Initial	0.60 M	1.00 M	~0

x mol/L HF dissociates to reach equilibrium

Change	$-x$	$\rightarrow$	$+x$	$+x$
Equil.	$0.60 - x$		$1.00 + x$	x

$$K_a = 7.2 \times 10^{-4} = \frac{[F^-][H^+]}{[HF]} = \frac{(1.00 + x)(x)}{(0.60 - x)} \approx \frac{1.00(x)}{0.60} \quad \text{(assuming } x \ll 0.60)$$

$x = [H^+] = 0.60 \times (7.2 \times 10^{-4}) = 4.3 \times 10^{-4} \, M$;   Assumptions good ($x$ is 7.2×10^{-2}% of 0.60).

$pH = -\log (4.3 \times 10^{-4}) = 3.37$

35.   Major species after NaOH added: HNO_2, NO_2^-, Na^+ and OH^-. The OH^- from the strong base will react with the best acid present (HNO_2). Any reaction involving a strong base is assumed to go to completion. Since all species present are in the same volume of solution, we can use molarity units to do the stoichiometry part of the problem (instead of moles). The stoichiometry problem is:

$$OH^- + HNO_2 \rightarrow NO_2^- + H_2O$$

	OH^-	HNO_2		NO_2^-	H_2O	
Before	0.10 mol/1.00 L	1.00 M		1.00 M		
Change	-0.10 M	-0.10 M	$\rightarrow$	$+0.10$ M		Reacts completely
After	0	0.90		1.10		

After all the OH⁻ reacts, we are left with a solution containing a weak acid (HNO_2) and its conjugate base (NO_2^-). This is what we call a buffer problem. We will solve this buffer problem using the K_a equilibrium reaction.

$$HNO_2 \rightleftharpoons NO_2^- + H^+$$

	HNO_2		NO_2^-	H^+
Initial	0.90 M		1.10 M	~0

x mol/L HNO_2 dissociates to reach equilibrium

| Change | -x | $\rightarrow$ | +x | +x |
| Equil. | 0.90 - x | | 1.10 + x | x |

$K_a = 4.0 \times 10^{-4} = \dfrac{(1.10 + x)(x)}{(0.90 - x)} \approx \dfrac{1.10(x)}{0.90}$, $x = [H^+] = 3.3 \times 10^{-4}$ M; pH = 3.48; Assumptions good.

Note: The added NaOH to this buffer solution changes the pH only from 3.40 to 3.48. If the NaOH were added to 1.0 L of pure water, the pH would change from 7.00 to 13.00.

Major species after HCl added: HNO_2, NO_2^-, H^+, Na^+, Cl^-; The added H^+ from the strong acid will react completely with the best base present (NO_2^-).

$$H^+ + NO_2^- \rightarrow HNO_2$$

	H^+		NO_2^-		HNO_2	
Before	$\dfrac{0.20 \text{ mol}}{1.00 \text{ L}}$		1.00 M		1.00 M	
Change	-0.20 M		-0.20 M	$\rightarrow$	+0.20 M	Reacts completely
After	0		0.80		1.20	

After all the H^+ has reacted, we have a buffer solution (a solution containing a weak acid and its conjugate base). Solving the buffer problem:

$$HNO_2 \rightleftharpoons NO_2^- + H^+$$

	HNO_2	NO_2^-	H^+
Initial	1.20 M	0.80 M	0
Equil.	1.20 - x	0.80 + x	+x

$K_a = 4.0 \times 10^{-4} = \dfrac{(0.80 + x)(x)}{1.20 - x} \approx \dfrac{0.80(x)}{1.20}$, $x = [H^+] = 6.0 \times 10^{-4}$ M; pH = 3.22; Assumptions good.

Note: The added HCl to this buffer solution changes the pH only from 3.40 to 3.22. If the HCl were added to 1.0 L of pure water, the pH would change from 7.00 to 0.70.

36.   When NaOH is added, the OH⁻ reacts completely with the best acid present, HF.

$$OH^- + HF \rightarrow F^- + H_2O$$

	OH^-		HF		F^-		H_2O	
Before	$\dfrac{0.10 \text{ mol}}{1.00 \text{ L}}$		0.60 M		1.00 M			
Change	-0.10 M		-0.10 M	$\rightarrow$	+0.10 M			Reacts completely
After	0		0.50		1.10			

We now have a new buffer problem. Solving the equilibrium part of the problem:

$$HF \quad \rightleftharpoons \quad F^- \quad + \quad H^+$$

Initial	0.50 M	1.10 M	~0
Equil.	0.50 - x	1.10 + x	x

$$K_a = 7.2 \times 10^{-4} = \frac{(1.10 + x)(x)}{(0.50 - x)} \approx \frac{1.10(x)}{(0.50)}, \quad x = [H^+] = 3.3 \times 10^{-4} \, M$$

pH = 3.48;   Assumptions good.

When HCl is added, H^+ reacts to completion with the best base (F^-) present.

$$H^+ \quad + \quad F^- \quad \rightarrow \quad HF$$

Before	0.20 M	1.00 M	0.60 M
After	0	0.80 M	0.80 M

After reaction, we have a buffer problem:

$$HF \quad \rightleftharpoons \quad F^- \quad + \quad H^+$$

Initial	0.80 M	0.80 M	0
Equil.	0.80 - x	0.80 + x	x

$$K_a = 7.2 \times 10^{-4} = \frac{(0.80 + x)(x)}{0.80 - x} \approx x, \quad x = [H^+] = 7.2 \times 10^{-4} \, M; \quad pH = 3.14; \quad \text{Assumptions good.}$$

37.   a. $$HC_2H_3O_2 \quad \rightleftharpoons \quad H^+ \quad + \quad C_2H_3O_2^- \qquad\qquad K_a = 1.8 \times 10^{-5}$$

Initial	0.10 M	~0	0.25 M

x mol/L $HC_2H_3O_2$ dissociates to reach equilibrium

Change	-x $\rightarrow$	+x	+x
Equil.	0.10 - x	x	0.25 + x

$$1.8 \times 10^{-5} = \frac{x(0.25 + x)}{(0.10 - x)} \approx \frac{x(0.25)}{0.10} \quad \text{(assuming } 0.25 + x \approx 0.25 \text{ and } 0.10 - x \approx 0.10)$$

$$x = [H^+] = 7.2 \times 10^{-6} \, M; \quad pH = 5.14 \qquad \text{Assumptions good by the 5\% rule.}$$

Alternatively, we can use the Henderson-Hasselbalch equation:

$$pH = pK_a + \log \frac{[\text{Base}]}{[\text{Acid}]} \quad \text{where } pK_a = -\log (1.8 \times 10^{-5}) = 4.74$$

$$pH = 4.74 + \log \frac{(0.25)}{(0.10)} = 4.74 + 0.40 = 5.14$$

The Henderson-Hasselbalch equation will be valid when assumptions of the type $0.10 - x \approx 0.10$ that we just made are valid. From a practical standpoint, this will almost always be true for useful buffer solutions. Note: The Henderson-Hasselbalch equation can only be used to solve for the pH of buffer solutions.

b. $pH = 4.74 + \log \dfrac{(0.10)}{(0.25)} = 4.74 + (-0.40) = 4.34$

c. $pH = 4.74 + \log \dfrac{(0.25)}{(0.25)} = 4.74 + 0.00 = 4.74$ ($pH = pK_a$ since [acid] = [base])

d. $pH = pK_a + \log \dfrac{[base]}{[acid]}$; $[base] = [C_2H_5NH_2] = 0.50\ M$; $[acid] = [C_2H_5NH_3^+] = 0.25\ M$

$K_a = \dfrac{K_w}{K_b} = \dfrac{1.0 \times 10^{-14}}{5.6 \times 10^{-4}} = 1.8 \times 10^{-11}$

$pH = -\log (1.8 \times 10^{-11}) + \log \left(\dfrac{0.50\ M}{0.25\ M} \right) = 10.74 + 0.30 = 11.04$

e. $pH = 10.74 + \log \left(\dfrac{0.50\ M}{0.50\ M} \right) = 10.74 + 0.00 = 10.74$

38. $50.0\ g\ NH_4Cl \times \dfrac{1\ mol\ NH_4Cl}{53.49\ g\ NH_4Cl} = 0.935\ mol\ NH_4Cl$ added to $1.00\ L$; $[NH_4^+] = 0.935\ M$

$pH = pK_a + \log \dfrac{[NH_3]}{[NH_4^+]} = -\log (5.6 \times 10^{-10}) + \log \left(\dfrac{0.75}{0.935} \right) = 9.25 - 0.096 = 9.15$

39. $[H^+]$ added $= \dfrac{0.010\ mol}{0.25\ L} = 0.040\ M$; The added H^+ reacts completely with NH_3 to form NH_4^+.

a.

	NH_3	$+$	H^+	$\rightarrow$	NH_4^+	
Before	0.050 M		0.040 M		0.15 M	
Change	-0.040		-0.040	$\rightarrow$	+0.040	Reacts completely
After	0.010		0		0.19	

A buffer solution still exists after H^+ reacts completely. Using the Henderson-Hasselbalch equation:

$pH = pK_a + \log \dfrac{[NH_3]}{[NH_4^+]} = -\log (5.6 \times 10^{-10}) + \log \left(\dfrac{0.010}{0.19} \right) = 9.25 + (-1.28) = 7.97$

b. NH_3 + H^+ $\rightarrow$ NH_4^+

Before	0.50 M	0.040 M		1.50 M
Change	-0.040	-0.040	$\rightarrow$	+0.040
After	0.46	0		1.54

Reacts completely

A buffer solution still exists. $pH = pK_a + \log \dfrac{[NH_3]}{[NH_4^+]} = 9.25 + \log\left(\dfrac{0.46}{1.54}\right) = 8.73$

The two buffers differ in their capacity and not in pH (both buffers had an initial pH = 8.77). Solution b has the greater capacity since it has the largest concentration of weak acid and conjugate base. Buffers with greater capacities will be able to absorb more H^+ or OH^- added.

40.   a. pK_b for $C_6H_5NH_2 = -\log(3.8 \times 10^{-10}) = 9.42$; pK_a for $C_6H_5NH_3^+ = 14.00 - 9.42 = 4.58$

$pH = pK_a + \log \dfrac{[C_6H_5NH_2]}{[C_6H_5NH_3^+]}$, $4.20 = 4.58 + \log \dfrac{0.50\,M}{[C_6H_5NH_3^+]}$

$-0.38 = \log \dfrac{0.50\,M}{[C_6H_5NH_3^+]}$, $[C_6H_5NH_3^+] = [C_6H_5NH_3Cl] = 1.2\,M$

b. $4.0\text{ g NaOH} \times \dfrac{1\text{ mol NaOH}}{40.00\text{ g}} \times \dfrac{1\text{ mol OH}^-}{\text{mol NaOH}} = 0.10\text{ mol OH}^-$, $[OH^-] = \dfrac{0.10\text{ mol}}{1.0\text{ L}} = 0.10\,M$

$C_6H_5NH_3^+$ + OH^- $\rightarrow$ $C_6H_5NH_2$ + H_2O

Before	1.2 M	0.10 M		0.50 M
Change	-0.10	-0.10	$\rightarrow$	+0.10
After	1.1	0		0.60

A buffer solution exists. $pH = 4.58 + \log\left(\dfrac{0.60}{1.1}\right) = 4.32$

41. $pH = pK_a + \log \dfrac{[C_2H_3O_2^-]}{[HC_2H_3O_2]}$; $pK_a = -\log(1.8 \times 10^{-5}) = 4.74$

Since the buffer components, $C_2H_3O_2^-$ and $HC_2H_3O_2$, are both in the same volume of water, the concentration ratio of $[C_2H_3O_2^-]/[HC_2H_3O_2]$ will equal the mol ratio of mol $C_2H_3O_2^-$/mol $HC_2H_3O_2$.

$5.00 = 4.74 + \log \dfrac{\text{mol } C_2H_3O_2^-}{\text{mol } HC_2H_3O_2}$; $\text{mol } HC_2H_3O_2 = 0.5000\text{ L} \times \dfrac{0.200\text{ mol}}{\text{L}} = 0.100\text{ mol}$

$0.26 = \log \dfrac{\text{mol } C_2H_3O_2^-}{0.100\text{ mol}}$, $\dfrac{\text{mol } C_2H_3O_2^-}{0.100} = 10^{0.26} = 1.8$,   mol $C_2H_3O_2^- = 0.18$ mol

mass $NaC_2H_3O_2 = 0.18\text{ mol } NaC_2H_3O_2 \times \dfrac{82.03\text{ g}}{\text{mol}} = 15\text{ g } NaC_2H_3O_2$

42. $pH = pK_a + \log \dfrac{[NO_2^-]}{[HNO_2]}$, $3.55 = \log(4.0 \times 10^{-4}) + \log \dfrac{[NO_2^-]}{[HNO_2]}$

$3.55 = 3.40 + \log \dfrac{[NO_2^-]}{[HNO_2]}$, $\dfrac{[NO_2^-]}{[HNO_2]} = 10^{0.15} = 1.4$

Let x = volume (L) HNO_2 solution needed, then $1.00 - x$ = volume of $NaNO_2$ solution needed to form this buffer solution.

$$\dfrac{[NO_2^-]}{[HNO_2]} = 1.4 = \dfrac{(1.00 - x) \times \dfrac{0.50 \text{ mol NaNO}_2}{L}}{x \times \dfrac{0.50 \text{ mol HNO}_2}{L}} = \dfrac{0.50 - 0.50x}{0.50x}$$

$0.70\, x = 0.50 - 0.50\, x$, $1.20\, x = 0.50$, $x = 0.42$ L

We need 0.42 L of 0.50 M HNO_2 and $1.00 - 0.42 = 0.58$ L of 0.50 M $NaNO_2$ to form a pH = 3.55 buffer solution.

43. $C_5H_5NH^+ \rightleftharpoons H^+ + C_5H_5N$ $K_a = \dfrac{K_w}{K_b} = \dfrac{1.0 \times 10^{-14}}{1.7 \times 10^{-9}} = 5.9 \times 10^{-6}$; $pK_a = -\log(5.9 \times 10^{-6}) = 5.23$

We will use the Henderson-Hasselbalch equation to calculate the concentration ratio necessary for each buffer.

$pH = pK_a + \log \dfrac{[base]}{[acid]}$, $pH = 5.23 + \log \dfrac{[C_5H_5N]}{[C_5H_5NH^+]}$

a. $4.50 = 5.23 + \log \dfrac{[C_5H_5N]}{[C_5H_5NH^+]}$

$\log \dfrac{[C_5H_5N]}{[C_5H_5NH^+]} = -0.73$

$\dfrac{[C_5H_5N]}{[C_5H_5NH^+]} = 10^{-0.73} = 0.19$

b. $5.00 = 5.23 + \log \dfrac{[C_5H_5N]}{[C_5H_5NH^+]}$

$\log \dfrac{[C_5H_5N]}{[C_5H_5NH^+]} = -0.23$

$\dfrac{[C_5H_5N]}{[C_5H_5NH^+]} = 10^{-0.23} = 0.59$

c. $5.23 = 5.23 + \log \dfrac{[C_5H_5N]}{[C_5H_5NH^+]}$

$\dfrac{[C_5H_5N]}{[C_5H_5NH^+]} = 10^{0.0} = 1.0$

d. $5.50 = 5.23 + \log \dfrac{[C_5H_5N]}{[C_5H_5NH^+]}$

$\dfrac{[C_5H_5N]}{[C_5H_5NH^+]} = 10^{0.27} = 1.9$

44. a. $pH = pK_a + \log \dfrac{[\text{Base}]}{[\text{Acid}]}$, $7.40 = -\log(4.3 \times 10^{-7}) + \log \dfrac{[HCO_3^-]}{[H_2CO_3]} = 6.37 + \log \dfrac{[HCO_3^-]}{[H_2CO_3]}$

$\dfrac{[HCO_3^-]}{[H_2CO_3]} = 10^{1.03} = 11$; $\dfrac{[H_2CO_3]}{[HCO_3^-]} = \dfrac{[CO_2]}{[HCO_3^-]} = \dfrac{1}{11} = 0.091$

b. $7.15 = -\log(6.2 \times 10^{-8}) + \log \dfrac{[HPO_4^{2-}]}{[H_2PO_4^-]}$, $7.15 = 7.21 + \log \dfrac{[HPO_4^{2-}]}{[H_2PO_4^-]}$

$\dfrac{[HPO_4^{2-}]}{[H_2PO_4^-]} = 10^{-0.06} = 0.9$, $\dfrac{[H_2PO_4^-]}{[HPO_4^{2-}]} = \dfrac{1}{0.9} = 1.1 \approx 1$

c. A best buffer has approximately equal concentrations of weak acid and conjugate base so that $pH \approx pK_a$ for a best buffer. The pK_a value for a $H_3PO_4/H_2PO_4^-$ buffer is $-\log(7.5 \times 10^{-3}) = 2.12$. A pH of 7.1 is too high for a $H_3PO_4/H_2PO_4^-$ buffer to be effective. At this high a pH, there would be so little H_3PO_4 present that we could hardly consider it a buffer. This solution would not be effective in resisting pH changes, especially when a strong base is added.

45. A best buffer has large and equal quantities of weak acid and conjugate base. Since [acid] = [base] for a best buffer, then $pH = pK_a + \log \dfrac{[\text{base}]}{[\text{acid}]} = pK_a + 0 = pK_a$ ($pH = pK_a$).

The best acid choice for a pH = 7.00 buffer would be the weak acid with a pK_a close to 7.0 or $K_a \approx 1 \times 10^{-7}$. HOCl is the best choice in Table 14.2 ($K_a = 3.5 \times 10^{-8}$; $pK_a = 7.46$). To make this buffer, we need to calculate the [base]/[acid] ratio.

$7.00 = 7.46 + \log \dfrac{[\text{base}]}{[\text{acid}]}$, $\dfrac{[OCl^-]}{[HOCl]} = 10^{-0.46} = 0.35$

Any OCl^-/HOCl buffer in a concentration ratio of 0.35:1 will have a pH = 7.00. One possibility is [NaOCl] = 0.35 M and [HOCl] = 1.0 M.

46. For a pH = 5.00 buffer, we want an acid with a pK_a close to 5.0. For a conjugate acid-base pair, $14.00 = pK_a + pK_b$. So for a pH = 5.00 buffer, we want the base to have a pK_b close to 14.0 - 5.0 = 9.0 or a K_b close to 1×10^{-9}. The best choice in Table 14.3 is pyridine, C_5H_5N, with $K_b = 1.7 \times 10^{-9}$.

$pH = pK_a + \log \dfrac{[\text{base}]}{[\text{acid}]}$; $pK_a = \dfrac{K_w}{K_b} = \dfrac{1.0 \times 10^{-14}}{1.7 \times 10^{-9}} = 5.9 \times 10^{-6}$

$5.00 = -\log(5.9 \times 10^{-6}) + \log \dfrac{[\text{base}]}{[\text{acid}]}$, $\dfrac{[C_5H_5N]}{[C_5H_5NH^+]} = 10^{-0.23} = 0.59$

There are several possibilities for this buffer. One possibility is a solution of $[C_5H_5N]$ = 0.59 M and $[C_5H_5NHCl]$ = 1.0 M. The pH of this solution will be 5.00 since the base to acid concentration ratio is 0.59:1.

47.   The reaction $OH^- + CH_3NH_3^+ \rightarrow CH_3NH_2 + H_2O$ goes to completion for solutions a, c and d (no reaction occurs between the species in solution b since both species are bases). After the OH^- reacts completely, there must be both $CH_3NH_3^+$ and CH_3NH_2 in solution for it to be a buffer. The important components of each solution (after the OH^- reacts completely) is/are:

a. $0.05\ M$ CH_3NH_2 (no $CH_3NH_3^+$ remains, no buffer)

b. $0.05\ M$ OH^- and $0.1\ M$ CH_3NH_2 (two bases present, no buffer)

c. $0.05\ M$ OH^- and $0.05\ M$ CH_3NH_2 (too much OH^- added, no $CH_3NH_3^+$ remains, no buffer)

d. $0.05\ M$ CH_3NH_2 and $0.05\ M$ $CH_3NH_3^+$ (a buffered solution results)

Only the combination in mixture d results in a buffer. Note that the concentrations are halved from the initial values. This is because equal volumes of two solutions were added together, which halves the concentrations.

48.   a.   No; A solution of a strong acid (HNO_3) and its conjugate base (NO_3^-) is not generally considered a buffered solution.

b.   No; Two acids are present (HNO_3 and HF), so it is not a buffered solution.

c. H^+ reacts completely with F^-. Since equal volumes are mixed, the initial concentrations in the mixture are $0.10\ M$ HNO_3 and $0.20\ M$ NaF.

	H^+	+	F^-	$\rightarrow$	HF	
Before	$0.10\ M$		$0.20\ M$		0	
Change	-0.10		-0.10	$\rightarrow$	$+0.10$	Reacts completely
After	0		0.10		0.10	

After H^+ reacts completely, a buffered solution results, i.e., a weak acid (HF) and its conjugate base (F^-) are both present in solution in large quantities.

d.   No; A strong acid (HNO_3) and a strong base ($NaOH$) do not form buffered solutions. They will neutralize each other to form H_2O.

49.   When OH^- is added, it converts $HC_2H_3O_2$ into $C_2H_3O_2^-$: $HC_2H_3O_2 + OH^- \rightarrow C_2H_3O_2^- + H_2O$

From this reaction, the moles of $C_2H_3O_2^-$ produced <u>equal</u> the moles of OH^- added. Also, the total concentration of acetic acid plus acetate ion must equal $2.0\ M$ (assuming no volume change on addition of NaOH). Summarizing for each solution:

$[C_2H_3O_2^-] + [HC_2H_3O] = 2.0\ M$ and $[C_2H_3O_2^-]$ produced = $[OH^-]$ added

a. $pH = pK_a + \log \dfrac{[C_2H_3O_2^-]}{[HC_2H_3O_2]}$; For $pH = pK_a$, $\log \dfrac{[C_2H_3O_2^-]}{[HC_2H_3O_2]} = 0$

Therefore, $\dfrac{[C_2H_3O_2^-]}{[HC_2H_3O_2]} = 1.0$ and $[C_2H_3O_2^-] = [HC_2H_3O_2]$

Since $[C_2H_3O_2^-] + [HC_2H_3O_2] = 2.0\ M$, then $[C_2H_3O_2^-] = [HC_2H_3O_2] = 1.0\ M = [OH^-]$ added

To produce a $1.0\ M\ C_2H_3O_2^-$ solution, we need to add 1.0 mol of NaOH to 1.0 L of the $2.0\ M\ HC_2H_3O_2$ solution. The resultant solution will have pH = pK_a = 4.74.

b. $4.00 = 4.74 + \log \dfrac{[C_2H_3O_2^-]}{[HC_2H_3O_2]}$, $\dfrac{[C_2H_3O_2^-]}{[HC_2H_3O_2]} = 10^{-0.74} = 0.18$

$[C_2H_3O_2^-] = 0.18\,[HC_2H_3O_2]$ or $[HC_2H_3O_2] = 5.6\,[C_2H_3O_2^-]$; Since $[C_2H_3O_2^-] + [HC_2H_3O_2] = 2.0\ M$, then:

$[C_2H_3O_2^-] + 5.6\,[C_2H_3O_2^-] = 2.0\ M,\ [C_2H_3O_2^-] = \dfrac{2.0}{6.6} = 0.30\ M = [OH^-]$ added

We need to add 0.30 mol of NaOH to 1.0 L of $2.0\ M\ HC_2H_3O_2$ solution to produce $0.30\ M$ $C_2H_3O_2^-$. The resultant solution will have pH = 4.00.

c. $5.00 = 4.74 + \log \dfrac{[C_2H_3O_2^-]}{[HC_2H_3O_2]}$, $\dfrac{[C_2H_3O_2^-]}{[HC_2H_3O_2]} = 10^{0.26} = 1.8$

$1.8\,[HC_2H_3O_2] = [C_2H_3O_2^-]$ or $[HC_2H_3O_2] = 0.56\,[C_2H_3O_2^-]$; Since $[HC_2H_3O_2] + [C_2H_3O_2^-] = 2.0\ M$, then:

$1.56\,[C_2H_3O_2^-] = 2.0\ M,\ [C_2H_3O_2^-] = 1.3\ M = [OH^-]$ added

We need to add 1.3 mol of NaOH to 1.0 L of $2.0\ M\ HC_2H_3O_2$ to produce a solution with pH = 5.00.

50.   When H^+ is added, it converts $C_2H_3O_2^-$ into $HC_2H_3O_2$: $C_2H_3O_2^- + H^+ \rightarrow HC_2H_3O_2$. From this reaction, the moles of $HC_2H_3O_2$ produced must equal the moles of H^+ added and the total concentration of acetate ion + acetic acid must equal $1.0\ M$ (assuming no volume change). Summarizing for each solution:

$[C_2H_3O_2^-] + [HC_2H_3O_2] = 1.0\ M$ and $[HC_2H_3O_2] = [H^+]$ added

a.   pH = pK_a + $\log \dfrac{[C_2H_3O_2^-]}{[HC_2H_3O_2]}$; For pH = pK_a, $[C_2H_3O_2^-] = [HC_2H_3O_2]$.

For this to be true, $[C_2H_3O_2^-] = [HC_2H_3O_2] = 0.50\ M = [H^+]$ added, which means that 0.50 mol of HCl must have been added to 1.0 L of the initial solution to produce a solution with pH = pK_a.

b. $4.20 = 4.74 + \log \dfrac{[C_2H_3O_2^-]}{[HC_2H_3O_2]}$, $\dfrac{[C_2H_3O_2^-]}{[HC_2H_3O_2]} = 10^{-0.54} = 0.29$

$[C_2H_3O_2^-] = 0.29\,[HC_2H_3O_2]$; $0.29\,[HC_2H_3O_2] + [HC_2H_3O_2] = 1.0\ M$

$[HC_2H_3O_2] = 0.78\ M = [H^+]$ added

0.78 mol of HCl must be added to produce a solution with pH = 4.20.

c. $5.00 = 4.74 + \log \dfrac{[C_2H_3O_2^-]}{[HC_2H_3O_2]}$, $\dfrac{[C_2H_3O_2^-]}{[HC_2H_3O_2]} = 10^{0.26} = 1.8$

$[C_2H_3O_2^-] = 1.8\,[HC_2H_3O_2]$; $1.8\,[HC_2H_3O_2] + [HC_2H_3O_2] = 1.0\ M$

$[HC_2H_3O_2] = 0.36\ M = [H^+]$ added

0.36 mol of HCl must be added to produce a solution with pH = 5.00.

Acid-Base Titrations

51.

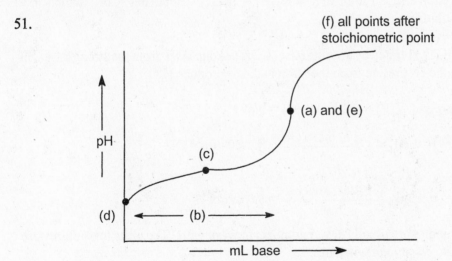

$HA + OH^- \rightarrow A^- + H_2O$;   Added OH^- from the strong base converts the weak acid, HA, into its conjugate base, A^-. Initially, before any OH^- is added (point d), HA is the dominant species present. After OH^- is added, both HA and A^- are present and a buffer solution results (region b). At the equivalence point (points a and e), exactly enough OH^- has been added to convert all of the weak acid, HA, into its conjugate base, A^-. Past the equivalence point (region f), excess OH^- is present. For the answer to part b, we included almost the entire buffer region. The maximum buffer region (or the region which is the best buffer solution) is around the halfway point to equivalence (point c). At this point, enough OH^- has been added to convert exactly one-half of the weak acid present initially into its conjugate base so $[HA] = [A^-]$ and $pH = pK_a$. A best buffer has about equal concentrations of weak acid and conjugate base present.

52.

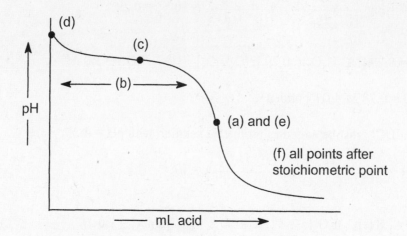

$B + H^+ \rightarrow BH^+$; Added H^+ from the strong acid converts the weak base, B, into its conjugate acid, BH^+. Initially, before any H^+ is added (point d), B is the dominant species present. After H^+ is added, both B and BH^+ are present and a buffered solution results (region b). At the equivalence point (points a and e), exactly enough H^+ has been added to convert all of the weak base present initially into its conjugate acid, BH^+. Past the equivalence point (region f), excess H^+ is present. For the answer to b, we included almost the entire buffer region. The maximum buffer region is around the halfway point to equivalence (point c) where $[B] = [BH^+]$. Here, $pH = pK_a$ which is a characteristic of a best buffer.

53.   This is a strong acid ($HClO_4$) titrated by a strong base (KOH). Added OH^- from the strong base will react completely with the H^+ present from the strong acid to produce H_2O.

   a.   Only strong acid present. $[H^+] = 0.200\ M$; pH = 0.699

   b.   mmol OH^- added $= 10.0\ mL \times \dfrac{0.100\ mmol\ OH^-}{mL} = 1.00\ mmol\ OH^-$

   mmol H^+ present $= 40.0\ mL \times \dfrac{0.200\ mmol\ H^+}{mL} = 8.0\ mmol\ H^+$

   Note: The units mmoles are usually easier numbers to work with. The units for molarity are moles/L but are also equal to mmoles/mL.

	H^+	+	OH^-	$\rightarrow$	H_2O
Before	8.00 mmol		1.00 mmol		
Change	-1.00 mmol		-1.00 mmol		Reacts completely
After	7.00 mmol		0		

   The excess H^+ determines pH. $[H^+]_{excess} = \dfrac{7.00\ mmol\ H^+}{40.0\ mL + 10.0\ mL} = 0.140\ M$; pH = 0.854

c. mmol OH⁻ added = 40.0 mL × 0.100 M = 4.00 mmol OH⁻

$$H^+ \quad + \quad OH^- \quad \rightarrow \quad H_2O$$

	H^+	OH^-
Before	8.00 mmol	4.00 mmol
After	4.00 mmol	0

$[H^+]_{excess} = \dfrac{4.00 \text{ mmol}}{(40.0 + 40.0) \text{ mL}} = 0.0500 \; M; \; pH = 1.301$

d. mmol OH⁻ added = 80.0 mL × 0.100 M = 8.00 mmol OH⁻; This is the equivalence point since we have added just enough OH⁻ to react with all the acid present. For a strong acid-strong base titration, pH = 7.00 at the equivalence point since only neutral species are present (K⁺, ClO₄⁻, H₂O).

e. mmol OH⁻ added = 100.0 mL × 0.100 M = 10.0 mmol OH⁻

$$H^+ \quad + \quad OH^- \quad \rightarrow \quad H_2O$$

	H^+	OH^-
Before	8.00 mmol	10.0 mmol
After	0	2.0 mmol

Past the equivalence point, the pH is determined by the excess OH⁻ present.

$[OH^-]_{excess} = \dfrac{2.0 \text{ mmol}}{(40.0 + 100.0) \text{ mL}} = 0.014 \; M; \; pOH = 1.85; \; pH = 12.15$

54. This is a strong base, Ba(OH)₂, titrated by a strong acid, HCl. The added strong acid will neutralize the OH⁻ from the strong base. As is always the case when a strong acid and/or strong base reacts, the reaction is assumed to go to completion.

a. Only a strong base is present, but it breaks up into two mol of OH⁻ ions for every mol of Ba(OH)₂. [OH⁻] = 2 × 0.100 M = 0.200 M; pOH = 0.699; pH = 13.301

b. mmol OH⁻ present = 80.0 mL × $\dfrac{0.100 \text{ mmol Ba(OH)}_2}{\text{mL}}$ × $\dfrac{2 \text{ mmol OH}^-}{\text{mmol Ba(OH)}_2}$ = 16.0 mmol OH⁻

mmol H⁺ added = 20.0 mL × $\dfrac{0.400 \text{ mmol H}^+}{\text{mL}}$ = 8.00 mmol H⁺

$$OH^- \quad + \quad H^+ \quad \rightarrow \quad H_2O$$

	OH^-	H^+	
Before	16.0 mmol	8.00 mmol	
Change	-8.00 mmol	-8.00 mmol	Reacts completely
After	8.0 mmol	0	

$[OH^-]_{excess} = \dfrac{8.0 \text{ mmol OH}^-}{80.0 \text{ mL} + 20.0 \text{ mL}} = 0.080 \; M; \; pOH = 1.10; \; pH = 12.90$

c. mmol H^+ added = 30.0 mL × 0.400 M = 12.0 mmol H^+

$$OH^- \quad + \quad H^+ \quad \rightarrow \quad H_2O$$

Before	16.0 mmol	12.0 mmol
After	4.0 mmol	0

$[OH^-]_{excess} = \dfrac{4.0 \text{ mmol } OH^-}{(80.0 + 30.0) \text{ mL}} = 0.036\ M$;　pOH = 1.44;　pH = 12.56

d. mmol H^+ added = 40.0 mL × 0.400 M = 16.0 mmol H^+; This is the equivalence point. Since the H^+ will exactly neutralize the OH^- from the strong base, all we have in solution is Ba^{2+}, Cl^- and H_2O. All are neutral species, so pH = 7.00.

e. mmol H^+ added = 80.0 mL × 0.400 M = 32.0 mmol H^+

$$OH^- \quad + \quad H^+ \quad \rightarrow \quad H_2O$$

Before	16.0 mmol	32.0 mmol
After	0	16.0 mmol

$[H^+]_{excess} = \dfrac{16.0 \text{ mmol}}{(80.0 + 80.0) \text{ mL}} = 0.100\ M$;　pH = 1.000

55. This is a weak acid ($HC_2H_3O_2$) titrated by a strong base (KOH).

a. Only a weak acid is present. Solving the weak acid problem:

$$HC_2H_3O_2 \quad \rightleftharpoons \quad H^+ \quad + \quad C_2H_3O_2^-$$

Initial	0.200 M	~0	0
	x mol/L $HC_2H_3O_2$ dissociates to reach equilibrium		
Change	$-x$	$\rightarrow$ $+x$	$+x$
Equil.	0.200 - x	x	x

$K_a = 1.8 \times 10^{-5} = \dfrac{x^2}{0.200 - x} = \dfrac{x^2}{0.200}$, $x = [H^+] = 1.9 \times 10^{-3}\ M$　pH = 2.72;　Assumptions good.

b. The added OH^- will react completely with the best acid present, $HC_2H_3O_2$.

mmol $HC_2H_3O_2$ present = 100.0 mL × $\dfrac{0.200 \text{ mmol } HC_2H_3O_2}{\text{mL}}$ = 20.0 mmol $HC_2H_3O_2$

mmol OH^- added = 50.0 mL × $\dfrac{0.100 \text{ mmol } OH^-}{\text{mL}}$ = 5.00 mmol OH^-

$$HC_2H_3O_2 \quad + \quad OH^- \quad \rightarrow \quad C_2H_3O_2^- \quad + \quad H_2O$$

Before	20.0 mmol	5.00 mmol	0	
Change	-5.00 mmol	-5.00 mmol	$\rightarrow$ +5.00 mmol	Reacts completely
After	15.0 mmol	0	5.00 mmol	

After reaction of all the strong base, we will have a buffered solution containing a weak acid ($HC_2H_3O_2$) and its conjugate base ($C_2H_3O_2^-$). We will use the Henderson-Hasselbalch equation to solve for the pH.

$$pH = pK_a + \log \frac{[C_2H_3O_2^-]}{[HC_2H_3O_2]} = -\log(1.8 \times 10^{-5}) + \log\left(\frac{5.00 \text{ mmol/V}_T}{15.0 \text{ mmol/V}_T}\right) \text{ where } V_T = \text{total volume}$$

$$pH = 4.74 + \log\left(\frac{5.00}{15.0}\right) = 4.74 + (-0.477) = 4.26$$

Note that the total volume cancels in the Henderson-Hasselbalch equation. For the [base]/[acid] term, the mole ratio equals the concentration ratio since the components of the buffer are always in the same volume of solution.

c.   mmol OH^- added = 100.0 mL × 0.100 mmol OH^-/mL = 10.0 mmol OH^-; The same amount (20.0 mmol) of $HC_2H_3O_2$ is present as before (it never changes). As before, let the OH^- react to completion, then see what remains in solution after the reaction.

	$HC_2H_3O_2$	+	OH^-	→	$C_2H_3O_2^-$	+	H_2O
Before	20.0 mmol		10.0 mmol		0		
After	10.0 mmol		0		10.0 mmol		

A buffered solution results after the reaction. Since $[C_2H_3O_2^-] = [HC_2H_3O_2] = 10.0$ mmol/total volume, then pH = pK_a. This is always true at the halfway point to equivalence for a weak acid/strong base titration, pH = pK_a.

$$pH = -\log(1.8 \times 10^{-5}) = 4.74$$

d.   mmol OH^- added = 150.0 mL × 0.100 M = 15.0 mmol OH^-. Added OH^- reacts completely with the weak acid.

	$HC_2H_3O_2$	+	OH^-	→	$C_2H_3O_2^-$	+	H_2O
Before	20.0 mmol		15.0 mmol		0		
After	5.0 mmol		0		15.0 mmol		

We have a buffered solution after all the OH^- reacts to completion. Using the Henderson-Hasselbalch equation:

$$pH = 4.74 + \log \frac{[C_2H_3O_2^-]}{[HC_2H_3O_2]} = 4.74 + \log\left(\frac{15.0 \text{ mmol}}{5.0 \text{ mmol}}\right) \quad \text{(Total volume cancels, so we can use mol ratios.)}$$

$$pH = 4.74 + 0.48 = 5.22$$

e.   mmol OH^- added = 200.00 mL × 0.100 M = 20.0 mmol OH^-; As before, let the added OH^- react to completion with the weak acid, then see what is in solution after this reaction.

$$HC_2H_3O_2 \quad + \quad OH^- \quad \rightarrow \quad C_2H_3O_2^- \quad + \quad H_2O$$

	$HC_2H_3O_2$	OH^-	$C_2H_3O_2^-$
Before	20.0 mmol	20.0 mmol	0
After	0	0	20.0 mmol

This is the equivalence point. Enough OH^- has been added to exactly neutralize all the weak acid present initially. All that remains that affects the pH at the equivalence point is the conjugate base of the weak acid, $C_2H_3O_2^-$. This is a weak base equilibrium problem.

$$C_2H_3O_2^- + H_2O \rightleftharpoons HC_2H_3O_2 + OH^- \quad K_b = \frac{K_w}{K_a} = \frac{1.0 \times 10^{-14}}{1.8 \times 10^{-5}} = 5.6 \times 10^{-10}$$

	$C_2H_3O_2^-$	$HC_2H_3O_2$	OH^-
Initial	20.0 mmol/300.0 mL	0	0

x mol/L $C_2H_3O_2^-$ reacts with H_2O to reach equilibrium

	$C_2H_3O_2^-$		$HC_2H_3O_2$	OH^-
Change	$-x$	$\rightarrow$	$+x$	$+x$
Equil.	$0.0667 - x$		x	x

$$K_b = 5.6 \times 10^{-10} = \frac{x^2}{0.0667 - x} \approx \frac{x^2}{0.0667}, \quad x = [OH^-] = 6.1 \times 10^{-6} \, M$$

pOH = 5.21; pH = 8.79; Assumptions good.

f. mmol OH^- added = 250.0 mL × 0.100 M = 25.0 mmol OH^-

$$HC_2H_3O_2 \quad + \quad OH^- \quad \rightarrow \quad C_2H_3O_2^- \quad + \quad H_2O$$

	$HC_2H_3O_2$	OH^-	$C_2H_3O_2^-$
Before	20.0 mmol	25.0 mmol	0
After	0	5.0 mmol	20.0 mmol

After the titration reaction, we will have a solution containing excess OH^- and a weak base, $C_2H_3O_2^-$. When a strong base and a weak base are both present, assume the amount of OH^- added from the weak base will be minimal, i.e., the pH past the equivalence point will be determined by the amount of excess strong base.

$$[OH^-]_{excess} = \frac{5.0 \, mmol}{100.0 \, mL + 250.0 \, mL} = 0.014 \, M; \quad pOH = 1.85; \quad pH = 12.15$$

56. This is a weak base (H_2NNH_2) titrated by a strong acid (HNO_3). To calculate the pH at the various points, let the strong acid react completely with the weak base present, then see what is in solution.

a. Only a weak base is present. Solve the weak base equilibrium problem.

$$H_2NNH_2 \quad + \quad H_2O \quad \rightleftharpoons \quad H_2NNH_3^+ \quad + \quad OH^-$$

	H_2NNH_2	$H_2NNH_3^+$	OH^-
Initial	0.100 M	0	~0
Equil.	$0.100 - x$	x	x

$$K_b = 3.0 \times 10^{-6} = \frac{x^2}{0.100 - x} \approx \frac{x^2}{0.100}, \quad x = [OH^-] = 5.5 \times 10^{-4} \, M$$

pOH = 3.26; pH = 10.74; Assumptions good.

b.   mmol H_2NNH_2 present = 100.0 mL $\times \dfrac{0.100 \text{ mmol } H_2NNH_2}{\text{mL}}$ = 10.0 mmol H_2NNH_2

mmol H^+ added = 20.0 mL $\times \dfrac{0.200 \text{ mmol } H^+}{\text{mL}}$ = 4.00 mmol H^+

	H_2NNH_2	+	H^+	$\rightarrow$	$H_2NNH_3^+$	
Before	10.0 mmol		4.00 mmol		0	
Change	-4.00 mmol		-4.00 mmol	$\rightarrow$	+4.00 mmol	Reacts completely
After	6.0 mmol		0		4.00 mmol	

A buffered solution results after the titration reaction. Solving using the Henderson-Hasselbalch equation:

$$pH = pK_a + \log \dfrac{[\text{base}]}{[\text{acid}]}, \quad K_a = \dfrac{K_w}{K_b} = \dfrac{1.0 \times 10^{-14}}{3.0 \times 10^{-6}} = 3.3 \times 10^{-9}$$

$$pH = -\log (3.3 \times 10^{-9}) + \log \left(\dfrac{6.0 \text{ mmol}/V_T}{4.00 \text{ mmol}/V_T} \right) \text{ where } V_T = \text{total volume, which cancels}$$

$$pH = 8.48 + \log (1.5) = 8.48 + 0.18 = 8.66$$

c.   mmol H^+ added = 25.0 mL $\times$ 0.200 M = 5.00 mmol H^+

	H_2NNH_2	+	H^+	$\rightarrow$	$H_2NNH_3^+$
Before	10.0 mmol		5.00 mmol		0
After	5.0 mmol		0		5.00 mmol

This is the halfway point to equivalence where $[H_2NNH_3^+] = [H_2NNH_2]$. At this point, $pH = pK_a$ (which is characteristic of the halfway point for any weak base/strong acid titration).

$$pH = -\log (3.3 \times 10^{-9}) = 8.48$$

d.   mmol H^+ added = 40.0 mL $\times$ 0.200 M = 8.00 mmol H^+

	H_2NNH_2	+	H^+	$\rightarrow$	$H_2NNH_3^+$
Before	10.0 mmol		8.00 mmol		0
After	2.0 mmol		0		8.00 mmol

A buffered solution results.

$$pH = pK_a + \log \dfrac{[\text{base}]}{[\text{acid}]} = 8.48 + \log \left(\dfrac{2.0 \text{ mmol}/V_T}{8.00 \text{ mmol}/V_T} \right) = 8.48 + (-0.60) = 7.88$$

e.   mmol H^+ added = 50.0 mL $\times$ 0.200 M = 10.0 mmol H^+

$$H_2NNH_2 \quad + \quad H^+ \quad \rightarrow \quad H_2NNH_3^+$$

Before	10.0 mmol	10.0 mmol	0
After	0	0	10.0 mmol

As is always the case in a weak base/strong acid titration, the pH at the equivalence point is acidic because only a weak acid ($H_2NNH_3^+$) is present. Solving the weak acid equilibrium problem:

$$H_2NNH_3^+ \quad \rightleftharpoons \quad H^+ \quad + \quad H_2NNH_2$$

Initial	10.0 mmol/150.0 mL	0	0
Equil.	0.0667 - x	x	x

$$K_a = 3.3 \times 10^{-9} = \frac{x^2}{0.0667 - x} \approx \frac{x^2}{0.0667}, \quad x = [H^+] = 1.5 \times 10^{-5} \, M$$

pH = 4.82; Assumptions good.

f.   mmol H^+ added = 100.0 mL $\times$ 0.200 M = 20.0 mmol H^+

$$H_2NNH_2 \quad + \quad H^+ \quad \rightarrow \quad H_2NNH_3^+$$

Before	10.0 mmol	20.0 mmol	0
After	0	10.0 mmol	10.0 mmol

Two acids are present past the equivalence point, but the excess H^+ will determine the pH of the solution since $H_2NNH_3^+$ is a weak acid.

$$[H^+]_{excess} = \frac{10.0 \text{ mmol}}{100.0 \text{ mL} + 100.0 \text{ mL}} = 0.0500 \, M; \quad pH = 1.301$$

57.   We will do sample calculations for the various parts of the titration. All results are summarized in Table 15.1 at the end of Exercise 15.60.

At the beginning of the titration, only the weak acid $HC_3H_5O_3$ is present.

$$HLac \quad \rightleftharpoons \quad H^+ \quad + \quad Lac^- \qquad K_a = 10^{-3.86} = 1.4 \times 10^{-4} \quad HLac = HC_3H_5O_3$$
$$Lac^- = C_3H_5O_3^-$$

Initial	0.100 M	~0	0
	x mol/L HLac dissociates to reach equilibrium		
Change	-x $\rightarrow$	+x	+x
Equil.	0.100 - x	x	x

$$1.4 \times 10^{-4} = \frac{x^2}{0.100 - x} = \frac{x^2}{0.100}, \quad x = [H^+] = 3.7 \times 10^{-3} \, M; \quad pH = 2.43 \quad \text{Assumptions good.}$$

Up to the stoichiometric point, we calculate the pH using the Henderson-Hasselbalch equation. This is the buffer region. For example, at 4.0 mL of NaOH added:

$$\text{initial mmol HLac present} = 25.0 \text{ mL} \times \frac{0.100 \text{ mmol}}{\text{mL}} = 2.50 \text{ mmol HLac}$$

$$\text{mmol OH}^- \text{ added} = 4.0 \text{ mL} \times \frac{0.100 \text{ mmol}}{\text{mL}} = 0.40 \text{ mmol OH}^-$$

Note: The units mmol are usually easier numbers to work with. The units for molarity are moles/L but are also equal to mmoles/mL.

The 0.40 mmol added OH$^-$ convert 0.40 mmoles HLac to 0.40 mmoles Lac$^-$ according to the equation:

$$\text{HLac} + \text{OH}^- \rightarrow \text{Lac}^- + \text{H}_2\text{O} \qquad \text{Reacts completely}$$

mmol HLac remaining = 2.50 - 0.40 = 2.10 mmol; mmol Lac$^-$ produced = 0.40 mmol

We have a buffered solution. Using the Henderson-Hasselbalch equation where pK$_a$ = 3.86:

$$\text{pH} = \text{pK}_a + \log \frac{[\text{Lac}^-]}{[\text{HLac}]} = 3.86 + \log \frac{(0.40)}{(2.10)} \qquad \begin{array}{l} \text{(Total volume cancels, so we can use} \\ \text{the ratio of moles or mmoles.)} \end{array}$$

$$\text{pH} = 3.86 - 0.72 = 3.14$$

Other points in the buffer region are calculated in a similar fashion. Perform a stoichiometry problem first, followed by a buffer problem. The buffer region includes all points up to 24.9 mL OH$^-$ added.

At the stoichiometric point (25.0 mL OH$^-$ added), we have added enough OH$^-$ to convert all of the HLac (2.50 mmol) into its conjugate base, Lac$^-$. All that is present is a weak base. To determine the pH, we perform a weak base calculation.

$$[\text{Lac}^-]_o = \frac{2.50 \text{ mmol}}{25.0 \text{ mL} + 25.0 \text{ mL}} = 0.0500 \; M$$

$$\text{Lac}^- + \text{H}_2\text{O} \rightleftharpoons \text{HLac} + \text{OH}^- \qquad K_b = \frac{1.0 \times 10^{-14}}{1.4 \times 10^{-4}} = 7.1 \times 10^{-11}$$

	Lac$^-$ + H$_2$O	→	HLac	+	OH$^-$
Initial	0.0500 M		0		0
	x mol/L Lac$^-$ reacts with H$_2$O to reach equilibrium				
Change	$-x$	→	$+x$		$+x$
Equil.	0.0500 - x		x		x

$$K_b = \frac{x^2}{0.0500 - x} \approx \frac{x^2}{0.0500} = 7.1 \times 10^{-11}$$

$x = [\text{OH}^-] = 1.9 \times 10^{-6} \; M$; pOH = 5.72; pH = 8.28   Assumptions good.

Past the stoichiometric point, we have added more than 2.50 mmol of NaOH. The pH will be determined by the excess OH^- ion present. An example of this calculation follows.

At 25.1 mL: OH^- added $= 25.1 \text{ mL} \times \dfrac{0.100 \text{ mmol}}{\text{mL}} = 2.51$ mmol OH^-; 2.50 mmol OH^- neutralizes all the weak acid present. The remainder is excess OH^-. excess $OH^- = 2.51 - 2.50 = 0.01$ mmol

$$[OH^-]_{excess} = \frac{0.01 \text{ mmol}}{(25.0 + 25.1) \text{ mL}} = 2 \times 10^{-4} M; \ pOH = 3.7; \ pH = 10.3$$

All results are listed in Table 15.1 at the end of the solution to Exercise 15.60.

58. Results for all points are summarized in Table 15.1 at the end of the solution to Exercise 15.60. At the beginning of the titration, we have a weak acid problem:

$$HOPr \rightleftharpoons H^+ + OPr^-$$

$HOPr = HC_3H_5O_2$
$OPr^- = C_3H_5O_2^-$

	HOPr		H^+	OPr^-
Initial	0.100 M		~0	0
	x mol/L HOPr acid dissociates to reach equilibrium			
Change	$-x$	$\rightarrow$	$+x$	$+x$
Equil.	$0.100 - x$		x	x

$$K_a = \frac{[H^+][OPr^-]}{[HOPr]} = 1.3 \times 10^{-5} = \frac{x^2}{0.100 - x} \approx \frac{x^2}{0.100}$$

$x = [H^+] = 1.1 \times 10^{-3} M; \ pH = 2.96$ Assumptions good.

The buffer region is from 4.0 - 24.9 mL of OH^- added. We will do a sample calculation at 24.0 mL OH^- added.

$$\text{initial mmol HOPr present} = 25.0 \text{ mL} \times \frac{0.100 \text{ mmol}}{\text{mL}} = 2.50 \text{ mmol HOPr}$$

$$\text{mmol } OH^- \text{ added} = 24.0 \text{ mL} \times \frac{0.100 \text{ mmol}}{\text{mL}} = 2.40 \text{ mmol } OH^-$$

The added strong base converts HOPr into OPr^-.

	HOPr	+	OH^-	$\rightarrow$	OPr^-	+	H_2O	
Before	2.50 mmol		2.40 mmol		0			
Change	-2.40		-2.40	$\rightarrow$	$+2.40$			Reacts completely
After	0.10 mmol		0		2.40 mmol			

A buffer solution results. Using the Henderson-Hasselbalch equation where $pK_a = -\log(1.3 \times 10^{-5}) = 4.89$:

$$pH = pK_a + \log\frac{[Base]}{[Acid]} = 4.89 + \log\frac{[OPr^-]}{[HOPr]}$$

$$pH = 4.89 + \log\left(\frac{2.40}{0.10}\right) = 4.89 + 1.38 = 6.27 \quad \text{(Volume cancels, so we can use the mol ratio.)}$$

All points in the buffer region, 4.0 mL to 24.9 mL, are calculated this way. See Table 15.1 at the end of Exercise 15.60 for all the results.

At the stoichiometric point, only a weak base (Opr⁻) is present:

$$OPr^- \quad + \quad H_2O \quad \rightleftharpoons \quad OH^- \quad + \quad HOPr$$

Initial $\quad\dfrac{2.50 \text{ mmol}}{50.0 \text{ mL}} = 0.0500 \ M \qquad\qquad 0 \qquad\qquad 0$

x mol/L OPr⁻ reacts with H_2O to reach equilibrium

Change $\quad\quad -x \qquad\qquad\qquad\qquad \rightarrow \quad +x \qquad\quad +x$

Equil. $\quad\quad 0.0500 - x \qquad\qquad\qquad\qquad x \qquad\qquad x$

$$K_b = \frac{[OH^-][HOPr]}{[OPr^-]} = \frac{K_w}{K_a} = 7.7 \times 10^{-10} = \frac{x^2}{0.0500 - x} \approx \frac{x^2}{0.0500}$$

$$x = 6.2 \times 10^{-6} \ M = [OH^-], \quad pOH = 5.21, \quad pH = 8.79 \qquad \text{Assumptions good.}$$

Beyond the stoichiometric point, the pH is determined by the excess strong base added. The results are the same as those in Exercise 15.57 (see Table 15.1).

For example at 26.0 mL NaOH added:

$$[OH^-] = \frac{2.60 \text{ mmol} - 2.50 \text{ mmol}}{(25.0 + 26.0) \text{ mL}} = 2.0 \times 10^{-3} \ M; \quad pOH = 2.70; \quad pH = 11.30$$

59. At beginning of the titration, only the weak base NH_3 is present. As always, solve for the pH using the K_b reaction for NH_3.

$$NH_3 \ + \ H_2O \quad \rightleftharpoons \quad NH_4^+ \ + \ OH^- \qquad K_b = 1.8 \times 10^{-5}$$

Initial $\quad 0.100 \ M \qquad\qquad\qquad 0 \qquad\quad \sim 0$

Equil. $\quad 0.100 - x \qquad\qquad\qquad x \qquad\quad x$

$$K_b = \frac{x^2}{0.100 - x} = \frac{x^2}{0.100} = 1.8 \times 10^{-5}$$

$$x = [OH^-] = 1.3 \times 10^{-3} \ M; \quad pOH = 2.89; \quad pH = 11.11 \quad \text{Assumptions good.}$$

In the buffer region (4.0 - 24.9 mL), we can use the Henderson-Hasselbalch equation:

$$K_a = \frac{1.0 \times 10^{-14}}{1.8 \times 10^{-5}} = 5.6 \times 10^{-10}; \quad pK_a = 9.25; \quad pH = 9.25 + \log\frac{[NH_3]}{[NH_4^+]}$$

We must determine the amounts of NH_3 and NH_4^+ present after the added H^+ reacts completely with the NH_3. For example, after 8.0 mL HCl are added:

$$\text{initial mmol } NH_3 \text{ present} = 25.0 \text{ mL} \times \frac{0.100 \text{ mmol}}{\text{mL}} = 2.50 \text{ mmol } NH_3$$

$$\text{mmol } H^+ \text{ added} = 8.0 \text{ mL} \times \frac{0.100 \text{ mmol}}{\text{mL}} = 0.80 \text{ mmol } H^+$$

Added H^+ reacts with NH_3 to completion: $NH_3 + H^+ \rightarrow NH_4^+$

mmol NH_3 remaining = 2.50 - 0.80 = 1.70 mmol; mmol NH_4^+ produced = 0.80 mmol

$pH = 9.25 + \log \frac{1.70}{0.80} = 9.58$ (Mole ratios can be used since the total volume cancels.)

Other points in the buffer region are calculated in similar fashion. Results are summarized in Table 15.1 at the end of Exercise 15.60.

At the stoichiometric point (25.0 mL H^+ added), just enough HCl has been added to convert all of the weak base (NH_3) into its conjugate acid (NH_4^+). Perform a weak acid calculation. $[NH_4^+]_o = 2.50$ mmol/50.0 mL = 0.0500 M

$$NH_4^+ \rightleftharpoons H^+ + NH_3 \qquad K_a = 5.6 \times 10^{-10}$$

Initial	0.0500 M	0	0
Equil.	0.0500 - x	x	x

$5.6 \times 10^{-10} = \frac{x^2}{0.0500 - x} = \frac{x^2}{0.0500}$, $x = [H^+] = 5.3 \times 10^{-6} M$; $pH = 5.28$ Assumptions good.

Beyond the stoichiometric point, the pH is determined by the excess H^+. For example, at 28.0 mL of H^+ added:

$$\text{mmol } H^+ \text{ added} = 28.0 \text{ mL} \times \frac{0.100 \text{ mmol}}{\text{mL}} = 2.80 \text{ mmol } H^+$$

Excess mmol H^+ = 2.80 mmol - 2.50 mmol = 0.30 mmol excess H^+

$$[H^+]_{excess} = \frac{0.30 \text{ mmol}}{(25.0 + 28.0) \text{ mL}} = 5.7 \times 10^{-3} M; \; pH = 2.24$$

All results are summarized in Table 15.1.

60.   Initially, a weak base problem:

$$py + H_2O \rightleftharpoons Hpy^+ + OH^- \qquad \text{py is pyridine}$$

Initial	0.100 M	0	~0
Equil.	0.100 - x	x	x

$$K_b = \frac{[Hpy^+][OH^-]}{[py]} = \frac{x^2}{0.100 - x} = \frac{x^2}{0.100} = 1.7 \times 10^{-9}$$

$x = [OH^-] = 1.3 \times 10^{-5} M$; pOH = 4.89; pH = 9.11   Assumptions good.

Buffer region (4.0 - 24.5 mL): Added H^+ reacts completely with py: $py + H^+ \rightarrow Hpy^+$. Determine the moles (or mmoles) of py and Hpy^+ after the reaction, and use the Henderson-Hasselbalch equation to solve for the pH.

$$K_a = \frac{K_w}{K_b} = \frac{1.0 \times 10^{-14}}{1.7 \times 10^{-9}} = 5.9 \times 10^{-6}; \ pK_a = 5.23; \ pH = 5.23 + \log \frac{[py]}{[Hpy^+]}$$

Results in the buffer region are summarized in Table 15.1 that follows this problem. See Exercise 15.59 for a similar sample calculation.

At the stoichiometric point (25.0 mL H^+ added), this is a weak acid problem since just enough H^+ has been added to convert all of the weak base into its conjugate acid. The initial concentration of Hpy^+ = 0.0500 M.

	Hpy^+	$\rightleftharpoons$	py	+	H^+	$K_a = 5.9 \times 10^{-6}$
Initial	0.0500 M		0		0	
Equil.	0.0500 - x		x		x	

$5.9 \times 10^{-6} = \dfrac{x^2}{0.0500 - x} = \dfrac{x^2}{0.0500}$, $x = [H^+] = 5.4 \times 10^{-4} M$; pH = 3.27   Assumptions good.

Beyond the equivalence point, the pH determination is made by calculating the concentration of excess H^+. See Exercise 15.59 for an example. All results are summarized in Table 15.1.

Table 15.1: Summary of pH Results for Exercises 15.57 - 15.60   (Graph follows)

Titrant mL	Exercise 15.57	Exercise 15.58	Exercise 15.59	Exercise 15.60
0.0	2.43	2.96	11.11	9.11
4.0	3.14	4.17	9.97	5.95
8.0	3.53	4.56	9.58	5.56
12.5	3.86	4.89	9.25	5.23
20.0	4.46	5.49	8.65	4.63
24.0	5.24	6.27	7.87	3.85
24.5	5.6	6.6	7.6	3.5
24.9	6.3	7.3	6.9	-
25.0	8.28	8.79	5.28	3.27
25.1	10.3	10.3	3.7	-
26.0	11.30	11.30	2.71	2.71
28.0	11.75	11.75	2.24	2.24
30.0	11.96	11.96	2.04	2.04

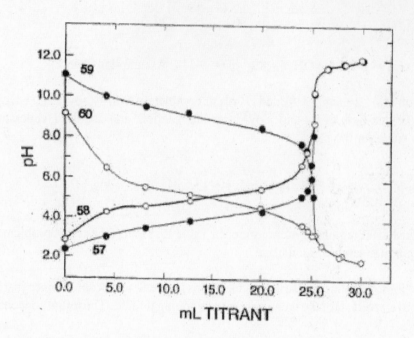

61.   a.   This is a weak acid/strong base titration. At the halfway point to equivalence, [weak acid] = [conjugate base], so pH = pK_a (always true for a weak acid/strong base titration).

pH = -log (6.4×10^{-5}) = 4.19

mmol $HC_7H_5O_2$ present = 100. mL × 0.10 M = 10. mmol $HC_7H_5O_2$. For the equivalence point, 10. mmol of OH^- must be added. The volume of OH^- added to reach the equivalence point is:

$$10.\ \text{mmol}\ OH^- \times \frac{1\ \text{mL}}{0.10\ \text{mmol}\ OH^-} = 1.0 \times 10^2\ \text{mL}\ OH^-$$

At the equivalence point, 10. mmol of $HC_7H_5O_2$ is neutralized by 10. mmol of OH^- to produce 10. mmol of $C_7H_5O_2^-$. This is a weak base. The total volume of the solution is 100.0 mL + 1.0 × 10^2 mL = 2.0 × 10^2 mL. Solving the weak base equilibrium problem:

$$C_7H_5O_2^- + H_2O \rightleftharpoons HC_7H_5O_2 + OH^- \quad K_b = \frac{K_w}{K_a} = \frac{1.0 \times 10^{-14}}{6.4 \times 10^{-5}} = 1.6 \times 10^{-10}$$

Initial   10. mmol/2.0 × 10^2 mL   0   0
Equil.   0.050 - x x x

$$K_b = 1.6 \times 10^{-10} = \frac{x^2}{0.050 - x} \approx \frac{x^2}{0.050}, \quad x = [OH^-] = 2.8 \times 10^{-6}\ M$$

pOH = 5.55; pH = 8.45   Assumptions good.

b.   At the halfway point to equivalence for a weak base/strong acid titration, pH = pK_a since [weak base] = [conjugate acid].

$$K_a = \frac{K_w}{K_b} = \frac{1.0 \times 10^{-14}}{5.6 \times 10^{-4}} = 1.8 \times 10^{-11}; \quad pH = pK_a = -\log(1.8 \times 10^{-11}) = 10.74$$

For the equivalence point (mmol acid added = mmol base present):

mmol $C_2H_5NH_2$ present = 100.0 mL $\times$ 0.10 M = 10. mmol $C_2H_5NH_2$

mL HNO_3 added = 10. mmol $H^+ \times \dfrac{1\ mL}{0.20\ mmol\ H^+}$ = 50. mL H^+

The strong acid added completely converts the weak base into its conjugate acid. Therefore, at the equivalence point, $[C_2H_5NH_3^+]_o$ = 10. mmol/(100.0 + 50.) mL = 0.067 M. Solving the weak acid equilibrium problem:

$$C_2H_5NH_3^+ \rightleftharpoons H^+ + C_2H_5NH_2$$

Initial   0.067 M 0   0
Equil.   0.067 - x x x

$$K_a = 1.8 \times 10^{-11} = \frac{x^2}{0.067 - x} \approx \frac{x^2}{0.067}, \quad x = [H^+] = 1.1 \times 10^{-6}\ M$$

pH = 5.96; Assumption good.

c. In a strong acid/strong base titration, the halfway point has no special significance other than exactly one-half of the original amount of acid present has been neutralized.

mmol H^+ present = 100.0 mL $\times$ 0.50 M = 50. mmol H^+

mL OH^- added = 25. mmol $OH^- \times \dfrac{1\ mL}{0.25\ mmol}$ = 1.0 $\times$ 10^2 mL OH^-

$$H^+ + OH^- \rightarrow H_2O$$

Before   50. mmol   25 mmol
After   25 mmol   0

$$[H^+]_{excess} = \frac{25\ mmol}{(100.0 + 1.0 \times 10^2)\ mL} = 0.13\ M; \quad pH = 0.89$$

At the equivalence point of a strong acid/strong base titration, only neutral species are present (Na^+, Cl^-, H_2O), so the pH = 7.00.

62.   50.0 mL $\times$ 1.0 M = 50. mmol CH_3NH_2 present initially; $CH_3NH_2 + H^+ \rightarrow CH_3NH_3^+$

a.   50.0 mL $\times$ 0.50 M = 25. mmol HCl added. The added H^+ will convert half of the CH_3NH_2 into $CH_3NH_3^+$. This is the halfway point to equivalence where $[CH_3NH_2] = [CH_3NH_3^+]$.

$$pH = pK_a + \log \frac{[CH_3NH_2]}{[CH_3NH_3^+]} = pK_a; \quad K_a = \frac{1.0 \times 10^{-14}}{4.4 \times 10^{-4}} = 2.3 \times 10^{-11}$$

$$pH = pK_a = -\log(2.3 \times 10^{-11}) = 10.64$$

b. It will take 100. mL of HCl solution to reach the stoichiometric point. Here the added H^+ will convert all of the CH_3NH_2 into its conjugate acid, $CH_3NH_3^+$.

$$[CH_3NH_3^+]_o = \frac{50.\ mmol}{150.\ mL} = 0.33\ M$$

	$CH_3NH_3^+$	$\rightleftharpoons$	H^+	$+$	CH_3NH_2	$K_a = \frac{K_w}{K_b} = 2.3 \times 10^{-11}$
Initial	0.33 M		0		0	
Equil.	0.33 - x		x		x	

$$2.3 \times 10^{-11} = \frac{x^2}{0.33 - x} \approx \frac{x^2}{0.33}, \quad x = [H^+] = 2.8 \times 10^{-6}\ M; \quad pH = 5.55 \quad \text{Assumptions good.}$$

63.   At equivalence point: $16.00\ mL \times 0.125\ mmol/mL = 2.00\ mmol\ OH^-$ added. There must be 2.00 mmol HX present initially.

2.00 mL NaOH added = $2.00\ mL \times 0.125\ mmol/mL = 0.250\ mmol\ OH^-$; 0.250 mmol of OH^- added will convert 0.250 mmol HX into 0.250 mmol X^-. Remaining HX = 2.00 - 0.250 = 1.75 mmol HX; This is a buffer solution where $[H^+] = 10^{-6.912} = 1.22 \times 10^{-7}\ M$. Let V_T = total volume of solution (which cancels). Assuming the initial and equilibrium HX and X^- concentrations are the same (a very good assumption for buffer solutions):

$$K_a = \frac{[H^+][X^-]}{[HX]} = \frac{1.22 \times 10^{-7}\,(0.250/V_T)}{1.75/V_T} = \frac{1.22 \times 10^{-7}(0.250)}{1.75} = 1.74 \times 10^{-8}$$

Note: We could also solve for K_a using the Henderson-Hasselbach equation

64.   a. 500.0 mL of HCl added represents the halfway point to equivalence. At the halfway point to equivalence in a weak base/strong acid titration, $pH = pK_a$. So, $pH = pK_a = 5.00$ and $K_a = 1.0 \times 10^{-5}$. Since $K_a = 1.0 \times 10^{-5}$ for the acid, then K_b for the conjugate base, A^-, equals 1.0×10^{-9} ($K_a \times K_b = K_w = 1.0 \times 10^{-14}$).

b. Added H^+ converts A^- into HA. At the equivalence point, only HA is present. The moles of HA present equal the moles of H^+ added (= $1.00\ L \times 0.100\ mol/L = 0.100\ mol$). The concentration of HA at the equivalence point is:

$$[HA]_o = \frac{0.100\ mol}{1.10\ L} = 0.0909\ M$$

Solving the weak acid equilibrium problem:

$$HA \quad \rightleftharpoons \quad H^+ \quad + \quad A^- \qquad\qquad K_a = 1.0 \times 10^{-5}$$

Initial	0.0909 M	0	0
Equil.	0.0909 - x	x	x

$$1.0 \times 10^{-5} = \frac{x^2}{0.0909 - x} \approx \frac{x^2}{0.0909}$$

$x = 9.5 \times 10^{-4}\ M = [H^+]$; pH = 3.02   Assumption good.

Indicators

65. $HIn \rightleftharpoons In^- + H^+ \quad K_a = \dfrac{[In^-][H^+]}{[HIn]} = 1.0 \times 10^{-9}$

a.   In a very acid solution, the HIn form dominates, so the solution will be yellow.

b.   The color change occurs when the concentration of the more dominant form is approximately ten times as great as the less dominant form of the indicator.

$\dfrac{[HIn]}{[In^-]} = \dfrac{10}{1}$; $K_a = 1.0 \times 10^{-9} = \left(\dfrac{1}{10} \right) [H^+]$, $[H^+] = 1 \times 10^{-8}\ M$; pH = 8.0 at color change

c.   This is way past the equivalence point (100.0 mL OH$^-$ added), so the solution is very basic and the In$^-$ form of the indicator dominates. The solution will be blue.

66.   The color of the indicator will change over the approximate range of pH = $pK_a \pm 1 = 5.3 \pm 1$. Therefore, the useful pH range of methyl red where it changes color would be about: 4.3 (red) - 6.3 (yellow). Note that at pH < 4.3, the HIn form of the indicator dominates, and the color of the solution is the color of HIn (red). At pH > 6.3, the In$^-$ form of the indicator dominates, and the color of the solution is the color of In$^-$ (yellow). In titrating a weak acid with a strong base, we start off with an acidic solution with pH < 4.3 so the color would change from red to orange at pH ~ 4.3. In titrating a weak base with a strong acid, the color change would be from yellow to orange at pH ~ 6.3. Only a weak base/strong acid titration would have an acidic pH at the equivalence point, so only in this type of titration would the color change of methyl red indicate the approximate endpoint.

67.   When choosing an indicator, we want the color change of the indicator to occur approximately at the pH of the equivalence point. Since the pH generally changes very rapidly at the equivalence point, we don't have to be exact. This is especially true for strong acid/strong base titrations. Some choices where color change occurs at about the pH of the equivalence point are:

Exercise	pH at eq. pt.	Indicator
15.53	7.00	bromthymol blue or phenol red
15.55	8.79	o-cresolphthalein or phenolphthalein

68.

Exercise	pH at eq. pt.	Indicator
15.54	7.00	bromthymol blue or phenol red
15.56	4.82	bromcresol green

69.

Exercise	pH at eq. pt.	Indicator
15.57	8.28	phenolphthalein
15.59	5.28	bromcresol green

70.

Exercise	pH at eq. pt.	Indicator
15.58	8.79	phenolphthalein
15.60	3.27	2,4-dinitrophenol

The titration in 15.60 is not feasible. The pH break at the equivalence point is too small.

71.   pH > 5 for bromcresol green to be blue. pH < 8 for thymol blue to be yellow. The pH is between 5 and 8.

72.   a. yellow   b. green (Both yellow and blue forms are present.)   c. yellow   d. blue

Solubility Equilibria

73.   a. $AgC_2H_3O_2(s) \rightleftharpoons Ag^+(aq) + C_2H_3O_2^-(aq)$ $K_{sp} = [Ag^+][C_2H_3O_2^-]$

b. $Al(OH)_3(s) \rightleftharpoons Al^{3+}(aq) + 3 OH^-(aq)$ $K_{sp} = [Al^{3+}][OH^-]^3$

c. $Ca_3(PO_4)_2(s) \rightleftharpoons 3 Ca^{2+}(aq) + 2 PO_4^{3-}(aq)$ $K_{sp} = [Ca^{2+}]^3[PO_4^{3-}]^2$

74.   a. $Ag_2CO_3(s) \rightleftharpoons 2 Ag^+(aq) + CO_3^{2-}(aq)$ $K_{sp} = [Ag^+]^2[CO_3^{2-}]$

b. $Ce(IO_3)_3(s) \rightleftharpoons Ce^{3+}(aq) + 3 IO_3^-(aq)$ $K_{sp} = [Ce^{3+}][IO_3^-]^3$

c. $BaF_2(s) \rightleftharpoons Ba^{2+}(aq) + 2 F^-(aq)$ $K_{sp} = [Ba^{2+}][F^-]^2$

75.   In our setup, s = solubility of the ionic solid in mol/L. This is defined as the maximum amount of a salt which can dissolve. Since solids do not appear in the K_{sp} expression, we do not need to worry about their initial and equilibrium amounts.

a.

	$CaC_2O_4(s)$	$\rightleftharpoons$	$Ca^{2+}(aq)$	+	$C_2O_4^{2-}(aq)$
Initial			0		0
	s mol/L of $CaC_2O_4(s)$ dissolves to reach equilibrium				
Change	$-s$	$\rightarrow$	$+s$		$+s$
Equil.			s		s

From the problem, $s = 4.8 \times 10^{-5}$ mol/L.

$K_{sp} = [Ca^{2+}] [C_2O_4^{2-}] = (s)(s) = s^2$, $K_{sp} = (4.8 \times 10^{-5})^2 = 2.3 \times 10^{-9}$

b. $BiI_3(s)$ $\rightleftharpoons$ $Bi^{3+}(aq)$ $+$ $3\ I^-(aq)$

Initial		0	0

s mol/L of $BiI_3(s)$ dissolves to reach equilibrium

Change	$-s$	$\rightarrow$	$+s$	$+3s$
Equil.			s	$3s$

$K_{sp} = [Bi^{3+}] [I^-]^3 = (s)(3s)^3 = 27\ s^4$, $K_{sp} = 27(1.32 \times 10^{-5})^4 = 8.20 \times 10^{-19}$

76.   a. $Pb_3(PO_4)_2(s)$ $\rightleftharpoons$ $3\ Pb^{2+}(aq)$ $+$ $2\ PO_4^{3-}(aq)$

Initial	0	0

s mol/L of $Pb_3(PO_4)_2(s)$ dissolves to reach equilibrium = molar solubility

Change	$-s$ $\rightarrow$	$+3s$	$+2s$
Equil.		$3s$	$2s$

$K_{sp} = [Pb^{2+}]^3 [PO_4^{3-}]^2 = (3s)^3(2s)^2 = 108\ s^5$, $K_{sp} = 108(6.2 \times 10^{-12})^5 = 9.9 \times 10^{-55}$

b. $Li_2CO_3(s)$ $\rightleftharpoons$ $2\ Li^+(aq)$ $+$ $CO_3^{2-}(aq)$

Initial	s = solubility (mol/L)	0	0
Equil.		$2s$	s

$K_{sp} = [Li^+]^2 [CO_3^{2-}] = (2s)^2(s) = 4s^3$, $K_{sp} = 4(7.4 \times 10^{-2})^3 = 1.6 \times 10^{-3}$

77. $PbBr_2(s)$ $\rightleftharpoons$ $Pb^{2+}(aq)$ $+$ $2\ Br^-(aq)$

Initial	0	0

s mol/L of $PbBr_2(s)$ dissolves to reach equilibrium

Change	$-s$	$\rightarrow$	$+s$	$+2s$
Equil.			s	$2s$

From the problem, $s = [Pb^{2+}] = 2.14 \times 10^{-2}\ M$. So:

$K_{sp} = [Pb^{2+}] [Br^-]^2 = s(2s)^2 = 4s^3$, $K_{sp} = 4(2.14 \times 10^{-2})^3 = 3.92 \times 10^{-5}$

78. $Ag_2C_2O_4(s)$ $\rightleftharpoons$ $2\ Ag^+(aq)$ $+$ $C_2O_4^{2-}(aq)$

Initial	s = solubility (mol/L)	0	0
Equil.		$2s$	s

From problem, $[Ag^+] = 2s = 2.2 \times 10^{-4}$ M, $s = 1.1 \times 10^{-4}$ M

$K_{sp} = [Ag^+]^2 [C_2O_4^{2-}] = (2s)^2(s) = 4s^3 = 4(1.1 \times 10^{-4})^3 = 5.3 \times 10^{-12}$

79.   In our setups, s = solubility in mol/L. Since solids do not appear in the K_{sp} expression, we do not need to worry about their initial or equilibrium amounts.

a. $Ag_3PO_4(s) \rightleftharpoons 3\ Ag^+(aq) + PO_4^{3-}(aq)$

Initial   0   0
                    s mol/L of $Ag_3PO_4(s)$ dissolves to reach equilibrium
Change $-s$ $\rightarrow$ $+3s$ $+s$
Equil. $3s$ s

$K_{sp} = 1.8 \times 10^{-18} = [Ag^+]^3 [PO_4^{3-}] = (3s)^3(s) = 27\ s^4$

$27\ s^4 = 1.8 \times 10^{-18}$, $s = (6.7 \times 10^{-20})^{1/4} = 1.6 \times 10^{-5}$ mol/L = molar solubility

b. $CaCO_3(s) \rightleftharpoons Ca^{2+}(aq) + CO_3^{2-}(aq)$
Initial s = solubility (mol/L)   0   0
Equil. s s

$K_{sp} = 8.7 \times 10^{-9} = [Ca^{2+}] [CO_3^{2-}] = s^2$, $s = 9.3 \times 10^{-5}$ mol/L

c. $Hg_2Cl_2(s) \rightleftharpoons Hg_2^{2+}(aq) + 2\ Cl^-(aq)$

Initial s = solubility (mol/L)   0   0
Equil. s $2s$

$K_{sp} = 1.1 \times 10^{-18} = [Hg_2^{2+}] [Cl^-]^2 = (s)(2s)^2 = 4s^3$, $s = 6.5 \times 10^{-7}$ mol/L

80.   a. $PbI_2(s) \rightleftharpoons Pb^{2+}(aq) + 2\ I^-(aq)$

Initial s = solubility (mol/L)   0   0
Equil. s $2s$

$K_{sp} = 1.4 \times 10^{-8} = [Pb^{2+}] [I^-]^2 = s(2s)^2 = 4s^3$

$s = (1.4 \times 10^{-8}/4)^{1/3} = 1.5 \times 10^{-3}$ mol/L = molar solubility

b. $CdCO_3(s) \rightleftharpoons Cd^{2+}(aq) + CO_3^{2-}(aq)$

Initial s = solubility (mol/L)   0   0
Equil. s s

$K_{sp} = 5.2 \times 10^{-12} = [Cd^{2+}] [CO_3^{2-}] = s^2$, $s = 2.3 \times 10^{-6}$ mol/L

c. $Sr_3(PO_4)_2(s)$ $\rightleftharpoons$ $3\ Sr^{2+}(aq)\ +\ 2\ PO_4^{3-}(aq)$

Initial s = solubility (mol/L)   0   0
Equil. $3s$ $2s$

$K_{sp} = 1 \times 10^{-31} = [Sr^{2+}]^3\ [PO_4^{3-}]^2 = (3s)^3(2s)^2 = 108\ s^5,\ s = 2 \times 10^{-7}$ mol/L

81.   Let s = solubility of $Co(OH)_3$ in mol/L. Note: Since solids do not appear in the K_{sp} expression, we do not need to worry about their initial or equilibrium amounts.

$Co(OH)_3(s)$ $\rightleftharpoons$ $Co^{3+}(aq)$ + $3\ OH^-(aq)$

Initial   0 $1.0 \times 10^{-7}\ M$ from water
              s mol/L of $Co(OH)_3(s)$ dissolves to reach equilibrium = molar solubility
Change $-s$ $\rightarrow$ $+s$ $+3s$
Equil. s $1.0 \times 10^{-7} + 3s$

$K_{sp} = 2.5 \times 10^{-43} = [Co^{3+}]\ [OH^-]^3 = (s)(1.0 \times 10^{-7} + 3s)^3 \approx s(1.0 \times 10^{-7})^3$

$s = \dfrac{2.5 \times 10^{-43}}{1.0 \times 10^{-21}} = 2.5 \times 10^{-22}$ mol/L   Assumption good $(1.0 \times 10^{-7} + 3s \approx 1.0 \times 10^{-7})$.

82. $Mg(OH)_2(s)$ $\rightleftharpoons$ $Mg^{2+}(aq)$ + $2\ OH^-(aq)$

Initial s = solubility (mol/L)   0 $1.0 \times 10^{-7}\ M$
Equil. s $1.0 \times 10^{-7} + 2s$

$K_{sp} = [Mg^{2+}]\ [OH^-]^2 = s(1.0 \times 10^{-7} + 2s)^2$; Assume that $1.0 \times 10^{-7} + 2s \approx 2s$, then:

$K_{sp} = 8.9 \times 10^{-12} = s(2s)^2 = 4s^3,\ s = 1.3 \times 10^{-4}$ mol/L

Assumption is good $(1.0 \times 10^{-7}$ is 0.04% of $2s)$. Molar solubility $= 1.3 \times 10^{-4}$ mol/L

83.   a.   Since both solids dissolve to produce 3 ions in solution, we can compare values of K_{sp} to determine relative molar solubility. Since the K_{sp} for CaF_2 is smaller, $CaF_2(s)$ is less soluble (in mol/L).

b.   We must calculate molar solubilities since each salt yields a different number of ions when it dissolves.

$Ca_3(PO_4)_2(s)$ $\rightleftharpoons$ $3\ Ca^{2+}(aq)\ +\ 2\ PO_4^{3-}(aq)$ $K_{sp} = 1.3 \times 10^{-32}$

Initial s = solubility (mol/L)   0   0
Equil. $3s$ $2s$

$K_{sp} = [Ca^{2+}]^3\ [PO_4^{3-}]^2 = (3s)^3(2s)^2 = 108\ s^5,\ s = (1.3 \times 10^{-32}/108)^{1/5} = 1.6 \times 10^{-7}$ mol/L

$$FePO_4(s) \rightleftharpoons Fe^{3+}(aq) + PO_4^{3-}(aq) \quad K_{sp} = 1.0 \times 10^{-22}$$

Initial	s = solubility (mol/L)	0	0
Equil.		s	s

$$K_{sp} = [Fe^{3+}][PO_4^{3-}] = s^2, \quad s = \sqrt{1.0 \times 10^{-22}} = 1.0 \times 10^{-11} \text{ mol/L}$$

Since the molar solubility of $FePO_4$ is smaller, $FePO_4$ is less soluble (in mol/L).

84. a.

$$FeC_2O_4(s) \rightleftharpoons Fe^{2+}(aq) + C_2O_4^{2-}(aq)$$

Equil.	s = solubility (mol/L)	s	s

$$K_{sp} = 2.1 \times 10^{-7} = [Fe^{2+}][C_2O_4^{2-}] = s^2, \quad s = 4.6 \times 10^{-4} \text{ mol/L}$$

$$Cu(IO_4)_2(s) \rightleftharpoons Cu^{2+}(aq) + 2\,IO_4^-(aq)$$

Equil.		s	$2s$

$$K_{sp} = 1.4 \times 10^{-7} = [Cu^{2+}][IO_4^-]^2 = s(2s)^2 = 4s^3, \quad s = (1.4 \times 10^{-7}/4)^{1/3} = 3.3 \times 10^{-3} \text{ mol/L}$$

By comparing calculated molar solubilities, $FeC_2O_4(s)$ is less soluble (in mol/L).

 b. Since each salt produces 3 ions in solution, we can compare K_{sp} values to determine relative molar solubilities. Therefore, $Mn(OH)_2(s)$ will be less soluble (in mol/L) since it has a smaller K_{sp} value.

85. a.

$$Fe(OH)_3(s) \rightleftharpoons Fe^{3+}(aq) + 3\,OH^-(aq)$$

Initial		0	$1 \times 10^{-7}\,M$ from water

s mol/L of $Fe(OH)_3(s)$ dissolves to reach equilibrium = molar solubility

Change	$-s$ $\rightarrow$	$+s$	$+3s$
Equil.		s	$1 \times 10^{-7} + 3s$

$$K_{sp} = 4 \times 10^{-38} = [Fe^{3+}][OH^-]^3 = (s)(1 \times 10^{-7} + 3s)^3 \approx s(1 \times 10^{-7})^3$$

$s = 4 \times 10^{-17}$ mol/L Assumption good ($3s \ll 1 \times 10^{-7}$).

 b.

$$Fe(OH)_3(s) \rightleftharpoons Fe^{3+}(aq) + 3\,OH^-(aq) \quad pH = 5.0 \text{ so } [OH^-] = 1 \times 10^{-9}\,M$$

Initial		0	$1 \times 10^{-9}\,M$ (buffered)

s mol/L dissolves to reach equilibrium

Change	$-s$ $\rightarrow$	$+s$	----- (assume no pH change in buffer)
Equil.		s	1×10^{-9}

$$K_{sp} = 4 \times 10^{-38} = [Fe^{3+}][OH^-]^3 = (s)(1 \times 10^{-9})^3, \quad s = 4 \times 10^{-11} \text{ mol/L} = \text{molar solubility}$$

c. $Fe(OH)_3(s) \rightleftharpoons Fe^{3+}(aq) + 3\ OH^-(aq)$ $pH = 11.0$ so $[OH^-] = 1 \times 10^{-3}\ M$

Initial		0	0.001 M (buffered)

s mol/L dissolves to reach equilibrium

Change	$-s$	$\rightarrow$	$+s$	----- (assume no pH change)
Equil.			s	0.001

$K_{sp} = 4 \times 10^{-38} = [Fe^{3+}]\,[OH^-]^3 = (s)(0.001)^3$, $s = 4 \times 10^{-29}$ mol/L = molar solubility

Note: As $[OH^-]$ increases, solubility decreases. This is the common ion effect.

86.   a. $Ag_2SO_4(s) \rightleftharpoons 2\ Ag^+(aq) + SO_4^{2-}(aq)$

Initial	s = solubility (mol/L)	0	0
Equil.		$2s$	s

$K_{sp} = 1.2 \times 10^{-5} = [Ag^+]^2\,[SO_4^{2-}] = (2s)^2 s = 4s^3$, $s = 1.4 \times 10^{-2}$ mol/L

b. $Ag_2SO_4(s) \rightleftharpoons 2\ Ag^+(aq) + SO_4^{2-}(aq)$

Initial	s = solubility (mol/L)	0.10 M	0
Equil.		$0.10 + 2s$	s

$K_{sp} = 1.2 \times 10^{-5} = (0.10 + 2s)^2(s) \approx (0.10)^2(s)$, $s = 1.2 \times 10^{-3}$ mol/L; Assumption good.

c. $Ag_2SO_4(s) \rightleftharpoons 2\ Ag^+(aq) + SO_4^{2-}(aq)$

Initial	s = solubility (mol/L)	0	0.20 M
Equil		$2s$	$0.20 + s$

$1.2 \times 10^{-5} = (2s)^2(0.20 + s) \approx 4s^2(0.20)$, $s = 3.9 \times 10^{-3}$ mol/L; Assumption good.

Note: Comparing the solubilities of parts b and c to that of part a illustrates that the solubility of a salt decreases when a common ion is present.

87. $Ca_3(PO_4)_2(s) \rightleftharpoons 3\ Ca^{2+}(aq) + 2\ PO_4^{3-}(aq)$

Initial		0	0.20 M

s mol/L of $Ca_3(PO_4)_2(s)$ dissolves to reach equilibrium

Change	$-s$	$\rightarrow$	$+3s$	$+2s$
Equil.			$3s$	$0.20 + 2s$

$K_{sp} = 1.3 \times 10^{-32} = [Ca^{2+}]^3\,[PO_4^{3-}]^2 = (3s)^3(0.20 + 2s)^2$

Assuming $0.20 + 2s \approx 0.20$: $1.3 \times 10^{-32} = (3s)^3(0.20)^2 = 27\,s^3(0.040)$

s = molar solubility = 2.3×10^{-11} mol/L; Assumption good.

88. $Ce(IO_3)_3(s) \rightleftharpoons Ce^{3+}(aq) + 3 IO_3^-(aq)$

Initial s = solubility (mol/L)   0   0.20 M
Equil. s $0.20 + 3s$

$K_{sp} = [Ce^{3+}] [IO_3^-]^3 = s(0.20 + 3s)^3$

From the problem, $s = 4.4 \times 10^{-8}$ mol/L; Solving for K_{sp}:

$K_{sp} = (4.4 \times 10^{-8}) \times [0.20 + 3(4.4 \times 10^{-8})]^3 = 3.5 \times 10^{-10}$

89.   If the anion in the salt can act as a base in water, the solubility of the salt will increase as the solution becomes more acidic. Added H^+ will react with the base, forming the conjugate acid. As the basic anion is removed, more of the salt will dissolve to replenish the basic anion. The salts with basic anions are Ag_3PO_4, $CaCO_3$, $CdCO_3$ and $Sr_3(PO_4)_2$. Hg_2Cl_2 and PbI_2 do not have any pH dependence since Cl^- and I^- are terrible bases (the conjugate bases of a strong acids).

$$Ag_3PO_4(s) + H^+(aq) \rightarrow 3 Ag^+(aq) + HPO_4^{2-}(aq) \xrightarrow{\text{excess } H^+} 3 Ag^+(aq) + H_3PO_4(aq)$$

$$CaCO_3(s) + H^+ \rightarrow Ca^{2+} + HCO_3^- \xrightarrow{\text{excess } H^+} Ca^{2+} + H_2CO_3 \ [H_2O(l) + CO_2(g)]$$

$$CdCO_3(s) + H^+ \rightarrow Cd^{2+} + HCO_3^- \xrightarrow{\text{excess } H^+} Cd^{2+} + H_2CO_3 \ [H_2O(l) + CO_2(g)]$$

$$Sr_3(PO_4)_2(s) + 2 H^+ \rightarrow 3 Sr^{2+} + 2 HPO_4^{2-} \xrightarrow{\text{excess } H^+} 3 Sr^{2+} + 2 H_3PO_4$$

90.   a. AgF   b. $Pb(OH)_2$ c. $Sr(NO_2)_2$ d. $Ni(CN)_2$

All the above salts have anions that are bases. The anions of the other choices are conjugate bases of strong acids. They have no basic properties in water and, therefore, do not have solubilities which depend on pH.

91.   Potentially, $BaSO_4(s)$ could form if Q is greater than K_{sp}.

$BaSO_4(s) \rightleftharpoons Ba^{2+}(aq) + SO_4^{2-}(aq)$ $K_{sp} = 1.5 \times 10^{-9}$

To calculate Q, we need the initial concentrations of Ba^{2+} and SO_4^{2-}.

$$[Ba^{2+}]_o = \frac{\text{mmoles } Ba^{2+}}{\text{total mL solution}} = \frac{75.0 \text{ mL} \times \dfrac{0.020 \text{ mmol } Ba^{2+}}{\text{mL}}}{75.0 \text{ mL} \times 125 \text{ mL}} = 0.0075 \ M$$

$$[SO_4^{2-}]_o = \frac{\text{mmoles } SO_4^{2-}}{\text{total mL solution}} = \frac{125 \text{ mL} \times \dfrac{0.040 \text{ mmol } SO_4^{2-}}{\text{mL}}}{200. \text{ mL}} = 0.025 \ M$$

$Q = [Ba^{2+}]_o[SO_4^{2-}]_o = (0.0075\ M)(0.025\ M) = 1.9 \times 10^{-4}$

$Q > K_{sp}$ (1.5×10^{-9}) so $BaSO_4(s)$ will form.

92.   The formation of $Mg(OH)_2(s)$ is the only possible precipitate. $Mg(OH)_2(s)$ will form if $Q > K_{sp}$.

$Mg(OH)_2(s) \rightleftharpoons Mg^{2+}(aq) + 2\ OH^-(aq)$ $K_{sp} = [Mg^{2+}][OH^-]^2 = 8.9 \times 10^{-12}$

$$[Mg^{2+}]_o = \frac{100.0\ mL \times 4.0 \times 10^{-4}\ mmol\ Mg^{2+}/mL}{100.0\ mL + 100.0\ mL} = 2.0 \times 10^{-4}\ M$$

$$[OH^-]_o = \frac{100.0\ mL \times 2.0 \times 10^{-4}\ mmol\ OH^-/mL}{200.0\ mL} = 1.0 \times 10^{-4}\ M$$

$Q = [Mg^{2+}]_o[OH^-]_o^2 = (2.0 \times 10^{-4}\ M)(1.0 \times 10^{-4})^2 = 2.0 \times 10^{-12}$

Since $Q < K_{sp}$, then $Mg(OH)_2(s)$ will not precipitate, so no precipitate forms.

93.   The concentrations of ions are large, so Q will be greater than K_{sp} and $BaC_2O_4(s)$ will form. To solve this problem, we will assume that the precipitation reaction goes to completion; then we will solve an equilibrium problem to get the actual ion concentrations.

$$100.\ mL \times \frac{0.200\ mmol\ K_2C_2O_4}{mL} = 20.0\ mmol\ K_2C_2O_4$$

$$150.\ mL \times \frac{0.250\ mmol\ BaBr_2}{mL} = 37.5\ mmol\ BaBr_2$$

$$Ba^{2+}(aq) \quad + \quad C_2O_4^{2-}(aq) \quad \rightarrow \quad BaC_2O_4(s) \qquad K = 1/K_{sp} \gg 1$$

	Ba²⁺	C₂O₄²⁻		BaC₂O₄	
Before	37.5 mmol	20.0 mmol		0	
Change	-20.0	-20.0	$\rightarrow$	+20.0	Reacts completely (K is large)
After	17.5	0		20.0	

New initial concentrations (after complete precipitation) are: $[Ba^{2+}] = \dfrac{17.5\ mmol}{250.\ mL} = 7.00 \times 10^{-2}\ M$

$[K^+] = \dfrac{2(20.0\ mmol)}{250.\ mL} = 0.160\ M;\ \ [Br^-] = \dfrac{2(37.5\ mmol)}{250.\ mL} = 0.300\ M$

For K^+ and Br^-, these are also the final concentrations. For Ba^{2+} and $C_2O_4^{2-}$, we need to perform an equilibrium calculation.

$$BaC_2O_4(s) \rightleftharpoons Ba^{2+}(aq) + C_2O_4^{2-}(aq) \qquad K_{sp} = 2.3 \times 10^{-8}$$

Initial   0.0700 M 0

s mol/L of $BaC_2O_4(s)$ dissolves to reach equilibrium

Equil.   0.0700 + s s

$$K_{sp} = 2.3 \times 10^{-8} = [Ba^{2+}][C_2O_4^{2-}] = (0.0700 + s)(s) \approx 0.0700\, s$$

$$s = [C_2O_4^{2-}] = 3.3 \times 10^{-7}\ \text{mol/L}; \quad [Ba^{2+}] = 0.0700\ M \quad \text{Assumption good } (s \ll 0.0700).$$

94. $50.0\ \text{mL} \times 0.10\ M = 5.0\ \text{mmol Pb}^{2+}$; $50.0\ \text{mL} \times 1.0\ M = 50.\ \text{mmol Cl}^-$. For this solution, $Q > K_{sp}$, so $PbCl_2$ precipitates. Assume precipitation of $PbCl_2(s)$ is complete. 5.0 mmol Pb^{2+} requires 10. mmol of Cl^- for complete precipitation, which leaves 40. mmol Cl^- in excess. Now, let some of the $PbCl_2(s)$ redissolve to establish equilibrium

$$PbCl_2(s) \rightleftharpoons Pb^{2+}(aq) + 2\,Cl^-(aq)$$

Initial   0 $\dfrac{40.\ \text{mmol}}{100.0\ \text{mL}} = 0.40\ M$

s mol/L of $PbCl_2(s)$ dissolves to reach equilibrium

Equil. s $0.40 + 2s$

$$K_{sp} = [Pb^{2+}][Cl^-]^2, \quad 1.6 \times 10^{-5} = s(0.40 + 2s)^2 \approx s(0.40)^2$$

$$s = 1.0 \times 10^{-4}\ \text{mol/L}; \quad \text{Assumption good.}$$

At equilibrium: $[Pb^{2+}] = s = 1.0 \times 10^{-4}\ \text{mol/L}; \quad [Cl^-] = 0.40 + 2s, \quad 0.40 + 2(1.0 \times 10^{-4}) = 0.40\ M$

95. $Ag_3PO_4(s) \rightleftharpoons 3\,Ag^+(aq) + PO_4^{3-}(aq)$; When Q is greater than K_{sp}, precipitation will occur. We will calculate the $[Ag^+]_o$ necessary for $Q = K_{sp}$. Any $[Ag^+]_o$ greater than this calculated number will cause precipitation of $Ag_3PO_4(s)$. In this problem, $[PO_4^{3-}]_o = [Na_3PO_4]_o = 1.0 \times 10^{-5}\ M$.

$$K_{sp} = 1.8 \times 10^{-18}; \quad Q = 1.8 \times 10^{-18} = [Ag^+]_o^3\,[PO_4^{3-}]_o = [Ag^+]_o^3\,(1.0 \times 10^{-5}\ M)$$

$$[Ag^+]_o = \left(\frac{1.8 \times 10^{-18}}{1.0 \times 10^{-5}} \right)^{1/3}, \quad [Ag^+]_o = 5.6 \times 10^{-5}\ M$$

When $[Ag^+]_o = [AgNO_3]_o$ is greater than $5.6 \times 10^{-5}\ M$, precipitation of $Ag_3PO_4(s)$ will occur.

96.   From Table 15.4, K_{sp} for $NiCO_3 = 1.4 \times 10^{-7}$ and K_{sp} for $CuCO_3 = 2.5 \times 10^{-10}$. From the K_{sp} values, $CuCO_3$ will precipitate first since it has the smaller K_{sp} value and will be the least soluble. For $CuCO_3(s)$, precipitation begins when:

$$[CO_3^{2-}] = \frac{K_{sp,\,CuCO_3}}{[Cu^{2+}]} = \frac{2.5 \times 10^{-10}}{0.25\ M} = 1.0 \times 10^{-9}\ M\ CO_3^{2-}$$

For $NiCO_3(s)$ to precipitate:

$$[CO_3^{2-}] = \frac{K_{sp, NiCO_3}}{[Ni^{2+}]} = \frac{1.4 \times 10^{-7}}{0.25\ M} = 5.6 \times 10^{-7}\ M\ CO_3^{2-}$$

Determining the $[Cu^{2+}]$ when $NiCO_3(s)$ begins to precipitate:

$$[Cu^{2+}] = \frac{K_{sp, CuCO_3}}{[CO_3^{2-}]} = \frac{2.5 \times 10^{-10}}{5.6 \times 10^{-7}\ M} = 4.5 \times 10^{-4}\ M\ Cu^{2+}$$

For successful separation, 1% Cu^{2+} or less of the initial amount of Cu^{2+} (0.25 M) must be present before $NiCO_3(s)$ begins to precipitate. The percent of Cu^{2+} present when $NiCO_3(s)$ begins to precipitate is:

$$\frac{4.5 \times 10^{-4}\ M}{0.25\ M} \times 100 = 0.18\%\ Cu^{2+}$$

Since less than 1% of the initial amount of Cu^{2+} remains, the metals can be separated through slow addition of $Na_2CO_3(aq)$.

Complex Ion Equilibria

97.   a.

$$Co^{2+} + NH_3 \rightleftharpoons CoNH_3^{2+} \qquad\qquad K_1$$
$$CoNH_3^{2+} + NH_3 \rightleftharpoons Co(NH_3)_2^{2+} \qquad K_2$$
$$Co(NH_3)_2^{2+} + NH_3 \rightleftharpoons Co(NH_3)_3^{2+} \qquad K_3$$
$$Co(NH_3)_3^{2+} + NH_3 \rightleftharpoons Co(NH_3)_4^{2+} \qquad K_4$$
$$Co(NH_3)_4^{2+} + NH_3 \rightleftharpoons Co(NH_3)_5^{2+} \qquad K_5$$
$$Co(NH_3)_5^{2+} + NH_3 \rightleftharpoons Co(NH_3)_6^{2+} \qquad K_6$$

$$\overline{\qquad Co^{2+} + 6\ NH_3 \rightleftharpoons Co(NH_3)_6^{2+} \qquad\qquad K_f = K_1K_2K_3K_4K_5K_6 \qquad}$$

Note: The various K's are included for your information. Each NH_3 adds with a corresponding K value associated with that reaction. The overall formation constant, K_f, for the overall reaction is equal to the product of all the stepwise K values.

   b.

$$Ag^+ + NH_3 \rightleftharpoons AgNH_3^+ \qquad\qquad K_1$$
$$AgNH_3^+ + NH_3 \rightleftharpoons Ag(NH_3)_2^+ \qquad K_2$$

$$\overline{\qquad Ag^+ + 2\ NH_3 \rightleftharpoons Ag(NH_3)_2^+ \qquad\qquad K_f = K_1K_2 \qquad}$$

98.   a.

$$Ni^{2+} + CN^- \rightleftharpoons NiCN^+ \qquad\qquad K_1$$
$$NiCN^+ + CN^- \rightleftharpoons Ni(CN)_2 \qquad K_2$$
$$Ni(CN)_2 + CN^- \rightleftharpoons Ni(CN)_3^- \qquad K_3$$
$$Ni(CN)_3^- + CN^- \rightleftharpoons Ni(CN)_4^{2-} \qquad K_4$$

$$\overline{\qquad Ni^{2+} + 4\ CN^- \rightleftharpoons Ni(CN)_4^{2-} \qquad\qquad K_f = K_1K_2K_3K_4 \qquad}$$

b. $Mn^{2+} + C_2O_4^{2-} \rightleftharpoons MnC_2O_4$ K_1
      $MnC_2O_4 + C_2O_4^{2-} \rightleftharpoons Mn(C_2O_4)_2^{2-}$ K_2
      _____

      $Mn^{2+} + 2\ C_2O_4^{2-} \rightleftharpoons Mn(C_2O_4)_2^{2-}$ $K_f = K_1K_2$

99. $Hg^{2+}(aq) + 2\ I^-(aq) \rightarrow HgI_2(s)$, orange ppt.; $HgI_2(s) + 2\ I^-(aq) \rightarrow HgI_4^{2-}(aq)$, soluble complex ion

100. $Ag^+(aq) + Cl^-(aq) \rightleftharpoons AgCl(s)$, white ppt.; $AgCl(s) + 2\ NH_3(aq) \rightleftharpoons Ag(NH_3)_2^+(aq) + Cl^-(aq)$

      $Ag(NH_3)_2^+(aq) + Br^-(aq) \rightleftharpoons AgBr(s) + 2\ NH_3(aq)$, pale yellow ppt.

      $AgBr(s) + 2\ S_2O_3^{2-}(aq) \rightleftharpoons Ag(S_2O_3)_2^{3-}(aq) + Br^-(aq)$

      $Ag(S_2O_3)_2^{3-}(aq) + I^-(aq) \rightleftharpoons AgI(s) + 2\ S_2O_3^{2-}(aq)$, yellow ppt.

      The least soluble salt (smallest K_{sp} value) must be AgI since it forms in the presence of Cl^- and Br^-. The most soluble salt (largest K_{sp} value) must be AgCl since it forms initially, but never reforms. The order of K_{sp} values are: K_{sp} (AgCl) > K_{sp} (AgBr) > K_{sp} (AgI)

101. The formation constant for HgI_4^{2-} is an extremely large number. Because of this, we will let the Hg^{2+} and I^- ions present initially react to completion, and then solve an equilibrium problem to determine the Hg^{2+} concentration.

      | | Hg^{2+} | $+$ | $4\ I^-$ | $\rightleftharpoons$ | HgI_4^{2-} | $K = 1.0 \times 10^{30}$ |

      | Before | 0.010 M | 0.78 M | | 0 | |
      | Change | -0.010 | -0.040 | $\rightarrow$ | +0.010 | Reacts completely (K large) |
      | After | 0 | 0.74 | | 0.010 | New initial |

      x mol/L HgI_4^{2-} dissociates to reach equilibrium

      | Change | $+x$ | $+4x$ | $\leftarrow$ | $-x$ | |
      | Equil. | x | $0.74 + 4x$ | | $0.010 - x$ | |

      $K = 1.0 \times 10^{30} = \dfrac{[HgI_4^{2-}]}{[Hg^{2+}][I^-]^4} = \dfrac{(0.010 - x)}{(x)(0.74 + 4x)^4}$; Making normal assumptions:

      $1.0 \times 10^{30} = \dfrac{(0.010)}{(x)(0.74)^4}$, $x = [Hg^{2+}] = 3.3 \times 10^{-32}\ M$ Assumptions good.

      Note: 3.3×10^{-32} mol/L corresponds to one Hg^{2+} ion per 5×10^7 L. It is very reasonable to approach the equilibrium in two steps. The reaction does essentially go to completion.

102. $[X^-]_0 = 5.00\ M$ and $[Cu^+]_0 = 1.0 \times 10^{-3}\ M$ since equal volumes of each reagent are mixed.

      Since the K values are much greater than 1, assume the reaction goes completely to CuX_3^{2-}, and then solve an equilibrium problem.

$$\begin{array}{ccccc} Cu^+ & + & 3\,X^- & \rightleftharpoons & CuX_3{}^{2-} \end{array} \quad K = K_1 \times K_2 \times K_3 = 1.0 \times 10^9$$

Before	$1.0 \times 10^{-3}\,M$	$5.00\ M$	0	
After	0	$5.00 - 3(10^{-3}) \approx 5.00$	1.0×10^{-3}	Reacts completely
Equil.	x	$5.00 + 3x$	$1.0 \times 10^{-3} - x$	

$$K = \frac{(1.0 \times 10^{-3} - x)}{x(5.00 + 3x)^3} = 1.0 \times 10^9 \approx \frac{1.0 \times 10^{-3}}{x(5.00)^3}, \quad x = [Cu^+] = 8.0 \times 10^{-15}\,M \quad \text{Assumptions good.}$$

$$[CuX_3{}^{2-}] = 1.0 \times 10^{-3} - 8.0 \times 10^{-15} = 1.0 \times 10^{-3}\,M$$

$$K_3 = \frac{[CuX_3{}^{2-}]}{[CuX_2{}^-]\,[X^-]} = 1.0 \times 10^3 = \frac{(1.0 \times 10^{-3})}{[CuX_2{}^-]\,(5.00)}, \quad [CuX_2{}^-] = 2.0 \times 10^{-7}\,M$$

Summarizing:

$[CuX_3{}^{2-}] = 1.0 \times 10^{-3}\,M$ (answer a)
$[CuX_2{}^-]\ \ = 2.0 \times 10^{-7}\,M$ (answer b)
$[Cu^{2+}]\ \ \ = 8.0 \times 10^{-15}\,M$ (answer c)

103.   a. $$AgI(s) \rightleftharpoons Ag^+(aq) + I^-(aq) \quad K_{sp} = [Ag^+][I^-] = 1.5 \times 10^{-16}$$

Initial	s = solubility (mol/L)	0	0
Equil.		s	s

$$K_{sp} = 1.5 \times 10^{-16} = s^2, \quad s = 1.2 \times 10^{-8}\ \text{mol/L}$$

b.
$$\begin{array}{ll} AgI(s) \rightleftharpoons Ag^+ + I^- & K_{sp} = 1.5 \times 10^{-16} \\ Ag^+ + 2\,NH_3 \rightleftharpoons Ag(NH_3)_2{}^+ & K_f = 1.7 \times 10^7 \end{array}$$

$$\rule{9cm}{0.4pt}$$

$$AgI(s) + 2\,NH_3(aq) \rightleftharpoons Ag(NH_3)_2{}^+(aq) + I^-(aq) \quad K = K_{sp} \times K_f = 2.6 \times 10^{-9}$$

$$\begin{array}{ccccc} AgI(s) & + & 2\,NH_3 & \rightleftharpoons & Ag(NH_3)_2{}^+ + I^- \end{array}$$

Initial	$3.0\ M$	0	0
	s mol/L of AgBr(s) dissolves to reach equilibrium = molar solubility		
Equil.	$3.0 - 2s$	s	s

$$K = \frac{[Ag(NH_3)_2{}^+][I^-]}{[NH_3]^2} = \frac{s^2}{(3.0 - 2s)^2} = 2.6 \times 10^{-9} \approx \frac{s^2}{(3.0)^2}, \quad s = 1.5 \times 10^{-4}\ \text{mol/L}$$

Assumption good.

c.   The presence of NH_3 increases the solubility of AgI. Added NH_3 removes Ag^+ from solution by forming the complex ion, $Ag(NH_3)_2{}^+$. As Ag^+ is removed, more AgI(s) will dissolve to replenish the Ag^+ concentration.

104.

$$AgBr(s) \rightleftharpoons Ag^+ + Br^- \qquad K_{sp} = 5.0 \times 10^{-13}$$
$$Ag^+ + 2 S_2O_3^{2-} \rightleftharpoons Ag(S_2O_3)_2^{3-} \qquad K_f = 2.9 \times 10^{13}$$

$$AgBr(s) + 2 S_2O_3^{2-} \rightleftharpoons Ag(S_2O_3)_2^{3-} + Br^- \qquad K = K_{sp} \times K_f = 14.5 \quad \text{(Carry extra sig. figs.)}$$

$$AgBr(s) \quad + \quad 2 S_2O_3^{2-} \quad \rightleftharpoons \quad Ag(S_2O_3)_2^{3-} \quad + \quad Br^-$$

Initial	0.500 M	0	0

s mol/L AgBr(s) dissolves to reach equilibrium

Change	$-s$	$-2s$ $\rightarrow$	$+s$	$+s$
Equil.		0.500 - 2s	s	s

$$K = \frac{s^2}{(0.500 - 2s)^2} = 14.5; \text{ Taking the square root of both sides:}$$

$$\frac{s}{0.500 - 2s} = 3.81, \quad s = 1.91 - 7.62\ s, \quad s = 0.222 \text{ mol/L}$$

$$1.00 \text{ L} \times \frac{0.222 \text{ mol AgBr}}{\text{L}} \times \frac{187.8 \text{ g AgBr}}{\text{mol AgBr}} = 41.7 \text{ g AgBr} = 42 \text{ g AgBr}$$

105. Test tube 1: added Cl$^-$ reacts with Ag$^+$ to form a silver chloride precipitate. The net ionic equation is Ag$^+$(aq) + Cl$^-$(aq) $\rightarrow$ AgCl(s). Test tube 2: added NH$_3$ reacts with Ag$^+$ ions to form a soluble complex ion, Ag(NH$_3$)$_2^+$. As this complex ion forms, Ag$^+$ is removed from the solution, which causes the AgCl(s) to dissolve. When enough NH$_3$ is added, all of the silver chloride precipitate will dissolve. The equation is AgCl(s) + 2 NH$_3$(aq) $\rightarrow$ Ag(NH$_3$)$_2^+$(aq) + Cl$^-$(aq). Test tube 3: added H$^+$ reacts with the weak base, NH$_3$, to form NH$_4^+$. As NH$_3$ is removed from the Ag(NH$_3$)$_2^+$ complex ion, Ag$^+$ ions are released to solution and can then react with Cl$^-$ to reform AgCl(s). The equations are Ag(NH$_3$)$_2^+$(aq) + 2 H$^+$(aq) $\rightarrow$ Ag$^+$(aq) + 2 NH$_4^+$(aq) and Ag$^+$(aq) + Cl$^-$(aq) $\rightarrow$ AgCl(s).

106. In NH$_3$, Cu^{2+} forms the soluble complex ion, Cu(NH$_3$)$_4^{2+}$. This increases the solubility of Cu(OH)$_2$(s) since added NH$_3$ removes Cu^{2+} from the equilibrium causing more Cu(OH)$_2$(s) to dissolve. In HNO$_3$, H$^+$ removes OH$^-$ from the K$_{sp}$ equilibrium causing more Cu(OH)$_2$(s) to dissolve. Any salt with basic anions will be more soluble in an acid solution. AgC$_2$H$_3$O$_2$(s) will be more soluble in either NH$_3$ or HNO$_3$. This is because Ag$^+$ forms the complex ion Ag(NH$_3$)$_2^+$, and C$_2$H$_3$O$_2^-$ is a weak base, so it will react with added H$^+$. AgCl(s) will be more soluble only in NH$_3$ due to Ag(NH$_3$)$_2^+$ formation. In acid, Cl$^-$ is a horrible base, so it doesn't react with added H$^+$. AgCl(s) will not be more soluble in HNO$_3$.

Additional Exercises

107. NH$_3$ + H$_2$O $\rightleftharpoons$ NH$_4^+$ + OH$^-$ $K_b = \dfrac{[NH_4^+][OH^-]}{[NH_3]}$; Taking the -log of the K$_b$ expression:

$$- \log K_b = - \log [OH^-] - \log \frac{[NH_4^+]}{[NH_3]}, \quad - \log [OH^-] = - \log K_b + \log \frac{[NH_4^+]}{[NH_3]}$$

$$pOH = pK_b + \log \frac{[NH_4^+]}{[NH_3]} \text{ or } pOH = pK_b + \log \frac{[Acid]}{[Base]}$$

108.   a. $pH = pK_a = -\log (6.4 \times 10^{-5}) = 4.19$ since $[HBz] = [Bz^-]$ where $HBz = C_6H_5CO_2H$ and $[Bz^-] = C_6H_5CO_2^-$.

b. $[Bz^-]$ will increase to $0.120\ M$, and $[HBz]$ will decrease to $0.080\ M$ after OH^- reacts completely with HBz. The Henderson-Hasselbalch equation is derived from the K_a equilibrium reaction.

$$pH = pK_a + \log \frac{[Bz^-]}{[HBz]}, \quad pH = 4.19 + \log \frac{(0.120)}{(0.080)} = 4.37$$

c.

	Bz^-	$+$	H_2O	$\rightleftharpoons$	HBz	$+$	OH^-
Initial	$0.120\ M$				$0.080\ M$		0
Equil.	$0.120 - x$				$0.080 + x$		x

$$K_b = \frac{K_w}{K_a} = \frac{1.0 \times 10^{-14}}{6.4 \times 10^{-5}} = \frac{(0.080 + x)\,(x)}{(0.120 - x)} \approx \frac{(0.080)\,(x)}{0.120}$$

$x = [OH^-] = 2.34 \times 10^{-10}\ M$ (carrying extra sig. figs.); Assumptions good.

$pOH = 9.63; \quad pH = 4.37$

d.   We get the same answer. Both equilibria involve the two major species, benzoic acid and benzoate anion. Both equilibria must hold true. K_b is related to K_a by K_w and $[OH^-]$ is related to $[H^+]$ by K_w, so all constants are interrelated.

109.   A best buffer is when $pH \approx pK_a$ since these solutions have about equal concentrations of weak acid and conjugate base. Therefore, choose combinations that yield a buffer where $pH \approx pK_a$, i.e., look at the acids available and choose the one whose pK_a is closest to the pH.

a.   Potassium fluoride + HCl will yield a buffer consisting of HF ($pK_a = 3.14$) and F^-.

b.   Benzoic acid + NaOH will yield a buffer consisting of benzoic acid ($pK_a = 4.19$) and benzoate anion.

c.   Sodium acetate + acetic acid ($pK_a = 4.74$) is the best choice for $pH = 5.0$ buffer since acetic acid has a pK_a value closest to 5.0.

d.   HOCl and NaOH: This is the best choice to produce a conjugate acid/base pair with $pH = 7.0$. This mixture would yield a buffer consisting of HOCl ($pK_a = 7.46$) and OCl^-. Actually, the best choice for a $pH = 7.0$ buffer is an equimolar mixture of ammonium chloride and sodium acetate. NH_4^+ is a weak acid ($K_a = 5.6 \times 10^{-10}$) and $C_2H_3O_2^-$ is a weak base ($K_b = 5.6 \times 10^{-10}$). A mixture of the two will give a buffer at $pH = 7.0$ since the weak acid and weak base are the same strengths (K_a for $NH_4^+ = K_b$ for $C_2H_3O_2^-$). $NH_4C_2H_3O_2$ is commercially available, and its solutions are used for $pH = 7.0$ buffers.

e.   Ammonium chloride + NaOH will yield a buffer consisting of NH_4^+ ($pK_a = 9.26$) and NH_3.

110. a. The optimum pH for a buffer is when pH = pK_a. At this pH, a buffer will have equal neutralization capacity for both added acid and base. As shown below, since the pK_a for TRISH$^+$ is about 8, the optimal buffer pH is about 8.

$$K_b = 1.19 \times 10^{-6}; \quad K_a = K_w/K_b = 8.40 \times 10^{-9}; \quad pK_a = -\log(8.40 \times 10^{-9}) = 8.076$$

b. $pH = pK_a + \log \dfrac{[TRIS]}{[TRISH^+]}, \quad 7.00 = 8.076 + \log \dfrac{[TRIS]}{[TRISH^+]}$

$\dfrac{[TRIS]}{[TRISH^+]} = 10^{-1.08} = 0.083$ (at pH = 7.00)

$9.00 = 8.076 + \log \dfrac{[TRIS]}{[TRISH^+]}, \quad \dfrac{[TRIS]}{[TRISH^+]} = 10^{0.92} = 8.3$ (at pH = 9.00)

c. $\dfrac{50.0 \text{ g TRIS}}{2.0 \text{ L}} \times \dfrac{1 \text{ mol}}{121.14 \text{ g}} = 0.206\,M = 0.21\,M = [TRIS]$

$\dfrac{65.0 \text{ g TRISHCl}}{2.0 \text{ L}} \times \dfrac{1 \text{ mol}}{157.60 \text{ g}} = 0.206\,M = 0.21\,M = [TRISHCl] = [TRISH^+]$

$pH = pK_a + \log \dfrac{[TRIS]}{[TRISH^+]} = 8.076 + \log \dfrac{(0.21)}{(0.21)} = 8.08$

The amount of H$^+$ added from HCl is: $0.50 \times 10^{-3} \text{ L} \times \dfrac{12 \text{ mol}}{\text{L}} = 6.0 \times 10^{-3} \text{ mol H}^+$

The H$^+$ from HCl will convert TRIS into TRISH$^+$. The reaction is:

	TRIS	+	H$^+$	→	TRISH$^+$	
Before	0.21 M		$\dfrac{6.0 \times 10^{-3}}{0.2005} = 0.030\,M$		0.21 M	
Change	-0.030		-0.030	→	+0.030	Reacts completely
After	0.18		0		0.24	

Now use the Henderson-Hasselbalch equation to solve the buffer problem.

$$pH = 8.076 + \log \left(\dfrac{0.18}{0.24} \right) = 7.95$$

111. a. $HC_2H_3O_2 + OH^- \rightleftharpoons C_2H_3O_2^- + H_2O$

$$K_{eq} = \dfrac{[C_2H_3O_2^-]}{[HC_2H_3O_2][OH^-]} \times \dfrac{[H^+]}{[H^+]} = \dfrac{K_{a, HC_2H_3O_2}}{K_w} = \dfrac{1.8 \times 10^{-5}}{1.0 \times 10^{-14}} = 1.8 \times 10^{9}$$

b. $C_2H_3O_2^- + H^+ \rightleftharpoons HC_2H_3O_2 \quad K_{eq} = \dfrac{[HC_2H_3O_2]}{[H^+][C_2H_3O_2^-]} = \dfrac{1}{K_{a, HC_2H_3O_2}} = 5.6 \times 10^{4}$

c. $HCl + NaOH \rightarrow NaCl + H_2O$

Net ionic equation is: $H^+ + OH^- \rightleftharpoons H_2O$; $K_{eq} = \dfrac{1}{K_w} = 1.0 \times 10^{14}$

112. a.   Since all acids are the same initial concentration, the pH curve with the highest pH at 0 mL of NaOH added will correspond to the titration of the weakest acid. This is pH curve f.

b.   The pH curve with the lowest pH at 0 mL of NaOH added will correspond to the titration of the strongest acid. This is pH curve a.

  The best point to look at to differentiate a strong acid from a weak acid titration (if initial concentrations are not known) is the equivalence point pH. If the pH = 7.00, the acid titrated is a strong acid; if the pH is greater than 7.00, the acid titrated is a weak acid.

c.   For a weak acid-strong base titration, the pH at the halfway point to equivalence is equal to the pK_a value. The pH curve, which represents the titration of an acid with $K_a = 1.0 \times 10^{-6}$, will have a pH $= -\log(1 \times 10^{-6}) = 6.0$ at the halfway point. The equivalence point, from the plots, occurs at 50 mL NaOH added, so the halfway point is 25 mL. Plot d has a pH ~6.0 at 25 mL of NaOH added, so the acid titrated in this pH curve (plot d) has $K_a \sim 1 \times 10^{-6}$.

113. In the final solution: $[H^+] = 10^{-2.15} = 7.1 \times 10^{-3}\ M$

Beginning mmol HCl = 500.0 mL $\times$ 0.200 mmol/mL = 100. mmol HCl

Amount of HCl that reacts with NaOH = 1.50×10^{-2} mmol/mL $\times$ V

$$\frac{7.1 \times 10^{-3}\ \text{mmol}}{\text{mL}} = \frac{\text{final mmol H}^+}{\text{total volume}} = \frac{100. - 0.0150\ V}{500.0 + V}$$

$3.6 + 7.1 \times 10^{-3}\ V = 100. - 1.50 \times 10^{-2}\ V$, $2.21 \times 10^{-2}\ V = 100. - 3.6$

$V = 4.36 \times 10^3$ mL = 4.36 L = 4.4 L NaOH

114. For a titration of a strong acid with a strong base, the added OH^- reacts completely with the H^+ present. To determine the pH, we calculate the concentration of excess H^+ or OH^- after the neutralization reaction and then calculate the pH.

0 mL: $[H^+] = 0.100\ M$ from HNO_3; pH = 1.000

4.0 mL: initial mmol H^+ present = 25.0 mL $\times \dfrac{0.100\ \text{mmol H}^+}{\text{mL}} = 2.50$ mmol H^+

  mmol OH^- added = 4.0 mL $\times \dfrac{0.100\ \text{mmol OH}^-}{\text{mL}} = 0.40$ mmol OH^-

  0.40 mmol OH^- reacts completely with 0.40 mmol H^+: $OH^- + H^+ \rightarrow H_2O$

  $[H^+]_{\text{excess}} = \dfrac{(2.50 - 0.40)\ \text{mmol}}{(25.0 + 4.0)\ \text{mL}} = 7.24 \times 10^{-2}\ M$; pH = 1.140

We follow the same procedure for the remaining calculations.

8.0 mL: $[H^+]_{excess} = \dfrac{(2.50 - 0.80)\ \text{mmol}}{33.0\ \text{mL}} = 5.15 \times 10^{-2}\ M$; pH = 1.288

12.5 mL: $[H^+]_{excess} = \dfrac{(2.50 - 1.25)\ \text{mmol}}{37.5\ \text{mL}} = 3.33 \times 10^{-2}\ M$; pH = 1.478

20.0 mL: $[H^+]_{excess} = \dfrac{(2.50 - 2.00)\ \text{mmol}}{45.0\ \text{mL}} = 1.1 \times 10^{-2}\ M$; pH = 1.96

24.0 mL: $[H^+]_{excess} = \dfrac{(2.50 - 2.40)\ \text{mmol}}{49.0\ \text{mL}} = 2.0 \times 10^{-3}\ M$; pH = 2.70

24.5 mL: $[H^+]_{excess} = \dfrac{(2.50 - 2.45)\ \text{mmol}}{49.5\ \text{mL}} = 1 \times 10^{-3}\ M$; pH = 3.0

24.9 mL: $[H^+]_{excess} = \dfrac{(2.50 - 2.49)\ \text{mmol}}{49.9\ \text{mL}} = 2 \times 10^{-4}\ M$; pH = 3.7

25.0 mL: Equivalence point; We have a neutral solution since there is no excess H^+ or OH^- remaining after the neutralization reaction. pH = 7.00

25.1 mL: base in excess, $[OH^-]_{excess} = \dfrac{(2.51 - 2.50)\ \text{mmol}}{50.1\ \text{mL}} = 2 \times 10^{-4}\ M$; pOH = 3.7; pH = 10.3

26.0 mL: $[OH^-]_{excess} = \dfrac{(2.60 - 2.50)\ \text{mmol}}{51.0\ \text{mL}} = 2.0 \times 10^{-3}\ M$; pOH = 2.70; pH = 11.30

28.0 mL: $[OH^-]_{excess} = \dfrac{(2.80 - 2.50)\ \text{mmol}}{53.0\ \text{mL}} = 5.7 \times 10^{-3}\ M$; pOH = 2.24; pH = 11.76

30.0 mL: $[OH^-]_{excess} = \dfrac{(3.00 - 2.50)\ \text{mmol}}{55.0\ \text{mL}} = 9.1 \times 10^{-3}\ M$; pOH = 2.04; pH = 11.96

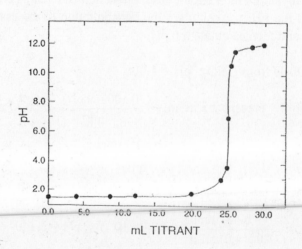

115. $HA + OH^- \rightarrow A^- + H_2O$ where HA = acetylsalicylic acid (assuming a monoprotic acid)

mmol HA present $= 27.36$ mL $OH^- \times \dfrac{0.5106 \text{ mmol } OH^-}{\text{mL } OH^-} \times \dfrac{1 \text{ mmol HA}}{\text{mmol } OH^-} = 13.97$ mmol HA

Molar mass of HA $= \dfrac{\text{grams}}{\text{mol}} = \dfrac{2.51 \text{ g HA}}{13.97 \times 10^{-3} \text{ mol HA}} = 180. \text{ g/mol}$

To determine the K_a value, use the pH data. After complete neutralization of acetylsalicylic acid by OH^-, we have 13.97 mmol of A^- produced from the neutralization reaction. A^- will react completely with the added H^+ and reform acetylsalicylic acid, HA.

mmol H^+ added $= 13.68$ mL $\times \dfrac{0.5106 \text{ mmol } H^+}{\text{mL}} = 6.985$ mmol H^+

	A^-	$+$	H^+	$\rightarrow$	HA	
Before	13.97 mmol		6.985 mmol		0	
Change	-6.985		-6.985	$\rightarrow$	+6.985	Reacts completely
After	6.985 mmol		0		6.985 mmol	

We have back titrated this solution to the halfway point to equivalence where pH = pK_a (assuming HA is a weak acid). We know this because after H^+ reacts completely, equal mmol of HA and A^- are present, which only occurs at the halfway point to equivalence. Assuming acetylsalicylic acid is a weak acid, then pH = pK_a = 3.48. $K_a = 10^{-3.48} = 3.3 \times 10^{-4}$

116. $HC_2H_3O_2 \rightleftharpoons H^+ + C_2H_3O_7^-$; Let C_o = initial concentration of $HC_2H_3O_2$

From normal weak acid setup: $K_a = 1.8 \times 10^{-5} = \dfrac{[H^+][C_2H_3O_2^-]}{[HC_2H_3O_2]} = \dfrac{[H^+]^2}{C_o - [H^+]}$

$[H^+] = 10^{-2.68} = 2.1 \times 10^{-3} M$; $1.8 \times 10^{-5} = \dfrac{(2.1 \times 10^{-3})^2}{C_o - 2.1 \times 10^{-3}}$, $C_o = 0.25 M$

25.0 mL $\times$ 0.25 mmol/mL = 6.3 mmol $HC_2H_3O_2$

Need 6.3 mmol KOH $= V_{KOH} \times 0.0975$ mmol/mL, $V_{KOH} = 65$ mL

117.   50.0 mL $\times$ 0.100 M = 5.00 mmol NaOH initially

at pH = 10.50, pOH = 3.50, $[OH^-] = 10^{-3.50} = 3.2 \times 10^{-4} M$

mmol OH^- remaining $= 3.2 \times 10^{-4}$ mmol/mL $\times$ 73.75 mL $= 2.4 \times 10^{-2}$ mmol

mmol OH^- that reacted = 5.00 - 0.024 = 4.98 mmol

Since the weak acid is monoprotic, 23.75 mL of the weak acid solution contains 4.98 mmol HA.

$$[HA]_o = \frac{4.98 \text{ mmol}}{23.75 \text{ mL}} = 0.210 \, M$$

118. a.
$$Cu(OH)_2 \rightleftharpoons Cu^{2+} + 2\,OH^- \qquad\qquad K_{sp} = 1.6 \times 10^{-19}$$
$$Cu^{2+} + 4\,NH_3 \rightleftharpoons Cu(NH_3)_4^{2+} \qquad\qquad K_f = 1.0 \times 10^{13}$$

$$\overline{Cu(OH)_2(s) + 4\,NH_3(aq) \rightleftharpoons Cu(NH_3)_4^{2+}(aq) + 2\,OH^-(aq) \qquad K = K_{sp}K_f = 1.6 \times 10^{-6}}$$

b.
$$Cu(OH)_2(s) \;+\; 4\,NH_3 \;\rightleftharpoons\; Cu(NH_3)_4^{2+} \;+\; 2\,OH^- \qquad K = 1.6 \times 10^{-6}$$

Initial	5.0 M	0	0.0095 M

s mol/L Cu(OH)$_2$ dissolves to reach equilibrium

Equil.	5.0 - 4s	s	0.0095 + 2s

$$K = 1.6 \times 10^{-6} = \frac{[Cu(NH_3)_4^{2+}][OH^-]^2}{[NH_3]^4} = \frac{s(0.0095 + 2s)^2}{(5.0 - 4s)^4}$$

If s is small: $1.6 \times 10^{-6} = \dfrac{s(0.0095)^2}{(5.0)^4}$, $s = 11.$ mol/L

Assumptions are not good. We will solve the problem by successive approximations.

$$s_{calc} = \frac{1.6 \times 10^{-6}\,(5.0 - 4s_{guess})^4}{(0.0095 + 2s_{guess})^2}, \text{ The results from six trials are:}$$

s_{guess}:   0.10, 0.050, 0.060, 0.055, 0.056

s_{calc}: 1.6×10^{-2}, 0.071, 0.049, 0.058, 0.056

Thus, the solubility of Cu(OH)$_2$ is 0.056 mol/L in 5.0 M NH$_3$.

119.
$$Ca_5(PO_4)_3OH(s) \;\rightleftharpoons\; 5\,Ca^{2+} \;+\; 3\,PO_4^{3-} \;+\; OH^-$$

Initial	s = solubility (mol/L)	0	0	1.0×10^{-7} from water
Equil.		5s	3s	$s + 1.0 \times 10^{-7} \approx s$

$$K_{sp} = 6.8 \times 10^{-37} = [Ca^{2+}]^5\,[PO_4^{3-}]^3\,[OH^-] = (5s)^5(3s)^3(s)$$

$6.8 \times 10^{-37} = (3125)(27)s^9$, $s = 2.7 \times 10^{-5}$ mol/L   Assumption is good.

The solubility of hydroxyapatite will increase as the solution gets more acidic since both phosphate and hydroxide can react with H$^+$.

$$Ca_5(PO_4)_3F(s) \rightleftharpoons 5\ Ca^{2+} + 3\ PO_4^{3-} + F^-$$

Initial	s = solubility (mol/L)	0	0	0
Equil.		$5s$	$3s$	s

$$K_{sp} = 1 \times 10^{-60} = (5s)^5(3s)^3(s) = (3125)(27)s^9, \quad s = 6 \times 10^{-8}\ \text{mol/L}$$

The hydroxyapatite in tooth enamel is converted to the less soluble fluorapatite by fluoride-treated water. The less soluble fluorapatite is more difficult to remove, making teeth less susceptible to decay.

120. $K_{sp} = 6.4 \times 10^{-9} = [Mg^{2+}]\ [F^-]^2, \quad 6.4 \times 10^{-9} = (0.00375 - y)(0.0625 - 2y)^2$

This is a cubic equation. No simplifying assumptions can be made since y is relatively large. Solving cubic equations is difficult unless you have a graphing calculator. However, if you don't have a graphing calculator, one way to solve this problem is to make the simplifying assumption to run the precipitation reaction to completion. This assumption is made because of the very small value for K, indicating that the ion concentrations are very small. Once this assumption is made, the problem becomes much easier to solve.

121. a. $$Pb(OH)_2(s) \rightleftharpoons Pb^{2+}(aq) + 2\ OH^-(aq)$$

Initial	s = solubility (mol/L)	0	$1.0 \times 10^{-7}\ M$ from water
Equil.		s	$1.0 \times 10^{-7} + 2s$

$$K_{sp} = 1.2 \times 10^{-15} = [Pb^{2+}]\ [OH^-]^2 = s(1.0 \times 10^{-7} + 2s)^2 \approx s(2s^2) = 4s^3$$

$s = [Pb^{2+}] = 6.7 \times 10^{-6}\ M$;   Assumption to ignore OH^- from water is good by the 5% rule.

b. $$Pb(OH)_2(s) \rightleftharpoons Pb^{2+}(aq) + 2\ OH^-(aq)$$

Initial		0	$0.10\ M$ pH = 13.00, $[OH^-] = 0.10\ M$
	s mol/L $Pb(OH)_2(s)$ dissolves to reach equilibrium		
Equil.		s	0.10 (buffered solution)

$$1.2 \times 10^{-15} = (s)(0.10)^2, \quad s = [Pb^{2+}] = 1.2 \times 10^{-13}\ M$$

c. We need to calculate the Pb^{2+} concentration in equilibrium with $EDTA^{4-}$. Since K is large for the formation of $PbEDTA^{2-}$, let the reaction go to completion; then solve an equilibrium problem to get the Pb^{2+} concentration.

$$Pb^{2+} + EDTA^{4-} \rightleftharpoons PbEDTA^{2-} \qquad K = 1.1 \times 10^{18}$$

	Pb^{2+}	$EDTA^{4-}$	$PbEDTA^{2-}$	
Before	0.010 M	0.050 M	0	
	0.010 mol/L Pb^{2+} reacts completely (large K)			
Change	-0.010	-0.010 $\rightarrow$	+0.010	Reacts completely
After	0	0.040	0.010	New initial
	x mol/L $PbEDTA^{2-}$ dissociates to reach equilibrium			
Equil.	x	0.040 + x	0.010 - x	

$$1.1 \times 10^{18} = \frac{(0.010 - x)}{(x)(0.040 + x)} \approx \frac{(0.010)}{x(0.040)}, \quad x = [Pb^{2+}] = 2.3 \times 10^{-19}\,M \quad \text{Assumptions good.}$$

Now calculate the solubility quotient for $Pb(OH)_2$ to see if precipitation occurs. The concentration of OH^- is $0.10\,M$ since we have a solution buffered at $pH = 13.00$.

$$Q = [Pb^{2+}]_o [OH^-]_o^2 = (2.3 \times 10^{-19})(0.10)^2 = 2.3 \times 10^{-21} < K_{sp}\,(1.2 \times 10^{-15})$$

$Pb(OH)_2(s)$ will not form since Q is less than K_{sp}.

Challenge Problems

122.   At 4.0 mL NaOH added: $\left|\dfrac{\Delta pH}{\Delta mL}\right| = \left|\dfrac{2.43 - 3.14}{0 - 4.0}\right| = 0.18$

The other points are calculated in a similar fashion. The results are summarized and plotted below. As can be seen from the plot, the advantage of this approach is that it is much easier to accurately determine the location of the equivalence point.

mL	pH	$\left|\Delta pH/\Delta mL\right|$
0	2.43	–
4.0	3.14	0.18
8.0	3.53	0.098
12.5	3.86	0.073
20.0	4.46	0.080
24.0	5.24	0.20
24.5	5.6	0.7
24.9	6.3	2
25.0	8.28	20
25.1	10.3	20
26.0	11.30	1
28.0	11.75	0.23
30.0	11.96	0.11

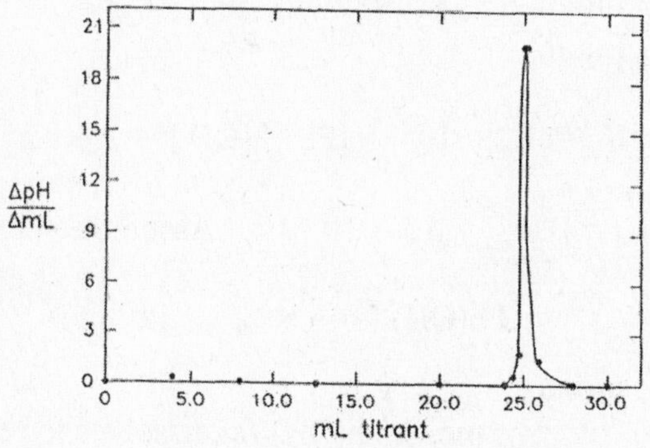

123.   mmol $HC_3H_5O_2$ present initially = $45.0\ mL \times \dfrac{0.750\ mmol}{mL} = 33.8\ mmol\ HC_3H_5O_2$

mmol $C_3H_5O_2^-$ present initially = $55.0\ mL \times \dfrac{0.700\ mmol}{mL} = 38.5\ mmol\ C_3H_5O_2^-$

The initial pH of the buffer is:

$$pH = pK_a + \log \frac{[C_3H_5O_2^-]}{[HC_3H_5O_2]} = -\log(1.3 \times 10^{-5}) + \log \frac{\dfrac{38.5\ mmol}{100.0\ mL}}{\dfrac{33.8\ mmol}{100.0\ mL}} = 4.89 + \log \frac{38.5}{33.8} = 4.95$$

Note: Since the buffer components are in the same volume of solution, we can use the mol (or mmol) ratio in the Henderson-Hasselbalch equation to solve for pH instead of using the concentration ratio of $[C_3H_5O_2^-]/[HC_3H_5O_2]$. The total volume always cancels for buffer solutions.

When NaOH is added, the pH will increase, and the added OH⁻ will convert $HC_3H_5O_2$ into $C_3H_5O_2^-$. The pH after addition of OH⁻ increases by 2.5%, so the resulting pH is:

$$4.95 + 0.025\,(4.95) = 5.07$$

At this pH, a buffer solution still exists and the mmol ratio between $C_3H_5O_2^-$ and $HC_3H_5O_2$ is:

$$pH = pK_a + \log \frac{\text{mmol}\,C_3H_5O_2^-}{\text{mmol}\,HC_3H_5O_2}, \quad 5.07 = 4.89 + \log \frac{\text{mmol}\,C_3H_5O_2^-}{\text{mmol}\,HC_3H_5O_2}$$

$$\frac{\text{mmol}\,C_3H_5O_2^-}{\text{mmol}\,HC_3H_5O_2} = 10^{0.18} = 1.5$$

Let x = mmol OH⁻ added to increase pH to 5.07. Since OH⁻ will essentially react to completion with $HC_3H_5O_2$, the setup for the problem using mmol is:

	$HC_3H_5O_2$	+	OH⁻	→	$C_3H_5O_2^-$	
Before	33.8 mmol		x mmol		38.5 mmol	
Change	$-x$		$-x$	→	$+x$	Reacts completely
After	33.8 - x		0		38.5 + x	

Solving for x:

$$\frac{\text{mmol}\,C_3H_5O_2^-}{\text{mmol}\,HC_3H_5O_2} = 1.5 = \frac{38.5 + x}{33.8 - x}, \quad 1.5\,(33.8 - x) = 38.5 + x, \quad x = 4.9 \text{ mmol OH⁻ added}$$

The volume of NaOH necessary to raise the pH by 2.5% is:

$$4.9 \text{ mmol NaOH} \times \frac{1 \text{ mL}}{0.10 \text{ mmol NaOH}} = 49 \text{ mL}$$

49 mL of 0.10 M NaOH must be added to increase the pH by 2.5%.

124. 0.400 mol/L × V_{NH_3} = mol NH_3 = mol NH_4^+ after reaction with HCl at the equivalence point.

At the equivalence point: $[NH_4^+]_o = \dfrac{\text{mol}\,NH_4^+}{\text{total volume}} = \dfrac{0.400 \times V_{NH_3}}{1.50 \times V_{NH_3}} = 0.267\,M$

	NH_4^+	⇌	H^+	+	NH_3
Initial	0.267 M		0		0
Equil.	0.267 - x		x		x

$$K_a = \frac{K_w}{K_b} = \frac{1.0 \times 10^{-14}}{1.8 \times 10^{-5}} = 5.6 \times 10^{-10} = \frac{x^2}{0.267 - x} \approx \frac{x^2}{0.267}$$

$x = [H^+] = 1.2 \times 10^{-5}\ M;\quad pH = 4.92$ Assumption good.

125.   For HOCl, $K_a = 3.5 \times 10^{-8}$ and $pK_a = -\log(3.5 \times 10^{-8}) = 7.46$; This will be a buffer solution since the pH is close to the pK_a value.

$$pH = pK_a + \log \frac{[OCl^-]}{[HOCl]}, \quad 8.00 = 7.46 + \log \frac{[OCl^-]}{[HOCl]}, \quad \frac{[OCl^-]}{[HOCl]} = 10^{0.54} = 3.5$$

$1.00\ L \times 0.0500\ M = 0.0500$ mol HOCl initially. Added OH^- converts HOCl into OCl^-. The total moles of OCl^- and HOCl must equal 0.0500 mol. Solving where n = moles:

$$n_{OCl^-} + n_{HOCl} = 0.0500 \text{ and } n_{OCl^-} = 3.5\ n_{HOCl}$$

$4.5\ n_{HOCl} = 0.0500,\quad n_{HOCl} = 0.011$ mol; $n_{OCl^-} = 0.039$ mol

We need to add 0.039 mol NaOH to produce 0.039 mol OCl^-.

0.039 mol $OH^- = V \times 0.0100\ M,\ V = 3.9\ L$ NaOH

126. $50.0\ mL \times 0.100\ M = 5.00$ mmol H_2SO_4; $30.0\ mL \times 0.100\ M = 3.00$ mmol HOCl

$25.0\ mL \times 0.200\ M = 5.00$ mmol NaOH; $10.0\ mL \times 0.150\ M = 1.50$ mmol KOH

$25.0\ mL \times 0.100\ M = 2.50$ mmol $Ba(OH)_2 = 5.00$ mmol OH^-

We've added 11.50 mmol OH^- total.

Let the OH^- react with the best acid present. This is H_2SO_4, which is a diprotic acid. For H_2SO_4, $K_{a_1} \gg 1$ and $K_{a_2} = 1.2 \times 10^{-2}$. The reaction is:

10.00 mmol $OH^- + 5.00$ mmol $H_2SO_4 \rightarrow 10.00$ mmol $H_2O + 5.00$ mmol SO_4^{2-}

OH^- still remains, and it reacts with the next best acid, HOCl ($K_a = 3.5 \times 10^{-8}$). The remaining 1.50 mmol OH^- will convert 1.50 mmol HOCl into 1.50 mmol OCl^-, resulting in a solution containing 1.50 mmol OCl^- and (3.00 - 1.50 =) 1.50 mmol HOCl. Major species at this point: HOCl, OCl^-, SO_4^{2-}, H_2O plus cations that don't affect pH. SO_4^{2-} is an extremely weak base ($K_b = 8.3 \times 10^{-13}$). Major equilibrium affecting pH: $HOCl \rightleftharpoons H^+ + OCl^-$. Since [HOCl] = [$OCl^-$], then:

$[H^+] = K_a = 3.5 \times 10^{-8}\ M;\quad pH = 7.46$ Assumptions good.

127.   The first titration plot (from 0-100.0 mL) corresponds to the titration of H_2A by OH^-. The reaction is $H_2A + OH^- \rightarrow HA^- + H_2O$. After all of the H_2A has been reacted, the second titration (from 100.0 - 200.0 mL) corresponds to the titration of HA^- by OH^-. The reaction is $HA^- + OH^- \rightarrow A^{2-} + H_2O$.

a.   At 100.0 mL of NaOH, just enough OH^- has been added to react completely with all of the H_2A present (mol OH^- added = mol H_2A present initially). From the balanced equation, the mol of HA^- produced will equal the mol of H_2A present initially. Since mol HA^- present at 100.0 mL OH^- added equals the mol of H_2A present initially, exactly 100.0 mL more of NaOH must be added to react with all of the HA^-. The volume of NaOH added to reach the second equivalence point equals 100.0 mL + 100.0 mL = 200.0 mL.

b. $H_2A + OH^- \rightarrow HA^- + H_2O$ is the reaction occurring from 0-100.0 mL NaOH added.

   i.   Since no reaction has taken place, H_2A and H_2O are the major species.

   ii.   Adding OH^- converts H_2A into HA^-. The major species up to 100.0 mL NaOH added are H_2A, HA^-, H_2O, and Na^+.

   iii.   At 100.0 mL NaOH added, mol of OH^- = mol H_2A, so all of the H_2A present initially has been converted into HA^-. The major species are HA^-, H_2O, and Na^+.

   iv.   Between 100.0 and 200.0 mL NaOH added, the OH^- converts HA^- into A^{2-}. The major species are HA^-, A^{2-}, H_2O, and Na^+.

   v.   At the second equivalence point (200.0 mL), just enough OH^- has been added to convert all of the HA^- into A^{2-}. The major species are A^{2-}, H_2O, and Na^+.

   vi.   Past 200.0 mL NaOH added, excess OH^- is present. The major species are OH^-, A^{2-}, H_2O, and Na^+.

c.   50.0 mL of NaOH added correspond to the first halfway point to equivalence. Exactly one-half of the H_2A present initially has been converted into its conjugate base HA^-, so $[H_2A] = [HA^-]$ in this buffer solution.

$$H_2A \rightleftharpoons HA^- + H^+ \qquad K_{a_1} = \frac{[HA^-][H^+]}{[H_2A]}$$

When $[HA^-] = [H_2A]$, then $K_{a_1} = [H^+]$ or $pK_{a_1} = pH$.

Here, pH = 4.0 so $pK_{a_1} = 4.0$ and $K_{a_1} = 10^{-4.0} = 1 \times 10^{-4}$.

150.0 mL of NaOH added correspond to the second halfway point to equivalence where $[HA^-] = [A^{2-}]$ in this buffer solution.

$$HA^- \rightleftharpoons A^{2-} + H^+ \qquad K_{a_2} = \frac{[A^{2-}][H^+]}{[HA^-]}$$

When $[A^{2-}] = [HA^-]$, then $K_{a_2} = [H^+]$ or $pK_{a_2} = pH$.

Here, $pH = 8.0$ so $pK_{a_2} = 8.0$ and $K_{a_2} = 10^{-8.0} = 1 \times 10^{-8}$.

128.   An indicator changes color at $pH \approx pK_a \pm 1$. The results from each indicator tells us something about the pH. The conclusions are summarized below:

Results from	pH
bromphenol blue	$\geq \sim 5.0$
bromcresol purple	$\leq \sim 5.0$
bromcresol green *	$pH \sim pK_a \sim 4.8 \sim 5.0$
alizarin	$\leq \sim 5.5$

*For bromcresol green, the resultant color is green.
This is a combination of the extremes (yellow and blue).
This occurs when $pH \sim pK_a$ of the indicator.

From the indicator results, the pH of the solution is about 5.0. We solve for K_a by setting up the typical weak acid problem.

$$HX \quad \rightleftharpoons \quad H^+ \ + \ X^-$$

Initial   1.0 M ~0   0
Equil.   1.0 - x x x

$$K_a = \frac{[H^+][X^-]}{[HX]} = \frac{x^2}{1.0 - x}; \quad \text{Since pH} \sim 5.0, \text{ then } [H^+] = x \approx 1 \times 10^{-5} \ M.$$

$$K_a \approx \frac{(1 \times 10^{-5})^2}{1.0 - 1 \times 10^{-5}} \approx 1 \times 10^{-10}$$

129.   a. $SrF_2(s) \quad \rightleftharpoons \quad Sr^{2+}(aq) \ + \ 2\,F^-(aq)$

Initial   0   0
                s mol/L SrF$_2$ dissolves to reach equilibrium
Equil. s $2s$

$$[Sr^{2+}][F^-]^2 = K_{sp} = 7.9 \times 10^{-10} = 4s^3, \quad s = 5.8 \times 10^{-4} \text{ mol/L}$$

b. Greater, because some of the F$^-$ would react with water:

$$F^- + H_2O \rightleftharpoons HF + OH^- \quad K_b = \frac{K_w}{K_a(HF)} = 1.4 \times 10^{-11}$$

This lowers the concentration of F$^-$, forcing more SrF$_2$ to dissolve.

c. $SrF_2(s) \rightleftharpoons Sr^{2+} + 2\,F^- \quad K_{sp} = 7.9 \times 10^{-10} = [Sr^{2+}]\,[F^-]^2$

Let s = solubility = [Sr^{2+}], then $2s$ = total F$^-$ concentration.

Since F$^-$ is a weak base, some of the F$^-$ is converted into HF. Therefore:

total F$^-$ concentration = $2s$ = [F$^-$] + [HF].

$$HF \rightleftharpoons H^+ + F^- \quad K_a = 7.2 \times 10^{-4} = \frac{[H^+]\,[F^-]}{[HF]} = \frac{1.0 \times 10^{-2}\,[F^-]}{[HF]} \quad \text{(since pH = 2.00 buffer)}$$

$7.2 \times 10^{-2} = \dfrac{[F^-]}{[HF]}$, [HF] = 14 [F$^-$]; Solving:

[Sr^{2+}] = s; $2s$ = [F$^-$] + [HF] = [F$^-$] + 14 [F$^-$], $2s$ = 15 [F$^-$], [F$^-$] = $2s/15$

$$7.9 \times 10^{-10} = [Sr^{2+}]\,[F^-]^2 = (s)\left(\frac{2s}{15}\right)^2, \quad s = 3.5 \times 10^{-3}\ mol/L$$

130. $M_3X_2(s) \quad \rightarrow \quad 3\,M^{2+}(aq) \quad + \quad 2\,X^{3-}(aq) \qquad K_{sp} = [M^{2+}]^3[X^{3-}]^2$

Initial s = solubility (mol/L)   0   0
Equil. $3s$ $2s$

$K_{sp} = (3s)^3(2s)^2 = 108\,s^5$; Total ion concentration = $3s + 2s = 5s$

$$\pi = iMRT, \ iM = \text{total ion concentration} = \frac{\pi}{RT} = \frac{2.64 \times 10^{-2}\ atm}{0.08206\ L\ atm/K \cdot mol \times 298\ K}$$

$$= 1.08 \times 10^{-3}\ mol/L$$

$5s = 1.08 \times 10^{-3}$ mol/L, $s = 2.16 \times 10^{-4}$ mol/L; $K_{sp} = 108\,s^5 = 108(2.16 \times 10^{-4})^5 = 5.08 \times 10^{-17}$

CHAPTER SIXTEEN

SPONTANEITY, ENTROPY, AND FREE ENERGY

Questions

7. a. A spontaneous process is one that occurs without any outside intervention.

    b. Entropy is a measure of disorder or randomness.

    c. The system is the portion of the universe in which we are interested.

    d. The surroundings are everything else in the universe besides the system.

8. a. Entropy increases; there is greater volume accessible to the randomly moving gas molecules which leads to an increase in disorder.

    b. The positional entropy doesn't change. There is no change in volume and thus no change in the numbers of positions of the molecules. The total entropy (ΔS_{univ}) increases because the increase in temperature increases the energy disorder (ΔS_{surr}).

    c. Entropy decreases because the volume decreases (P and V are inversely related).

9. Living organisms need an external source of energy to carry out these processes. Green plants use the energy from sunlight to produce glucose from carbon dioxide and water by photosynthesis. In the human body, the energy released from the metabolism of glucose helps drive the synthesis of proteins. For all processes combined, ΔS_{univ} must be greater than zero (second law).

10. No; When using ΔG_f° values in Appendix 4, we have specified a temperature of 25 °C. Further, if gases or solutions are involved, we have specified partial pressures of 1 atm and solute concentrations of 1 molar. At other temperatures and compositions, the reaction may not be spontaneous. A negative ΔG° value means the reaction is spontaneous under <u>standard conditions</u>.

11. Dispersion increases the entropy of the universe since the more widely something is dispersed, the greater the disorder. We must do work to overcome this disorder. In terms of the second law, it would be more advantageous to prevent contamination of the environment than to clean it up later. As a substance disperses, we have a much larger area that must be decontaminated.

12.   All thermodynamic functions depend on temperature. However, ΔH and ΔS are the least dependent on temperature and are commonly assumed to be temperature independent. ΔG has the strongest temperature dependence. To determine ΔG, the temperature must be known in order to use the equation $\Delta G = \Delta H - T\Delta S$.

13. $w_{max} = \Delta G$; When ΔG is negative, the magnitude of ΔG is equal to the maximum possible useful work obtainable from the process (at constant T and P). When ΔG is positive, the magnitude of ΔG is equal to the minimum amount of work that must be expended to make the process spontaneous. Due to waste energy (heat) in any real process, the amount of useful work obtainable from a spontaneous process is always less than w_{max} and, for a nonspontaneous reaction, an amount of work greater than w_{max} must be applied to make the process spontaneous.

14.   The rate of a reaction is directly related to temperature. As the temperature increases, the rate of a reaction increases. Spontaneity, however, does not necessarily have a direct relationship to temperature. The temperature dependence of spontaneity depends on the signs of the values for ΔH and ΔS (see Table 16.5 of the text). For example, when ΔH and ΔS are both negative, the reaction becomes more favorable thermodynamically (ΔG becomes more negative) with decreasing temperature. This is just the opposite of the kinetics dependence on temperature. Other sign combinations of ΔH and ΔS have different spontaneity temperature dependence.

Exercises

Spontaneity, Entropy, and the Second Law of Thermodynamics: Free Energy

15.   a, b and c; From our own experiences, salt water, colored water and rust form without any outside intervention. A bedroom, however, spontaneously gets cluttered. It takes an outside energy source to clean a bedroom.

16.   c and d; It takes an outside energy source to build a house and to launch and keep a satellite in orbit.

17.   We draw all of the possible arrangements of the two particles in the three levels.

2 kJ	__	__	x	__	x	xx
1 kJ	__	x	__	xx	x	__
0 kJ	xx	x	x	__	__	__

Total E =   0 kJ   1 kJ   2 kJ   2 kJ   3 kJ   4 kJ

The most likely total energy is 2 kJ.

18.

2 kJ	—	—	AB	—	—	B	A	B	A
1 kJ	—	AB	—	B	A	—	—	A	B
0 kJ	AB	—	—	A	B	A	B	—	—
E_T =	0 kJ	2 kJ	4 kJ	1 kJ	1 kJ	2 kJ	2 kJ	3 kJ	3 kJ

The most likely total energy is 2 kJ.

19. a. H_2 at 100°C and 0.5 atm; Higher temperature and lower pressure means greater volume and hence, greater positional entropy.

  b. N_2 at STP has the greater volume. c. $H_2O(l)$ is more disordered than $H_2O(s)$.

20. Of the three phases (solid, liquid, and gas), solids are the most ordered and gases are the most disordered. Thus, a, b, and f (melting, sublimation, and boiling) involve an increase in the entropy of the system since going from a solid to a liquid or a solid to a gas or a liquid to a gas increases disorder. For freezing (process c), a substance goes from the more disordered liquid state to the more ordered solid state, hence, entropy decreases. Process d (mixing) involves an increase in disorder (entropy) while separation increases order (decreases the entropy of the system). So of all the processes, a, b, d, and f involve an increase in the entropy of the system.

21. a. To boil a liquid requires heat. Hence, this is an endothermic process. All endothermic processes decrease the entropy of the surroundings (ΔS_{surr} is negative).

  b. This is an exothermic process. Heat is released when gas molecules slow down enough to form the solid. In exothermic processes, the entropy of the surroundings increases (ΔS_{surr} is positive).

22. a. $\Delta S_{surr} = \dfrac{-\Delta H}{T} = \dfrac{-(-2221 \text{ kJ})}{298 \text{ K}} = 7.45 \text{ kJ/K} = 7.45 \times 10^3 \text{ J/K}$

  b. $\Delta S_{surr} = \dfrac{-\Delta H}{T} = \dfrac{-112 \text{ kJ}}{298 \text{ K}} = -0.376 \text{ kJ/K} = -376 \text{ J/K}$

23. $\Delta G = \Delta H - T\Delta S$; When ΔG is negative, the process will be spontaneous.

  a. $\Delta G = \Delta H - T\Delta S = 25 \times 10^3 \text{ J} - (300. \text{ K})(5.0 \text{ J/K}) = 24,000 \text{ J}$, Not spontaneous

  b. $\Delta G = 25,000 \text{ J} - (300. \text{ K})(100. \text{ J/K}) = -5000 \text{ J}$, Spontaneous

  c. Without calculating ΔG, we know this reaction will be spontaneous at all temperatures. ΔH is negative and ΔS is positive ($-T\Delta S < 0$). ΔG will always be less than zero with these sign combinations for ΔH and ΔS.

  d. $\Delta G = -1.0 \times 10^4 \text{ J} - (200. \text{ K})(-40. \text{ J/K}) = -2000 \text{ J}$, Spontaneous

24. $\Delta G = \Delta H - T\Delta S$; A process is spontaneous when $\Delta G < 0$. For the following, assume ΔH and ΔS are temperature independent.

   a.   When ΔH and ΔS are both negative, ΔG will be negative below a certain temperature where the favorable ΔH term dominates. When $\Delta G = 0$, then $\Delta H = T\Delta S$. Solving for this temperature:

$$T = \frac{\Delta H}{\Delta S} = \frac{-18,000 \text{ J}}{-60. \text{ J/K}} = 3.0 \times 10^2 \text{ K}$$

   At $T < 3.0 \times 10^2$ K, this process will be spontaneous ($\Delta G < 0$).

   b.   When ΔH and ΔS are both positive, ΔG will be negative above a certain temperature where the favorable ΔS term dominates.

$$T = \frac{\Delta H}{\Delta S} = \frac{18,000 \text{ J}}{60. \text{ J/K}} = 3.0 \times 10^2 \text{ K}$$

   At $T > 3.0 \times 10^2$ K, this process will be spontaneous ($\Delta G < 0$).

   c.   When ΔH is positive and ΔS is negative, this process can never be spontaneous at any temperature since ΔG can never be negative.

   d.   When ΔH is negative and ΔS is positive, this process is spontaneous at all temperatures since ΔG will always be negative.

25.   At the boiling point, $\Delta G = 0$ so $\Delta H = T\Delta S$.

$$\Delta S = \frac{\Delta H}{T} = \frac{27.5 \text{ kJ/mol}}{(273 + 35) \text{ K}} = 8.93 \times 10^{-2} \text{ kJ/K} \bullet \text{mol} = 89.3 \text{ J/K} \bullet \text{mol}$$

26.   At the boiling point, $\Delta G = 0$ so $\Delta H = T\Delta S$. $T = \dfrac{\Delta H}{\Delta S} = \dfrac{58.51 \times 10^3 \text{ J/mol}}{92.92 \text{ J/K} \bullet \text{mol}} = 629.7$ K

27.   a. $NH_3(s) \rightarrow NH_3(l)$; $\Delta G = \Delta H - T\Delta S = 5650$ J/mol $- 200.$ K $(28.9$ J/K$\bullet$mol$)$

   $\Delta G = 5650$ J/mol $- 5780$ J/mol $= -130$ J/mol

   Yes, NH_3 will melt since $\Delta G < 0$ at this temperature.

   b.   At the melting point, $\Delta G = 0$ so $T = \dfrac{\Delta H}{\Delta S} = \dfrac{5650 \text{ J/mol}}{28.9 \text{ J/K} \bullet \text{mol}} = 196$ K.

28.   At the melting point, $\Delta G = 0$ so $\Delta H = T\Delta S$. $\Delta S = \dfrac{\Delta H}{T} = \dfrac{35.2 \times 10^3 \text{ J/mol}}{3680 \text{ K}} = 9.57$ J/K$\bullet$mol

Chemical Reactions: Entropy Changes and Free Energy

29.   a.   Decrease in disorder; $\Delta S°(-)$ b.   Increase in disorder; $\Delta S°(+)$

c.   Decrease in disorder $(\Delta n < 0)$; $\Delta S°(-)$ d.   Increase in disorder $(\Delta n > 0)$; $\Delta S°(+)$

For c and d, concentrate on the gaseous products and reactants. When there are more gaseous product molecules than gaseous reactant molecules $(\Delta n > 0)$, then $\Delta S°$ will be positive (disorder increases). When Δn is negative, then $\Delta S°$ is negative (disorder decreases).

30.   a.   Decrease in disorder $(\Delta n < 0)$; $\Delta S°(-)$ b.   Decrease in disorder $(\Delta n < 0)$; $\Delta S°(-)$

c.   Increase in disorder; $\Delta S°(+)$ d.   Increase in disorder; $\Delta S°(+)$

31.   a. $C_{graphite}(s)$; Diamond is a more ordered structure than graphite.

b. $C_2H_5OH(g)$; The gaseous state is more disordered than the liquid state.

c. $CO_2(g)$; The gaseous state is more disordered than the solid state.

32.   a.   He (10 K); S = 0 at 0 K   b. N_2O; More complicated molecule

c. $H_2O(l)$: The liquid state is more disordered than the solid state.

33.   a. $2 H_2S(g) + SO_2(g) \rightarrow 3 S_{rhombic}(s) + 2 H_2O(g)$; Since there are more molecules of reactant gases compared to product molecules of gas $(\Delta n = 2 - 3 < 0)$, then $\Delta S°$ will be negative.
        $\Delta S° = \Sigma n_p S°_{products} - \Sigma n_r S°_{reactants}$

$\Delta S° = [3 \text{ mol } S_{rhombic}(s) (32 \text{ J/K}\bullet mol) + 2 \text{ mol } H_2O(g) (189 \text{ J/K}\bullet mol)]$

$- [2 \text{ mol } H_2S(g) (206 \text{ J/K}\bullet mol) + 1 \text{ mol } SO_2(g) (248 \text{ J/K}\bullet mol)]$

$\Delta S° = 474 \text{ J/K} - 660. \text{ J/K} = -186 \text{ J/K}$

b. $2 SO_3(g) \rightarrow 2 SO_2(g) + O_2(g)$; Since Δn of gases is positive $(\Delta n = 3-2)$, $\Delta S°$ will be positive.

$\Delta S = 2 \text{ mol}(248 \text{ J/K} \bullet mol) + 1 \text{ mol}(205 \text{ J/K}\bullet mol) - [2 \text{ mol}(257 \text{ J/K}\bullet mol)] = 187 \text{ J/K}$

c. $Fe_2O_3(s) + 3 H_2(g) \rightarrow 2 Fe(s) + 3 H_2O(g)$; Since Δn of gases = 0 $(\Delta n = 3 - 3)$, we can't easily predict if $\Delta S°$ will be positive of negative.

$\Delta S = 2 \text{ mol}(27 \text{ J/K}\bullet mol) + 3 \text{ mol}(189 \text{ J/K}\bullet mol) - [1 \text{ mol}(90. \text{ J/K}\bullet mol) + 3 \text{ mol}(131 \text{ J/K}\bullet mol)]$

$\Delta S = 138 \text{ J/K}$

34.   a. $H_2(g) + 1/2 \, O_2(g) \rightarrow H_2O(l)$; Since Δn of gases is negative, ΔS° will be negative.

$\Delta S^\circ = 1 \text{ mol } H_2O(l)(70. \text{ J/K•mol}) - [1 \text{ mol } H_2(g)(131 \text{ J/K•mol}) + 1/2 \text{ mol } O_2(g)(205 \text{ J/K•mol})]$

$\Delta S^\circ = 70. \text{ J/K} - 234 \text{ J/K} = -164 \text{ J/K}$

   b. $N_2(g) + 3 \, H_2(g) \rightarrow 2 \, NH_3(g)$; Since Δn of gases is negative, ΔS° will be negative.

$\Delta S^\circ = 2(193) - [1(192) + 3(131)] = -199 \text{ J/K}$

   c. $HCl(g) \rightarrow H^+(aq) + Cl^-(aq)$; The gaseous state dominates predictions of ΔS°. Here the gaseous state is more disordered than the ions in solution, so ΔS° will be negative.

$\Delta S^\circ = 1 \text{ mol } H^+(0) + 1 \text{ mol } Cl^-(57 \text{ J/K•mol}) - 1 \text{ mol } HCl(187 \text{ J/K•mol}) = -130. \text{ J/K}$

35. $C_2H_2(g) + 4 \, F_2(g) \rightarrow 2 \, CF_4(g) + H_2(g)$; $\Delta S^\circ = 2 S^\circ_{CF_4} + S^\circ_{H_2} - [S^\circ_{C_2H_2} + 4 S^\circ_{F_2}]$

$-358 \text{ J/K} = (2 \text{ mol}) S^\circ_{CF_4} + 131 \text{ J/K} - [201 \text{ J/K} + 4(203 \text{ J/K})]$, $S^\circ_{CF_4} = 262 \text{ J/K•mol}$

36. $-144 \text{ J/K} = (2 \text{ mol}) S^\circ_{AlBr_3} - [2(28 \text{ J/K}) + 3(152 \text{ J/K})]$, $S^\circ_{AlBr_3} = 184 \text{ J/K•mol}$

37.   a. $S_{rhombic} \rightarrow S_{monoclinic}$; This phase transition is spontaneous ($\Delta G < 0$) at temperatures above $95^\circ C$. $\Delta G = \Delta H - T\Delta S$; For ΔG to be negative only above a certain temperature, ΔH is positive and ΔS is positive (see Table 16.5 of text).

   b.   Since ΔS is positive, $S_{rhombic}$ is the more ordered crystalline structure.

38.   Enthalpy is not favorable, so ΔS must provide the driving force for the change. Thus, ΔS is positive. There is an increase in disorder, so the original enzyme has the more ordered structure.

39.   a.   When a bond is formed, energy is released, so ΔH is negative. Since there are more reactant molecules of gas than product molecules of gas ($\Delta n < 0$), ΔS will be negative.

   b. $\Delta G = \Delta H - T\Delta S$; For this reaction to be spontaneous ($\Delta G < 0$), the favorable enthalpy term must dominate. The reaction will be spontaneous at low temperatures where the ΔH term dominates.

40.   Since there are more product gas molecules than reactant gas molecules ($\Delta n > 0$), ΔS will be positive. From the signs of ΔH and ΔS, this reaction is spontaneous at all temperatures. It will cost money to heat the reaction mixture. Since there is no thermodynamic reason to do this, the purpose of the elevated temperature must be to increase the rate of the reaction, i.e., kinetic reasons.

41.   a.

	$CH_4(g)$	+	$2\,O_2(g)$	→	$CO_2(g)$	+	$2\,H_2O(g)$
ΔH_f°	-75 kJ/mol		0		-393.5		-242
ΔG_f°	-51 kJ/mol		0		-394		-229
S°	186 J/K•mol		205		214		189

Data from Appendix 4

$$\Delta H^\circ = \Sigma n_p \Delta H^\circ_{f,\,products} - \Sigma n_r \Delta H^\circ_{f,\,reactants}\,;\ \ \Delta S^\circ = \Sigma n_p S^\circ_{products} - \Sigma n_r S^\circ_{reactants}$$

$$\Delta H^\circ = 2\ mol(-242\ kJ/mol) + 1\ mol(-393.5\ kJ/mol) - [1\ mol(-75\ kJ/mol)] = -803\ kJ$$

$$\Delta S^\circ = 2\ mol(189\ J/K\bullet mol) + 1\ mol(214\ J/K\bullet mol)$$

$$- [1\ mol(186\ J/K\bullet mol) + 2\ mol(205\ J/K\bullet mol)] = -4\ J/K$$

There are two ways to get ΔG°. We can use $\Delta G^\circ = \Delta H^\circ - T\Delta S^\circ$ (be careful of units):

$$\Delta G^\circ = \Delta H^\circ - T\Delta S^\circ = -803 \times 10^3\ J - (298\ K)(-4\ J/K) = -8.018 \times 10^5\ J = -802\ kJ$$

or we can use ΔG_f° values where $\Delta G^\circ = \Sigma n_p \Delta G^\circ_{f,\,products} - \Sigma n_r \Delta G^\circ_{f,\,reactants}$:

$$\Delta G^\circ = 2\ mol(-229\ kJ/mol) + 1\ mol(-394\ kJ/mol) - [1\ mol(-51\ kJ/mol)]$$

$$\Delta G^\circ = -801\ kJ\ \ \text{(Answers are the same within round-off error.)}$$

b.

	$6\,CO_2(g)$	+	$6\,H_2O(l)$	→	$C_6H_{12}O_6(s)$	+	$6\,O_2(g)$
ΔH_f°	-393.5 kJ/mol		-286		-1275		0
S°	214 J/K•mol		70.		212		205

$$\Delta H^\circ = -1275 - [6(-286) + 6(-393.5)] = 2802\ kJ$$

$$\Delta S^\circ = 6(205) + 212 - [6(214) + 6(70.)] = -262\ J/K$$

$$\Delta G^\circ = 2802\ kJ - (298\ K)(-0.262\ kJ/K) = 2880.\ kJ$$

c.

	$P_4O_{10}(s)$	+	$6\,H_2O(l)$	→	$4\,H_3PO_4(s)$
ΔH_f° (kJ/mol)	-2984		-286		-1279
S° (J/K•mol)	229		70.		110.

$\Delta H° = 4 \text{ mol}(-1279 \text{ kJ/mol}) - [1 \text{ mol}(-2984 \text{ kJ/mol}) + 6 \text{ mol}(-286 \text{ kJ/mol})] = -416 \text{ kJ}$

$\Delta S° = 4(110.) - [229 + 6(70.)] = -209 \text{ J/K}$

$\Delta G° = \Delta H° - T\Delta S° = -416 \text{ kJ} - (298 \text{ K})(-0.209 \text{ kJ/K}) = -354 \text{ kJ}$

d.

	$HCl(g)$	$+$ $NH_3(g)$	$\rightarrow$ $NH_4Cl(s)$
$\Delta H_f°$ (kJ/mol)	-92	-46	-314
$S°$ (J/K•mol)	187	193	96

$\Delta H° = -314 - [-92 - 46] = -176 \text{ kJ}$; $\Delta S° = 96 - [187 + 193] = -284 \text{ J/K}$

$\Delta G° = \Delta H° - T\Delta S° = -176 \text{ kJ} - (298 \text{ K})(-0.284 \text{ kJ/K}) = -91 \text{ kJ}$

42. $\Delta G° = -58.03 \text{ kJ} - (298 \text{ K})(-0.1766 \text{ kJ/K}) = -5.40 \text{ kJ}$

$$\Delta G° = 0 = \Delta H° - T\Delta S°, \quad T = \frac{\Delta H°}{\Delta S°} = \frac{-58.03 \text{ kJ}}{-0.1766 \text{ kJ/K}} = 328.6 \text{ K}$$

$\Delta G°$ is negative below 328.6 K where the favorable $\Delta H°$ term dominates.

43. $CH_4(g) + CO_2(g) \rightarrow CH_3CO_2H(l)$

$\Delta H° = -484 - [-75 + (-393.5)] = -16 \text{ kJ}$; $\Delta S° = 160 - [186 + 214] = -240. \text{ J/K}$

$\Delta G° = \Delta H° - T\Delta S° = -16 \text{ kJ} - (298 \text{ K})(-0.240 \text{ kJ/K}) = 56 \text{ kJ}$

This reaction is spontaneous only at temperatures below $T = \Delta H°/\Delta S° = 67 \text{ K}$ (where the favorable $\Delta H°$ term will dominate, giving a negative $\Delta G°$ value). This is not practical. Substances will be in condensed phases, and rates will be very slow at this extremely low temperature.

$CH_3OH(g) + CO(g) \rightarrow CH_3CO_2H(l)$

$\Delta H° = -484 - [-110.5 + (-201)] = -173 \text{ kJ}$; $\Delta S° = 160 - [198 + 240.] = -278 \text{ J/K}$

$\Delta G° = -173 \text{ kJ} - (298 \text{ K})(-0.278 \text{ kJ/K}) = -90. \text{ kJ}$

This reaction also has a favorable enthalpy and an unfavorable entropy term. This reaction is spontaneous at temperatures below $T = \Delta H°/\Delta S° = 622 \text{ K}$. The reaction of CH_3OH and CO will be preferred. It is spontaneous at high enough temperatures that the rates of reaction should be reasonable.

44. $C_2H_4(g) + H_2O(g) \rightarrow CH_3CH_2OH(l)$

$\Delta H° = -278 - (52 - 242) = -88$ kJ; $\Delta S° = 161 - (219 + 189) = -247$ J/K

When $\Delta G° = 0$, $\Delta H° = T\Delta S°$, $T = \dfrac{\Delta H°}{\Delta S°} = \dfrac{-88 \times 10^3 \text{ J}}{-247 \text{ J/K}} = 360$ K

Since the signs of $\Delta H°$ and $\Delta S°$ are both negative, this reaction will be spontaneous at temperatures below 360 K (where the favorable $\Delta H°$ term will dominate).

$C_2H_6(g) + H_2O(g) \rightarrow CH_3CH_2OH(l) + H_2(g)$

$\Delta H° = -278 - (-84.7 - 242) = 49$ kJ; $\Delta S° = 131 + 161 - (229.5 + 189) = -127$ J/K

This reaction can never be spontaneous because of the signs of $\Delta H°$ and $\Delta S°$.

Thus, the reaction $C_2H_4(g) + H_2O(g) \rightarrow C_2H_5OH(l)$ would be preferred.

45. $SO_3(g) \rightarrow SO_2(g) + 1/2\ O_2(g)$ $\Delta G° = -1/2\ (-142$ kJ$)$
        $S(s) + 3/2\ O_2(g) \rightarrow SO_3(g)$ $\Delta G° = -371$ kJ
_____

        $S(s) + O_2(g) \rightarrow SO_2(g)$ $\Delta G° = 71$ kJ $- 371$ kJ $= -300.$ kJ

46. $6\ C(s) + 6\ O_2(g) \rightarrow 6\ CO_2(g)$ $\Delta G° = 6(-394$ kJ$)$
        $3\ H_2(g) + 3/2\ O_2(g) \rightarrow 3\ H_2O(l)$ $\Delta G° = 3(-237$ kJ$)$
        $6\ CO_2(g) + 3\ H_2O(l) \rightarrow C_6H_6(l) + 15/2\ O_2(g)$ $\Delta G° = -1/2\ (-6399$ kJ$)$
_____

        $6\ C(s) + 3\ H_2(g) \rightarrow C_6H_6(l)$ $\Delta G° = 125$ kJ

47. $\Delta G° = \Sigma n_p \Delta G°_{f\ products} - \Sigma n_r \Delta G°_{f\ reactants}$, -374 kJ $= -1105$ kJ $- \Delta G°_{f\ SF_4}$, $\Delta G°_{f\ SF_4} = -731$ kJ/mol

48. $-5490.$ kJ $= 8(-394$ kJ$) + 10(-237$ kJ$) - 2\ \Delta G°_{f\ C_4H_{10}}$, $\Delta G°_{f\ C_4H_{10}} = -16$ kJ/mol

49. $\Delta G° = \Sigma n_p \Delta G°_{f\ products} - \Sigma n_r \Delta G°_{f\ reactants}$

$\Delta G° = 1$ mol$(-2698$ kJ/mol$) + 6$ mol$(-237$ kJ/mol$) - [4$ mol$(13$ kJ/mol$) + 8$ mol$(0)] = -4172$ kJ

50. a. $\Delta G° = 2(-270.$ kJ$) - 2(-502$ kJ$) = 464$ kJ

b. Since $\Delta G°$ is positive, this reaction is not spontaneous at standard conditions at 298 K.

c. $\Delta G° = \Delta H° - T\Delta S°$, $\Delta H° = \Delta G° + T\Delta S° = 464$ kJ $+ 298$ K$(0.179$ kJ/K$) = 517$ kJ

We need to solve for the temperature when $\Delta G° = 0$:

$\Delta G° = 0 = \Delta H° - T\Delta S°$, $T = \dfrac{\Delta H°}{\Delta S°} = \dfrac{517 \text{ kJ}}{0.179 \text{ kJ/K}} = 2890$ K

This reaction will be spontaneous at standard conditions ($\Delta G° < 0$) at T > 2890 K, where the favorable entropy term will dominate.

Free Energy: Pressure Dependence and Equilibrium

51. $\Delta G = \Delta G° + RT \ln Q$; For this reaction: $\Delta G = \Delta G° + RT \ln \dfrac{P_{NO_2} \times P_{O_2}}{P_{NO} \times P_{O_3}}$

$\Delta G° = 1 \text{ mol}(52 \text{ kJ/mol}) + 1 \text{ mol}(0) - [1 \text{ mol}(87 \text{ kJ/mol}) + 1 \text{ mol}(163 \text{ kJ/mol})] = -198 \text{ kJ}$

$\Delta G = -198 \text{ kJ} + \dfrac{8.3145 \text{ J/K} \bullet \text{mol}}{1000 \text{ J/kJ}} (298 \text{ K}) \ln \dfrac{(1.00 \times 10^{-7})\,(1.00 \times 10^{-3})}{(1.00 \times 10^{-6})\,(2.00 \times 10^{-6})}$

$\Delta G = -198 \text{ kJ} + 9.69 \text{ kJ} = -188 \text{ kJ}$

52. $\Delta G° = 3(0) + 2(-229) - [2(-34) + 1(-300.)] = -90. \text{ kJ}$

$\Delta G = \Delta G° + RT \ln \dfrac{P_{H_2O}^2}{P_{H_2S}^2 \times P_{SO_2}} = -90. \text{ kJ} + \dfrac{(8.3145)\,(298)}{1000} \text{ kJ} \left[\ln \dfrac{(0.030)^2}{(1.0 \times 10^{-4})^2\,(0.010)} \right]$

$\Delta G = -90. \text{ kJ} + 39.7 \text{ kJ} = -50. \text{ kJ}$

53. $\Delta G = \Delta G° + RT \ln Q = \Delta G° + RT \ln \dfrac{P_{N_2O_4}}{P_{NO_2}^2}$

$\Delta G° = 1 \text{ mol}(98 \text{ kJ/mol}) - 2 \text{ mol}(52 \text{ kJ/mol}) = -6 \text{ kJ}$

a.   These are standard conditions, so $\Delta G = \Delta G°$ since Q = 1 and ln Q = 0. Since $\Delta G°$ is negative, then the forward reaction is spontaneous. The reaction shifts right to reach equilibrium.

b. $\Delta G = -6 \times 10^3 \text{ J} + 8.3145 \text{ J/K} \bullet \text{mol} (298 \text{ K}) \ln \dfrac{0.50}{(0.21)^2}$

$\Delta G = -6 \times 10^3 \text{ J} + 6.0 \times 10^3 \text{ J} = 0$

Since $\Delta G = 0$, this reaction is at equilibrium (no shift).

c. $\Delta G = -6 \times 10^3 \text{ J} + 8.3145 \text{ J/K} \bullet \text{mol} (298 \text{ K}) \ln \dfrac{1.6}{(0.29)^2}$

$\Delta G = -6 \times 10^3 \text{ J} + 7.3 \times 10^3 \text{ J} = 1.3 \times 10^3 \text{ J} = 1 \times 10^3 \text{ J}$

Since ΔG is positive, the reverse reaction is spontaneous, so the reaction shifts to the left to reach equilibrium.

54. $\Delta H° = 2\ \Delta H°_{f\ NH_3} = 2(-46) = -92$ kJ; $\Delta G° = 2\ \Delta G°_{f\ NH_3} = 2(-17) = -34$ kJ

$\Delta S° = 2(193$ J/K$) - [192$ J/K $+ 3(131$ J/K$)] = -199$ J/K; $\Delta G° = -RT \ln K$

$$K = \exp \frac{-\Delta G°}{RT} = \exp\left(\frac{-(-34{,}000\ \text{J})}{(8.3145\ \text{J/K}\bullet\text{mol})\,(298\ \text{K})}\right) = e^{13.72} = 9.1 \times 10^5$$

Note: When determining exponents, we will round off after the calculation is complete. This helps eliminate excessive round-off error.

a. $\Delta G = \Delta G° + RT \ln \dfrac{P^2_{NH_3}}{P_{N_2} \times P^3_{H_2}} = -34\ \text{kJ} + \dfrac{(8.3145\ \text{J/K}\bullet\text{mol})\,(298\ \text{K})}{1000\ \text{J/kJ}} \ln \dfrac{(50.)^2}{(200.)\,(200.)^3}$

$\Delta G = -34$ kJ $- 33$ kJ $= -67$ kJ

b. $\Delta G = -34\ \text{kJ} + \dfrac{(8.3145\ \text{J/K}\bullet\text{mol})\,(298\ \text{K})}{1000\ \text{J/kJ}} \ln \dfrac{(200.)^2}{(200.)\,(600.)^3}$

$\Delta G = -34$ kJ $- 34.4$ kJ $= -68$ kJ

55.   At 25.0°C: $\Delta G° = \Delta H° - T\Delta S° = -58.03 \times 10^3$ J/mol $- (298.2$ K$)(-176.6$ J/K$\bullet$mol$)$

$= -5.37 \times 10^3$ J/mol

$\Delta G° = -RT \ln K,\ \ln K = \dfrac{-\Delta G°}{RT} = \dfrac{-(-5.37 \times 10^3\ \text{J/mol})}{(8.3145\ \text{J/K}\bullet\text{mol})\,(298.2\ \text{K})} = 2.166,\ \ K = e^{2.166} = 8.72$

At 100.0°C: $\Delta G° = -58.03 \times 10^3$ J/mol $- (373.2$ K$)(-176.6$ J/K$\bullet$mol$) = 7.88 \times 10^3$ J/mol

$\ln K = \dfrac{-(7.88 \times 10^3\ \text{J/mol})}{(8.3145\ \text{J/K}\bullet\text{mol})\,(373.2\ \text{K})} = -2.540,\ \ K = e^{-2.540} = 0.0789$

Note: When determining exponents, we will round off after the calculation is complete. This helps eliminate excessive round-off error.

56.   a. $\Delta H° = 2$ mol$(-92$ kJ/mol$) - [1$ mol$(0) + 1$ mol$(0)] = -184$ kJ

$\Delta S° = 2$ mol$(187$ J/K$\bullet$mol$) - [1$ mol$(131$ J/K$\bullet$mol$) + 1$ mol$(223$ J/K$\bullet$mol$)] = 20.$ J/K

$\Delta G° = \Delta H° - T\Delta S° = -184 \times 10^3$ J $- 298$ K $(20.$ J/K$) = -1.90 \times 10^5$ J $= -190.$ kJ

$\Delta G° = -RT \ln K,\ \ln K = \dfrac{-\Delta G°}{RT} = \dfrac{-(-1.90 \times 10^5\ \text{J})}{8.3145\ \text{J/K}\bullet\text{mol}\,(298\ \text{K})} = 76.683,\ \ K = e^{76.683} = 2.01 \times 10^{33}$

b.   These are standard conditions, so $\Delta G = \Delta G° = -190.$ kJ. Since ΔG is negative, the forward reaction is spontaneous, so the reaction shifts right to reach equilibrium.

57.   **a.** $\Delta G° = -RT \ln K = -\dfrac{8.3145 \text{ J}}{\text{K mol}}$ (298 K) $\ln (1.00 \times 10^{-14}) = 7.99 \times 10^4$ J $= 79.9$ kJ/mol

      b. $\Delta G°_{313} = -RT \ln K = -\dfrac{8.3145 \text{ J}}{\text{K mol}}$ (313 K) $\ln (2.92 \times 10^{-14}) = 8.11 \times 10^4$ J $= 81.1$ kJ/mol

58.   **a.**

	$\Delta H°_f$ (kJ/mol)	S° (J/K•mol)
$NH_3(g)$	-46	193
$O_2(g)$	0	205
$NO(g)$	90.	211
$H_2O(g)$	-242	189
$NO_2(g)$	34	240.
$HNO_3(l)$	-174	156
$H_2O(l)$	-286	70.

$4 NH_3(g) + 5 O_2(g) \rightarrow 4 NO(g) + 6 H_2O(g)$

$\Delta H° = 6(-242) + 4(90.) - [4(-46)] = -908$ kJ

$\Delta S° = 4(211) + 6(189) - [4(193) + 5(205)] = 181$ J/K

$\Delta G° = -908$ kJ $- 298$ K (0.181 kJ/K) $= -962$ kJ

$\Delta G° = -RT \ln K, \ \ln K = \dfrac{-\Delta G°}{RT} = \left(\dfrac{-(-962 \times 10^3 \text{ J})}{8.3145 \text{ J/K•mol} \times 298 \text{ K}} \right) = 388$

$\ln K = 2.303 \log K, \ \log K = 168, \ K = 10^{168}$ (an extremely large value)

$2 NO(g) + O_2(g) \rightarrow 2 NO_2(g)$

$\Delta H° = 2(34) - [2(90.)] = -112$ kJ; $\ \Delta S° = 2(240.) - [2(211) + (205)] = -147$ J/K

$\Delta G° = -112$ kJ $- (298$ K$)(-0.147$ kJ/K$) = -68$ kJ

$K = \exp \dfrac{-\Delta G°}{RT} = \exp\left(\dfrac{-(-68,000 \text{ J})}{8.3145 \text{ J/K•mol} (298 \text{ K})} \right) = e^{27.44} = 8.3 \times 10^{11}$

Note: When determining exponents, we will round off after the calculation is complete.

$3 NO_2(g) + H_2O(l) \rightarrow 2 HNO_3(l) + NO(g)$

$\Delta H° = 2(-174) + (90.) - [3(34) + (-286)] = -74$ kJ

$\Delta S° = 2(156) + (211) - [3(240.) + (70.)] = -267$ J/K

$\Delta G° = -74$ kJ $- (298$ K$)(-0.267$ kJ/K$) = 6$ kJ

$K = \exp \dfrac{-\Delta G°}{RT} = \exp\left(\dfrac{-6000 \text{ J}}{8.3145 \text{ J/K•mol} (298 \text{ K})} \right) = e^{-2.4} = 9 \times 10^{-2}$

b. $\Delta G° = -RT \ln K$; $T = 825°C = (825 + 273)K = 1098$ K; We must determine $\Delta G°$ at 1098 K.

$$\Delta G°_{1098} = \Delta H° - T\Delta S° = -908 \text{ kJ} - (1098 \text{ K})(0.181 \text{ kJ/K}) = -1107 \text{ kJ}$$

$$K = \exp\frac{-\Delta G°}{RT} = \exp\left(\frac{-(-1.107 \times 10^6 \text{ J})}{8.3145 \text{ J/K} \cdot \text{mol}(1098 \text{ K})}\right) = e^{121.258} = 4.589 \times 10^{52}$$

c. There is no thermodynamic reason for the elevated temperature since $\Delta H°$ is negative and $\Delta S°$ is positive. Thus, the purpose of the high temperature must be to increase the rate of the reaction.

59. $$K = \frac{P^2_{NF_3}}{P_{N_2} \times P^3_{F_2}} = \frac{(0.48)^2}{0.021(0.063)^3} = 4.4 \times 10^4$$

$$\Delta G°_{800} = -RT \ln K = -8.3145 \text{ J/K} \cdot \text{mol} (800. \text{ K}) \ln (4.4 \times 10^4) = -7.1 \times 10^4 \text{ J/mol} = -71 \text{ kJ/mol}$$

60. $2 SO_2(g) + O_2(g) \rightarrow 2 SO_3(g)$; $\Delta G° = 2(-371 \text{ kJ}) - [2(-300. \text{ kJ})] = -142 \text{ kJ}$

$$\Delta G° = -RT \ln K, \ln K = \frac{-\Delta G°}{RT} = \frac{-(-142,000 \text{ J})}{8.3145 \text{ J/K} \cdot \text{mol} (298 \text{ K})} = 57.311, K = e^{57.311} = 7.76 \times 10^{24}$$

$$K = 7.76 \times 10^{24} = \frac{P^2_{SO_3}}{P^2_{SO_2} \times P_{O_2}} = \frac{(2.0)^2}{P^2_{SO_2} \times (0.50)}, \ P_{SO_2} = 1.0 \times 10^{-12} \text{ atm}$$

From the negative value of $\Delta G°$, this reaction is spontaneous at standard conditions. Since there are more molecules of reactant gases than product gases, $\Delta S°$ will be negative (unfavorable). Therefore, this reaction must be exothermic ($\Delta H° < 0$). When $\Delta H°$ and $\Delta S°$ are both negative, the reaction will be spontaneous at relatively low temperatures where the favorable $\Delta H°$ term dominates.

61. The equation $\ln K = \frac{-\Delta H°}{R}\left(\frac{1}{T}\right) + \frac{\Delta S°}{R}$ is in the form of a straight line equation ($y = mx + b$). A

graph of $\ln K$ vs. $1/T$ will yield a straight line with slope $= m = -\Delta H°/R$ and a y-intercept $= b = \Delta S°/R$.

From the plot:

$$\text{slope} = \frac{\Delta y}{\Delta x} = \frac{0 - 40.}{3.0 \times 10^{-3} \text{ K}^{-1} - 0} = -1.3 \times 10^4 \text{ K}$$

$-1.3 \times 10^4 \text{ K} = -\Delta H°/R, \ \Delta H° = 1.3 \times 10^4 \text{ K} \times 8.3145 \text{ J/K} \cdot \text{mol} = 1.1 \times 10^5 \text{ J/mol}$

y-intercept $= 40. = \Delta S°/R, \ \Delta S° = 40. \times 8.3145 \text{ J/K} \cdot \text{mol} = 330 \text{ J/K} \cdot \text{mol}$

As seen here, when $\Delta H°$ is positive, the slope of the $\ln K$ vs. $1/T$ plot is negative. When $\Delta H°$ is negative, as in an exothermic process, the slope of the $\ln K$ vs. $1/T$ plot will be positive (slope $= -\Delta H°/R$).

62.   A graph of ln K vs. 1/T will yield a straight line with slope equal to $-\Delta H°/R$ and y-intercept equal to $\Delta S°/R$.

Temp (°C)	T(K)	1000/T (K^{-1})	K$_w$	ln K$_w$
0	273	3.66	1.14×10^{-15}	-34.408
25	298	3.36	1.00×10^{-14}	-32.236
35	308	3.25	2.09×10^{-14}	-31.499
40.	313	3.19	2.92×10^{-14}	-31.165
50.	323	3.10	5.47×10^{-14}	-30.537

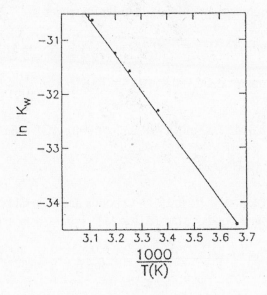

The straight line equation (from a calculator) is: $\ln K = -6.91 \times 10^3 \left(\dfrac{1}{T}\right) - 9.09$

$\text{Slope} = -6.91 \times 10^3 \text{ K} = \dfrac{-\Delta H°}{R}$, $\Delta H° = -(-6.91 \times 10^3 \text{ K} \times 8.3145 \text{ J/K·mol}) = 5.75 \times 10^4 \text{ J/mol}$

$\text{y-intercept} = -9.09 = \dfrac{\Delta S°}{R}$, $\Delta S° = -9.09 \times 8.3145 \text{ J/K·mol} = -75.6 \text{ J/K·mol}$

Additional Exercises

63.   It appears the sum of the two processes has no net change. This is not so. By the second law of thermodynamics, ΔS_{univ} must have increased even though it looks as if we have gone through a cyclic process.

64.   When an ionic solid dissolves, one would expect the disorder of the system to increase, so ΔS_{sys} is positive. Since temperature increased as the solid dissolved, this is an exothermic process and ΔS_{surr} is positive ($\Delta S_{surr} = -\Delta H/T$). Since the solid did dissolve, the dissolving process is spontaneous, so ΔS_{univ} is positive (as it must be when ΔS_{sys} and ΔS_{surr} are both positive).

65.   The introduction of mistakes is an effect of entropy. The purpose of redundant information is to provide a control to check the "correctness" of the transmitted information.

66.   At boiling point, $\Delta G = 0$ so $\Delta S = \dfrac{\Delta H_{vap}}{T}$; For methane: $\Delta S = \dfrac{8.20 \times 10^3 \text{ J/mol}}{112 \text{ K}} = 73.2 \text{ J/mol} \cdot \text{K}$

For hexane: $\Delta S = \dfrac{28.9 \times 10^3 \text{ J/mol}}{342 \text{ K}} = 84.5 \text{ J/mol} \cdot \text{K}$

$V_{met} = \dfrac{nRT}{P} = \dfrac{1.00 \text{ mol} (0.08206)(112 \text{ K})}{1.00 \text{ atm}} = 9.19 \text{ L};$ $V_{hex} = \dfrac{nRT}{P} = R(342 \text{ K}) = 28.1 \text{ L}$

Hexane has the larger molar volume at the boiling point, so hexane should have the larger entropy. As the volume of a gas increases, positional disorder increases.

67.   S (monoclinic) $\rightarrow$ S (rhombic); $\Delta H° = 0 - 0.30 = -0.30 \text{ kJ};$ $\Delta S° = 31.73 - 32.55 = -0.82 \text{ J/K}$

At the conversion temperature: $\Delta G° = 0$ so $\Delta H° = T\Delta S°;$ $T = \dfrac{\Delta H°}{\Delta S°} = \dfrac{-3.0 \times 10^2 \text{ J}}{-0.82 \text{ J/K}} = 370 \text{ K}$

68.   a. $\Delta G° = -RT \ln K = -(8.3145 \text{ J/K} \cdot \text{mol})(298 \text{ K}) \ln 0.090,$ $\Delta G° = 6.0 \times 10^3 \text{ J/mol} = 6.0 \text{ kJ/mol}$

b.   H–O–H + Cl–O–Cl $\rightarrow$ 2 H–O–Cl

On each side of the reaction, there are 2 H–O bonds and 2 O–Cl bonds. Both sides have the same number and type of bonds. Thus, $\Delta H \approx \Delta H° \approx 0$.

c. $\Delta G° = \Delta H° - T\Delta S°,$ $\Delta S° = \dfrac{\Delta H° - \Delta G°}{T} = \dfrac{0 - 6.0 \times 10^3 \text{ J}}{298 \text{ K}} = -20. \text{ J/K}$

d.   For $H_2O(g)$, $\Delta H_f° = -242 \text{ kJ/mol}$ and $S° = 189 \text{ J/K} \cdot \text{mol}$

$\Delta H° = 0 = 2 \, \Delta H_{f \text{ HOCl}}° - [1 \text{ mol}(-242 \text{ kJ/mol}) + 1 \text{ mol}(80.3 \text{ J/K/mol})],$ $\Delta H_{f \text{ HOCl}}° = -81 \text{ kJ/mol}$

$-20. \text{ J/K} = 2 \, S_{HOCl}° - [1 \text{ mol}(189 \text{ J/K} \cdot \text{mol}) + 1 \text{ mol}(266.1 \text{ J/K} \cdot \text{mol})],$ $S_{HOCl}° = 218 \text{ J/K} \cdot \text{mol}$

e.   Assuming $\Delta H°$ and $\Delta S°$ are T independent: $\Delta G_{500}° = 0 - (500. \text{ K})(-20. \text{ J/K}) = 1.0 \times 10^4 \text{ J}$

$\Delta G° = -RT \ln K,$ $K = \exp\left(\dfrac{-\Delta G°}{RT}\right) = \exp\left(\dfrac{-1.0 \times 10^4}{(8.3145)(500.)}\right) = e^{-2.41} = 0.090$

f. $\Delta G = \Delta G° + RT \ln \dfrac{P_{HOCl}^2}{P_{H_2O} \times P_{Cl_2O}}$; From part a, $\Delta G° = 6.0 \text{ kJ/mol}$.

We should express all P's in atm. However, because we perform the pressure conversion the same number of times in the numerator and denominator, the factors of 760 torr/atm will all cancel. Thus, we can use the pressures in units of torr.

$\Delta G = 6.0 \text{ kJ/mol} + \dfrac{(8.3145 \text{ J/K} \cdot \text{mol})(298 \text{ K})}{1000 \text{ J/kJ}} \ln\left(\dfrac{(0.10)^2}{(18)(2.0)}\right) = 6.0 - 20. = -14 \text{ kJ/mol}$

69. $HgbO_2 \quad \rightarrow Hgb + O_2 \qquad \Delta G° = -(-70 \text{ kJ})$
$Hgb + CO \rightarrow HgbCO \qquad \Delta G° = -80 \text{ kJ}$

$$\overline{HgbO_2 + CO \rightarrow HgbCO + O_2 \quad \Delta G° = -10 \text{ kJ}}$$

$$\Delta G° = -RT\ln K, \quad K = \exp\left(\frac{-\Delta G°}{RT}\right) = \exp\left(\frac{-(-10 \times 10^3 \text{ J})}{(8.3145 \text{ J/K} \cdot \text{mol})(298 \text{ K})}\right) = 60$$

70. $Ba(NO_3)_2(s) \rightleftharpoons Ba^{2+}(aq) + 2 NO_3^-(aq) \quad K = K_{sp}; \quad \Delta G° = -561 + 2(-109) - (-797) = 18 \text{ kJ}$

$$\Delta G° = -RT \ln K_{sp}, \quad \ln K_{sp} = \frac{-\Delta G°}{RT} = \frac{-18,000 \text{ J}}{8.3145 \text{ J/K} \cdot \text{mol} (298 \text{ K})} = -7.26, \quad K_{sp} = e^{-7.26} = 7.0 \times 10^{-4}$$

71. $HF(aq) \rightleftharpoons H^+(aq) + F^-(aq); \quad \Delta G = \Delta G° + RT \ln \dfrac{[H^+][F^-]}{[HF]}$

$$\Delta G° = -RT \ln K = -(8.3145 \text{ J/K} \cdot \text{mol})(298 \text{ K}) \ln (7.2 \times 10^{-4}) = 1.8 \times 10^4 \text{ J/mol}$$

a.   The concentrations are all at standard conditions, so $\Delta G = \Delta G° = 1.8 \times 10^4 \text{ J/mol}$ (since $Q = 1.0$ and $\ln Q = 0$). Since $\Delta G°$ is positive, then the reaction shifts left to reach equilibrium.

b. $\Delta G = 1.8 \times 10^4 \text{ J/mol} + (8.3145 \text{ J/K} \cdot \text{mol})(298 \text{ K}) \ln \dfrac{(2.7 \times 10^{-2})^2}{0.98}$

$\Delta G = 1.8 \times 10^4 \text{ J/mol} - 1.8 \times 10^4 \text{ J/mol} = 0$

Since $\Delta G = 0$, then the reaction is at equilibrium (no shift).

c. $\Delta G = 1.8 \times 10^4 + 8.3145 (298) \ln \dfrac{(1.0 \times 10^{-5})^2}{1.0 \times 10^{-5}} = -1.1 \times 10^4 \text{ J/mol}; \text{ shifts right}$

d. $\Delta G = 1.8 \times 10^4 + 8.3145 (298) \ln \dfrac{7.2 \times 10^{-4}(0.27)}{0.27} = 1.8 \times 10^4 - 1.8 \times 10^4 = 0; \text{ at equilibrium}$

e. $\Delta G = 1.8 \times 10^4 + 8.3145 (298) \ln \dfrac{1.0 \times 10^{-3}(0.67)}{0.52} = 2 \times 10^3 \text{ J/mol}; \text{ shifts left}$

72. $K^+ \text{(blood)} \rightleftharpoons K^+ \text{(muscle)} \quad \Delta G° = 0; \quad \Delta G = RT \ln \left(\dfrac{[K^+]_m}{[K^+]_b}\right); \quad \Delta G = w_{max}$

$$\Delta G = \frac{8.3145 \text{ J}}{K \text{ mol}} (310. \text{ K}) \ln \left(\frac{0.15}{0.0050}\right), \quad \Delta G = 8.8 \times 10^3 \text{ J/mol} = 8.8 \text{ kJ/mol}$$

At least 8.8 kJ of work must be applied.

Other ions will have to be transported in order to maintain electroneutrality. Either anions must be transported into the cells, or cations (Na^+) in the cell must be transported to the blood. The latter is what happens: $[Na^+]$ in blood is greater than $[Na^+]$ in cells as a result of this pumping.

73. a. $\Delta G° = - RT \ln K$, $K = \exp(-\Delta G°/RT) = \exp\left(\dfrac{-(-30,500\ J)}{8.3145\ J/K\bullet mol \times 298\ K}\right) = 2.22 \times 10^5$

b. $C_6H_{12}O_6(s) + 6\ O_2(g) \rightarrow 6\ CO_2(g) + 6\ H_2O(l)$

$\Delta G° = 6\ mol(-394\ kJ/mol) + 6\ mol(-237\ kJ/mol) - 1\ mol(-911\ kJ/mol) = -2875\ kJ$

$\dfrac{2875\ kJ}{mol\ glucose} \times \dfrac{1\ mol\ ATP}{30.5\ kJ} = \dfrac{94.3\ mol\ ATP}{mol\ glucose}$; 94.3 molecules ATP/molecule glucose

This is an overstatement. The assumption that all of the free energy goes into this reaction is false. Actually, only 38 moles of ATP are produced by the metabolism of one mole of glucose.

74. a. $\Delta G° = -RT \ln K$

$\ln K = \dfrac{-\Delta G°}{RT} = \dfrac{-14,000\ J}{(8.3145\ J/K\bullet mol)\ (298\ K)} = -5.65$, $K = e^{-5.65} = 3.5 \times 10^{-3}$

b.

Glutamic acid + NH$_3$ → Glutamine + H$_2$O	$\Delta G° = 14\ kJ$
ATP + H$_2$O → ADP + H$_2$PO$_4^-$	$\Delta G° = -30.5\ kJ$

Glutamic acid + ATP + NH$_3$ → Glutamine + ADP + H$_2$PO$_4^-$ $\Delta G° = 14 - 30.5 = -17\ kJ$

$\ln K = \dfrac{-\Delta G°}{RT} = \dfrac{-(-17,000\ J)}{8.3145\ J/K\bullet mol\ (298\ K)} = 6.86$, $K = e^{6.86} = 9.5 \times 10^2$

75. ΔS is more favorable for reaction two than for reaction one, resulting in $K_2 > K_1$. In reaction one, seven particles in solution are forming one particle. In reaction two, four particles form one particle, which results in a smaller decrease in disorder than for reaction one.

76. $\Delta G° = -RT \ln K$; When $K = 1.00$, $\Delta G° = 0$ since $\ln 1.00 = 0$. $\Delta G° = 0 = \Delta H° - T\Delta S°$, $\Delta H° = T\Delta S°$

$\Delta H° = 3(-242\ kJ) - [-826\ kJ] = 100.\ kJ$; $\Delta S° = 2(27\ J/K) + 3(189\ J/K) - [90.\ J/K + 3(131\ J/K)]$

$= 138\ J/K$

$\Delta H° = T\Delta S°$, $T = \dfrac{\Delta H°}{\Delta S°} = \dfrac{100.\ kJ}{0.138\ kJ/K} = 725\ K$

Challenge Problems

77. $3\ O_2(g) \rightleftharpoons 2\ O_3(g)$; $\Delta H° = 2(143\ kJ) = 286\ kJ$; $\Delta G° = 2(163\ kJ) = 326\ kJ$

$\ln K = \dfrac{-\Delta G°}{RT} = \dfrac{-326 \times 10^3\ J}{(8.3145\ J/K\bullet mol)\ (298\ K)} = -131.573$, $K = e^{-131.573} = 7.22 \times 10^{-58}$

We need the value of K at 230. K. From Section 16.8 of the text: $\ln K = \dfrac{-\Delta H°}{RT} + \dfrac{\Delta S°}{R}$

For two sets of K and T:

$$\ln K_1 = \frac{-\Delta H°}{R}\left(\frac{1}{T_1}\right) + \frac{\Delta S°}{R}; \quad \ln K_2 = \frac{-\Delta H°}{R}\left(\frac{1}{T_2}\right) + \frac{\Delta S°}{R}$$

Subtracting the first expression from the second:

$$\ln K_2 - \ln K_1 = \frac{\Delta H°}{R}\left(\frac{1}{T_1} - \frac{1}{T_2}\right) \text{ or } \ln\frac{K_2}{K_1} = \frac{\Delta H°}{R}\left(\frac{1}{T_1} - \frac{1}{T_2}\right)$$

Let $K_2 = 7.22 \times 10^{-58}$, $T_2 = 298$; $K_1 = K_{230}$, $T_1 = 230.$ K; $\Delta H° = 286 \times 10^3$ J

$$\ln\frac{7.22 \times 10^{-58}}{K_{230}} = \frac{286 \times 10^3}{8.3145}\left(\frac{1}{230.} - \frac{1}{298}\right) = 34.13$$

$$\frac{7.22 \times 10^{-58}}{K_{230}} = e^{34.13} = 6.6 \times 10^{14}, \quad K_{230} = 1.1 \times 10^{-72}$$

$$K_{230} = 1.1 \times 10^{-72} = \frac{P_{O_3}^2}{P_{O_2}^3} = \frac{P_{O_3}^2}{(1.0 \times 10^{-3} \text{ atm})^3}, \quad P_{O_3} = 3.3 \times 10^{-41} \text{ atm}$$

The volume occupied by one molecule of ozone is:

$$V = \frac{nRT}{P} = \frac{(1/6.022 \times 10^{23} \text{ mol}) (0.08206 \text{ L atm/mol·K}) (230. \text{ K})}{(3.3 \times 10^{-41} \text{ atm})}, \quad V = 9.5 \times 10^{17} \text{ L}$$

Equilibrium is probably not maintained under these conditions. When only two ozone molecules are in a volume of 9.5×10^{17} L, the reaction is not at equilibrium. Under these conditions, $Q > K$ and the reaction shifts left. But with only 2 ozone molecules in this huge volume, it is extremely unlikely they will collide with each other. In these conditions, the concentration of ozone is not large enough to maintain equilibrium.

78.   Arrangement I and V: $S = k \ln W$; $W = 1$; $S = k \ln 1 = 0$

       Arrangement II and IV: $W = 4$; $S = k \ln 4 = 1.38 \times 10^{-23}$ J/K $\ln 4$, $S = 1.91 \times 10^{-23}$ J/K

       Arrangement III: $W = 6$; $S = k \ln 6 = 2.47 \times 10^{-23}$ J/K

79.   a.   From the plot, the activation energy of the reverse reaction is $E_a + (-\Delta G°) = E_a - \Delta G°$ ($\Delta G°$ is a negative number as drawn in the diagram).

$$k_f = A \exp\left(\frac{-E_a}{RT}\right) \text{ and } k_r = A \exp\left(\frac{-(E_a - \Delta G°)}{RT}\right), \quad \frac{k_f}{k_r} = \frac{A \exp\left(\frac{-E_a}{RT}\right)}{A \exp\left(\frac{-(E_a - \Delta G°)}{RT}\right)}$$

If the A factors are equal: $\dfrac{k_f}{k_r} = \exp\left(\dfrac{-E_a}{RT} + \dfrac{(E_a - \Delta G^\circ)}{RT} \right) = \exp\left(\dfrac{-\Delta G^\circ}{RT} \right)$

From $\Delta G^\circ = -RT \ln K$, $K = \exp\left(\dfrac{-\Delta G^\circ}{RT} \right)$; Since K and $\dfrac{k_f}{k_r}$ are both equal to the same expression, then $K = \dfrac{k_f}{k_r}$.

b. A catalyst will lower the activation energy for both the forward and reverse reaction (but not change ΔG°). Therefore, a catalyst must increase the rate of both the forward and reverse reactions.

80. At equilibrium:

$$P_{H_2} = \frac{nRT}{V} = \frac{\left(\dfrac{1.10 \times 10^{13} \text{ molecules}}{6.022 \times 10^{23} \text{ molecules/mol}} \right) \left(\dfrac{0.08206 \text{ L atm}}{\text{mol K}} \right) (298 \text{ K})}{1.00 \text{ L}} = 4.47 \times 10^{-10} \text{ atm}$$

The pressure of H_2 decreased from 1.00 atm to 4.47×10^{-10} atm. Essentially all of the H_2 and Br_2 has reacted. Therefore, $P_{HBr} = 2.00$ atm since there is a 2:1 mol ratio between HBr and H_2 in the balanced equation. Since we began with equal moles of H_2 and Br_2, then we will have equal moles of H_2 and Br_2 at equilibrium. Therefore, $P_{H_2} = P_{Br_2} = 4.47 \times 10^{-10}$ atm.

$$K = \frac{P_{HBr}^2}{P_{H_2} \times P_{Br_2}} = \frac{(2.00)^2}{(4.47 \times 10^{-10})^2} = 2.00 \times 10^{19} \quad \text{Assumptions good.}$$

$\Delta G^\circ = -RT \ln K = -(8.3145 \text{ J/K} \cdot \text{mol})(298 \text{ K}) \ln (2.00 \times 10^{19}) = -1.10 \times 10^5$ J/mol

$$\Delta S^\circ = \frac{\Delta H^\circ - \Delta G^\circ}{T} = \frac{-103,800 \text{ J/mol} - (-1.10 \times 10^5 \text{ J/mol})}{298 \text{ K}} = 20 \text{ J/K} \cdot \text{mol}$$

81. a. $\Delta G^\circ = G_B^\circ - G_A^\circ = 11{,}718 - 8996 = 2722$ J

$$K = \exp\left(\frac{-\Delta G^\circ}{RT} \right) = \exp\left(\frac{-2722 \text{ J}}{(8.3145 \text{ J/K} \cdot \text{mol})(298 \text{ K})} \right) = 0.333$$

b. Since Q = 1.00 > K, reaction shifts left. Let x = atm of B(g) which reacts to reach equilibrium.

$$A(g) \quad \rightleftharpoons \quad B(g)$$

Initial	1.00 atm	1.00 atm
Equil.	1.00 + x	1.00 - x

$K = \dfrac{P_B}{P_A} = \dfrac{1.00 - x}{1.00 + x} = 0.333$, $1.00 - x = 0.333 + 0.333\,x$, $x = 0.50$ atm

$P_B = 1.00 - 0.50 = 0.50$ atm; $P_A = 1.00 + 0.50 = 1.50$ atm

c. $\Delta G = \Delta G° + RT \ln Q = \Delta G° + RT \ln (P_B/P_A)$

$\Delta G = 2722\ J + (8.3145)(298) \ln (0.50/1.50) = 2722\ J - 2722\ J = 0$ (carrying extra sig. figs.)

82.   From Exercise 16.61, $\ln K = \dfrac{-\Delta H°}{RT} + \dfrac{\Delta S°}{R}$. For K at two temperatures, T_1 and T_2, the equation can be manipulated to give (see Exercise 16.77): $\ln \dfrac{K_2}{K_1} = \dfrac{\Delta H°}{R}\left(\dfrac{1}{T_1} - \dfrac{1}{T_2}\right)$

$\ln\left(\dfrac{3.25 \times 10^{-2}}{8.84}\right) = \dfrac{\Delta H°}{8.3145\ \text{J/K·mol}}\left(\dfrac{1}{298\ K} - \dfrac{1}{348\ K}\right)$

$-5.61 = (5.8 \times 10^{-5}\ \text{mol/J}) (\Delta H°),\ \ \Delta H° = -9.7 \times 10^4\ \text{J/mol}$

For K = 8.84 at T = 25°C:

$\ln 8.84 = \dfrac{-(-9.7 \times 10^4\ \text{J/mol})}{(8.3145\ \text{J/K·mol})\ (298\ K)} + \dfrac{\Delta S°}{8.3145\ \text{J/K·mol}},\ \dfrac{\Delta S°}{8.3145} = -37,\ \Delta S° = -310\ \text{J/K·mol}$

We get the same value for $\Delta S°$ using $K = 3.25 \times 10^{-2}$ at T = 348 K data. $\Delta G° = -RT \ln K$. When K = 1.00, then $\Delta G° = 0$ since $\ln 1.00 = 0$. $\Delta G° = 0 = \Delta H° - T\Delta S°$. Assuming $\Delta H°$ and $\Delta S°$ do not depend on temperature:

$\Delta H° = T\Delta S°,\ \ T = \dfrac{\Delta H°}{\Delta S°} = \dfrac{-9.7 \times 10^4\ \text{J/mol}}{-310\ \text{J/K·mol}} = 310\ K$

83. $K_p = P_{CO_2}$;   To insure Ag_2CO_3 from decomposing, P_{CO_2} should be greater than K_p.

From Exercise 16.61, $\ln K = \dfrac{-\Delta H°}{RT} + \dfrac{\Delta S°}{R}$. For two conditions of K and T, the equation is:

$\ln \dfrac{K_2}{K_1} = \dfrac{\Delta H°}{R}\left(\dfrac{1}{T_1} - \dfrac{1}{T_2}\right)$

Let $T_1 = 25°C = 298\ K$, $K_1 = 6.23 \times 10^{-3}$ torr; $T_2 = 110.°C = 383\ K$, $K_2 = ?$

$\ln \dfrac{K_2}{6.23 \times 10^{-3}\ \text{torr}} = \dfrac{79.14 \times 10^3\ \text{J/mol}}{8.3145\ \text{J/K·mol}}\left(\dfrac{1}{298\ K} - \dfrac{1}{383\ K}\right)$

$\ln \dfrac{K_2}{6.23 \times 10^{-3}} = 7.1,\ \dfrac{K_2}{6.23 \times 10^{-3}} = e^{7.1} = 1.2 \times 10^3,\ K_2 = 7.5\ \text{torr}$

To prevent decomposition of Ag_2CO_3, the partial pressure of CO_2 should be greater than 7.5 torr.

84.   From the problem, $\chi^L_{C_6H_6} = \chi^L_{CCl_4} = 0.500$. We need the pure vapor pressures ($P°$) in order to calculate the vapor pressure of the solution. Using the thermodynamic data:

$$C_6H_6(l) \rightleftharpoons C_6H_6(g) \quad K = P_{C_6H_6} = P^\circ_{C_6H_6} \text{ at } 25°C$$

$$\Delta G^\circ_{rxn} = \Delta G^\circ_{f\,C_6H_6(g)} - \Delta G^\circ_{f\,C_6H_6(l)} = 129.66 \text{ kJ/mol} - 124.50 \text{ kJ/mol} = 5.16 \text{ kJ/mol}$$

$$\Delta G^\circ = -RT \ln K, \quad \ln K = \frac{-\Delta G^\circ}{RT} = \frac{-5.16 \times 10^3 \text{ J/mol}}{(8.3145 \text{ J/K} \cdot \text{mol})(298 \text{ K})} = -2.08$$

$$K = P^\circ_{C_6H_6} = e^{-2.08} = 0.125 \text{ atm}$$

For CCl_4: $\Delta G^\circ_{rxn} = \Delta G^\circ_{f\,CCl_4(g)} - \Delta G^\circ_{f\,CCl_4(l)} = -60.59 \text{ kJ/mol} - (-65.21 \text{ kJ/mol}) = 4.62 \text{ kJ/mol}$

$$K = P^\circ_{CCl_4} = \exp\left(\frac{-\Delta G^\circ}{RT}\right) = \exp\left(\frac{-4620 \text{ J/mol}}{8.3145 \text{ J/K} \cdot \text{mol} \times 298 \text{ K}}\right) = 0.155 \text{ atm}$$

$$P_{C_6H_6} = \chi^L_{C_6H_6} P^\circ_{C_6H_6} = 0.500 \,(0.125 \text{ atm}) = 0.0625 \text{ atm}; \quad P_{CCl_4} = 0.500 \,(0.155 \text{ atm}) = 0.0775 \text{ atm}$$

$$\chi^V_{C_6H_6} = \frac{P_{C_6H_6}}{P_{tot}} = \frac{0.0625 \text{ atm}}{0.0625 \text{ atm} + 0.0775 \text{ atm}} = \frac{0.0625}{0.1400} = 0.446; \quad \chi^V_{CCl_4} = 1.000 - 0.446 = 0.554$$

85.   Use the thermodynamic data to calculate the boiling point of the solvent.

At boiling point, $\Delta G = 0 = \Delta H - T\Delta S$, $\Delta H = T\Delta S$, $T = \dfrac{\Delta H}{\Delta S} = \dfrac{33.90 \times 10^3 \text{ J/mol}}{95.95 \text{ J/K} \cdot \text{mol}} = 353.3 \text{ K}$

$\Delta T = K_b m$, $(355.4 \text{ K} - 353.3 \text{ K}) = 2.5 \text{ K kg/mol} \,(m)$, $m = \dfrac{2.1}{2.5} = 0.84 \text{ mol/kg}$

$$\text{mass solvent} = 150. \text{ mL} \times \frac{0.879 \text{ g}}{\text{mL}} \times \frac{1 \text{ kg}}{1000 \text{ g}} = 0.132 \text{ kg}$$

$$\text{mass solute} = 0.132 \text{ kg solvent} \times \frac{0.84 \text{ mol solute}}{\text{kg solvent}} \times \frac{142 \text{ g}}{\text{mol}} = 15.7 \text{ g} = 16 \text{ g solute}$$

CHAPTER SEVENTEEN

ELECTROCHEMISTRY

Review of Oxidation - Reduction Reactions

13. Oxidation: increase in oxidation number; loss of electrons

    Reduction: decrease in oxidation number; gain of electrons

14. See Table 4.2 in Chapter 4 of the text for rules for assigning oxidation numbers.

    a. H (+1), O (-2), N (+5)　　　　b. Cl (-1), Cu (+2)

    c. O (0)　　　　　　　　　　　　d. H (+1), O (-1)

    e. Mg (+2), O (-2), S (+6)　　　　f. Ag (0)

    g. Pb (+2), O (-2), S (+6)　　　　h. O (-2), Pb (+4)

    i. Na (+1), O (-2), C (+3)　　　　j. O (-2), C (+4)

    k. $(NH_4)_2Ce(SO_4)_3$ contains NH_4^+ ions and SO_4^{2-} ions. Thus, cerium exists as the Ce^{4+} ion. H (+1), N (-3), Ce (+4), S (+6), O (-2)

    l. O (-2), Cr (+3)

15. The species oxidized shows an increase in oxidation numbers and is called the reducing agent. The species reduced shows a decrease in oxidation numbers and is called the oxidizing agent. The pertinent oxidation numbers are listed by the substance oxidized and the substance reduced.

	Redox?	Ox. Agent	Red. Agent	Substance Oxidized	Substance Reduced
a.	Yes	H_2O	CH_4	CH_4 (C, -4 → +2)	H_2O (H, +1 → 0)
b.	Yes	$AgNO_3$	Cu	Cu (0 → +2)	$AgNO_3$ (Ag, +1 → 0)
c.	Yes	HCl	Zn	Zn (0 → +2)	HCl (H, +1 → 0)

    d. No; There is no change in any of the oxidation numbers.

16.   See Chapter 4.10 of the text for rules on balancing oxidation-reduction reactions.

a. $Cr \rightarrow Cr^{3+} + 3\ e^-$

$$NO_3^- \rightarrow NO$$
$$4\ H^+ + NO_3^- \rightarrow NO + 2\ H_2O$$
$$3\ e^- + 4\ H^+ + NO_3^- \rightarrow NO + 2\ H_2O$$

$$Cr \rightarrow Cr^{3+} + 3\ e^-$$
$$3\ e^- + 4\ H^+ + NO_3^- \rightarrow NO + 2\ H_2O$$

$$4\ H^+(aq) + NO_3^-(aq) + Cr(s) \rightarrow Cr^{3+}(aq) + NO(g) + 2\ H_2O(l)$$

b. $(Al \rightarrow Al^{3+} + 3\ e^-) \times 5$

$$MnO_4^- \rightarrow Mn^{2+}$$
$$8\ H^+ + MnO_4^- \rightarrow Mn^{2+} + 4\ H_2O$$
$$(5\ e^- + 8\ H^+ + MnO_4^- \rightarrow Mn^{2+} + 4\ H_2O) \times 3$$

$$5\ Al \rightarrow 5\ Al^{3+} + 15\ e^-$$
$$15\ e^- + 24\ H^+ + 3\ MnO_4^- \rightarrow 3\ Mn^{2+} + 12\ H_2O$$

$$24\ H^+(aq) + 3\ MnO_4^-(aq) + 5\ Al(s) \rightarrow 5\ Al^{3+}(aq) + 3\ Mn^{2+}(aq) + 12\ H_2O(l)$$

c. $(Ce^{4+} + e^- \rightarrow Ce^{3+}) \times 6$

$$CH_3OH \rightarrow CO_2$$
$$H_2O + CH_3OH \rightarrow CO_2 + 6\ H^+$$
$$H_2O + CH_3OH \rightarrow CO_2 + 6\ H^+ + 6\ e^-$$

$$6\ Ce^{4+} + 6\ e^- \rightarrow 6\ Ce^{3+}$$
$$H_2O + CH_3OH \rightarrow CO_2 + 6\ H^+ + 6\ e^-$$

$$H_2O(l) + CH_3OH(aq) + 6\ Ce^{4+}(aq) \rightarrow 6\ Ce^{3+}(aq) + CO_2(g) + 6\ H^+(aq)$$

d. $PO_3^{3-} \rightarrow PO_4^{3-}$
$(H_2O + PO_3^{3-} \rightarrow PO_4^{3-} + 2\ H^+ + 2\ e^-) \times 3$

$$MnO_4^- \rightarrow MnO_2$$
$$(3\ e^- + 4\ H^+ + MnO_4^- \rightarrow MnO_2 + 2\ H_2O) \times 2$$

$$3\ H_2O + 3\ PO_3^{3-} \rightarrow 3\ PO_4^{3-} + 6\ H^+ + 6\ e^-$$
$$6\ e^- + 8\ H^+ + 2\ MnO_4^- \rightarrow 2\ MnO_2 + 4\ H_2O$$

$$2\ H^+ + 2\ MnO_4^- + 3\ PO_3^{3-} \rightarrow 3\ PO_4^{3-} + 2\ MnO_2 + H_2O$$

Now convert to a basic solution by adding 2 OH⁻ to <u>both</u> sides. $2\ H^+ + 2\ OH^- \rightarrow 2\ H_2O$ on the reactant side. After converting H^+ to OH^-, simplify the overall equation by crossing off 1 H_2O on each side of the reaction. The overall balanced equation is:

$$H_2O(l) + 2\ MnO_4^-(aq) + 3\ PO_3^{3-}(aq) \rightarrow 3\ PO_4^{3-}(aq) + 2\ MnO_2(s) + 2\ OH^-(aq)$$

e.
$$Mg \rightarrow Mg(OH)_2 \qquad\qquad\qquad OCl^- \rightarrow Cl^-$$
$$2\,H_2O + Mg \rightarrow Mg(OH)_2 + 2\,H^+ + 2\,e^- \qquad 2\,e^- + 2\,H^+ + OCl^- \rightarrow Cl^- + H_2O$$

$$2\,H_2O + Mg \rightarrow Mg(OH)_2 + 2\,H^+ + 2\,e^-$$
$$2\,e^- + 2\,H^+ + OCl^- \rightarrow Cl^- + H_2O$$

$$\overline{OCl^-(aq) + H_2O(l) + Mg(s) \rightarrow Mg(OH)_2(s) + Cl^-(aq)}$$

The final overall reaction does not contain H^+, so we are done.

f.
$$H_2CO \rightarrow HCO_3^- \qquad\qquad\qquad Ag(NH_3)_2^+ \rightarrow Ag + 2\,NH_3$$
$$2\,H_2O + H_2CO \rightarrow HCO_3^- + 5\,H^+ + 4\,e^- \qquad (e^- + Ag(NH_3)_2^+ \rightarrow Ag + 2\,NH_3) \times 4$$

$$2\,H_2O + H_2CO \rightarrow HCO_3^- + 5\,H^+ + 4\,e^-$$
$$4\,e^- + 4\,Ag(NH_3)_2^+ \rightarrow 4\,Ag + 8\,NH_3$$

$$\overline{4\,Ag(NH_3)_2^+ + 2\,H_2O + H_2CO \rightarrow HCO_3^- + 5\,H^+ + 4\,Ag + 8\,NH_3}$$

Convert to a basic solution by adding 5 OH^- to both sides (5 H^+ + 5 $OH^- \rightarrow$ 5 H_2O). Then, cross off 2 H_2O on both sides, which gives the overall balanced equation:

$$5\,OH^-(aq) + 4\,Ag(NH_3)_2^+(aq) + H_2CO(aq) \rightarrow HCO_3^-(aq) + 3\,H_2O(l) + 4\,Ag(s) + 8\,NH_3(aq)$$

Questions

17.   In a galvanic cell, a spontaneous redox reaction occurs which produces an electric current. In an electrolytic cell, electricity is used to force a redox reaction that is not spontaneous to occur.

18.   The salt bridge allows counter ions to flow into the two cell compartments to maintain electrical neutrality. Without a salt bridge, no sustained electron flow can occur.

19.   a.   Cathode: The electrode at which reduction occurs.

   b.   Anode: The electrode at which oxidation occurs.

   c.   Oxidation half-reaction: The half-reaction in which electrons are products. In a galvanic cell, the oxidation half-reaction always occurs at the anode.

   d.   Reduction half-reaction: The half-reaction in which electrons are reactants. In a galvanic cell, the reduction half-reaction always occurs at the cathode.

20.   a.   Purification by electrolysis is called electrorefining. See the text for a discussion of the electrorefining of copper. Electrorefining is possible because of the selectivity of electrode reactions. The anode is made up of the impure metal. A potential is applied so just the metal of interest and all more easily oxidized metals are oxidized at the anode. The metal of interest is the only metal plated at the cathode due to the careful control of the potential applied. The metal ions that could plate out at the cathode in preference to the metal we are purifying will not be in solution, because these metals were not oxidized at the anode.

    b.   A more easily oxidized metal is placed in electrical contact with the metal we are trying to protect. It is oxidized in preference to the protected metal. The protected metal becomes the cathode electrode, thus, cathodic protection.

21.   As a battery discharges, E_{cell} decreases, eventually reaching zero. A charged battery is not at equilibrium. At equilibrium, $E_{cell} = 0$ and $\Delta G = 0$. We get no work out of an equilibrium system. A battery is useful to us because it can do work as it approaches equilibrium.

22.   Standard reduction potentials are an intensive property, i.e., they do not depend on how many times the reaction occurs. As long as the concentrations of ions and gases are 1 M or 1 atm, standard reduction potentials and standard oxidation potentials are a constant and not dependent on the coefficients in the balanced equation.

23.   Both fuel cells and batteries are galvanic cells. However, fuel cells, unlike batteries, have the reactants continuously supplied and can produce a current indefinitely.

24.   Moisture must be present to act as a medium for ion flow between the anodic and cathodic regions. Salt provides ions necessary to complete the electrical circuit in the corrosion process. Together, salt and water make up the salt bridge in this spontaneous electrochemical process.

     Three methods discussed in the text to prevent corrosion are galvanizing, alloying and cathodic protection. Galvanizing coats the metal of interest (usually iron) with zinc, which is an easily oxidized metal. Alloying mixes in metals that form durable, effective oxide coatings over the metal of interest. Cathodic protection connects, via a wire, a more easily oxidized metal to the metal we are trying to protect. The more active metal is preferentially oxidized, thus protecting our metal object from corrosion.

Exercises

Galvanic Cells, Cell Potentials, Standard Reduction Potentials, and Free Energy

25.   A typical galvanic cell diagram is:

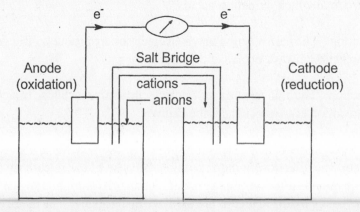

     The diagram for all cells will look like this. The contents of each half-cell compartment will be identified for each reaction, with all solute concentrations at 1.0 M and all gases at 1.0 atm. For Exercises 17.25 and 17.26, the flow of ions through the salt bridge was not asked for in the questions.

If asked, however, cations always flow into the cathode compartment, and anions always flow into the anode compartment. This is required to keep each compartment electrically neutral.

a. Table 17.1 of the text lists balanced reduction half-reactions for many substances. For this overall reaction, we need the Cl_2 to Cl^- reduction half-reaction and the Cr^{3+} to $Cr_2O_7^{2-}$ oxidation half-reaction. Manipulating these two half-reactions gives the overall balanced equation.

$$(Cl_2 + 2\ e^- \rightarrow 2\ Cl^-) \times 3$$
$$\underline{7\ H_2O + 2\ Cr^{3+} \rightarrow Cr_2O_7^{2-} + 14\ H^+ + 6\ e^-}$$
$$7\ H_2O(l) + 2\ Cr^{3+}(aq) + 3\ Cl_2(g) \rightarrow Cr_2O_7^{2-}(aq) + 6\ Cl^-(aq) + 14\ H^+(aq)$$

The contents of each compartment are:

Cathode: Pt electrode; Cl_2 bubbled into solution, Cl^- in solution

Anode: Pt electrode; Cr^{3+}, H^+, and $Cr_2O_7^{2-}$ in solution

We need a nonreactive metal to use as the electrode in each case, since all the reactants and products are in solution. Pt is a common choice. Another possibility is graphite.

b. $\quad\quad$ $Cu^{2+} + 2\ e^- \rightarrow Cu$
$$\underline{Mg \rightarrow Mg^{2+} + 2e^-}$$

$Cu^{2+}(aq) + Mg(s) \rightarrow Cu(s) + Mg^{2+}(aq)$

Cathode: Cu electrode; Cu^{2+} in solution;   Anode: Mg electrode; Mg^{2+} in solution

26. Reference Exercise 17.25 for a typical galvanic cell diagram. The contents of each half-cell compartment are identified below with all solute concentrations at 1.0 M and all gases at 1.0 atm.

a. Reference Table 17.1 for the balanced half-reactions.

$$5\ e^- + 6\ H^+ + IO_3^- \rightarrow 1/2\ I_2 + 3\ H_2O$$
$$\underline{(Fe^{2+} \rightarrow Fe^{3+} + e^-) \times 5}$$

$$6\ H^+ + IO_3^- + 5\ Fe^{2+} \rightarrow 5\ Fe^{3+} + 1/2\ I_2 + 3\ H_2O$$

or $12\ H^+(aq) + 2\ IO_3^-(aq) + 10\ Fe^{2+}(aq) \rightarrow 10\ Fe^{3+}(aq) + I_2(aq) + 6\ H_2O(l)$

Cathode: Pt electrode; IO_3^-, I_2 and H_2SO_4 (H^+ source) in solution.

Note: $I_2(s)$ would make a poor electrode since it sublimes.

Anode: Pt electrode; Fe^{2+} and Fe^{3+} in solution

b.
$$(Ag^+ + e^- \rightarrow Ag) \times 2$$
$$Zn \rightarrow Zn^{2+} + 2\,e^-$$

$$Zn(s) + 2\,Ag^+(aq) \rightarrow 2\,Ag(s) + Zn^{2+}(aq)$$

Cathode: Ag electrode; Ag^+ in solution;   Anode: Zn electrode; Zn^{2+} in solution

27.   To determine $E°$ for the overall cell reaction, we must add the standard reduction potential to the standard oxidation potential ($E°_{cell} = E°_{red} + E°_{ox}$). Reference Table 17.1 for values of standard reduction potentials. Remember that $E°_{ox} = -E°_{red}$ and that standard potentials are not multiplied by the integer used to obtain the overall balanced equation.

25a. $E°_{cell} = E°_{Cl_2 \rightarrow Cl^-} + E°_{Cr^{3+} \rightarrow Cr_2O_7^{2-}} = 1.36\text{ V} + (-1.33\text{ V}) = 0.03\text{ V}$

25b. $E°_{cell} = E°_{Cu^{2+} \rightarrow Cu} + E°_{Mg \rightarrow Mg^{2+}} = 0.34\text{ V} + 2.37\text{ V} = 2.71\text{ V}$

28.   26a. $E°_{cell} = E°_{IO_3^- \rightarrow I_2} + E°_{Fe^{2+} \rightarrow Fe^{3+}} = 1.20\text{ V} + (-0.77\text{ V}) = 0.43\text{ V}$

26b. $E°_{cell} = E°_{Ag^+ \rightarrow Ag} + E°_{Zn \rightarrow Zn^{2+}} = 0.80\text{ V} + 0.76\text{ V} = 1.56\text{ V}$

29.   Reference Exercise 17.25 for a typical galvanic cell design. The contents of each half-cell compartment are identified below with all solute concentrations at 1.0 M and all gases at 1.0 atm. For each pair of half-reactions, the half-reaction with the largest standard reduction potential will be the cathode reaction, and the half-reaction with the smallest reduction potential will be reversed to become the anode reaction. Only this combination gives a spontaneous overall reaction, i.e., a reaction with a positive overall standard cell potential. Note that in a galvanic cell as illustrated in Exercise 17.25, the cations in the salt bridge migrate to the cathode, and the anions migrate to the anode.

a.
$$Cl_2 + 2\,e^- \rightarrow 2\,Cl^- \qquad E° = 1.36\text{ V}$$
$$2\,Br^- \rightarrow Br_2 + 2\,e^- \qquad -E° = -1.09\text{ V}$$

$$Cl_2(g) + 2\,Br^-(aq) \rightarrow Br_2(aq) + 2\,Cl^-(aq) \qquad E°_{cell} = 0.27\text{ V}$$

The contents of each compartment are:

Cathode: Pt electrode; $Cl_2(g)$ bubbled in, Cl^- in solution

Anode: Pt electrode; Br_2 and Br^- in solution

b.
$$(2\,e^- + 2\,H^+ + IO_4^- \rightarrow IO_3^- + H_2O) \times 5 \qquad E° = 1.60\text{ V}$$
$$(4\,H_2O + Mn^{2+} \rightarrow MnO_4^- + 8\,H^+ + 5\,e^-) \times 2 \qquad -E° = -1.51\text{ V}$$

$$10\,H^+ + 5\,IO_4^- + 8\,H_2O + 2\,Mn^{2+} \rightarrow 5\,IO_3^- + 5\,H_2O + 2\,MnO_4^- + 16\,H^+ \quad E°_{cell} = 0.09\text{ V}$$

This simplifies to:

$$3 H_2O(l) + 5 IO_4^-(aq) + 2 Mn^{2+}(aq) \rightarrow 5 IO_3^-(aq) + 2 MnO_4^-(aq) + 6 H^+(aq) \quad E^\circ_{cell} = 0.09 \text{ V}$$

Cathode: Pt electrode; IO_4^-, IO_3^-, and H_2SO_4 (as a source of H^+) in solution

Anode: Pt electrode; Mn^{2+}, MnO_4^- and H_2SO_4 in solution

30.   Reference Exercise 17.25 for a typical galvanic cell design. The contents of each half-cell compartment are identified below, with all solute concentrations at 1.0 M and all gases at 1.0 atm.

a. $H_2O_2 + 2 H^+ + 2 e^- \rightarrow 2 H_2O$ $E^\circ = 1.78$ V

           $H_2O_2 \rightarrow O_2 + 2 H^+ + 2 e^-$ $-E^\circ = -0.68$ V

           $2 H_2O_2(aq) \rightarrow 2 H_2O(l) + O_2(g)$ $E^\circ_{cell} = 1.10$ V

Cathode: Pt electrode; H_2O_2 and H^+ in solution

Anode: Pt electrode; $O_2(g)$ bubbled in, H_2O_2 and H^+ in solution

b. $(Fe^{3+} + 3 e^- \rightarrow Fe) \times 2$ $E^\circ = -0.036$ V

           $(Mn \rightarrow Mn^{2+} + 2 e^-) \times 3$ $-E^\circ = 1.18$ V

           $2 Fe^{3+}(aq) + 3 Mn(s) \rightarrow 2 Fe(s) + 3 Mn^{2+}(aq)$ $E^\circ_{cell} = 1.14$ V

Cathode: Fe electrode; Fe^{3+} in solution;   Anode: Mn electrode; Mn^{2+} in solution

31.   In standard line notation, the anode is listed first and the cathode is listed last. A double line separates the two compartments. By convention, the electrodes are on the ends with all solutes and gases towards the middle. A single line is used to indicate a phase change. We also include all concentrations.

25a. $Pt \mid Cr^{3+}$ (1.0 M), $Cr_2O_7^{2-}$ (1.0 M), H^+ (1.0 M) $\mid\mid Cl_2$ (1.0 atm)$\mid Cl^-$ (1.0 M) $\mid$ Pt

25b. $Mg \mid Mg^{2+}$ (1.0 M) $\mid\mid Cu^{2+}$ (1.0 M) $\mid$ Cu

29a. $Pt \mid Br^-$ (1.0 M), Br_2 (1.0 M) $\mid\mid Cl_2$ (1.0 atm) $\mid Cl^-$ (1.0 M) $\mid$ Pt

29b. $Pt \mid Mn^{2+}$ (1.0 M), MnO_4^- (1.0 M), H^+ (1.0 M) $\mid\mid IO_4^-$ (1.0 M), H^+ (1.0 M), IO_3^- (1.0 M) $\mid$ Pt

32.   26a. $Pt \mid Fe^{2+}$ (1.0 M), Fe^{3+} (1.0 M) $\mid\mid IO_3^-$ (1.0 M), H^+ (1.0 M), I_2 (1.0 M)$\mid$ Pt

26b. $Zn \mid Zn^{2+}$ (1.0 M) $\mid\mid Ag^+$ (1.0 M) $\mid$ Ag

30a. $Pt \mid H_2O_2$ (1.0 M), H^+ (1.0 M) $\mid O_2$ (1.0 atm)$\mid\mid H_2O_2$ (1.0 M), H^+ (1.0 M)$\mid$Pt

30b. $Mn \mid Mn^{2+}$ (1.0 M) $\mid\mid Fe^{3+}$ (1.0 M) $\mid$ Fe

33.   Locate the pertinent half-reactions in Table 17.1, and then figure which combination will give a
       positive standard cell potential. In all cases, the anode compartment contains the species with the
       smallest standard reduction potential. For part a, the copper compartment is the anode, and in part
       b, the cadmium compartment is the anode.

   a. $Au^{3+} + 3\ e^- \rightarrow Au$ $E° = 1.50\ V$
                       $(Cu^+ \rightarrow Cu^{2+} + e^-) \times 3$ $-E° = -0.16\ V$
       _____

                $Au^{3+}(aq) + 3\ Cu^+(aq) \rightarrow Au(s) + 3\ Cu^{2+}(aq)$ $E°_{cell} = 1.34\ V$

   b. $(VO_2^+ + 2\ H^+ + e^- \rightarrow VO^{2+} + H_2O) \times 2$ $E° = 1.00\ V$
                       $Cd \rightarrow Cd^{2+} + 2e^-$ $-E° = 0.40\ V$
       _____

       $2\ VO_2^+(aq) + 4\ H^+(aq) + Cd(s) \rightarrow 2\ VO^{2+}(aq) + 2\ H_2O(l) + Cd^{2+}(aq)$ $E°_{cell} = 1.40\ V$

34. a. $(H_2O_2 + 2\ H^+ + 2\ e^- \rightarrow 2\ H_2O) \times 3$ $E° = 1.78\ V$
                   $2\ Cr^{3+} + 7\ H_2O \rightarrow Cr_2O_7^{2-} + 14\ H^+ + 6\ e^-$ $-E° = -1.33\ V$
       _____

       $3\ H_2O_2(aq) + 2\ Cr^{3+}(aq) + H_2O(l) \rightarrow Cr_2O_7^{2-}(aq) + 8\ H^+(aq)$ $E°_{cell} = 0.45\ V$

   b. $(2\ H^+ + 2\ e^- \rightarrow H_2) \times 3$ $E° = 0.00\ V$
                    $(Al \rightarrow Al^{3+} + 3\ e^-) \times 2$ $-E° = 1.66\ V$
       _____

       $6\ H^+(aq) + 2\ Al(s) \rightarrow 3\ H_2(g) + 2\ Al^{3+}(aq)$ $E°_{cell} = 1.66\ V$

35. a. $2\ Ag^+ + 2\ e^- \rightarrow 2\ Ag$ $E° = 0.80\ V$
          $Cu \rightarrow Cu^{2+} + 2\ e^-$ $-E° = -0.34\ V$
       _____

       $2\ Ag^+ + Cu \rightarrow Cu^{2+} + 2\ Ag$ $E°_{cell} = 0.46\ V$ Spontaneous at standard conditions
                                                                            $(E°_{cell} > 0)$.

   b. $Zn^{2+} + 2\ e^- \rightarrow Zn$ $E° = -0.76\ V$
        $Ni \rightarrow Ni^{2+} + 2\ e^-$ $-E° = 0.23\ V$
       _____

       $Zn^{2+} + Ni \rightarrow Zn + Ni^{2+}$ $E°_{cell} = -0.53\ V$ Not spontaneous at standard conditions
                                                                             $(E°_{cell} < 0)$.

36. a. $(5\ e^- + 8\ H^+ + MnO_4^- \rightarrow Mn^{2+} + 4\ H_2O) \times 2$ $E° = 1.51\ V$
          $(2\ I^- \rightarrow I_2 + 2\ e^-) \times 5$ $-E° = -0.54\ V$
       _____

       $16\ H^+ + 2\ MnO_4^- + 10\ I^- \rightarrow 5\ I_2 + 2\ Mn^{2+} + 8\ H_2O$ $E°_{cell} = 0.97\ V$ Spontaneous

   b. $(5\ e^- + 8\ H^+ + MnO_4^- \rightarrow Mn^{2+} + 4\ H_2O) \times 2$ $E° = 1.51\ V$
        $(2\ F^- \rightarrow F_2 + 2\ e^-) \times 5$ $-E° = -2.87\ V$
       _____

       $16\ H^+ + 2\ MnO_4^- + 10\ F^- \rightarrow 5\ F_2 + 2\ Mn^{2+} + 8\ H_2O$ $E°_{cell} = -1.36\ V$ Not spontaneous

37.
$$Cl_2 + 2\,e^- \rightarrow 2\,Cl^- \qquad\qquad E° = 1.36\ V$$
$$(ClO_2^- \rightarrow ClO_2 + e^-) \times 2 \qquad -E° = -0.954\ V$$

$$2\ ClO_2^-(aq) + Cl_2(g) \rightarrow 2\ ClO_2(aq) + 2\ Cl^-(aq) \qquad E°_{cell} = 0.41\ V = 0.41\ J/C$$

$$\Delta G° = -nFE°_{cell} = -(2\ mol\ e^-)(96{,}485\ C/mol\ e^-)(0.41\ J/C) = -7.9 \times 10^4\ J = -79\ kJ$$

38.   a.
$$(4\ H^+ + NO_3^- + 3\ e^- \rightarrow NO + 2\ H_2O) \times 2 \qquad\qquad E° = 0.96\ V$$
$$(Mn \rightarrow Mn^{2+} + 2\ e^-) \times 3 \qquad\qquad -E° = 1.18\ V$$

$$3\ Mn(s) + 8\ H^+(aq) + 2\ NO_3^-(aq) \rightarrow 2\ NO(g) + 4\ H_2O(l) + 3\ Mn^{2+}(aq) \qquad E°_{cell} = 2.14\ V$$

$$(2\ e^- + 2\ H^+ + IO_4^- \rightarrow IO_3^- + H_2O) \times 5 \qquad\qquad E° = 1.60\ V$$
$$(Mn^{2+} + 4\ H_2O \rightarrow MnO_4^- + 8\ H^+ + 5\ e^-) \times 2 \qquad\qquad -E° = -1.51\ V$$

$$5\ IO_4^-(aq) + 2\ Mn^{2+}(aq) + 3\ H_2O(l) \rightarrow 5\ IO_3^-(aq) + 2\ MnO_4^-(aq) + 6\ H^+(aq) \quad E°_{cell} = 0.09\ V$$

b. Nitric acid oxidation (see above for $E°_{cell}$):

$$\Delta G° = -nFE°_{cell} = -(6\ mol\ e^-)(96{,}485\ C/mol\ e^-)(2.14\ J/C) = -1.24 \times 10^6\ J = -1240\ kJ$$

Periodate oxidation (see above for $E°_{cell}$):

$$\Delta G° = -(10\ mol\ e^-)(96{,}485\ C/mol\ e^-)(0.09\ J/C)(1\ kJ/1000\ J) = -90\ kJ$$

39. Since the cells are at standard conditions, $w_{max} = \Delta G = \Delta G° = -nFE°_{cell}$. See Exercise 17.33 for the balanced overall equations and for $E°_{cell}$.

33a. $w_{max} = -(3\ mol\ e^-)(96{,}485\ C/mol\ e^-)(1.34\ J/C) = -3.88 \times 10^5\ J = -388\ kJ$

33b. $w_{max} = -(2\ mol\ e^-)(96{,}485\ C/mol\ e^-)(1.40\ J/C) = -2.70 \times 10^5\ J = -270.\ kJ$

40. Since the cells are at standard conditions, $w_{max} = \Delta G = \Delta G° = -nFE°_{cell}$. See Exercise 17.34 for the balanced overall equations and for $E°_{cell}$.

34a. $w_{max} = -(6\ mol\ e^-)(96{,}485\ C/mol\ e^-)(0.45\ J/C) = -2.6 \times 10^5\ J = -260\ kJ$

34b. $w_{max} = -(6\ mol\ e^-)(96{,}485\ C/mol\ e^-)(1.66\ J/C) = -9.61 \times 10^5\ J = -961\ kJ$

41. $CH_3OH(l) + 3/2\ O_2(g) \rightarrow CO_2(g) + 2\ H_2O(l) \qquad \Delta G° = 2(-237) + (-394) - [-166] = -702\ kJ$

The balanced half-reactions are:

$$H_2O + CH_3OH \rightarrow CO_2 + 6\ H^+ + 6\ e^- \quad and \quad O_2 + 4\ H^+ + 4\ e^- \rightarrow 2\ H_2O$$

For $3/2$ mol O_2, 6 moles of electrons will be transferred (n = 6).

$$\Delta G° = -nFE°, \quad E° = \frac{-\Delta G°}{nF} = \frac{-(-702{,}000\ J)}{(6\ mol\ e^-)\,(96{,}485\ C/mol\ e^-)} = 1.21\ J/C = 1.21\ V$$

42. $Fe^{2+} + 2 e^- \rightarrow Fe$ $E° = -0.44$ V $= -0.44$ J/C

$\Delta G° = -nFE° = -(2 \text{ mol e}^-)(96,485 \text{ C/mol e}^-)(-0.44 \text{ J/C})(1 \text{ kJ}/1000 \text{ J}) = 85$ kJ

$85 \text{ kJ} = 0 - [\Delta G°_{f\,Fe^{2+}} + 0]$, $\Delta G°_{f\,Fe^{2+}} = -85$ kJ

We can get $\Delta G°_{f\,Fe^{3+}}$ two ways. Consider: $Fe^{3+} + e^- \rightarrow Fe^{2+}$ $E° = 0.77$ V

$\Delta G° = -(1 \text{ mol e})(96,485 \text{ C/mol e}^-)(0.77 \text{ J/C}) = -74,300 \text{ J} = -74$ kJ

$Fe^{2+} \rightarrow Fe^{3+} + e^-$	$\Delta G° = 74$ kJ
$Fe \rightarrow Fe^{2+} + 2 e^-$	$\Delta G° = -85$ kJ

$Fe \rightarrow Fe^{3+} + 3 e^-$ $\Delta G° = -11$ kJ, $\Delta G°_{f\,Fe^{3+}} = -11$ kJ/mol

or consider: $Fe^{3+} + 3 e^- \rightarrow Fe$ $E° = -0.036$ V

$\Delta G° = -(3 \text{ mol e}^-)(96,485 \text{ C/mol e}^-)(-0.036 \text{ J/C}) = 10,400 \text{ J} \approx 10.$ kJ

10. kJ $= 0 - [\Delta G°_{f\,Fe^{3+}} + 0]$, $\Delta G°_{f\,Fe^{3+}} = -10.$ kJ/mol; Round-off error explains the 1 kJ discrepancy.

43.   Good oxidizing agents are easily reduced. Oxidizing agents are on the left side of the reduction half-reactions listed in Table 17.1. We look for the largest, most positive standard reduction potentials to correspond to the best oxidizing agents. The ordering from worst to best oxidizing agents is:

	K^+	<	H_2O	<	Cd^{2+}	<	I_2	<	$AuCl_4^-$	<	IO_3^-
$E°(V)$	-2.92		-0.83		-0.40		0.54		0.99		1.20

44.   Good reducing agents are easily oxidized. The reducing agents are on the right side of the reduction half-reactions listed in Table 17.1. The best reducing agents have the most negative standard reduction potentials ($E°$) or the most positive standard oxidation potentials, $E°_{ox}$ $(= -E°)$.

	F^-	<	Cr^{3+}	<	Fe^{2+}	<	H_2	<	Zn	<	Li
$-E°(V)$	-2.87		-1.33		-0.77		0.00		0.76		3.05

45.   a. $2 H^+ + 2 e^- \rightarrow H_2$ $E° = 0.00$ V; $Cu \rightarrow Cu^{2+} + 2 e^-$ $-E° = -0.34$ V

$E°_{cell} = -0.34$ V; No, H^+ cannot oxidize Cu to Cu^{2+} at standard conditions ($E°_{cell} < 0$).

b. $Fe^{3+} + e^- \rightarrow Fe^{2+}$ $E° = 0.77$ V; $2 I^- \rightarrow I_2 + 2 e^-$ $-E° = -0.54$ V

$E°_{cell} = 0.77 - 0.54 = 0.23$ V; Yes, Fe^{3+} can oxidize I^- to I_2.

c. $H_2 \rightarrow 2 H^+ + 2 e^-$ $-E° = 0.00$ V; $Ag^+ + e^- \rightarrow Ag$ $E° = 0.80$ V

$E°_{cell} = 0.80$ V; Yes, H_2 can reduce Ag^+ to Ag at standard conditions ($E°_{cell} > 0$).

d. $Fe^{2+} \rightarrow Fe^{3+} + e^-$ $-E° = -0.77$ V; $Cr^{3+} + e^- \rightarrow Cr^{2+}$ $E° = -0.50$ V

$E°_{cell} = -0.50 - 0.77 = -1.27$ V; No, Fe^{2+} cannot reduce Cr^{3+} to Cr^{2+} at standard conditions.

46. $Cl_2 + 2 e^- \rightarrow 2 Cl^-$ $E° = 1.36$ V $Ag^+ + e^- \rightarrow Ag$ $E° = 0.80$ V
     $Pb^{2+} + 2 e^- \rightarrow Pb$ $E° = -0.13$ V $Zn^{2+} + 2 e^- \rightarrow Zn$ $E° = -0.76$ V
     $Na^+ + e^- \rightarrow Na$ $E° = -2.71$ V

a. Oxidizing agents (species reduced) are on the left side of the above reduction half-reactions. Of the species available, Ag^+ would be the best oxidizing agent since it has the largest $E°$ value. Note that Cl_2 is a better oxidizing agent than Ag^+, but it is not one of the choices listed.

b. Reducing agents (species oxidized) are on the right side of the reduction half-reactions. Of the species available, Zn would be the best reducing agent since it has the largest $-E°$ value.

c. $SO_4^{2-} + 4 H^+ + 2 e^- \rightarrow H_2SO_3 + H_2O$ $E° = 0.20$ V; SO_4^{2-} can oxidize Pb and Zn at standard conditions. When SO_4^{2-} is coupled with these reagents, $E°_{cell}$ is positive.

d. $Al \rightarrow Al^{3+} + 3 e^-$ $-E° = 1.66$ V; Al can oxidize Ag^+ and Zn^{2+} at standard conditions since $E°_{cell} > 0$.

47. a. $2 Br^- \rightarrow Br_2 + 2 e^-$ $-E° = -1.09$ V; $2 Cl^- \rightarrow Cl_2 + 2 e^-$ $-E° = -1.36$ V; $E° > 1.09$ V to oxidize Br^-; $E° < 1.36$ V to not oxidize Cl^-; $Cr_2O_7^{2-}$, O_2, MnO_2, and IO_3^- are all possible since when all of these oxidizing agents are coupled with Br^-, $E°_{cell} > 0$, and when coupled with Cl^-, $E°_{cell} < 0$ (assuming standard conditions).

b. $Mn \rightarrow Mn^{2+} + 2 e^-$ $-E° = 1.18$; $Ni \rightarrow Ni^{2+} + 2 e^-$ $-E° = 0.23$ V; Any oxidizing agent with -0.23 V $> E° > -1.18$ V will work. $PbSO_4$, Cd^{2+}, Fe^{2+}, Cr^{3+}, Zn^{2+} and H_2O will be able to oxidize Mn but not Ni (assuming standard conditions).

48. a. $Cu^{2+} + 2 e^- \rightarrow Cu$ $E° = 0.34$ V; $Cu^{2+} + e^- \rightarrow Cu^+$ $E° = 0.16$ V; To reduce Cu^{2+} to Cu but not reduce Cu^{2+} to Cu^+, the reducing agent must have a standard oxidation potential ($E°_{ox} = -E°$) between -0.34 V and -0.16 V (so $E°_{cell}$ is positive only for the Cu^{2+} to Cu reduction). The reducing agents (species oxidized) are on the right side of the half-reactions in Table 17.1. The reagents at standard conditions which have $E°_{ox}$ ($= -E°$) between -0.34 V and -0.16 V are Ag (in 1.0 M Cl^-) and H_2SO_3.

b. $Br_2 + 2 e^- \rightarrow 2 Br^-$ $E° = 1.09$ V; $I_2 + 2 e^- \rightarrow 2 I^-$ $E° = 0.54$ V; From Table 17.1, VO^{2+}, Au (in 1.0 M Cl^-), NO, ClO_2^-, Hg_2^{2+}, Ag, Hg, Fe^{2+}, H_2O_2 and MnO_4^- are all capable at standard conditions of reducing Br_2 to Br^- but not reducing I_2 to I^-. When these reagents are coupled with Br_2, $E°_{cell} > 0$, and when coupled with I_2, $E°_{cell} < 0$.

49. $ClO^- + H_2O + 2 e^- \rightarrow 2 OH^- + Cl^-$ $E° = 0.90$ V
         $2 NH_3 + 2 OH^- \rightarrow N_2H_4 + 2 H_2O + 2 e^-$ $-E° = 0.10$ V
         _____
         $ClO^-(aq) + 2 NH_3(aq) \rightarrow Cl^-(aq) + N_2H_4(aq) + H_2O(l)$ $E°_{cell} = 1.00$ V

Since E°_{cell} is positive for this reaction, then at standard conditions ClO^- can spontaneously oxidize NH_3 to the somewhat toxic N_2H_4.

50. $$Tl^{3+} + 2\ e^- \rightarrow Tl^+ \qquad\qquad E^\circ = 1.25\ V$$
$$3\ I^- \rightarrow I_3^- + 2\ e^- \qquad\qquad -E^\circ = -0.55\ V$$

$$Tl^{3+} + 3\ I^- \rightarrow Tl^+ + I_3^- \qquad E^\circ_{cell} = 0.70\ V$$

In solution, Tl^{3+} can oxidize I^- to I_3^-. Thus, we expect TlI_3 to be thallium(I) triiodide.

The Nernst Equation

51. $$H_2O_2 + 2\ H^+ + 2\ e^- \rightarrow 2\ H_2O \qquad\qquad E^\circ = 1.78\ V$$
$$(Ag \rightarrow Ag^+ + e^-) \times 2 \qquad\qquad -E^\circ = -0.80\ V$$

$$H_2O_2(aq) + 2\ H^+(aq) + 2\ Ag(s) \rightarrow 2\ H_2O(l) + 2\ Ag^+(aq) \qquad E^\circ_{cell} = 0.98\ V$$

a. A galvanic cell is based on spontaneous chemical reactions. At standard conditions, this reaction produces a voltage of 0.98 V. Any change in concentration that increases the tendency of the forward reaction to occur will increase the cell potential. Conversely, any change in concentration that decreases the tendency of the forward reaction to occur (increases the tendency of the reverse reaction to occur) will decrease the cell potential. Using Le Chatelier's principle, increasing the reactant concentrations of H_2O_2 and H^+ from 1.0 M to 2.0 M will drive the forward reaction further to right (will further increase the tendency of the forward reaction to occur). Therefore, E_{cell} will be greater than E°_{cell}.

b. Here, we decreased the reactant concentration of H^+ and increased the product concentration of Ag^+ from the standard conditions. This decreases the tendency of the forward reaction to occur which will decrease E_{cell} as compared to E°_{cell} ($E_{cell} < E^\circ_{cell}$).

52.   The concentrations of Fe^{2+} in the two compartments are now 0.01 M and $1 \times 10^{-7}\ M$. The driving force for this reaction is to equalize the Fe^{2+} concentrations in the two compartments. This occurs if the compartment with $1 \times 10^{-7}\ M\ Fe^{2+}$ becomes the anode (Fe will be oxidized to Fe^{2+}) and the compartment with the 0.01 $M\ Fe^{2+}$ becomes the cathode (Fe^{2+} will be reduced to Fe). Electron flow, as always for galvanic cells, goes from the anode to the cathode, so electron flow will go from the right compartment ($[Fe^{2+}] = 1 \times 10^{-7}\ M$) to the left compartment ($[Fe^{2+}] = 0.01\ M$).

53.   For concentration cells, the driving force for the reaction is the difference in ion concentrations between the anode and cathode. In order to equalize the ion concentrations, the anode always has the smaller ion concentration. The general setup for this concentration cell is:

Cathode: $Ag^+(x\ M) + e^- \rightarrow Ag \qquad\qquad E^\circ = 0.80\ V$
Anode: $Ag \rightarrow Ag^+ (y\ M) + e^- \qquad\qquad -E^\circ = -0.80\ V$

$$Ag^+(\text{cathode},\ x\ M) \rightarrow Ag^+ (\text{anode},\ y\ M) \qquad E^\circ_{cell} = 0.00\ V$$

$$E_{cell} = E^\circ_{cell} - \frac{0.0591}{n} \log Q = \frac{-0.0591}{1} \log \frac{[Ag^+]_{anode}}{[Ag^+]_{cathode}}$$

For each concentration cell, we will calculate the cell potential using the above equation. Remember that the anode always has the smaller ion concentration.

a. Since both compartments are at standard conditions ($[Ag^+] = 1.0\,M$) then $E_{cell} = E^\circ_{cell} = 0\,V$. No voltage is produced since no reaction occurs. Concentration cells only produce a voltage when the ion concentrations are <u>not</u> equal.

b. Cathode = $2.0\,M\,Ag^+$; Anode = $1.0\,M\,Ag^+$; Electron flow is always from the anode to the cathode, so electrons flow to the right in the diagram.

$$E_{cell} = \frac{-0.0591}{n} \log \frac{[Ag^+]_{anode}}{[Ag^+]_{cathode}} = \frac{-0.0591}{1} \log \frac{1.0}{2.0} = 0.018\,V$$

c. Cathode = $1.0\,M\,Ag^+$; Anode = $0.10\,M\,Ag^+$; Electrons flow to the left in the diagram.

$$E_{cell} = \frac{-0.0591}{n} \log \frac{[Ag^+]_{anode}}{[Ag^+]_{cathode}} = \frac{-0.0591}{1} \log \frac{0.10}{1.0} = 0.059\,V$$

d. Cathode = $1.0\,M\,Ag^+$; Anode = $4.0 \times 10^{-5}\,M\,Ag^+$; Electrons flow to the left in the diagram.

$$E_{cell} = \frac{-0.0591}{1} \log \frac{4.0 \times 10^{-5}}{1.0} = 0.26\,V$$

e. Since the ion concentrations are the same, then $\log ([Ag^+]_{anode}/[Ag^+]_{cathode}) = \log (1.0) = 0$ and $E_{cell} = 0$. No electron flow occurs.

54.   As is the case for all concentration cells, $E^\circ_{cell} = 0$, and the smaller ion concentration is always in the anode compartment. The general Nernst equation for the $Ni \mid Ni^{2+} (x\,M) \mid\mid Ni^{2+}(y\,M) \mid Ni$ concentration cell is:

$$E_{cell} = E^\circ_{cell} - \frac{0.0591}{n} \log Q = \frac{-0.0591}{2} \log \frac{[Ni^{2+}]_{anode}}{[Ni^{2+}]_{cathode}}$$

a. Since both compartments are at standard conditions ($[Ni^{2+}] = 1.0\,M$), then $E_{cell} = E^\circ_{cell} = 0\,V$. No electron flow occurs.

b. Cathode = $2.0\,M\,Ni^{2+}$; Anode = $1.0\,M\,Ni^{2+}$; Electron flow is always from the anode to the cathode, so electrons flow to the right in the diagram.

$$E_{cell} = \frac{-0.0591}{2} \log \frac{[Ni^{2+}]_{anode}}{[Ni^{2+}]_{cathode}} = \frac{-0.0591}{2} \log \frac{1.0}{2.0} = 8.9 \times 10^{-3}\,V$$

c. Cathode = $1.0\,M\,Ni^{2+}$; Anode = $0.10\,M\,Ni^{2+}$; Electrons flow to the left in the diagram.

$$E_{cell} = \frac{-0.0591}{2} \log \frac{0.10}{1.0} = 0.030\,V$$

d. Cathode = $1.0\,M\,Ni^{2+}$; Anode = $4.0 \times 10^{-5}\,M\,Ni^{2+}$; Electrons flow to the left in the diagram.

$$E_{cell} = \frac{-0.0591}{2} \log \frac{4.0 \times 10^{-5}}{1.0} = 0.13\,V$$

e.   Since both concentrations are equal, $\log (2.5/2.5) = \log 1.0 = 0$ and $E_{cell} = 0$. No electron flow
     occurs.

55.
$$5\ e^- + 8\ H^+ + MnO_4^- \rightarrow Mn^{2+} + 4\ H_2O \qquad\qquad E° = 1.51\ V$$
$$(Fe^{2+} \rightarrow Fe^{3+} + e^-) \times 5 \qquad\qquad -E° = -0.77\ V$$

$$8\ H^+(aq) + MnO_4^-(aq) + 5\ Fe^{2+}(aq) \rightarrow 5\ Fe^{3+}(aq) + Mn^{2+}(aq) + 4\ H_2O(l) \qquad E°_{cell} = 0.74\ V$$

$$E_{cell} = E°_{cell} - \frac{0.0591}{n} \log Q = 0.74\ V - \frac{0.0591}{5} \log \frac{[Fe^{3+}]^5\ [Mn^{2+}]}{[Fe^{2+}]^5\ [MnO_4^-]\ [H^+]^8}; \quad \text{pH} = 4.0 \text{ so } H^+ = 1 \times 10^{-4}\ M$$

$$E_{cell} = 0.74 - \frac{0.0591}{5} \log \frac{(1 \times 10^{-6})^5\ (1 \times 10^{-6})}{(1 \times 10^{-3})^5\ (1 \times 10^{-2})\ (1 \times 10^{-4})^8}$$

$$E_{cell} = 0.74 - \frac{0.0591}{5} \log (1 \times 10^{13}) = 0.74\ V - 0.15\ V = 0.59\ V = 0.6\ V \text{ (1 sig. fig. due to concentrations)}$$

Yes, $E_{cell} > 0$ so the reaction will occur as written.

56. $n = 2$ for this reaction (lead goes from $Pb \rightarrow Pb^{2+}$ in $PbSO_4$).

$$E = E° - \frac{0.0591}{2} \log \frac{1}{[H^+]^2 [HSO_4^-]^2} = 2.04\ V - \frac{0.0591}{2} \log \frac{1}{(4.5)^2\ (4.5)^2}$$

$$E = 2.04\ V + 0.077\ V = 2.12\ V$$

57.
$$Cu^{2+} + 2\ e^- \rightarrow Cu \qquad\qquad E° = 0.34\ V$$
$$Zn \rightarrow Zn^{2+} + 2\ e^- \qquad\qquad -E° = 0.76\ V$$

$$Cu^{2+}(aq) + Zn(s) \rightarrow Zn^{2+}(aq) + Cu(s) \qquad E°_{cell} = 1.10\ V$$

Since Zn^{2+} is a product in the reaction, the Zn^{2+} concentration increases from $1.00\ M$ to $1.20\ M$. This
means that the reactant concentration of Cu^{2+} must decrease from $1.00\ M$ to $0.80\ M$ (from the 1:1
mol ratio in the balanced reaction).

$$E_{cell} = E°_{cell} - \frac{0.0591}{n} \log Q = 1.10\ V - \frac{0.0591}{2} \log \frac{[Zn^{2+}]}{[Cu^{2+}]}$$

$$E_{cell} = 1.10\ V - \frac{0.0591}{2} \log \frac{1.20}{0.80} = 1.10\ V - 0.0052\ V = 1.09\ V$$

58.
$$(Pb^{2+} + 2\ e^- \rightarrow Pb) \times 3 \qquad\qquad E° = -0.13\ V$$
$$(Al \rightarrow Al^{3+} + 3\ e^-) \times 2 \qquad\qquad -E° = 1.66\ V$$

$$3\ Pb^{2+}(aq) + 2\ Al(s) \rightarrow 3\ Pb(s) + 2\ Al^{3+}(aq) \qquad E°_{cell} = 1.53\ V$$

From the balanced reaction, when the Al^{3+} has increased by 0.60 mol/L (Al^{3+} is a product in the
spontaneous reaction), then the Pb^{2+} concentration has decreased by 3/2 (0.60 mol/L) = 0.90 M.

$$E_{cell} = 1.53 \text{ V} - \frac{0.0591}{6} \log \frac{[Al^{3+}]^2}{[Pb^{2+}]^3} = 1.53 - \frac{0.0591}{6} \log \frac{(1.60)^2}{(0.10)^3}$$

$$E_{cell} = 1.53 \text{ V} - 0.034 \text{ V} = 1.50 \text{ V}$$

59. $Cu^{2+}(aq) + H_2(g) \rightarrow 2 H^+(aq) + Cu(s)$ $E^\circ_{cell} = 0.34 \text{ V} - 0.00\text{V} = 0.34 \text{ V}$; n = 2 mol electrons

Since $P_{H_2} = 1.0$ atm and $[H^+] = 1.0\ M$: $E_{cell} = E^\circ_{cell} - \frac{0.0591}{2} \log \frac{1}{[Cu^{2+}]}$

a. $E_{cell} = 0.34 \text{ V} - \frac{0.0591}{2} \log \frac{1}{2.5 \times 10^{-4}} = 0.34 \text{ V} - 0.11\text{V} = 0.23 \text{ V}$

b. $0.195 \text{ V} = 0.34 \text{ V} - \frac{0.0591}{2} \log \frac{1}{[Cu^{2+}]}$, $\log \frac{1}{[Cu^{2+}]} = 4.91$, $[Cu^{2+}] = 10^{-4.91} = 1.2 \times 10^{-5}\ M$

Note: When determining exponents, we will carry extra significant figures.

60. $3 Ni^{2+}(aq) + 2 Al(s) \rightarrow 2 Al^{3+}(aq) + 3 Ni(s)$ $E^\circ_{cell} = -0.23 + 1.66 = 1.43 \text{ V}$; n = 6 mol electrons

a. $E_{cell} = 1.43 \text{ V} - \frac{0.0591}{6} \log \frac{[Al^{3+}]^2}{[Ni^{2+}]^3} = 1.43 - \frac{0.0591}{6} \log \frac{(7.2 \times 10^{-3})^2}{(1.0)^3}$

$E_{cell} = 1.43 \text{ V} - (-0.042 \text{ V}) = 1.47 \text{ V}$

b. $1.62 \text{ V} = 1.43 \text{ V} - \frac{0.0591}{6} \log \frac{[Al^{3+}]^2}{(1.0)^3}$, $\log [Al^{3+}]^2 = -19.29$

$[Al^{3+}]^2 = 10^{-19.29}$, $[Al^{3+}] = 2.3 \times 10^{-10}\ M$

61. $Cu^{2+}(aq) + H_2(g) \rightarrow 2 H^+(aq) + Cu(s)$ $E^\circ_{cell} = 0.34 \text{ V} - 0.00 \text{ V} = 0.34 \text{ V}$; n = 2

Since $P_{H_2} = 1.0$ atm and $[H^+] = 1.0\ M$: $E_{cell} = E^\circ_{cell} - \frac{0.0591}{2} \log \frac{1}{[Cu^{2+}]}$

Use the K_{sp} expression to calculate the Cu^{2+} concentration in the cell.

$Cu(OH)_2(s) \rightleftharpoons Cu^{2+}(aq) + 2 OH^-(aq)$ $K_{sp} = 1.6 \times 10^{-19} = [Cu^{2+}][OH^-]^2$

From problem, $[OH^-] = 0.10\ M$, so: $[Cu^{2+}] = \frac{1.6 \times 10^{-19}}{(0.10)^2} = 1.6 \times 10^{-17}\ M$

$E_{cell} = E^\circ_{cell} - \frac{0.0591}{2} \log \frac{1}{[Cu^{2+}]} = 0.34 \text{ V} - \frac{0.0591}{2} \log \frac{1}{1.6 \times 10^{-17}} = 0.34 - 0.50 = -0.16 \text{ V}$

Since $E_{cell} < 0$, the forward reaction is not spontaneous, but the reverse reaction is spontaneous. The Cu electrode becomes the anode and $E_{cell} = 0.16$ V for the reverse reaction. The cell reaction is:
$2 H^+(aq) + Cu(s) \rightarrow Cu^{2+}(aq) + H_2(g)$.

62. $3\ Ni^{2+}(aq) + 2\ Al(s) \rightarrow 2\ Al^{3+}(aq) + 3\ Ni(s)$ $E^{\circ}_{cell} = -0.23\ V + 1.66\ V = 1.43\ V;\ n = 6$

$$E_{cell} = E^{\circ}_{cell} - \frac{0.0591}{n} \log \frac{[Al^{3+}]^2}{[Ni^{2+}]^3},\quad 1.82\ V = 1.43\ V - \frac{0.0591}{6} \log \frac{[Al^{3+}]^2}{(1.0)^3}$$

$\log [Al^{3+}]^2 = -39.59,\quad [Al^{3+}]^2 = 10^{-39.59},\quad [Al^{3+}] = 1.6 \times 10^{-20}\ M$

$Al(OH)_3(s) \rightleftharpoons Al^{3+}(aq) + 3\ OH^-(aq)$ $K_{sp} = [Al^{3+}][OH^-]^3$; From the problem, $[OH^-] = 1.0 \times 10^{-4}\ M$.

$K_{sp} = (1.6 \times 10^{-20})(1.0 \times 10^{-4})^3 = 1.6 \times 10^{-32}$

63.   See Exercises 17.25, 17.27 and 17.29 for balanced reactions and standard cell potentials. Balanced reactions are necessary to determine n, the moles of electrons transferred.

25a. $7\ H_2O + 2\ Cr^{3+} + 3\ Cl_2 \rightarrow Cr_2O_7^{2-} + 6\ Cl^- + 14\ H^+$ $E^{\circ}_{cell} = 0.03\ V = 0.03\ J/C$

$\Delta G^{\circ} = -nFE^{\circ}_{cell} = -(6\ mol\ e^-)(96,485\ C/mol\ e^-)(0.03\ J/C) = -1.7 \times 10^4\ J = -20\ kJ$

$E_{cell} = E^{\circ}_{cell} - \frac{0.0591}{n} \log Q$: At equilibrium, $E_{cell} = 0$ and $Q = K$, so $E^{\circ}_{cell} = \frac{0.0591}{n} \log K$

$\log K = \frac{nE^{\circ}}{0.0591} = \frac{6(0.03)}{0.0591} = 3.05,\quad K = 10^{3.05} = 1 \times 10^3$

Note: When determining exponents, we will round off to the correct number of significant figures after the calculation is complete in order to help eliminate excessive round-off error.

25b. $\Delta G^{\circ} = -(2\ mol\ e^-)(96,485\ C/mol\ e^-)(2.71\ J/C) = -5.23 \times 10^5\ J = -523\ kJ$

$\log K = \frac{2(2.71)}{0.0591} = 91.709,\quad K = 5.12 \times 10^{91}$

29a. $\Delta G^{\circ} = -(2\ mol\ e^-)(96,485\ C/mol\ e^-)(0.27\ J/C) = -5.21 \times 10^4\ J = -52\ kJ$

$\log K = \frac{2(0.27)}{0.0591} = 9.14,\quad K = 1.4 \times 10^9$

29b. $\Delta G^{\circ} = -(10\ mol\ e^-)(96,485\ C/mol\ e^-)(0.09\ J/C) = -8.7 \times 10^4\ J = -90\ kJ$

$\log K = \frac{10(0.09)}{0.0591} = 15.23,\quad K = 2 \times 10^{15}$

64. $\Delta G^{\circ} = -nFE^{\circ}_{cell};\quad E^{\circ}_{cell} = \frac{0.0591}{n} \log K,\quad \log K = \frac{nE^{\circ}}{0.0591}$

26a. $\Delta G^{\circ} = -(10\ mol\ e^-)(96,485\ C/mol\ e^-)(0.43\ J/C) = -4.1 \times 10^5\ J = -410\ kJ$

$\log K = \frac{10(0.43)}{0.0591} = 72.76,\quad K = 10^{72.76} = 5.8 \times 10^{72}$

26b. $\Delta G° = -(2 \text{ mol } e^-)(96,485 \text{ C/mol } e^-)(1.56 \text{ J/C}) = -3.01 \times 10^5 \text{ J} = -301 \text{ kJ}$

$\log K = \dfrac{2(1.56)}{0.0591} = 52.792, \ K = 6.19 \times 10^{52}$

30a. $\Delta G° = -(2 \text{ mol } e^-)(96,485 \text{ C/mol } e^-)(1.10 \text{ J/C}) = -2.12 \times 10^5 \text{ J} = -212 \text{ kJ}$

$\log K = \dfrac{2(1.10)}{0.0591} = 37.225, \ K = 1.68 \times 10^{37}$

30b. $\Delta G° = -(6 \text{ mol } e^-)(96,485 \text{ C/mol } e^-)(1.14 \text{ J/C}) = -6.60 \times 10^5 \text{ J} = -660. \text{ kJ}$

$\log K = \dfrac{6(1.14)}{0.0591} = 115.736, \ K = 5.45 \times 10^{115}$

65.   a.   Possible reaction: $I_2(s) + 2 \text{ Cl}^-(aq) \to 2 \text{ I}^-(aq) + Cl_2(g)$ $E°_{cell} = 0.54 \text{ V} - 1.36 \text{ V} = -0.82 \text{ V}$
This reaction is not spontaneous at standard conditions since $E°_{cell} < 0$. No reaction occurs.

b.   Possible reaction: $Cl_2(g) + 2 \text{ I}^-(aq) \to I_2(s) + 2 \text{ Cl}^-(aq)$ $E°_{cell} = 0.82 \text{ V}$; This reaction is spontaneous at standard conditions since $E°_{cell} > 0$. The reaction will occur.

$Cl_2(g) + 2 \text{ I}^-(aq) \to I_2(s) + 2 \text{ Cl}^-(aq)$ $E°_{cell} = 0.82 \text{ V} = 0.82 \text{ J/C}$

$\Delta G° = -nFE°_{cell} = -(2 \text{ mol } e^-)(96,485 \text{ C/mol } e^-)(0.82 \text{ J/C}) = -1.6 \times 10^5 \text{ J} = -160 \text{ kJ}$

$E° = \dfrac{0.0591}{n} \log K, \ \log K = \dfrac{nE°}{0.0591} = \dfrac{2(0.82)}{0.0591} = 27.75, \ K = 10^{27.75} = 5.6 \times 10^{27}$

c.   Possible reaction: $2 \text{ Ag}(s) + Cu^{2+}(aq) \to Cu(s) + 2 \text{ Ag}^+(aq)$ $E°_{cell} = -0.46 \text{ V}$; No reaction occurs.

d. Fe^{2+} can be oxidized or reduced. The other species present are H^+, SO_4^{2-}, H_2O, and O_2 from air. Only O_2 in the presence of H^+ has a large enough standard reduction potential to oxidize Fe^{2+} to Fe^{3+} (resulting in $E°_{cell} > 0$). All other combinations, including the possible reduction of Fe^{2+}, give negative cell potentials. The spontaneous reaction is:

$4 \text{ Fe}^{2+}(aq) + 4 \text{ H}^+(aq) + O_2(g) \to 4 \text{ Fe}^{3+}(aq) + 2 \text{ H}_2O(l)$ $E°_{cell} = 1.23 - 0.77 = 0.46 \text{ V}$

$\Delta G° = -nFE°_{cell} = -(4 \text{ mol } e^-)(96,485 \text{ C/mol } e^-)(0.46 \text{ J/C})(1 \text{ kJ}/1000 \text{ J}) = -180 \text{ kJ}$

$\log K = \dfrac{4(0.46)}{0.0591} = 31.13, \ K = 1.3 \times 10^{31}$

66.   a. $Cu^+ + e^- \to Cu$ $\qquad\qquad$ $E° = 0.52 \text{ V}$
$\qquad Cu^+ \to Cu^{2+} + e^-$ $\qquad\qquad$ $-E° = -0.16 \text{ V}$

$\rule{8cm}{0.4pt}$

$2 \text{ Cu}^+(aq) \to Cu^{2+}(aq) + Cu(s)$ $\qquad$ $E°_{cell} = 0.36 \text{ V}$; Spontaneous

$\Delta G° = -nFE°_{cell} = -(1 \text{ mol } e^-)(96,485 \text{ C/mol } e)(0.36 \text{ J/C}) = -34,700 \text{ J} = -35 \text{ kJ}$

$E°_{cell} = \dfrac{0.0591}{n} \log K, \ \log K = \dfrac{nE°}{0.0591} = \dfrac{1(0.36)}{0.0591} = 6.09, \ K = 10^{6.09} = 1.2 \times 10^6$

b.

$$Fe^{2+} + 2\ e^- \rightarrow Fe \qquad\qquad E° = -0.44\ V$$
$$(Fe^{2+} \rightarrow Fe^{3+} + e^-) \times 2 \qquad -E° = -0.77\ V$$

$$3\ Fe^{2+}(aq) \rightarrow 2\ Fe^{3+}(aq) + Fe(s) \qquad E°_{cell} = -1.21\ V; \ Not\ spontaneous$$

c.

$$HClO_2 + 2\ H^+ + 2\ e^- \rightarrow HClO + H_2O \qquad\qquad E° = 1.65\ V$$
$$HClO_2 + H_2O \rightarrow ClO_3^- + 3\ H^+ + 2\ e^- \qquad\qquad -E° = -1.21\ V$$

$$2\ HClO_2(aq) \rightarrow ClO_3^-(aq) + H^+(aq) + HClO(aq) \quad E°_{cell} = 0.44\ V; \ Spontaneous$$

$$\Delta G° = -nFE°_{cell} = -(2\ mol\ e^-)(96,485\ C/mol\ e^-)(0.44\ J/C) = -84,900\ J = -85\ kJ$$

$$\log K = \frac{nE°}{0.0591} = \frac{2(0.44)}{0.0591} = 14.89, \ K = 7.8 \times 10^{14}$$

67. a.

$$Au^{3+} + 3\ e^- \rightarrow Au \qquad\qquad E° = 1.50\ V$$
$$(Tl \rightarrow Tl^+ + e^-) \times 3 \qquad -E° = 0.34\ V$$

$$Au^{3+}(aq) + 3\ Tl(s) \rightarrow Au(s) + 3\ Tl^+(aq) \qquad E°_{cell} = 1.84\ V$$

b. $\Delta G° = -nFE°_{cell} = -(3\ mol\ e^-)(96,485\ C/mol\ e^-)(1.84\ J/C) = -5.33 \times 10^5\ J = -533\ kJ$

$$\log K = \frac{nE°}{0.0591} = \frac{3(1.84)}{0.0591} = 93.401, \ K = 10^{93.401} = 2.52 \times 10^{93}$$

c. $E_{cell} = 1.84\ V - \dfrac{0.0591}{3} \log \dfrac{[Tl^+]^3}{[Au^{3+}]} = 1.84 - \dfrac{0.0591}{3} \log \dfrac{(1.0 \times 10^{-4})^3}{1.0 \times 10^{-2}}$

$E_{cell} = 1.84 - (-0.20) = 2.04\ V$

68.

$$(Cr^{2+} \rightarrow Cr^{3+} + e^-) \times 2$$
$$Co^{2+} + 2\ e^- \rightarrow Co$$

$$2\ Cr^{2+} + Co^{2+} \rightarrow 2\ Cr^{3+} + Co$$

$$E°_{cell} = \frac{0.0591}{n} \log K = \frac{0.0591}{2} \log (2.79 \times 10^7) = 0.220\ V$$

$$E = E° - \frac{0.0591}{n} \log \frac{[Cr^{3+}]^2}{[Cr^{2+}]^2[Co^{2+}]} = 0.220\ V - \frac{0.0591}{2} \log \frac{(2.0)^2}{(0.30)^2(0.20)} = 0.151\ V$$

$$\Delta G = -nFE = -(2\ mol\ e^-)(96,485\ C/mol\ e^-)(0.151\ J/C) = -2.91 \times 10^4\ J = -29.1\ kJ$$

69. $CdS + 2 e^- \rightarrow Cd + S^{2-}$ $E° = -1.21$ V
      $Cd \rightarrow Cd^{2+} + 2 e^-$ $-E° = 0.402$ V

$CdS(s) \rightarrow Cd^{2+}(aq) + S^{2-}(aq)$ $E°_{cell} = -0.81$ V $K_{sp} = ?$

$\log K_{sp} = \dfrac{nE°}{0.0591} = \dfrac{2(-0.81)}{0.0591} = -27.41$, $K_{sp} = 10^{-27.41} = 3.9 \times 10^{-28}$

70. $Al^{3+} + 3 e^- \rightarrow Al$ $E° = -1.66$ V
      $Al + 6 F^- \rightarrow AlF_6^{3-} + 3 e^-$ $-E° = 2.07$ V

$Al^{3+}(aq) + 6 F^-(aq) \rightarrow AlF_6^{3-}(aq)$ $E°_{cell} = 0.41$ V $K = ?$

$\log K = \dfrac{nE°}{0.0591} = \dfrac{3(0.41)}{0.0591} = 20.81$, $K = 10^{20.81} = 6.5 \times 10^{20}$

71. $Ag^+ + e^- \rightarrow Ag$ $E° = 0.80$ V
      $Ag + 2 S_2O_3^{2-} \rightarrow Ag(S_2O_3)_2^{3-} + e^-$ $-E° = -0.017$ V

$Ag^+(aq) + 2 S_2O_3^{2-}(aq) \rightarrow Ag(S_2O_3)_2^{3-}(aq)$ $E°_{cell} = 0.78$ V $K = ?$

For this overall reaction, $E°_{cell} = \dfrac{0.0591}{n} \log K$

$\log K = \dfrac{nE°}{0.0591} = \dfrac{(1)(0.78)}{0.0591} = 13.20$, $K = 10^{13.20} = 1.6 \times 10^{13}$

72. $CuI + e^- \rightarrow Cu + I^-$ $E°_{CuI} = ?$
      $Cu \rightarrow Cu^+ + e^-$ $-E° = -0.52$ V

$CuI(s) \rightarrow Cu^+(aq) + I^-(aq)$ $E°_{cell} = E°_{CuI} - 0.52$ V

For this overall reaction, $K = K_{sp} = 1.1 \times 10^{-12}$:

$E°_{cell} = \dfrac{0.0591}{n} \log K_{sp} = \dfrac{0.0591}{1} \log (1.1 \times 10^{-12}) = -0.71$ V

$E°_{cell} = -0.71$ V $= E°_{CuI} - 0.52$, $E°_{CuI} = -0.19$ V

Electrolysis

73. a. $Al^{3+} + 3 e^- \rightarrow Al$; 3 mol e^- are needed to produce 1 mol Al from Al^{3+}.

1.0×10^3 g Al $\times \dfrac{1\,mol\,Al}{26.98\,g\,Al} \times \dfrac{3\,mol\,e^-}{mol\,Al} \times \dfrac{96{,}485\,C}{mol\,e^-} \times \dfrac{1\,s}{100.0\,C} = 1.07 \times 10^5$ s $= 30.$ hours

   b. 1.0 g Ni $\times \dfrac{1\,mol\,Ni}{58.69\,g\,Ni} \times \dfrac{2\,mol\,e^-}{mol\,Ni} \times \dfrac{96{,}485\,C}{mol\,e^-} \times \dfrac{1\,s}{100.0\,C} = 33$ s

c. $5.0 \text{ mol Ag} \times \dfrac{1 \text{ mol e}^-}{\text{mol Ag}} \times \dfrac{96{,}485 \text{ C}}{\text{mol e}^-} \times \dfrac{1 \text{ s}}{100.0 \text{ C}} = 4.8 \times 10^3 \text{ s} = 1.3 \text{ hours}$

74. The oxidation state of bismuth in BiO^+ is +3 because oxygen has a -2 oxidation state in this ion. Therefore, 3 moles of electrons are required to reduce the bismuth in BiO^+ to $Bi(s)$.

$10.0 \text{ g Bi} \times \dfrac{1 \text{ mol Bi}}{209.0 \text{ g Bi}} \times \dfrac{3 \text{ mol e}^-}{\text{mol Bi}} \times \dfrac{96{,}485 \text{ C}}{\text{mol e}^-} \times \dfrac{1 \text{ s}}{25.0 \text{ C}} = 554 \text{ s} = 9.23 \text{ min}$

75. $15 \text{ A} = \dfrac{15 \text{ C}}{\text{s}} \times \dfrac{60 \text{ s}}{\text{min}} \times \dfrac{60 \text{ min}}{\text{h}} = 5.4 \times 10^4 \text{ C of charge passed in 1 hour}$

a. $5.4 \times 10^4 \text{ C} \times \dfrac{1 \text{ mol e}^-}{96{,}485 \text{ C}} \times \dfrac{1 \text{ mol Co}}{2 \text{ mol e}^-} \times \dfrac{58.93 \text{ g Co}}{\text{mol Co}} = 16 \text{ g Co}$

b. $5.4 \times 10^4 \text{ C} \times \dfrac{1 \text{ mol e}^-}{96{,}485 \text{ C}} \times \dfrac{1 \text{ mol Hf}}{4 \text{ mol e}^-} \times \dfrac{178.5 \text{ g Hf}}{\text{mol Hf}} = 25 \text{ g Hf}$

c. $2 \text{ I}^- \rightarrow I_2 + 2 \text{ e}^-$; $5.4 \times 10^4 \text{ C} \times \dfrac{1 \text{ mol e}^-}{96{,}485 \text{ C}} \times \dfrac{1 \text{ mol I}_2}{2 \text{ mol e}^-} \times \dfrac{253.8 \text{ g I}_2}{\text{mol I}_2} = 71 \text{ g I}_2$

d. $CrO_3(l) \rightarrow Cr^{6+} + 3 \text{ O}^{2-}$; 6 mol e$^-$ are needed to produce 1 mol Cr from molten CrO_3.

$5.4 \times 10^4 \text{ C} \times \dfrac{1 \text{ mol e}^-}{96{,}485 \text{ C}} \times \dfrac{1 \text{ mol Cr}}{6 \text{ mol e}^-} \times \dfrac{52.00 \text{ g Cr}}{\text{mol Cr}} = 4.9 \text{ g Cr}$

76. Al is in the +3 oxidation in Al_2O_3, so 3 mol e$^-$ are needed to convert Al^{3+} into $Al(s)$.

$2.00 \text{ h} \times \dfrac{60 \text{ min}}{\text{h}} \times \dfrac{60 \text{ s}}{\text{min}} \times \dfrac{1.00 \times 10^6 \text{ C}}{\text{s}} \times \dfrac{1 \text{ mol e}^-}{96{,}485 \text{ C}} \times \dfrac{1 \text{ mol Al}}{3 \text{ mol e}^-} \times \dfrac{26.98 \text{ g Al}}{\text{mol Al}} = 6.71 \times 10^5 \text{ g}$

77. $1397 \text{ s} \times \dfrac{6.50 \text{ C}}{\text{s}} \times \dfrac{1 \text{ mol e}^-}{96{,}485 \text{ C}} \times \dfrac{1 \text{ mol M}}{3 \text{ mol e}^-} = 3.14 \times 10^{-2} \text{ mol M where M = unknown metal}$

Molar mass $= \dfrac{1.41 \text{ g M}}{3.14 \times 10^{-2} \text{ mol M}} = \dfrac{44.9 \text{ g}}{\text{mol}}$; The element is scandium. Sc forms 3+ ions.

78. Alkaline earth metals form +2 ions, so 2 mol of e$^-$ are transferred to form the metal, M.

$\text{mol M} = 748 \text{ s} \times \dfrac{5.00 \text{ C}}{\text{s}} \times \dfrac{1 \text{ mol e}^-}{96{,}485 \text{ C}} \times \dfrac{1 \text{ mol M}}{2 \text{ mol e}^-} = 1.94 \times 10^{-2} \text{ mol M}$

$\text{molar mass of M} = \dfrac{0.471 \text{ g M}}{1.94 \times 10^{-2} \text{ mol M}} = 24.3 \text{ g/mol}$; $MgCl_2$ was electrolyzed.

79. F_2 is produced at the anode: $2\,F^- \rightarrow F_2 + 2\,e^-$

$$2.00\ h \times \frac{60\ min}{h} \times \frac{60\ s}{min} \times \frac{10.0\ C}{s} \times \frac{1\ mol\ e^-}{96,485\ C} = 0.746\ mol\ e^-$$

$$0.746\ mol\ e^- \times \frac{1\ mol\ F_2}{2\ mol\ e^-} = 0.373\ mol\ F_2;\quad PV = nRT,\quad V = \frac{nRT}{P}$$

$$V = \frac{(0.373\ mol)\,(0.08206\ L\cdot atm/K\cdot mol)\,(298\ K)}{1.00\ atm} = 9.12\ L\ F_2$$

K is produced at the cathode: $K^+ + e^- \rightarrow K$

$$0.746\ mol\ e^- \times \frac{1\ mol\ K}{mol\ e^-} \times \frac{39.10\ g\ K}{mol\ K} = 29.2\ g\ K$$

80.   The half-reactions for the electrolysis of water are:

$$(2\,e^- + 2\,H_2O \rightarrow H_2 + 2\,OH^-) \times 2$$
$$2\,H_2O \rightarrow 4\,H^+ + O_2 + 4\,e^-$$
$$\overline{}$$
$$2\,H_2O(l) \rightarrow 2\,H_2(g) + O_2(g)$$

Note: $4\,H^+ + 4\,OH^- \rightarrow 4\,H_2O$ and n = 4 for this reaction as it is written.

$$15.0\ min \times \frac{60\ s}{min} \times \frac{2.50\ C}{s} \times \frac{1\ mol\ e^-}{96,485\ C} \times \frac{2\ mol\ H_2}{4\ mol\ e^-} = 1.17 \times 10^{-2}\ mol\ H_2$$

At STP, 1 mole of an ideal gas occupies a volume of 22.42 L (see Chapter 5 of the text).

$$1.17 \times 10^{-2}\ mol\ H_2 \times \frac{22.42\ L}{mol\ H_2} = 0.262\ L = 262\ mL\ H_2$$

$$1.17 \times 10^{-2}\ mol\ H_2 \times \frac{1\ mol\ O_2}{2\ mol\ H_2} \times \frac{22.42\ L}{mol\ O_2} = 0.131\ L = 131\ mL\ O_2$$

81. $$\frac{150. \times 10^3\ g\ C_6H_8N_2}{h} \times \frac{1\ h}{60\ min} \times \frac{1\ min}{60\ s} \times \frac{1\ mol\ C_6H_8N_2}{108.14\ g\ C_6H_8N_2} \times \frac{2\ mol\ e^-}{mol\ C_6H_8N_2} \times \frac{96,485\ C}{mol\ e^-}$$

$$= 7.44 \times 10^4\ C/s\ \text{or a current of}\ 7.44 \times 10^4\ A$$

82. $Al^{3+} + 3\,e^- \rightarrow Al$; 3 mol e^- are needed to produce Al from Al^{3+}

$$2000\ lb\ Al \times \frac{453.6\ g}{lb} \times \frac{1\ mol\ Al}{26.98\ g} \times \frac{3\ mol\ e^-}{mol\ Al} \times \frac{96,485\ C}{mol\ e^-} = 1 \times 10^{10}\ C\ \text{of electricity needed}$$

$$\frac{1 \times 10^{10}\ C}{24\ h} \times \frac{1\ h}{60\ min} \times \frac{1\ min}{60\ s} = 1 \times 10^5\ C/s = 1 \times 10^5\ A$$

83. $2.30 \text{ min} \times \dfrac{60 \text{ s}}{\text{min}} = 138 \text{ s}$; $138 \text{ s} \times \dfrac{2.00 \text{ C}}{\text{s}} \times \dfrac{1 \text{ mol e}^-}{96{,}485 \text{ C}} \times \dfrac{1 \text{ mol Ag}}{\text{mol e}^-} = 2.86 \times 10^{-3} \text{ mol Ag}$

$[\text{Ag}^+] = 2.86 \times 10^{-3} \text{ mol Ag}^+/0.250 \text{ L} = 1.14 \times 10^{-2} \, M$

84. $0.50 \text{ L} \times 0.010 \text{ mol Pt}^{4+}/\text{L} = 5.0 \times 10^{-3} \text{ mol Pt}^{4+}$

To plate out 99% of the Pt^{4+}, we will produce $0.99 \times 5.0 \times 10^{-3}$ mol Pt.

$0.99 \times 5.0 \times 10^{-3} \text{ mol Pt} \times \dfrac{4 \text{ mol e}^-}{\text{mol Pt}} \times \dfrac{96{,}485 \text{ C}}{\text{mol e}^-} \times \dfrac{1 \text{ s}}{4.00 \text{ C}} = 480 \text{ s}$

85. | | | | |
     |---|---|---|---|
     | $\text{Au}^{3+} + 3 \text{ e}^- \rightarrow \text{Au}$ | $E° = 1.50 \text{ V}$ | $\text{Ni}^{2+} + 2 \text{ e}^- \rightarrow \text{Ni}$ | $E° = -0.23 \text{ V}$ |
     | $\text{Ag}^+ + \text{e}^- \rightarrow \text{Ag}$ | $E° = 0.80 \text{ V}$ | $\text{Cd}^{2+} + 2 \text{ e}^- \rightarrow \text{Cd}$ | $E° = -0.40 \text{ V}$ |

$2 \text{ H}_2\text{O} + 2\text{e}^- \rightarrow \text{H}_2 + 2 \text{ OH}^-$ $E° = -0.83 \text{ V}$

Au(s) will plate out first since it has the most positive reduction potential, followed by Ag(s), which is followed by Ni(s), and finally Cd(s) will plate out last since it has the most negative reduction potential of the metals listed.

86. Species present: Fe^{2+}, SO_4^{2-}, H^+ and H_2O. The possible cathode reactions are:

$\text{SO}_4^{2-} + 4 \text{ H}^+ + 2 \text{ e}^- \rightarrow \text{H}_2\text{SO}_3 + \text{H}_2\text{O}$	$E° = 0.20 \text{ V}$
$2 \text{ H}^+ + 2 \text{ e}^- \rightarrow \text{H}_2$	$E° = 0.00 \text{ V}$
$\text{Fe}^{2+} + 2 \text{ e}^- \rightarrow \text{Fe}$	$E° = -0.44 \text{ V}$
$2 \text{ H}_2\text{O} + 2 \text{ e}^- \rightarrow \text{H}_2 + 2 \text{ OH}^-$	$E° = -0.83 \text{ V}$

Reduction of SO_4^{2-} will occur at the cathode since $E°_{\text{SO}_4^{2-}}$ is most positive. The possible anode reactions are:

$\text{Fe}^{2+} \rightarrow \text{Fe}^{3+} + \text{e}^-$	$-E° = -0.77 \text{ V}$
$2 \text{ H}_2\text{O} \rightarrow \text{O}_2 + 4 \text{ H}^+ + 4 \text{ e}^-$	$-E° = -1.23 \text{ V}$

Oxidation of Fe^{2+} will occur at the anode since $-E°_{\text{Fe}^{2+}}$ is most positive.

87. Reduction occurs at the cathode, and oxidation occurs at the anode. First, determine all the species present, then look up pertinent reduction and/or oxidation potentials in Table 17.1 for all these species. The cathode reaction will be the reaction with the most positive reduction potential, and the anode reaction will be the reaction with the most positive oxidation potential.

a. Species present: Ni^{2+} and Br^-; Ni^{2+} can be reduced to Ni, and Br^- can be oxidized to Br_2 (from Table 17.1). The reactions are:

Cathode:	$\text{Ni}^{2+} + 2\text{e}^- \rightarrow \text{Ni}$	$E° = -0.23 \text{ V}$
Anode:	$2 \text{ Br}^- \rightarrow \text{Br}_2 + 2 \text{ e}^-$	$-E° = -1.09 \text{ V}$

b. Species present: Al^{3+} and F^-; Al^{3+} can be reduced, and F^- can be oxidized. The reactions are:

Cathode: $Al^{3+} + 3\ e^- \rightarrow Al$ $E° = -1.66$ V
Anode: $2\ F^- \rightarrow F_2 + 2\ e^-$ $-E° = -2.87$ V

c. Species present: Mn^{2+} and I^-; Mn^{2+} can be reduced, and I^- can be oxidized. The reactions are:

Cathode: $Mn^{2+} + 2\ e^- \rightarrow Mn$ $E° = -1.18$ V
Anode: $2\ I^- \rightarrow I_2 + 2\ e^-$ $-E° = -0.54$ V

88. These are all in aqueous solutions, so we must also consider the reduction and oxidation of H_2O in addition to the potential redox reactions of the ions present. For the cathode reaction, the species with the most positive reduction potential will be reduced, and for the anode reaction, the species with the most positive oxidation potential will be oxidized.

a. Species present: Ni^{2+}, Br^- and H_2O. Possible cathode reactions are:

$Ni^{2+} + 2e^- \rightarrow Ni$ $E° = -0.23$ V
$2\ H_2O + 2\ e^- \rightarrow H_2 + 2\ OH^-$ $E° = -0.83$ V

Since it is easier to reduce Ni^{2+} than H_2O (assuming standard conditions), Ni^{2+} will be reduced by the above cathode reaction.

Possible anode reactions are:

$2\ Br^- \rightarrow Br_2 + 2\ e^-$ $-E° = -1.09$ V
$2\ H_2O \rightarrow O_2 + 4\ H^+ + 4\ e^-$ $-E° = -1.23$ V

Since Br^- is easier to oxidize than H_2O (assuming standard conditions), then Br^- will be oxidized by the above anode reaction.

b. Species present: Al^{3+}, F^- and H_2O; Al^{3+} and H_2O can be reduced. The reduction potentials are $E° = -1.66$ V for Al^{3+} and $E° = -0.83$ V for H_2O (assuming standard conditions). H_2O will be reduced at the cathode ($2\ H_2O + 2\ e^- \rightarrow H_2 + 2\ OH^-$).

F^- and H_2O can be oxidized. The oxidation potentials are $-E° = -2.87$ V for F^- and $-E° = -1.23$ V for H_2O (assuming standard conditions). From the potentials, we would predict H_2O to be oxidized at the anode ($2\ H_2O \rightarrow O_2 + 4\ H^+ + 4\ e^-$).

c. Species present: Mn^{2+}, I^- and H_2O; Mn^{2+} and H_2O can be reduced. The possible cathode reactions are:

$Mn^{2+} + 2\ e^- \rightarrow Mn$ $E° = -1.18$ V
$2\ H_2O + 2\ e^- \rightarrow H_2 + 2\ OH^-$ $E° = -0.83$ V

Reduction of H_2O will occur at the cathode since $E°_{H_2O}$ is most positive.

I^- and H_2O can be oxidized. The possible anode reactions are:

$$2\ I^- \rightarrow I_2 + 2\ e^- \qquad\qquad -E° = -0.54\ V$$
$$2\ H_2O \rightarrow O_2 + 4\ H^+ + 4\ e^- \qquad -E° = -1.23\ V$$

Oxidation of I^- will occur at the anode since $-E_{I^-}°$ is most positive.

Additional Exercises

89.   The half-reaction for the SCE is:

$$Hg_2Cl_2 + 2\ e^- \rightarrow 2\ Hg + 2\ Cl^- \qquad E_{SCE} = 0.242\ V$$

For a spontaneous reaction to occur, E_{cell} must be positive. Using the standard reduction potentials in Table 17.1 and the given SCE potential, deduce which combination will produce a positive overall cell potential.

a. $Cu^{2+} + 2\ e^- \rightarrow Cu \qquad E° = 0.34\ V$

E_{cell} = 0.34 - 0.242 = 0.10 V; SCE is the anode.

b. $Fe^{3+} + e^- \rightarrow Fe^{2+} \qquad E° = 0.77\ V$

E_{cell} = 0.77 - 0.242 = 0.53 V; SCE is the anode.

c. $AgCl + e^- \rightarrow Ag + Cl^- \quad E° = 0.22\ V$

E_{cell} = 0.242 - 0.22 = 0.02 V; SCE is the cathode.

d. $Al^{3+} + 3\ e^- \rightarrow Al \qquad E° = -1.66\ V$

E_{cell} = 0.242 + 1.66 = 1.90 V; SCE is the cathode.

e. $Ni^{2+} + 2\ e^- \rightarrow Ni \qquad E° = -0.23\ V$

E_{cell} = 0.242 + 0.23 = 0.47 V; SCE is the cathode.

90.   The potential oxidizing agents are NO_3^- and H^+. Hydrogen ion cannot oxidize Pt under either condition. Nitrate cannot oxidize Pt unless there is Cl^- in the solution. Aqua regia has both Cl^- and NO_3^-. The overall reaction is:

$$(NO_3^- + 4\ H^+ + 3\ e^- \rightarrow NO + 2\ H_2O) \times 2 \qquad\qquad E° = \ \ 0.96\ V$$
$$(4\ Cl^- + \ Pt \rightarrow PtCl_4^{2-} + 2\ e^-) \times 3 \qquad\qquad -E° = -0.755\ V$$

$$12\ Cl^-(aq) + 3\ Pt(s) + 2\ NO_3^-(aq) + 8\ H^+(aq) \rightarrow 3\ PtCl_4^{2-}(aq) + 2\ NO(g) + 4\ H_2O(l) \qquad E_{cell}° = 0.21\ V$$

91. $2 Ag^+(aq) + Cu(s) \rightarrow Cu^{2+}(aq) + 2 Ag(s)$ $E°_{cell} = 0.80 - 0.34 V = 0.46 V$; A galvanic cell produces a voltage as the forward reaction occurs. Any stress that increases the tendency of the forward reaction to occur will increase the cell potential, while a stress that decreases the tendency of the forward reaction to occur will decrease the cell potential.

a.   Added Cu^{2+} (a product ion) will decrease the tendency of the forward reaction to occur, which will decrease the cell potential.

b.   Added NH_3 removes Cu^{2+} in the form of $Cu(NH_3)_4^{2+}$. Removal of a product ion will increase the tendency of the forward reaction to occur, which will increase the cell potential.

c.   Added Cl^- removes Ag^+ in the form of $AgCl(s)$. Removal of a reactant ion will decrease the tendency of the forward reaction to occur, which will decrease the cell potential.

d. $Q_1 = \dfrac{[Cu^{2+}]_o}{[Ag^+]_o^2}$; As the volume of solution is doubled, each concentration is halved.

$$Q_2 = \frac{1/2\,[Cu^{2+}]_o}{(1/2\,[Ag^+]_o)^2} = \frac{2[Cu^{2+}]_o}{[Ag^+]_o^2} = 2\,Q_1$$

The reaction quotient is doubled as the concentrations are halved. Since reactions are spontaneous when $Q < K$ and since Q increases when the solution volume doubles, the reaction is closer to equilibrium, which will decrease the cell potential.

e.   Since $Ag(s)$ is not a reactant in this spontaneous reaction, and since solids do not appear in the reaction quotient expressions, replacing the silver electrode with a platinum electrode will have no effect on the cell potential.

92.
$$\begin{array}{ll}
(Al^{3+} + 3\,e^- \rightarrow Al) \times 2 & E° = -1.66 V \\
(M \rightarrow M^{2+} + 2\,e^-) \times 3 & -E° = ?
\end{array}$$

$3 M(s) + 2 Al^{3+}(aq) \rightarrow 2 Al(s) + 3 M^{2+}(aq)$ $E°_{cell} = -E° - 1.66 V$

$\Delta G° = -nFE°_{cell}$, $-411 \times 10^3 J = -(6 \text{ mol } e^-)(96,485 \text{ C/mol } e^-)(E°_{cell})$, $E°_{cell} = 0.71 V$

$E°_{cell} = -E° - 1.66 V = 0.71 V$, $-E° = 2.37$ or $E° = -2.37$

From table 17.1, the reduction potential for $Mg^{2+} + 2 e^- \rightarrow Mg$ is $-2.37 V$, which fits the data. Hence, the metal is magnesium.

93.   a. $\Delta G° = \Sigma n_p \Delta G°_{f,\,products} - \Sigma n_r \Delta G°_{f,\,reactants} = 2(-480.) + 3(86) - [3(-40.)] = -582 \text{ kJ}$

From oxidation numbers, $n = 6$. $\Delta G° = -nFE°$, $E° = \dfrac{-\Delta G°}{nF} = \dfrac{-(-582,000 \text{ J})}{6(96,485) \text{ C}} = 1.01 V$

$\log K = \dfrac{nE°}{0.0591} = \dfrac{6(1.01)}{0.0591} = 102.538$, $K = 10^{102.538} = 3.45 \times 10^{102}$

b.
$$2\ e^- + Ag_2S \rightarrow 2\ Ag + S^{2-}) \times 3 \qquad E^{\circ}_{Ag_2S} = ?$$
$$(Al \rightarrow Al^{3+} + 3\ e^-) \times 2 \qquad -E^{\circ} = 1.66\ V$$

$$3\ Ag_2S(s) + 2\ Al(s) \rightarrow 6\ Ag(s) + 3\ S^{2-}(aq) + 2\ Al^{3+}(aq) \quad E^{\circ}_{cell} = 1.01\ V = E^{\circ}_{Ag_2S} + 1.66V$$

$$E^{\circ}_{Ag_2S} = 1.01\ V - 1.66\ V = -0.65\ V$$

94. $Zn \rightarrow Zn^{2+} + 2\ e^- \quad -E^{\circ} = 0.76\ V; \quad Fe \rightarrow Fe^{2+} + 2\ e^- \quad -E^{\circ} = 0.44\ V$

It is easier to oxidize Zn than Fe, so the Zn would be oxidized, protecting the iron of the *Monitor's* hull.

95.   From Exercise 17.27a: $3\ Cl_2(g) + 2\ Cr^{3+}(aq) + 7\ H_2O(l) \rightleftharpoons 14\ H^+(aq) + Cr_2O_7^{2-}(aq) + 6\ Cl^-(aq)$
$E^{\circ}_{cell} = 0.03\ V$

$$E_{cell} = E^{\circ}_{cell} - \frac{0.0591}{6} \log \frac{[Cr_2O_7^{2-}]\ [H^+]^{14}[Cl^-]^6}{[Cr^{3+}]^2\ P^3_{Cl_2}}$$

When $K_2Cr_2O_7$ and Cl^- are added to concentrated H_2SO_4, Q becomes a large number due to $[H^+]^{14}$ term. The log of a large number is positive. E_{cell} becomes negative, which means the reverse reaction becomes spontaneous. The pungent fumes were $Cl_2(g)$.

96.   Aluminum has the ability to form a durable oxide coating over its surface. Once the HCl dissolves this oxide coating, Al is exposed to H^+ and is easily oxidized to Al^{3+}, i.e., the Al foil disappears after the oxide coating is dissolved.

97.   Consider the strongest oxidizing agent combined with the strongest reducing agent from Table 17.1:

$$F_2 + 2\ e^- \rightarrow 2\ F^- \qquad E^{\circ} = 2.87\ V$$
$$(Li \rightarrow Li^+ + e^-) \times 2 \qquad -E^{\circ} = 3.05\ V$$

$$F_2(g) + 2\ Li(s) \rightarrow 2\ Li^+(aq) + 2\ F^-(aq) \qquad E^{\circ}_{cell} = 5.92\ V$$

The claim is impossible. The strongest oxidizing agent and reducing agent when combined only give an E°_{cell} value of about 6 V.

98. $2\ H_2(g) + O_2(g) \rightarrow 2\ H_2O(l)$; Oxygen goes from the zero oxidation state to the -2 oxidation state in H_2O. Since two mol O appear in the balanced reaction, then n = 4 mol electrons transferred.

a. $E^{\circ}_{cell} = \dfrac{0.0591}{n} \log K = \dfrac{0.0591}{4} \log (1.28 \times 10^{83})$, $E^{\circ}_{cell} = 1.23\ V$

$\Delta G^{\circ} = -nFE^{\circ}_{cell} = -(4\ mol\ e^-)(96,485\ C/mol\ e^-)(1.23\ J/C) = -4.75 \times 10^5\ J = -475\ kJ$

b.   Since mol of gas decrease as reactants are converted into products, then ΔS° will be negative (unfavorable). Since the value of ΔG° is negative, then ΔH° must be negative (ΔG° = ΔH° - $T\Delta S^{\circ}$).

c. $\Delta G = w_{max} = \Delta H - T\Delta S$. Since ΔS is negative, then as T increases, ΔG becomes more positive (closer to zero). Therefore, w_{max} will decrease as T increases.

99.   a. $O_2 + 2\ H_2O + 4\ e^- \rightarrow 4\ OH^-$ $E° = 0.40\ V$

   $(H_2 + 2\ OH^- \rightarrow 2\ H_2O + 2\ e^-) \times 2$ $-E° = 0.83\ V$

   _____

   $2\ H_2(g) + O_2(g) \rightarrow 2\ H_2O(l)$ $E°_{cell} = 1.23\ V = 1.23\ J/C$

Since standard conditions are assumed, then $w_{max} = \Delta G°$ for 2 mol H_2O produced.

$\Delta G° = -nFE°_{cell} = -(4\ mol\ e^-)(96,485\ C/mol\ e^-)(1.23\ J/C) = -475,000\ J = -475\ kJ$

For 1.00×10^3 g H_2O produced, w_{max} is:

$$1.00 \times 10^3 \text{ g } H_2O \times \frac{1\ mol\ H_2O}{18.02\ g\ H_2O} \times \frac{-475\ kJ}{2\ mol\ H_2O} = -13,200\ kJ = w_{max}$$

The work done can be no larger than the free energy change. The best that could happen is that all of the free energy released would go into doing work, but this does not occur in any real process since there is always waste energy in a real process. Fuel cells are more efficient in converting chemical energy into electrical energy; they are also less massive. The major disadvantage is that they are expensive. In addition, $H_2(g)$ and $O_2(g)$ are an explosive mixture if ignited; much more so than fossil fuels.

100.   Cadmium goes from the zero oxidation state to the +2 oxidation state in $Cd(OH)_2$. Since one mol of Cd appears in the balanced reaction, then n = 2 mol electrons transferred. At standard conditions:

$w_{max} = \Delta G° = -nFE°$, $w_{max} = -(2\ mol\ e^-)(96,485\ C/mol\ e^-)(1.10\ J/C) = -2.12 \times 10^5\ J = -212\ kJ$

101. $(CO + O^{2-} \rightarrow CO_2 + 2\ e^-) \times 2$

   $O_2 + 4\ e^- \rightarrow 2\ O^{2-}$

   _____

   $2\ CO + O_2 \rightarrow 2\ CO_2$

$\Delta G = -nFE$, $E = \dfrac{-\Delta G}{nF} = \dfrac{-(-380 \times 10^3\ J)}{(4\ mol\ e^-)(96,485\ C/mol\ e^-)} = 0.98\ V$

102.   In the electrolysis of aqueous sodium chloride, H_2O is reduced in preference to Na^+, and Cl^- is oxidized in preference to H_2O. The anode reaction is $2\ Cl^- \rightarrow Cl_2 + 2\ e^-$, and the cathode reaction is $2\ H_2O + 2\ e^- \rightarrow H_2 + 2\ OH^-$. The overall reaction is:

$2\ H_2O(l) + 2\ Cl^-(aq) \rightarrow Cl_2(g) + H_2(g) + 2\ OH^-(aq)$.

From the 1:1 mol ratio between Cl_2 and H_2 in the overall balanced reaction, if 257 L of $Cl_2(g)$ are produced, then 257 L of $H_2(g)$ will also be produced since moles and volume of gas are directly proportional at constant T and P (see Chapter 5 of text).

103. $\text{mol e}^- = 50.0 \text{ min} \times \dfrac{60 \text{ s}}{\text{min}} \times \dfrac{2.50 \text{ C}}{\text{s}} \times \dfrac{1 \text{ mol e}^-}{96{,}485 \text{ C}} = 7.77 \times 10^{-2} \text{ mol e}^-$

   $\text{mol Ru} = 2.618 \text{ g Ru} \times \dfrac{1 \text{ mol Ru}}{101.1 \text{ g Ru}} = 2.590 \times 10^{-2} \text{ mol Ru}$

   $\dfrac{\text{mol e}^-}{\text{mol Ru}} = \dfrac{7.77 \times 10^{-2} \text{ mol e}^-}{2.590 \times 10^{-2} \text{ mol Ru}} = 3.00;$ The charge on the ruthenium ions is $+3$ ($Ru^{3+} + 3 \text{ e}^- \rightarrow Ru$).

104. $15 \text{ kWh} = \dfrac{15000 \text{ J h}}{\text{s}} \times \dfrac{60 \text{ s}}{\text{min}} \times \dfrac{60 \text{ min}}{\text{h}} = 5.4 \times 10^7 \text{ J or } 5.4 \times 10^4 \text{ kJ}$ (Hall process)

   To melt 1.0 kg Al requires: $1.0 \times 10^3 \text{ g Al} \times \dfrac{1 \text{ mol Al}}{26.98 \text{ g}} \times \dfrac{10.7 \text{ kJ}}{\text{mol Al}} = 4.0 \times 10^2 \text{ kJ}$

   It is feasible to recycle Al by melting the metal because, in theory, it takes less than 1% of the energy required to produce the same amount of Al by the Hall process.

Challenge Problems

105. $\Delta G^\circ = -nFE^\circ = \Delta H^\circ - T\Delta S^\circ, \quad E^\circ = \dfrac{T\Delta S^\circ}{nF} - \dfrac{\Delta H^\circ}{nF}$

   If we graph E° vs. T we should get a straight line ($y = mx + b$). The slope of the line is equal to $\Delta S^\circ/nF$, and the y-intercept is equal to $-\Delta H^\circ/nF$. From the equation above, E° will have a small temperature dependence when ΔS° is close to zero.

106.   a.   We can calculate ΔG° from $\Delta G^\circ = \Delta H^\circ - T\Delta S^\circ$ and then E° from $\Delta G^\circ = -nFE^\circ$; or we can use the equation derived in Exercise 17.105. For this reaction, $n = 2$ (from oxidation states).

   $E^\circ_{-20} = \dfrac{T\Delta S^\circ - \Delta H^\circ}{nF} = \dfrac{(253 \text{ K})(263.5 \text{ J/K}) + 315.9 \times 10^3 \text{ J}}{(2 \text{ mol e}^-)(96{,}485 \text{ C/mol e}^-)} = 1.98 \text{ J/C} = 1.98 \text{ V}$

   b. $E_{-20} = E^\circ_{-20} - \dfrac{RT}{nF} \ln Q = 1.98 \text{ V} - \dfrac{RT}{nF} \ln \dfrac{1}{[H^+]^2 [HSO_4^-]^2}$

   $E_{-20} = 1.98 \text{ V} - \dfrac{(8.3145 \text{ J/K} \bullet \text{mol})(253 \text{ K})}{(2 \text{ mol e}^-)(96{,}485 \text{ C/mol e}^-)} \ln \dfrac{1}{(4.5)^2 (4.5)^2} = 1.98 \text{ V} + 0.066 \text{ V} = 2.05 \text{ V}$

   c.   From Exercise 17.56, $E = 2.12 \text{ V}$ at 25°C. As the temperature decreases, the cell potential decreases. Also, oil becomes more viscous at lower temperatures, which adds to the difficulty of starting an engine on a cold day. The combination of these two factors results in batteries failing more often on cold days than on warm days.

107. $(Ag^+ + e^- \rightarrow Ag) \times 2$ $E^\circ = 0.80 \text{ V}$

   $\underline{\quad\quad Pb \rightarrow Pb^{2+} + 2 e^- \quad\quad -E^\circ = -(-0.13) \quad\quad}$

   $2 Ag^+ + Pb \rightarrow 2 Ag + Pb^{2+}$ $E^\circ_{cell} = 0.93 \text{ V}$

$$E = E° - \frac{0.0591}{n} \log \frac{[Pb^{2+}]}{[Ag^+]^2}, \quad 0.83 \text{ V} = 0.93 \text{ V} - \frac{0.0591}{2} \log \frac{(1.8)}{[Ag^+]^2}$$

$$\log \frac{(1.8)}{[Ag^+]^2} = \frac{0.10(2)}{0.0591} = 3.4, \quad \frac{(1.8)}{[Ag^+]^2} = 10^{3.4}, \quad [Ag^+] = 0.027 \, M$$

$$Ag_2SO_4(s) \rightleftharpoons 2 \, Ag^+(aq) + SO_4^{2-}(aq) \quad K_{sp} = [Ag^+]^2[SO_4^{2-}]$$

Initial s = solubility (mol/L)   0   0
Equil.   2s   s

From problem: $2s = 0.027 \, M$, $s = 0.027/2$

$K_{sp} = (2s)^2(s) = (0.027)^2(0.027/2) = 9.8 \times 10^{-6}$

108.   a. $Zn(s) + Cu^{2+}(aq) \rightarrow Zn^{2+}(aq) + Cu(s)$ $E°_{cell} = 1.10 \text{ V}$; $E_{cell} = 1.10 \text{ V} - \frac{0.0591}{2} \log \frac{[Zn^{2+}]}{[Cu^{2+}]}$

$$E_{cell} = 1.10 \text{ V} - \frac{0.0591}{2} \log \frac{0.10}{2.50} = 1.10 \text{ V} + 0.041 \text{ V} = 1.14 \text{ V}$$

b. $10.0 \text{ h} \times \frac{60 \text{ min}}{\text{h}} \times \frac{60 \text{ s}}{\text{min}} \times \frac{10.0 \text{ C}}{\text{s}} \times \frac{1 \text{ mol e}^-}{96,485 \text{ C}} \times \frac{1 \text{ mol Cu}}{2 \text{ mol e}^-} = 1.87 \text{ mol Cu produced}$

The Cu^{2+} concentration will decrease by 1.87 mol/L, and the Zn^{2+} concentration will increase by 1.87 mol/L.

$[Cu^{2+}] = 2.50 - 1.87 = 0.63 \, M$; $[Zn^{2+}] = 0.10 + 1.87 = 1.97 \, M$

$$E_{cell} = 1.10 \text{ V} - \frac{0.0591}{2} \log \frac{1.97}{0.63} = 1.10 \text{ V} - 0.015 \text{ V} = 1.09 \text{ V}$$

c. $1.87 \text{ mol Zn consumed} \times \frac{65.38 \text{ g Zn}}{\text{mol Zn}} = 122 \text{ g Zn}$; Mass of electrode = 200. - 122 = 78 g Zn

$1.87 \text{ mol Cu formed} \times \frac{63.55 \text{ g Cu}}{\text{mol Cu}} = 119 \text{ g Cu}$; Mass of electrode = 200. + 119 = 319 g Cu

d.   Three things could possibly cause this battery to go dead:

   1.   All of the Zn is consumed.
   2.   All of the Cu^{2+} is consumed.
   3.   Equilibrium is reached ($E_{cell} = 0$).

We began with 2.50 mol Cu^{2+} and 200. g Zn × 1 mol Zn/65.38 g Zn = 3.06 mol Zn. Cu^{2+} is the limiting reagent and will run out first. To react all the Cu^{2+} requires:

$$2.50 \text{ mol Cu}^{2+} \times \frac{2 \text{ mol e}^-}{\text{mol Cu}^{2+}} \times \frac{96,485 \text{ C}}{\text{mol e}^-} \times \frac{1 \text{ s}}{10.0 \text{ C}} \times \frac{1 \text{ h}}{3600 \text{ s}} = 13.4 \text{ h}$$

For equilibrium to be reached: $E = 0 = 1.10 \text{ V} - \dfrac{0.0591}{2} \log \dfrac{[Zn^{2+}]}{[Cu^{2+}]}$

$\dfrac{[Zn^{2+}]}{[Cu^{2+}]} = K = 10^{2(1.10)/0.0591} = 1.68 \times 10^{37}$

This is such a large equilibrium constant that virtually all of the Cu^{2+} must react to reach equilibrium. So, the battery will go dead in 13.4 hours.

109.

$$2\,H^+ + 2\,e^- \rightarrow H_2 \qquad\qquad\qquad E° = 0.000 \text{ V}$$
$$\underline{ Fe \rightarrow Fe^{2+} + 2\,e^- \qquad\qquad -E° = -(-0.440\text{V})}$$

$$2\,H^+(aq) + Fe(s) \rightarrow H_2(g) + Fe^{2+}(aq) \qquad E°_{cell} = 0.440 \text{ V}$$

$E_{cell} = E°_{cell} - \dfrac{0.0591}{n} \log Q$, where $n = 2$ and $Q = \dfrac{P_{H_2} \times [Fe^{2+}]}{[H^+]^2}$

To determine K_a for the weak acid, first use the electrochemical data to determine the H^+ concentration in the half-cell containing the weak acid.

$0.333 \text{ V} = 0.440 \text{ V} - \dfrac{0.0591}{2} \log \dfrac{1.00\,(1.00 \times 10^{-3})}{[H^+]^2}$

$\dfrac{0.107(2)}{0.0591} = \log \dfrac{1.00 \times 10^{-3}}{[H^+]^2}, \quad \dfrac{1.00 \times 10^{-3}}{[H^+]^2} = 10^{3.621} = 4.18 \times 10^3, \quad [H^+] = 4.89 \times 10^{-4} \, M$

Now we can solve for the K_a value of the weak acid HA through the normal setup for a weak acid problem.

$$HA \quad\rightleftharpoons\quad H^+ \quad + \quad A^- \qquad K_a = \dfrac{[H^+][A^-]}{[HA]}$$

Initial	1.00 *M*	~0	0
Equil.	1.00 - *x*	*x*	*x*

$K_a = \dfrac{x^2}{1.00 - x}$ where $x = [H^+] = 4.89 \times 10^{-4} \, M$, $K_a = \dfrac{(4.89 \times 10^{-4})^2}{1.00 - 4.89 \times 10^{-4}} = 2.39 \times 10^{-7}$

110.   a.   Nonreactive anions are present in each half-cell to balance the cation charges.

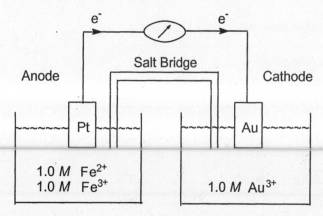

b. $Au^{3+}(aq) + 3\ Fe^{2+}(aq) \rightarrow 3\ Fe^{3+}(aq) + Au(s)$ $E^{\circ}_{cell} = 1.50 - 0.77 = 0.73\ V$

$$E_{cell} = E^{\circ}_{cell} - \frac{0.0591}{n}\ \log Q = 0.73\ V - \frac{0.0591}{3}\ \log \frac{[Fe^{3+}]^3}{[Au^{3+}]\,[Fe^{2+}]^3}$$

Since $[Fe^{3+}] = [Fe^{2+}] = 1.0\ M$: $0.31\ V = 0.73\ V - \dfrac{0.0591}{3}\ \log \dfrac{1}{[Au^{3+}]}$

$$\frac{3(-0.42)}{0.0591} = -\log \frac{1}{[Au^{3+}]}, \quad \log[Au^{3+}] = -21.32, \quad [Au^{3+}] = 10^{-21.32} = 4.8 \times 10^{-22}\ M$$

$Au^{3+} + 4\ Cl^- \rightleftharpoons AuCl_4^-$; Since the equilibrium Au^{3+} concentration is so small, assume $[AuCl_4^-] \approx [Au^{3+}]_o \approx 1.0\ M$, i.e., assume K is large, so the reaction essentially goes to completion.

$$K = \frac{[AuCl_4^-]}{[Au^{3+}]\,[Cl^-]^4} = \frac{1.0}{(4.8 \times 10^{-22})\,(0.10)^4} = 2.1 \times 10^{25}; \quad \text{Assumption good (K is large).}$$

111.   a. $E_{cell} = E_{ref} + 0.05916\ pH$, $0.480\ V = 0.250\ V + 0.05916\ pH$

$$pH = \frac{0.480 - 0.250}{0.05916} = 3.888; \quad \text{Uncertainty} = \pm 1\ mV = \pm 0.001\ V$$

$$pH_{max} = \frac{0.481 - 0.250}{0.05916} = 3.905; \quad pH_{min} = \frac{0.479 - 0.250}{0.05916} = 3.871$$

So, if the uncertainty in potential is $\pm 0.001\ V$, the uncertainty in pH is ± 0.017 or about ± 0.02 pH units. For this measurement, $[H^+] = 10^{-3.888} = 1.29 \times 10^{-4}\ M$. For an error of $+1\ mV$, $[H^+] = 10^{-3.905} = 1.24 \times 10^{-4}\ M$. For an error of $-1\ mV$, $[H^+] = 10^{-3.871} = 1.35 \times 10^{-4}\ M$. So, the uncertainty in $[H^+]$ is $\pm 0.06 \times 10^{-4}\ M = \pm 6 \times 10^{-6}\ M$.

b.   From part a, we will be within ± 0.02 pH units if we measure the potential to the nearest ± 0.001 V (1 mV).

112.   a.   From Table 17.1: $2\ H_2O + 2\ e^- \rightarrow H_2 + 2\ OH^-$ $E^{\circ} = -0.83\ V$

$$E^{\circ}_{cell} = E^{\circ}_{H_2O} - E^{\circ}_{Zr} = -0.83\ V + 2.36\ V = 1.53\ V$$

Yes, the reduction of H_2O to H_2 by Zr is spontaneous at standard conditions since $E^{\circ}_{cell} > 0$.

b. $(2\ H_2O + 2\ e^- \rightarrow H_2 + 2\ OH^-) \times 2$
       $Zr + 4\ OH^- \rightarrow ZrO_2{\cdot}H_2O + H_2O + 4\ e^-$

$$\overline{3\ H_2O(l) + Zr(s) \rightarrow 2\ H_2(g) + ZrO_2{\cdot}H_2O(s)}$$

c. $\Delta G^{\circ} = -nFE^{\circ} = -(4 \text{ mol e}^-)(96{,}485 \text{ C/mol e}^-)(1.53 \text{ J/C}) = -5.90 \times 10^5 \text{ J} = -590. \text{ kJ}$

$E = E^{\circ} - \dfrac{0.0591}{n} \log Q; \quad \text{At equilibrium, } E = 0 \text{ and } Q = K.$

$E^{\circ} = \dfrac{0.0591}{n} \log K, \quad \log K = \dfrac{4(1.53)}{0.0591} = 104, \quad K \approx 10^{104}$

d. $1.00 \times 10^3 \text{ kg Zr} \times \dfrac{1000 \text{ g}}{\text{kg}} \times \dfrac{1 \text{ mol Zr}}{91.22 \text{ g Zr}} \times \dfrac{2 \text{ mol H}_2}{\text{mol Zr}} = 2.19 \times 10^4 \text{ mol H}_2$

$2.19 \times 10^4 \text{ mol H}_2 \times \dfrac{2.016 \text{ g H}_2}{\text{mol H}_2} = 4.42 \times 10^4 \text{ g H}_2$

$V = \dfrac{nRT}{P} = \dfrac{(2.19 \times 10^4 \text{ mol}) \, (0.08206 \text{ L atm/mol}\bullet\text{K}) \, (1273 \text{ K})}{1.0 \text{ atm}} = 2.3 \times 10^6 \text{ L H}_2$

e. Probably yes; Less radioactivity overall was released by venting the H_2 than what would have been released if the H_2 had exploded inside the reactor (as happened at Chernobyl). Neither alternative is pleasant, but venting the radioactive hydrogen is the less unpleasant of the two alternatives.

113. a.
$$(Ag^+ + e^- \rightarrow Ag) \times 2 \qquad\qquad E^{\circ} = 0.80 \text{ V}$$
$$Cu \rightarrow Cu^{2+} + 2 e^- \qquad\qquad -E^{\circ} = -0.34 \text{ V}$$

$$2 \, Ag^+(aq) + Cu(s) \rightarrow 2 \, Ag(s) + Cu^{2+}(aq) \quad E^{\circ}_{cell} = 0.46 \text{ V}$$

$$E_{cell} = E^{\circ}_{cell} - \dfrac{0.0591}{n} \log Q \quad \text{where } n = 2 \text{ and } Q = \dfrac{[Cu^{2+}]}{[Ag^+]^2}$$

To calculate E_{cell}, we need to use the K_{sp} data to determine $[Ag^+]$.

$$AgCl(s) \quad \rightleftharpoons \quad Ag^+(aq) \quad + \quad Cl^-(aq) \quad K_{sp} = 1.6 \times 10^{-10} = [Ag^+] \, [Cl^-]$$

Initial s = solubility (mol/L)   0   0
Equil. s s

$K_{sp} = 1.6 \times 10^{-10} = s^2, \quad s = [Ag^+] = 1.3 \times 10^{-5} \text{ mol/L}$

$E_{cell} = 0.46 \text{ V} - \dfrac{0.0591}{2} \log \dfrac{2.0}{(1.3 \times 10^{-5})^2} = 0.46 \text{ V} - 0.30 = 0.16 \text{ V}$

b. $Cu^{2+}(aq) + 4\,NH_3(aq) \rightleftharpoons Cu(NH_4)_4^{2+}(aq)$ $K = 1.0 \times 10^{13} = \dfrac{[Cu(NH_3)_4^{2+}]}{[Cu^{2+}][NH_3]^4}$

Since K is very large for the formation of $Cu(NH_3)_4^{2+}$, the forward reaction is dominant. At equilibrium, essentially all of the 2.0 $M\,Cu^{2+}$ will react to form 2.0 $M\,Cu(NH_3)_4^{2+}$. This reaction requires 8.0 $M\,NH_3$ to react with all of the Cu^{2+} in the balanced equation. Therefore, the mol of NH_3 added to 1.0 L solution will be larger than 8.0 mol since some NH_3 must be present at equilibrium. In order to calculate how much NH_3 is present at equilibrium, we need to use the electrochemical data to determine the Cu^{2+} concentration.

$$E_{cell} = E_{cell}^{\circ} - \frac{0.0591}{n}\log Q,\quad 0.52\,V = 0.46\,V - \frac{0.0591}{2}\log \frac{[Cu^{2+}]}{(1.3 \times 10^{-5})^2}$$

$$\log \frac{[Cu^{2+}]}{(1.3 \times 10^{-5})^2} = \frac{-0.06(2)}{0.0591} = -2.03,\quad \frac{[Cu^{2+}]}{(1.3 \times 10^{-5})^2} = 10^{-2.03} = 9.3 \times 10^{-3}$$

$[Cu^{2+}] = 1.6 \times 10^{-12} = 2 \times 10^{-12}\,M$ (We carried extra significant figures in the calculation.)

Note: Our assumption that the 2.0 $M\,Cu^{2+}$ essentially reacts to completion is excellent as only $2 \times 10^{-12}\,M\,Cu^{2+}$ remains after this reaction. Now we can solve for the equilibrium $[NH_3]$.

$$K = 1.0 \times 10^{13} = \frac{[Cu(NH_3)_4^{2+}]}{[Cu^{2+}][NH_3]^4} = \frac{(2.0)}{(2 \times 10^{-12})[NH_3]^4},\quad [NH_3] = 0.6\,M$$

Since 1.0 L of solution is present, then 0.6 mol NH_3 remains at equilibrium. The total mol of NH_3 added is 0.6 mol plus the 8.0 mol NH_3 necessary to form 2.0 $M\,Cu(NH_3)_4^{2+}$. Therefore, 8.0 + 0.6 = 8.6 mol NH_3 were added.

CHAPTER EIGHTEEN

THE NUCLEUS: A CHEMIST'S VIEW

Questions

1. Fission: Splitting of a heavy nucleus into two (or more) lighter nuclei.

   Fusion: Combining two light nuclei to form a heavier nucleus.

   The maximum binding energy per nucleon occurs at Fe. Nuclei smaller than Fe become more stable by fusing to form heavier nuclei closer in mass to Fe. Nuclei larger than Fe form more stable nuclei by splitting to form lighter nuclei closer in mass to Fe.

2. Characteristic frequencies of energies emitted in a nuclear reaction suggest that discrete energy levels exist in the nucleus. The extra stability of certain numbers of nucleons and the predominance of nuclei with even numbers of nucleons suggest that the nuclear structure might be described by using quantum numbers.

3. The assumptions are that the ^{14}C level in the atmosphere is constant or that the ^{14}C level at the time the plant died can be calculated. A constant ^{14}C level is a poor assumption, and accounting for variation is complicated. Another problem is that some of the material must be destroyed to determine the ^{14}C level.

4. A nonradioactive substance can be put in equilibrium with a radioactive substance. The two materials can then be checked to see if all the radioactivity remains in the original material or if it has been scrambled by the equilibrium.

5. No, coal-fired power plants also pose risks. A partial list of risks is:

Coal	Nuclear
Air pollution	Radiation exposure to workers
Coal mine accidents	Disposal of wastes
Health risks to miners	Meltdown
(black lung disease)	Terrorists
	Public fear

6.   Even though gamma rays penetrate human tissue very deeply, they are very small and cause only occasional ionization of biomolecules. Alpha particles, because they are much more massive, are very effective at causing ionization of biomolecules and produce a dense trail of damage once they get inside an organism.

7.   For fusion reactions, a collision of sufficient energy must occur between two positively charged particles to initiate the reaction. This requires high temperatures. In fission, an electrically neutral neutron collides with the positively charged nucleus. This has a much lower activation energy.

8.   Moderator: Slows the neutrons to increase the efficiency of the fission reaction.

      Control rods: Absorbs neutrons to slow or halt the fission reaction.

Exercises

Radioactive Decay and Nuclear Transformations

9.   All nuclear reactions must be charge balanced and mass balanced. To charge balance, balance the sum of the atomic numbers on each side of the reaction, and to mass balance, balance the sum of the mass numbers on each side of the reaction.

   a. $^{51}_{24}Cr + ^{\ 0}_{-1}e \rightarrow ^{51}_{23}V$

   b. $^{131}_{\ 53}I \rightarrow ^{\ 0}_{-1}e + ^{131}_{\ 54}Xe$

10.   a. $^{32}_{15}P \rightarrow ^{\ 0}_{-1}e + ^{32}_{16}S$

   b. $^{235}_{\ 92}U \rightarrow ^{4}_{2}He + ^{231}_{\ 90}Th$

11.   a. $^{68}_{31}Ga + ^{\ 0}_{-1}e \rightarrow ^{68}_{30}Zn$ b. $^{62}_{29}Cu \rightarrow ^{\ 0}_{+1}e + ^{62}_{28}Ni$

   c. $^{212}_{\ 87}Fr \rightarrow ^{4}_{2}He + ^{208}_{\ 85}At$ d. $^{129}_{\ 51}Sb \rightarrow ^{\ 0}_{-1}e + ^{129}_{\ 52}Te$

12.   a. $^{73}_{31}Ga \rightarrow ^{73}_{32}Ge + ^{\ 0}_{-1}e$ b. $^{192}_{\ 78}Pt \rightarrow ^{188}_{\ 76}Os + ^{4}_{2}He$

   c. $^{205}_{\ 83}Bi \rightarrow ^{205}_{\ 82}Pb + ^{\ 0}_{+1}e$ d. $^{241}_{\ 96}Cm + ^{\ 0}_{-1}e \rightarrow ^{241}_{\ 95}Am$

13. $^{247}_{97}Bk \rightarrow ^{207}_{82}Pb + ?\ ^{4}_{2}He + ?\ ^{0}_{-1}e$; The change in mass number (247 - 207 = 40) is due exclusively to the alpha particles. A change in mass number of 40 requires 10 $^{4}_{2}He$ particles to be produced. The atomic number only changes by 97 - 82 = 15. The 10 alpha particles change the atomic number by 20, so 5 $^{0}_{-1}e$ (5 beta particles) are produced in the decay series of ^{247}Bk to ^{207}Pb.

14. a. $^{241}_{95}Am \rightarrow ^{4}_{2}He + ^{237}_{93}Np$

     b. $^{241}_{95}Am \rightarrow 8\ ^{4}_{2}He + 4\ ^{0}_{-1}e + ^{209}_{83}Bi$; The final product is $^{209}_{83}Bi$.

     c. $^{241}_{95}Am \rightarrow ^{237}_{93}Np + \alpha \rightarrow ^{233}_{91}Pa + \alpha \rightarrow ^{233}_{92}U + \beta \rightarrow ^{229}_{90}Th + \alpha \rightarrow ^{225}_{88}Ra + \alpha$

$^{213}_{84}Po + \beta \leftarrow ^{213}_{83}Bi + \alpha \leftarrow ^{217}_{85}At + \alpha \leftarrow ^{221}_{87}Fr + \alpha \leftarrow ^{225}_{89}Ac + \beta$

$^{209}_{82}Pb + \alpha \rightarrow ^{209}_{83}Bi + \beta$

The intermediate radionuclides are:

$^{237}_{93}Np$, $^{233}_{91}Pa$, $^{233}_{92}U$, $^{229}_{90}Th$, $^{225}_{88}Ra$, $^{225}_{89}Ac$, $^{221}_{87}Fr$, $^{217}_{85}At$, $^{213}_{83}Bi$, $^{213}_{84}Po$, and $^{209}_{82}Pb$.

15. $^{53}_{26}Fe$ has too many protons. It will undergo either positron production, electron capture and/or alpha particle production. $^{59}_{26}Fe$ has too many neutrons and will undergo beta particle production. (See Table 18.2 of the text.)

16. Reference Table 18.2 of the text for potential radioactive decay processes. ^{17}F and ^{18}F contain too many protons or too few neutrons. Electron capture and positron production are both possible decay mechanisms that increase the neutron to proton ratio. Alpha particle production also increases the neutron to proton ratio, but it is not likely for these light nuclei. ^{21}F contains too many neutrons or too few protons. Beta production lowers the neutron to proton ratio, so we expect ^{21}F to be a β-emitter.

17. a. $^{249}_{98}Cf + ^{18}_{8}O \rightarrow ^{263}_{106}Sg + 4\ ^{1}_{0}n$ b. $^{259}_{104}Rf$; $^{263}_{106}Sg \rightarrow ^{4}_{2}He + ^{259}_{104}Rf$

18. a. $^{240}_{95}Am + ^{4}_{2}He \rightarrow ^{243}_{97}Bk + ^{1}_{0}n$ b. $^{238}_{92}U + ^{12}_{6}C \rightarrow ^{244}_{98}Cf + 6\ ^{1}_{0}n$

     c. $^{249}_{98}Cf + ^{15}_{7}N \rightarrow ^{260}_{105}Db + 4\ ^{1}_{0}n$ d. $^{249}_{98}Cf + ^{10}_{5}B \rightarrow ^{257}_{103}Lr + 2\ ^{1}_{0}n$

Kinetics of Radioactive Decay

19.   All radioactive decay follows first-order kinetics where $t_{1/2} = (\ln 2)/k$.

$$t_{1/2} = \frac{\ln 2}{k} = \frac{0.693}{1.0 \times 10^{-3} \, h^{-1}} = 690 \, h$$

20. $$k = \frac{\ln 2}{t_{1/2}} = \frac{0.69315}{432.2 \, yr} \times \frac{1 \, yr}{365 \, d} \times \frac{1 \, d}{24 \, h} \times \frac{1 \, hr}{3600 \, s} = 5.086 \times 10^{-11} \, s^{-1}$$

$$\text{Rate} = kN = 5.086 \times 10^{-11} \, s^{-1} \times 5.00 \, g \times \frac{1 \, mol}{241 \, g} \times \frac{6.022 \times 10^{23} \, nuclei}{mol} = 6.35 \times 10^{11} \, decays/s$$

6.35×10^{11} alpha particles are emitted each second from a 5.00 g ^{241}Am sample.

21.   Kr-81 is most stable since it has the longest half-life while Kr-73 is hottest (least stable) since it has the shortest half-life.

12.5% of each isotope will remain after 3 half-lives:

$$100\% \xrightarrow{t_{1/2}} 50\% \xrightarrow{t_{1/2}} 25\% \xrightarrow{t_{1/2}} 12.5\%$$

For Kr-73: t = 3(27 s) = 81 s

For Kr-74: t = 3(11.5 min) = 34.5 min

For Kr-76: t = 3(14.8 h) = 44.4 h

For Kr-81: t = 3(2.1 × 10^5 yr) = 6.3 × 10^5 yr

22.   a. $$k = \frac{\ln 2}{t_{1/2}} = \frac{0.6931}{12.8 \, d} \times \frac{1 \, d}{24 \, h} \times \frac{1 \, h}{3600 \, s} = 6.27 \times 10^{-7} \, s^{-1}$$

b. $$\text{Rate} = kN = 6.27 \times 10^{-7} \, s^{-1} \times \left(28.0 \times 10^{-3} \, g \times \frac{1 \, mol}{64.0 \, g} \times \frac{6.022 \times 10^{23} \, nuclei}{mol} \right)$$

$$\text{Rate} = 1.65 \times 10^{14} \, decays/s$$

c.   25% of the ^{64}Cu will remain after 2 half-lives (100% decays to 50% after one half-life, which decays to 25% after a second half-life). Hence, 2(12.8 days) = 25.6 days is the time frame for the experiment.

23.   Units for N and N_o are usually the number of nuclei but can also be grams if the units are the same for both N and N_o. In this problem m = the mass of ^{32}P that remains.

$$175 \, mg \, Na_3 \, ^{32}PO_4 \times \frac{32.0 \, mg \, ^{32}P}{165.0 \, mg \, Na_3 \, ^{32}PO_4} = 33.9 \, mg \, ^{32}P \text{ initially}; k = \frac{\ln 2}{t_{1/2}}$$

$$\ln\left(\frac{N}{N_o}\right) = -kt = \frac{-0.6931\,t}{t_{1/2}}, \quad \ln\left(\frac{m}{33.9\,mg}\right) = \frac{-0.6931\,(35.0\,d)}{14.3\,d};\ \text{Carrying extra sig. figs.:}$$

$$\ln(m) = -1.696 + 3.523 = 1.827, \quad m = e^{1.827} = 6.22\ mg\ {}^{32}P\ \text{remains}$$

24.   a. $0.0100\ Ci \times \dfrac{3.7 \times 10^{10}\ decays/s}{Ci} = 3.7 \times 10^{8}\ decays/s;\quad k = \dfrac{\ln 2}{t_{1/2}}$

   Rate = kN, $\dfrac{3.7 \times 10^{8}\ decays}{s} = \left(\dfrac{0.6931}{2.87\,h} \times \dfrac{1\,h}{3600\,s}\right) \times N,\quad N = 5.5 \times 10^{12}$ atoms of ^{38}S

   5.5×10^{12} atoms $^{38}S \times \dfrac{1\ mol\ ^{38}S}{6.02 \times 10^{23}\ atoms} \times \dfrac{1\ mol\ Na_2{}^{38}SO_4}{mol\ ^{38}S} = 9.1 \times 10^{-12}\ mol\ Na_2{}^{38}SO_4$

   $9.1 \times 10^{-12}\ mol\ Na_2{}^{38}SO_4 \times \dfrac{148.0\ g\ Na_2{}^{38}SO_4}{mol\ Na_2{}^{38}SO_4} = 1.3 \times 10^{-9}\ g = 1.3\ ng\ Na_2{}^{38}SO_4$

   b.   99.99% decays, 0.01% left; $\ln\left(\dfrac{0.01}{100}\right) = -kt = \dfrac{-0.6931\,t}{2.87\,h},\quad t = 38.1\ \text{hours} \approx 40\ \text{hours}$

25. $t = 58.0\ yr;\quad k = \dfrac{\ln 2}{t_{1/2}};\quad \ln\left(\dfrac{N}{N_o}\right) = -kt = \dfrac{-0.6931 \times 58.0\ yr}{28.8\ yr} = -1.40,\quad \left(\dfrac{N}{N_o}\right) = e^{-1.40} = 0.247$

24.7% of the ^{90}Sr remains as of July 16, 2003.

26. $\ln(N/N_o) = -kt;\quad N = 0.010\ N_o;\quad t_{1/2} = (\ln 2)/k$

   $\ln(0.010) = \dfrac{-(\ln 2)\,t}{t_{1/2}} = \dfrac{-0.693\,t}{8.1\,d},\quad t = 54\ \text{days}$

27. $k = \dfrac{\ln 2}{t_{1/2}};\quad \ln\left(\dfrac{N}{N_o}\right) = -kt = \dfrac{-0.6931\,t}{t_{1/2}},\quad \ln\left(\dfrac{N}{13.6}\right) = \dfrac{-0.693\,(15,000\ yr)}{5730\ yr} = -1.8$

   $\dfrac{N}{13.6} = e^{-1.8} = 0.17,\quad N = 13.6 \times 0.17 = 2.3$ counts per minute per g of C

If we had 10. mg C, we would see:

   $10.\ mg \times \dfrac{1\ g}{1000\ mg} \times \dfrac{2.3\ counts}{min\ g} = \dfrac{0.023\ counts}{min}$

It would take roughly 40 min to see a single disintegration. This is too long to wait, and the background radiation would probably be much greater than the ^{14}C activity. Thus, ^{14}C dating is not practical for very small samples.

28. $t_{1/2} = 5730$ yr; $k = (\ln 2)/t_{1/2}$; $\ln (N/N_o) = -kt$; $\ln \dfrac{15.1}{15.3} = \dfrac{-(\ln 2)\,t}{5730\ \text{yr}}$, $t = 109$ yr

No; From ^{14}C dating, the painting was produced (at the earliest) during the late 1800s.

29.   Assuming 1.000 g ^{238}U present in a sample, then 0.688 g ^{206}Pb is present. Since 1 mol ^{206}Pb is produced per mol ^{238}U decayed, then:

$$^{238}\text{U decayed} = 0.688\ \text{g Pb} \times \frac{1\,\text{mol Pb}}{206\ \text{g Pb}} \times \frac{1\,\text{mol U}}{\text{mol Pb}} \times \frac{238\ \text{g U}}{\text{mol U}} = 0.795\ \text{g}\ ^{238}\text{U}$$

Original mass ^{238}U present = 1.000 g + 0.795 g = 1.795 g ^{238}U

$$\ln\!\left(\frac{N}{N_o}\right) = -kt = \frac{-(\ln 2)\,t}{t_{1/2}},\ \ \ln\!\left(\frac{1.000\ \text{g}}{1.795\ \text{g}}\right) = \frac{-0.693\ (t)}{4.5 \times 10^9\ \text{yr}},\ \ t = 3.8 \times 10^9\ \text{yr}$$

30.   a.   The decay of ^{40}K is not the sole source of ^{40}Ca.

       b.   Decay of ^{40}K is the sole source of ^{40}Ar and that no ^{40}Ar is lost over the years.

       c. $\dfrac{0.95\ \text{g}\ ^{40}\text{Ar}}{1.00\ \text{g}\ ^{40}\text{K}}$ = current mass ratio

            0.95 g of ^{40}K decayed to ^{40}Ar. 0.95 g of ^{40}K is only 10.7% of the total ^{40}K that decayed, or:

               0.107 (m) = 0.95 g, m = 8.9 g = total mass of ^{40}K that decayed

            Mass of ^{40}K when the rock was formed was 1.00 g + 8.9 g = 9.9 g.

            $$\ln\!\left(\frac{1.00\ \text{g}\ ^{40}\text{K}}{9.9\ \text{g}\ ^{40}\text{K}}\right) = -kt = \frac{-(\ln 2)\,t}{t_{1/2}} = \frac{-0.6931\,t}{1.27 \times 10^9\ \text{yr}},\ \ t = 4.2 \times 10^9\ \text{years old}$$

       d.   If some ^{40}Ar escaped, then the measured ratio of ^{40}Ar/^{40}K is less than it should be. We would calculate the age of the rocks to be less than it actually is.

Energy Changes in Nuclear Reactions

31. $\Delta E = \Delta mc^2$, $\Delta m = \dfrac{\Delta E}{c^2} = \dfrac{3.9 \times 10^{23}\ \text{kg m}^2/\text{s}^2}{(3.00 \times 10^8\ \text{m/s})^2} = 4.3 \times 10^6$ kg

The sun loses 4.3×10^6 kg of mass each second. Note: 1 J = 1 kg m^2/s^2

32. $$\frac{1.8 \times 10^{14} \text{ kJ}}{\text{s}} \times \frac{1000 \text{ J}}{\text{kJ}} \times \frac{3600 \text{ s}}{\text{h}} \times \frac{24 \text{ h}}{\text{day}} = 1.6 \times 10^{22} \text{ J/day}$$

$$\Delta E = \Delta mc^2, \ \Delta m = \frac{\Delta E}{c^2} = \frac{1.6 \times 10^{22} \text{ J}}{(3.00 \times 10^8 \text{ m/s})^2} = 1.8 \times 10^5 \text{ kg of solar material provides 1 day of solar energy to the earth.}$$

$$1.6 \times 10^{22} \text{ J} \times \frac{1 \text{ kJ}}{1000 \text{ J}} \times \frac{1 \text{ g}}{32 \text{ kJ}} \times \frac{1 \text{ kg}}{1000 \text{ g}} = 5.0 \times 10^{14} \text{ kg of coal is needed to provide the same amount of energy.}$$

33. We need to determine the mass defect, Δm, between the mass of the nucleus and the mass of the individual parts that make up the nucleus. Once Δm is known, we can then calculate ΔE (the binding energy) using $E = mc^2$. Note: $1 \text{ J} = 1 \text{ kg m}^2/\text{s}^2$.

For $^{232}_{94}\text{Pu}$ (94 e, 94 p, 138 n):

mass of ^{232}Pu nucleus = 3.85285×10^{-22} g - mass of 94 electrons

mass of ^{232}Pu nucleus = 3.85285×10^{-22} g - $94(9.10939 \times 10^{-28})$ g = 3.85199×10^{-22} g

$\Delta m = 3.85199 \times 10^{-22}$ g - (mass of 94 protons + mass of 138 neutrons)

$\Delta m = 3.85199 \times 10^{-22}$ g - $[94(1.67262 \times 10^{-24}) + 138(1.67493 \times 10^{-24})]$ g = -3.168×10^{-24} g

For 1 mol of nuclei: $\Delta m = -3.168 \times 10^{-24}$ g/nuclei $\times 6.0221 \times 10^{23}$ nuclei/mol = -1.908 g/mol

$\Delta E = \Delta mc^2 = (-1.908 \times 10^{-3} \text{ kg/mol})(2.9979 \times 10^8 \text{ m/s})^2 = -1.715 \times 10^{14} = \text{J/mol}$

For $^{231}_{91}\text{Pa}$ (91 e, 91 p, 140 n):

mass of ^{231}Pa nucleus = 3.83616×10^{-22} g - $91(9.10939 \times 10^{-28})$ g = 3.83533×10^{-22} g

$\Delta m = 3.83533 \times 10^{-22}$ g - $[91(1.67262 \times 10^{-24}) + 140(1.67493 \times 10^{-24})]$ g = -3.166×10^{-24} g

$$\Delta E = \Delta mc^2 = \frac{-3.166 \times 10^{-27} \text{ kg}}{\text{nuclei}} \times \frac{6.0221 \times 10^{23} \text{ nuclei}}{\text{mol}} \times \left(\frac{2.9979 \times 10^8 \text{ m}}{\text{s}}\right)^2$$

$$= -1.714 \times 10^{14} \text{ J/mol}$$

34. From the table at the back of the text, the mass of a proton = 1.00728 amu, the mass of a neutron = 1.00866 amu, and the mass of an electron = 5.486×10^{-4} amu.

Mass of nucleus = mass of atom - mass of electrons = 55.9349 - 26(0.0005486) = 55.9206 amu

$26\,^1_1\text{H} + 30\,^1_0\text{n} \rightarrow\, ^{56}_{26}\text{Fe}; \ \Delta m = 55.9206 \text{ amu} - [26(1.00728) + 30(1.00866)] \text{ amu} = -0.5285 \text{ amu}$

$$\Delta E = \Delta mc^2 = -0.5285 \text{ amu} \times \frac{1.6605 \times 10^{-27} \text{ kg}}{\text{amu}} \times (2.9979 \times 10^8 \text{ m/s})^2 = -7.887 \times 10^{-11} \text{ J}$$

$$\frac{\text{binding energy}}{\text{nucleon}} = \frac{7.887 \times 10^{-11} \text{ J}}{56 \text{ nucleons}} = 1.408 \times 10^{-12} \text{ J/nucleon}$$

35.   Let m_e = mass of electron; For ^{12}C (6e, 6p, 6n): mass defect = Δm = mass of ^{12}C nucleus -[mass of 6 protons + mass of 6 neutrons]. Note: the atomic masses of the elements given include the mass of the electrons.

$\Delta m = 12.0000 \text{ amu} - 6 \, m_e - [6(1.00782 - m_e) + 6(1.00866)]$; Mass of electrons cancel.

$\Delta m = 12.0000 - [6(1.00782) + 6(1.00866)] = -0.0989 \text{ amu}$

$$\Delta E = \Delta mc^2 = -0.0989 \text{ amu} \times \frac{1.6605 \times 10^{-27} \text{ kg}}{\text{amu}} \times (2.9979 \times 10^8 \text{ m/s})^2 = -1.48 \times 10^{-11} \text{ J}$$

$$\frac{\text{BE}}{\text{nucleon}} = \frac{1.48 \times 10^{-11} \text{ J}}{12 \text{ nucleons}} = 1.23 \times 10^{-12} \text{ J/nucleon}$$

For ^{235}U (92e, 92p, 143n):

$\Delta m = 235.0439 - 92 \, m_e - [92(1.00782 - m_e) + 143(1.00866)] = -1.9139 \text{ amu}$

$$\Delta E = \Delta mc^2 = -1.9139 \text{ amu} \times \frac{1.66054 \times 10^{-27} \text{ kg}}{\text{amu}} \times (2.99792 \times 10^8 \text{ m/s})^2 = -2.8563 \times 10^{-10} \text{ J}$$

$$\frac{\text{BE}}{\text{nucleon}} = \frac{2.8563 \times 10^{-10} \text{ J}}{235 \text{ nucleons}} = 1.2154 \times 10^{-12} \text{ J/nucleon}$$

Since ^{56}Fe is the most stable known nucleus, the binding energy per nucleon for ^{56}Fe (1.408×10^{-12} J/nucleon) will be larger than that for ^{12}C or ^{235}U (see Figure 18.9 of the text).

36.   For ^2_1H: mass defect = Δm = mass of ^2_1H nucleus - mass of proton - mass of neutron; Let's determine the mass defect in a slightly different way than in Exercise 18.35. Instead of using the atomic mass of hydrogen-1, we will use the mass of the electron and the mass of the proton. The mass of the ^2H nucleus will equal the atomic mass of ^2H minus the mass of the electron in a ^2H atom. From the back of the text, the pertinent masses are: $m_e = 5.49 \times 10^{-4}$ amu, $m_p = 1.00728$ amu, $m_n = 1.00866$ amu.

$\Delta m = 2.01410 \text{ amu} - 0.000549 \text{ amu} - [1.00728 \text{ amu} + 1.00866 \text{ amu}] = -2.39 \times 10^{-3} \text{ amu}$

$$\Delta E = \Delta mc^2 = -2.39 \times 10^{-3} \text{ amu} \times \frac{1.6605 \times 10^{-27} \text{ kg}}{\text{amu}} \times (2.998 \times 10^8 \text{ m/s})^2 = -3.57 \times 10^{-13} \text{ J}$$

$$\frac{\text{BE}}{\text{nucleon}} = \frac{3.57 \times 10^{-13} \text{ J}}{2 \text{ nucleons}} = 1.79 \times 10^{-13} \text{ J/nucleon}$$

For $_{1}^{3}$H: $\Delta m = 3.01605 - 0.000549 - [1.00728 + 2(1.00866)] = -9.10 \times 10^{-3}$ amu

$$\Delta E = -9.10 \times 10^{-3} \text{ amu} \times \frac{1.6605 \times 10^{-27} \text{ kg}}{\text{amu}} \times (2.998 \times 10^8 \text{ m/s})^2 = -1.36 \times 10^{-12} \text{ J}$$

$$\frac{BE}{\text{nucleon}} = \frac{1.36 \times 10^{-12} \text{ J}}{3 \text{ nucleons}} = 4.53 \times 10^{-13} \text{ J/nucleon}$$

37. $_{1}^{1}$H + $_{1}^{1}$H → $_{1}^{2}$H + $_{+1}^{0}$e; $\Delta m = (2.01410 \text{ amu} - m_e + m_e) - 2(1.00782 \text{ amu} - m_e)$

$\Delta m = 2.01410 - 2(1.00782) + 2(0.000549) = -4.4 \times 10^{-4}$ amu for two protons reacting

When two mol of protons undergo fusion, $\Delta m = -4.4 \times 10^{-4}$ g.

$\Delta E = \Delta mc^2 = -4.4 \times 10^{-7} \text{ kg} \times (3.00 \times 10^8 \text{ m/s})^2 = -4.0 \times 10^{10}$ J

$$\frac{-4.0 \times 10^{10} \text{ J}}{2 \text{ mol protons}} \times \frac{1 \text{ mol}}{1.01 \text{ g}} = -2.0 \times 10^{10} \text{ J/g of hydrogen nuclei}$$

38. $_{1}^{2}$H + $_{1}^{3}$H → $_{2}^{4}$He + $_{0}^{1}$n; Using atomic masses, the masses of the electrons cancel when determining Δm for this nuclear reaction.

$\Delta m = [4.00260 + 1.00866 - (2.01410 + 3.01605)]$ amu $= -1.889 \times 10^{-2}$ amu

For the production of one mol of $_{2}^{4}$He: $\Delta m = -1.889 \times 10^{-2}$ g $= -1.889 \times 10^{-5}$ kg

$\Delta E = \Delta mc^2 = -1.889 \times 10^{-5} \text{ kg} \times (2.9979 \times 10^8 \text{ m/s})^2 = -1.698 \times 10^{12}$ J/mol

For 1 nucleus of $_{2}^{4}$He:

$$\frac{-1.698 \times 10^{12} \text{ J}}{\text{mol}} \times \frac{1 \text{ mol}}{6.0221 \times 10^{23} \text{ nuclei}} = -2.820 \times 10^{-12} \text{ J/nucleus}$$

Detection, Uses, and Health Effects of Radiation

39.   The Geiger-Müller tube has a certain response time. After the gas in the tube ionizes to produce a "count," some time must elapse for the gas to return to an electrically neutral state. The response of the tube levels off because, at high activities, radioactive particles are entering the tube faster than the tube can respond to them.

40.   Not all of the emitted radiation enters the Geiger-Müller tube. The fraction of radiation entering the tube must be constant.

41.   All evolved oxygen in O_2 comes from water and not from carbon dioxide.

42.   Water is produced in this reaction by removing an OH group from one substance and H from the other substance. There are two ways to do this:

i. $CH_3C(=O)$—(OH + H)—$^{18}OCH_3$ $\longrightarrow$ $CH_3C(=O)$—$^{18}OCH_3$ + HO—H

ii. $CH_3CO(=O)$—(H + H^{18}O)—CH_3 $\longrightarrow$ $CH_3CO(=O)$—CH_3 + H—^{18}OH

Since the water produced is not radioactive, methyl acetate forms by the first reaction in which all the oxygen-18 ends up in methyl acetate.

43.   Release of Sr is probably more harmful. Xe is chemically unreactive. Strontium is in the same family as calcium and could be absorbed and concentrated in the body in a fashion similar to Ca. This puts the radioactive Sr in the bones: red blood cells are produced in bone marrow. Xe would not be readily incorporated into the body.

The chemical properties determine where a radioactive material may be concentrated in the body or how easily it may be excreted. The length of time of exposure and what is exposed to radiation significantly affects the health hazard. (See exercise 18.44 for a specific example.)

44.   i) and ii) mean that Pu is not a significant threat outside the body. Our skin is sufficient to keep out the α particles. If Pu gets inside the body, it is easily oxidized to Pu^{4+} (iv), which is chemically similar to Fe^{3+} (iii). Thus, Pu^{4+} will concentrate in tissues where Fe^{3+} is found. One of these is the bone marrow where red blood cells are produced. Once inside the body, α particles cause considerable damage.

Additional Exercises

45.   The most abundant isotope is generally the most stable isotope. The periodic table predicts that the most stable isotopes for exercises a - d are ^{39}K, ^{56}Fe, ^{23}Na and ^{204}Tl. (Reference Table 18.2 of the text for potential decay processes.)

a.   Unstable; ^{45}K has too many neutrons and will undergo beta particle production.

b.   Stable

c.   Unstable; ^{20}Na has too few neutrons and will most likely undergo electron capture or positron production. Alpha particle production makes too severe of a change to be a likely decay process for the relatively light ^{20}Na nuclei. Alpha particle production usually occurs for heavy nuclei.

d.   Unstable; ^{194}Tl has too few neutrons and will undergo electron capture, positron production and/or alpha particle production.

46. $\ln(N/N_o) = -kt$; $k = (\ln 2)/t_{1/2}$; $N = 0.001 \times N_o$

$$\ln\left(\frac{0.001 \times N_o}{N_o}\right) = \frac{-(\ln 2)\,t}{24,100 \text{ yr}}, \quad \ln 0.001 = -2.88 \times 10^{-5}\,t, \quad t = 2 \times 10^5 \text{ yr} = 200,000 \text{ yr (!)}$$

47. $\ln\left(\dfrac{N}{N_o}\right) = -kt = \dfrac{-(\ln 2)\,t}{12.3 \text{ yr}}, \quad \ln\left(\dfrac{0.17 \times N_o}{N_o}\right) = -5.64 \times 10^{-2}\,t, \quad t = 31.4 \text{ yr}$

It takes 31.4 yr for the tritium to decay to 17% of the original amount. Hence, the watch stopped fluorescing enough to be read in 1975 (1944 + 31.4).

48. $\Delta m = -2(5.486 \times 10^{-4} \text{ amu}) = -1.097 \times 10^{-3} \text{ amu}$

$$\Delta E = \Delta mc^2 = -1.097 \times 10^{-3} \text{ amu} \times \frac{1.6605 \times 10^{-27} \text{ kg}}{\text{amu}} \times (2.9979 \times 10^8 \text{ m/s})^2 = -1.637 \times 10^{-13} \text{ J}$$

$E_{photon} = 1/2(1.637 \times 10^{-13} \text{ J}) = 8.185 \times 10^{-14} \text{ J} = hc/\lambda$

$$\lambda = \frac{hc}{E} = \frac{6.6261 \times 10^{-34} \text{ J s} \times 2.9979 \times 10^8 \text{ m/s}}{8.185 \times 10^{-14} \text{ J}} = 2.427 \times 10^{-12} \text{ m} = 2.427 \times 10^{-3} \text{ nm}$$

49. $20{,}000 \text{ ton TNT} \times \dfrac{4 \times 10^9 \text{ J}}{\text{ton TNT}} \times \dfrac{1 \text{ mol } {}^{235}\text{U}}{2 \times 10^{13} \text{ J}} \times \dfrac{235 \text{ g } {}^{235}\text{U}}{\text{mol } {}^{235}\text{U}} = 940 \text{ g } {}^{235}\text{U} \approx 900 \text{ g } {}^{235}\text{U}$

This assumes that all of the ^{235}U undergoes fission.

50.   In order to sustain a nuclear chain reaction, the neutrons produced by the fission must be contained within the fissionable material so that they can go on to cause other fissions. The fissionable material must be closely packed together to ensure that neutrons are not lost to the outside. The critical mass is the mass of material in which exactly one neutron from each fission event causes another fission event so that the process sustains itself. A supercritical situation occurs when more than one neutron from each fission event causes another fission event. In this case, the process rapidly escalates and the heat build up causes a violent explosion.

Challenge Problems

51.   Assuming that the radionuclide is long lived enough such that no significant decay occurs during the time of the experiment, the total counts of radioactivity injected are:

$$0.10 \text{ mL} \times \frac{5.0 \times 10^3 \text{ cpm}}{\text{mL}} = 5.0 \times 10^2 \text{ cpm}$$

Assuming that the total activity is uniformly distributed only in the rat's blood, the blood volume is:

$$V \times \frac{48 \text{ cpm}}{\text{mL}} = 5.0 \times 10^2 \text{ cpm}, \quad V = 10.4 \text{ mL} = 10. \text{ mL}$$

52.   a.   From Table 17.1: $2 H_2O + 2 e^- \rightarrow H_2 + 2 OH^-$ $E° = -0.83$ V

$E°_{cell} = E°_{H_2O} - E°_{Zr} = -0.83$ V $+ 2.36$ V $= 1.53$ V

Yes, the reduction of H_2O to H_2 by Zr is spontaneous at standard conditions since $E°_{cell} > 0$.

   b. $(2 H_2O + 2 e^- \rightarrow H_2 + 2 OH^-) \times 2$
         $Zr + 4 OH^- \rightarrow ZrO_2 \bullet H_2O + H_2O + 4 e^-$
         _____
         $3 H_2O(l) + Zr(s) \rightarrow 2 H_2(g) + ZrO_2 \bullet H_2O(s)$

   c. $\Delta G° = -nFE° = -(4$ mol $e^-)(96{,}485$ C/mol $e^-)(1.53$ J/C$) = -5.90 \times 10^5$ J $= -590.$ kJ

$E = E° - \dfrac{0.0591}{n} \log Q$; At equilibrium, $E = 0$ and $Q = K$.

$E° = \dfrac{0.0591}{n} \log K$, $\log K = \dfrac{4(1.53)}{0.0591} = 104$, $K \approx 10^{104}$

   d. 1.00×10^3 kg Zr $\times \dfrac{1000\,g}{kg} \times \dfrac{1\,mol\,Zr}{91.22\,g\,Zr} \times \dfrac{2\,mol\,H_2}{mol\,Zr} = 2.19 \times 10^4$ mol H_2

2.19×10^4 mol $H_2 \times \dfrac{2.016\,g\,H_2}{mol\,H_2} = 4.42 \times 10^4$ g H_2

$V = \dfrac{nRT}{P} = \dfrac{(2.19 \times 10^4\,mol)\,(0.08206\,L\,atm/mol \bullet K)\,(1273\,K)}{1.0\,atm} = 2.3 \times 10^6$ L H_2

   e.   Probably yes; Less radioactivity overall was released by venting the H_2 than what would have been released if the H_2 had exploded inside the reactor (as happened at Chernobyl). Neither alternative is pleasant, but venting the radioactive hydrogen is the less unpleasant of the two alternatives.

53.   a. ^{12}C; It takes part in the first step of the reaction but is regenerated in the last step. ^{12}C is not consumed, so it is not a reactant.

   b. ^{13}N, ^{13}C, ^{14}N, ^{15}O, and ^{15}N are the intermediates.

   c. $4\,{}^{1}_{1}H \rightarrow {}^{4}_{2}He + 2\,{}^{0}_{+1}e$; $\Delta m = 4.00260$ amu $- 2\,m_e + 2\,m_e - [4(1.00782$ amu $- m_e)]$

$\Delta m = 4.00260 - 4(1.00782) + 4(0.000549) = -0.02648$ amu for 4 protons reacting

For 4 mol of protons, $\Delta m = -0.02648$ g and ΔE for the reaction is:

$\Delta E = \Delta mc^2 = -2.648 \times 10^{-5}$ kg $\times (2.9979 \times 10^8$ m/s$)^2 = -2.380 \times 10^{12}$ J

For 1 mol of protons reacting: $\dfrac{-2.380 \times 10^{12}\,J}{4\,mol\,{}^{1}H} = -5.950 \times 10^{11}$ J/mol ^{1}H

54.   a. $^{238}_{92}U \rightarrow ^{222}_{86}Rn + ? \ ^{4}_{2}He + ? \ ^{0}_{-1}e$; To account for the mass number change, 4 alpha particles

are needed.   To balance the number of protons, 2 beta particles are needed.

$^{222}_{86}Rn \rightarrow ^{4}_{2}He + ^{218}_{84}Po$;   Polonium-84 is produced when ^{222}Rn decays.

b.   Alpha particles cause significant ionization damage when inside a living organism. Since the half-life of ^{222}Rn is relatively short, a significant number of alpha particles will be produced when ^{222}Rn is present (even for a short period of time) in the lungs.

c. $^{222}_{86}Rn \rightarrow ^{4}_{2}He + ^{218}_{84}Po$; $^{218}_{84}Po \rightarrow ^{4}_{2}He + ^{214}_{82}Pb$; Polonium-218 is produced when radon-222

decays. ^{218}Po is a more potent alpha particle producer since it has a much shorter half-life than ^{222}Rn. In addition, ^{218}Po is a solid, so it can get trapped in the lung tissue once it is produced. Once trapped, the alpha particles produced from polonium-218 (with its very short half-life) can cause significant ionization damage.

d.   Rate $= kN$;   rate $= \dfrac{4.0 \text{ pCi}}{L} \times \dfrac{1 \times 10^{-12} \text{ Ci}}{\text{pCi}} \times \dfrac{3.7 \times 10^{10} \text{ decays/sec}}{\text{Ci}} = 0.15$ decays/sec • L

$k = \dfrac{\ln 2}{t_{1/2}} = \dfrac{0.6931}{3.82 \text{ d}} \times \dfrac{1 \text{ d}}{24 \text{ hr}} \times \dfrac{1 \text{ hr}}{3600 \text{ s}} = 2.10 \times 10^{-6} \text{ s}^{-1}$

$N = \dfrac{\text{rate}}{k} \times \dfrac{0.15 \text{ decay/sec} \cdot L}{2.10 \times 10^{-6} \text{ s}^{-1}} = 7.1 \times 10^{4} \ ^{222}Rn$ atoms/L

$\dfrac{7.1 \times 10^{4} \ ^{222}Rn \text{ atoms}}{L} \times \dfrac{1 \text{ mol } ^{222}Rn}{6.02 \times 10^{23} \text{ atoms}} = 1.2 \times 10^{-19}$ mol ^{222}Rn/L

55.   mol I $= \dfrac{33 \text{ counts}}{\text{min}} \times \dfrac{1 \text{ mol I} \cdot \text{min}}{5.0 \times 10^{11} \text{ counts}} = 6.6 \times 10^{-11}$ mol I

$[I^-] = \dfrac{6.6 \times 10^{-11} \text{ mol I}^-}{0.150 \text{ L}} = 4.4 \times 10^{-10}$ mol/L

$$Hg_2I_2(s) \rightarrow Hg_2^{2+}(aq) \quad + \quad 2\ I^-(aq) \qquad K_{sp} = [Hg_2^{2+}][I^-]^2$$

Initial s = solubility (mol/L)   0   0

Equil. s $2s$

From the problem, $2s = 4.4 \times 10^{-10}$ mol/L, $s = 2.2 \times 10^{-10}$ mol/L

$K_{sp} = (s)(2s)^2 = (2.2 \times 10^{-10})(4.4 \times 10^{-10})^2 = 4.3 \times 10^{-29}$

CHAPTER NINETEEN

THE REPRESENTATIVE ELEMENTS: GROUPS 1A THROUGH 4A

Questions

1.   The gravity of the earth is not strong enough to keep the light H_2 molecules in the atmosphere.

2.   Ammonia production; Hydrogenation of vegetable oils

3.   Ionic, covalent, and metallic (or interstitial); The ionic and covalent hydrides are true compounds obeying the law of definite proportions and differ from each other in the type of bonding. The interstitial hydrides are more like solid solutions of hydrogen with a transition metal and do not obey the law of definite proportions.

4.   The small size of the Li^+ cation results in a much greater attraction to water. The attraction to water is not as great for the other alkali metal ions. Thus, lithium salts tend to absorb water.

5.   Hydrogen forms many compounds in which the oxidation state is +1, as do the Group 1A elements. For example, H_2SO_4 and HCl compare to Na_2SO_4 and $NaCl$. On the other hand, hydrogen forms diatomic H_2 molecules and is a nonmetal, while the Group 1A elements are metals. Hydrogen also forms compounds with a -1 oxidation state, which is not characteristic of Group 1A metals, e.g., NaH.

6. $4 KO_2(s) + 2 CO_2(g) \rightarrow 2 K_2CO_3(s) + 3 O_2(g)$; Potassium superoxide can react with exhaled CO_2 to produce O_2, which then can be breathed.

7.   Group 1A and 2A metals are all easily oxidized. They must be produced in the absence of materials (H_2O, O_2) that are capable of oxidizing them.

8.   The acidity decreases. Solutions of Be^{2+} are acidic, while solutions of the other M^{2+} ions are neutral.

9.   In graphite, planes of carbon atoms slide easily along each other. In addition, graphite is not volatile. The lubricant will not be lost when used in a high-vacuum environment.

10.   Quartz: Crystalline, long range order; The structure is an ordered arrangement of 12- membered rings, each containing 6 – Si and 6 – O atoms.

      Amorphous SiO_2:   No long range order; Irregular arrangement that contains many different ring sizes. See Chapter 10.5 of the text.

11.   Group 3A elements have one fewer valence electron than Si or Ge. A p-type semiconductor would form.

12.   Size decreases from left to right and increases going down the periodic table. So, going one element right and one element down would result in a similar size for the two elements diagonal to each other. The ionization energies will be similar for the diagonal elements since the periodic trends also oppose each other. Electron affinities are harder to predict, but atoms with similar size and ionization energy should also have similar electron affinities.

Exercises

Group 1A Elements

13.   a. $\Delta H^\circ = -110.5 - [-75 + (-242)] = 207$ kJ; $\Delta S^\circ = 198 + 3(131) - [186 + 189] = 216$ J/K

   b. $\Delta G^\circ = \Delta H^\circ - T\Delta S^\circ$; $\Delta G^\circ = 0$ when $T = \dfrac{\Delta H^\circ}{\Delta S^\circ} = \dfrac{207 \times 10^3 \text{ J}}{216 \text{ J/K}} = 958$ K

   At T > 958 K and standard pressures, the favorable ΔS° term dominates, and the reaction is spontaneous ($\Delta G^\circ < 0$).

14.   a. $\Delta H^\circ = 2(-46$ kJ$) = -92$ kJ; $\Delta S^\circ = 2(193$ J/K$) - [3(131$ J/K$) + 192$ J/K$] = -199$ J/K;

   $\Delta G^\circ = \Delta H^\circ - T\Delta S^\circ = -92$ kJ $- 298$ K$(-0.199$ kJ/K$) = -33$ kJ

   b.   Since ΔG° is negative, then this reaction is spontaneous at standard conditions.

   c. $\Delta G^\circ = 0$ when $T = \dfrac{\Delta H^\circ}{\Delta S^\circ} = \dfrac{-92 \text{ kJ}}{-0.199 \text{ kJ/K}} = 460$ K

   At T < 460 K and standard pressures, the favorable ΔH° term dominates and the reaction is spontaneous ($\Delta G^\circ < 0$).

15.   a.   lithium oxide;   b.   potassium superoxide;   c.   sodium peroxide

16.   a. Li_3N b. K_2CO_3 c.   RbOH   d.   NaH

17.   a. $Li_2O(s) + H_2O(l) \rightarrow 2$ LiOH(aq)   b. $Na_2O_2(s) + 2 H_2O(l) \rightarrow 2$ NaOH(aq) $+ H_2O_2$(aq)

   c.   LiH(s) $+ H_2O(l) \rightarrow H_2(g) +$ LiOH(aq)   d. $2 KO_2(s) + 2H_2O(l) \rightarrow 2$ KOH(aq) $+ O_2(g) + H_2O_2$(aq)

18. $4 \, Li(s) + O_2(g) \rightarrow 2 \, Li_2O(s)$

$16 \, Li(s) + S_8(s) \rightarrow 8 \, Li_2S(s); \quad 2 \, Li(s) + Cl_2(g) \rightarrow 2 \, LiCl(s)$

$12 \, Li(s) + P_4(s) \rightarrow 4 \, Li_3P(s); \quad 2 \, Li(s) + H_2(g) \rightarrow 2 \, LiH(s)$

$2 \, Li(s) + 2 \, H_2O(l) \rightarrow 2 \, LiOH(aq) + H_2(g); \quad 2 \, Li(s) + 2 \, HCl(aq) \rightarrow 2 \, LiCl(aq) + H_2(g)$

19. $2 \, Li(s) + 2 \, C_2H_2(g) \rightarrow 2 \, LiC_2H(s) + H_2(g)$;   This is an oxidation-reduction reaction.

20.   We need another reactant beside NaCl(aq) since oxygen and hydrogen are in some of the products. The obvious choice is H_2O.

$$2 \, NaCl(aq) + 2 \, H_2O(l) \rightarrow Cl_2(g) + H_2(g) + 2 \, NaOH(aq)$$

Note that hydrogen is reduced and chlorine is oxidized in this electrolysis process.

Group 2A Elements

21.   a. magnesium carbonate   b. barium sulfate   c. strontium hydroxide

22.   a. Ca_3N_2 b. $BeCl_2$ c. BaH_2

23. $CaCO_3(s) + H_2SO_4(aq) \rightarrow CaSO_4(aq) + H_2O(l) + CO_2(g)$

24. $2 \, Sr(s) + O_2(g) \rightarrow 2 \, SrO(s); \quad 8 \, Sr(s) + S_8(s) \rightarrow 8 \, SrS(s)$

$Sr(s) + Cl_2(g) \rightarrow SrCl_2(s); \quad 6 \, Sr(s) + P_4(s) \rightarrow 2 \, Sr_3P_2(s)$

$Sr(s) + H_2(g) \rightarrow SrH_2(s); \quad Sr(s) + 2 \, H_2O(l) \rightarrow Sr(OH)_2(aq) + H_2(g)$

$Sr(s) + 2 \, HCl(aq) \rightarrow SrCl_2(aq) + H_2(g)$

25.   In the gas phase, linear molecules would exist.

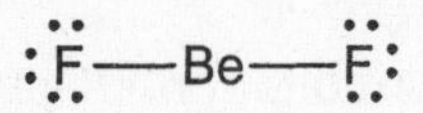

$$:\ddot{F}\!-\!Be\!-\!\ddot{F}:$$

In the solid state, BeF_2 has the following extended structure:

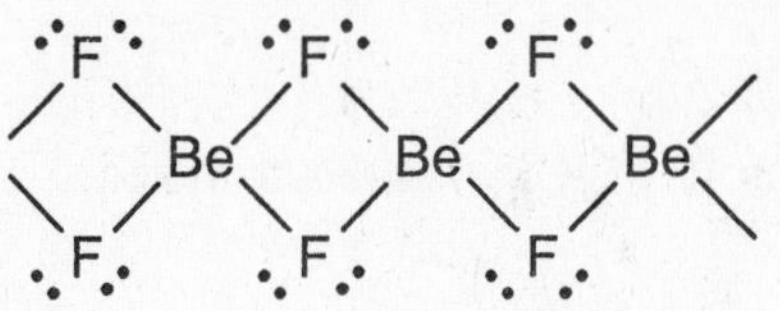

26. $BeCl_2$, with only four valence electrons, needs four more electrons to satisfy the octet rule. NH_3 has a lone pair of electrons on the N atom. Therefore, $BeCl_2$ will react with two NH_3 molecules in order to achieve the octet rule, making $BeCl_2(NH_3)_2$ the likely product in excess ammonia. $BeCl_2(NH_3)_2$ has $2 + 2(7) + 2(5) + 6(1) = 32$ valence electrons.

27. $$\frac{1 \text{ mg F}^-}{L} \times \frac{1 \text{ g}}{1000 \text{ mg}} \times \frac{1 \text{ mol F}^-}{19.00 \text{ g F}^-} = 5.3 \times 10^{-5} \, M\,F^- = 5 \times 10^{-5} \, M\,F^-$$

$CaF_2(s) \rightleftharpoons Ca^{2+}(aq) + 2\,F^-(aq)$ $K_{sp} = [Ca^{2+}][F^-]^2 = 4.0 \times 10^{-11}$; Precipitation will occur when $Q > K_{sp}$. Let's calculate $[Ca^{2+}]$ so that $Q = K_{sp}$.

$$Q = 4.0 \times 10^{-11} = [Ca^{2+}]_o [F^-]_o^2 = [Ca^{2+}]_o (5 \times 10^{-5})^2, \quad [Ca^{2+}]_o = 2 \times 10^{-2} \, M$$

$CaF_2(s)$ will precipitate when $[Ca^{2+}]_o > 2 \times 10^{-2} \, M$. Therefore, hard water should have a calcium ion concentration of less than $2 \times 10^{-2} \, M$ in order to avoid $CaF_2(s)$ formation.

28.
	$CaCO_3(s)$	$\rightleftharpoons$	$Ca^{2+}(aq)$	+	$CO_3^{2-}(aq)$
Initial	s = solubility (mol/L)		0		0
Equil.			s		s

$K_{sp} = 8.7 \times 10^{-9} = [Ca^{2+}][CO_3^{2-}] = s^2, \quad s = 9.3 \times 10^{-5}$ mol/L

29. $Ba^{2+} + 2\,e^- \rightarrow Ba$; $6.00 \text{ hr} \times \dfrac{60 \text{ min}}{\text{hr}} \times \dfrac{60 \text{ s}}{\text{min}} \times \dfrac{2.50 \times 10^5 \text{ C}}{\text{s}} \times \dfrac{1 \text{ mol e}^-}{96,485 \text{ C}} \times \dfrac{1 \text{ mol Ba}}{2 \text{ mol e}^-}$

$$\times \frac{137.3 \text{ g Ba}}{\text{mol Ba}} = 3.84 \times 10^6 \text{ g Ba}$$

30.   Alkaline earth metals form +2 ions so 2 mol of e^- are transferred to form the metal, M.

$$\text{mol M} = 748 \text{ s} \times \frac{5.00 \text{ C}}{\text{s}} \times \frac{1 \text{ mol e}^-}{96,485 \text{ C}} \times \frac{1 \text{ mol M}}{2 \text{ mol e}^-} = 1.94 \times 10^{-2} \text{ mol M}$$

$$\text{molar mass of M} = \frac{0.471 \text{ g M}}{1.94 \times 10^{-2} \text{ mol M}} = 24.3 \text{ g/mol}; \quad MgCl_2 \text{ was electrolyzed.}$$

Group 3A Elements

31.   a. AlN   b. GaF_3 c. Ga_2S_3

32. Tl_2O_3, thallium(III) oxide; Tl_2O, thallium(I) oxide; $InCl_3$, indium(III) chloride;
      InCl, indium(I) chloride

33. $B_2H_6(g) + 3\ O_2(g) \rightarrow 2\ B(OH)_3(s)$

34. $B_2O_3(s) + 3\ Mg(s) \rightarrow 3\ MgO(s) + 2\ B(s)$

35. $Ga_2O_3(s) + 6\ H^+(aq) \rightarrow 2\ Ga^{3+}(aq) + 3\ H_2O(l)$

      $Ga_2O_3(s) + 2\ OH^-(aq) + 3\ H_2O(l) \rightarrow 2\ Ga(OH)_4^-(aq)$

      $In_2O_3(s) + 6\ H^+(aq) \rightarrow 2\ In^{3+}(aq) + 3\ H_2O(l);\ In_2O_3(s) + OH^-(aq) \rightarrow$ no reaction

36. $Al(OH)_3(s) + 3\ H^+(aq) \rightarrow Al^{3+}(aq) + 3\ H_2O(l)$

      $Al(OH)_3(s) + OH^-(aq) \rightarrow Al(OH)_4^-(aq)$

37. $2\ Ga(s) + 3\ F_2(g) \rightarrow 2\ GaF_3(s);\ 4\ Ga(s) + 3\ O_2(g) \rightarrow 2\ Ga_2O_3(s)$

      $16\ Ga(s) + 3\ S_8(s) \rightarrow 8\ Ga_2S_3(s);\ 2\ Ga(s) + N_2(g) \rightarrow 2\ GaN(s)$

      Note: GaN would be predicted, but in practice, this reaction does not occur.

      $2\ Ga(s) + 6\ HCl(aq) \rightarrow 2\ GaCl_3(aq) + 3\ H_2(g)$

38. $2\ Al(s) + 2\ NaOH(aq) + 6\ H_2O(l) \rightarrow 2\ Al(OH)_4^-(aq) + 2\ Na^+(aq) + 3\ H_2(g)$

Group 4A Elements

39. CF_4, $4 + 4(7) = 32\ e^-$ GeF_4, $4 + 4(7) = 32\ e^-$ GeF_6^{2-}, $4 + 6(7) + 2 = 48\ e^-$

      tetrahedral; 109.5°; sp^3 tetrahedral; 109.5°; sp^3 octahedral; 90°; d^2sp^3

      In order to form CF_6^{2-}, carbon would have to expand its octet of electrons. Carbon compounds do
      not expand their octet because of the small atomic size of carbon and because no low energy d-
      orbitals are available for carbon to accommodate the extra electrons.

40. CS_2 has $4 + 2(6) = 16$ valence electrons. C_3S_2 has $3(4) + 2(6) = 24$ valence electrons.

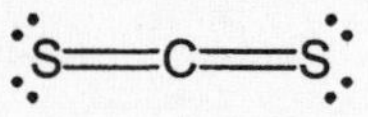

$$\ddot{S}=C=\ddot{S} \quad \text{linear;} \qquad \ddot{S}=C=C=C=\ddot{S} \quad \text{linear}$$

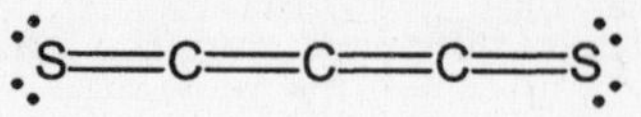

41.   a. $SiO_2(s) + 2\,C(s) \rightarrow Si(s) + 2\,CO(g)$

b. $SiCl_4(l) + 2\,Mg(s) \rightarrow Si(s) + 2\,MgCl_2(s)$

c. $Na_2SiF_6(s) + 4\,Na(s) \rightarrow Si(s) + 6\,NaF(s)$

42. $Sn(s) + 2\,Cl_2(g) \rightarrow SnCl_4(s);\; Sn(s) + O_2(g) \rightarrow SnO_2(s)$

$Sn(s) + 2\,HCl(aq) \rightarrow SnCl_2(aq) + H_2(g)$

43.   Lead is very toxic. As the temperature of the water increases, the solubility of lead will increase. Drinking hot tap water from pipes containing lead solder could result in higher lead concentrations in the body.

44.

	$Pb(OH)_2(s)$	$\rightleftharpoons$	Pb^{2+}	+	$2\,OH^-$	$K_{sp} = 1.2 \times 10^{-15} = [Pb^{2+}][OH^-]^2$
Initial	s = solubility (mol/L)		0		$1.0 \times 10^{-7}\,M$	
Equil.			s		$1.0 \times 10^{-7} + 2s \approx 2s$	

$K_{sp} = (s)(2s)^2 = 1.2 \times 10^{-15}$, $4s^3 = 1.2 \times 10^{-15}$, $s = 6.7 \times 10^{-6}$ mol/L; Assumption good.

$Pb(OH)_2(s)$ is more soluble in acidic solutions. Added H^+ reacts with OH^- to form H_2O. As OH^- is removed through this reaction, more $Pb(OH)_2(s)$ will dissolve.

45. $C_6H_{12}O_6(aq) \rightarrow 2\,C_2H_5OH(aq) + 2\,CO_2(g)$

46. $Sn(s) + 2F_2(g) \rightarrow SnF_4(s)$, tin(IV) fluoride; $Sn(s) + F_2(g) \rightarrow SnF_2(s)$, tin(II) fluoride

47.   The π electrons are free to move in graphite, thus giving it greater conductivity (lower resistance). The electrons in graphite have the greatest mobility within sheets of carbon atoms, resulting in a lower resistance in the basal plane. Electrons in diamond are not mobile (high resistance). The structure of diamond is uniform in all directions; thus, resistivity has no directional dependence in diamond.

48.   SiC would have a covalent network structure similar to diamond.

Additional Exercises

49. $2 \, K(s) + 2 \, H_2O(l) \rightarrow 2 \, KOH(aq) + H_2(g)$, $\Delta H° = 2(-481 \text{ kJ}) - 2(-286 \text{ kJ}) = -390. \text{ kJ}$

$$5.00 \text{ g K} \times \frac{1 \text{ mol K}}{39.10 \text{ g K}} \times \frac{-390. \text{ kJ}}{2 \text{ mol K}} = -24.9 \text{ kJ of heat released upon reaction of 5.00 g of potassium.}$$

$$24{,}900 \text{ J} = \frac{4.18 \text{ J}}{\text{g} \,°\text{C}} \times (1.00 \times 10^3 \text{ g}) \times \Delta T, \; \Delta T = \frac{24{,}900}{4.18 \times 1.00 \times 10^3} = 5.96°\text{C}$$

Final temperature = $24.0 + 5.96 = 30.0°\text{C}$

50.   For 589.0 nm: $\nu = \dfrac{c}{\lambda} = \dfrac{2.9979 \times 10^8 \text{ m/s}}{589.0 \times 10^{-9} \text{ m}} = 5.090 \times 10^{14} \text{ s}^{-1}$

$E = h\nu = 6.6261 \times 10^{-34} \text{ J s} \times 5.090 \times 10^{14} \text{ s}^{-1} = 3.373 \times 10^{-19} \text{ J}$

For 589.6 nm: $\nu = c/\lambda = 5.085 \times 10^{14} \text{ s}^{-1}$; $E = h\nu = 3.369 \times 10^{-19} \text{ J}$

The energies in kJ/mol are:

$$3.373 \times 10^{-19} \text{ J} \times \frac{1 \text{ kJ}}{1000 \text{ J}} \times \frac{6.0221 \times 10^{23}}{\text{mol}} = 203.1 \text{ kJ/mol}$$

$$3.369 \times 10^{-19} \text{ J} \times \frac{1 \text{ kJ}}{1000 \text{ J}} \times \frac{6.0221 \times 10^{23}}{\text{mol}} = 202.9 \text{ kJ/mol}$$

51.   Strontium and calcium are both alkaline earth metals, so they have similar chemical properties. Since milk is a good source of calcium, strontium could replace some calcium in milk without much difficulty.

52.   The Be^{2+} ion is a Lewis acid and has a strong affinity for the lone pairs of electrons on oxygen in water. Thus, the compound is not dehydrated easily. The ion in solution is $Be(H_2O)_4^{2+}$. The acidic solution results from the reaction: $Be(H_2O)_4^{2+}(aq) \rightleftharpoons Be(H_2O)_3(OH)^+(aq) + H^+(aq)$

53.   The "inert pair effect" refers to the difficulty of removing the pair of s electrons from some of the elements in the fifth and sixth periods of the periodic table. As a result, multiple oxidation states are exhibited for the heavier elements of Groups 3A and 4A. In^+, In^{3+}, Tl^+ and Tl^{3+} oxidation states are all important to the chemistry of In and Tl.

54.   Element 113 would fall below Tl in the periodic table. Element 113: $[Rn] \, 7s^2 5f^{14} 6d^{10} 7p^1$.

55.   Major species present: $Al(H_2O)_6^{3+}$ ($K_a = 1.4 \times 10^{-5}$), NO_3^- (neutral) and H_2O ($K_w = 1.0 \times 10^{-14}$); $Al(H_2O)_6^{3+}$ is a stronger acid than water so it will be the dominant H^+ producer.

$$Al(H_2O)_6^{3+} \quad\rightleftharpoons\quad Al(H_2O)_5(OH)^{2+} \quad + \quad H^+$$

Initial	0.050 M	0	~0

x mol/L $Al(H_2O)_6^{3+}$ dissociates to reach equilibrium

Change	$-x$	$\rightarrow$	$+x$	$+x$
Equil.	0.050 - x		x	x

$$K_a = 1.4 \times 10^{-5} = \frac{[Al(H_2O)_5(OH)^{2+}][H^+]}{[Al(H_2O)_6^{3+}]} = \frac{x^2}{0.050-x} \approx \frac{x^2}{0.050}$$

$x = 8.4 \times 10^{-4}\ M = [H^+]$; pH = $-\log(8.4 \times 10^{-4})$ = 3.08; Assumptions good.

56.

$$Tl^{3+} + 2\ e^- \rightarrow Tl^+ \qquad\qquad E° = 1.25\ V$$
$$3\ I^- \rightarrow I_3^- + 2\ e^- \qquad\qquad -E° = -0.55\ V$$
$$\overline{\phantom{Tl^{3+} + 3\ I^- \rightarrow Tl^+ + I_3^- \qquad\qquad E°_{cell} = 0.70\ V}}$$
$$Tl^{3+} + 3\ I^- \rightarrow Tl^+ + I_3^- \qquad\qquad E°_{cell} = 0.70\ V$$

In solution, Tl^{3+} can oxidize I^- to I_3^-. Thus, we expect TlI_3 to be thallium(I) triiodide.

57.   Ga(I): $[Ar]4s^2 3d^{10}$, no unpaired e^-; Ga(III): $[Ar]3d^{10}$, no unpaired e^-

Ga(II): $[Ar]4s^1 3d^{10}$, 1 unpaired e^-; Note: s electrons are lost before the d electrons.

If the compound contained Ga(II), it would be paramagnetic, and if the compound contained Ga(I) and Ga(III), it would be diamagnetic. This can be determined easily by measuring the mass of a sample in the presence and in the absence of a magnetic field. Paramagnetic compounds will have an apparent greater mass in a magnetic field.

58.   a.   Out of 100.0 g of compound there are:

$$44.4\ g\ Ca \times \frac{1\ mol}{40.08\ g} = 1.11\ mol\ Ca; \quad 20.0\ g\ Al \times \frac{1\ mol}{26.98\ g} = 0.741\ mol\ Al$$

$$35.6\ g\ O \times \frac{1\ mol}{16.00\ g} = 2.23\ mol\ O$$

$$\frac{1.11}{0.741} = 1.50; \quad \frac{0.741}{0.741} = 1.00; \quad \frac{2.23}{0.741} = 3.01; \quad \text{Empirical formula is } Ca_3Al_2O_6.$$

   b. $Ca_9Al_6O_{18}$

   c.   There are covalent bonds between Al and O atoms in the $Al_6O_{18}^{18-}$ anion; sp^3 hybrid orbitals on aluminum overlap with sp^3 hybrid orbitals on oxygen to form the sigma bonds.

59. $750. \text{ mL grape juice} \times \dfrac{12 \text{ mL } C_2H_5OH}{100. \text{ mL juice}} \times \dfrac{0.79 \text{ g } C_2H_5OH}{\text{mL}} \times \dfrac{1 \text{ mol } C_2H_5OH}{46.07 \text{ g}} \times \dfrac{2 \text{ mol } CO_2}{2 \text{ mol } C_2H_5OH}$

$$= 1.54 \text{ mol } CO_2 \quad \text{(carry extra significant figure)}$$

$1.54 \text{ mol } CO_2 = \text{total mol } CO_2 = \text{mol } CO_2(g) + \text{mol } CO_2(aq) = n_g + n_{aq}$

$$P_{CO_2} = \frac{n_g RT}{V} = \frac{n_g \left(\dfrac{0.08206 \text{ L atm}}{\text{mol K}} \right) (298 \text{ K})}{75 \times 10^{-3} \text{ L}} = 326 \, n_g; \quad P_{CO_2} = \frac{C}{k} = \frac{\dfrac{n_{aq}}{0.750 \text{ L}}}{\dfrac{3.1 \times 10^{-2} \text{ mol}}{\text{L atm}}} = 43.0 \, n_{aq}$$

$P_{CO_2} = 326 \, n_g = 43.0 \, n_{aq}$ and from above $n_{aq} = 1.54 - n_g$; Solving:

$326 \, n_g = 43.0(1.54 - n_g), \quad 369 \, n_g = 66.2, \quad n_g = 0.18 \text{ mol}$

$P_{CO_2} = 326(0.18) = 59 \text{ atm in gas phase}$

$C = k P_{CO_2} = \dfrac{3.1 \times 10^{-2} \text{ mol}}{\text{L atm}} \times 59 \text{ atm}, \quad C = 1.8 \text{ mol } CO_2/\text{L in wine}$

60. Pb_3O_4: We assign -2 for the oxidation state of O. The sum of the oxidation states of Pb must be +8. We get this if two of the lead atoms are Pb(II) and one is Pb(IV). Therefore, the mole ratio of lead(II) to lead(IV) is 2:1.

61. $Pb(NO_3)_2(aq) + H_3AsO_4(aq) \rightarrow PbHAsO_4(s) + 2 \, HNO_3(aq)$

Note: The insecticide used is $PbHAsO_4$ and is commonly called lead arsenate. This is not the correct name, however. Correctly, lead arsenate would be $Pb_3(AsO_4)_2$ and $PbHAsO_4$ should be named lead hydrogen arsenate.

Challenge Problems

62.   a. $K^+ \text{ (blood)} \rightleftharpoons K^+ \text{ (muscle)} \quad \Delta G° = 0; \quad \Delta G = RT \ln \left(\dfrac{[K^+]_m}{[K^+]_b} \right); \quad \Delta G = w_{max}$

$$\Delta G = \frac{8.3145 \text{ J}}{\text{K mol}} (310. \text{ K}) \ln \left(\frac{0.15}{0.0050} \right), \quad \Delta G = 8.8 \times 10^3 \text{ J/mol} = 8.8 \text{ kJ/mol}$$

At least 8.8 kJ of work must be applied to transport 1 mol K^+.

b. Other ions will have to be transported in order to maintain electroneutrality. Either anions must be transported into the cells, or cations (Na^+) in the cell must be transported to the blood. The latter is what happens: $[Na^+]$ in blood is greater than $[Na^+]$ in cells as a result of this pumping.

c. $\Delta G° = -RT \ln K = -(8.3145 \text{ J/K} \cdot \text{mol})(310. \text{ K}) \ln 1.7 \times 10^5 = -3.1 \times 10^4 \text{ J/mol} = -31 \text{ kJ/mol}$

The hydrolysis of ATP (at standard conditions) provides 31 kJ/mol of energy to do work. We must add 8.8 kJ of work to transport 1.0 mol of K^+.

$$8.8 \text{ kJ} \times \frac{1 \text{ mol ATP}}{31 \text{ kJ}} = 0.28 \text{ mol ATP must be hydrolyzed.}$$

63. $$Pb^{2+} + H_2EDTA^{2-} \rightleftharpoons PbEDTA^{2-} + 2 H^+$$

Before	0.0010 M	0.050 M	0	$1.0 \times 10^{-6} M$	(buffer, $[H^+]$ constant)
Change	-0.0010	-0.0010	$\rightarrow$ +0.0010	No change	Reacts completely
After	0	0.049	0.0010	1.0×10^{-6}	New initial conditions

x mol/L $PbEDTA^{2-}$ dissociates to reach equilibrium

Change	+x	+x	$\leftarrow$ -x	---	
Equil.	x	0.049 + x	0.0010 - x	1.0×10^{-6}	(buffer)

$$K = 1.0 \times 10^{23} = \frac{[PbEDTA^{2-}][H^+]^2}{[Pb^{2+}][H_2EDTA^{2-}]} = \frac{(0.0010 - x)(1.0 \times 10^{-6})^2}{(x)(0.049 + x)}$$

$$1.0 \times 10^{23} \approx \frac{(0.0010)(1.0 \times 10^{-12})}{x(0.049)}, \quad x = [Pb^{2+}] = 2.0 \times 10^{-37} M \text{ Assumptions good.}$$

64. $SiCl_4(l) + 2 H_2O(l) \rightarrow SiO_2(s) + 4 H^+(aq) + 4 Cl^-(aq)$

$\Delta H° = -911 + 4(0) + 4(-167) - [-687 + 2(-286)] = -320. \text{ kJ}$

$\Delta S° = 42 + 4(0) + 4(57) - [240. + 2(70.)] = -110. \text{ J/K}; \quad \Delta G° = \Delta H° - T\Delta S°$

$\Delta G° = 0$ when $T = \Delta H°/\Delta S° = -320. \times 10^3 \text{ J}/(-110. \text{ J/K}) = 2910 \text{ K}$

Due to the favorable $\Delta H°$ term, this reaction is spontaneous at temperatures below 2910 K.

The corresponding reaction for CCl_4 is:

$CCl_4(l) + 2 H_2O(l) \rightarrow CO_2(g) + 4 H^+(aq) + 4 Cl^-(aq)$

$\Delta H° = -393.5 + 4(0) + 4(-167) - [-135 + 2(-286)] = -355 \text{ kJ}$

$\Delta S° = 214 + 4(0) + 4(57) - [216 + 2(70.)] = 86 \text{ J/K}$

Thermodynamics predict that this reaction would be spontaneous at any temperature.

The answer must lie with kinetics. $SiCl_4$ reacts because an activated complex can form by a water molecule attaching to silicon in $SiCl_4$. The activated complex requires silicon to form a fifth bond. Silicon has low energy 3d orbitals available to expand the octet. Carbon will not break the octet rule; therefore, CCl_4 cannot form this activated complex. CCl_4 and H_2O require a different pathway to get to products. The different pathway has a higher activation energy and, in turn, the reaction is much slower. (See Exercise 19.65.)

65.   Carbon cannot form the fifth bond necessary for the transition state because of the small atomic size of carbon and because carbon doesn't have low energy d orbitals available to expand the octet.

66.   White tin is stable at normal temperatures. Gray tin is stable at temperatures below 13.2°C. Thus for the phase change: Sn(gray) → Sn(white), ΔG is (-) at T > 13.2°C and ΔG is (+) at T < 13.2°C. This is only possible if ΔH is (+) and ΔS is (+). Thus, gray tin has the more ordered structure.

CHAPTER TWENTY

THE REPRESENTATIVE ELEMENTS: GROUPS 5A THROUGH 8A

Questions

1. $N_2(g) + 3 H_2(g) \rightleftharpoons 2 NH_3(g) + heat$

    a.   This reaction is exothermic, so an increase in temperature will decrease the value of K (see Table 13.3 of text.) This has the effect of lowering the amount of $NH_3(g)$ produced at equilibrium. The temperature increase, therefore, must be for kinetics reasons. When the temperature increases, the reaction reaches equilibrium much faster. At low temperatures, this reaction is very slow, too slow to be of any use.

    b.   As $NH_3(g)$ is removed, the reaction shifts right to produce more $NH_3(g)$.

    c.   A catalyst has no effect on the equilibrium position. The purpose of a catalyst is to speed up a reaction so it reaches equilibrium more quickly.

    d.   When the pressure of reactants and products is high, the reaction shifts to the side that has fewer gas molecules. Since the product side contains 2 molecules of gas compared to 4 molecules of gas on the reactant side, the reaction shifts right to products at high pressures of reactants and products.

2.   White phosphorus consists of discrete tetrahedral P_4 molecules. The bond angles in the P_4 tetrahedrons are only 60°, which makes P_4 very reactive, especially towards oxygen. Red and black phosphorus are covalent network solids. In red phosphorus, the P_4 tetrahedra are bonded to each other in chains, making them less reactive than white phosphorus. They need a source of energy to react with oxygen, such as when one strikes a match. Black phosphorus is crystalline, with the P atoms tightly bonded to each other in the crystal, and is fairly unreactive towards oxygen.

3.   The pollution provides nitrogen and phosphorous nutrients so the algae can grow. The algae consume oxygen, causing fish to die.

4.   In the upper atmosphere, O_3 acts as a filter for UV radiation: $O_3 \xrightarrow{h\nu} O_2 + O$

      O_3 is also a powerful oxidizing agent. It irritates the lungs and eyes, and, at high concentration, it is toxic. The smell of a "fresh spring day" is O_3 formed during lightning discharges. Toxic materials don't necessarily smell bad. For example, HCN smells like almonds.

5.   Plastic sulfur consists of long S_n chains of sulfur atoms. As plastic sulfur becomes brittle, the long chains break down into S_8 rings.

6.   In the presence of S^{2-}, sulfur forms polysulfide ions, S_n^{2-}, which are soluble like most species with charges, e.g., $S_8 + S^{2-} \rightleftharpoons S_9^{2-}$. Nitric acid oxidizes S^{2-} to S, which causes sulfur to precipitate out of solution.

7.   Fluorine is the most reactive of the halogens because it is the most electronegative atom and the bond in the F_2 molecule is very weak.

8.   One reason is that the H—F bond is stronger than the other hydrohalides, making it more difficult to form H^+ and F^-. The main reason HF is a weak acid is entropy. When $F^-(aq)$ forms from the dissociation of HF, there is a high degree of ordering that takes place as water molecules hydrate this small ion. Entropy is considerably more unfavorable for the formation of hydrated F^- than for the formation of the other hydrated halides. The result of the more unfavorable $\Delta S°$ term is a positive $\Delta G°$ value which leads to a K_a value less than one.

9.   Helium is unreactive and doesn't combine with any other elements. It is a very light gas and would easily escape the earth's gravitational pull as the planet was formed.

10.   In Mendeleev's time none of the noble gases were known. Since an entire family was missing, no gaps seemed to appear in the periodic arrangement. Mendeleev had no evidence to predict the existence of such a family. The heavier members of the noble gases are not really inert. Xe and Kr have been shown to react and form compounds with other elements.

Exercises

Group 5A Elements

11. NO_4^{3-}

Both NO_4^{3-} and PO_4^{3-} have 32 valence electrons, so both have similar Lewis structures. From the Lewis structure for NO_4^{3-}, the central N atom has a tetrahedral arrangement of electron pairs. N is small. There is probably not enough room for all 4 oxygen atoms around N. P is larger, thus, PO_4^{3-} is stable.

PO_3^-

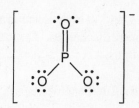

PO_3^- and NO_3^- each have 24 valence electrons so both have similar Lewis structures. From the Lewis structure for PO_3^-, PO_3^- has a trigonal planar arrangement of electron pairs about the central P atom (two single bonds and one double bond). $P=O$ bonds are not particularly stable, while $N=O$ bonds are stable. Thus, NO_3^- is stable.

12.   a. PF_5;   N is too small and doesn't have low energy d-orbitals to expand its octet to form NF_5.

       b. AsF_5;   I is too large to fit 5 atoms of I around As.

       c. NF_3;   N is too small for three large bromine atoms to fit around it.

13.   a. $NH_4NO_3(s) \xrightarrow{heat} N_2O(g) + 2 H_2O(g)$

       b. $2 N_2O_5(g) \rightarrow 4 NO_2(g) + O_2(g)$

       c. $2 K_3P(s) + 6 H_2O(l) \rightarrow 2 PH_3(g) + 6 KOH(aq)$

       d. $PBr_3(l) + 3 H_2O(l) \rightarrow H_3PO_3(aq) + 3 HBr(aq)$

       e. $2 NH_3(aq) + NaOCl(aq) \rightarrow N_2H_4(aq) + NaCl(aq) + H_2O(l)$

14. $4 As(s) + 3 O_2(g) \rightarrow As_4O_6(s)$; $4 As(s) + 5 O_2(g) \rightarrow As_4O_{10}(s)$

       $As_4O_6(s) + 6 H_2O(l) \rightarrow 4 H_3AsO_3(aq)$; $As_4O_{10}(s) + 6 H_2O(l) \rightarrow 4 H_3AsO_4(aq)$

15.   Unbalanced equation:

       $CaF_2\cdot3Ca_3(PO_4)_2(s) + H_2SO_4(aq) \rightarrow H_3PO_4(aq) + HF(aq) + CaSO_4\cdot2H_2O(s)$

       Balancing Ca^{2+}, F^-, and PO_4^{3-}:

       $CaF_2\cdot3Ca_3(PO_4)_2(s) + H_2SO_4(aq) \rightarrow 6 H_3PO_4(aq) + 2 HF(aq) + 10 CaSO_4\cdot2H_2O(s)$

On the righthand side, there are 20 extra hydrogen atoms, 10 extra sulfates, and 20 extra water molecules. We can balance the hydrogen and sulfate with 10 sulfuric acid molecules. The extra waters came from the water in the sulfuric acid solution. The balanced equation is:

$CaF_2\cdot3Ca_3(PO_4)_2(s) + 10 H_2SO_4(aq) + 20 H_2O(l) \rightarrow 6 H_3PO_4(aq) + 2 HF(aq) + 10 CaSO_4\cdot2H_2O(s)$

16.   a. NO_2, $5 + 2(6) = 17$ e⁻ N_2O_4, $2(5) + 4(6) = 34$ e⁻

          plus other resonance structures   plus other resonance structures

       b. BF_3, $3 + 3(7) = 24$ e⁻ NH_3, $5 + 3(1) = 8$ e⁻

       BF_3NH_3, $24 + 8 = 32$ e⁻

In reaction a, NO_2 has an odd number of electrons, so it is impossible to satisfy the octet rule. By dimerizing to form N_2O_4, the odd electron on two NO_2 molecules can pair up, giving a species whose Lewis structure can satisfy the octet rule. In general, odd electron species are very reactive.   In reaction b, BF_3 can be considered electron deficient. Boron has only six electrons around it. By forming BF_3NH_3, the boron atom satisfies the octet rule by accepting a lone pair of electrons from NH_3 to form a fourth bond.

17. $2 NaN_3(s) \rightarrow 2 Na(s) + 3 N_2(g)$

$$n_{N_2} = \frac{PV}{RT} = \frac{1.00 \text{ atm} \times 70.0 \text{ L}}{\dfrac{0.08206 \text{ L atm}}{\text{mol K}} \times 273 \text{ K}} = 3.12 \text{ mol } N_2 \text{ needed to fill air bag.}$$

$$\text{mol } NaN_3 \text{ reacted} = 3.12 \text{ mol } N_2 \times \frac{2 \text{ mol } NaN_3}{3 \text{ mol } N_2} = 2.08 \text{ mol } NaN_3$$

18.   For ammonia (in one minute):

$$n_{NH_3} = \frac{PV}{RT} = \frac{90. \text{ atm} \times 500. \text{ L}}{\dfrac{0.08206 \text{ L atm}}{\text{mol K}} \times 496 \text{ K}} = 1.1 \times 10^3 \text{ mol } NH_3$$

NH_3 flows into the reactor at a rate of 1.1×10^3 mol/min.

For CO_2 (in one minute):

$$n_{CO_2} = \frac{PV}{RT} = \frac{45 \text{ atm} \times 600. \text{ L}}{\dfrac{0.08206 \text{ L atm}}{\text{mol K}} \times 496 \text{ K}} = 6.6 \times 10^2 \text{ mol } CO_2$$

CO_2 flows into the reactor at 6.6×10^2 mol/min.

To react completely with 1.1×10^3 mol NH_3/min, we need:

$$\frac{1.1 \times 10^3 \text{ mol } NH_3}{\text{min}} \times \frac{1 \text{ mol } CO_2}{2 \text{ mol } NH_3} = 5.5 \times 10^2 \text{ mol } CO_2/\text{min}$$

Since 660 mol CO_2/min are present, ammonia is the limiting reagent.

$$\frac{1.1 \times 10^3 \text{ mol } NH_3}{\text{min}} \times \frac{1 \text{ mol urea}}{2 \text{ mol } NH_3} \times \frac{60.06 \text{ g urea}}{\text{mol urea}} = 3.3 \times 10^4 \text{ g urea/min}$$

19.

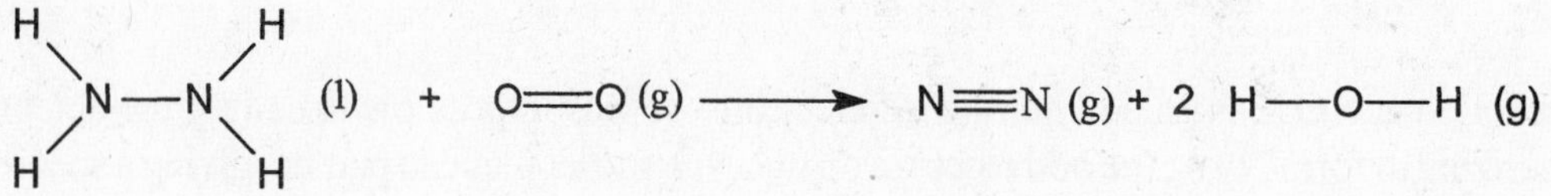

Bonds broken:

    1 N – N (160. kJ/mol)
    4 N – H (391 kJ/mol)
    1 O = O (495 kJ/mol)

Bonds formed:

    1 N≡ N (941 kJ/mol)
    2 × 2 O – H (467 kJ/mol)

$\Delta H = 160. + 4(391) + 495 - [941 + 4(467)] = 2219 \text{ kJ} - 2809 \text{ kJ} = -590. \text{ kJ}$

20. $5 N_2O_4(l) + 4 N_2H_3CH_3(l) \rightarrow 12 H_2O(g) + 9 N_2(g) + 4 CO_2(g)$

$$\Delta H° = \left[12 \text{ mol}\left(\frac{-242 \text{ kJ}}{\text{mol}}\right) + 4 \text{ mol}\left(\frac{-393.5 \text{ kJ}}{\text{mol}}\right)\right]$$

$$- \left[5 \text{ mol}\left(\frac{-20. \text{ kJ}}{\text{mol}}\right) + 4 \text{ mol}\left(\frac{54 \text{ kJ}}{\text{mol}}\right)\right] = -4594 \text{ kJ}$$

Using bond energies, $\Delta H = -5.0 \times 10^3$ kJ (from Sample Exercise 20.2). When using bond energies to calculate ΔH, the enthalpy change is assumed to be due only to the difference in bond strength between reactants and products. Bond energies generally give a very good estimate for ΔH for gas phase reactions. However, when solids and liquids are present, ΔH estimates from bond energy differences is not as good. This is because the difference in the strength of the intermolecular forces between reactants and products is not considered when using bond energies. Here, the reactants are in the liquid phase. The loss in strength of the intermolecular forces as the liquid reactants are converted to the gaseous products was not considered when using bond energies in Sample Exercise 20.2; hence, the large difference between the two calculated ΔH values.

21. $1/2\ N_2(g) + 1/2\ O_2(g) \rightarrow NO(g)$ $\Delta G° = \Delta G°_{f,NO} = 87$ kJ/mol; By definition, $\Delta G°_f$ for a compound equals the free energy change that would accompany the formation of 1 mol of that compound from its elements in their standard states. NO (and some other oxides of nitrogen) have weaker bonds as compared to the triple bond of N_2 and the double bond of O_2. Because of this, NO (and some other oxides of nitrogen) have higher (positive) standard free energies of formation as compared to the relatively stable N_2 and O_2 molecules.

22. $\Delta H° = 2(90.\ \text{kJ}) - [0 + 0] = 180.\ \text{kJ}$; $\Delta S° = 2(211\ \text{J/K}) - [192 + 205] = 25\ \text{J/K}$

$\Delta G° = 2(87\ \text{kJ}) - [0] = 174\ \text{kJ}$

At the high temperatures in automobile engines, the reaction $N_2 + O_2 \rightarrow 2\ NO$ becomes spontaneous since the favorable $\Delta S°$ term will become dominate. In the atmosphere, even though $2\ NO \rightarrow N_2 + O_2$ is spontaneous at the cooler temperatures of the atmosphere, it doesn't occur because the rate is slow. Therefore, higher concentrations of NO are present in the atmosphere as compared to what is predicted by thermodynamics.

23.   M.O. model:

NO^+: $(\sigma_{2s})^2(\sigma_{2s}{}^*)^2(\pi_{2p})^4(\sigma_{2p})^2$, Bond order $= (8 - 2)/2 = 3$, 0 unpaired e^- (diamagnetic)

NO: $(\sigma_{2s})^2(\sigma_{2s}{}^*)^2(\pi_{2p})^4(\sigma_{2p})^2(\pi_{2p}{}^*)^1$, B.O. $= 2.5$, 1 unpaired e^- (paramagnetic)

NO^-: $(\sigma_{2s})^2(\sigma_{2s}{}^*)^2(\pi_{2p})^4(\sigma_{2p})^2(\pi_{2p}{}^*)^2$, B.O. $= 2$, 2 unpaired e^- (paramagnetic)

Lewis structures: NO^+: $\left[\ :N\equiv O:\ \right]^+$

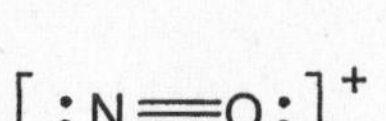

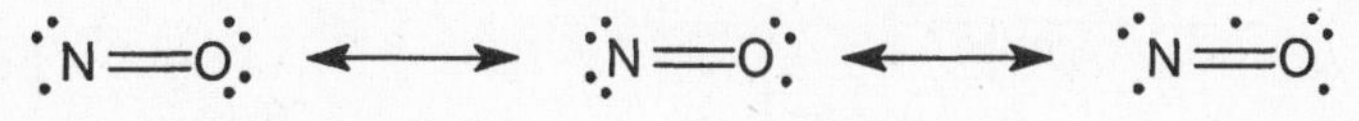

NO: $\ddot{N}=\ddot{O}\ \longleftrightarrow\ :\ddot{N}=\ddot{O}:\ \longleftrightarrow\ \dot{N}\equiv O:$

NO^-: $\left[\ :\ddot{N}=\ddot{O}:\ \right]^-$

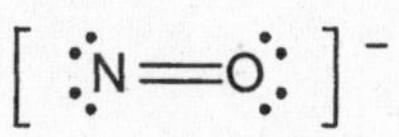

The two models give the same results only for NO^+ (a triple bond with no unpaired electrons).

Lewis structures are not adequate for NO and NO^-. The M.O. model gives a better representation for all three species. For NO, Lewis structures are poor for odd electron species. For NO^-, both models predict a double bond, but only the MO model correctly predicts that NO^- is paramagnetic.

24.   For $NCl_3 \rightarrow NCl_2 + Cl$, only the N – Cl bond is broken. For $O=N-Cl \rightarrow NO + Cl$, the NO bond gets stronger (bond order increases from 2.0 to 2.5) when the N – Cl bond is broken. This makes ΔH for the reaction smaller than just the energy necessary to break the N – Cl bond.

25.   a. $H_3PO_4 > H_3PO_3$; The strongest acid has the most oxygen atoms.

       b. $H_3PO_4 > H_2PO_4^- > HPO_4^{2-}$; Acid strength decreases as protons are removed.

26.   TSP = Na_3PO_4; PO_4^{3-} is the conjugate base of the weak acid HPO_4^{2-} ($K_a = 4.8 \times 10^{-13}$). All conjugate bases of weak acids are effective bases ($K_b = K_w/K_a = 1.0 \times 10^{-14}/4.8 \times 10^{-13} = 2.1 \times 10^{-2}$). The weak base reaction of PO_4^{3-} with H_2O is: $PO_4^{3-} + H_2O \rightleftharpoons HPO_4^{2-} + OH^-$ $K_b = 2.1 \times 10^{-2}$.

27.   The acidic protons are attached to oxygen.

$H_4P_2O_6$ (50 valence e$^-$): $H_4P_2O_5$ (44 valence e$^-$):

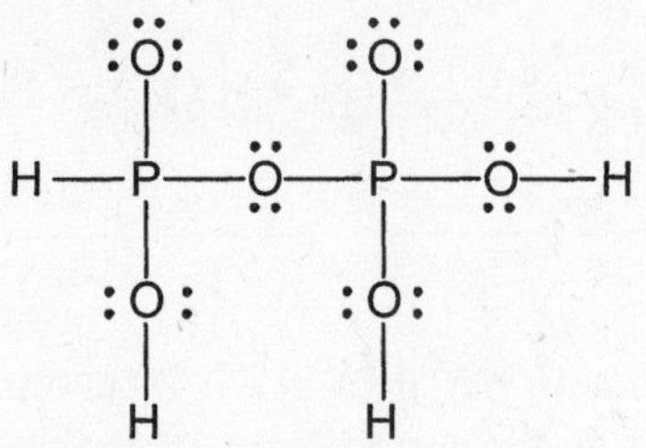

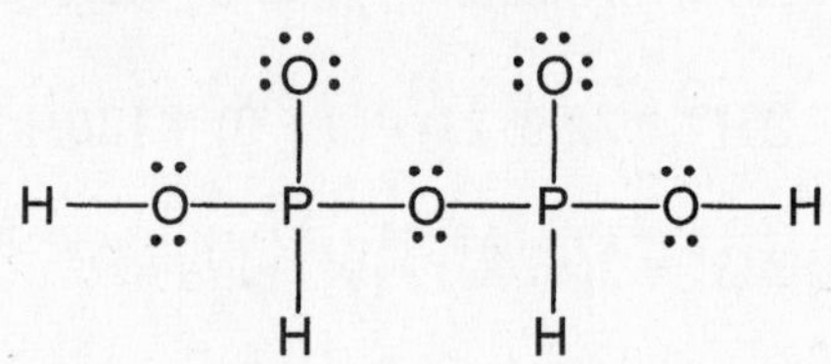

28.   a. SbF_5, 40 valence e$^-$ HSO_3F, 32 valence e$^-$ $H_2SO_3F^+$, 32 valence e$^-$

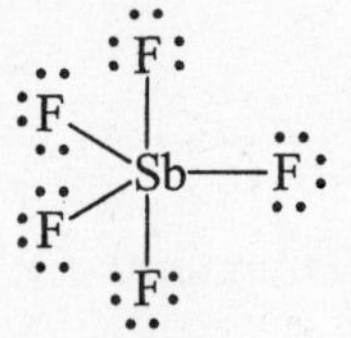

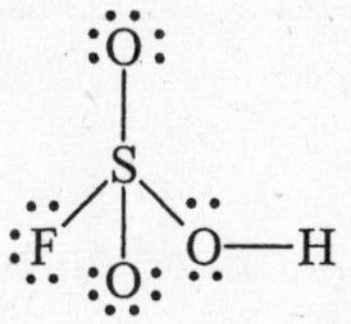

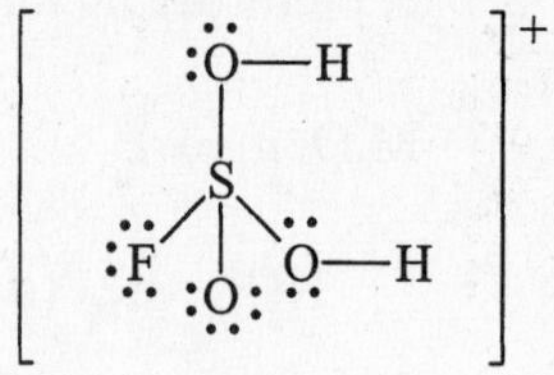

                dsp^3 sp^3 sp^3

        F_5SbOSO_2FH, 72 valence e$^-$ $F_5SbOSO_2F^-$, 72 valence e$^-$

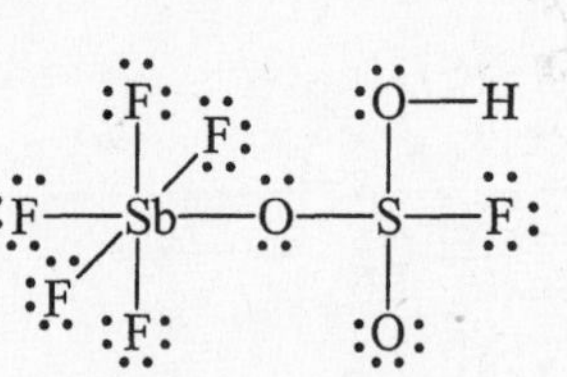

        Sb: d^{2}sp^3; S: sp^3 Sb: d^{2}sp^3; S: sp^3

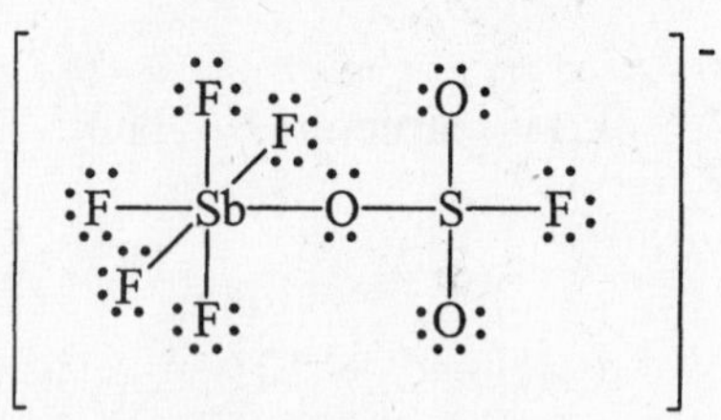

       b. The active protonating species is $H_2SO_3F^+$, the species with two OH bonds.

Group 6A Elements

29. $O = O - O \rightarrow O = O + O$

Break $O - O$ bond: $\Delta H = \dfrac{146 \text{ kJ}}{\text{mol}} \times \dfrac{1 \text{ mol}}{6.022 \times 10^{23}} = 2.42 \times 10^{-22} \text{ kJ} = 2.42 \times 10^{-19} \text{ J}$

A photon of light must contain at least 2.42×10^{-19} J to break one $O - O$ bond.

$E_{photon} = \dfrac{hc}{\lambda}$, $\lambda = \dfrac{hc}{E} = \dfrac{(6.626 \times 10^{-34} \text{ J s}) (2.998 \times 10^{8} \text{ m/s})}{2.42 \times 10^{-19} \text{ J}} = 8.21 \times 10^{-7} \text{ m} = 821 \text{ nm}$

30.   From Figure 7.2 in the text, light from violet to green will work.

31.   a. $2 SO_2(g) + O_2(g) \rightarrow 2 SO_3(g)$ b. $SO_3(g) + H_2O(l) \rightarrow H_2SO_4(aq)$

c. $2 Na_2S_2O_3(aq) + I_2(aq) \rightarrow Na_2S_4O_6(aq) + 2 NaI(aq)$

d. $Cu(s) + 2 H_2SO_4(aq) \rightarrow CuSO_4(aq) + 2 H_2O(l) + SO_2(aq)$

32. $H_2SeO_4(aq) + 3 SO_2(g) \rightarrow Se(s) + 3 SO_3(g) + H_2O(l)$

33.   a. SO_3^{2-}, $6 + 3(6) + 2 = 26$ e⁻   b. O_3, $3(6) = 18$ e⁻

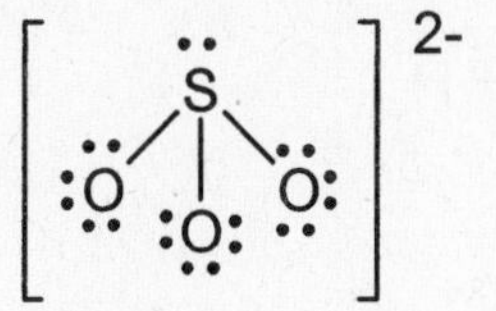

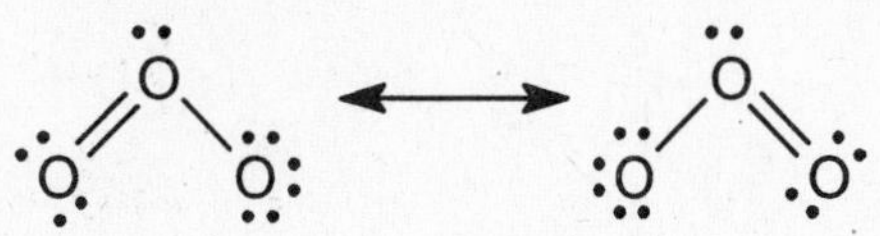

trigonal pyramid; $\approx 109.5°$; sp^3 V-shaped; $\approx 120°$; sp^2

c. SCl_2, $6 + 2(7) = 20$ e⁻   d. $SeBr_4$, $6 + 4(7) = 34$ e⁻

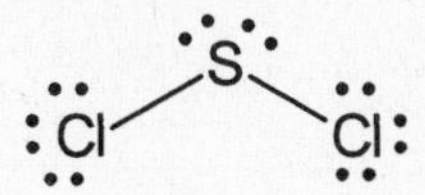

V-shaped; $\approx 109.5°$; sp^3

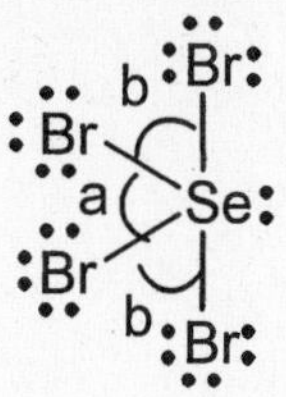

see-saw; a $\approx 120°$, b $\approx 90°$; dsp^3

e. TeF_6, $6 + 6(7) = 48$ e⁻

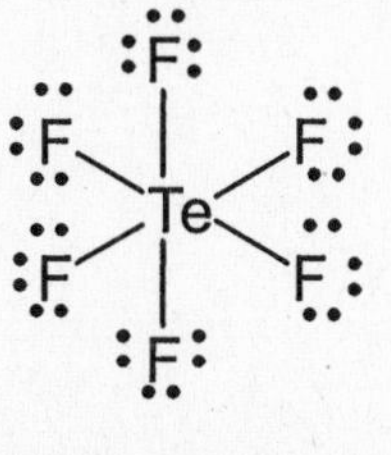

octahedral; $90°$; d^2sp^3

34. S_2N_2 has $2(6) + 2(5) = 22$ valence electrons.

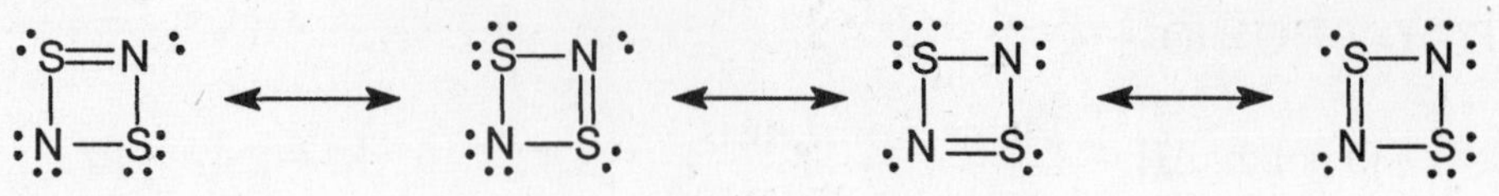

Group 7A Elements

35. O_2F_2 has $2(6) + 2(7) = 26$ valence e⁻; From the following Lewis structure, each oxygen atom has a tetrahedral arrangement of electron pairs. Therefore, bond angles ≈ 109.5° and each O is sp^3 hybridized.

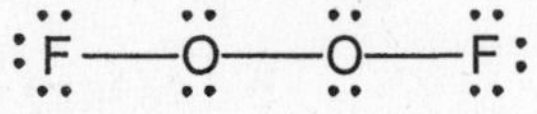

| Formal Charge | 0 | 0 | 0 | 0 |
| Oxid. Number | -1 | +1 | +1 | -1 |

Oxidation numbers are more useful. We are forced to assign +1 as the oxidation number for oxygen. Oxygen is very electronegative, and +1 is not a stable oxidation state for this element.

36.   a. CCl_2F_2, $4 + 2(7) + 2(7) = 32$ e⁻   b. $HClO_4$, $1 + 7 + 4(6) = 32$ e⁻

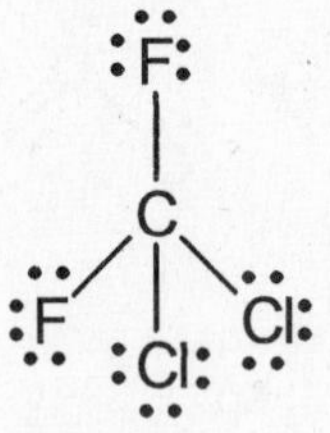

tetrahedral; 109.5°; sp^3

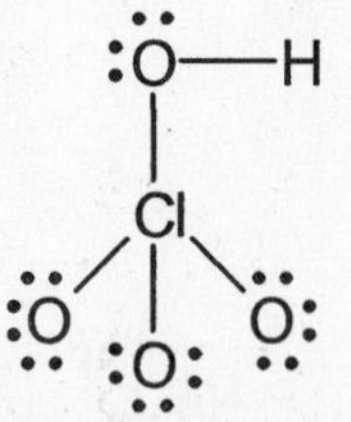

About Cl: tetrahedral; 109.5°; sp^3
About O: V-shaped; ≈ 109.5°; sp^3

c. ICl_3, $7 + 3(7) = 28$ e⁻   d. BrF_5, $7 + 5(7) = 42$ e⁻

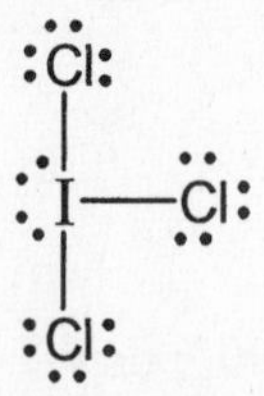

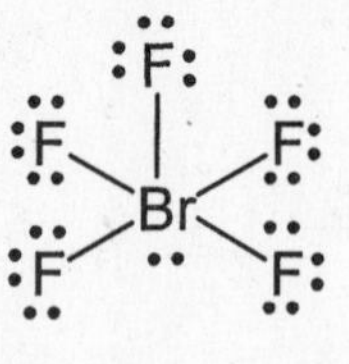

T-shaped; ≈ 90°; dsp^3 square pyramid; ≈ 90°; d^2sp^3

37.   a. $BaCl_2(s) + H_2SO_4(aq) \rightarrow BaSO_4(s) + 2\ HCl(g)$

         b. $BrF(s) + H_2O(l) \rightarrow HF(aq) + HOBr(aq)$

         c. $SiO_2(s) + 4\ HF(aq) \rightarrow SiF_4(g) + 2\ H_2O(l)$

38.   a. $F_2 + H_2O \rightarrow HOF + HF$; $2\,HOF \rightarrow 2\,HF + O_2$; $HOF + H_2O \rightarrow HF + H_2O_2$ (dilute acid)

In dilute base, HOF exists as OF^- and HF exists as F^-. The balanced reaction is:

$$(2e^- + H_2O + OF^- \rightarrow F^- + 2\,OH^-) \times 2$$
$$4\,OH^- \rightarrow O_2 + 2\,H_2O + 4e^-$$

$$\overline{}$$

$$2\,OF^- \rightarrow O_2 + 2\,F^-$$

b.   HOF: Assign +1 to H and -1 to F. The oxidation state of oxygen is then zero. Oxygen is very electronegative. A zero oxidation state is not very stable since oxygen is a very good oxidizing agent.

39. $\quad ClO^- + H_2O + 2\,e^- \rightarrow 2\,OH^- + Cl^- \qquad\qquad E° = 0.90\ V$
$\quad 2\,NH_3 + 2\,OH^- \rightarrow N_2H_4 + 2\,H_2O + 2\,e^- \qquad -E° = 0.10\ V$

$$\overline{}$$

$ClO^-(aq) + 2\,NH_3(aq) \rightarrow Cl^-(aq) + N_2H_4(aq) + H_2O(l) \qquad E°_{cell} = 1.00\ V$

Since $E°_{cell}$ is positive for this reaction, then at standard conditions ClO^- can spontaneously oxidize NH_3 to the somewhat toxic N_2H_4.

40.   A disproportion reaction is an oxidation-reduction reaction in which one species will act as both the oxidizing agent and reducing agent. The species reacts with itself, forming products with higher and lower oxidation states. For example, $2\,Cu^+ \rightarrow Cu + Cu^{2+}$ is a disproportion reaction.

$HClO_2$ will disproportionate at standard conditions since $E°_{cell} > 0$:

$\quad HClO_2 + 2\,H^+ + 2\,e^- \rightarrow HClO + H_2O \qquad\qquad E° = 1.65\ V$
$\quad HClO_2 + H_2O \rightarrow ClO_3^- + 3\,H^+ + 2\,e^- \qquad\quad -E° = -1.21\ V$

$$\overline{}$$

$2\,HClO_2(aq) \rightarrow HClO(aq) + ClO_3^-(aq) + H^+(aq) \qquad E°_{cell} = 0.44\ V$

Group 8A Elements

41.   Xe has one more valence electron than I. Thus, the isoelectric species will have I plus one extra electron substituted for Xe, giving a species with a net minus one charge.

a. IO_4^- b. IO_3^- c. IF_2^- d. IF_4^- e. IF_6^-

42.   a. KrF_2, $8 + 2(7) = 22\ e^-$ b. KrF_4, $8 + 4(7) = 36\ e^-$

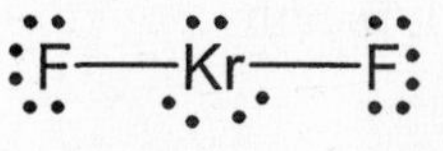

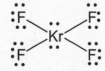

linear; 180°; dsp^3

square planar; 90°; d^2sp^3

c. XeO_2F_2, $8 + 2(6) + 2(7) = 34$ e⁻

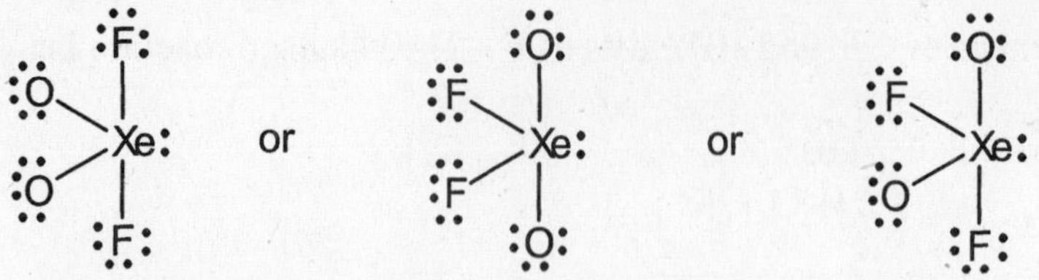

All are: see-saw; $\approx 90°$ and $\approx 120°$; dsp^3

d. XeO_2F_4, $8 + 2(6) + 4(7) = 48$ e⁻

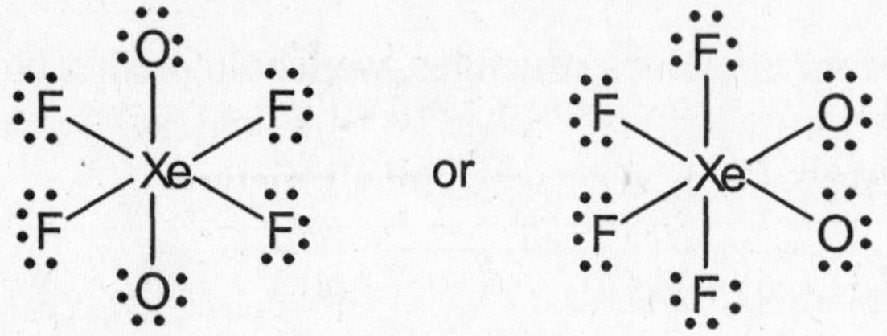

All are: octahedral; $90°$; d^2sp^3

43. XeF_2 can react with oxygen to produce explosive xenon oxides and oxyfluorides.

44. $10.0 \text{ m} \times 10.0 \text{ m} \times 10.0 \text{ m} = 1.00 \times 10^3 \text{ m}^3$; From Table 20.12, volume % Ar = 0.9%.

$$1.00 \times 10^3 \text{ m}^3 \times \left(\frac{10 \text{ dm}}{\text{m}}\right)^3 \times \frac{1 \text{ L}}{\text{dm}^3} \times \frac{0.9 \text{ L Ar}}{100 \text{ L air}} = 9 \times 10^3 \text{ L of Ar in the room}$$

$$PV = nRT, \quad n = \frac{PV}{RT} = \frac{(1.0 \text{ atm}) (9 \times 10^3 \text{ L})}{(0.08206 \text{ L atm/mol·K}) (298 \text{ K})} = 4 \times 10^2 \text{ mol Ar}$$

$$4 \times 10^2 \text{ mol Ar} \times \frac{39.95 \text{ g}}{\text{mol}} = 2 \times 10^4 \text{ g Ar in the room}$$

$$4 \times 10^2 \text{ mol Ar} \times \frac{6.022 \times 10^{23} \text{ atoms}}{\text{mol}} = 2 \times 10^{26} \text{ atoms Ar in the room}$$

A 2 L breath contains: $2 \text{ L air} \times \dfrac{0.9 \text{ L Ar}}{100 \text{ L air}} = 2 \times 10^{-2} \text{ L Ar}$

$$n = \frac{PV}{RT} = \frac{(1.0 \text{ atm}) (2 \times 10^{-2} \text{ L})}{(0.08206 \text{ L atm/mol·K}) (298 \text{ K})} = 8 \times 10^{-4} \text{ mol Ar}$$

$$8 \times 10^{-4} \text{ mol Ar} \times \frac{6.022 \times 10^{23} \text{ atoms}}{\text{mol}} = 5 \times 10^{20} \text{ atoms of Ar in a 2 L breath}$$

Since Ar and Rn are both noble gases, both species will be relatively unreactive. However, all nuclei of Rn are radioactive, unlike most nuclei of Ar. It is the radioactive decay products of Rn that can cause biological damage when inhaled.

Additional Exercises

45.   As the halogen atoms get larger, it becomes more difficult to fit three halogen atoms around the small nitrogen atom, and the NX_3 molecule becomes less stable.

46.   a.   The Lewis structures for NNO and NON are:

$$:N=N=O: \longleftrightarrow :N\equiv N-O: \longleftrightarrow :N-N\equiv O:$$

$$:N=O=N: \longleftrightarrow :N\equiv O-N: \longleftrightarrow :N-O\equiv N:$$

The NNO structure is correct. From the Lewis structures, we would predict both NNO and NON to be linear. However, we would predict NNO to be polar and NON to be nonpolar. Since experiments show N_2O to be polar, NNO is the correct structure.

b.   Formal charge = number of valence electrons of atoms - [(number of lone pair electrons) + 1/2 (number of shared electrons)].

$$:N=N=O: \longleftrightarrow :N\equiv N-O: \longleftrightarrow :N-N\equiv O:$$
$$\ \ -1\ \ \ +1\ \ \ 0 \qquad\qquad 0\ \ \ +1\ \ \ -1 \qquad\qquad -2\ \ \ +1\ \ \ +1$$

The formal charges for the atoms in the various resonance structures are below each atom. The central N is sp-hybridized in all of the resonance structures. We can probably ignore the third resonance structure on the basis of the relatively large formal charges compared to the first two resonance structures.

c.   The sp hybrid orbitals on the center N overlap with atomic orbitals (or hybrid orbitals) on the other two atoms to form the two sigma bonds. The remaining two unhybridized p orbitals on the center N overlap with two p orbitals on the peripheral N to form the two π bonds.

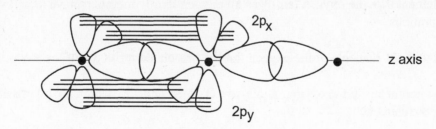

47.   OCN⁻ has $6 + 4 + 5 + 1 = 16$ valence electrons.

$$\left[:O=C=N:\right]^{-} \longleftrightarrow \left[:O-C\equiv N:\right]^{-} \longleftrightarrow \left[:O\equiv C-N:\right]^{-}$$

Formal
charge   0   0   -1   -1   0   0   +1   0   -2

Only the first two resonance structures should be important. The third places a positive formal charge on the most electronegative atom in the ion and a -2 formal charge on N.

CNO^-:

$$\left[\; :C\!\!=\!\!N\!\!=\!\!O:\; \right]^- \longleftrightarrow \left[\; :C\!\!\equiv\!\!N\!\!-\!\!\ddot{O}:\; \right]^- \longleftrightarrow \left[\; :\ddot{C}\!\!-\!\!N\!\!\equiv\!\!O:\; \right]^-$$

Formal
charge   -2   +1   0   -1   +1   -1   -3   +1   +1

All of the resonance structures for fulminate (CNO^-) involve greater formal charges than in cyanate (OCN^-), making fulminate more reactive (less stable).

48.   a.
$$(2e^- + NaBiO_3 \rightarrow BiO_3^{3-} + Na^+) \times 5$$
$$(4\,H_2O + Mn^{2+} \rightarrow MnO_4^- + 8\,H^+ + 5e^-) \times 2$$

$$8\,H_2O(l) + 2\,Mn^{2+}(aq) + 5\,NaBiO_3(s) \rightarrow 2\,MnO_4^-(aq) + 16\,H^+(aq) + 5\,BiO_3^{3-}(aq) + 5\,Na^+(aq)$$

b.   Bismuthate exists as a covalent network solid: $(BiO_3^-)_x$.

49. 1.0×10^4 kg waste $\times \dfrac{3.0\,\text{kg NH}_4^+}{100\,\text{kg waste}} \times \dfrac{1000\,\text{g}}{\text{kg}} \times \dfrac{1\,\text{mol NH}_4^+}{18.04\,\text{g NH}_4^+} \times \dfrac{1\,\text{mol C}_5\text{H}_7\text{O}_2\text{N}}{55\,\text{mol NH}_4^+}$

$$\times \dfrac{113.12\,\text{g C}_5\text{H}_7\text{O}_2\text{N}}{\text{mol C}_5\text{H}_7\text{O}_2\text{N}} = 3.4 \times 10^4 \text{ g tissue if all NH}_4^+ \text{ converted}$$

Since only 95% of the NH_4^+ ions react:

mass of tissue $= (0.95)\,(3.4 \times 10^4 \text{ g}) = 3.2 \times 10^4$ g or 32 kg bacterial tissue

50.   This element is in the oxygen family as all oxygen family members have ns^2np^4 valence electron configurations.

a.   As with all elements of the oxygen family, this element has 6 valence electrons.

b.   The nonmetals in the oxygen family are O, S, Se and Te, which are all possible identities for the element.

c.   Ions in the oxygen family are -2 charged in ionic compounds. Li_2X would be the formula between Li^+ and X^{2-} ions.

d.   In general, radii increase from right to left across the periodic table and increase going down a family. From this trend, the radius of the unknown element must be smaller than the Ba radius.

e.   The ioniation energy trend is the opposite of the radii trend indicated in the previous answer. From this trend, the unknown element will have a smaller ionization energy than fluorine.

51. TeF_5^- has $6 + 5(7) + 1 = 42$ valence electrons.

The lone pair of electrons around Te exerts a stronger repulsion than the bonding pairs, pushing the four square planar F's away from the lone pair and thus reducing the bond angles between the axial F atom and the square planar F atoms.

52.   a. $AgCl(s) \xrightarrow{hv} Ag(s) + Cl$; The reactive chlorine atom is trapped in the crystal. When light is removed, Cl reacts with silver atoms to reform AgCl, i.e., the reverse reaction occurs. In pure AgCl, the Cl atoms escape, making the reverse reaction impossible.

       b.   Over time, chlorine is lost and the dark silver metal is permanent.

53.   Release of Sr is probably more harmful. Xe is chemically unreactive. Strontium is in the same family as calcium and could be absorbed and concentrated in the body in a fashion similar to Ca. This puts the radioactive Sr in the bones: red blood cells are produced in bone marrow. Xe would not be readily incorporated in the body.

       The chemical properties determine where a radioactive material may be concentrated in the body or how easily it may be excreted. The length of time of exposure and what is exposed to radiation significantly affects the health hazard.

Challenge Problems

54.   In order to form a π bond, the d and p orbitals must overlap "side to side" instead of "head to head" as in sigma bonds. A representation of the "side to side" overlap follows.

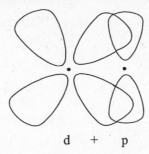

d   +   p

55.   For the reaction:

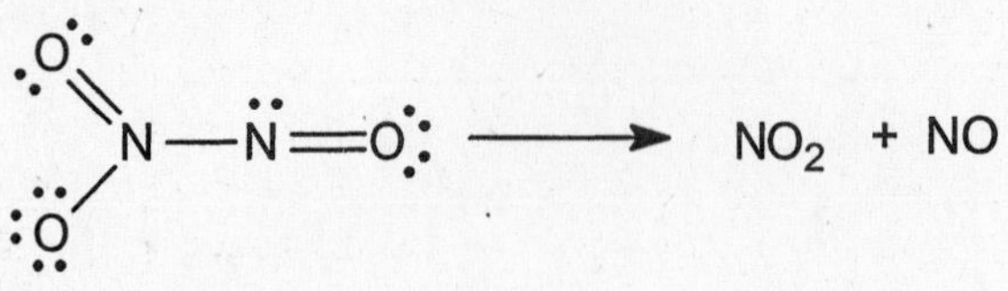

the activation energy must in some way involve breaking a nitrogen-nitrogen single bond.
For the reaction:

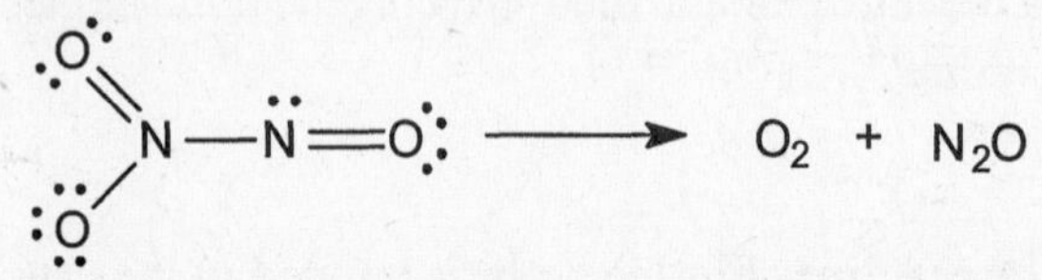

at some point nitrogen-oxygen bonds must be broken. N–N single bonds (160. kJ/mol) are weaker
than N–O single bonds (201 kJ/mol). In addition, resonance structures indicate that there is more
double bond character in the N–O bonds than in the N–N bond. Thus, NO_2 and NO are preferred
by kinetics because of the lower activation energy.

56. $Mg^{2+} + P_3O_{10}^{5-} \rightleftharpoons MgP_3O_{10}^{3-}$ $K = 4.0 \times 10^8$; $[Mg^{2+}]_o = \dfrac{50. \times 10^{-3} \text{ g}}{\text{L}} \times \dfrac{1 \text{ mol}}{24.31 \text{ g}} = 2.1 \times 10^{-3} \ M$

$[P_3O_{10}^{5-}]_o = \dfrac{40. \text{ g } Na_5P_3O_{10}}{\text{L}} \times \dfrac{1 \text{ mol}}{367.86 \text{ g}} = 0.11 \ M$

Assume the reaction goes to completion since K is large. Then solve the back equilibrium problem
to determine the small amount of Mg^{2+} present.

$$Mg^{2+} \quad + \quad P_3O_{10}^{5-} \quad \rightleftharpoons \quad MgP_3O_{10}^{3-}$$

	Mg^{2+}	$P_3O_{10}^{5-}$	$MgP_3O_{10}^{3-}$	
Before	$2.1 \times 10^{-3} \ M$	$0.11 \ M$	0	
Change	-2.1×10^{-3}	-2.1×10^{-3} $\rightarrow$	$+2.1 \times 10^{-3}$	React completely
After	0	0.11	2.1×10^{-3}	New initial condition

x mol/L $MgP_3O_{10}^{3-}$ dissociates to reach equilibrium

Change	$+x$	$+x$ $\leftarrow$	$-x$	
Equil.	x	$0.11 + x$	$2.1 \times 10^{-3} - x$	

$K = 4.0 \times 10^8 = \dfrac{[MgP_3O_{10}^{3-}]}{[Mg^{2+}][P_3O_{10}^{5-}]} = \dfrac{2.1 \times 10^{-3} - x}{x(0.11 + x)}$ (assume $x \ll 2.1 \times 10^{-3}$)

$4.0 \times 10^8 \approx \dfrac{2.1 \times 10^{-3}}{x(0.11)}$, $x = [Mg^{2+}] = 4.8 \times 10^{-11} \ M$; Assumptions good.

57.   a.   NO is the catalyst. NO is present in the first step of the mechanism on the reactant side, but it is not a reactant since it is regenerated in the second step and does not appear in the overall balanced equation.

b. NO_2 is an intermediate. Intermediates also never appear in the overall balanced equation. In a mechanism, intermediates always appear first on the product side while catalysts always appear first on the reactant side.

c. $k = A \exp(-E_a/RT)$; $\dfrac{k_{cat}}{k_{un}} = \dfrac{A \exp[-E_a(cat)/RT]}{A \exp[-E_a(un)/RT]} = \exp\left(\dfrac{E_a(un) - E_a(cat)}{RT}\right)$

$\dfrac{k_{cat}}{k_{un}} = \exp\left(\dfrac{2100 \text{ J/mol}}{8.3145 \text{ J/K}\bullet\text{mol} \times 298 \text{ K}}\right) = e^{0.85} = 2.3$

The catalyzed reaction is approximately 2.3 times faster than the uncatalyzed reaction at 25 °C.

d.   The mechanism for the chlorine-catalyzed destruction of ozone is:

$O_3(g) + Cl(g) \rightarrow O_2(g) + ClO(g)$ slow
$ClO(g) + O(g) \rightarrow O_2(g) + Cl(g)$ fast
_____

$O_3(g) + O(g) \rightarrow 2\ O_2(g)$

e.   Since the chlorine atom-catalyzed reaction has a lower activation energy, the Cl-catalyzed rate is faster. Hence, Cl is a more effective catalyst. Using the activation energy, we can estimate the efficiency that Cl atoms destroy ozone as compared to NO molecules.

At 25 °C: $\dfrac{k_{Cl}}{k_{NO}} = \exp\left(\dfrac{-E_a(Cl)}{RT} + \dfrac{E_a(NO)}{RT}\right) = \exp\left(\dfrac{(-2100 + 11,900) \text{ J/mol}}{(8.3145 \times 298) \text{ J/mol}}\right) = e^{3.96} = 52$

At 25 °C, the Cl catalyzed reaction is roughly 52 times faster (more efficient) than the NO catalyzed reaction, assuming the frequency factor A is the same for each reaction and assuming similar rate laws.

58. $3\ O_2(g) \rightleftharpoons 2\ O_3(g)$; $\Delta H° = 2(143 \text{ kJ}) = 286 \text{ kJ}$; $\Delta G° = 2(163 \text{ kJ}) = 326 \text{ kJ}$

$\ln K = \dfrac{-\Delta G°}{RT} = \dfrac{-326 \times 10^3 \text{ J}}{8.3145 \text{ J/K}\bullet\text{mol} \times 298 \text{ K}} = -131.573,\ \ K = e^{-131.573} = 7.22 \times 10^{-58}$

Note: We carried extra significant figures for the K calculation.

We need the value of K at 230. K. From Section 16.8 of the text: $\ln K = \dfrac{-\Delta H°}{RT} + \dfrac{\Delta S°}{R}$

For two sets of K and T:

$$\ln K_1 = \frac{-\Delta H^\circ}{R}\left(\frac{1}{T_1}\right) + \frac{\Delta S^\circ}{R}; \ \ln K_2 = \frac{-\Delta H^\circ}{R}\left(\frac{1}{T_2}\right) + \frac{\Delta S^\circ}{R}$$

Subtracting the first expression from the second:

$$\ln K_2 - \ln K_1 = \frac{\Delta H^\circ}{R}\left(\frac{1}{T_1} - \frac{1}{T_2}\right) \text{ or } \ln\frac{K_2}{K_1} = \frac{\Delta H^\circ}{R}\left(\frac{1}{T_1} - \frac{1}{T_2}\right)$$

Let $K_2 = 7.22 \times 10^{-58}$, $T_2 = 298$; $K_1 = K_{230}$, $T_1 = 230.$ K; $\Delta H^\circ = 286 \times 10^3$ J

$$\ln\frac{7.22 \times 10^{-58}}{K_{230}} = \frac{286 \times 10^3}{8.3145}\left(\frac{1}{230.} - \frac{1}{298}\right) = 34.13 \text{ (Carrying extra sig. figs.)}$$

$$\frac{7.22 \times 10^{-58}}{K_{230}} = e^{34.13} = 6.6 \times 10^{14}, \ K_{230} = 1.1 \times 10^{-72}$$

$$K_{230} = 1.1 \times 10^{-72} = \frac{P_{O_3}^2}{P_{O_2}^3} = \frac{P_{O_3}^2}{(1.0 \times 10^{-3})^3}, \ P_{O_3} = 3.3 \times 10^{-41} \text{ atm}$$

The volume occupied by one molecule of ozone is:

$$V = \frac{nRT}{P} = \frac{(1/6.022 \times 10^{23} \text{ mol}) \times 0.08206 \text{ L atm/mol•K} \times 230. \text{ K}}{3.3 \times 10^{-41} \text{ atm}}, \ V = 9.5 \times 10^{17} \text{ L}$$

Equilibrium is probably not maintained under these conditions. When only two ozone molecules are in a volume of 9.5×10^{17} L, the reaction is not at equilibrium. Under these conditions, Q > K and the reaction shifts left. But with only 2 ozone molecules in this huge volume, it is extremely unlikely that they will collide with each other. In these conditions, the concentration of ozone is not large enough to maintain equilibrium.

CHAPTER TWENTY-ONE

TRANSITION METALS AND COORDINATION CHEMISTRY

Questions

5. a. Ligand: Species that donates a pair of electrons to form a covalent bond to a metal ion. Ligands act as Lewis bases (electron pair donors).

    b. Chelate: Ligand that can form more than one bond to a metal ion.

    c. Bidentate: Ligand that forms two bonds to a metal ion.

    d. Complex ion: Metal ion plus ligands.

6. Since transition metals form bonds to species that donate lone pairs of electrons, transition metals are Lewis acids (electron pair acceptors). The Lewis bases in coordination compounds are the ligands that must have at least one unshared pair of electrons. The coordinate covalent bond between the ligand and the transition metal just indicates that both electrons in the bond originally came from one of the atoms in the bond.

7. a. Isomers: Species with the same formulas but different properties. See the text for examples of the following types of isomers.

    b. Structural isomers: Isomers that have one or more bonds that are different.

    c. Stereoisomers: Isomers that contain the same bonds but differ in how the atoms are arranged in space.

    d. Coordination isomers: Structural isomers that differ in the atoms that make up the complex ion.

    e. Linkage isomers: Structural isomers that differ in how one or more ligands are attached to the transition metal.

    f. Geometric isomers: (Cis - trans isomerism); Stereoisomers that differ in the positions of atoms with respect to a rigid ring, bond, or each other.

g. Optical isomers: Stereoisomers that are nonsuperimposable mirror images of each other; that is, they are different in the same way that our left and right hands are different.

8. a. Ligand that will give complex ions with the maximum number of unpaired electrons.

b. Ligand that will give complex ions with the minimum number of unpaired electrons.

c. Complex ion with a minimum number of unpaired electrons (low-spin = strong-field).

d. Complex ion with a maximum number of unpaired electrons (high-spin = weak-field).

9. Cu^{2+}: $[Ar]3d^9$; Cu^+: $[Ar]3d^{10}$; Cu(II) is d^9 and Cu(I) is d^{10}. Color is a result of the electron transfer between split d orbitals. This cannot occur for the filled d orbitals in Cu(I). Cd^{2+}, like Cu^+, is also d^{10}. We would not expect $Cd(NH_3)_4Cl_2$ to be colored since the d orbitals are filled in this Cd^{2+} complex.

10. Sc^{3+} has no electrons in d orbitals. Ti^{3+} and V^{3+} have d electrons present. Color of transition metal complexes results from electron transfer between split d orbitals. If no d electrons are present, no electron transfer can occur, and the compounds are not colored.

11. The d-orbital splitting in tetrahedral complexes is less than one-half the d-orbital splitting in octahedral complexes. There are no known ligands powerful enough to produce the strong-field case, hence all tetrahedral complexes are weak-field or high spin.

12. CN^- and CO form much stronger complexes with Fe(II) than O_2. Thus, O_2 cannot be transported by hemoglobin in the presence of CN^- or CO.

13. $Fe_2O_3(s) + 6 H_2C_2O_4(aq) \rightarrow 2 Fe(C_2O_4)_3{}^{3-}(aq) + 3 H_2O(l) + 6 H^+(aq)$; The oxalate anion forms a soluble complex ion with iron in rust (Fe_2O_3), which allows rust stains to be removed.

14. Most transition metals have unfilled d orbitals, which creates a large number of valence electrons that can be removed. Stable ions of the representative metals are determined by how many s and p valence electrons can be removed. In general, representative metals lose all of the s and p valence electrons to form their stable ions. Transition metals generally lose the s electron(s) to form +1 and +2 ions, but they can also lose some (or all) of the d electrons to form other oxidation states as well.

15. The lanthanide elements are located just before the 5d transition metals. The lanthanide contraction is the steady decrease in the atomic radii of the lanthanide elements when going from left to right across the periodic table. As a result of the lanthanide contraction, the sizes of the 4d and 5d elements are very similar (see Exercise 7.132). This leads to a greater similarity in the chemistry of the 4d and 5d elements in a given vertical group.

16. See text for examples.

a. Roasting: Converting sulfide minerals to oxides by heating in air below their melting points.

b. Smelting: Reducing metal ions to the free metal.

  c. Flotation: Separation of mineral particles in an ore from the unwanted impurities. This process depends on the greater wetability of the mineral particles as compared to the unwanted impurities.

  d. Leaching: The extraction of metals from ores using aqueous chemical solutions.

  e. Gangue: The impurities (such as clay, sand or rock) in an ore.

17. Advantages: cheap energy cost; less air pollution; Disadvantages: chemicals used in hydrometallurgy are expensive and sometimes toxic.

18. In zone refining, a bar of impure metal travels through a heater. The impurities present are more soluble in the molten metal than in the solid metal. As the molten zone moves down a metal, the impurities are swept along with the liquid, leaving behind relatively pure metal.

Exercises

Transition Metals and Coordination Compounds

19. a. Ni: $[Ar]4s^2 3d^8$ b. Cd: $[Kr]5s^2 4d^{10}$

  c. Zr: $[Kr]5s^2 4d^2$ d. Os: $[Xe]6s^2 4f^{14} 5d^6$

20. Transition metal ions lose the s electrons before the d electrons.

  a. Ni^{2+}: $[Ar]3d^8$ b. Cd^{2+}: $[Kr]4d^{10}$

  c. Zr^{3+}: $[Kr]4d^1$; Zr^{4+}: $[Kr]$ d. Os^{2+}: $[Xe]4f^{14} 5d^6$; Os^{3+}: $[Xe]4f^{14} 5d^5$

21. Transition metal ions lose the s electrons before the d electrons.

  a. Ti: $[Ar]4s^2 3d^2$ b. Re: $[Xe]6s^2 4f^{14} 5d^5$ c. Ir: $[Xe]6s^2 4f^{14} 5d^7$

    Ti^{2+}: $[Ar]3d^2$ Re^{2+}: $[Xe]4f^{14} 5d^5$ Ir^{2+}: $[Xe]4f^{14} 5d^7$

    Ti^{4+}: $[Ar]$ or $[Ne]3s^2 3p^6$ Re^{3+}: $[Xe]4f^{14} 5d^4$ Ir^{3+}: $[Xe]4f^{14} 5d^6$

22. Cr and Cu are exceptions to the normal filling order of electrons.

  a. Cr: $[Ar]4s^1 3d^5$ b. Cu: $[Ar]4s^1 3d^{10}$ c. V: $[Ar]4s^2 3d^3$

    Cr^{2+}: $[Ar]3d^4$ Cu^+: $[Ar]3d^{10}$ V^{2+}: $[Ar]3d^3$

    Cr^{3+}: $[Ar]3d^3$ Cu^{2+}: $[Ar]3d^9$ V^{3+}: $[Ar]3d^2$

23. $CoCl_2(s) + 6 H_2O(g) \rightleftharpoons CoCl_2 \cdot 6 H_2O(s)$; If rain were imminent, there would be a lot of water vapor in the air causing the reaction to shift to the right. The indicator would take on the color of $CoCl_2 \cdot 6 H_2O$, pink.

24. $H^+ + OH^- \rightarrow H_2O$; Sodium hydroxide (NaOH) will react with the H^+ on the product side of the reaction. This effectively removes H^+ from the equilibrium, which will shift the reaction to the right to produce more H^+ and CrO_4^{2-}. Since more CrO_4^{2-} is produced, the solution turns yellow.

25.   Test tube 1: added Cl^- reacts with Ag^+ to form a silver chloride precipitate. The net ionic equation is $Ag^+(aq) + Cl^-(aq) \rightarrow AgCl(s)$. Test tube 2: added NH_3 reacts with Ag^+ ions to form a soluble complex ion $Ag(NH_3)_2^+$. As this complex ion forms, Ag^+ is removed from solution, which causes the $AgCl(s)$ to dissolve. When enough NH_3 is added, all of the silver chloride precipitate will dissolve. The equation is $AgCl(s) + 2 NH_3(aq) \rightarrow Ag(NH_3)_2^+(aq) + Cl^-(aq)$. Test tube 3: added H^+ reacts with the weak base NH_3 to form NH_4^+. As NH_3 is removed from the $Ag(NH_3)_2^+$ complex ion, Ag^+ ions are released to solution and can then react with Cl^- to reform $AgCl(s)$.   The equations are $Ag(NH_3)_2^+(aq) + 2 H^+(aq) \rightarrow Ag^+(aq) + 2 NH_4^+(aq)$ and $Ag^+(aq) + Cl^-(aq) \rightarrow AgCl(s)$.

26. CN^- is a weak base, so OH^- ions are present that lead to precipitation of $Ni(OH)_2(s)$. As excess CN^- is added, the $Ni(CN)_4^{2-}$ complex ion forms. The two reactions are:

$Ni^{2+}(aq) + 2 OH^-(aq) \rightarrow Ni(OH)_2(s)$; The precipitate is $Ni(OH)_2(s)$.

$Ni(OH)_2(s) + 4 CN^-(aq) \rightarrow Ni(CN)_4^{2-}(aq) + 2 OH^-(aq)$; $Ni(CN)_4^{2-}$ is a soluble species.

27.   Since each compound contains an octahedral complex ion, the formulas for the compounds are $[Co(NH_3)_6]I_3$, $[Pt(NH_3)_4I_2]I_2$, $Na_2[PtI_6]$ and $[Cr(NH_3)_4I_2]I$. Note that in some cases, the I^- ions are ligands bound to the transition metal ion as required for a coordination number of 6, while in other cases the I^- ions are counter ions required to balance the charge of the complex ion. The $AgNO_3$ solution will only precipitate the I^- counter ions and will not precipitate the I^- ligands. Therefore, 3 mol of AgI will precipitate per mol of $[Co(NH_3)_6]I_3$, 2 mol of AgI will precipitate per mol of $[Pt(NH_3)_4I_2]I_2$, 0 mol of AgI will precipitate per mol of $Na_2[PtI_6]$, and 1 mol of AgI will precipitate per mol of $[Cr(NH_3)_4I_2]I$.

28. $BaCl_2$ gives no precipitate, so SO_4^{2-} must be in the coordination sphere. A precipitate with $AgNO_3$ means the Cl^- is not in the coordination sphere. Since there are only four ammonia molecules in the coordination sphere, the SO_4^{2-} must be acting as a bidentate ligand. The structure is:

29. To determine the oxidation state of the metal, you must know the charges of the various common ligands (see Table 21.13 of the text).

   a. pentaamminechlororuthenium(III) ion

   b. hexacyanoferrate(II) ion

   c. tris(ethylenediamine)manganese(II) ion

   d. pentaamminenitrocobalt(III) ion

30. a. tetracyanonicklate(II) ion

   b. tetraamminedichlorochromium(III) ion

   c. tris(oxalato)ferrate(III) ion

   d. tetraaquadithiocyanatocobalt(III) ion

31. a. hexaamminecobalt(II) chloride

   b. hexaaquacobalt(III) iodide

   c. potassium tetrachloroplatinate(II)

   d. potassium hexachloroplatinate(II)

   e. pentaamminechlorocobalt(III) chloride

   f. triamminetrinitrocobalt(III)

32. a. pentaaquabromochromium(III) bromide

   b. sodium hexacyanocobaltate(III)

   c. bis(ethylenediamine)dinitroiron(III) chloride

   d. tetraamminediiodoplatinum(IV) tetraiodoplatinate(II)

33. a. $K_2[CoCl_4]$

   b. $[Pt(H_2O)(CO)_3]Br_2$

   c. $Na_3[Fe(CN)_2(C_2O_4)_2]$

   d. $[Cr(NH_3)_3Cl(H_2NCH_2CH_2NH_2)]I_2$

34. a. $FeCl_4^-$

   b. $[Ru(NH_3)_5H_2O]^{3+}$

   c. $[Cr(CO)_4(OH)_2]^+$

   d. $[Pt(NH_3)Cl_3]^-$

35. a.

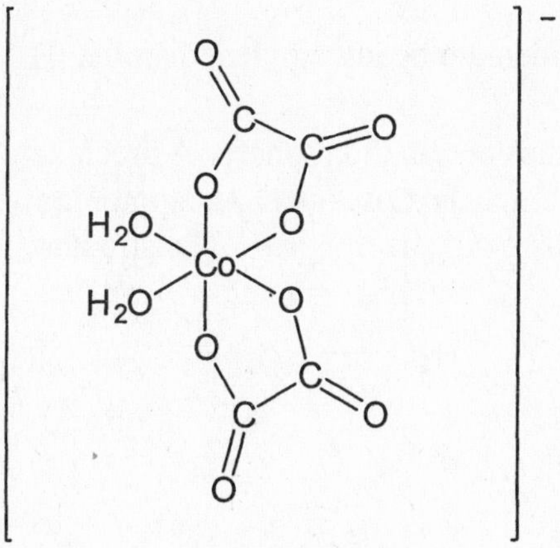

cis

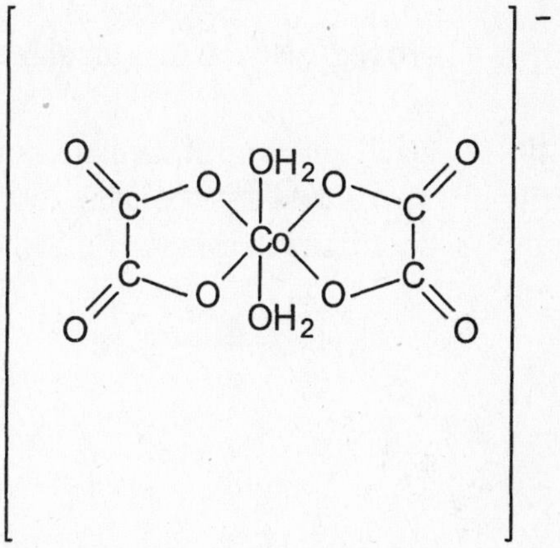

trans

Note: $C_2O_4^{2-}$ is a bidentate ligand. Bidentate ligands bond to the metal at two positions that are 90° apart from each other in octahedral complexes. Bidentate ligands do not bond to the metal at positions 180° apart.

b.

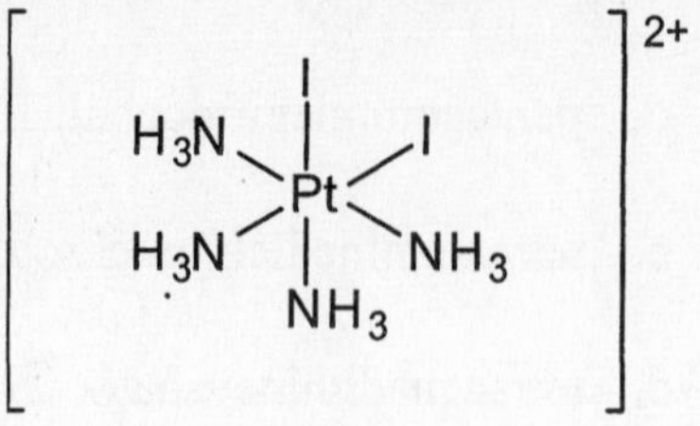

cis

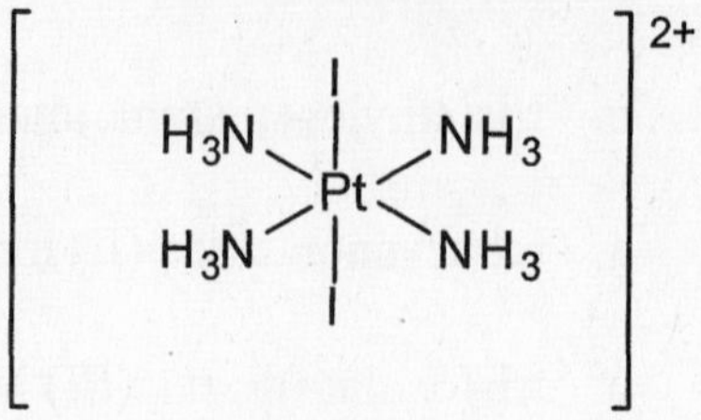

trans

c.

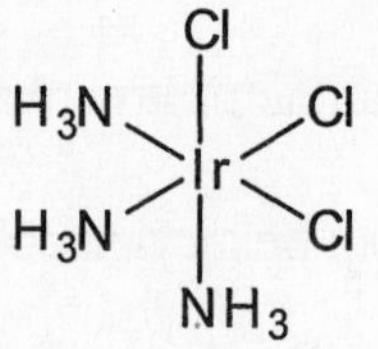

cis

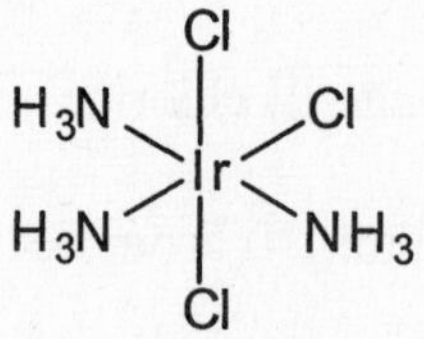

trans

d.

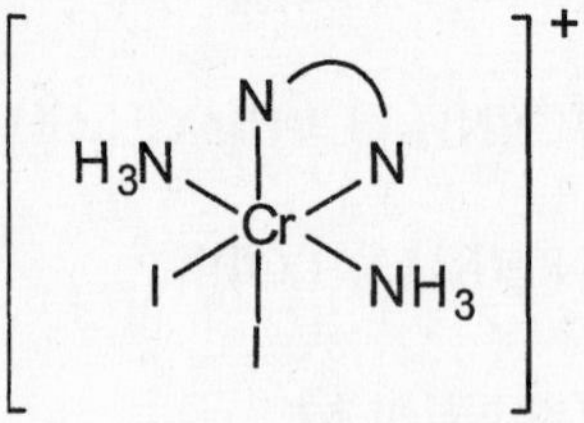

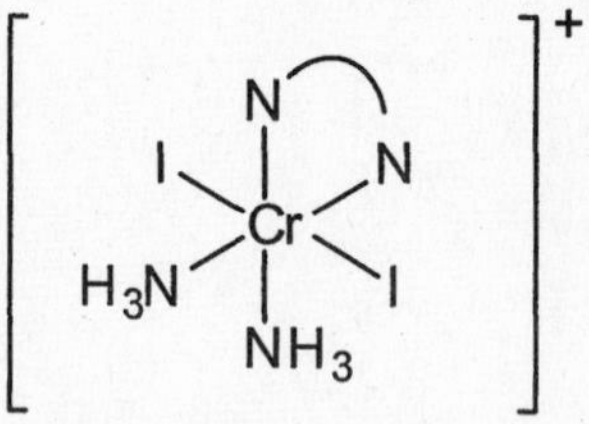

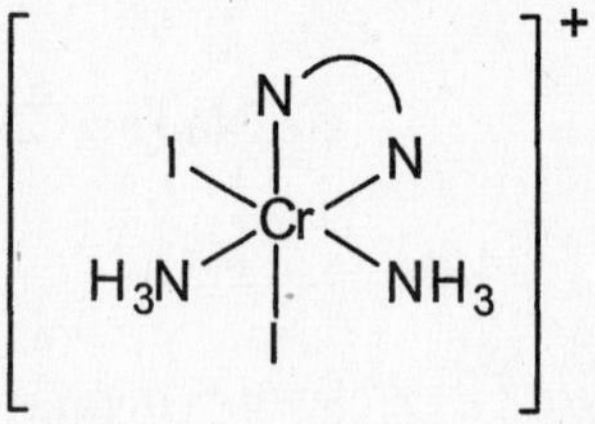

Note: N   N   is an abbreviation for the bidentate ligand ethylenediamine ($H_2NCH_2CH_2NH_2$).

36.   a.

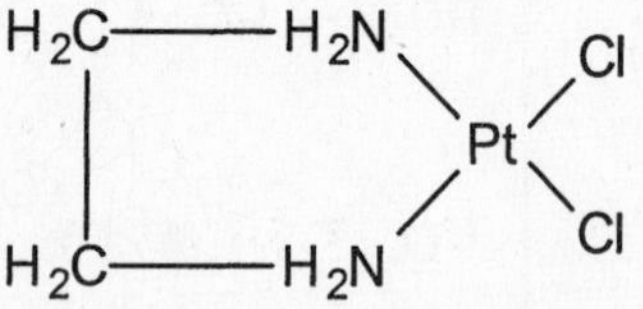

b.

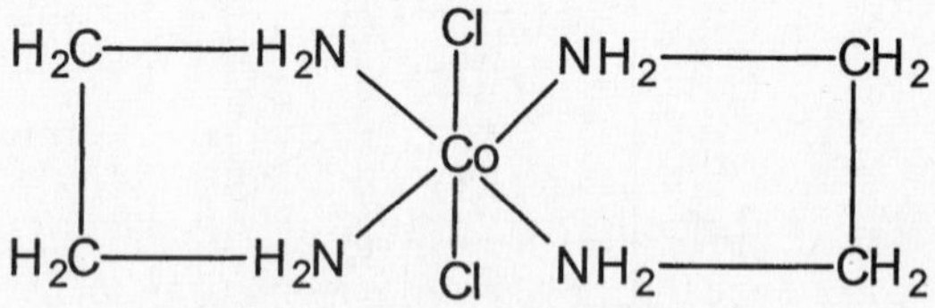

c.

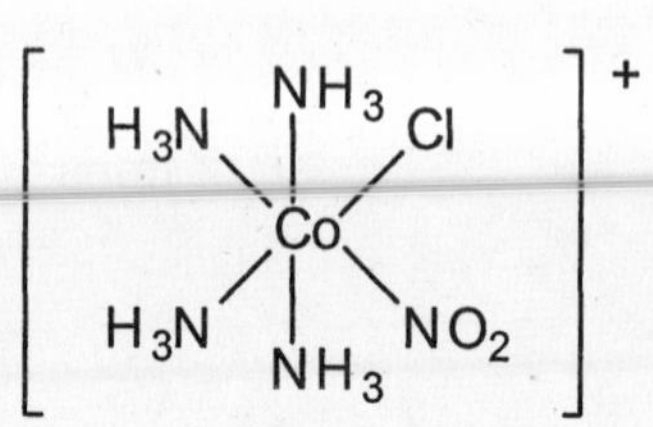

d.

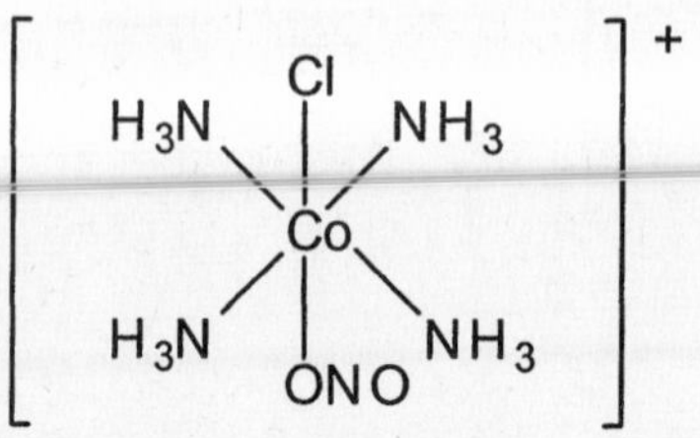

e.

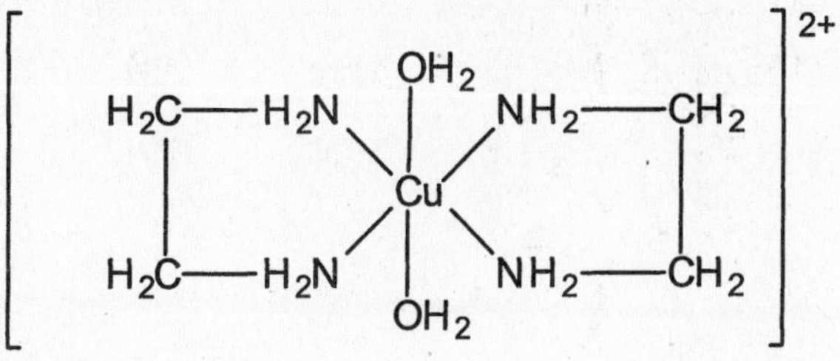

37.

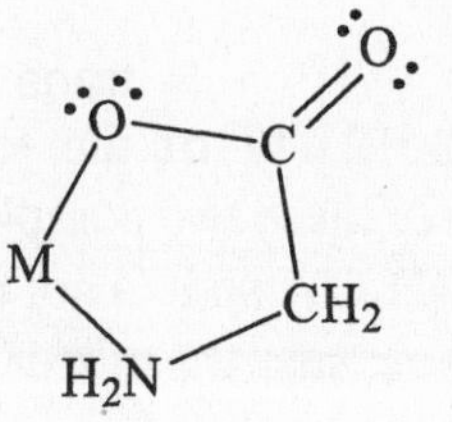

M = transition metal ion

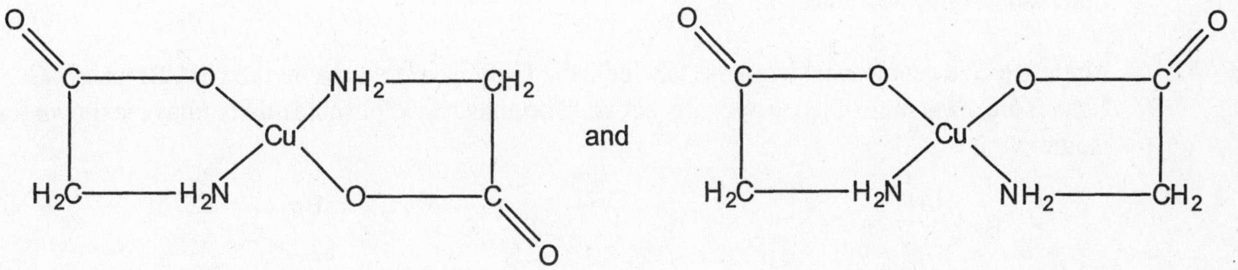

and

38.   M = metal ion

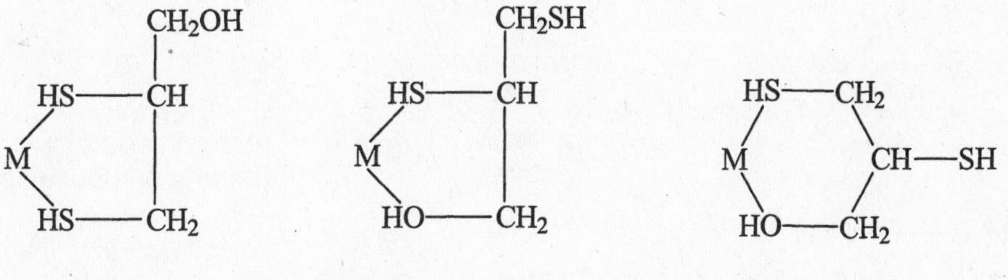

39.   Linkage isomers differ in the way the ligand bonds to the metal. SCN^- can bond through the sulfur or through the nitrogen atom. NO_2^- can bond through the nitrogen or through the oxygen atom. OCN^- can bond through the oxygen or through the nitrogen atom. N_3^-, $NH_2CH_2CH_2NH_2$ and I^- are not capable of linkage isomerism.

40.

(structures)

H_3N—Co with NO_2 (top), NH_3, NH_3, H_3N, H_3N, NO_2 (bottom)

O_2N—Co with NO_2 (top), NH_3, NH_3, H_3N, NH_3 (bottom)

H_3N—Co with ONO (top), NH_3, NH_3, H_3N, ONO (bottom)

ONO—Co with ONO (top), NH_3, H_3N, NH_3, NH_3 (bottom)

H_3N—Co with ONO (top), NH_3, NH_3, H_3N, NO_2 (bottom)

O_2N—Co with ONO (top), NH_3, NH_3, H_3N, NH_3 (bottom)

41.   Similar to the molecules discussed in Figures 21.16 and 21.17 of the text, $Cr(acac)_3$ and cis-$Cr(acac)_2(H_2O)_2$ are optically active. The mirror images of these two complexes are nonsuperimposable. There is a plane of symmetry in trans-$Cr(acac)_2(H_2O)_2$, so it is not optically active. A molecule with a plane of symmetry is never optically active because the mirror images are always superimposable. A plane of symmetry is a plane through a molecule where one side reflects the other side of the molecule.

42.   There are five geometrical isomers (labeled i-v). Only isomer v, where the CN^-, Br^- and H_2O ligands are cis to each other, is optically active. The nonsuperimposable mirror image is shown for isomer v.

i Br—Pt with CN (top), OH_2, OH_2, Br, CN (bottom)

ii Br—Pt with OH_2 (top), CN, CN, Br, OH_2 (bottom)

iii H_2O—Pt with Br (top), CN, CN, H_2O, Br (bottom)

iv Br—Pt with OH_2 (top), CN, Br, NC, OH_2 (bottom)

v Br—Pt with CN (top), CN, OH_2, Br, OH_2 (bottom)

optically active   mirror

NC—Pt with NC (top), Br, Br, H_2O, H_2O (bottom)

mirror image of v
(nonsuperimposable)

Bonding, Color, and Magnetism in Coordination Compounds

43.   **a.** Fe^{2+}: $[Ar]3d^6$

High spin, small Δ Low spin, large Δ

b. Fe^{3+}: $[Ar]3d^5$

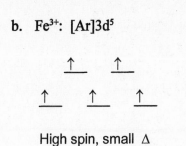

High spin, small Δ

c. Ni^{2+}: $[Ar]3d^8$

44.   a. Zn^{2+}: $[Ar]3d^{10}$

b. Co^{2+}: $[Ar]3d^7$

High spin, small Δ Low spin, large Δ

c. Ti^{3+}: $[Ar]3d^1$

45.   Since fluorine has a -1 charge as a ligand, chromium has a +2 oxidation state in CrF_6^{4-}. The electron configuration of Cr^{2+} is: $[Ar]3d^4$. For four unpaired electrons, this must be a weak-field (high-spin) case where the splitting of the d-orbitals is small and the number of unpaired electrons is maximized. The crystal field diagram for this ion is:

small Δ

46. NH_3 and H_2O are neutral ligands, so the oxidation states of the metals are Co^{3+} and Fe^{2+}. Both have six d electrons ($[Ar]3d^6$). To explain the magnetic properties, we must have a strong-field for $Co(NH_3)_6^{3+}$ and a weak-field for $Fe(H_2O)_6^{2+}$.

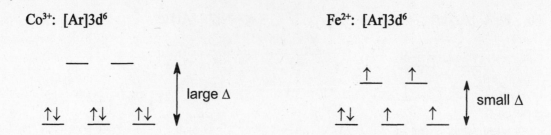

Co³⁺: [Ar]3d⁶   Fe²⁺: [Ar]3d⁶

Only this splitting of d-orbitals gives a diamagnetic $Co(NH_3)_6^{3+}$ (no unpaired electrons) and a paramagnetic $Fe(H_2O)_6^{2+}$ (unpaired electrons present).

47.   To determine the crystal field diagrams, you need to determine the oxidation state of the transition metal, which can only be determined if you know the charges of the ligands (see Table 21.13). The electron configurations and the crystal field diagrams follow.

a. Ru^{2+}: [Kr]4d⁶, no unpaired e⁻   b. Ni^{2+}: [Ar]3d⁸, 2 unpaired e⁻

Low spin, large Δ

c. V^{3+}: [Ar]3d², 2 unpaired e⁻

Note: Ni^{2+} must have 2 unpaired electrons, whether high-spin or low-spin, and V^{3+} must have 2 unpaired electrons, whether high-spin or low-spin.

48.   In both compounds, iron is in the +3 oxidation state with an electron configuration of [Ar]3d⁵. Fe^{3+} complexes have one unpaired electron when a strong-field case and five unpaired electrons when a weak-field case. $Fe(CN)_6^{2-}$ is a strong-field case and $Fe(SCN)_6^{3-}$ is a weak-field case. Therefore, cyanide, CN⁻, is a stronger field ligand than thiocyanate, SCN⁻.

49.   From Table 21.16 of the text, the violet complex ion absorbs yellow-green light (λ ~ 570 nm), the yellow complex ion absorbs blue light (λ ~ 450 nm), and the green complex ion absorbs red light (λ ~ 650 nm). The spectrochemical series shows that NH_3 is a stronger-field ligand than H_2O which is a stronger-field ligand than Cl⁻. Therefore, $Cr(NH_3)_6^{3+}$ will have the largest d-orbital splitting and will absorb the lowest wavelength electromagnetic radiation (λ ~ 450 nm) since energy and wavelength are inversely related (λ = hc/E). Thus, the yellow solution contains the $Cr(NH_3)_6^{3+}$ complex ion. Similarly, we would expect the $Cr(H_2O)_4Cl_2^+$ complex ion to have the smallest d-orbital splitting since it contains the weakest-field ligands. The green solution with the longest wavelength

of absorbed light contains the $Cr(H_2O)_4Cl_2^+$ complex ion. This leaves the violet solution, which contains the $Cr(H_2O)_6^{3+}$ complex ion. This makes sense as we would expect $Cr(H_2O)_6^{3+}$ to absorb light of a wavelength between that of $Cr(NH_3)_6^{3+}$ and $Cr(H_2O)_4Cl_2^+$.

50.   All these complex ions contain Co^{3+} bound to different ligands, so the difference in d-orbital splitting for each complex ion is due to the difference in ligands. The spectrochemical series indicates that CN^- is a stronger field ligand than NH_3 which is a stronger field ligand than F^-. Therefore, $Co(CN)_6^{3-}$ will have the largest d-orbital splitting and will absorb the lowest wavelength electromagnetic radiation ($\lambda = 290$ nm) since energy and wavelength are inversely related ($\lambda = hc/E$). $Co(NH_3)_6^{3+}$ will absorb 440 nm electromagnetic radiation while CoF_6^{3-} will absorb the longest wavelength electromagnetic radiation ($\lambda = 770$ nm) since F^- is the weakest field ligand present.

51. $CoBr_6^{4-}$ has an octahedral structure, and $CoBr_4^{2-}$ has a tetrahedral structure (as do most Co^{2+} complexes with four ligands). Coordination complexes absorb electromagnetic radiation (EMR) of energy equal to the energy difference between the split d-orbitals. Since the tetrahedral d-orbital splitting is less than one-half of the octahedral d-orbital splitting, tetrahedral complexes will absorb lower energy EMR, which corresponds to longer wavelength EMR ($E = hc/\lambda$). Therefore, $CoBr_6^{2-}$ will absorb EMR having a wavelength shorter than 3.4×10^{-6} m.

52. Co^{2+}: $[Ar]3d^7$;   The corresponding d-orbital splitting diagram for tetrahedral Co^{2+} complexes is:

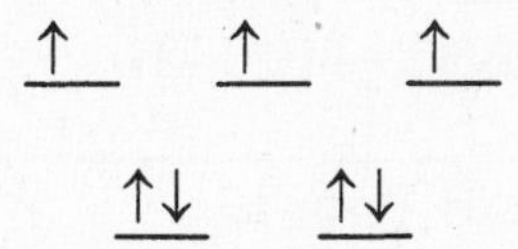

$$\uparrow \qquad \uparrow \qquad \uparrow$$

$$\uparrow\downarrow \qquad \uparrow\downarrow$$

All tetrahedral complexes are high-spin since the d-orbital splitting is small. Ions with 2 or 7 d electrons should give the most stable tetrahedral complexes since they have the greatest number of electrons in the lower energy orbitals compared to the number of electrons in the higher energy orbitals.

53.   Since the ligands are Cl^-, then iron is in the +3 oxidation state. Fe^{3+}: $[Ar]3d^5$

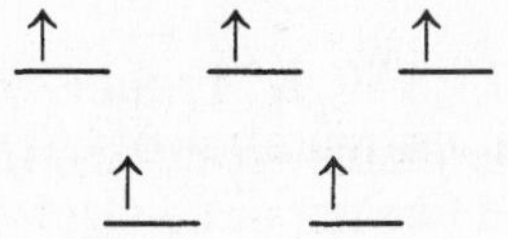

$$\uparrow \qquad \uparrow \qquad \uparrow$$

$$\uparrow \qquad \uparrow$$

Since all tetrahedral complexes are high-spin, there are 5 unpaired electrons in $FeCl_4^-$.

54.   Pd is in the +2 oxidation state in $PdCl_4^{2-}$; Pd^{2+}: $[Kr]4d^8$. If $PdCl_4^{2-}$ were a tetrahedral complex, then it would have 2 unpaired electrons and would be paramagnetic (see diagram below). Instead, $PdCl_4^{2-}$ has a square planar molecular structure with a d-orbital splitting diagram shown on the next page. Note that all electrons are paired in the square planar diagram, which explains the diamagnetic properties of $PdCl_4^{2-}$.

———

$\underline{\uparrow\downarrow} \quad \underline{\uparrow} \quad \underline{\uparrow}$ $\qquad\qquad$ $\underline{\uparrow\downarrow}$

$\underline{\uparrow\downarrow} \quad \underline{\uparrow\downarrow}$ $\qquad\qquad$ $\underline{\uparrow\downarrow}$

$\qquad\qquad\qquad\qquad\qquad$ $\underline{\uparrow\downarrow} \quad \underline{\uparrow\downarrow}$

$\qquad$ tetrahedral d^8 $\qquad\qquad\qquad$ square planar d^8

Metallurgy

55.   a.   To avoid fractions, let's first calculate ΔH for the reaction:

$$6 \text{ FeO(s)} + 6 \text{ CO(g)} \rightarrow 6 \text{ Fe(s)} + 6 \text{ CO}_2\text{(g)}$$

6 FeO + 2 CO$_2$ → 2 Fe$_3$O$_4$ + 2 CO	$\Delta H° = -2(18 \text{ kJ})$
2 Fe$_3$O$_4$ + CO$_2$ → 3 Fe$_2$O$_3$ + CO	$\Delta H° = -(-39 \text{ kJ})$
3 Fe$_2$O$_3$ + 9 CO → 6 Fe + 9 CO$_2$	$\Delta H° = 3(-23 \text{ kJ})$

$$6 \text{ FeO(s)} + 6 \text{ CO(g)} \rightarrow 6 \text{ Fe(s)} + 6 \text{ CO}_2\text{(g)} \qquad \Delta H° = -66 \text{ kJ}$$

So for: FeO(s) + CO(g) → Fe(s) + CO$_2$(g) $\qquad\qquad$ $\Delta H° = \dfrac{-66 \text{ kJ}}{6} = -11 \text{ kJ}$

b. $\Delta H° = 2(-110.5 \text{ kJ}) - [-393.5 \text{ kJ} + 0] = 172.5 \text{ kJ}$

$\Delta S° = 2(198 \text{ J/K}) - [214 \text{ J/K} + 6 \text{ J/K}] = 176 \text{ J/K}$

$\Delta G° = \Delta H° - T\Delta S°$, $\Delta G° = 0$ when $T = \dfrac{\Delta H°}{\Delta S°} = \dfrac{172.5 \text{ kJ}}{0.176 \text{ kJ/K}} = 980. \text{ K}$

Due to the favorable $\Delta S°$ term, this reaction is spontaneous at T > 980. K. From Figure 21.36 of the text, this reaction takes place in the blast furnace at temperatures greater than 980. K as required by thermodynamics.

56.   3 Fe + C → Fe$_3$C; $\Delta H° = 21 - [3(0) + 0] = 21 \text{ kJ}$

$\Delta S° = 108 - [3(27) + 6] = 21 \text{ J/K}$

$\Delta G° = \Delta H° - T\Delta S°$; When $\Delta H°$ and $\Delta S°$ are both positive, the reaction is spontaneous at high temperatures where the favorable $\Delta S°$ term becomes dominant. Thus, to incorporate carbon into steel, high temperatures are needed for thermodynamic reasons but will also be beneficial for kinetic reasons (as the temperature increases, the rate of the reaction will increase). The relative amount of Fe$_3$C (cementite) which remains in the steel depends on the cooling process. If the steel is cooled slowly, there is time for the equilibrium to shift back to the left; small crystals of carbon form giving a relatively ductile steel. If cooling is rapid, there is not enough time for the equilibrium to shift back to the left; Fe_3C is still present in the steel, and the steel is more brittle. Which cooling process occurs depends on the desired properties of the steel. The process of tempering fine tunes the steel to the desired properties by repeated heating and cooling.

Additional Exercises

57.   i. $0.0203 \text{ g } CrO_3 \times \dfrac{52.00 \text{ g Cr}}{100.0 \text{ g } CrO_3} = 0.0106 \text{ g Cr}; \quad \% \text{ Cr} = \dfrac{0.0106}{0.105} \times 100 = 10.1\% \text{ Cr}$

    ii. $32.93 \times 10^{-3} \text{ L HCl} \times \dfrac{0.100 \text{ mol HCl}}{\text{L}} \times \dfrac{1 \text{ mol } NH_3}{\text{mol HCl}} \times \dfrac{17.03 \text{ g } NH_3}{\text{mol}} = 0.0561 \text{ g } NH_3$

        $\% \, NH_3 = \dfrac{0.0561 \text{ g}}{0.341 \text{ g}} \times 100 = 16.5\% \, NH_3$

    iii. $73.53\% \text{ I} + 16.5\% \, NH_3 + 10.1\% \text{ Cr} = 100.1\%$; The compound must be composed of only Cr, NH_3, and I.

        Out of 100.00 g of compound:

        $10.1 \text{ g Cr} \times \dfrac{1 \text{ mol}}{52.00 \text{ g}} = 0.194$ $\dfrac{0.194}{0.194} = 1.00$

        $16.5 \text{ g } NH_3 \times \dfrac{1 \text{ mol}}{17.03 \text{ g}} = 0.969$ $\dfrac{0.969}{0.194} = 4.99$

        $73.53 \text{ g I} \times \dfrac{1 \text{ mol}}{126.9 \text{ g}} = 0.5794$ $\dfrac{0.5794}{0.194} = 2.99$

    $Cr(NH_3)_5I_3$ is the empirical formula. Cr(III) forms octahedral complexes. So, compound A is made of the octahedral $[Cr(NH_3)_5I]^{2+}$ complex ion and two I^- counter ions; the formula is $[Cr(NH_3)_5I]I_2$. Let's check this proposed formula using the freezing point data.

    iv. $\Delta T_f = iK_f m$; For $[Cr(NH_3)_5I]I_2$, i = 3.0 (assuming complete dissociation).

    $m = \dfrac{0.601 \text{ g complex}}{1.000 \times 10^{-2} \text{ kg } H_2O} \times \dfrac{1 \text{ mol complex}}{517.9 \text{ g complex}} = 0.116 \text{ molal}$

    $\Delta T_f = 3.0 \times 1.86°\text{C/molal} \times 0.116 \text{ molal} = 0.65°\text{C}$

    Since ΔT_f is close to the measured value, then this is consistent with the formula $[Cr(NH_3)_5I]I_2$.

58.   a.   Copper is both oxidized and reduced in this reaction, so, yes, this reaction is an oxidation-reduction reaction. The oxidation state of copper in $[Cu(NH_3)_4]Cl_2$ is +2, the oxidation state of copper in Cu is zero, and the oxidation state of copper in $[Cu(NH_3)_4]Cl$ is +1.

b. Total mass of copper used:

$$10{,}000 \text{ boards} \times \frac{(8.0 \text{ cm} \times 16.0 \text{ cm} \times 0.060 \text{ cm})}{\text{board}} \times \frac{8.96 \text{ g}}{\text{cm}^3} = 6.9 \times 10^5 \text{ g Cu}$$

Amount of Cu to be recovered $= 0.80 \times 6.9 \times 10^5 \text{ g} = 5.5 \times 10^5 \text{ g Cu}$

$$5.5 \times 10^5 \text{ g Cu} \times \frac{1 \text{ mol Cu}}{63.55 \text{ g Cu}} \times \frac{1 \text{ mol } [Cu(NH_3)_4]Cl_2}{\text{mol Cu}} \times \frac{202.59 \text{ g } [Cu(NH_3)_4]Cl_2}{\text{mol } [Cu(NH_3)_4]Cl_2}$$

$$= 1.8 \times 10^6 \text{ g } [Cu(NH_3)_4]Cl_2$$

$$5.5 \times 10^5 \text{ g Cu} \times \frac{1 \text{ mol Cu}}{63.55 \text{ g Cu}} \times \frac{4 \text{ mol NH}_3}{\text{mol Cu}} \times \frac{17.03 \text{ g NH}_3}{\text{mol NH}_3} = 5.9 \times 10^5 \text{ g NH}_3$$

59.   a. 2; Forms bonds through the lone pairs on the two oxygen atoms.

b. 3; Forms bonds through the lone pairs on the three nitrogen atoms.

c. 4; Forms bonds through the two nitrogen atoms and the two oxygen atoms.

d. 4; Forms bonds through the four nitrogen atoms.

60.   a. In the following structures, we omitted the 4 NH_3 ligands coordinated to the outside cobalt atoms.

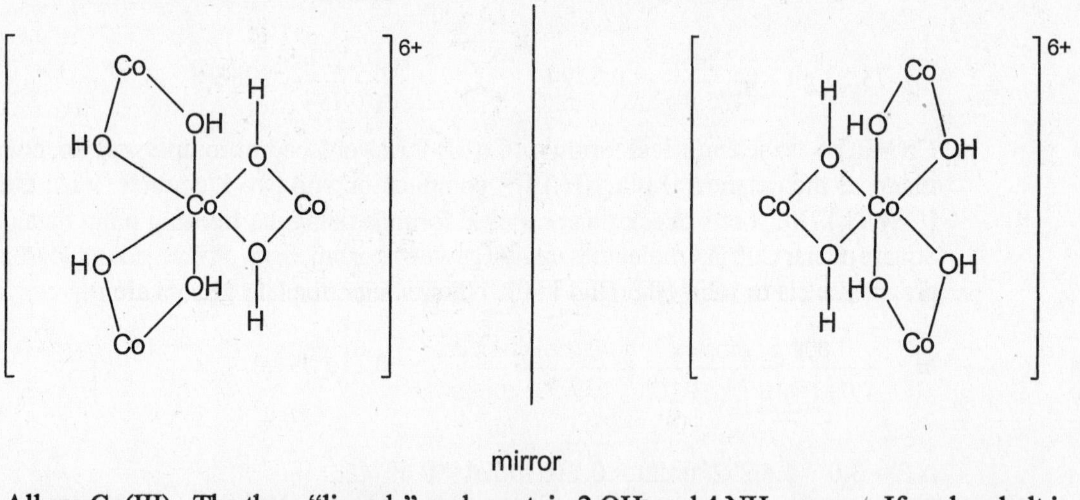

mirror

b. All are Co(III). The three "ligands" each contain 2 OH⁻ and 4 NH_3 groups. If each cobalt is in the +3 oxidation state, then each ligand has a +1 overall charge. The +3 charge from the three ligands, along with the +3 charge of the central cobalt atom, gives the overall complex a +6 charge. This is balanced by the -6 charge of the six Cl⁻ ions.

c. Co^{3+}: $[Ar]3d^6$; There are zero unpaired electrons if a low-spin (strong-field) case.

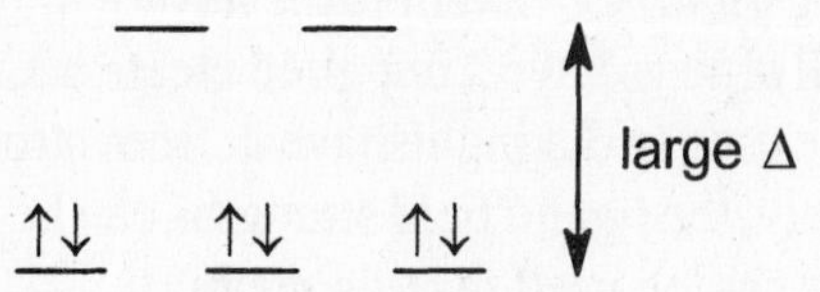

61. a. $Ru(phen)_3^{2+}$ exhibits optical isomerism [similar to $Co(en)_3^{3+}$ in Figure 21.16 of the text].

b. Ru^{2+}: $[Kr]4d^6$; Since there are no unpaired electrons, Ru^{2+} is a strong-field (low-spin) case.

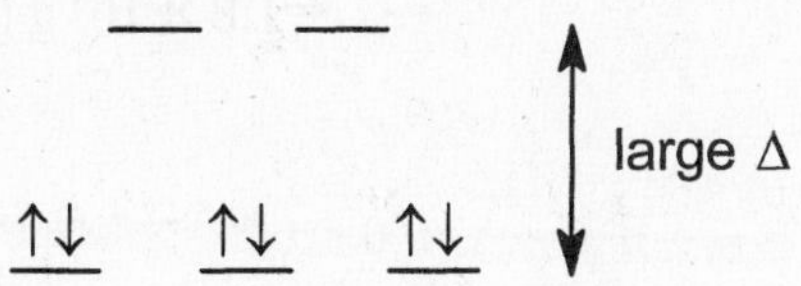

62. a. $Be(tfa)_2$ exhibits optical isomerism. A representation for the tetrahedral optical isomers are:

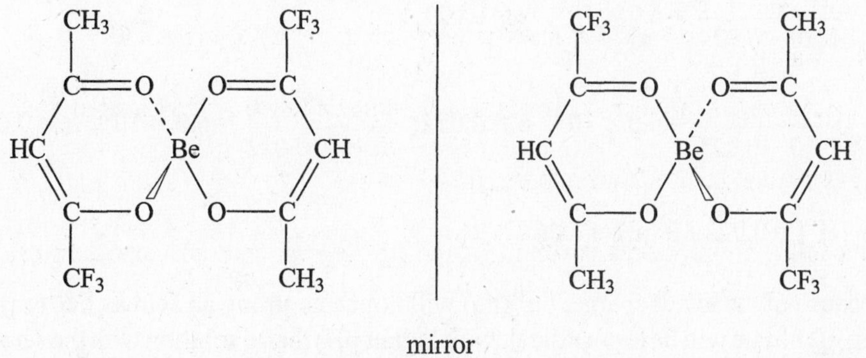

mirror

Note: The dotted line indicates a bond pointing into the plane of the paper, and the wedge indicates a bond pointing out of the plane of the paper.

b. Square planar $Cu(tfa)_2$ molecules exhibit geometric isomerism. In one geometric isomer, the CF_3 groups are cis to each other and in the other isomer, the CF_3 groups are trans.

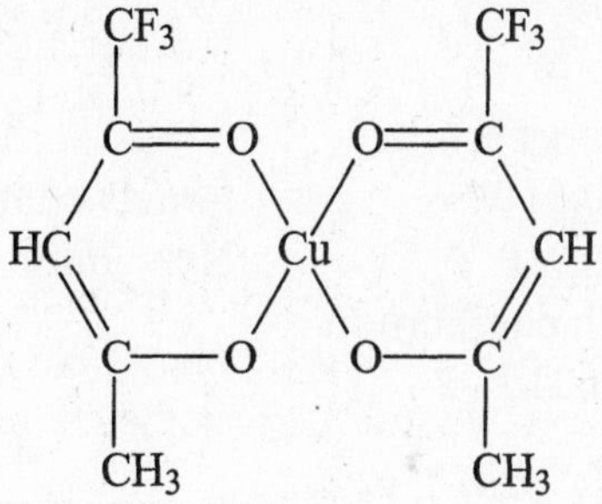

cis 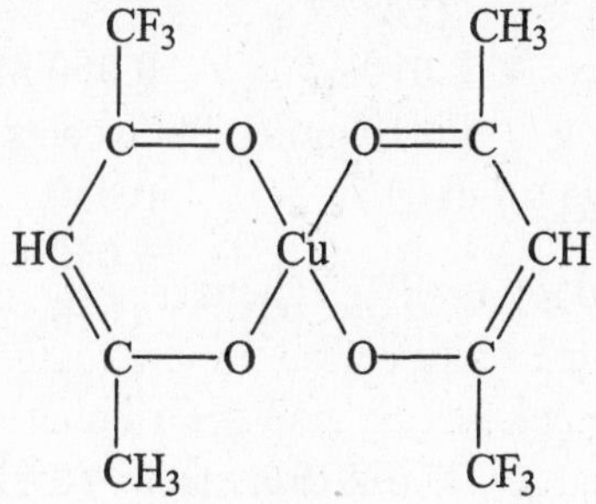   trans

63.   Octahedral Cr^{2+} complexes should be used. Cr^{2+}: $[Ar]3d^4$; High-spin (weak-field) Cr^{2+} complexes have 4 unpaired electrons and low-spin (strong-field) Cr^{2+} complexes have 2 unpaired electrons. Ni^{2+}: $[Ar]3d^8$; Octahedral Ni^{2+} complexes will always have 2 unpaired electrons, whether high or low-spin. Therefore, Ni^{2+} complexes cannot be used to distinguish weak from strong-field ligands by examining magnetic properties. Alternatively, the ligand field strengths can be measured using visible spectra. Either Cr^{2+} or Ni^{2+} complexes can be used for this method.

64.   a. $Fe(H_2O)_6^{3+} + H_2O \rightleftharpoons Fe(H_2O)_5(OH)^{2+} + H_3O^+$

Initial	0.10 M	0	~0
Equil.	0.10 - x	x	x

$$K_a = \frac{[Fe(H_2O)_5(OH)^{2+}][H_3O^+]}{[Fe(H_2O)_6^{3+}]} = 6.0 \times 10^{-3} = \frac{x^2}{0.10 - x} \approx \frac{x^2}{0.10}$$

$x = 2.4 \times 10^{-2}$; Assumption is poor (x is 24% of 0.10). Using successive approximations:

$$\frac{x^2}{0.10 - 0.024} = 6.0 \times 10^{-3},\ x = 0.021$$

$$\frac{x^2}{0.10 - 0.021} = 6.0 \times 10^{-3},\ x = 0.022; \quad \frac{x^2}{0.10 - 0.022} = 6.0 \times 10^{-3},\ x = 0.022$$

$x = [H^+] = 0.022\ M$; pH = 1.66

b.   Because of the lower charge, $Fe^{2+}(aq)$ will not be as strong an acid as $Fe^{3+}(aq)$. A solution of iron(II) nitrate will be less acidic (have a higher pH) than a solution with the same concentration of iron(III) nitrate.

65.   We need to calculate the Pb^{2+} concentration in equilibrium with $EDTA^{4-}$. Since K is large for the formation of $PbEDTA^{2-}$, let the reaction go to completion; then solve an equilibrium problem to get the Pb^{2+} concentration.

$$Pb^{2+} + EDTA^{4-} \rightleftharpoons PbEDTA^{2-} \quad K = 1.1 \times 10^{18}$$

Before	0.010 M	0.050 M	0	
	0.010 mol/L Pb^{2+} reacts completely (large K)			
Change	-0.010	-0.010	$\rightarrow$ +0.010	Reacts completely
After	0	0.040	0.010	New initial condition
	x mol/L $PbEDTA^{2-}$ dissociates to reach equilibrium			
Equil.	x	0.040 + x	0.010 - x	

$$1.1 \times 10^{18} = \frac{(0.010 - x)}{(x)(0.040 + x)} \approx \frac{(0.010)}{x(0.040)},\ x = [Pb^{2+}] = 2.3 \times 10^{-19}\ M \text{ Assumptions good.}$$

Now calculate the solubility quotient for $Pb(OH)_2$ to see if precipitation occurs. The concentration of OH^- is 0.10 M since we have a solution buffered at pH = 13.00.

$$Q = [Pb^{2+}]_o [OH^-]_o^2 = (2.3 \times 10^{-19})(0.10)^2 = 2.3 \times 10^{-21} < K_{sp} (1.2 \times 10^{-15})$$

$Pb(OH)_2(s)$ will not form since Q is less than K_{sp}.

66.   a.   In the lungs, there is a lot of O_2, and the equilibrium favors $Hb(O_2)_4$. In the cells, there is a deficiency of O_2, and the equilibrium favors HbH_4^{4+}.

      b. CO_2 is a weak acid, $CO_2 + H_2O \rightleftharpoons HCO_3^- + H^+$. Removing CO_2 essentially decreases H^+. $Hb(O_2)_4$ is then favored, and O_2 is not released by hemoglobin in the cells. Breathing into a paper bag increases $[CO_2]$ in the blood, thus increasing $[H^+]$ which shifts the reaction left.

      c. CO_2 builds up in the blood, and it becomes too acidic, driving the equilibrium to the left. Hemoglobin can't bind O_2 as strongly in the lungs. Bicarbonate ion acts as a base in water and neutralizes the excess acidity.

67. $HbO_2 \quad\quad \rightarrow Hb + O_2 \quad\quad \Delta G° = -(-70 \text{ kJ})$
      $Hb + CO \rightarrow HbCO \quad\quad \Delta G° = -80 \text{ kJ}$

      $HbO_2 + CO \rightarrow HbCO + O_2 \quad \Delta G° = -10 \text{ kJ}$

      $$\Delta G° = -RT \ln K, \quad K = \exp\left(\frac{-\Delta G°}{RT}\right) = \exp\left(\frac{-(-10 \times 10^3 \text{ J})}{(8.3145 \text{ J/K} \cdot \text{mol})(298 \text{ kJ})}\right) = 60$$

Challenge Problems

68.   a. $\Delta S°$ will be negative because there is a decrease in the number of moles of gas.

      b.   Since $\Delta S°$ is negative, $\Delta H°$ must be negative for the reaction to be spontaneous at some temperatures. Therefore, ΔS_{surr} is positive.

      c. $Ni(s) + 4 CO(g) \rightleftharpoons Ni(CO)_4(g)$

         $\Delta H° = -607 - [4(-110.5)] = -165 \text{ kJ}; \quad \Delta S° = 417 - [4(198) + (30.)] = -405 \text{ J/K}$

      d. $\Delta G° = 0 = \Delta H° - T\Delta S°, \quad T = \dfrac{\Delta H°}{\Delta S°} = \dfrac{-165 \times 10^3 \text{ J}}{-405 \text{ J/K}} = 407 \text{ K or } 134°C$

      e. $T = 50.°C + 273 = 323 \text{ K}$

         $\Delta G°_{323} = -165 \text{ kJ} - (323 \text{ K})(-0.405 \text{ kJ/K}) = -34 \text{ kJ}$

         $$\ln K = \frac{-\Delta G°}{RT} = \frac{-(-34,000 \text{ J})}{(8.3145 \text{ J/K} \cdot \text{mol})(323 \text{ K})} = 12.66, \quad K = e^{12.66} = 3.1 \times 10^5$$

f. $T = 227°C + 273 = 500. \text{ K}$

$\Delta G^°_{500} = -165 \text{ kJ} - (500. \text{ K})(-0.405 \text{ kJ/K}) = 38 \text{ kJ}$

$\ln K = \dfrac{-38,000}{(8.3145)(500.)} = -9.14, \quad K = e^{-9.14} = 1.1 \times 10^{-4}$

g.   The temperature change causes the value of the equilibrium constant to change from a large value favoring formation of $Ni(CO)_4$ to a small value favoring the decomposition of $Ni(CO)_4$ into pure Ni and CO. This is exactly what is wanted in order to purify a nickel sample.

69.   a.   Consider the following electrochemical cell:

$$Co^{3+} + e^- \rightarrow Co^{2+} \qquad\qquad E^° = 1.82 \text{ V}$$
$$Co(en)_3^{2+} \rightarrow Co(en)_3^{3+} + e^- \qquad -E^° = ?$$

$$\overline{Co^{3+} + Co(en)_3^{2+} \rightarrow Co^{2+} + Co(en)_3^{3+} \qquad E^°_{cell} = 1.82 - E^°}$$

The equilibrium constant for this overall reaction is:

$$Co^{3+} + 3 \text{ en} \rightarrow Co(en)_3^{3+} \qquad\qquad K_1 = 2.0 \times 10^{47}$$
$$Co(en)_3^{2+} \rightarrow Co^{2+} + 3 \text{ en} \qquad\qquad K_2 = 1/1.5 \times 10^{12}$$

$$\overline{Co^{3+} + Co(en)_3^{2+} \rightarrow Co(en)_3^{3+} + Co^{2+} \qquad K = K_1K_2 = \dfrac{2.0 \times 10^{47}}{1.5 \times 10^{12}} = 1.3 \times 10^{35}}$$

From the Nernst equation for the overall reaction:

$$E^°_{cell} = \dfrac{0.0591}{n} \log K = \dfrac{0.0591}{1} \log(1.3 \times 10^{35}), \quad E^°_{cell} = 2.08 \text{ V}$$

$$E^°_{cell} = 1.82 - E^° = 2.08 \text{ V}, \quad E^° = 1.82 \text{ V} - 2.08 \text{ V} = -0.26 \text{ V}$$

b.   The stronger oxidizing agent will be the more easily reduced species and will have the more positive standard reduction potential. From the reduction potentials, Co^{3+} ($E^° = 1.82$ V) is a much stronger oxidizing agent than $Co(en)_3^{3+}$ ($E^° = -0.26$ V).

c.   In aqueous solution, Co^{3+} forms the hydrated transition metal complex, $Co(H_2O)_6^{3+}$. In both complexes, $Co(H_2O)_6^{3+}$ and $Co(en)_3^{3+}$, cobalt exists as Co^{3+} which has 6 d electrons.

Assuming a strong-field case for each complex ion, the d-orbital splitting diagram for each is:

$$\underline{\quad} \quad \underline{\quad} \qquad\qquad e_g$$

$$\underline{\uparrow\downarrow} \quad \underline{\uparrow\downarrow} \quad \underline{\uparrow\downarrow} \qquad t_{2g}$$

When each complex gains an electron, the electron enters a higher energy e_g orbital. Since en is a stronger field ligand than H_2O, the d-orbital splitting is larger for $Co(en)_3^{3+}$, and it takes more energy to add an electron to $Co(en)_3^{3+}$ than to $Co(H_2O)_6^{3+}$. Therefore, it is more favorable for $Co(H_2O)_6^{3+}$ to gain an electron than for $Co(en)_3^{3+}$ to gain an electron.

70. $\overset{\text{II}}{(H_2O)_5Cr} - Cl - \overset{\text{III}}{Co(NH_3)_5} \rightarrow \overset{\text{III}}{(H_2O)_5Cr} - Cl - \overset{\text{II}}{Co(NH_3)_5} \rightarrow Cr(H_2O)_5Cl^{2+} + Co(II)$ complex

Yes; After the oxidation, the ligands on Cr(III) won't exchange. Since Cl^- is in the coordination sphere, it must have formed a bond to Cr(II) before the electron transfer occurred (as proposed through the formation of the intermediate).

71.   No; In all three cases, six bonds are formed between Ni^{2+} and nitrogen, so ΔH values should be similar. $\Delta S°$ for formation of the complex ion is most negative for 6 NH_3 molecules reacting with a metal ion (7 independent species become 1). For penten reacting with a metal ion, 2 independent species become 1, so $\Delta S°$ is least negative of all three of the reactions. Thus, the chelate effect occurs because the more bonds a chelating agent can form to the metal, the more favorable $\Delta S°$ is for the formation of the complex ion and the larger the formation constant.

72.

The $d_{x^2-y^2}$ and d_{xy} orbitals are in the plane of the three ligands and should be destabilized the most. The amount of destabilization should be about equal when all the possible interactions are considered. The d_{z^2} orbital has some electron density in the xy plane (the doughnut) and should be destabilized a lesser amount as compared to the $d_{x^2-y^2}$ and d_{xy} orbitals. The d_{xz} and d_{yz} orbitals have no electron density in the plane and should be lowest in energy.

73.

The d_{z^2} orbital will be destabilized much more than in the trigonal planar case (see Exercise 21.72). The d_{z^2} orbital has electron density on the z-axis directed at the two axial ligands. The $d_{x^2-y^2}$ and d_{xy} orbitals are in the plane of the three trigonal planar ligands and should be destabilized a lesser amount as compared to the d_{z^2} orbital; only a portion of the electron density in the $d_{x^2-y^2}$ and d_{xy} orbitals is directed at the ligands. The d_{xz} and d_{yz} orbitals will be destabilized the least since the electron density is directed between the ligands.

74.   a. $AgBr(s) \rightleftharpoons$ $Ag^+(aq)$ + $Br^-(aq)$ $K_{sp} = [Ag^+][Br^-] = 5.0 \times 10^{-13}$

Initial   s = solubility (mol/L)   0   0
Equil.   s   s

$K_{sp} = 5.0 \times 10^{-13} = s^2$, $s = 7.1 \times 10^{-7}$ mol/L

   b. $AgBr(s) \rightleftharpoons Ag^+ + Br^-$ $K_{sp} = 5.0 \times 10^{-13}$
                         $Ag^+ + 2 NH_3 \rightleftharpoons Ag(NH_3)_2^+$ $K_f = 1.7 \times 10^7$
   _____
         $AgBr(s) + 2 NH_3(aq) \rightleftharpoons Ag(NH_3)_2^+(aq) + Br^-(aq)$ $K = K_{sp} \times K_f = 8.5 \times 10^{-6}$

                    $AgBr(s)$ + $2 NH_3$ $\rightleftharpoons$ $Ag(NH_3)_2^+$ + Br^-

Initial   3.0 M 0   0
            s mol/L of AgBr(s) dissolves to reach equilibrium = molar solubility
Equil.   3.0 - 2s   s   s

$K = \dfrac{[Ag(NH_3)_2^+][Br^-]}{[NH_3]^2} = \dfrac{s^2}{(3.0 - 2s)^2} = 8.5 \times 10^{-6} \approx \dfrac{s^2}{(3.0)^2}$, $s = 8.7 \times 10^{-3}$ mol/L

Assumption good.

   c. The presence of NH_3 increases the solubility of AgBr. Added NH_3 removes Ag^+ from solution by forming the complex ion, $Ag(NH_3)_2^+$. As Ag^+ is removed, more AgBr(s) will dissolve to replenish the Ag^+ concentration.

   d. mass AgBr = $0.2500 \text{ L} \times \dfrac{8.7 \times 10^{-3} \text{ mol AgBr}}{\text{L}} \times \dfrac{187.8 \text{ g AgBr}}{\text{mol AgBr}} = 0.41$ g AgBr

   e. Added HNO_3 will have no effect on the AgBr(s) solubility in pure water. Neither H^+ nor NO_3^- react with Ag^+ or Br^- ions. Br^- is the conjugate base of the strong acid HBr, so it is a terrible base. However, added HNO_3 will reduce the solubility of AgBr(s) in the ammonia solution. NH_3 is a weak base ($K_b = 1.8 \times 10^{-5}$). Added H^+ will react with NH_3 to form NH_4^+. As NH_3 is removed, a smaller amount of the $Ag(NH_3)_2^+$ complex ion will form, resulting in a smaller amount of AgBr(s) that will dissolve.

CHAPTER TWENTY-TWO

ORGANIC AND BIOLOGICAL MOLECULES

Questions

1. There is only one chain of consecutive C-atoms in the molecule. They are not all in a true straight line since the bond angles at each carbon are the tetrahedral angles of 109.5°.

2. London dispersion (LD) forces are the primary intermolecular forces exhibited by hydrocarbons. The strength of the LD forces depends on the surface area contact among neighboring molecules. As branching increases, there is less surface area contact among neighboring molecules, leading to weaker LD forces and lower boiling points.

3. Resonance: All atoms are in the same position. Only the positions of π electrons are different.

    Isomerism: Atoms are in different locations in space.

    Isomers are distinctly different substances. Resonance is the use of more than one Lewis structure to describe the bonding in a single compound. Resonance structures are **not** isomers.

4. Structural isomers: Same formula but different bonding, either in the kinds of bonds present or the way in which the bonds connect atoms to each other.

    Geometrical isomers: Same formula and same bonds, but differ in the arrangement of atoms in space about a rigid bond or ring.

5. Substitution: An atom or group is replaced by another atom or group.

    e.g., H in benzene is replaced by Cl. $C_6H_6 + Cl_2 \xrightarrow{\text{catalyst}} C_6H_5Cl + HCl$

    Addition: Atoms or groups are added to a molecule.

    e.g., Cl_2 adds to ethene. $CH_2 = CH_2 + Cl_2 \rightarrow CH_2Cl - CH_2Cl$

6. a. Addition polymer: Polymer that forms by adding monomer units together (usually by reacting double bonds). Teflon, polyvinyl chloride and polyethylene are examples of addition polymers.

    b. Condensation polymer: Polymer that forms when two monomers combine by eliminating a small molecule (usually H_2O or HCl). Nylon and dacron are examples of condensation polymers.

c. Copolymer: Polymer formed from more than one type of monomer. Nylon and dacron are copolymers.

7.   A thermoplastic polymer can be remelted; a thermoset polymer cannot be softened once it is formed.

8.   The physical properties depend on the strengths of the intermolecular forces among adjacent polymer chains. These forces are affected by chain length and extent of branching.

     longer chains = stronger intermolecular forces; branched chains = weaker intermolecular forces

9.   Crosslinking makes a polymer more rigid by bonding adjacent polymer chains together.

10. The regular arrangement of the methyl groups in the isotactic chains allows adjacent polymer chains to pack together very closely. This leads to stronger intermolecular forces among chains as compared to atactic polypropylene where the packing of polymer chains is not as efficient.

11. In nylon, hydrogen bonding interactions occur due to the presence of N–H bonds in the polymer. For a given polymer chain length, there are more N–H groups in Nylon-46 as compared to Nylon-6. Hence, Nylon-46 forms a stronger polymer compared to Nylon-6 due to the increased hydrogen bonding interactions.

12. Polyvinyl chloride contains some polar C–Cl bonds compared to only nonpolar C–H bonds in polyethylene. The stronger intermolecular forces would be found in polyvinyl chloride since there are dipole-dipole forces present in PVC that are not present in polyethylene.

13. Primary: The amino acid sequence in the protein. Covalent bonds (peptide linkages) are the forces that link the various amino acids together in the primary structure.

     Secondary: Includes structural features known as α-helix or pleated sheet. Both are maintained mostly through hydrogen bonding interactions.

     Tertiary: The overall shape of a protein (long and narrow or globular) maintained by hydrophobic and hydrophilic interactions, such as salt linkages, hydrogen bonds, disulfide linkages and dispersion forces.

14. Denaturation changes the three-dimensional structure of a protein. Once the structure is affected, the function of the protein will also be affected.

15. All amino acids can act as both a weak acid and a weak base; this is the requirement for a buffer. The weak acid is the carboxylic end of the amino acid, and the weak base is the amine end of the amino acid.

16. A disaccharide is a carbohydrate formed by bonding two monosaccharides (simple sugars) together. In sucrose, the simple sugars are glucose and fructose, and the bond formed between these two monosaccharides is called a glycoside linkage.

17.   Hydrogen bonding interactions occur among the numerous –OH groups of starch and the water molecules.

18.   Optical isomers: The same formula and the same bonds, but the compounds are nonsuperimposable mirror images of each other. The key to identifying optical isomerism in organic compounds is to look for a tetrahedral carbon atom with four different substituents attached. When four different groups are bonded to a carbon atom, then a nonsuperimposable mirror image does exist.

19.   They all contain nitrogen atoms with lone pairs of electrons.

20.   DNA: Deoxyribose sugar; double stranded; Adenine, cytosine, guanine and thymine are the bases.

        RNA: Ribose sugar; single stranded; Adenine, cytosine, guanine and uracil are the bases.

21.   A deletion may change the entire code for a protein, thus giving an entirely different sequence of amino acids. A substitution will change only one single amino acid in a protein.

22.   When the two strands of a DNA molecule are compared, it is found that a given base in one strand is always found paired with a particular base in the other strand. Because of the shapes and side atoms along the rings of the nitrogen bases, only certain pairs are able to approach and hydrogen bond with each other in the double helix. Adenine is always found paired with thymine; cytosine is always found paired with guanine. When a DNA helix unwinds for replication during cell division, only the appropriate complementary bases are able to approach and bond to the nitrogen bases of each strand. For example, for a guanine-cytosine pair in the original DNA, when the two strands separate, only a new cytosine molecule can approach and bond to the original guanine, and only a new guanine molecule can approach and bond to the original cytosine.

Exercises

Hydrocarbons

23.   i.

        $CH_3-CH_2-CH_2-CH_2-CH_2-CH_3$

        ii.

        $CH_3-CH(CH_3)-CH_2-CH_2-CH_3$

        iii.

        $CH_3-CH_2-CH(CH_3)-CH_2-CH_3$

        iv.

        $CH_3-C(CH_3)(CH_3)-CH_2-CH_3$

        v.

        $CH_3-CH(CH_3)-CH(CH_3)-CH_3$

        All other possibilities are identical to one of these five compounds.

24.   See Exercise 22.23 for the structures. The names of structures i - v respectively, are: hexane (or n-hexane), 2-methylpentane, 3-methylpentane, 2,2-dimethylbutane and 2,3-dimethylbutane.

25.   A difficult task in this problem is recognizing different compounds from compounds that differ by rotations about one or more C–C bonds (called conformations). The best way to distinguish different compounds from conformations is to name them. Different name = different compound; same name = same compound, so it is not an isomer, but instead, is a conformation.

a. 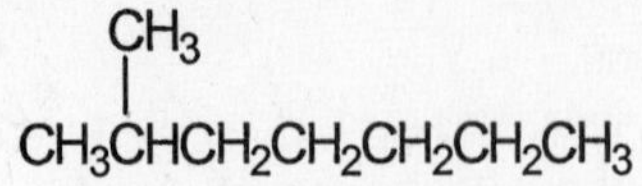
$$CH_3$$
$$|$$
$$CH_3CHCH_2CH_2CH_2CH_2CH_3$$
2-methylheptane

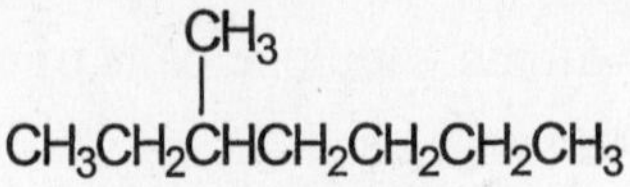

$$CH_3$$
$$|$$
$$CH_3CH_2CHCH_2CH_2CH_2CH_3$$
3-methylheptane

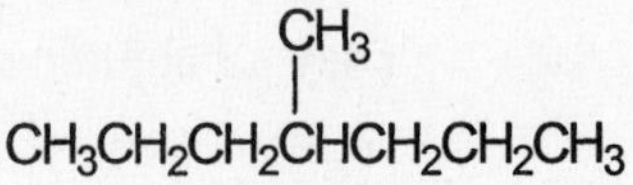

$$CH_3$$
$$|$$
$$CH_3CH_2CH_2CHCH_2CH_2CH_3$$
4-methylheptane

b. 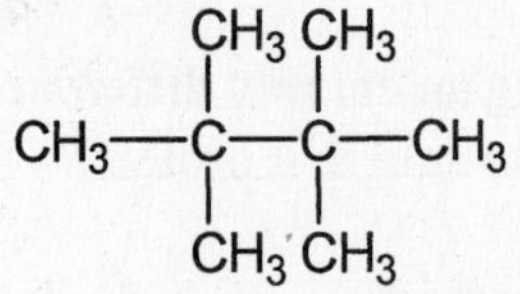
$$CH_3\ CH_3$$
$$|\quad\ |$$
$$CH_3—C—C—CH_3$$
$$|\quad\ |$$
$$CH_3\ CH_3$$
2,2,3,3-tetramethylbutane

26.   a. 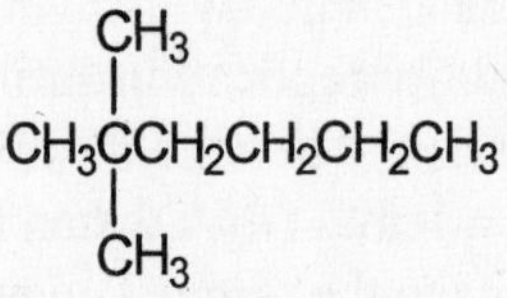
$$CH_3$$
$$|$$
$$CH_3CCH_2CH_2CH_2CH_3$$
$$|$$
$$CH_3$$
2,2-dimethylhexane

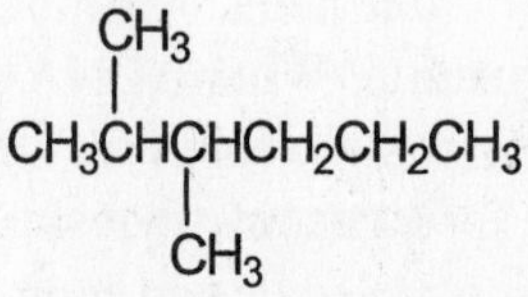

$$CH_3$$
$$|$$
$$CH_3CHCHCH_2CH_2CH_3$$
$$|$$
$$CH_3$$
2,3-dimethylhexane

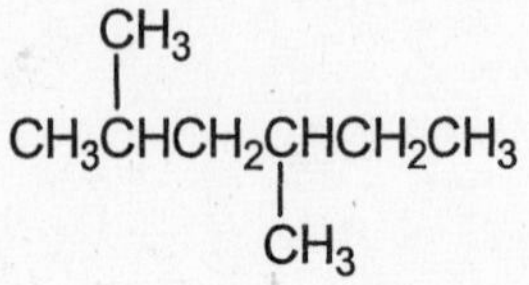

$$CH_3$$
$$|$$
$$CH_3CHCH_2CHCH_2CH_3$$
$$|$$
$$CH_3$$
2,4-dimethylhexane

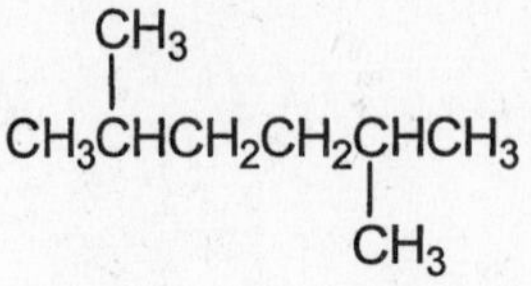

$$CH_3$$
$$|$$
$$CH_3CHCH_2CH_2CHCH_3$$
$$|$$
$$CH_3$$
2,5-dimethylhexane

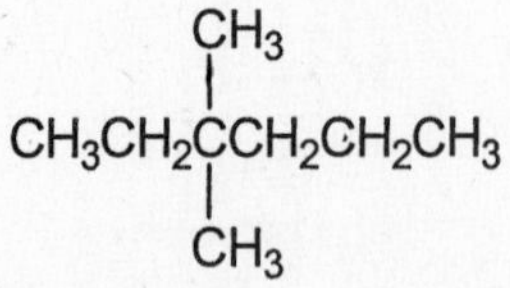

$$CH_3$$
$$|$$
$$CH_3CH_2CCH_2CH_2CH_3$$
$$|$$
$$CH_3$$
3,3-dimethylhexane

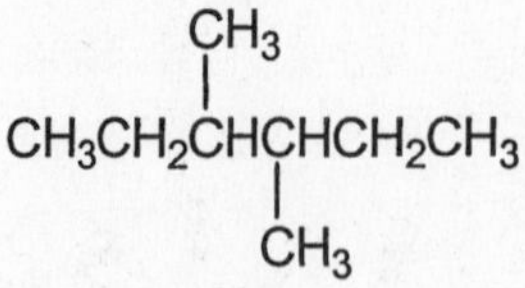

$$CH_3$$
$$|$$
$$CH_3CH_2CHCHCH_2CH_3$$
$$|$$
$$CH_3$$
3,4-dimethylhexane

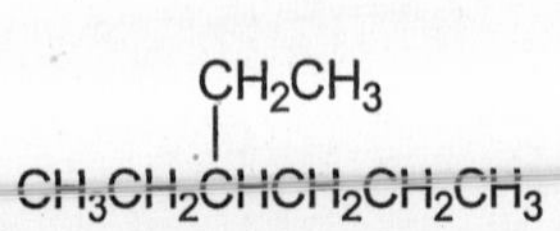

$$CH_2CH_3$$
$$|$$
$$CH_3CH_2CHCH_2CH_2CH_3$$
3-ethylhexane

b.

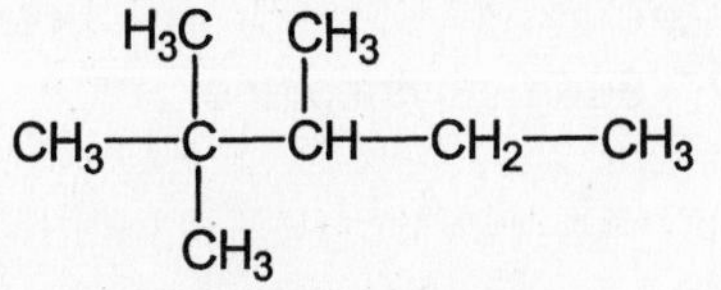

```
      H3C   CH3
        \   |
CH3 — C — CH — CH2 — CH3
        |
       CH3
```
2,2,3-trimethylpentane

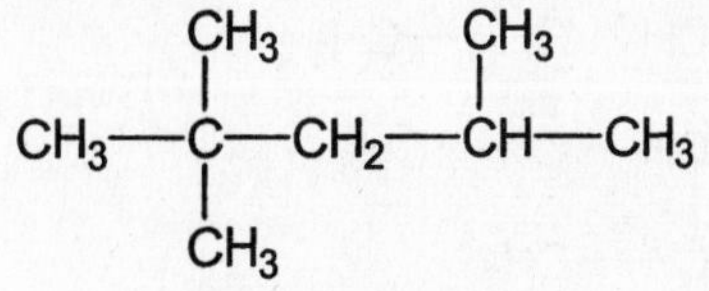

```
       CH3      CH3
        |        |
CH3 — C — CH2 — CH — CH3
        |
       CH3
```
2,2,4-trimethylpentane

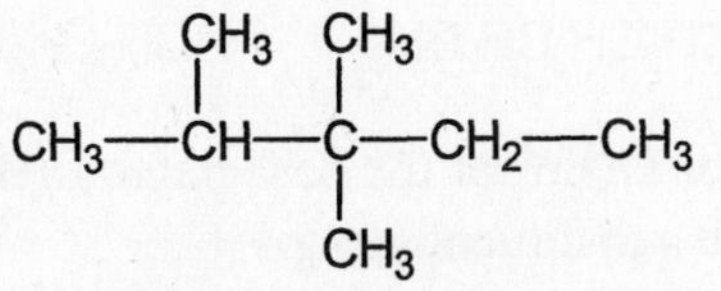

```
      CH3  CH3
       |    |
CH3 — CH — C — CH2 — CH3
            |
           CH3
```
2,3,3-trimethylpentane

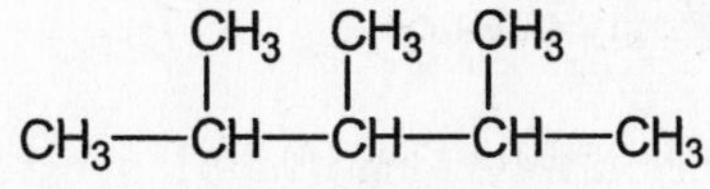

```
      CH3  CH3  CH3
       |    |    |
CH3 — CH — CH — CH — CH3
```
2,3,4-trimethylpentane

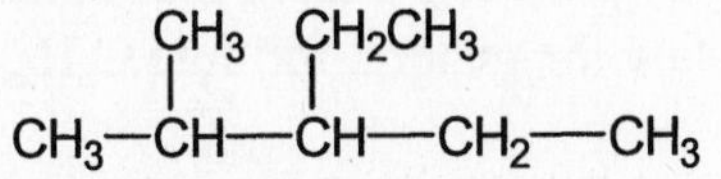

```
      CH3   CH2CH3
       |     |
CH3 — CH — CH — CH2 — CH3
```
3-ethyl-2-methylpentane

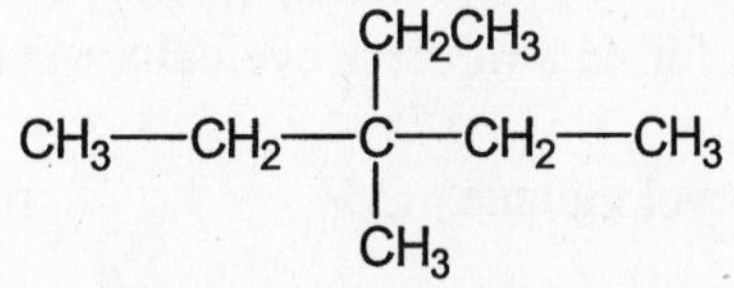

```
            CH2CH3
             |
CH3 — CH2 — C — CH2 — CH3
             |
            CH3
```
3-ethyl-3-methylpentane

27. a.

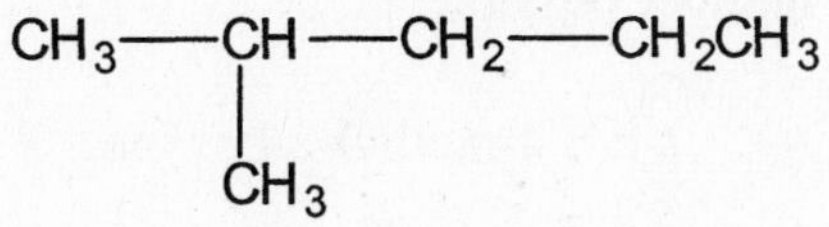

```
CH3 — CH — CH2 — CH2CH3
       |
      CH3
```

b.

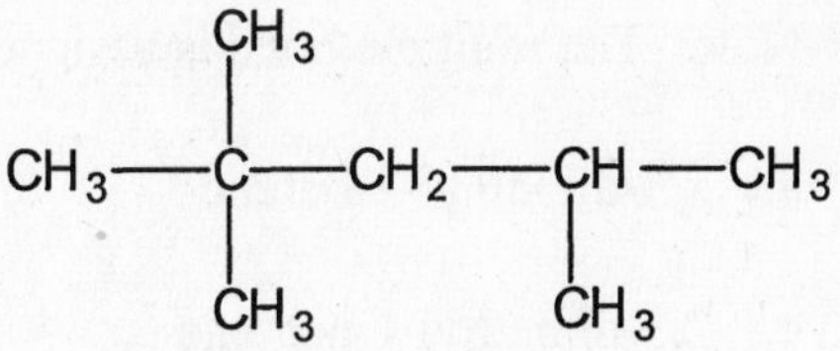

```
            CH3
             |
CH3 — C — CH2 — CH — CH3
       |          |
      CH3        CH3
```

c.

```
CH3 — CH — CH2CH2CH3
       |
CH3 — C — CH3
       |
      CH3
```

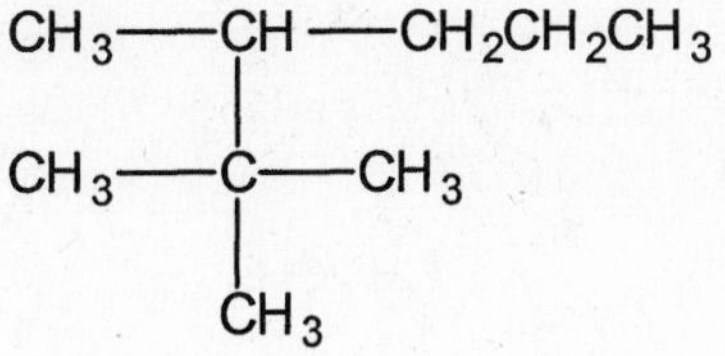

d. The longest chain is 6 carbons long.

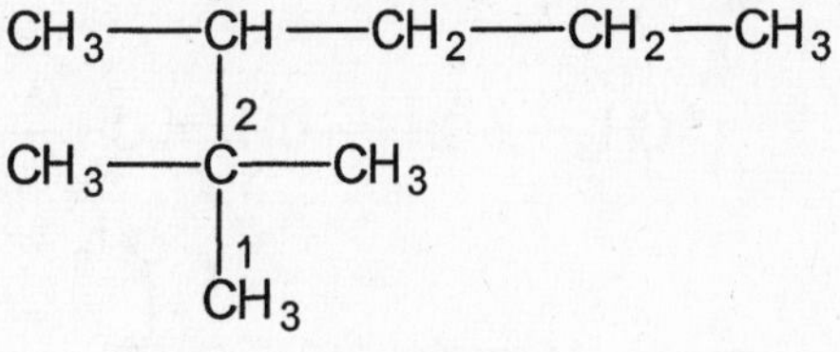

```
        3       4       5      6
CH3 — CH — CH2 — CH2 — CH3
       |2
CH3 — C — CH3
       |1
      CH3
```

2,2,3-trimethylhexane

28.

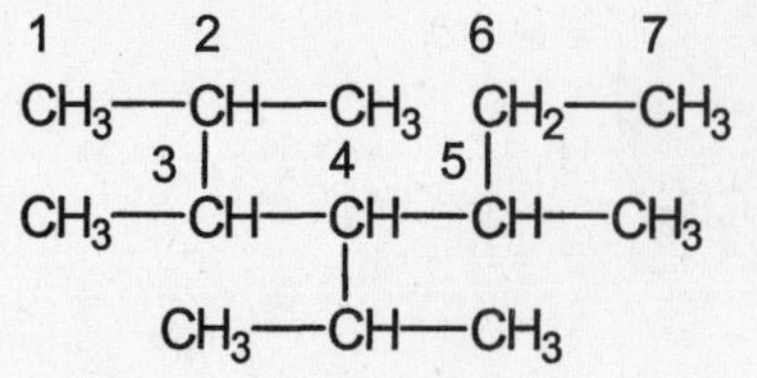

$$\begin{array}{ccccccc}
1 & 2 & & 6 & 7 \\
CH_3 & —CH—CH_3 & & CH_2 & —CH_3 \\
& |\ 3 & & |\ 4\ \ 5 \\
CH_3 & —CH—CH—CH—CH_3 \\
& | \\
& CH_3—CH—CH_3
\end{array}$$

4-isopropyl-2,3,5-trimethylheptane

29.   a. 2,2,4-trimethylhexane   b. 5-methylnonane   c. 2,2,4,4-tetramethylpentane

d. 3-ethyl-3-methyloctane

Note: For alkanes, always identify the longest carbon chain for the base name first, then number the carbons to give the lowest overall numbers for the substituent groups.

30.   The hydrogen atoms in ring compounds are commonly omitted. In organic compounds, carbon atoms satisfy the octet rule of electrons by forming four bonds to other atoms. Therefore, add C-H bonds to the carbon atoms in the ring in order to give each C atom four bonds. You can also determine the formula of these cycloalkanes by using the general formula C_nH_{2n}.

a. isopropylcyclobutane; C_7H_{14} b. 1-tert-butyl-3-methylcyclopentane; $C_{10}H_{20}$

c. 1,3-dimethyl-2-propylcyclohexane; $C_{11}H_{22}$

31.   a. 1-butene   b. 4-methyl-2-hexene   c. 2,5-dimethyl-3-heptene

Note: The multiple bond is assigned the lowest number possible.

32.   a. 2,3-dimethyl-2-butene   b. 4-methyl-2-hexyne

c. 2,3-dimethyl-1-pentene

33.   a. $CH_3–CH_2–CH{=}CH–CH_2–CH_3$ b. $CH_3–CH{=}CH–CH{=}CH–CH_2CH_3$

c.

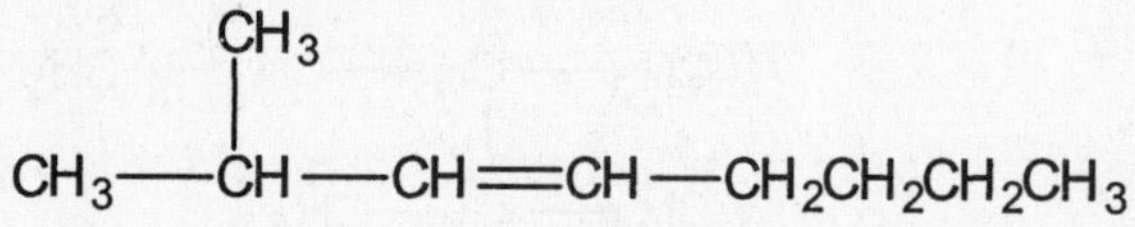

$$\begin{array}{c}
CH_3 \\
| \\
CH_3—CH—CH{=}CH—CH_2CH_2CH_2CH_3
\end{array}$$

34.

a. 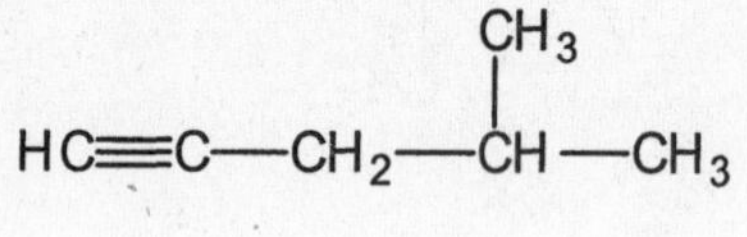 $HC{\equiv}C—CH_2—\overset{\displaystyle CH_3}{\underset{\displaystyle |}{CH}}—CH_3$ b. $H_2C{=}C—\overset{\displaystyle CH_3}{\underset{\displaystyle |}{\underset{\displaystyle CH_3}{C}}}—CH_2CH_2CH_3$ with $CH_3\ CH_3$ above

c. $CH_3CH_2—\overset{\displaystyle CH_2CH_3}{\underset{\displaystyle |}{CH}}—CH{=}CH—CH_2CH_2CH_2CH_2CH_3$

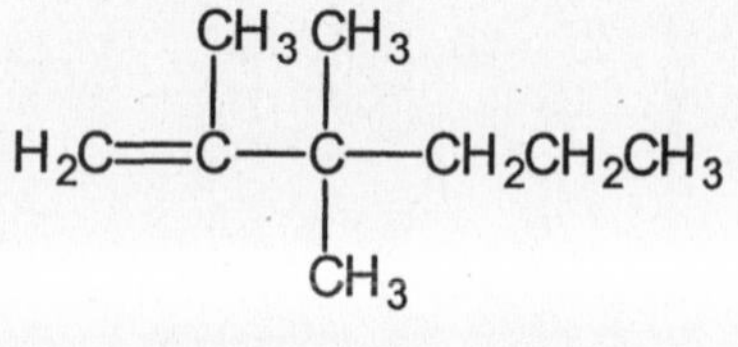

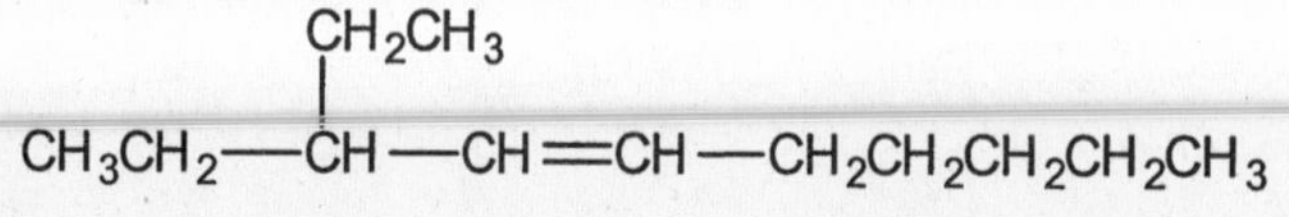

35.   a.

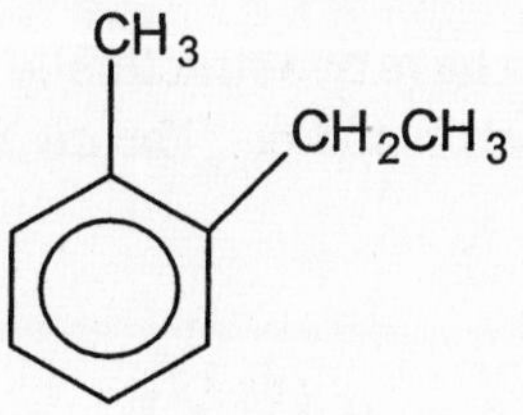

b.

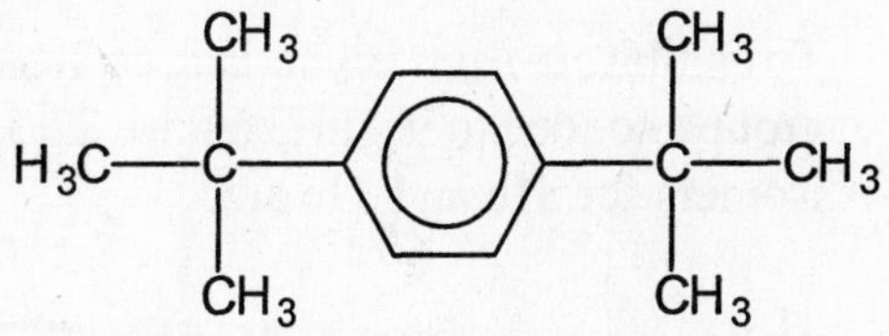

c.

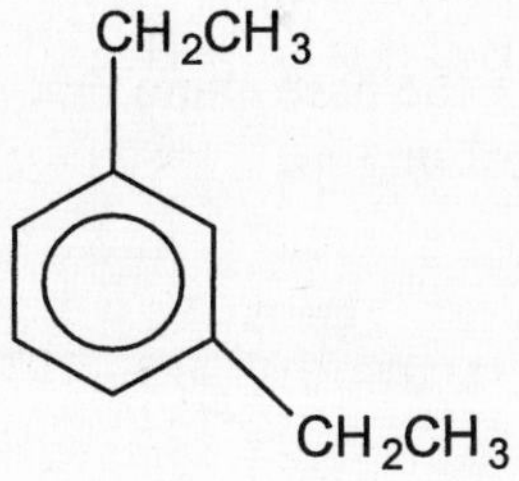

d.

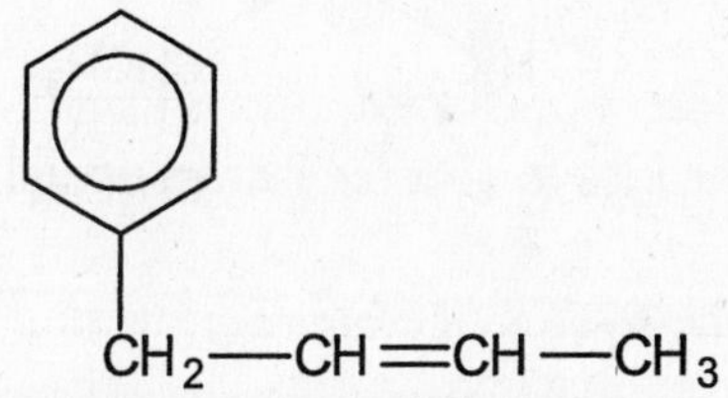

36.   isopropylbenzene or 2-phenylpropane

37.   a.   1,3-dichlorobutane   b.   1,1,1-trichlorobutane

      c.   2,3-dichloro-2,4-dimethylhexane   d.   1,2-difluoroethane

38.   a.   3-chloro-1-butene   b.   1-ethyl-3-methycyclopentene

      c.   3-chloro-4-propylcyclopentene   d.   1,2,4-trimethylcyclohexane

      e.   2-bromotoluene (or 1-bromo-2-methylbenzene) f.   1-bromo-2-methylcyclohexane

      g.   4-bromo-3-methylcyclohexene

Note: If the location of the double bond is not given in the name, it is assumed to be located between C_1 and C_2. Also, when the base name can be numbered in equivalent ways, give the first substituent group the lowest number, e.g., for part f, 1-bromo-2-methylcyclohexane is preferred to 2-bromo-1-methycyclohexane.

Isomerism

39.　To exhibit *cis-trans* isomerism, each carbon in the double bond must have two structurally different groups bonded to it. In Exercise 22.31, this occurs for compounds b and c. The *cis* and *trans* isomers for 31b and 31c are:

31 b.

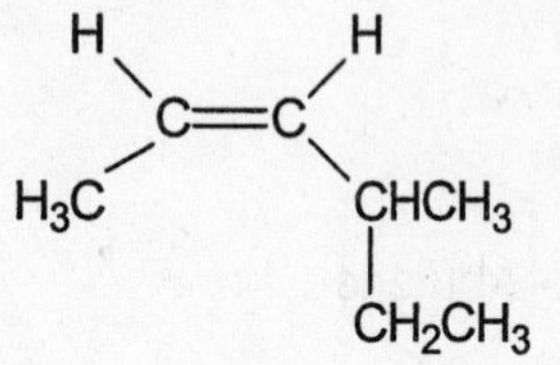

cis 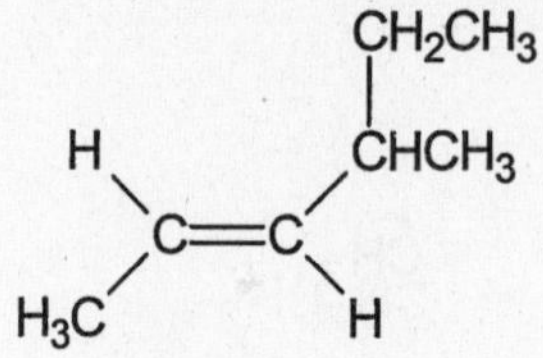　　　　　　　　　　　　　　　　　trans

31 c.

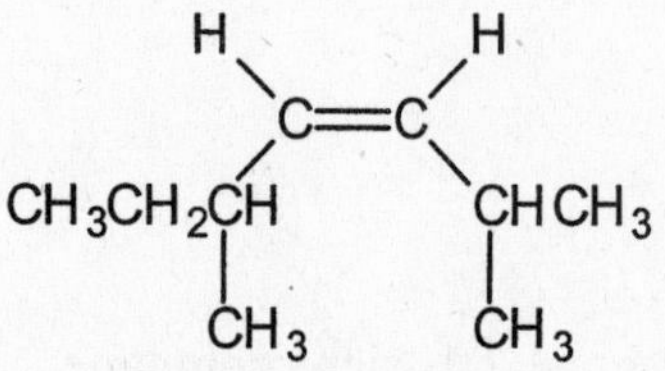

cis 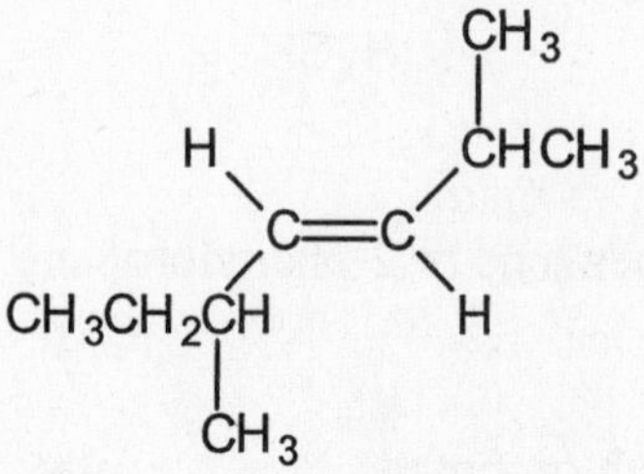　　　　　　　　　　　　　　　　　trans

Similarly, all the compounds in Exercise 22.33 also exhibit *cis-trans* isomerism.

In compound a of Exercise 22.31, the carbon in the double bond does **not** contain two different groups. The first carbon in the double bond contains two H atoms. To illustrate that this compound does not exhibit *cis-trans* isomerism, lets look at the potential *cis-trans* isomers.

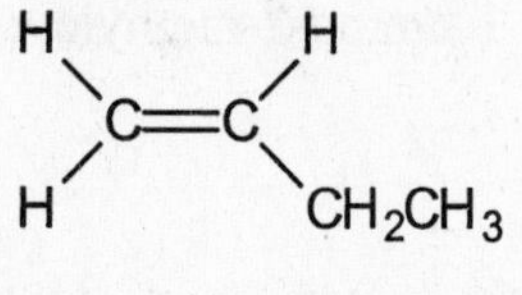

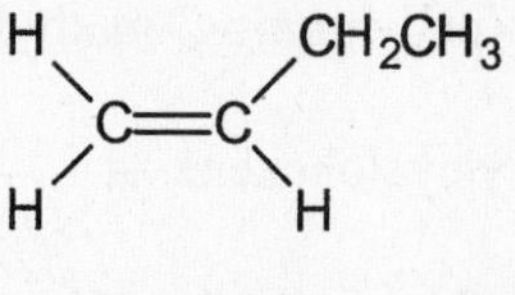

These are the same compounds; they only differ by a simple rotation of the molecule. Therefore, they are **not** isomers of each other, but instead are the same compound.

40.　In Exercise 22.32, none of the compounds can exhibit *cis-trans* isomerism since none of the carbons with the multiple bond have two different groups bonded to each. In Exercise 22.34, only 3-ethyl-4-decene can exhibit *cis-trans* isomerism since the fourth and fifth carbons each have two different groups bonded to the carbon atoms with the double bond.

41. C_5H_{10} has the general formula for alkenes, C_nH_{2n}. To distinguish the different isomers from each other, we will name them. Each isomer must have a different name.

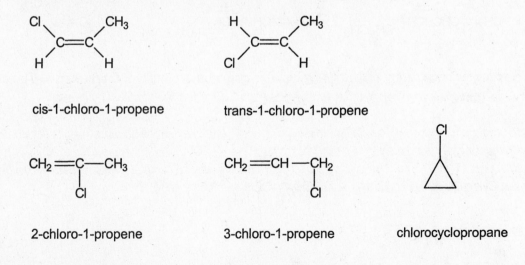

$CH_2\!=\!CHCH_2CH_2CH_3$

1-pentene

$CH_3CH\!=\!CHCH_2CH_3$

2-pentene

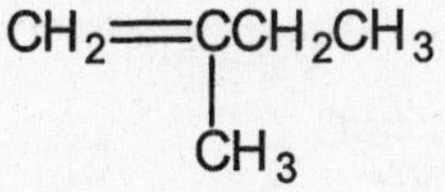

$CH_2\!=\!CCH_2CH_3$
         |
        CH_3

2-methyl-1-butene

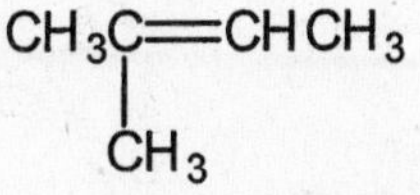

$CH_3C\!=\!CHCH_3$
       |
      CH_3

2-methyl-2-butene

$CH_3CHCH\!=\!CH_2$
       |
      CH_3

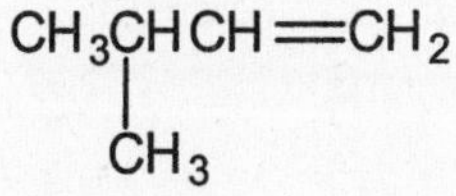

3-methyl-1-butene

42.   Only 2-pentene exhibits cis-trans isomerism. The isomers are:

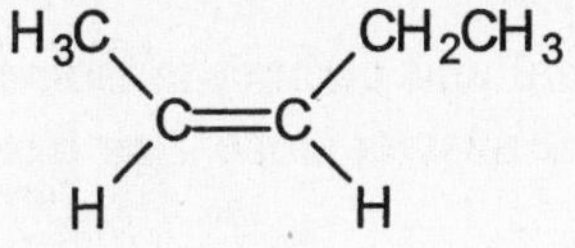

cis

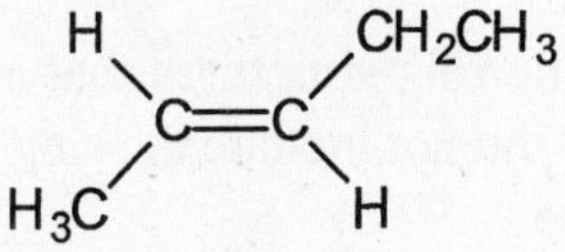

trans

The other isomers of C_5H_{10} do not contain carbons in the double bonds that have two different groups attached.

43.   To help distinguish the different isomers, we will name them.

cis-1-chloro-1-propene

trans-1-chloro-1-propene

$CH_2\!=\!C\!-\!CH_3$
          |
         Cl

2-chloro-1-propene

$CH_2\!=\!CH\!-\!CH_2$
               |
              Cl

3-chloro-1-propene

chlorocyclopropane

44. $HCBrCl-CH=CH_2$

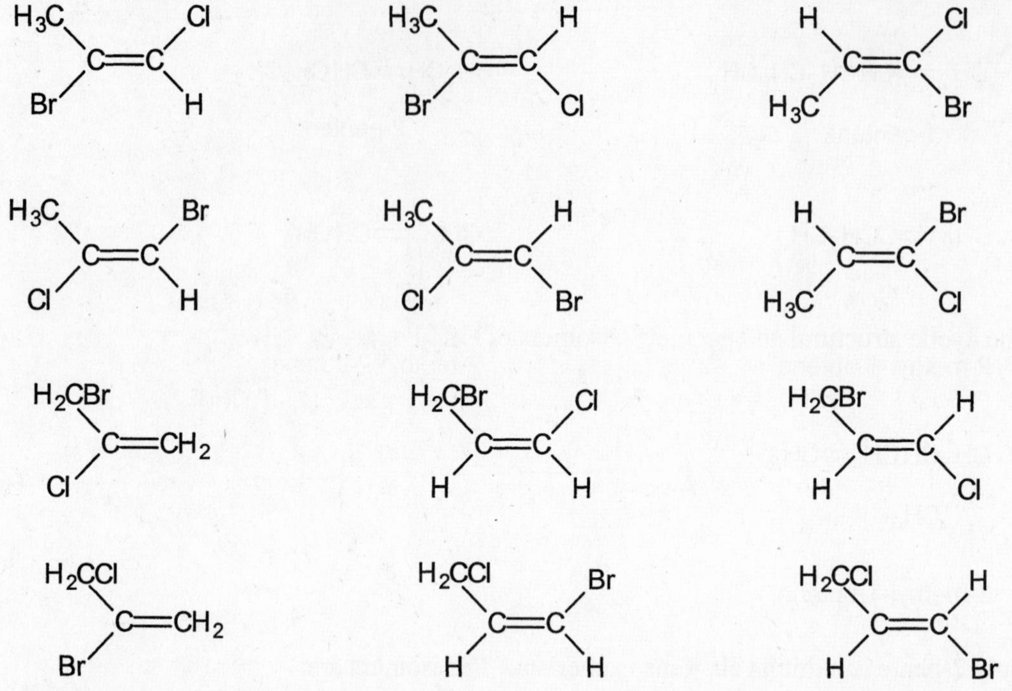

$$\begin{array}{ccc}
\underset{Br}{\overset{H_3C}{>}}C=C\underset{H}{\overset{Cl}{<}} &
\underset{Br}{\overset{H_3C}{>}}C=C\underset{Cl}{\overset{H}{<}} &
\underset{H_3C}{\overset{H}{>}}C=C\underset{Br}{\overset{Cl}{<}}
\end{array}$$

$$\begin{array}{ccc}
\underset{Cl}{\overset{H_3C}{>}}C=C\underset{H}{\overset{Br}{<}} &
\underset{Cl}{\overset{H_3C}{>}}C=C\underset{Br}{\overset{H}{<}} &
\underset{H_3C}{\overset{H}{>}}C=C\underset{Cl}{\overset{Br}{<}}
\end{array}$$

$$\begin{array}{ccc}
\underset{Cl}{\overset{H_2CBr}{>}}C=CH_2 &
\underset{H}{\overset{H_2CBr}{>}}C=C\underset{H}{\overset{Cl}{<}} &
\underset{H}{\overset{H_2CBr}{>}}C=C\underset{Cl}{\overset{H}{<}}
\end{array}$$

$$\begin{array}{ccc}
\underset{Br}{\overset{H_2CCl}{>}}C=CH_2 &
\underset{H}{\overset{H_2CCl}{>}}C=C\underset{H}{\overset{Br}{<}} &
\underset{H}{\overset{H_2CCl}{>}}C=C\underset{Br}{\overset{H}{<}}
\end{array}$$

Note: There are some ring structures that are also structural and geometric isomers of bromochloropropene. We did not include the ring structures in the answer since their base name is not bromochloropropene.

45.

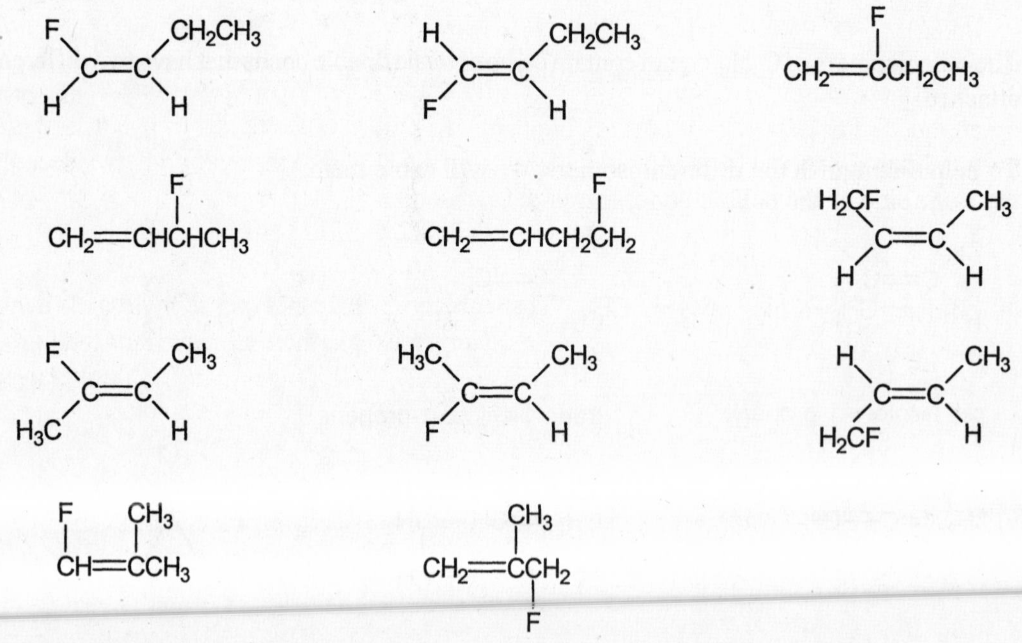

$$\begin{array}{ccc}
\underset{H}{\overset{F}{>}}C=C\underset{H}{\overset{CH_2CH_3}{<}} &
\underset{F}{\overset{H}{>}}C=C\underset{H}{\overset{CH_2CH_3}{<}} &
CH_2=\overset{\overset{F}{|}}{C}CH_2CH_3
\end{array}$$

$$\begin{array}{ccc}
CH_2=CH\overset{\overset{F}{|}}{C}HCH_3 &
CH_2=CHCH_2\overset{\overset{F}{|}}{C}H_2 &
\underset{H}{\overset{H_2CF}{>}}C=C\underset{H}{\overset{CH_3}{<}}
\end{array}$$

$$\begin{array}{ccc}
\underset{H_3C}{\overset{F}{>}}C=C\underset{H}{\overset{CH_3}{<}} &
\underset{F}{\overset{H_3C}{>}}C=C\underset{H}{\overset{CH_3}{<}} &
\underset{H_2CF}{\overset{H}{>}}C=C\underset{H}{\overset{CH_3}{<}}
\end{array}$$

$$\begin{array}{cc}
\overset{\overset{F}{|}}{C}H=\overset{\overset{CH_3}{|}}{C}CH_3 &
CH_2=\overset{\overset{CH_3}{|}}{\underset{\underset{F}{|}}{C}}CH_2
\end{array}$$

46.   The *cis* isomer has the CH_3 groups on the same side of the ring. The *trans* isomer has the CH_3 groups on opposite sides of the ring.

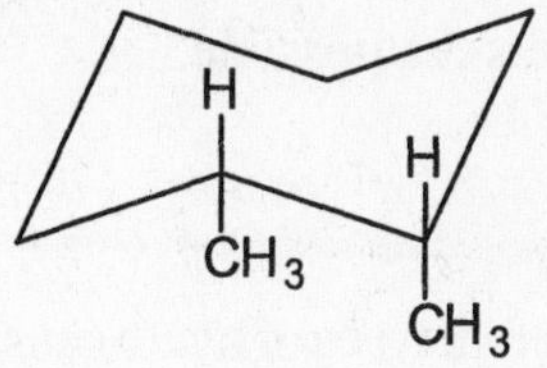

cis 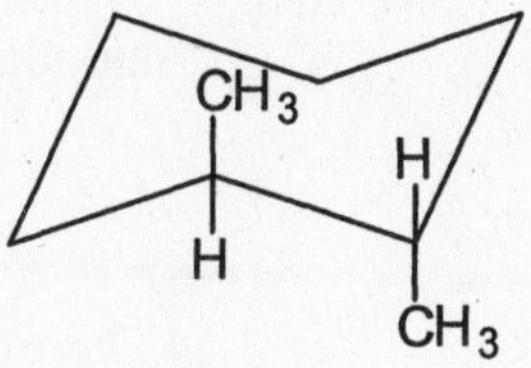   trans

The cyclic structural and geometric isomers of C_4H_7F are:

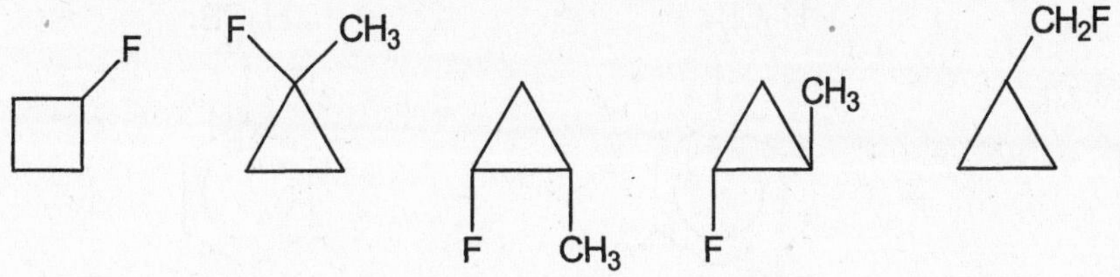

cis   trans

47.   a. 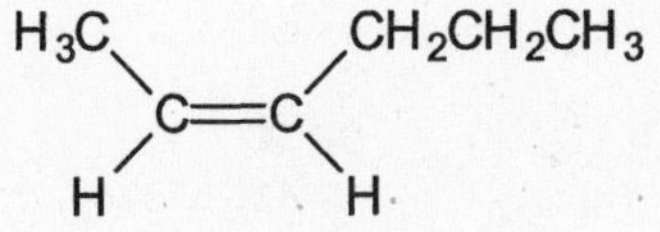   b. 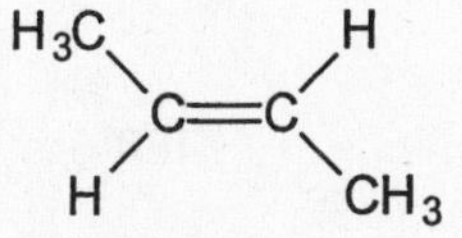   c.

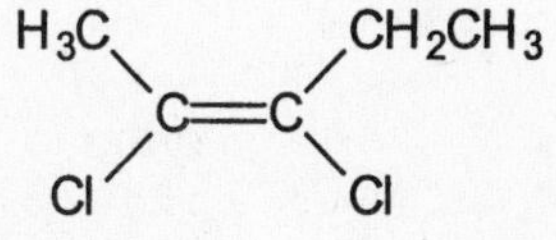

48.   a. cis-1-bromo-1-propene   b. cis-4-ethyl-3-methyl-3-heptene

      c. trans-1,4-diiodo-2-propyl-1-pentene

      Note: In general, cis-trans designations refer to the relative positions of the largest groups. In compound b, the largest group off the first carbon in the double bond is CH_2CH_3, and the largest group off the second carbon in the double bond is $CH_2CH_2CH_3$. Since their relative placement is on the same side of the double bond, this is the cis isomer.

49.   a. $CH_3^*\!-\!CH_2^*\!-\!CH_2^*\!-\!CH_2\!-\!CH_3$ There are three different types of hydrogens in n-pentane (see asterisks). Thus there are three mono-chloro isomers of n-pentane (1-chloropentane, 2-chloropentane and 3-chloropentane).

      b. 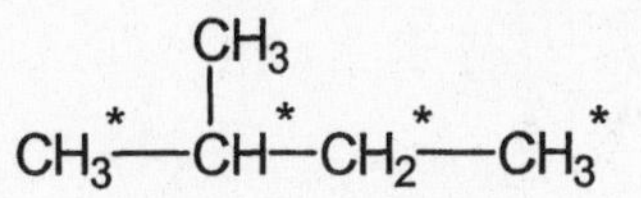   There are four different types of hydrogens in 2-methyl-butane, so four monochloro isomers of 2-methylbutane are possible.

c.

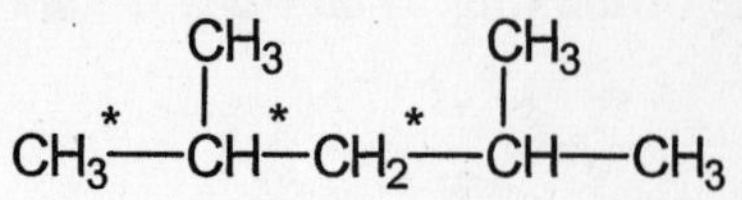

CH₃   CH₃
 | |
CH₃—*CH—*CH₂—*CH—CH₃

There are three different types of hydrogens, so three monochloro isomers are possible.

d.

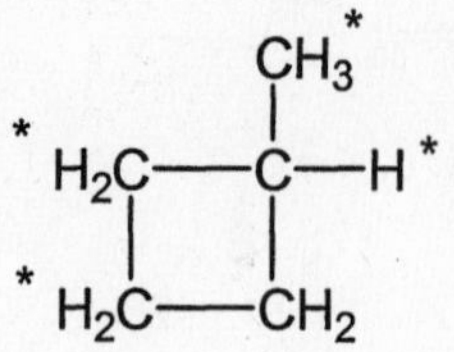

            CH₃ *
             |
   * H₂C——C——H *
        |
   * H₂C——CH₂

There are four different types of hydrogens, so four monochloro isomers are possible.

50.   a.

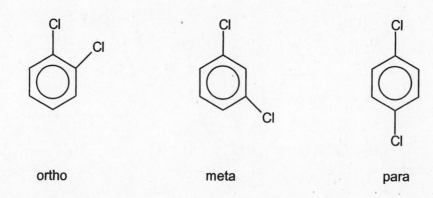

ortho   meta   para

b.   There are three trichlorobenzenes (1,2,3-trichlorobenzene, 1,2,4-trichlorobenzene and 1,3,5-trichlorobenzene).

c.   The meta isomer will be very difficult to synthesize.

d.   1,3,5-trichlorobenzene will be the most difficult to synthesize since all Cl groups are meta to each other in this compound.

Functional Groups

51.   Reference Table 22.5 for the common functional groups.

a.   ketone   b.   aldehyde   c.   carboxylic acid   d.   amine

52.   a.

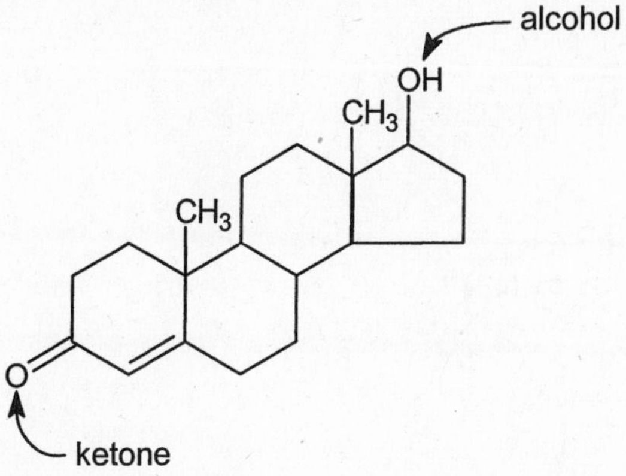

b.

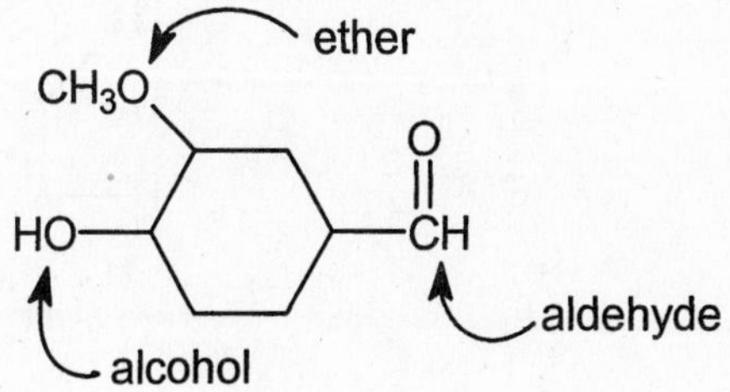

c.

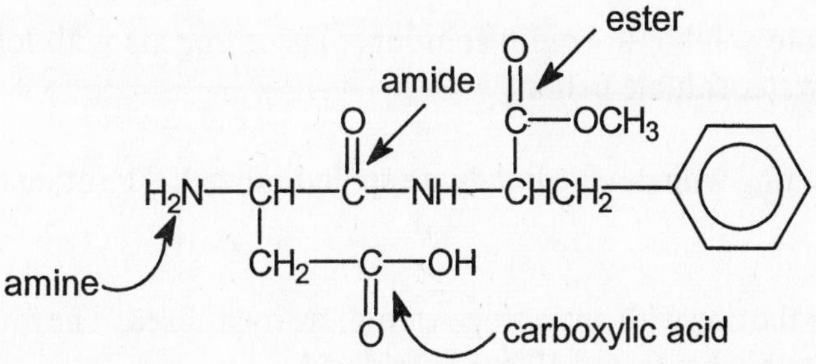

Note: the amide functional group $\left(\begin{array}{c} R-C-N-R'' \end{array} \right)$ is not covered in Chapter 22 of the text. We point it out for your information.

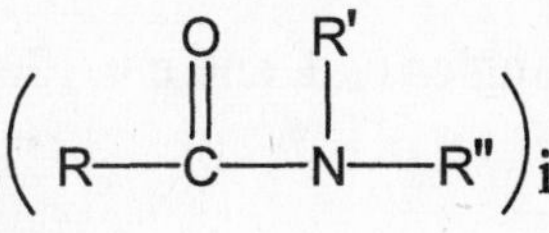

53.   a.

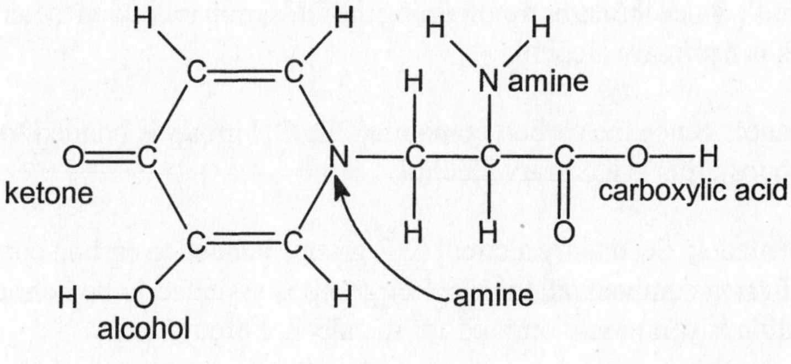

b.   5 carbons in the ring and the carbon in $-CO_2H$: sp^2; the other two carbons: sp^3

c.   24 sigma bonds; 4 pi bonds

54.   Hydrogen atoms are usually omitted from ring structures. In organic compounds, the carbon atoms generally form four bonds. With this in mind, the following structure has the missing hydrogen atoms included in order to give each carbon atom the four bond requirement.

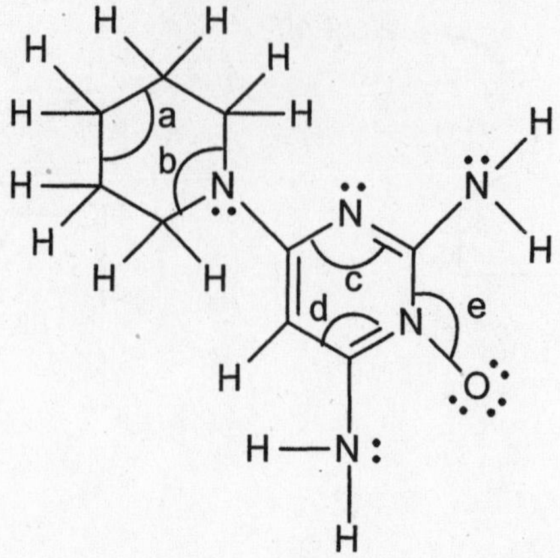

a. Minoxidil would be more soluble in acidic solution. The nitrogens with lone pairs can be protonated, forming a water soluble cation.

b. The two nitrogens in the ring with double bonds are sp^2 hybridized. The other three N's are sp^3 hybridized.

c. The five carbon atoms in the ring with one nitrogen are all sp^3 hybridized. The four carbon atoms in the other ring with double bonds are all sp^2 hybridized.

d. Angles a and b $\approx 109.5°$; Angles c, d, and e $\approx 120°$

e. 31 sigma bonds

f. 3 pi bonds

55. a. 3-chloro-1-butanol: Since the carbon containing the OH group is bonded to just 1 other carbon (1 R group), this is a primary alcohol.

b. 3-methyl-3-hexanol; Since the carbon containing the OH group is bonded to three other carbons (3 R groups), this is a tertrary alcohol

c. 2-methylcyclopentanol; Secondary alcohol (2 R groups bonded to carbon containing the OH group); Note: In ring compounds, the alcohol group is assumed to be bonded to C_1, so the number designation is commonly omitted for the alcohol group.

56.

a. 
$$\overset{\text{OH}}{\underset{|}{\text{CH}_2}}\text{—CH}_2\text{—CH}_2\text{—CH}_3 \qquad \text{primary alcohol}$$

b. 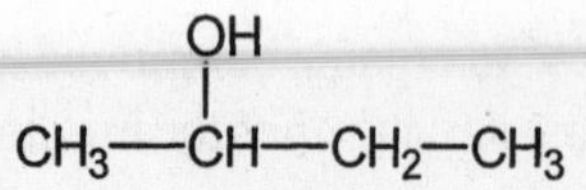
$$\text{CH}_3\text{—}\overset{\text{OH}}{\underset{|}{\text{CH}}}\text{—CH}_2\text{—CH}_3 \qquad \text{secondary alcohol}$$

c.

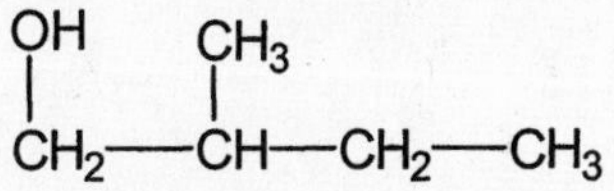

$$\underset{CH_2}{\overset{OH}{|}} - \underset{CH}{\overset{CH_3}{|}} - CH_2 - CH_3$$ primary alcohol

d.

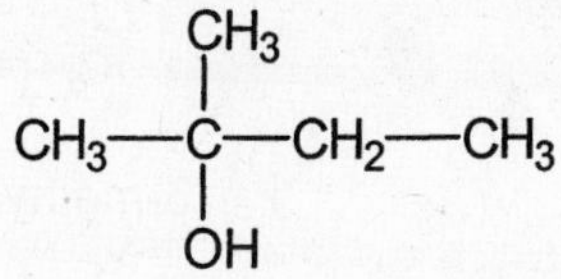

$$CH_3 - \underset{\underset{OH}{|}}{\overset{\overset{CH_3}{|}}{C}} - CH_2 - CH_3$$ tertiary alcohol

57.

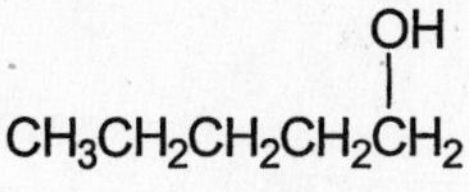

$$\underset{CH_3CH_2CH_2CH_2CH_2}{\overset{OH}{|}}$$

1-pentanol

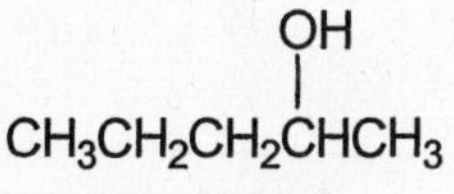

$$CH_3CH_2CH_2\underset{\overset{|}{|}}{\overset{OH}{C}}HCH_3$$

2-pentanol

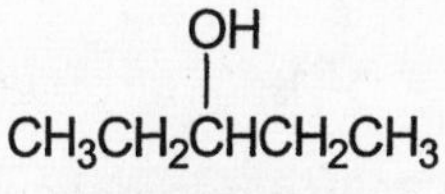

$$CH_3CH_2\underset{}{\overset{OH}{|}}CHCH_2CH_3$$

3-pentanol

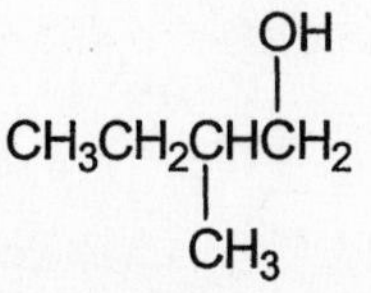

$$CH_3CH_2\underset{\underset{CH_3}{|}}{\overset{\overset{OH}{|}}{C}}HCH_2$$

2-methyl-1-butanol

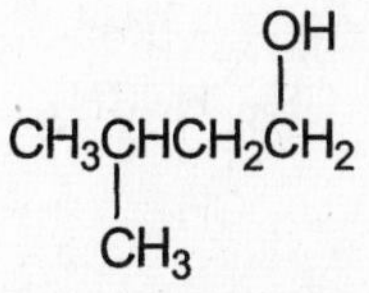

$$CH_3\underset{\underset{CH_3}{|}}{C}HCH_2CH_2 \overset{OH}{|}$$

3-methyl-1-butanol

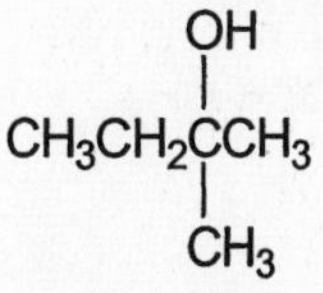

$$CH_3CH_2\underset{\underset{CH_3}{|}}{\overset{\overset{OH}{|}}{C}}CH_3$$

2-methyl-2-butanol

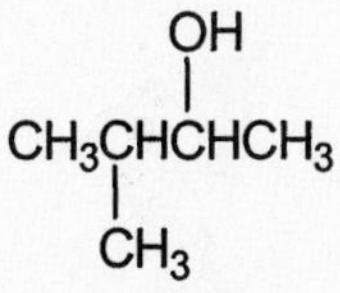

$$CH_3\underset{\underset{CH_3}{|}}{\overset{\overset{OH}{|}}{C}}HCHCH_3$$

3-methyl-2-butanol

$$CH_3 - \underset{\underset{CH_3}{|}}{\overset{\overset{CH_3}{|}}{C}} - \overset{OH}{CH_2}$$

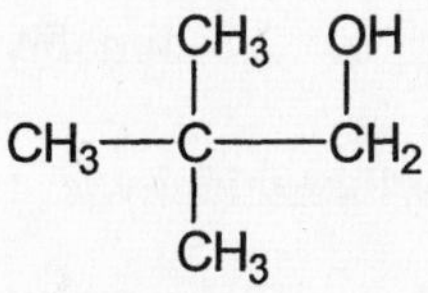

2,2-dimethyl-1-propanol

There are six isomeric ethers with formula $C_5H_{12}O$. The structures follow.

$$CH_3 - O - CH_2CH_2CH_2CH_3$$

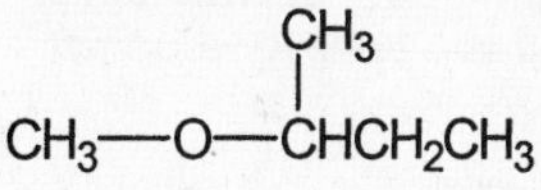

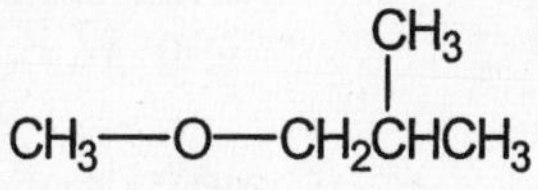

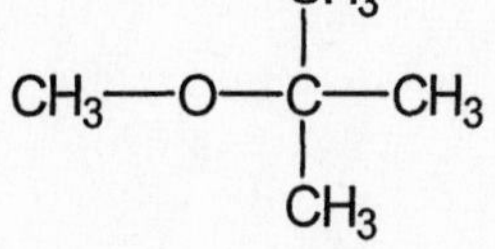

$$CH_3 - O - \underset{\underset{|}{|}}{\overset{\overset{CH_3}{|}}{C}}HCH_2CH_3$$

$$CH_3 - O - CH_2\underset{\underset{CH_3}{|}}{\overset{\overset{CH_3}{|}}{C}}HCH_3$$

$$CH_3 - O - \underset{\underset{CH_3}{|}}{\overset{\overset{CH_3}{|}}{C}} - CH_3$$

$$CH_3CH_2 - O - CH_2CH_2CH_3$$

$$CH_3CH_2 - O - \underset{\underset{CH_3}{|}}{\overset{\overset{CH_3}{|}}{C}}H$$

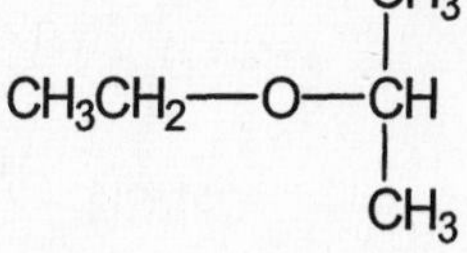

58.   There are four aldehydes and three ketones with formula $C_5H_{10}O$. The structures follow.

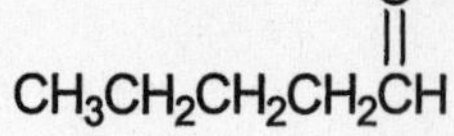

$$CH_3CH_2CH_2CH_2\overset{\overset{\displaystyle O}{\|}}{C}H$$

pentanal

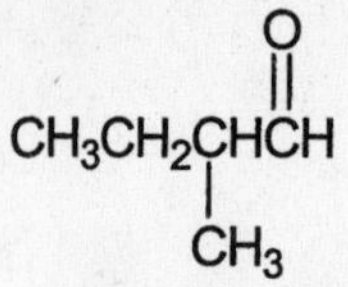

$$CH_3CH_2\underset{\underset{\displaystyle CH_3}{|}}{C}H\overset{\overset{\displaystyle O}{\|}}{C}H$$

2-methylbutanal

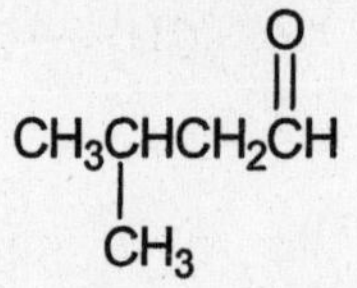

$$CH_3\underset{\underset{\displaystyle CH_3}{|}}{C}HCH_2\overset{\overset{\displaystyle O}{\|}}{C}H$$

3-methylbutanal

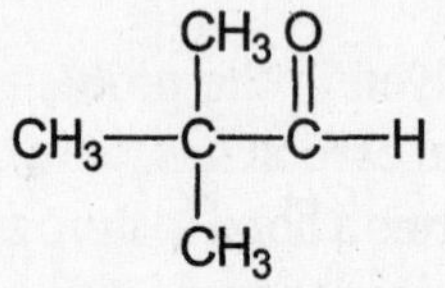

$$CH_3\overset{\overset{\displaystyle CH_3}{|}}{\underset{\underset{\displaystyle CH_3}{|}}{C}}\overset{\overset{\displaystyle O}{\|}}{C}-H$$

2,2-dimethylpropanal

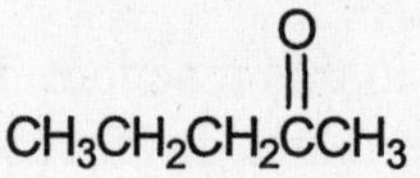

$$CH_3CH_2CH_2\overset{\overset{\displaystyle O}{\|}}{C}CH_3$$

2-pentanone

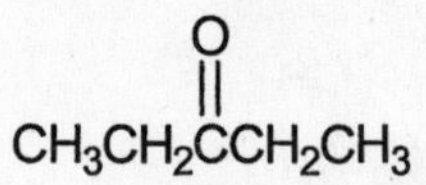

$$CH_3CH_2\overset{\overset{\displaystyle O}{\|}}{C}CH_2CH_3$$

3-pentanone

$$CH_3\underset{\underset{\displaystyle CH_3}{|}}{C}H\overset{\overset{\displaystyle O}{\|}}{C}CH_3$$

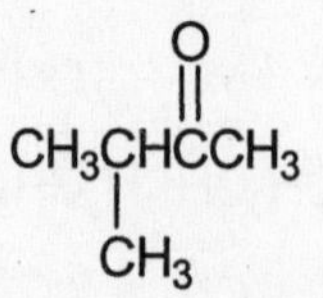

3-methyl-2-butanone

59.   a. 4,5-dichloro-3-hexanone   b. 2,3-dimethylpentanal

       c. 3-methylbenzaldehyde or m-methylbenzaldehyde

60.   a. 4-chlorobenzoic acid or p-chlorobenzoic acid

       b. 3-ethyl-2-methylhexanoic acid

       c. methanoic acid (common name = formic acid)

Reactions of Organic Compounds

61.   a. 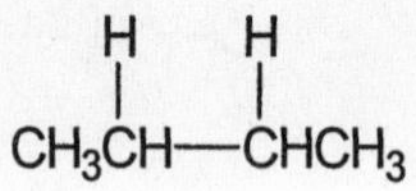 $CH_3\overset{\overset{\displaystyle H}{|}}{C}H-\overset{\overset{\displaystyle H}{|}}{C}HCH_3$ b. $\overset{\overset{\displaystyle Cl}{|}}{C}H_2-\overset{\overset{\displaystyle Cl}{|}}{C}H\overset{\overset{\displaystyle Cl}{|}}{C}H-\overset{\overset{\displaystyle Cl}{|}}{C}H$
                                                                                                          with CH_3 and CH_3 below

       c.   [benzene ring]—Cl + HCl

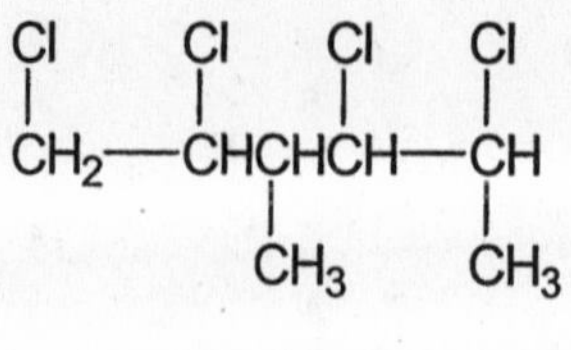

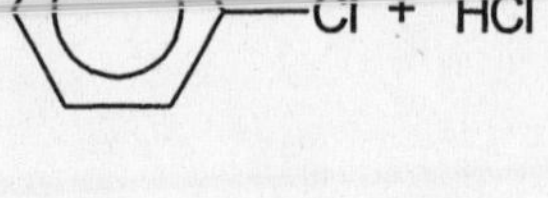

d. $C_4H_8(g) + 6\,O_2(g) \;\rightarrow\; 4\,CO_2(g) + 4\,H_2O(g)$

62.   a. The two possible products for the addition of HOH to this alkene are:

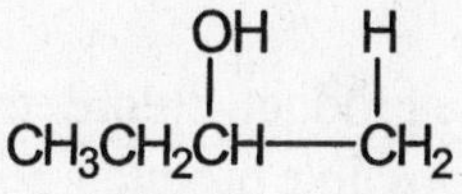

major product 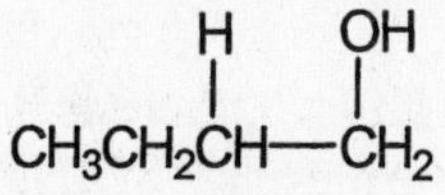   minor product

We would get both products in this reaction. Using the rule given in the problem, the first compound listed is the major product. In the reactant, the terminal carbon has more hydrogens bonded to it (2 vs. 1), so H forms a bond to this carbon, and OH forms a bond to the other carbon in the double bond for the major product. We will list only the major product for the remaining parts to this problem.

b. 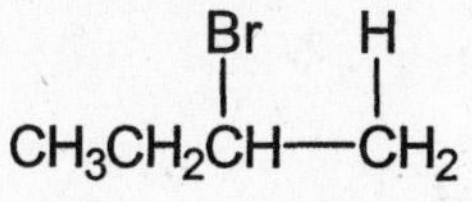   c.

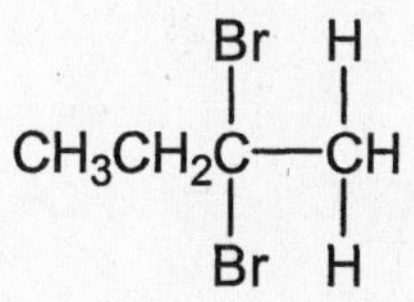

d. 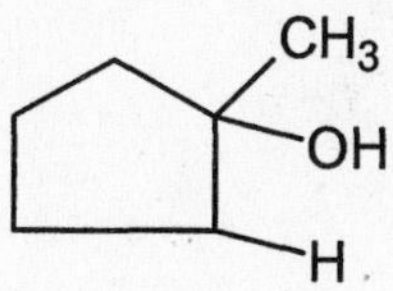   e.

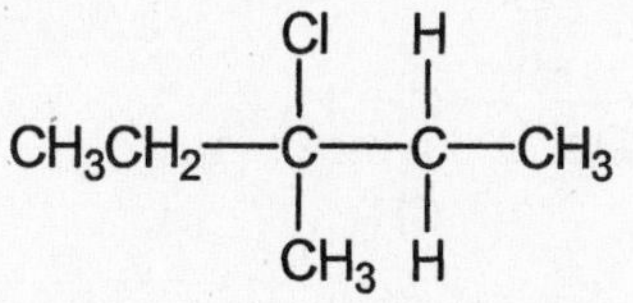

63.

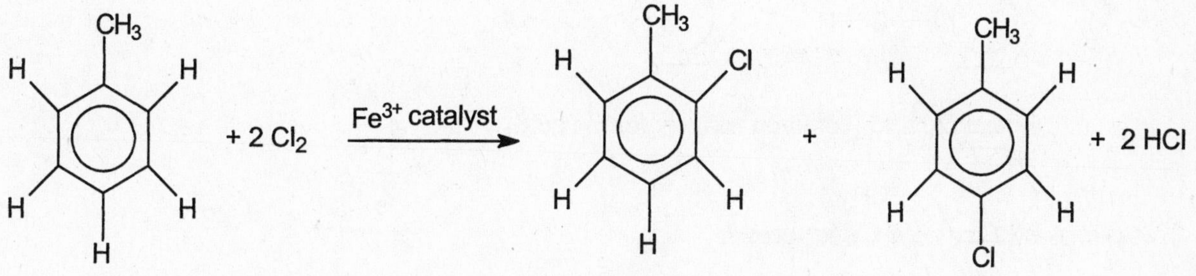

ortho   para

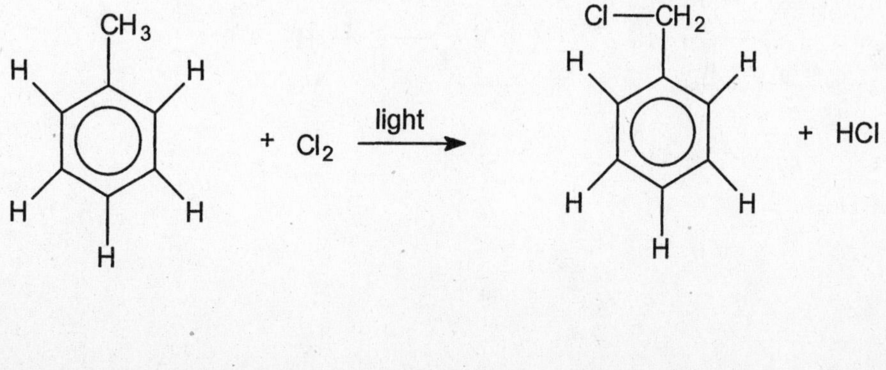

To substitute for the benzene ring hydrogens, an iron(III) catalyst must be present. Without this special iron catalyst, the benzene ring hydrogens are unreactive. To substitute for an alkane hydrogen, light must be present. For toluene, the light-catalyzed reaction substitutes a chlorine for a hydrogen in the methyl group attached to the benzene ring.

64.   When $CH_2=CH_2$ reacts with HCl, there is only one possible product, chloroethane. When Cl_2 is reacted with CH_3CH_3 (in the presence of light), there are six possible products because any number of the six hydrogens in ethane can be substituted for by Cl. The light-catalyzed substitution reaction is very difficult to control, hence, it is not a very efficient method of producing monochlorinated alkanes.

65.   Primary alcohols (a, d and f) are oxidized to aldehydes, which can be oxidized further to carboxylic acids. Secondary alcohols (b, e and f) are oxidized to ketones, and tertiary alcohols (c and f) do not undergo this type of oxidation reaction. Note that compound f contains a primary, secondary and tertiary alcohol. For the primary alcohols (a, d and f), we listed both the aldehyde and the carboxylic acid as possible products.

a.

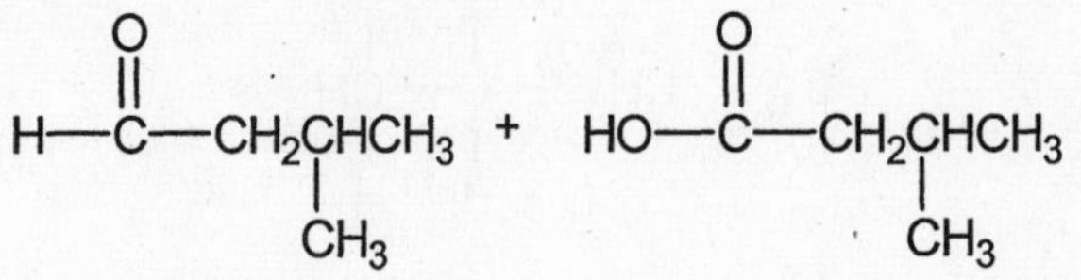

b. 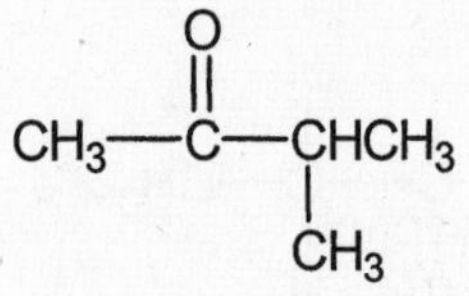   c.   No reaction

d.

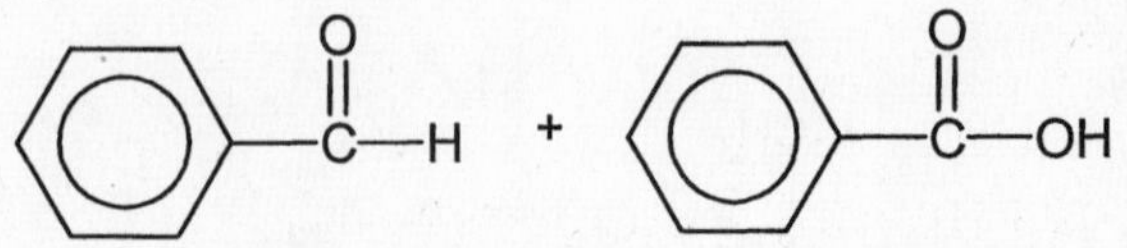

e.

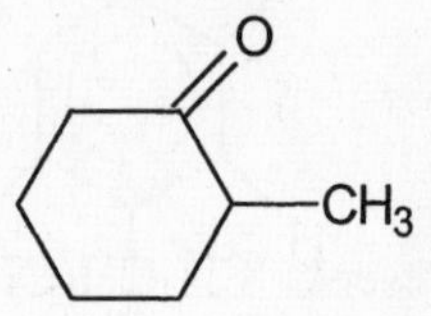

f.

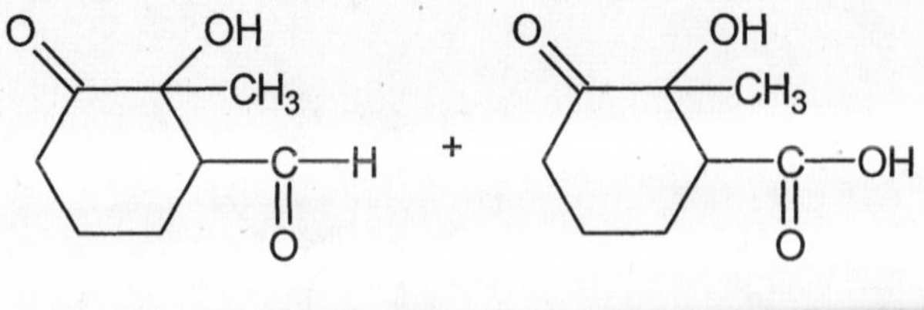

66.   **a.**

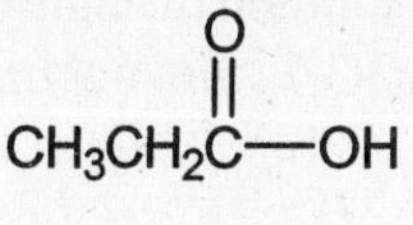

$$CH_3CH_2\overset{\displaystyle O}{\overset{\|}{C}}-OH$$

b.

$$CH_3CH_2\underset{\underset{\displaystyle CH_3}{|}}{\overset{\overset{\displaystyle CH_3}{|}}{CHCH}}-\overset{\displaystyle O}{\overset{\|}{C}}-OH$$

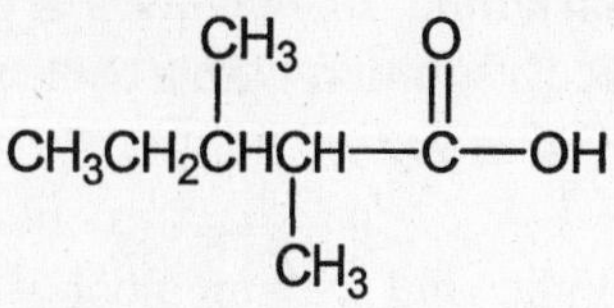

c.

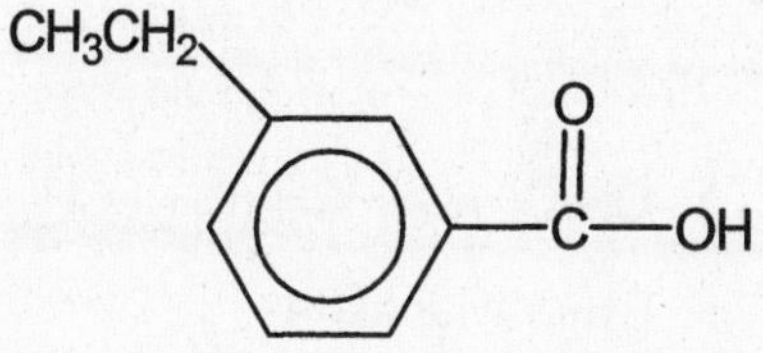

67.   **a.** $CH_3CH=CH_2 + Br_2 \rightarrow CH_3CHBrCH_2Br$ (Addition reaction of Br_2 with propene)

b.

$$CH_3-\underset{\underset{\displaystyle OH}{|}}{CH}-CH_3 \xrightarrow{\text{oxidation}} CH_3-\overset{\displaystyle O}{\overset{\|}{C}}-CH_3$$

Oxidation of 2-propanol yields acetone (2-propanone).

c.

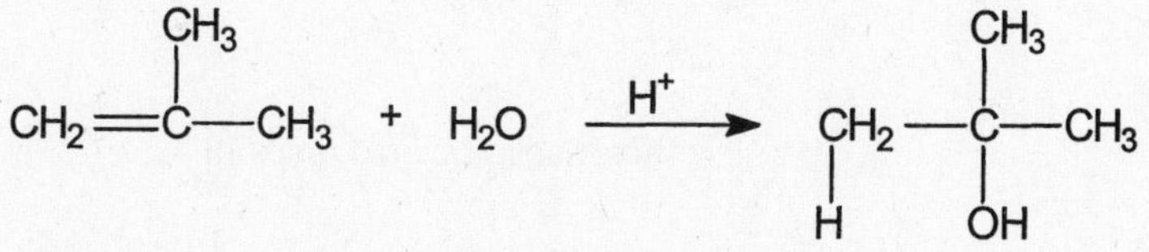

$$CH_2=\underset{}{\overset{\overset{\displaystyle CH_3}{|}}{C}}-CH_3 \ + \ H_2O \xrightarrow{H^+} CH_2-\underset{\underset{\displaystyle OH}{|}}{\overset{\overset{\displaystyle CH_3}{|}}{C}}-CH_3$$
$$H$$

Addition of H_2O to 2-methylpropene would yield tert-butyl alcohol (2-methyl-2-propanol) as the major product.

d. $CH_3CH_2CH_2OH \xrightarrow{KMnO_4} CH_3CH_2\overset{\displaystyle O}{\overset{\|}{C}}-OH$

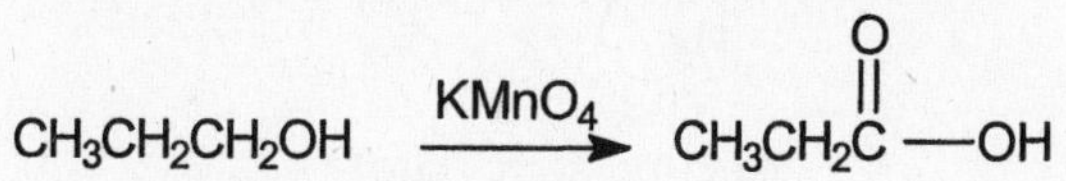

Oxidation of 1-propanol would eventually yield propanoic acid. Propanal is produced first in this reaction and is then oxidized to propanoic acid.

68.   **a.** $CH_2=CHCH_2CH_3$ will react with Cl_2 without any catalyst present. $CH_3CH_2CH_2CH_3$ only reacts with Cl_2 when ultraviolet light is present.

b. $CH_3CH_2CH_2\overset{\displaystyle O}{\overset{\|}{C}}OH$ is an acid, so this compound should react positively with a base like $NaHCO_3$. The other compound is a ketone, which will not react with a base.

c. $CH_3CH_2CH_2OH$ can be oxidized with $KMnO_4$ to propanoic acid. 2-propanone (a ketone) will not react with $KMnO_4$.

d. $CH_3CH_2NH_2$ is an amine, so it behaves as a base in water. Dissolution of some of this base in water will produce a solution with a basic pH. The ether, CH_3OCH_3, will not produce a basic pH when dissolved in water.

69.   Reaction of a carboxylic acid with an alcohol can produce these esters.

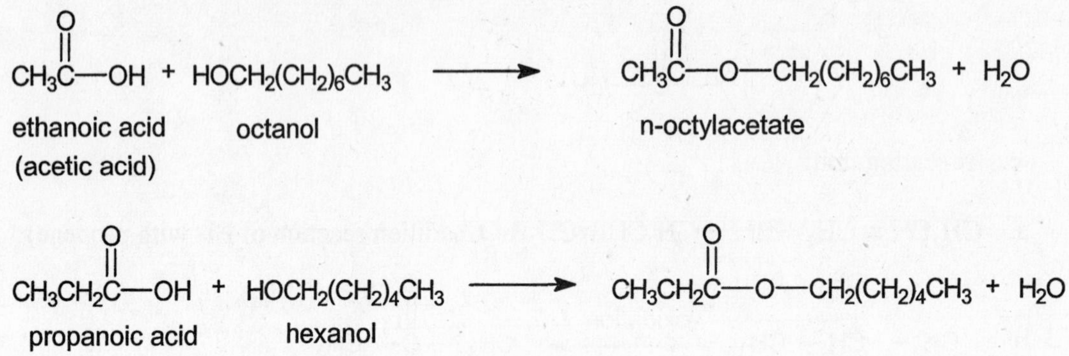

ethanoic acid   octanol   n-octylacetate
(acetic acid)

propanoic acid   hexanol

70.

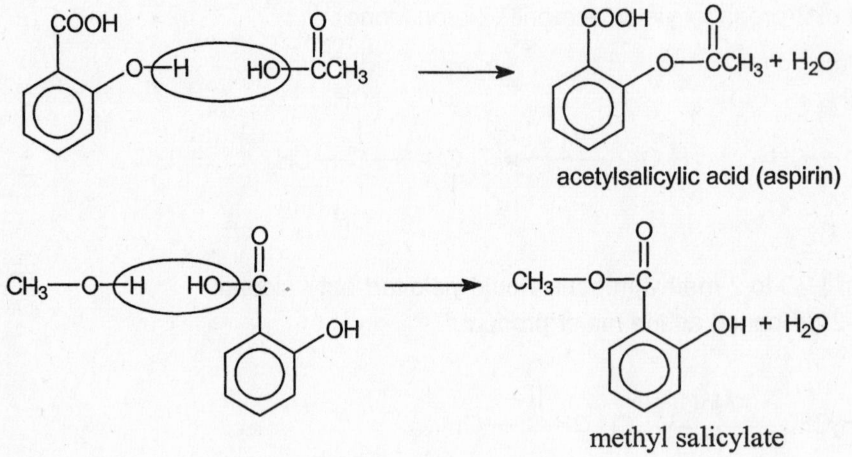

acetylsalicylic acid (aspirin)

methyl salicylate

Polymers

71.   The backbone of the polymer contains only carbon atoms, which indicates that Kel-F is an addition polymer. The smallest repeating unit of the polymer and the monomer used to produce this polymer are:

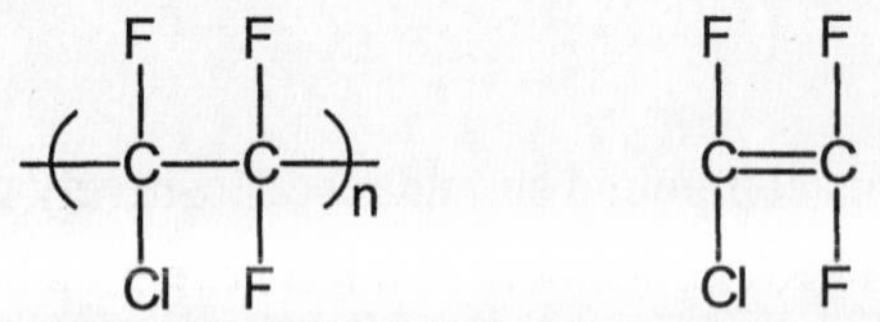

Note: Condensation polymers generally have O or N atoms in the backbone of the polymer.

72.   a.   repeating unit: $-(-CHF-CH_2-)_n$ monomer: $CHF=CH_2$

b.

repeating unit: 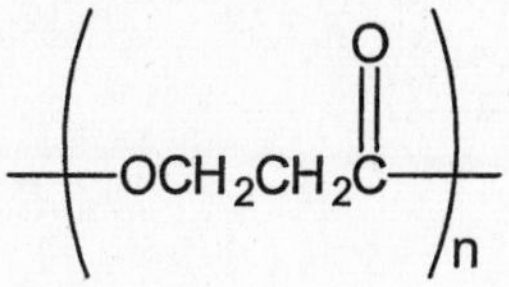   monomer: $HO-CH_2CH_2-CO_2H$

c.   repeating unit:

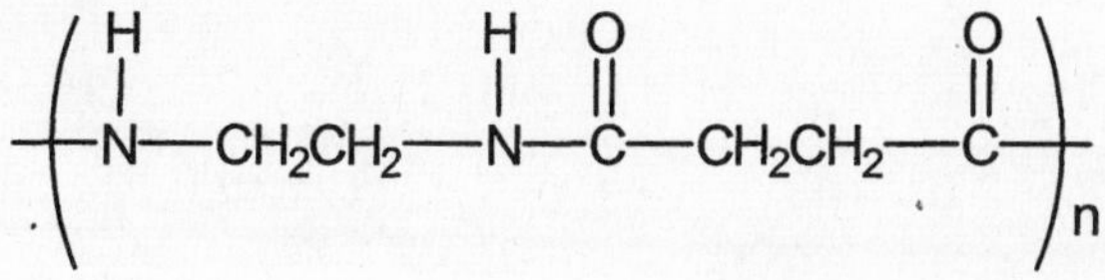

copolymer of: $H_2NCH_2CH_2NH_2$ and $HO_2CCH_2CH_2CO_2H$

d.   monomer: e.   monomer:

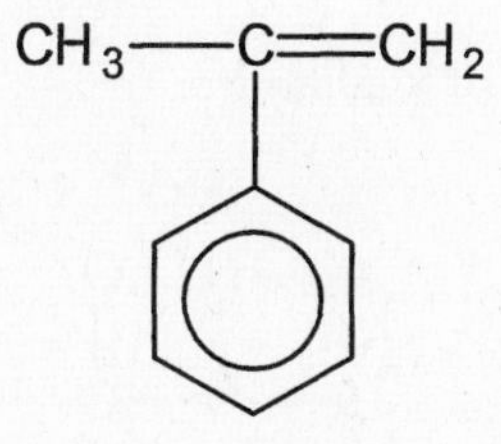

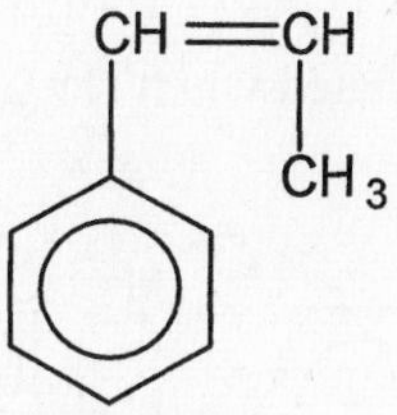

f.   monomer: $CClF=CF_2$

g.   copolymer of:

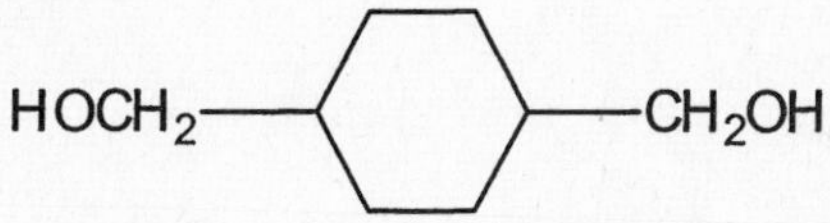

and

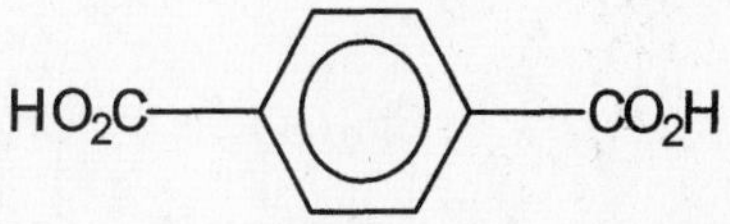

Addition polymers: a, d, e and f; Condensation polymers: b, c and g; Copolymer: c and g

73.

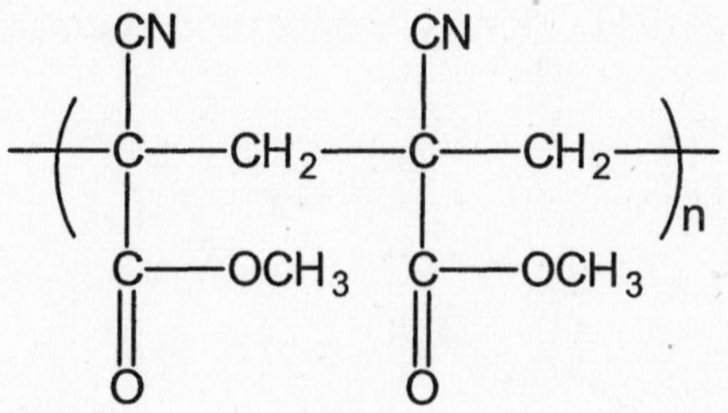

Super glue is an addition polymer formed by reaction of the C=C bond in methyl cyanoacrylate.

74.   a.   2-methyl-1,3-butadiene

b.

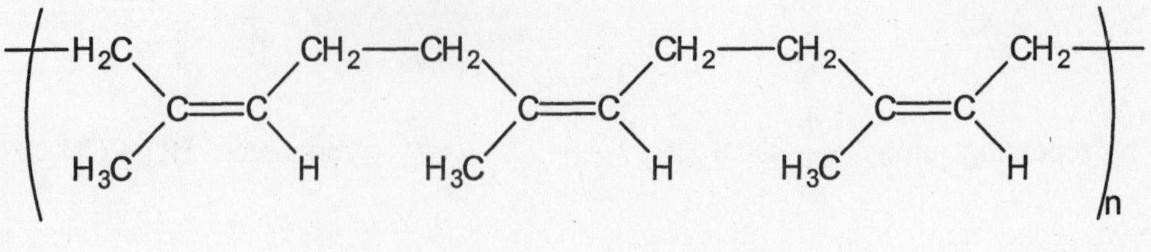

cis-polyisoprene (natural rubber)

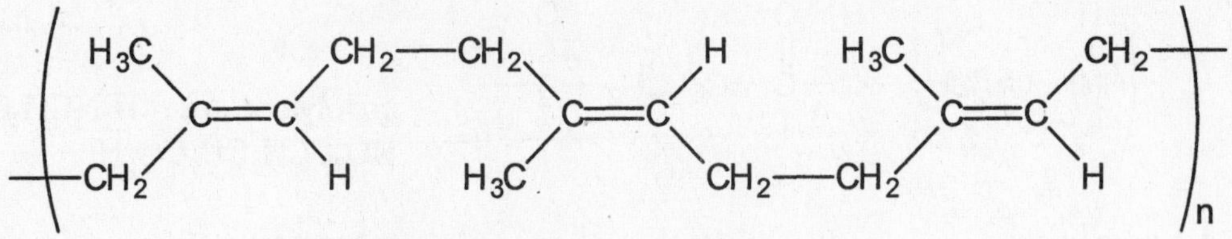

trans-polyisoprene (gutta percha)

75. H_2O is eliminated when Kevlar forms. Two repeating units of Kevlar are:

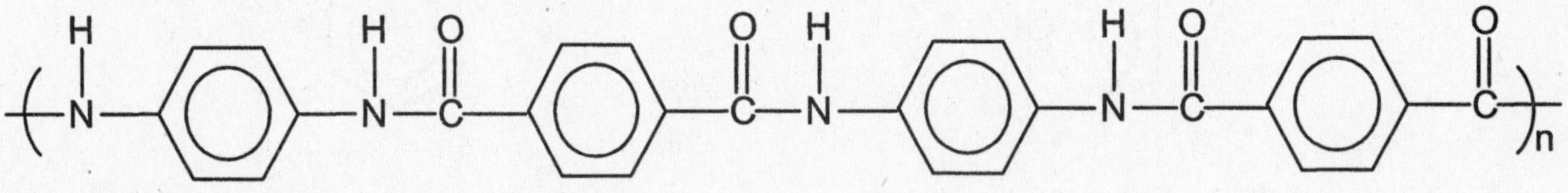

76.   This condensation polymer forms by elimination of water. The ester functional group repeats, hence the term, polyester.

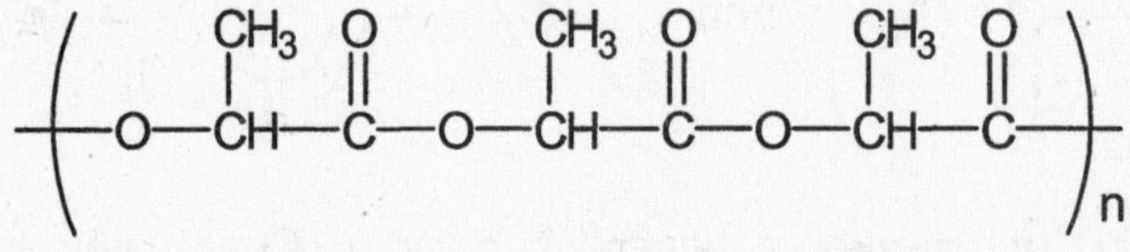

77.   This is a condensation polymer where two molecules of H_2O form when the monomers link together.

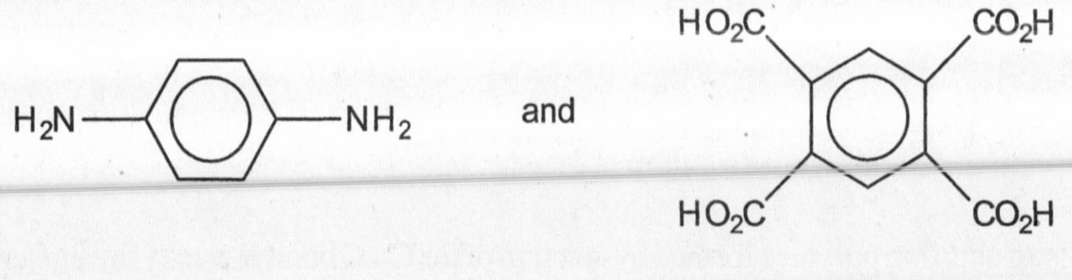

78.

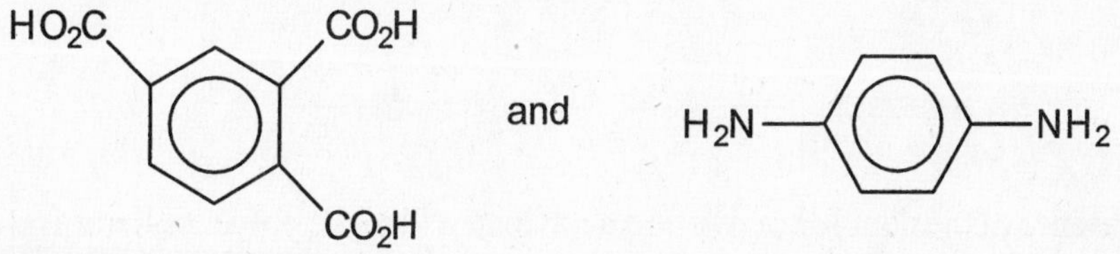

and

79. Divinylbenzene has two reactive double bonds that are both used when divinylbenzene inserts itself into two adjacent polymer chains. The chains cannot move past each other because the crosslinks bond adjacent polymer chains together, making the polymer more rigid.

80.   a.

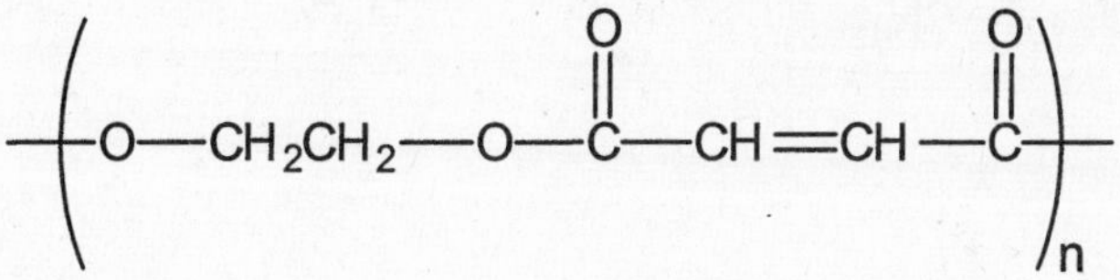

b.

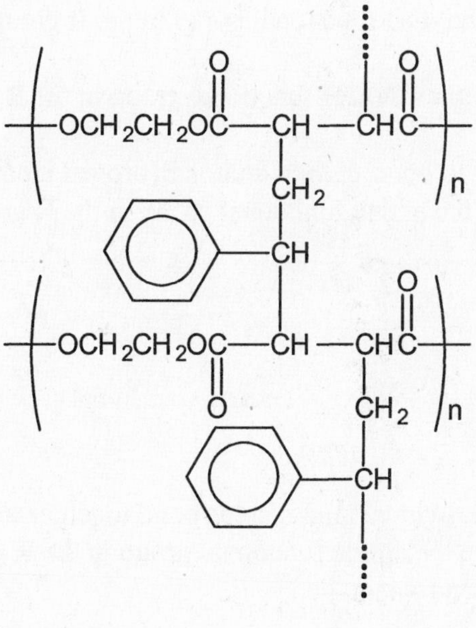

81   a. The polymer formed using 1,2-diaminoethane will exhibit relatively strong hydrogen bonding interactions between adjacent polymer chains. Since hydrogen bonding is not present in the ethylene glycol polymer (a polyester polymer forms), the 1,2-diaminoethane polymer will be stronger.

   b. The presence of rigid groups (benzene rings or multiple bonds) makes the polymer stiffer. Hence, the monomer with the benzene ring will produce the more rigid polymer.

c. Polyacetylene will have a double bond in the carbon backbone of the polymer.

$$n \; HC \equiv CH \longrightarrow \left(CH = CH \right)_n$$

The presence of the double bond in polyacetylene will make polyacetylene a more rigid polymer than polyethylene. Polyethylene doesn't have $C=C$ bonds in the backbone of the polymer (the double bonds in the monomers react to form the polymer).

82.   At low temperatures, the polymer is coiled into balls. The forces between poly(lauryl methacrylate) and oil molecules will be minimal, and the effect on viscosity will be minimal. At higher temperatures, the chains of the polymer will unwind and become tangled with the oil molecules, increasing the viscosity of the oil. Thus, the presence of the polymer counteracts the temperature effect, and the viscosity of the oil remains relatively constant.

Natural Polymers

83.   a. Serine, tyrosine and threonine contain the -OH functional group in the R group.

   b. Aspartic acid and glutamic acid contain the -COOH functional group in the R group.

   c. An amine group has a nitrogen bonded to other carbon and/or hydrogen atoms. Histidine, lysine, arginine and tryptophan contain the amine functional group in the R group.

   d. The amide functional group is:

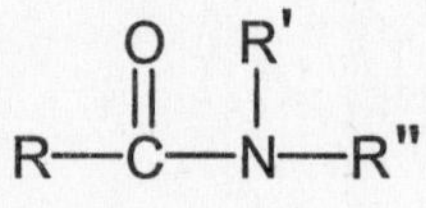

$$R - \overset{\overset{\textstyle O}{\|}}{C} - \overset{\overset{\textstyle R'}{|}}{N} - R''$$

This functional group is formed when individual amino acids bond together to form the peptide linkage. Glutamine and asparagine have the amide functional group in the R group.

84.   Crystalline amino acids exist as zwitterions, $^+H_3NCRHCOO^-$, held together by ionic forces. The ionic interparticle forces are strong. Before the temperature gets high enough to melt the solid, the amino acid decomposes.

85   a.   Aspartic acid and phenylalanine make up aspartame.

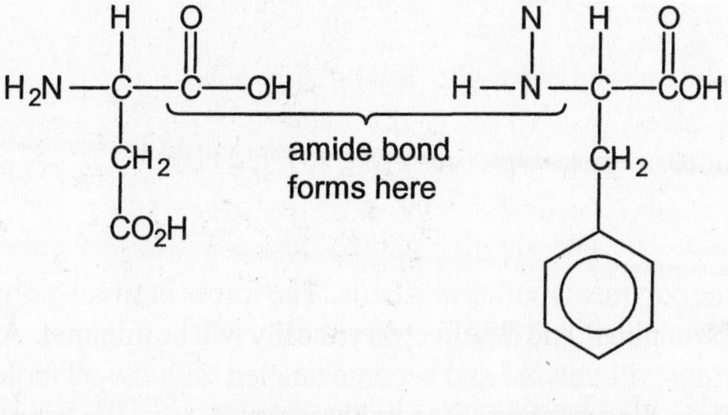

amide bond
forms here

b.   Aspartame contains the methyl ester of phenylalanine. This ester can hydrolyze to form
methanol:

$$R-CO_2CH_3 + H_2O \rightleftharpoons RCO_2H + HOCH_3$$

86.

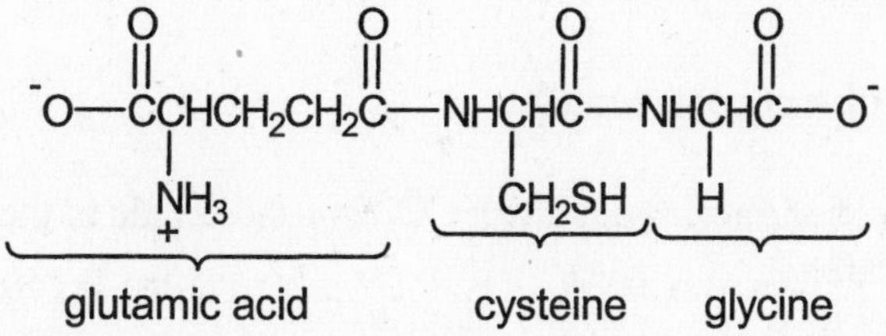

glutamic acid   cysteine   glycine

Glutamic acid, cysteine and glycine are the three amino acids in glutathione. Glutamic acid
uses the -COOH functional group in the R group to bond to cysteine instead of the carboxylic
acid group bonded to the α-carbon. The cysteine-glycine bond is the typical peptide linkage.

87.

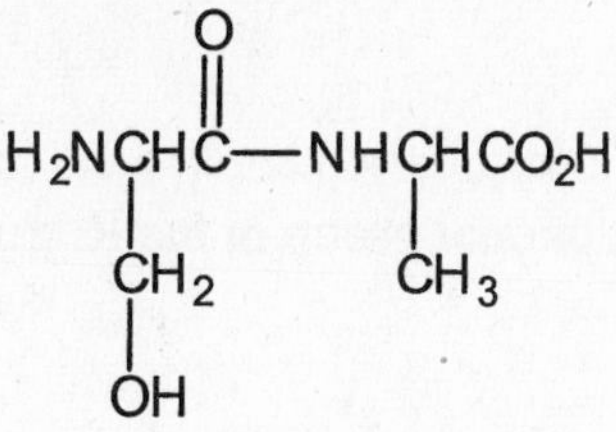

ser - ala 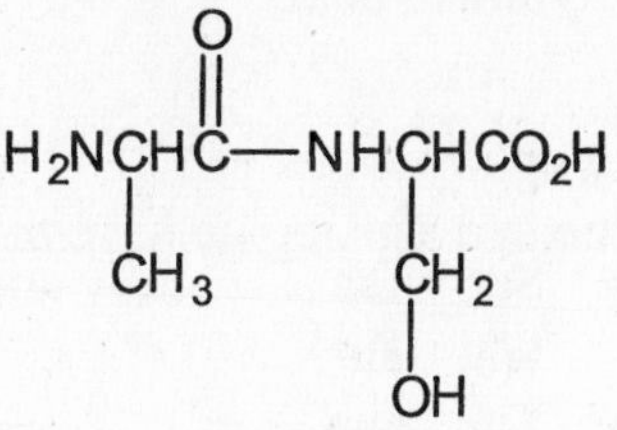   ala - ser

88.

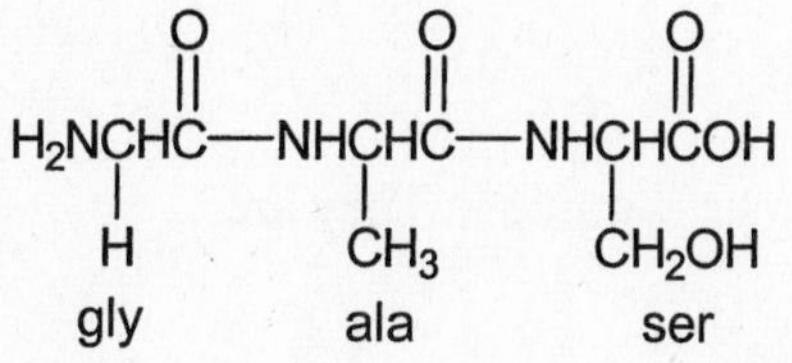

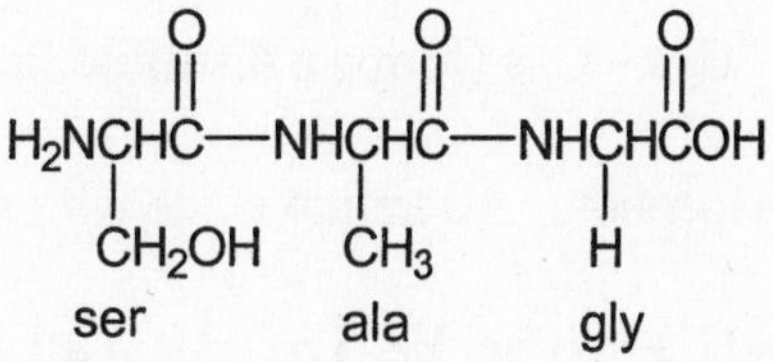

gly   ala   ser   ser   ala   gly

There are six possible tripeptides with gly, ala and ser. The other four tripeptides are gly-ser-ala, ser-
gly-ala, ala-gly-ser and ala-ser-gly.

89.   a. Six tetrapeptides are possible. From NH_2 to CO_2H end:

                phe-phe-gly-gly, gly-gly-phe-phe, gly-phe-phe-gly,

                phe-gly-gly-phe, phe-gly-phe-gly, gly-phe-gly-phe

        b. Twelve tetrapeptides are possible. From NH_2 to CO_2H end:

                phe-phe-gly-ala, phe-phe-ala-gly, phe-gly-phe-ala,

                phe-gly-ala-phe, phe-ala-phe-gly, phe-ala-gly-phe,

                gly-phe-phe-ala, gly-phe-ala-phe, gly-ala-phe-phe

                ala-phe-phe-gly, ala-phe-gly-phe, ala-gly-phe-phe

90.   There are 5 possibilities for the first amino acid, 4 possibilities for the second amino acid, 3 possibilities for the third amino acid, 2 possibilities for the fourth amino acid and 1 possibility for the last amino acid. The number of possible sequences is:

            $5 \times 4 \times 3 \times 2 \times 1 = 5! = 120$ different pentapeptides

91.   a. Ionic: Need NH_2 on side chain of one amino acid with CO_2H on side chain of the other amino acid. The possibilities are:

                NH_2 on side chain = His, Lys or Arg; CO_2H on side chain = Asp or Glu

        b. Hydrogen bonding: Need N–H or O–H bond present in side chain. The hydrogen bonding interaction occurs between the X–H bond and a carbonyl group from any amino acid.

                    X–H $\cdots\cdots$ O = C (carbonyl group)

                    Ser   Asn   Any amino acid
                    Glu   Thr
                    Tyr   Asp
                    His   Gln
                    Arg   Lys

        c. Covalent: Cys – Cys (forms a disulfide linkage)

        d. London dispersion: All amino acids with nonpolar R groups. They are:

                Gly, Ala, Pro, Phe, Ile, Trp, Met, Leu and Val

        e. Dipole-dipole: Need side chain with OH group. Tyr, Thr and Ser all could form this specific dipole-dipole force with each other since all contain an OH group in the side chain.

92.   Reference Exercise 22.91 for a more detailed discussion of these various interactions.

      a. Covalent   b. Hydrogen bonding

      c. Ionic   d. London dispersion

93.   Glutamic acid: $R = -CH_2CH_2CO_2H$; Valine: $R = -CH(CH_3)_2$; A polar side chain is replaced by a nonpolar side chain. This could affect the tertiary structure of hemoglobin and the ability of hemoglobin to bind oxygen.

94.   Glutamic acid: $R = -CH_2CH_2COOH$; Glutamine: $R = -CH_2CH_2CONH_2$; The R groups only differ by OH vs NH_2. Both of these groups are capable of forming hydrogen bonding interactions, so the change in intermolecular forces is minimal. Thus, this change is not critical because the secondary and tertiary structures of hemoglobin should not be greatly affected.

95.   See Figures 22.29 and 22.30 of the text for examples of the cyclization process.

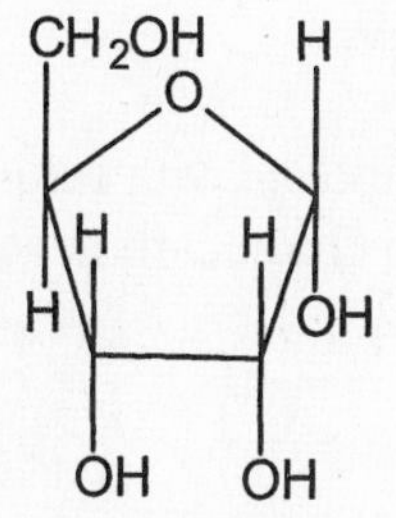

D-Ribose 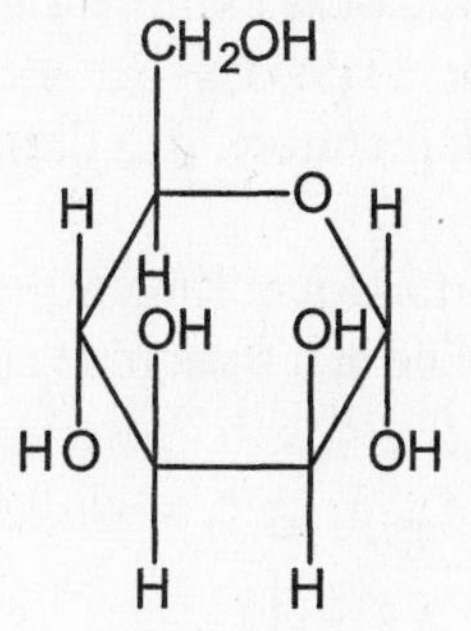   D-Mannose

96.   The chiral carbon atoms are marked with asterisks. A chiral carbon atom has four different substituent groups attached.

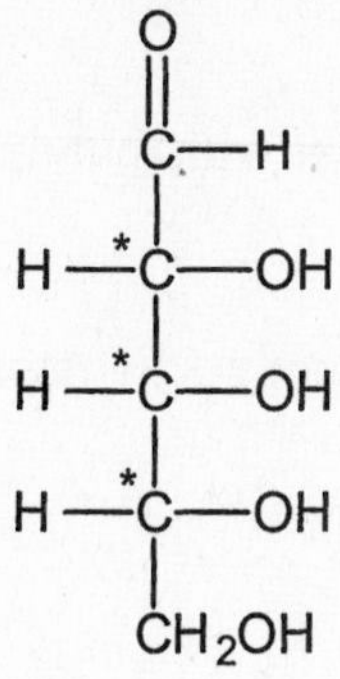

D-Ribose

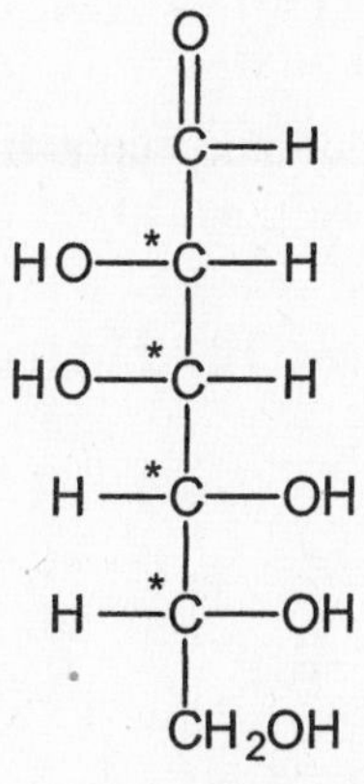

D-Mannose

97.   The aldohexoses contain 6 carbons and the aldehyde functional group. Glucose, mannose and galactose are aldohexoses. Ribose and arabinose are aldopentoses since they contain 5 carbons with the aldehyde functional group. The ketohexose (6 carbons + ketone functional group) is fructose and the ketopentose (5 carbons + ketone functional group) is ribulose.

98.   This is an example of Le Chatelier's principle at work. For the equilibrium reactions among the various forms of glucose, reference Figure 22.30 of the text. The chemical tests involve reaction of the aldehyde group found only in the open-chain structure. As the aldehyde group is reacted, the equilibrium between the cyclic forms of glucose and the open-chain structure will shift to produce more of the open-chain structure. This process continues until either the glucose or the chemicals used in the tests run out.

99.   The α and β forms of glucose differ in the orientation of a hydroxy group on one specific carbon in the cyclic forms (see Figure 22.30 of the text). Starch is a polymer composed of only α-D-glucose, and cellulose is a polymer composed of only β-D-glucose.

100.   Humans do not possess the necessary enzymes to break the β-glycosidic linkages found in cellulose. Cows, however, do possess the necessary enzymes to break down cellulose into the β-D-glucose monomers and, therefore, can derive nutrition from cellulose.

101.   A chiral carbon has four different groups attached to it. A compound with a chiral carbon is optically active. Isoleucine and threonine contain more than the one chiral carbon atom (see asterisks).

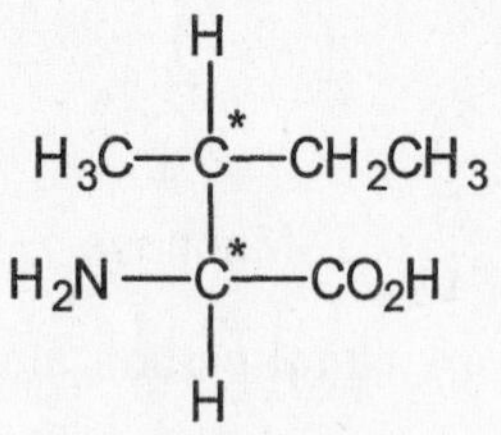

isoleucine 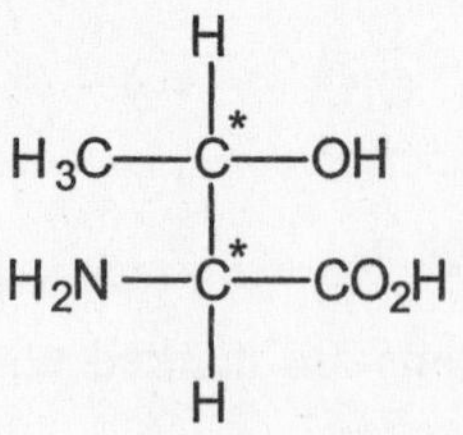   threonine

102.   There is no chiral carbon atom in glycine since it contains no carbon atoms with four different groups bonded to it.

103.   Only one of the isomers is optically active. The chiral carbon in this optically active isomer is marked with an asterisk.

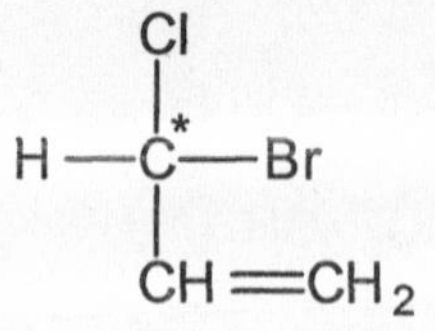

104.

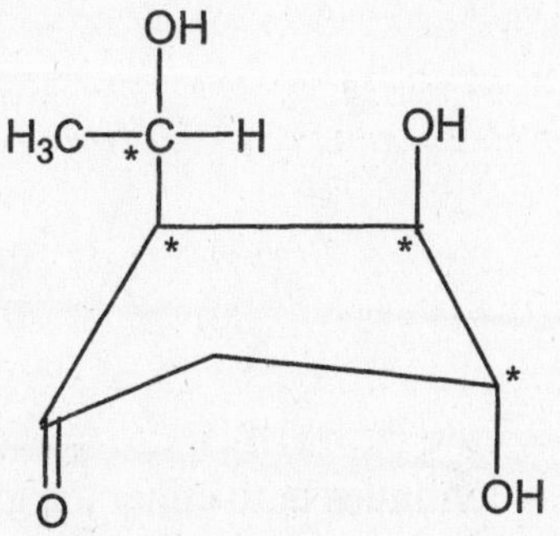

The compound has four chiral carbon atoms. The fourth group bonded to the three chiral carbon atoms in the ring is a hydrogen atom.

105.   The complimentary base pairs in DNA are cytosine (C) and guanine (G), and thymine (T) and adenine (A). The complimentary sequence is: C-C-A-G-A-T-A-T-G

106.   For each letter, there are 4 choices; A, T, G, or C. Hence, the total number of codons is $4 \times 4 \times 4 = 64$.

107.   Uracil will hydrogen bond to adenine.

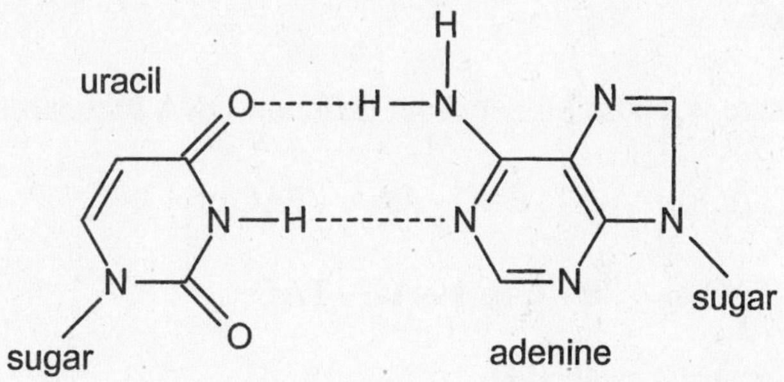

108.   The tautomer could hydrogen bond to guanine, forming a G–T base pair instead of A–T.

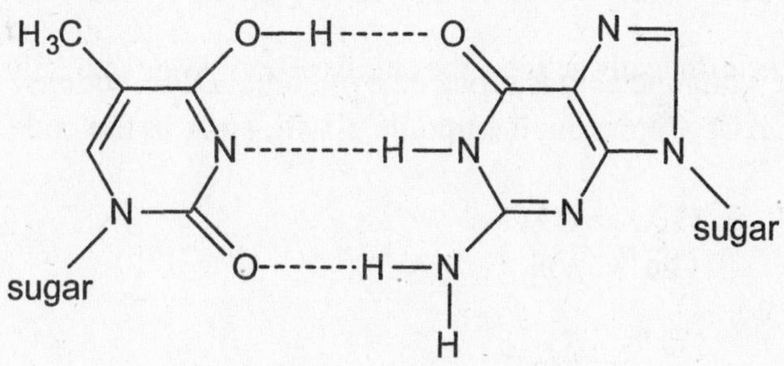

109.   Base pair:

RNA   DNA

A . . . . T

G C

C . . . . G

U . . . . A

a.   Glu: CTT, CTC   Val: CAA, CAG, CAT, CAC

Met: TAC   Trp: ACC

Phe: AAA, AAG   Asp: CTA, CTG

b.   DNA sequence for trp-glu-phe-met:

ACC - CTT - AAA - TAC
       or   or
       CTC   AAG

c.   Due to glu and phe, there is a possibility of four different DNA sequences. They are:

ACC - CTT - AAA - TAC or ACC - CTC - AAA - TAC or

ACC - CTT - AAG - TAC or ACC - CTC - AAG - TAC

d.   T—A—C—C—T—G—A—A—G

met   asp   phe

e.   TAC - CTA - AAG; TAC - CTA - AAA; TAC - CTG - AAA

110.   In sickle cell anemia, glutamic acid is replaced by valine. DNA codons: Glu: CTT, CTC; Val: CAA, CAG, CAT, CAC; Replacing the middle T with an A in the code for Glu will code for Val.

CTT → CAT or CTC → CAC
Glu   Val   Glu   Val

Additional Exercises

111. CH_2Cl-CH_2Cl, 1-2-dichloroethane: There is free rotation about the C–C single bond that doesn't lead to different compounds. $CHCl=CHCl$, 1,2-dichloroethene: There is no rotation about the C=C double bond. This creates the cis and trans isomers, which are different compounds.

112.   a.   Only one monochlorination product can form (1-chloro-2,2-dimethylpropane). The other possibilities differ from this compound by a simple rotation, so they are not different compounds.

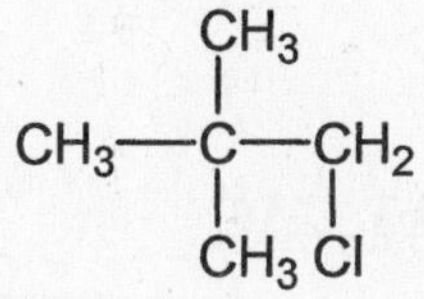

   b.   Three different monochlorination products are possible (ignoring cis-trans isomers).

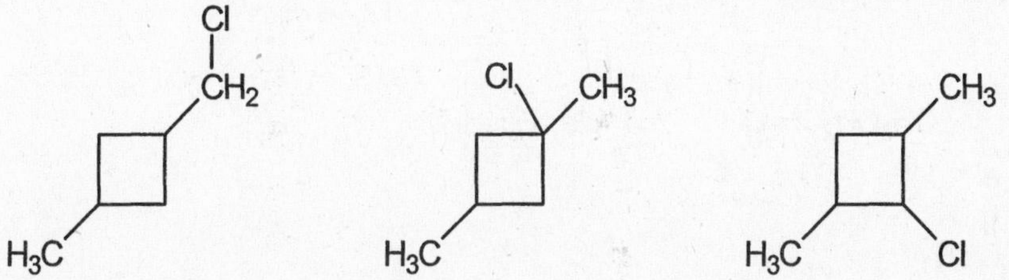

   c.   Two different monochlorination products are possible (the other possibilities differ by a simple rotation of one of these two compounds).

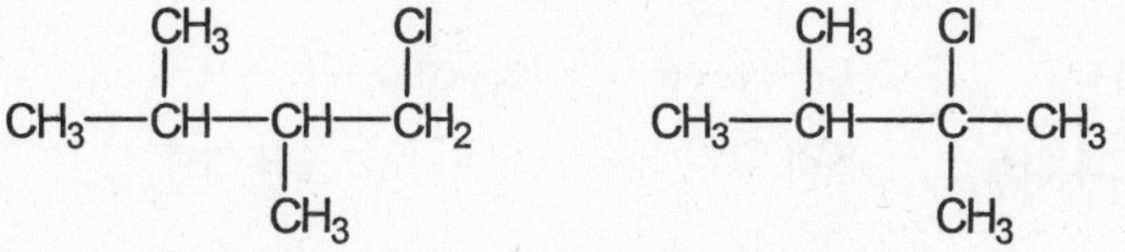

113.

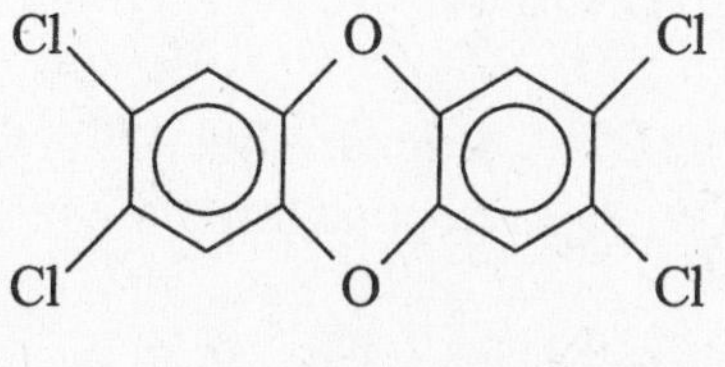

   There are many possibilities for isomers. Any structure with four chlorines replacing four hydrogens in any four of the numbered positions would be an isomer, i.e., 1,2,3,4-tetrachloro-dibenzo-p-dioxin is a possible isomer.

114.

a.   trans-2-butene:

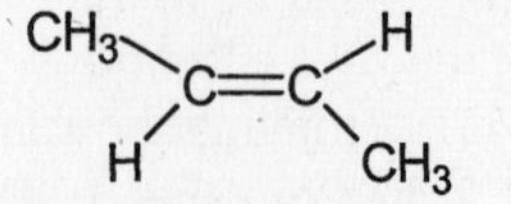

$$CH_3 \diagdown C = C \diagup H$$
$$H \diagup \diagdown CH_3$$

, formula = C_4H_8

or

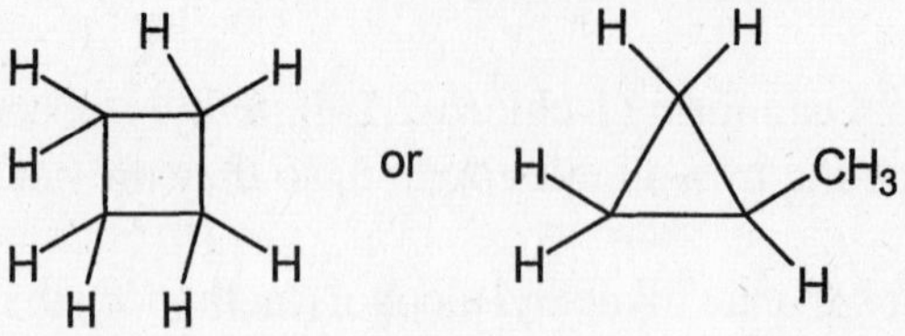

b.   propanoic acid: $CH_3CH_2\overset{\overset{O}{\|}}{C}—OH$,   formula = $C_3H_6O_2$

$CH_3\overset{\overset{O}{\|}}{C}—O—CH_3$ or $H\overset{\overset{O}{\|}}{C}—O—CH_2CH_3$

c.   butanal: $CH_3CH_2CH_2\overset{\overset{O}{\|}}{C}H$,   formula = C_4H_8O

$CH_3CH_2\overset{\overset{O}{\|}}{C}CH_3$

d.   butylamine: $CH_3CH_2CH_2CH_2NH_2$,   formula = $C_4H_{11}N$:

A secondary amine has two R groups bonded to N.

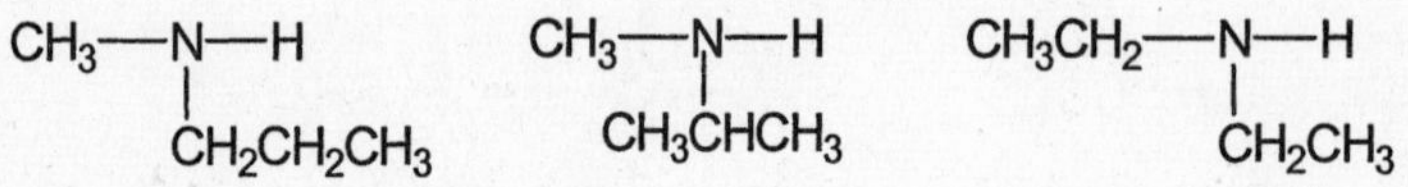

$CH_3—\overset{\overset{\displaystyle}{N}}{\underset{\displaystyle CH_2CH_2CH_3}{|}}—H$ $CH_3—\overset{}{\underset{CH_3CHCH_3}{N}}—H$ $CH_3CH_2—\overset{}{\underset{CH_2CH_3}{N}}—H$

e.   A tertiary amine has three R groups bonded to N. (See answer d for structure of butylamine.)

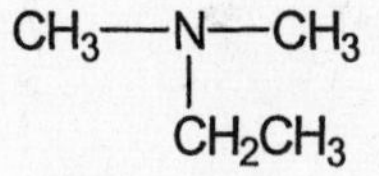

$CH_3—\overset{}{\underset{CH_2CH_3}{N}}—CH_3$

f.   2-methyl-2-propanol: $CH_3\overset{\overset{CH_3}{|}}{\underset{OH}{C}}CH_3$,   formula = $C_4H_{10}O$

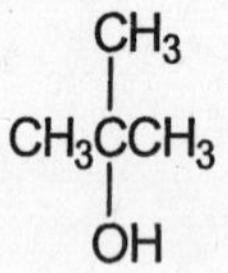

$CH_3—O—CH_2CH_2CH_3$ $CH_3—O—\overset{\overset{CH_3}{|}}{\underset{CH_3}{C}}H$ $CH_3CH_2—O—CH_2CH_3$

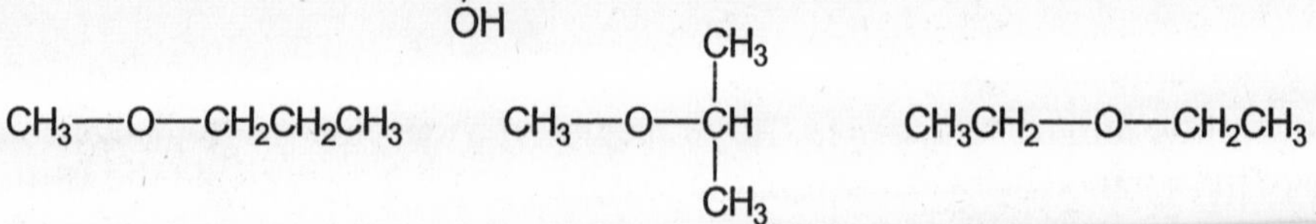

g.   A secondary alcohol has two R groups attached to the carbon bonded to the OH group. (See answer f for the structure of 2-methyl-2-propanol.)

$$\underset{\displaystyle CH_3CHCH_2CH_3}{\overset{\displaystyle OH}{\displaystyle |}}$$

115.   Alcohols consist of two parts, the polar OH group and the nonpolar hydrocarbon chain attached to the OH group. As the length of the nonpolar hydrocarbon chain increases, the solubility of the alcohol decreases in water, a very polar solvent. In methyl alcohol (methanol), the polar OH group overrides the effect of the nonpolar CH_3 group, and methyl alcohol is soluble in water. In stearyl alcohol, the molecule consists mostly of the long nonpolar hydrocarbon chain, so it is insoluble in water.

116. $CH_3CH_2CH_2CH_2CH_2CH_2CH_2COOH + OH^- \rightarrow CH_3–(CH_2)_6–COO^- + H_2O$; Octanoic acid is more soluble in 1 M NaOH. Added OH^- will remove the acidic proton from octanoic acid, creating a charged species. As is the case with any substance with an overall charge, solubility in water increases. When morphine is reacted with H^+, the amine group is protonated, creating a positive charge on morphine ($R_3N + H^+ \rightarrow R_3\overset{+}{N}H$). By treating morphine with HCl, an ionic compound results which is more soluble in water and in the blood stream than the neutral covalent form of morphine.

117.   The structures, the types of intermolecular forces exerted, and the boiling points for the compounds are:

$$CH_3CH_2CH_2\overset{\displaystyle O}{\overset{\displaystyle \|}{C}}OH \qquad\qquad\qquad CH_3CH_2CH_2CH_2CH_2OH$$

butanoic acid, 164°C   1-pentanol, 137°C
LD + dipole + H bonding   LD + H bonding

$$CH_3CH_2CH_2CH_2\overset{\displaystyle O}{\overset{\displaystyle \|}{C}}H \qquad\qquad\qquad CH_3CH_2CH_2CH_2CH_2CH_3$$

pentanal, 103°C   n-hexane, 69°C
LD + dipole   LD only

All these compounds have about the same molar mass. Therefore, the London dispersion (LD) forces in each are about the same. The other types of forces determine the boiling point order. Since butanoic acid and 1-pentanol both exhibit hydrogen bonding interactions, these two compounds will have the two highest boiling points. Butanoic acid has the highest boiling point since it exhibits H bonding along with dipole-dipole forces due to the polar $C = O$ bond.

118.   Water is produced in this reaction by removing an OH group from one substance and H from the other substance. There are two ways to do this:

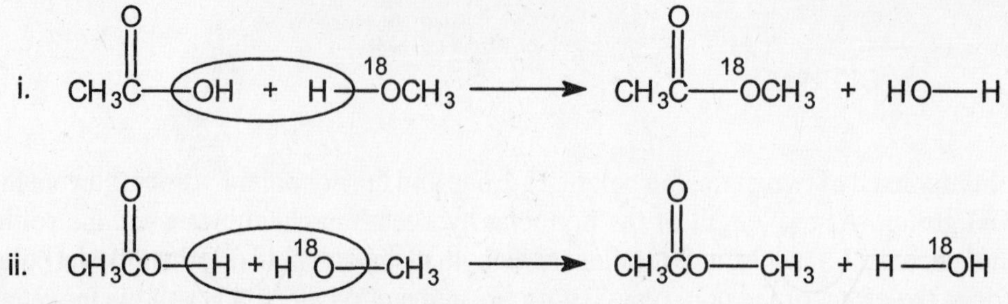

i.

ii.

Since the water produced is not radioactive, methyl acetate forms by the first reaction where all the oxygen-18 ends up in methyl acetate.

119. $85.63 \text{ g C} \times \dfrac{1 \text{ mol C}}{12.01 \text{ g C}} = 7.130 \text{ mol C}; \quad 14.37 \text{ g H} \times \dfrac{1 \text{ mol H}}{1.008 \text{ g H}} = 14.26 \text{ mol H}$

Since the mol H to mol C ratio is 2:1 (14.26/7.130 = 2.000), the empirical formula is CH_2. The empirical formula mass $\approx 12 + 2(1) = 14$. Since $4 \times 14 = 56$ puts the molar mass between 50 and 60, the molecular formula is C_4H_8.

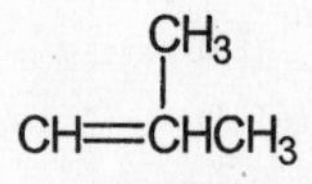

The isomers of C_4H_8 are:

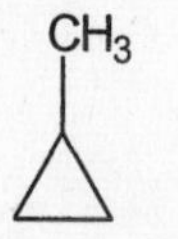

Only the alkenes will react with H_2O to produce alcohols, and only 1-butene will produce a secondary alcohol for the major product and a primary alcohol for the minor product.

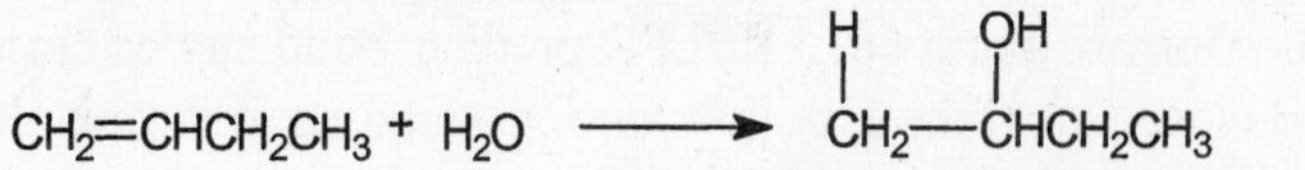

2-butene will produce only a secondary alcohol when reacted with H_2O, and 2-methyl-1-propene will produce a tertiary alcohol as the major product and a primary alcohol as the minor product.

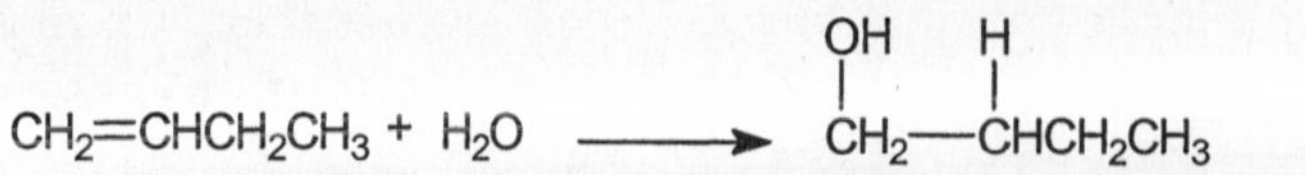

120.   The monomers for nitrile are $CH_2=CHCN$ (acrylonitrile) and $CH_2=CHCH=CH_2$ (butadiene). The structure of polymer nitrile is:

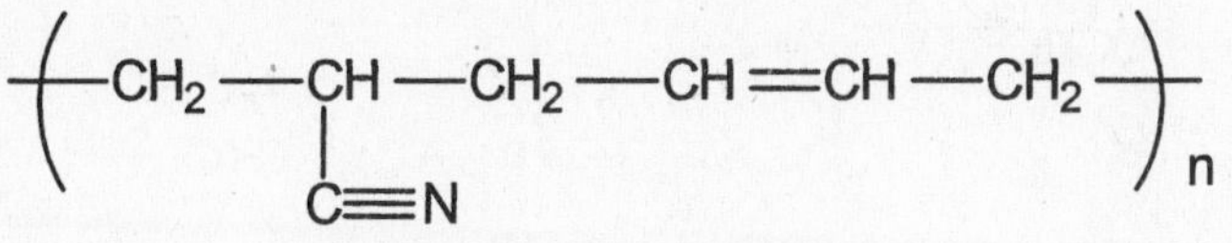

$$\left(CH_2-\underset{\underset{C\equiv N}{|}}{CH}-CH_2-CH=CH-CH_2\right)_n$$

121.   a. 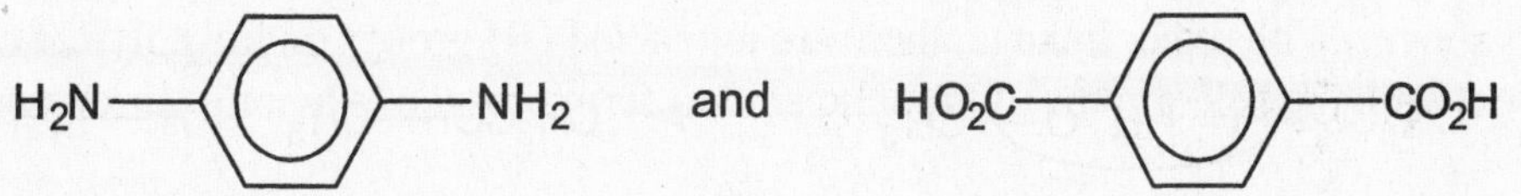 $H_2N-\langle\bigcirc\rangle-NH_2$ and $HO_2C-\langle\bigcirc\rangle-CO_2H$

       b.   Repeating unit:

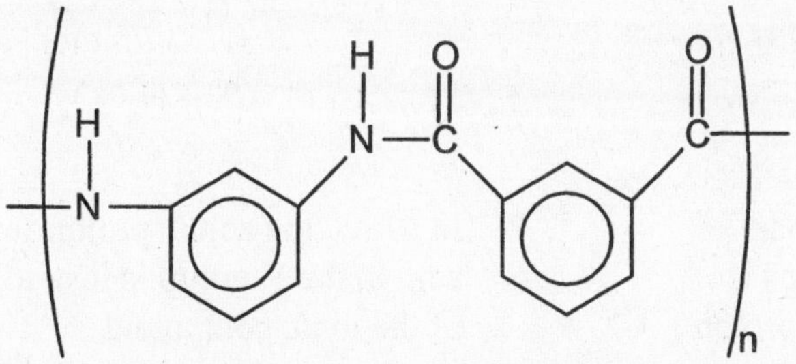

       The two polymers differ in the substitution pattern on the benzene rings. The Kevlar chain is straighter, and there is more efficient hydrogen bonding between Kevlar chains than between Nomex chains.

122.   Polyacrylonitrile:

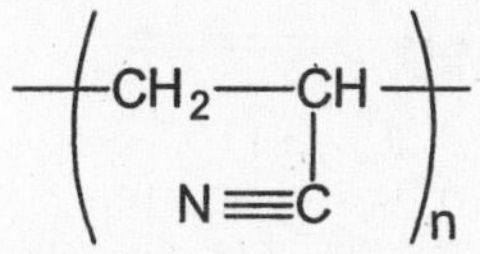

$$\left(CH_2-\underset{\underset{N\equiv C}{|}}{CH}\right)_n$$

       The CN triple bond is very strong and will not easily break in the combustion process. A likely combustion product is the toxic gas hydrogen cyanide, HCN(g).

123.   a.   The bond angles in the ring are about 60°. VSEPR predicts bond angles close to 109°. The bonding electrons are closer together than they prefer, resulting is strong electron-electron repulsions. Thus, ethylene oxide is unstable (reactive).

       b.   The ring opens up during polymerization; the monomers link together through the formation of O–C bonds.

$$\left(O-CH_2CH_2-O-CH_2CH_2-O-CH_2CH_2\right)_n$$

124.

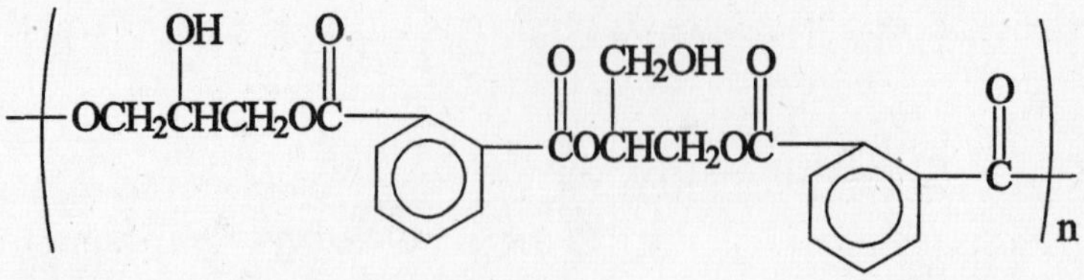

Two linkages are possible with glycerol. A possible repeating unit with both types of linkages is shown above. With either linkage, there are unreacted OH groups on the polymer chains. These can react with the acid groups of phthalic acid to form crosslinks among various polymer chains.

125. Glutamic acid:

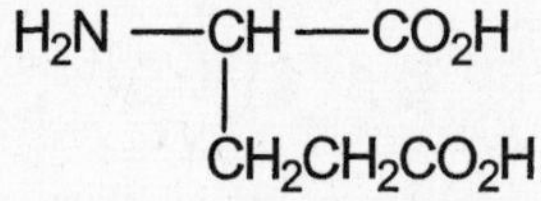

Monosodium glutamate:

One of the two acidic protons in the carboxylic acid groups is lost to form MSG. Which proton is lost is impossible for you to predict.

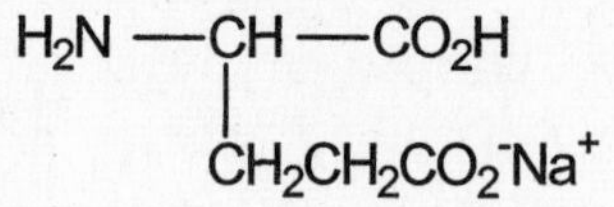

In MSG, the acidic proton from the carboxylic acid in the R group is lost, allowing formation of the ionic compound.

126. a.

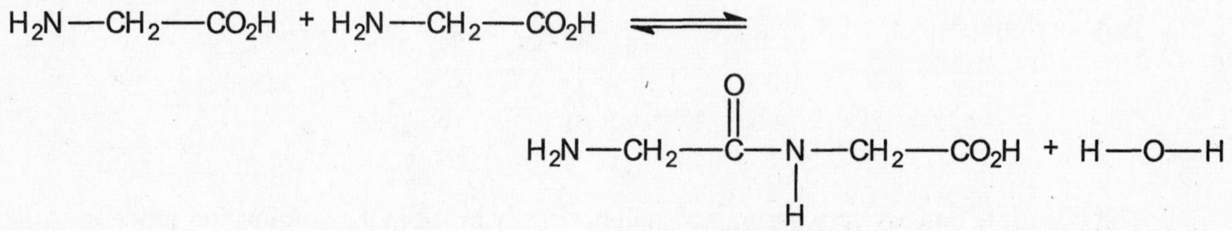

Bonds broken:

1 C – O (358 kJ/mol)
1 H – N (391 kJ/mol)

Bonds formed:

1 C – N (305 kJ/mol)
1 H – O (467 kJ/mol)

$\Delta H = 358 + 391 - (305 + 467) = -23$ kJ

b. ΔS for this process is negative (unfavorable) since order increases (disorder decreases).

c. $\Delta G = \Delta H - T\Delta S$; ΔG is positive because of the unfavorable entropy change. The reaction is not spontaneous.

127. $\Delta G = \Delta H - T\Delta S$; For the reaction, we break a P–O and O–H bond and form a P–O and O–H bond, so $\Delta H \approx 0$. ΔS for this process is negative since order increases. Thus, $\Delta G > 0$ and the reaction is not spontaneous.

128.   Both proteins and nucleic acids must form for life to exist. From the simple analysis, it looks as if life can't exist, an obviously incorrect assumption. A cell is not an isolated system. There is an external source of energy to drive the reactions. A photosynthetic plant uses sunlight, and animals use the carbohydrates produced by plants as sources of energy. When all processes are combined, ΔS_{univ} must be greater than zero as is dictated by the second law of thermodynamics.

129.   Alanine can be thought of as a diprotic acid. The first proton to leave comes from the carboxylic acid end with $K_a = 4.5 \times 10^{-3}$. The second proton to leave comes from the protonated amine end (K_a for $R-NH_3^+ = K_w/K_b = 1.0 \times 10^{-14}/7.4 \times 10^{-5} = 1.4 \times 10^{-10}$).

In $1.0\,M\,H^+$, both the carboxylic acid and the amine end will be protonated since H^+ is in excess. The protonated form of alanine is below. In $1.0\,M\,OH^-$, the dibasic form of alanine will be present since the excess OH^- will remove all acidic protons from alanine. The dibasic form of alanine follows.

$$1.0\,M\,H^+: \quad \underset{\text{protonated form}}{\overset{\overset{\displaystyle CH_3\quad O}{\displaystyle |\quad\ \ ||}}{H_3\overset{+}{N}-CH-C-OH}} \qquad\qquad 1.0\,M\,OH^-: \quad \underset{\text{dibasic form}}{\overset{\overset{\displaystyle CH_3\quad O}{\displaystyle |\quad\ \ ||}}{H_2N-CH-C-O^-}}$$

130.   The number of approximate base pairs in a DNA molecule is:

$$\frac{4.5 \times 10^9 \text{ g/mol}}{600 \text{ g/mol}} = 8 \times 10^6 \text{ base pairs}$$

The approximate number of complete turns in a DNA molecule is:

$$8 \times 10^6 \text{ base pairs} \times \frac{0.34 \text{ nm}}{\text{base pair}} \times \frac{1 \text{ turn}}{3.4 \text{ nm}} = 8 \times 10^5 \text{ turns}$$

131.   The Cl^- ions are lost upon binding to DNA. The dimension is just right for cisplatin to bond to two adjacent bases in one strand of the helix, which inhibits DNA synthesis.

132.   a. $^+H_3NCH_2COO^- + H_2O \rightleftharpoons H_2NCH_2CO_2^- + H_3O^+$

$$K_{eq} = K_a\,(-NH_3^+) = \frac{K_w}{K_b\,(-NH_2)} = \frac{1.0 \times 10^{-14}}{6.0 \times 10^{-5}} = 1.7 \times 10^{-10}$$

       b. $H_2NCH_2CO_2^- + H_2O \rightleftharpoons H_2NCH_2CO_2H + OH^-$

$$K_{eq} = K_b\,(-CO_2^-) = \frac{K_w}{K_a\,(-CO_2H)} = \frac{1.0 \times 10^{-14}}{4.3 \times 10^{-3}} = 2.3 \times 10^{-12}$$

c. $^+H_3NCH_2CO_2H \rightleftharpoons 2\ H^+ + H_2NCH_2CO_2^-$

$$K_{eq} = K_a(-CO_2H) \times K_a(-NH_3^+) = (4.3 \times 10^{-3})(1.7 \times 10^{-10}) = 7.3 \times 10^{-13}$$

Challenge Problems

133.   For the reaction:

$$^+H_3NCH_2CO_2H \rightleftharpoons 2\ H^+ + H_2NCH_2CO_2^- \qquad K_{eq} = 7.3 \times 10^{-13} = K_a\ (-CO_2H) \times K_a\ (-NH_3^+)$$

$$7.3 \times 10^{-13} = \frac{[H^+]^2[H_2NCH_2CO_2^-]}{[^+H_3NCH_2CO_2H]} = [H^+]^2, \quad [H^+] = (7.3 \times 10^{-13})^{1/2}$$

$[H^+] = 8.5 \times 10^{-7}$; pH = -log $[H^+]$ = 6.07 = isoelectric point

134.   a.   The new amino acid is most similar to methionine due to its $-CH_2CH_2SCH_3$ R group.

        b.   The new amino acid replaces methionine. The structure of the tetrapeptide is:

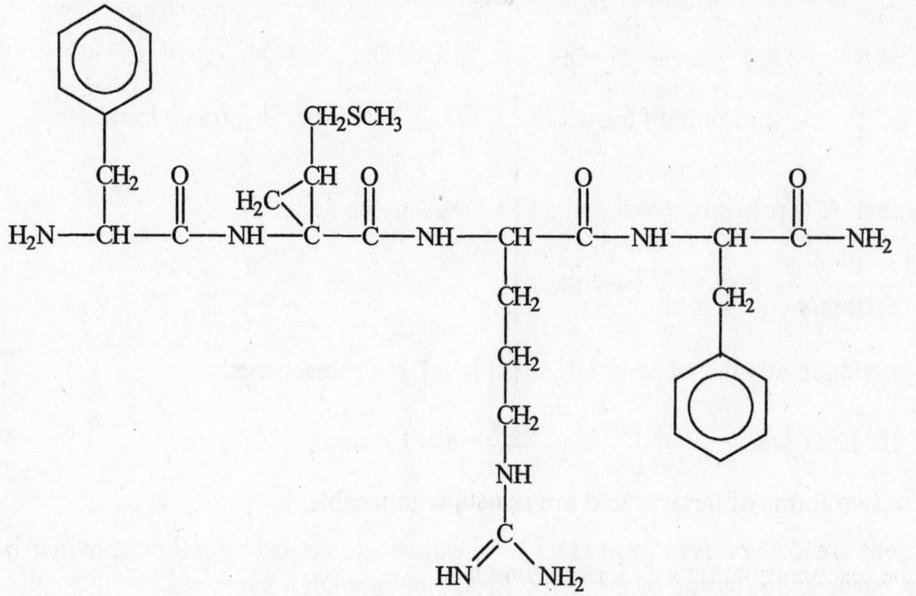

        c.   The chiral carbons are indicated with an asterisk.

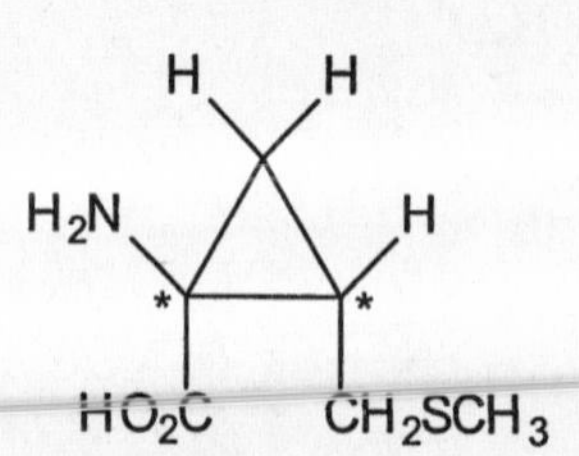

135. a. Even though this form of tartaric acid contains 2 chiral carbon atoms (see asterisks in the following structure), the mirror image of this form of tartaric acid is superimposable. Therefore, it is not optically active. An easier way to identify optical activity in molecules with two or more chiral carbon atoms is to look for a plane of symmetry in the molecule. If a molecule has a plane of symmetry, then it is never optically active. A plane of symmetry is a plane that bisects the molecule where one side exactly reflects on the other side.

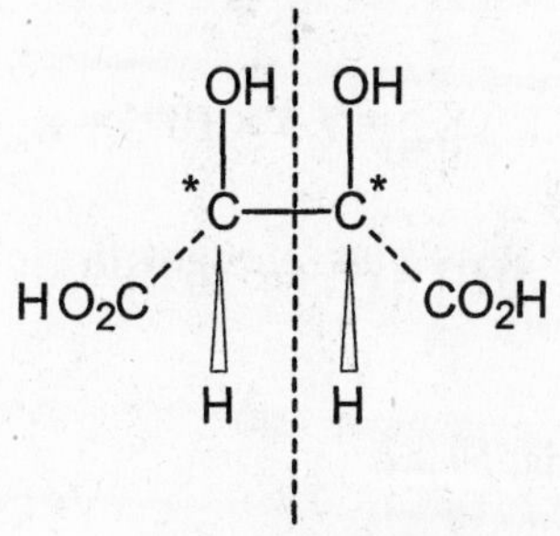

$$
\begin{array}{ccc}
 & OH \;\vdots\; OH & \\
 & \overset{*}{C}\!-\!\vdots\!-\!\overset{*}{C} & \\
HO_2C & \vdots & CO_2H \\
 & H \;\vdots\; H & \\
\end{array}
$$

symmetry plane

   b. The optically active forms of tartaric acid have no plane of symmetry. The structures of the optically active forms of tartaric acid are:

$$
\begin{array}{ccc}
OH \quad OH & \Big| & OH \quad OH \\
C\!-\!C & \Big| & C\!-\!C \\
H \quad\quad CO_2H & \Big| & HO_2C \quad\quad H \\
CO_2H \; H & \Big| & H \quad CO_2H \\
\end{array}
$$

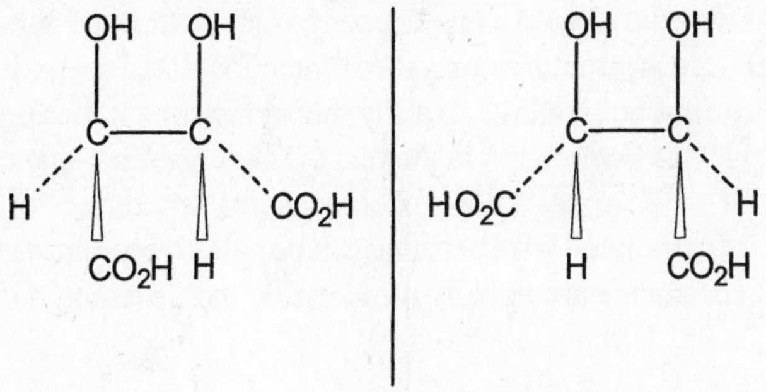

mirror

   These two forms of tartaric acid are nonsuperimposable.

136. One of the resonance structures for benzene is:

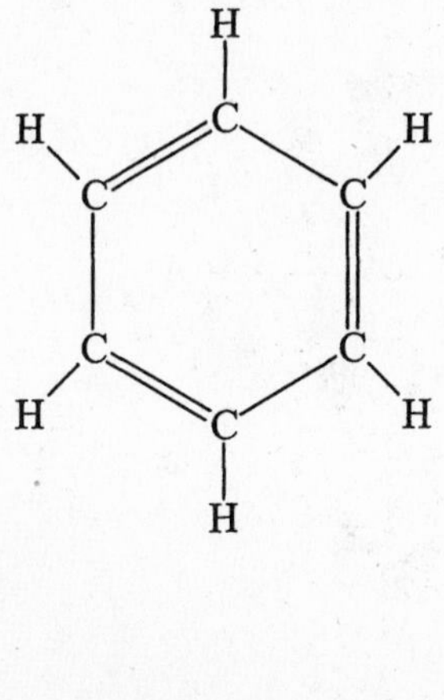

To break $C_6H_6(g)$ into $C(g)$ and $H(g)$ requires breaking 6 C–H bonds, 3 C=C bonds and 3 C–C bonds:

$$C_6H_6(g) \rightarrow 6\,C(g) + 6\,H(g) \quad \Delta H = 6\,D_{C-H} + 3\,D_{C=C} + 3\,D_{C-C}$$

$$\Delta H = 6(413\text{ kJ}) + 3(614\text{ kJ}) + 3(347\text{ kJ}) = 5361\text{ kJ}$$

The question asks for ΔH_f° for $C_6H_6(g)$, which is ΔH for the reaction:

$$6\,C(s) + 3\,H_2(g) \rightarrow C_6H_6(g) \quad \Delta H = \Delta H_{f,\,C_6H_6(g)}^\circ$$

To calculate ΔH for this reaction, we will use Hess's law along with the ΔH_f° value for $C(g)$ and the bond energy value for H_2 ($D_{H_2} = 432$ kJ/mol).

$$
\begin{aligned}
6\,C(g) + 6\,H(g) &\rightarrow C_6H_6(g) & \Delta H_1 &= -5361\text{ kJ} \\
6\,C(s) &\rightarrow 6\,C(g) & \Delta H_2 &= 6(717\text{ kJ}) \\
3\,H_2(g) &\rightarrow 6\,H(g) & \Delta H_3 &= 3(432\text{ kJ})
\end{aligned}
$$

$$6\,C(s) + 3\,H_2(g) \rightarrow C_6H_6(g) \quad \Delta H = \Delta H_1 + \Delta H_2 + \Delta H_3 = 237\text{ kJ}; \quad \Delta H_{f,\,C_6H_6(g)}^\circ = 237\text{ kJ/mol}$$

The experimental ΔH_f° for $C_6H_6(g)$ is more stable (lower in energy) by 154 kJ as compared to ΔH_f° calculated from bond energies (83 - 237 = -154 kJ). This extra stability is related to benzene's ability to exhibit resonance. Two equivalent Lewis structures can be drawn for benzene. The π bonding system implied by each Lewis structure consists of three localized π bonds. This is not correct as all C–C bonds in benzene are equivalent. We say the π electrons in benzene are delocalized over the entire surface of C_6H_6 (see Section 9.5 of the text). The large discrepancy between ΔH_f° values is due to the delocalized π electrons, whose effect was not accounted for in the calculated ΔH_f° value. The extra stability associated with benzene can be called resonance stabilization. In general, molecules that exhibit resonance are usually more stable than predicted using bond energies.

137.

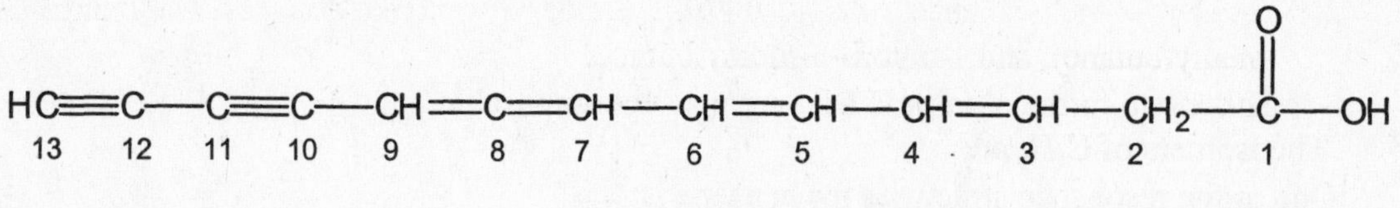

138.

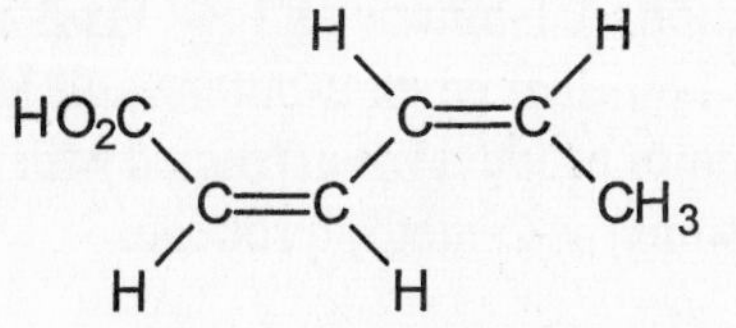

cis-2-cis-4-hexadienoic acid

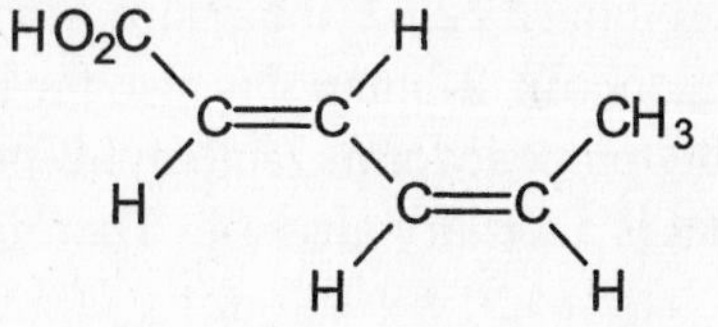

trans-2-cis-4-hexadienoic acid

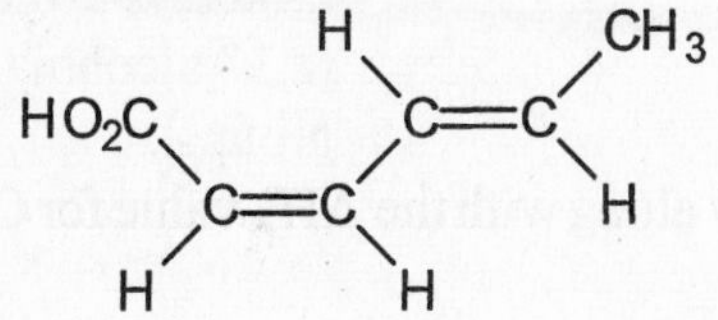

cis-2-trans-4-hexadienoic acid

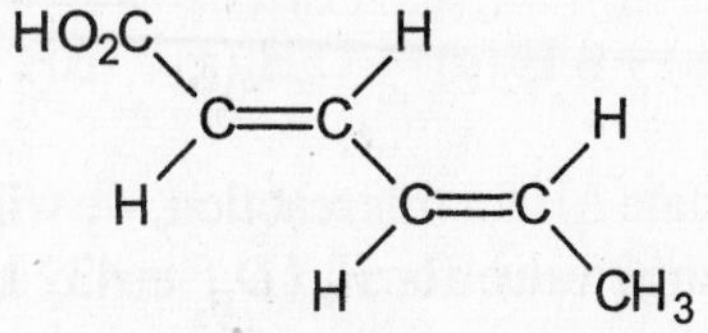

trans-2-trans-4-hexadienoic acid

139.   a.   The three structural isomers of C_5H_{12} are:

$CH_3CH_2CH_2CH_2CH_3$ $CH_3CHCH_2CH_3$
                                                  |
                                                  CH_3

n-pentane   2-methylbutane 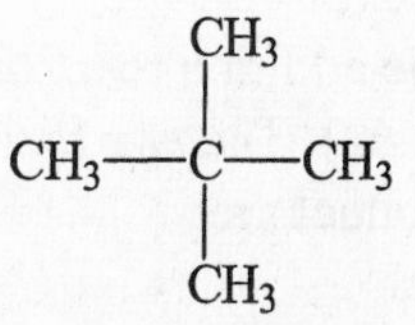   2,2-dimethylpropane

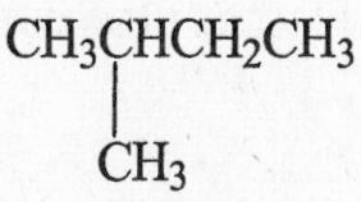

n-pentane will form three different monochlorination products: 1-chloropentane, 2-chloro-pentane and 3-chloropentane (the other possible monochlorination products differ by a simple rotation of the molecule; they are not different products from the ones listed). 2-2,dimethylpropane will only form one monochlorination product: 1-chloro-2,2-dimethyl-propane. 2-methylbutane is the isomer of C_5H_{12} that forms four different monochlorination products: 1-chloro-2-methylbutane, 2-chloro-2-methylbutane, 3-chloro-2-methylbutane (or we could name this compound 2-chloro-3-methylbutane), and 1-chloro-3-methylbutane.

b.   The isomers of C_4H_8 are:

CH_2=$CHCH_2CH_3$ CH_3CH=$CHCH_3$

1-butene   2-butene   2-methyl-1-propene or
                                                       2-methylpropene

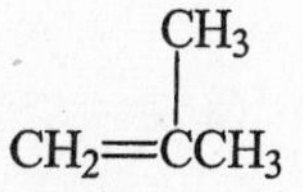

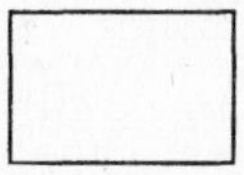

cyclobutane    methylcyclopropane

The cyclic structures will not react with H_2O; only the alkenes will add H_2O to the double bond. From Exercise 22.62, the major product of the reaction of 1-butene and H_2O is 2-butanol (a 2° alcohol). 2-butanol is also the major (and only) product when 2-butene and H_2O react. 2-methylpropene forms 2-methyl-2-propanol as the major product when reacted with H_2O; this product is a tertiary alcohol. Therefore, the C_4H_8 isomer is 2-methylpropene.

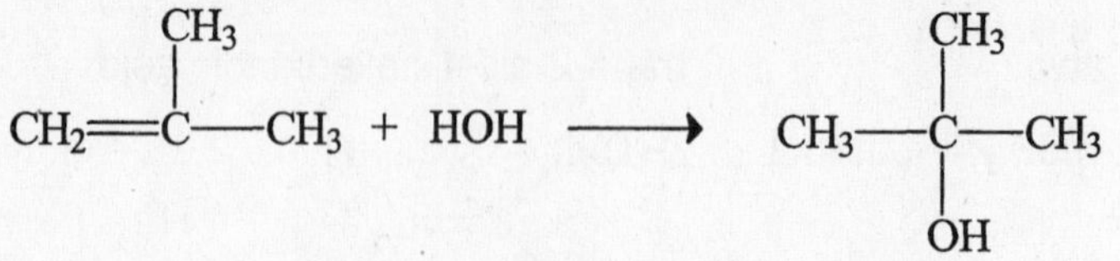

2-methyl-2-propanol (a 3° alcohol, 3 R groups)

c.   The structure of 1-chloro-1-methylcyclohexane is:

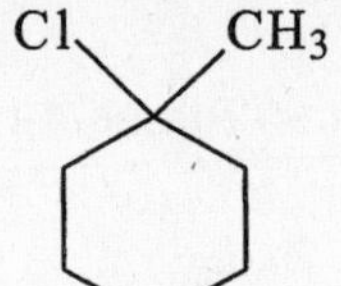

The addition reaction of HCl with an alkene is a likely choice for this reaction (see Exercise 22.62). The two isomers of C_7H_{12} that produce 1-chloro-1-methylcyclohexane as the major product are:

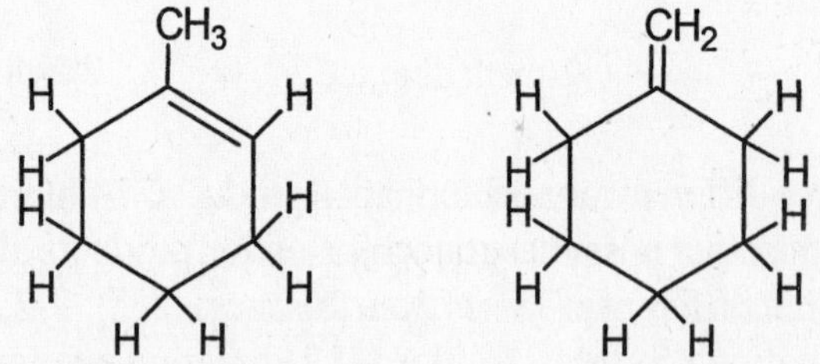

d.   Working backwards, 2° alcohols produce ketones when they are oxidized (1° alcohols produce aldehydes, then carboxylic acids). The easiest way to produce the 2° alcohol from a hydrocarbon is to add H_2O to an alkene. The alkene reacted is 1-propene (or propene).

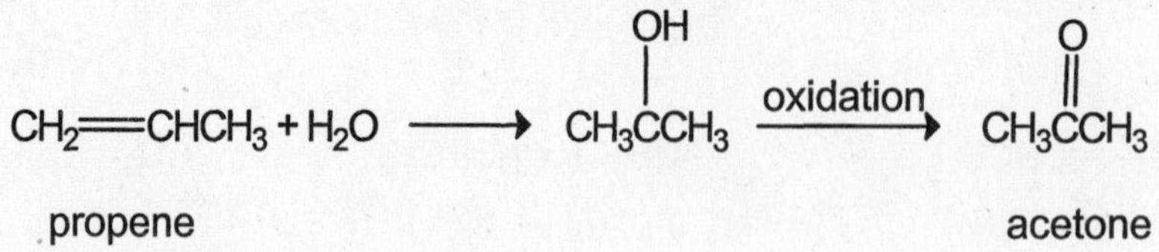

propene   acetone

e. The $C_5H_{12}O$ formula has too many hydrogens to be anything other than an alcohol (or an unreactive ether). 1° alcohols are first oxidized to aldehydes, then to carboxylic acids. Therefore, we want a 1° alcohol. The 1° alcohols with formula $C_5H_{12}O$ are:

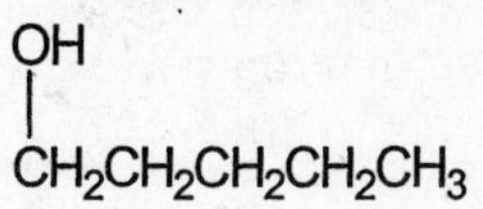

1-pentanol 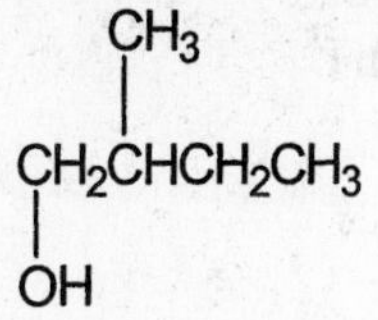   2-methyl-1-butanol 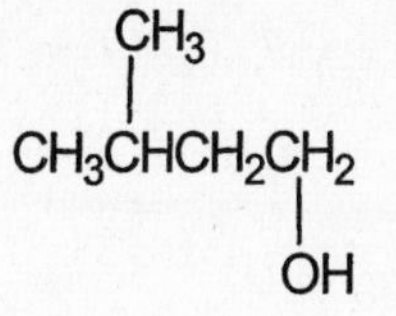   3-methyl-1-butanol 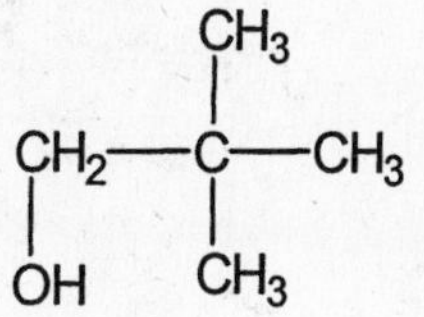   2,2-dimethyl-1-propanol

There are other alcohols with formula $C_5H_{12}O$, but they are all 2° or 3° alcohols, which do not produce carboxylic acids when oxidized.

140.   a.

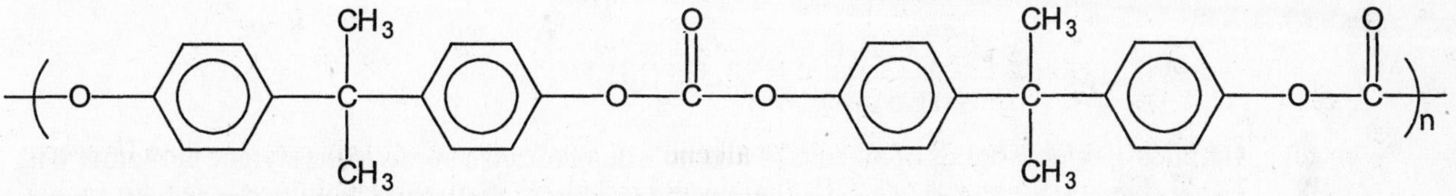

b. Condensation; HCl is eliminated when the polymer bonds form.

141.

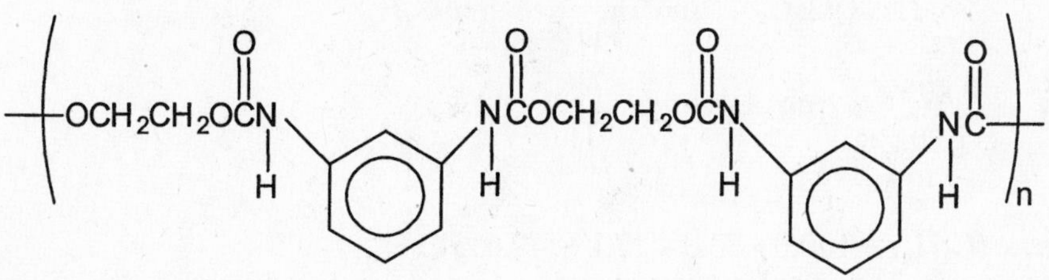

142.   a.

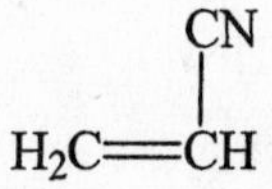

acrylonitrile    butadiene 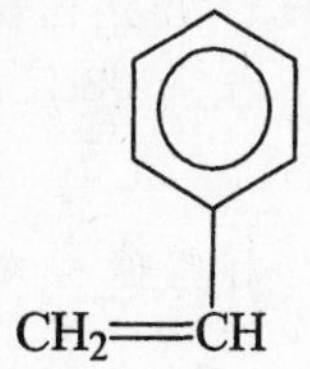   styrene

The structure of ABS plastic assuming a 1:1:1 mol ratio is:

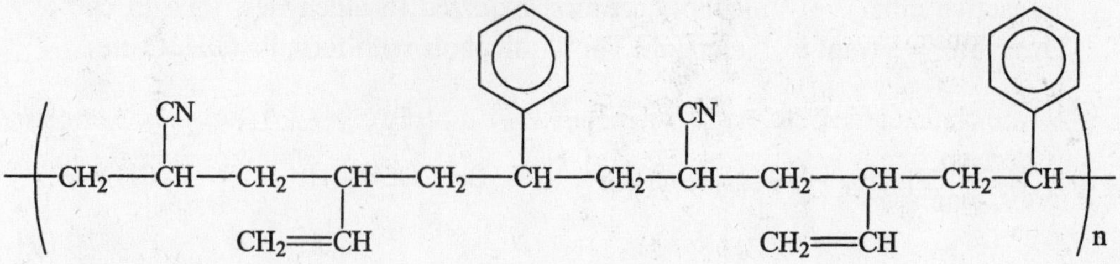

Note: Butadiene does not polymerize in a linear fashion in ABS plastic (unlike other butadiene polymers). There is no way for you to be able to predict this.

b. Only acrylonitrile contains nitrogen. If we have 100.00 g of polymer:

$$8.80 \text{ g N} \times \frac{1 \text{ mol } C_3H_3N}{14.01 \text{ g N}} \times \frac{53.06 \text{ g } C_3H_3N}{1 \text{ mol } C_3H_3N} = 33.3 \text{ g } C_3H_3N$$

$$\% \; C_3H_3N = \frac{33.3 \text{ g } C_3H_3N}{100.00 \text{ g polymer}} = 33.3\% \; C_3H_3N$$

Br_2 adds to double bonds of alkenes (benzene's delocalized π bonds in the styrene monomer will not react with Br_2 unless a special catalyst is present). Only butadiene in the polymer has a reactive double bond. From the polymer structure in part a, butadiene will react in a 1:1 mol ratio with Br_2.

$$0.605 \text{ g } Br_2 \times \frac{1 \text{ mol } Br_2}{159.8 \text{ g } Br_2} \times \frac{1 \text{ mol } C_4H_6}{\text{mol } Br_2} \times \frac{54.09 \text{ g } C_4H_6}{\text{mol } C_4H_6} = 0.205 \text{ g } C_4H_6$$

$$\% \; C_4H_6 = \frac{0.205 \text{ g}}{1.20 \text{ g}} \times 100 = 17.1\% \; C_4H_6$$

$$\% \text{ styrene } (C_8H_8) = 100.0 - 33.3 - 17.1 = 49.6\% \; C_8H_8.$$

c. If we have 100.0 g of polymer:

$$33.3 \text{ g } C_3H_3N \times \frac{1 \text{ mol } C_3H_3N}{53.06 \text{ g}} = 0.628 \text{ mol } C_3H_3N$$

$$17.1 \text{ g } C_4H_6 \times \frac{1 \text{ mol } C_4H_6}{54.09 \text{ g } C_4H_6} = 0.316 \text{ mol } C_4H_6$$

$$49.6 \text{ g } C_8H_8 \times \frac{1 \text{ mol } C_8H_8}{104.14 \text{ g } C_8H_8} = 0.476 \text{ mol } C_8H_8$$

Dividing by 0.316: $\frac{0.628}{0.316} = 1.99; \; \frac{0.316}{0.316} = 1.00; \; \frac{0.476}{0.316} = 1.51$

This is close to a mol ratio of 4:2:3. Thus, there are 4 acrylonitrile to 2 butadiene to 3 styrene molecules in this polymer sample; or $(A_4B_2S_3)_n$.

143.   a. The temperature of the rubber band increases when it is stretched.

b. Exothermic since heat is released.

c. As the chains are stretched, they line up more closely together, resulting in stronger London dispersion forces between the chains. Heat is released as the strength of the intermolecular forces increases.

d. Stretching is not spontaneous so, ΔG is positive. $\Delta G = \Delta H - T\Delta S$; Since ΔH is negative then ΔS must be negative in order to give a positive ΔG.

e.

unstretched   stretched

The structure of the stretched polymer is more ordered (lower S).